10th TSME-International Conference on Mechanical Engineering (TSME-ICoME2019)

IOP Conference Series: Materials Science and Engineering
Volume 886

Pattaya, Thailand
10 - 13 December 2019

ISBN: 978-1-7138-3622-3
ISSN: 1757-8981

Printed from e-media with permission by:

Curran Associates, Inc.
57 Morehouse Lane
Red Hook, NY 12571

Some format issues inherent in the e-media version may also appear in this print version.

Institute of Physics
Dirac House, Temple Back
Bristol BS1 6BE UK

Phone: 44 1 17 929 7481
Fax: 44 1 17 920 0979

techtracking@iop.org

Additional copies of this publication are available from:

Curran Associates, Inc.
57 Morehouse Lane
Red Hook, NY 12571 USA
Phone: 845-758-0400
Fax: 845-758-2633
Email: curran@proceedings.com
Web: www.proceedings.com

TABLE OF CONTENTS

COMPUTATION AND SIMULATION TECHNIQUES (CST)

DYNAMIC SYSTEMS, ROBOTICS AND CONTROLS (DRC)

THERMAL SYSTEM AND FLUID MECHANICS (TSF)

IOP Conf. Series: Materials Science and Engineering **886** (2020) 011001 doi:10.1088/1757-899X/886/1/011001

PREFACE

The 10th TSME International Conference on Mechanical Engineering (TSME-ICoME 2019) was hosted by the School of Mechanical Engineering, Institute of Engineering, Suranaree University of Technology, Nakhon Ratchasima, Thailand. The conference was held at A-One The Royal Cruise Hotel, Pattaya, Thailand, from 10th – 13th December 2019. The conference is a leading conference of the Thai Society of Mechanical Engineers (TSME) and represents one of the only exclusive events in Thailand where everyone "mechanically-engineering-related" can share their ideas and establish long-lasting collaborations and networks.

The scientific program covered a wide range of topics, including Alternative Energy and Combustion (AEC), Automotive, Aerospace and Marine Engineering (AME), Applied Mechanics, Materials and Manufacturing (AMM), Biomechanics and Bioengineering (BME), Computation and Simulation Techniques (CST), Dynamic Systems, Robotics and Controls (DRC), Engineering Education (EDU), Energy Technology and Management (ETM), and Thermal System and Fluid Mechanics (TSF)

This year conference brought together more than 200 delegates to exchange knowledge by in-depth discussions and to promote international and multi-disciplinary collaborations. The theme of this year's conference is "Emerging and Disruptive Technologies" and includes mechanical research fields, industrial panel sessions, keynote speeches, papers and poster presentation.

We would like to thank the program chairs, organization staff, and the members of the program committees for their work. Thanks also go to Editors and Reviewers from the Thailand Society of Mechanical Engineers (TSME), for their wonderful editorial and review service to this proceeding.

We grateful to all those who have contributed to success of TSME-ICoME 2019. We hope that all participants and other interested readers benefit scientifically from the proceedings and also find it stimulating in the process. Finally, we would like to wish you success in your technical presentations and social networking.

We hope you have a unique, rewarding and enjoyable week at TSME-ICoME 2019, Thailand.

With our warmest regards

The Organizing Committees

December 10th – 13th, 2019

Thailand

IOP Conf. Series: Materials Science and Engineering **886** (2020) 011001 doi:10.1088/1757-899X/886/1/011001

Address at the 10th Thai Society of Mechanical Engineers International Conference on Mechanical Engineering (TSME-ICoME 2019)

In 2010, TSME started to hold yearly international conferences under the name of the International Conference on Mechanical Engineering (ICoME) with collaboration from the American Society of Mechanical Engineers (ASME), the Japan Society of Mechanical Engineers (JSME), and the United Kingdom's Institute of Mechanical Engineers (IMechE). This year there is also cooperation with the Korean Society of Mechanical Engineers (KSME).

The objectives of the International Conference on Mechanical Engineering (ICoME) are to encourage students, researchers, and interested persons to present their works, to create researching networks, to publish the research works for industry sectors, to find the needs of public and private organizations, to fulfill these demands to enhance the growth of country, and to be a hub in the field of Mechanical Engineering for anyone with an interest.

The ICoME conference has been organized and popular for nearly 10 years. In this honor, I would like to express my congratulation and thankfulness to Suranaree University of Technology (SUT) for the dedication to organize the 10th TSME-ICoME (TSME-ICoME 2019) where will be held at A-One The Royal Cruise Hotel in Pattaya, Thailand – the gateway to Eastern Thailand from 10th – 13th December 2019. On behalf of the president of the association, I would like to thank you for your supporting team and all staff for hard working with the association's board to make this conference potential. I hope the conference to be successful for all purposes and believe that this conference will be a gathering of researchers in mechanical engineering.

Prof. Dr. Phadungsak Rattanadecho

President of Thai Society of Mechanical Engineers (TSME)

Peer review statement

All papers published in this volume of *IOP Conference Series: Materials Science and Engineering* have been peer reviewed through processes administered by the proceedings Editors. Reviews were conducted by expert referees to the professional and scientific standards expected of a proceedings journal published by IOP Publishing.

IOP Conf. Series: Materials Science and Engineering **886** (2020) 012001 doi:10.1088/1757-899X/886/1/012001

Effect of Equivalent Ratio Variation on a Two-Stage Distributed Combustion

Seekharin Komonhirun[1*], Pisit Yongyingsakthavon[1], Songrit Tanchatchawan[2] and Udomkiat Nonthakeaw[1]
[1]Department of Mechanical and Aerospace Engineering, King Mongkut's University of Technology North Bangkok
[2]Thailand Institute of Scientific and Technology Research, Khlong 5, Khlong luang, Pathum Thani, Thailand

* Corresponding Author: k.seekharin@gmail.com

Abstract
Distributed combustion is a recent technology that has a feasibility to apply on the industrial scale. It provides relatively low combustion temperature with rather uniform distribution and very low emission (NOx). The two-stage distributed combustion system is established to reduce the complexity of operating in the industrial application. The operation range of this combustion system is governed by the equivalent ratio of the first and second-stages of combustion. This research is attempted to study the effects of equivalent ratio and to identify the operation range of the combustion system. There are two stages in this experiment: 1. To set up using a diesel combustor (first-stage) 2. To use an LPG distributed combustor (second-stage). Part of the aim of this project is to improve the efficiency of the single-stage diesel combustor. The air to fuel ratio of the first-stage combustor is controlled to provide the 5% excess oxygen for the second-stage combustor. The equivalent ratio of the second-stage is varied as 1, 0.9, 0.8, 0.7 and 0.6. The results have shown that stoichiometric distributed combustion yields the highest outlet temperature and first and second law efficiency. The temperature is decreased when reducing the equivalent ratio. Finally, this research also provides the formula that can be used to determine the appropriate operating range in the term of A/F for the two-stage distributed combustion system conveniently applied to the single-stage available in the industrial.

Keywords: Distributed combustion, Thermal efficiency, Experiment

1. Introduction
The biodiesel is the vegetable oil blended with diesel fuel, which is commonly used in the industry to reduce the fuel price [1-5]. However, the operation of the biodiesel combustor around stoichiometric range can cause soot, carbon monoxide and nitrogen oxide (NOx). These gases are emissions which

causes pollution issues that directly impact the global environment. Lean combustion is a promising choice to reduce emissions especially soot. However, this combustion process provides lower combustion temperature and thermal efficiency. There are several methods to improve the combustion efficiency [6-11]. To date, there are few studies that have investigated the association between the distributed combustion and the first stage combustor. Currently, there are no data on suitable operation conditions for the distributed combustor [12]. Distributed combustion is the combustion that can reduce pollution especially NOx. The operating temperature range is around 800 C to 1000 C, which has potential in the industrial application [13]. At this temperature, the NOx reaction is inactivated due to the low activation energy [14-15]. The prominent point of this phenomenon is the invisible flame. In conventional combustion, the flame illuminates in yellow light. However, the distributed combustor emits the light in the infrared wavelength due to its low combustion temperature [16]. To achieve this phenomenon, it requires two conditions. The one is to lower the oxygen concentration at the inlet air of the distributed combustion stage from 21% in atmospheric condition to around 5% to 7%. The second condition requires the auto-ignition temperature to ignite and sustain combustion due to the lower inlet oxygen concentration that reduces the activation energy of air [17-18]. The design of combustor to achieve this condition can be developed by several methods such as air recirculation, exhaust gas extractor or preheated air inlet with first-stage combustor [19-22]. In addition, many works operate the disturbed combustion at stoichiometric. There are opportunities to apply distributed combustion as the second stage to take its advantages and also observe operation in lean combustion range. In this research, the biodiesel combustor is operated in lean combustion at the first-stage. The air is combusted with biodiesel to provide the high temperature and lower exhaust oxygen concentration before entering the second-stage distributed combustion. LPG fuel is injected into the distributed combustion chamber and ignited by the auto-ignition phenomenon.

$$C_{3.3}H_{8.6} + 5.74\,(O_2 + 19N_2) \Leftrightarrow 3.3\,CO_2 + 4.3\,H_2O + 109.06N_2 \quad (1)$$

$$\phi = \frac{(A/F)_{st}}{(A/F)_{operation}} \quad (2)$$

The stoichiometric reaction of the distributed combustion is shown in equation (1) with LPG fuel that contains 70% propane and 30% butane. The reaction is written in 5% excess oxygen from the first-stage. This oxygen concentration is the widely used value for the distributed combustion. Oxygen dilution increases the nitrogen concentration percentage and is drastically increased. The combustion in higher nitrogen concentration causes reducing combustion temperature because the energy absorbed by nitrogen gas [23]. In an investigation into the operation, the oxygen concentration of the first-stage biodiesel combustor and the equivalent ratio of the second-stage distributed combustor must be controlled mutually to achieve the desired operation. In this research, the effects of the equivalent ratio variation of the second-stage distributed combustion are chosen to study. The equivalent ratio (ϕ) is the ratio of the stoichiometric air/fuel and the operation air/fuel ratio, which is described in equation (2). The equivalent ratio is varied between 0.6 and 1. The first and second law efficiencies are investigated by experiment.

2. Methodology

The model of the second-stage combustor distributed combustor is applied from Hosseini et al. (2014) [24-26] due to its simplicity to apply with industrial application. The dynamic similitude is used for scaling the size of the combustor. Re_{air}/Re_{fuel} ratio is maintained during the resizing from a small lab-scale size to an industrial size, which provides 6-8 kW loads. During the design phase, the simulation with computational fluid dynamic (CFD) has been applied to observe combustion and flow phenomena inside the combustion chamber [27]. The shape of the combustion chamber is a single-cylinder cover with ceramic insulation. It contains one LPG fuel hole at the center and eight inlet holes connect to the first-stage biodiesel combustor and one exhaust outlet. Three K-type thermocouples are installed to measure the temperature distribution at 60, 120 and 180 cm. along the combustion chamber axis. The thermo couple and gas analyzer called "Testo-350" is installed at the exhaust outlet to measure temperature and concentration of oxygen, NOx and CO. In the first-stage combustor, the B7 diesel fuel is injected by the pressurized nitrogen and the atmospheric airflow is injected by the air blower. Air, biodiesel and LPG flow rates are measured by flow meters. The pressure swirl atomizer is used to inject biodiesel fuel into the first-stage combustion chamber. Biodiesel fuel is ignited with an electrical ignition system. The temperature outlet of the first-stage combustor is also measured with the thermocouple. After the design phase, the two-stage combustion system is constructed and tested.

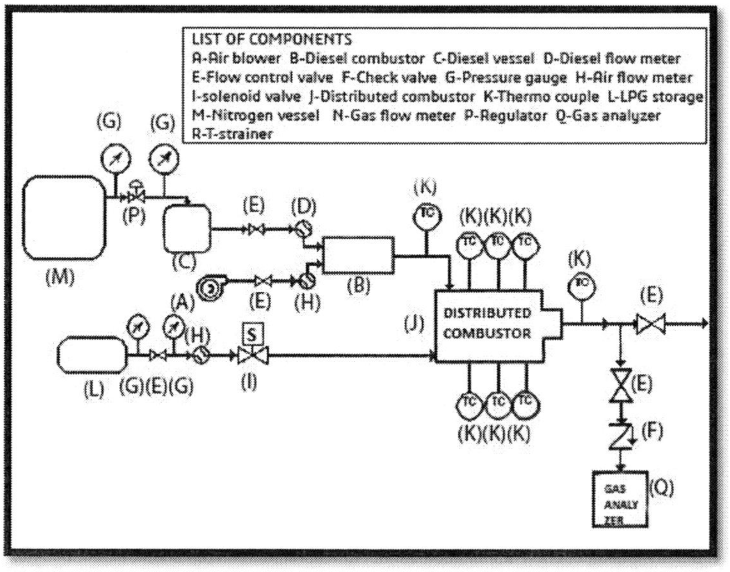

Figure 1. The two-stage combustion system diagram

2.1 A two-stage distributed combustion: setting up an experimental system

Before the experiment of the two-stage distributed combustion, the first-stage biodiesel combustor is tested and the temperature outlet is measured. The measurement data is used for calculating the first and the second law efficiencies. In this study, the operation of the biodiesel combustor can provide excess oxygen for the second-stage combustor in a range of 3% - 7%. Below 3% excess oxygen concentration, the soot occurred can cause damage to the measurement system while beyond 7%, the combustion blows off. Therefore, 5% excess oxygen concentration is most appropriate and selected for all experiment. The B7 fuel has a heating value of 42.53 MJ/kg, which corresponding to the flow rate of 19.88 g/s and the airflow rate is 374.33 g/min. This result in a steady outlet temperature is at 610 OC, which is above the auto-ignition temperature of the first-stage LPG fuel.

As shown in Table 1, when combined with the second-stage distributed combustor, the experiment is carried out in five cases of the equivalent ratio varied from 1 to 0.6 with 0.1 decrements. The LPG

IOP Conf. Series: Materials Science and Engineering **886** (2020) 012001 doi:10.1088/1757-899X/886/1/012001

supply is decreased and A/F ratio is increased when reducing the equivalent ratio. It contains 70% propane and 30% butane and has a heating value of 49.73 MJ/kg. The condition of the first-stage combustion is maintained constant for all cases. The auto-ignition of the second-stage is occurred by the preheated air from the first-stage diesel combustor. In a preliminary test, it is found that the distributed combustion can be initiated at around 500 OC of the exhaust gases from the first stage for all ranges of the equivalent ratio. One interesting finding is that there are no effects of equivalent ratio.

The temperature distribution along with the combustion chamber axis is measured and used for calculating the first and the second law efficiencies.

Table 1. Test conditions of the effects of the variation of the equivalent ratio

Test Conditions	Case I	Case II	Case III	Case IV	Case V
The first-stage combustion					
B7 mass flow rate (g/s)	19.88	19.88	19.88	19.88	19.88
Air flow rate (g/min)	374.33				
Resulted exhaust temperature (^{O}C)	610 ^{O}C	610 ^{O}C	610 ^{O}C	610 ^{O}C	610 ^{O}C
The second-stage combustion					
Equivalent ratio	1	0.9	0.8	0.7	0.6
Calculated A/F ratio	63.77	70.85	79.71	91.10	106.83
LPG mass flow rate (g/min)	5.87	5.28	4.70	4.11	3.52
LPG temperature (K)	300	300	300	300	300
LPG Composition	C_3H_8 70%, C_4H_{10} 30%				

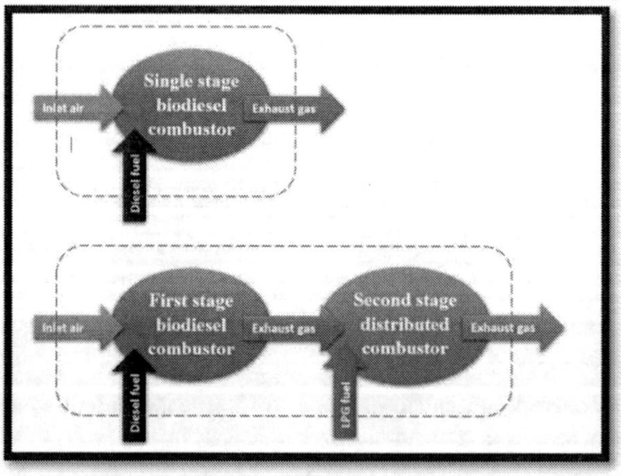

Figure 2. Schematic of the single-stage and two-stage combustion systems

2.2 Single-stage and two-stage distributed combustion efficiency analysis

The first law efficiency is calculated and the results of each case are compared. Figure 2 shows the control volumes of single-stage and two-stage combustion system. The first-law efficiency for both systems can be described as:

$$\eta_{th} = \frac{\dot{m}_{out}h_{out} - \dot{m}h_{air\ inlet} - \dot{m}h_{diesel\ inlet}}{(\dot{m}LHV_{Diesel})} \quad (3)$$

$$\eta_{th} = \frac{\dot{m}_{out}h_{out} - \dot{m}h_{air\ inlet} - \dot{m}h_{diesel\ inlet} - \dot{m}h_{LPG\ inlet}}{(\dot{m}LHV_{Diesel} + \dot{m}LHV_{LPG})} \quad (4)$$

The second law efficiency concept is the comparison between the actual conditions with the best theoretic possible condition. It can be calculated from the proportion of the exergy destroyed and exergy expense substituted from the unity [28] as described in equation (5).

$$\eta_{ex} = \frac{ex_{Outlet}}{ex_{Expense}} = 1 - \left(\frac{ex_{Destroyed}}{ex_{Expense}}\right) \quad (5)$$

While the exergy destroyed can be calculated from entropy generation in equation (6) for the single-stage system and equation (7) for the two-stage system.

$$ex_{Destroyed} = \frac{(T_r S_{gen})}{\square\square_{\square\square\square\square\square\square}} \dot{\square}_{\square\square\square\square\square\square} \quad (6)$$

$$ex_{Destroyed} = \frac{(T_{r1} S_{gen1})}{\square\square_{\square\square\square\square\square\square}} \dot{\square}_{\square\square\square\square\square\square} + \frac{(T_{r2} S_{gen2})}{\square\square_{\square\square\square}} \dot{\square}_{\square\square\square} \quad (7)$$

The entropy generation for both systems can be written in equations (8) and (9).

$$S_{gen} = S_{prod} - S_{react} + \frac{Q_{first-stage}}{T_r} \quad (8)$$

$$S_{gen} = S_{gen1} + S_{gen2} = (S_{prod1} - S_{react1}) + \frac{Q_{first-stage}}{T_{r1}} + (S_{prod2} - S_{react2}) + \frac{Q_{second-stage}}{T_{r2}} \quad (9)$$

The entropies of reactant and product are calculated from the summation of the entropy of each species, i, involved in combustion reaction by using the exact entropy analysis equation as shown in equation (10) where $\bar{s}_i^o$ is an absolute entropy [29].

$$S = \sum N_i \bar{s}_i = \sum N_i [\bar{s}_i^o(T, P_o) - R_u \ln(y_i \frac{P_m}{\square_0})] \quad (10)$$

The heat of combustion is written as Q_{out} in equation (11) and computed from the enthalpy difference between reactant and product species. The enthalpy of each species is the summation of the enthalpy of formation ($\bar{h}_f^o$) and the sensible enthalpy ($\bar{h} - \bar{h}^o$).

$$Q_{out} = \sum N_r(\bar{h}_f^o + \bar{h} - \bar{h}^o)_r - \sum N_p(\bar{h}_f^o + \bar{h} - \bar{h}^o)_p \quad (11)$$

Finally, the cost analysis is also performed to observe the advancing of the distributed combustion combination. Equation (12) and (13) is used to calculate the fuel cost for the single-stage and two-stage combustion systems, respectively, in kilowatts hour per baht.

$$\square\square\square\square_{\square\square\square\square\square} = \dot{\square}_{\square\square\square\square\square\square} \square\square\square\square_{\square\square\square\square\square\square} \quad (12)$$

$$\square\square\square\square_{\square\square\square\square\square} = \dot{\square}_{\square\square\square\square\square\square} \square\square\square\square_{\square\square\square\square\square\square} + \dot{\square}_{\square\square\square} \square\square\square\square_{\square\square\square} \quad (13)$$

where $\dot{\square}$ is volume flow rate.

3. Results and Discussion

3.1 The effects of the variation of equivalent ratio on the two-stage combustor

In the experiment, the first-stage combustion of biodiesel fuel with the atmospheric air supplied by the air blower is started at first. Then, the second-stage distributed combustor is ignited through the auto-ignition process when the temperature of the exhaust gas from the first-stage combustor reach above the auto-ignition temperature of the LPG fuel. The results of this study show that the distributed combustion can be attained at around 500OC of the first-stage exhaust. Surprisingly, the temperature of the auto-ignition of the butane and propane are 405 OC and 450 OC, respectively [15]. This can be noticed by the rapid increase of temperature inside the second-stage and provide the non-illuminated flame. Temperature and emissions are observed at the steady condition.

Figure 3 presents the measured temperatures at the exit and inside the second-stage combustion chamber. The results show that the temperature is highest in the reaction zone, which is near the fuel injection nozzle. The uniform-distribution temperature is shown through a slight decrease along the combustion chamber axis. This demonstrates the prominence of distributed combustion. Considering the influence of the equivalent ratio, the temperature is highest at the stoichiometric condition and decreases when lowering the equivalent ratio to 0.6 (the leanest in this experiment). The effects of the equivalent ratio on temperature distribution are also presented.

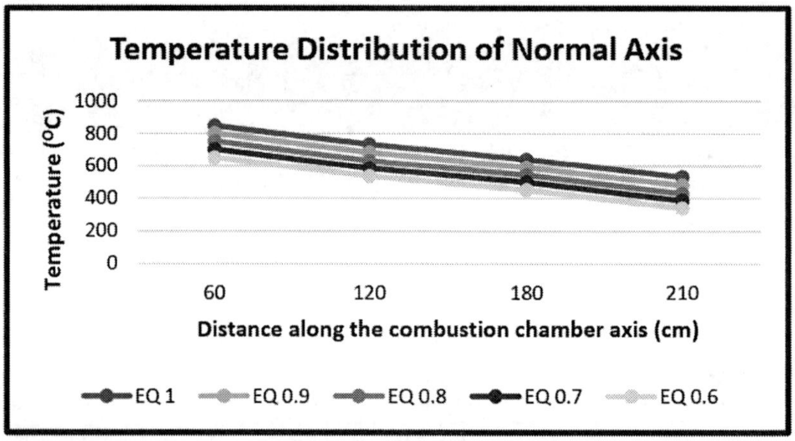

Figure 3. Temperature distribution of normal axis of combustion chamber for the equivalent ratio variation

As can be seen from Figure 4, NOx is incredibly low in all cases in the emissions measurement. The values are nearly zero as it was expected. The thermal NOx reaction is suppressed due to the low combustion temperature. For the investigation of CO, it shows the low level for all cases but not nearly zero like NOx. Both NOx and CO are maximized at the equivalent ratio of 1 and decreased when the equivalent ratio is reduced. Reduction of NOx can be explained by the relatively lower temperature of distributed combustion. In the case of CO, the decrease is from the lower feed of fuel as the supplied air is maintained constant.

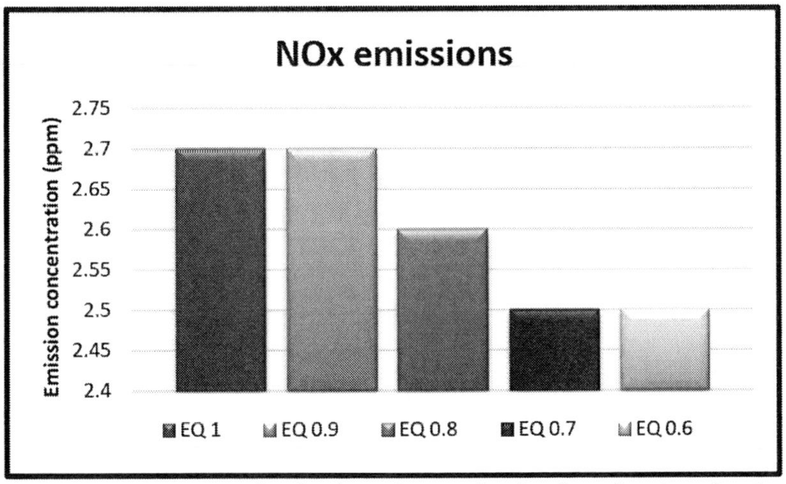

Figure 4. Temperature distribution of normal axis of combustion chamber for the equivalent ratio variation

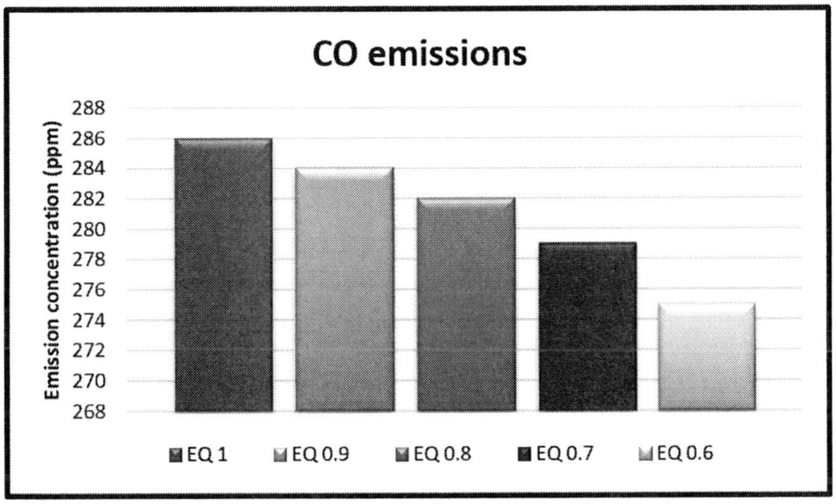

Figure 5. Temperature distribution of normal axis of combustion chamber for the equivalent ratio variation

3.2 The analysis of the first law

The first law efficiency for both single-stage and two-stage combustion systems are demonstrated in Figure 6. The results show that the two-stage combustion with 5% oxygen concentration has slightly higher efficiency than the single-stage combustion with excess 5% oxygen concentration, at stoichiometric. The first law efficiency tends to decreased when decreasing the equivalent ratio. It is obviously described by lower outlet enthalpy, as shown in equations (3) and (4). The lowest efficiency is found in the case of the 0.6 equivalent ratio which is about 5% dropped from the stoichiometric condition. It was suggested that there is an opportunity to improve the efficiency of the single-stage combustion by combining the distributed combustion as the second-stage. In this study, the distributed combustion is only 4.86 kW, which is just 34.73 % of the first-stage conventional combustion.

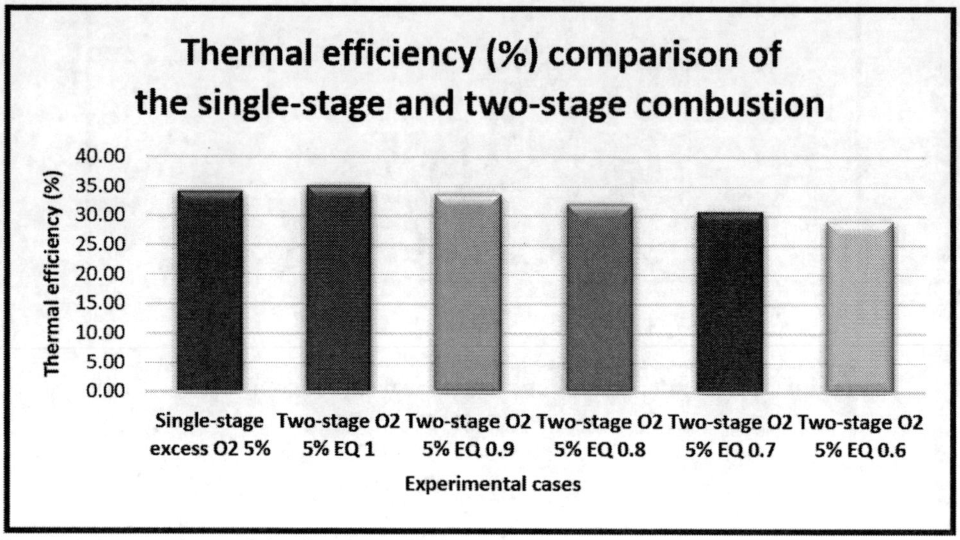

Figure 6. Comparison of the first law efficiency between the single-stage and the distributed combustor with the equivalent ratio variation

3.3 The analysis of the second law

Regarding the analysis of the second law, it yields the same trend. The highest efficiency is found at the stoichiometric point. Interestingly, the 0.9 equivalent ratio case also has slightly higher second law efficiency than the single-stage combustion. In other cases, the second law efficiency becomes lower when the equivalent ratio is decreased. This is due to the lower heat release rate. The heat is less released because of heat absorption by the more excess oxygen and produced nitrogen in distributed combustion.

As shown in Figure 8, the worthiness in kWh of output per fuel price is found in the case of the distributed combustion with 5% oxygen concentration and equivalent ratio of 1. The cost is worthy than using the single-stage combustion system. Also, the operation (nearly stoichiometric condition i.e. 0.9 equivalent ratios) yields lower cost. Another important finding was that the feasibility of using the two-stage combustion for industrial application i.e. the gas carburizing process [30]. However, this result of the 0.7 and 0.6 equivalent ratios are not appropriate for energy efficiency improvement

3.4 Formulated equations of A/F and equivalent ratio for the two-stage combustion

The distributed combustor is operated by two important control parameters which are the excess oxygen concentration from the first-stage and equivalent ratio of the distributed combustion stage. In the later stage of the study, the simply used equations of equivalent and A/F ratio are developed for the two-stage combustion system with any CxHy fuel. This is because it is expensive in the calculation to have the test conditions of A/F corresponding to require equivalent ratio. The equations are as follows:

IOP Conf. Series: Materials Science and Engineering **886** (2020) 012001 doi:10.1088/1757-899X/886/1/012001

$$\phi = \frac{\dfrac{\left(X+\left(\frac{Y}{4}\right)\right)}{\left(1-\frac{\varphi_{O_2\,out}}{\varphi_{O_2\,in}}\right)}}{X+\left[\left(\frac{Y}{4}\right)\left(1-\varphi_{O_2\,out}\right)\right]} \quad (14)$$

$$A/_F = \frac{X+\left[\left(\frac{Y}{4}\right)\left(1-\varphi_{O_2\,out}\right)\right]}{\varphi_{O_2\,in}-\varphi_{O_2\,out}} \quad (15)$$

X is the amount of carbon atom in fuel chemical formula
Y is the amount of hydrogen atom in fuel chemical formula
$\varphi_{O_2\,out}$ is the mole fraction of the outlet oxygen concentration
$\varphi_{O_2\,in}$ is the mole fraction of the inlet oxygen concentration
ϕ is the equivalent ratio

In equation (14), the $\varphi_{O_2\,out}$ is generally not equal to $\varphi_{O_2\,in}$ when there is a reaction in the second-stage. In equation (15) $\varphi_{O_2\,out}$ cannot be unity. A/F and equivalent ratio are related and controlled by X, Y, $\varphi_{O_2\,in}$ and $\varphi_{O_2\,out}$. When the fuel is selected, the controlling parameters are reduced to $\varphi_{O_2\,in}$ and $\varphi_{O_2\,out}$. These equations are limited to stoichiometric and lean combustion. $\varphi_{O_2\,in}$ controls inlet oxygen concentration for the distributed combustion, while $\varphi_{O_2\,out}$ controls the desired oxygen concentration at the outlet.

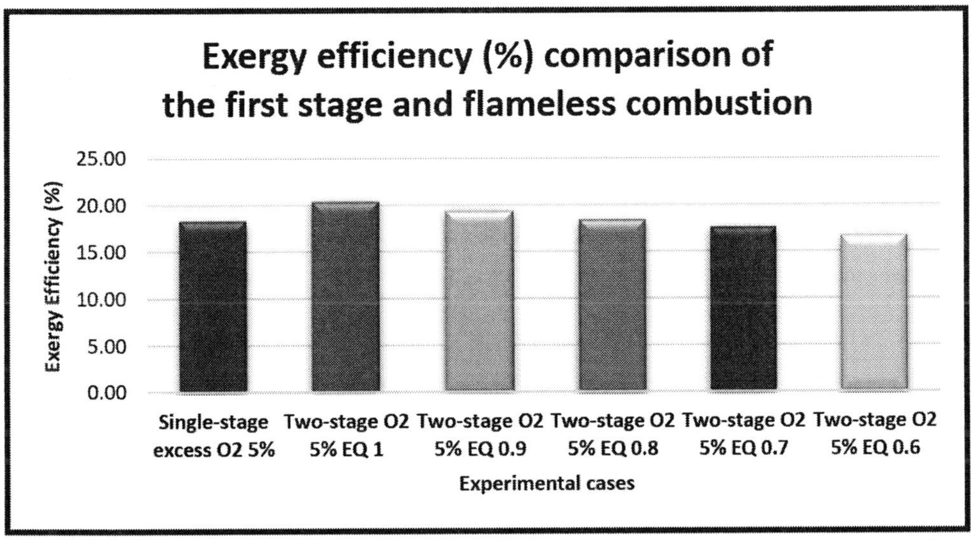

Figure 7. Comparison of the efficiency of exergy between the single-stage and the distributed combustor with the equivalent ratio variation

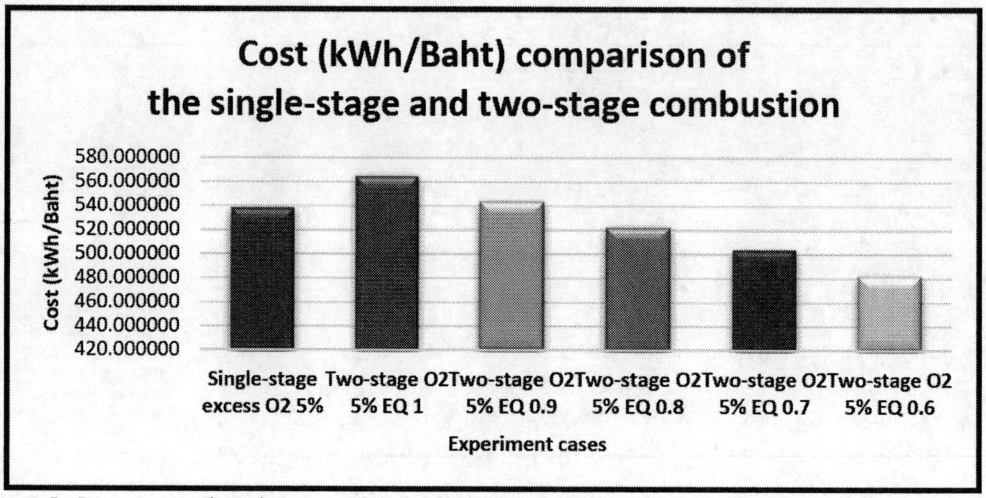

Figure 8. Cost comparison between the single-stage combustor and the distributed combustor with the equivalent ratio variation

4. Conclusion

This study set out to combine the LPG distributed combustor to the B7 biodiesel combustor to improve the first law and the second law efficiencies. This study has identified the advantage of using distributed combustion as second-stage. The research has also shown that emissions, NOx and CO are low values especially NOx concentration is nearly zero. The results of this research support the idea that the first law and the second law efficiencies are improved, and it can lead to lower energy cost at the same firing rate. These data suggest that the effects of the equivalent ratio are not recommended for energy cost saving because when the heat release is decreased, it can cause lower efficiency. The present study lays the groundwork for future research into applying the distributed combustion to the conventional single-stage combustion system.

Acknowledgement

The authors sincerely thank to the "Development of Machinery and Industrial Equipment (DMIE)" research center and Department of Mechanical and Aerospace Engineering (MAE), King Mongkut's University of Technology North Bangkok (KMUTNB) for the instrumental supports and facilities, and to Thailand Research Fund (TRF) and National research council of Thailand (NRCT) for giving both opportunity and funding for this research.

We also would like to thank Thailand Institute of Scientific and Technology Research (TISTR) for a student's scholarship in the partnership program in production of graduates in Master's degree and Doctoral degree between Thailand Institute of Scientific and Technology Research (TISTR) and educational institutions.

References

[1] Ferreira P Johson É Leão D A and Fernandes M 2019 *Physica A* **517** 86–96
[2] Ju K Su B Zhou D Zhou P and Zhang Y 2015 *Energy* **93** 1353-1360
[3] Zimmer A Cazarolli J Teixeira R M Viscardi S L C Cavalcanti E S H Gerbase A E and Bento F M 2013 *Fuel* **112** 153-162

[4] Graboski M S and McCormick R L 1998 *Progress in Energy Combustion and Science* **24** 125-164
[5] Wangwattana S 2001 *Proc. Conference on research direction and development of biodiesel for commerce (Bangkok)* (Thailand)
[6] Wang D Zhu G Li Z and Xia C 2019 *Fuel* **237**(July 2018) 833–839
[7] Han X Yang Z Wang M Tjong J and Zheng M 2017 *Applied Energy* **198** 347–359
[8] Shu J Fu J Liu J Ma Y Wang S Deng B and Zeng D 2019 *Applied Energy* **233** 182–195
[9] Liu F Yang Z Li Y and Wu H 2019 *Applied Thermal Engineering* **148** 1233–1245
[10] Wang Z Wu S Huang Y Huang S Shi S Cheng X and Huang R 2018 *Fuel* **231** 438–448
[11] Divekar P S Chen X Tjong J and Zheng M 2016 *Energy Conversion and Management* **112** 369–381
[12] Abuelnuor A A A Wahid M A Hosseini S E Saat A Saqr K M Sait H H and Osman M 2014 *Renew. Sustain. Energy Rev.* **33** 363–370
[13] Wünning J G 2005 Distributed Combustion and Its Applications 1–12
[14] Gao X Duan F Lim S C and Yip M S 2013 *Energy* **59** 559–569
[15] Turns S R 2012 *An introduction to combustion concept and application* (New York:McGraw-Hill) 170-172
[16] Arghode V K and Gupta A K 2010 *Applied Energy* **87**(5) 1631–1640
[17] Duwig C Li B Li Z S and Alden M 2012 *Combustion and Flame* **159** 306–316
[18] Tsuji H Gupta A Hasegawa T Katsuki M Kishimoto K and Morita M 2002 *High Temperature Air Combustion* (Boca Raton:CRC Press)
[19] Colourado A F Herrera B and Amell A 2010 *Bioresour. Technol.* **101**(7) 2443-2449
[20] Sanchez M Cadavid F and Amell A 2013 *Applied Energy* **111** 240–246
[21] Reddy V M Katoch A Roberts W L and Kumar S 2014 *Proc. Combust. Inst.*
[22] Reddy V M and Kumar S 2013 *Propulsion and Power Research* **2**(2) 139–147
[23] Rafidi N Blasiak W and Gupta A K 2008 *J. Engineering for Gas Turbine and Power, Trans. of ASME* **130** 023001-8
[24] Hosseini S E and Wahid M A 2014 *Energy Convers. Manag.* **86** 1154–1163
[25] Hosseini S E Bagheri G and Wahid M A 2014 *Energy Convers. Manag.* **81** 41–50
[26] Hosseini S E and Wahid M A 2013 *Energy Conversion and Management* **74** 426–432
[27] Komonhirun S Yongyingsakthavorn P and Nontakeaw U 2017 *Material Science and Engineering* **297** 012071
[28] Dincer I and Rosen M A 2012 *EXERGY: energy, environment and sustainable development* (Google eBook)
[29] Cengel Y A and Boles M A 2015 *Thermodynamic an engineering approach* (New York:McGraw-Hill) 760–793
[30] Holms T Gas Carburizing and Carbonitriding in Furnace Atmospheres (Sweden:AGA AB) **1** 3–18

IOP Conf. Series: Materials Science and Engineering **886** (2020) 012002 doi:10.1088/1757-899X/886/1/012002

The Potential Study of Coffee, Acacia Wood and Corn Cob Residues to produce Biomass Pellets Fuel

Benya Kasantiukl[1,*]

[1]Mechanical Engineering Department, Faculty o Engineering, Kasetsart University, Kamphaeng Saen Campus, Nakornpathom, 73140, Thailand

* Corresponding Author: benya.k@ku.th

Abstract

The objective of this research is to study the technical and economic feasibility of biomass pellet production from coffee, corncob and acacia wood residues for using as alternative energy. The biomass pellet in this research was produced by mixing three residues with the amount of tapioca starch not more than 20% by mass. The fuel properties of pellet were analyzed according to PFI and ASTM standards. The test results showed that the heating value of the pellet from coffee, acacia wood and corncob residues are 23.56, 20.43 and 19.46 MJ/kg, respectively. The moisture, ash, and durability index are 0.15-0.43, 0.954-0.957, and 95.37-98.52 %, respectively which are according to the standard. The bulk density is 221.04-277.93 kg/m3. The economic feasibility of pellet production was analyzed using Net Present Value (NPV), Reduction Rate (RR) and payback period (PB). The analysis results show that the production of biomass pellets from the residues of coffee, acacia wood and corncobs are economically feasible with RR = 27.56, 17.88, 14.61%, NPV=186,837 ฿, 91,684 ฿, 62,119 ฿, and payback period = 4 years and 7 months, 5 years and 4 months, 6 years and 8 months, respectively. Moreover, the sensitivity analysis results show that the most influential risk factors of the pellet production are fuel price and the quantity of fuel produced daily, followed by labor wages, production days a year, and price of machines.

1. Introduction

Energy is one of the important factors affecting today's lifestyles of people. Energy consumption in Thailand has continuously increased. In 2017, Thailand had final energy consumption of 83,691 ktoe. This includes traditional renewable energy (firewood wood and rice husk) at the total of about 6.23 per cent of the total final energy consumption and from renewable energy (agricultural waste and biomass) at the total of about 9.16 per cent of the total final energy consumption [1]. Of particular concern is the use of firewood-type fuel may reduce forest resources. At present, Thailand has only the forest area remaining about 30 per cent of the total land area [2]. Therefore, it is necessary to find alternative sources of energy to reduce such problems. A renewable energy source that is easy to find and has a high potential in Thailand is biomass energy from agricultural waste [3]. The use of agricultural residues to produce biomass can help reduce the use of firewood fuel and can help solve the problem of waste disposal and agricultural waste materials at the same time. In addition, the use of energy from biomass is also environmentally friendly without affecting global warming because the amount of carbon dioxide emissions throughout the life cycle is zero or carbon neutral [4]. The biomass is, therefore, an appropriate source of energy for sustainable development. The processing of agricultural waste which is popular to produce sources the renewable energy is extrusion into biomass pellet fuel. This is a method that is not complicated. It can be easily done in the community to get the fuel with more inflammable ignition point than firewood and charcoal and still be clean energy [5]. Various types of agricultural waste materials such as rice straw, tree branches, leaves and corncob [6], corncob and shell [7], coffee grounds and tea residue [8], rice husk, sugar cane bagasse and sawdust [9] have been investigated to produce as compressed biomass fuel. The properties of this compressed fuel are good to be used as the fuel to replace the firewood and charcoal.

In this study, therefore, such agricultural waste as spent coffee ground (SCG), acacia wood and sawdust mixed corncobs was taken to produce the biomass compressed pellet fuel to determine the feasibility consistent with the Pellet Fuel Institute (PFI) Standards [10], to study the properties to be used as renewable heat energy instead of using charcoal and to reduce the cost of buying fuel from the seller. This not only helps reduce environmental problems but also gets a benefit from agricultural waste and increases the value of the waste materials. Moreover, the biomass compressed pellet fuel is still considered as an alternative and sustainable source of renewable energy.

2. Materials and Methods

This research was divided into 4 steps as follows: analysis of raw materials used in pellet fuel compression, pellet fuel compression, analysis of the technical production of pellet fuel and the economic feasibility study of the production of compressed fuel from coffee grounds, acacia wood and corncobs mixed with sawdust. Each step has the following details.

2.1 Analyses of the Raw Materials Used in the Fuel Pellet Compression

To compress the fuel pellets in this research, an extrusion technique [11] through a screw compressor was used; furthermore, tapioca starch was used as a binder because it is high-calorific and able to stick the raw materials into a single material with good physical properties [12]. In this step, the properties of the raw materials used for compressed pellet fuel including coffee grounds, acacia wood and corncob mixed with sawdust together with tapioca starch as a binder were analyzed.

2.2 Fuel Pellet Compression

The process is as follows:

2.2.1 The biomass was chopped to a size of not more than 2 mm as shown in Figure 1a.

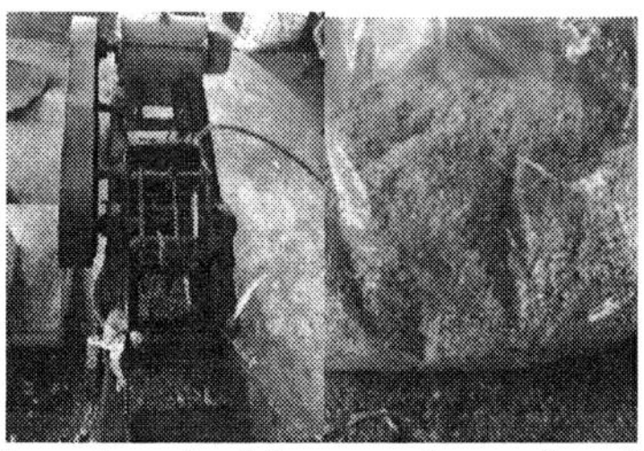
Figure 1a. Chopping the biomass through a 2-mm sieve

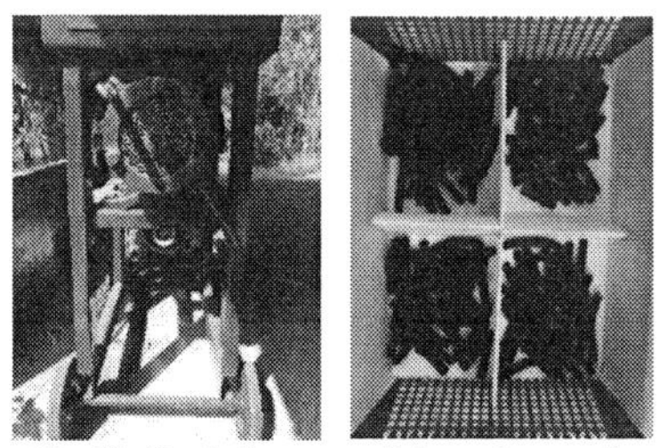
Figure 1b. Pelletizing machine & products

2.2.2 The biomass was formed into 3 samples of chopstick-like fuel pellets as follows:
Sample 1, coffee grounds: 10-kg coffee grounds 1-kg tapioca starch and 5-litre water
Sample 2, corncobs: 5-kg corncobs, 5-kg sawdust, 1-kg tapioca starch and 5-litre water

Sample 3, acacia wood: 10-kg acacia wood, 1-kg tapioca starch and 5-litre water

2.2.3 Then, the fuel pellets formed from the biomass (see Figure 2a) into were taken to be baked to wick moisture away to be more solid and not easily broken while packing or moving. Either drying under the sun (as shown in Figure 2a) or baking in the oven may be used (see Figure 2b).

Figure 2a. Fuel pellets drying under the sun

Figure 2b. Baking the fuel pellets at 103°C

2.2.4 In Fig.3 the fuel pellets were packed in bags to prevent moisture and waiting for the testing of the properties according to the standards.

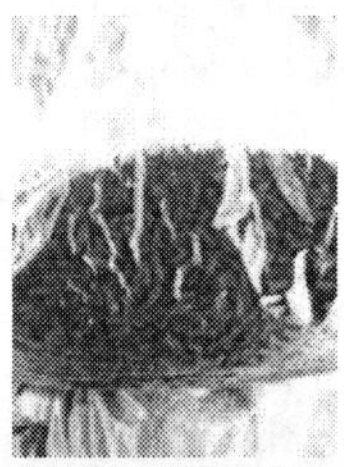

Figure 3. Packing the fuel pellet formed from the biomass

2.3 Analyze of Technical Suitability of the Fuel Pellet Production
According to the Pellet Fuel Institute (henceforth PFI) Standards and the ASTM, in each test, 13.04kg. of compressed fuel biomass is prescribed to be divided and tested according to the ASTM C 702-98 [13], as shown in Figure 4 below:

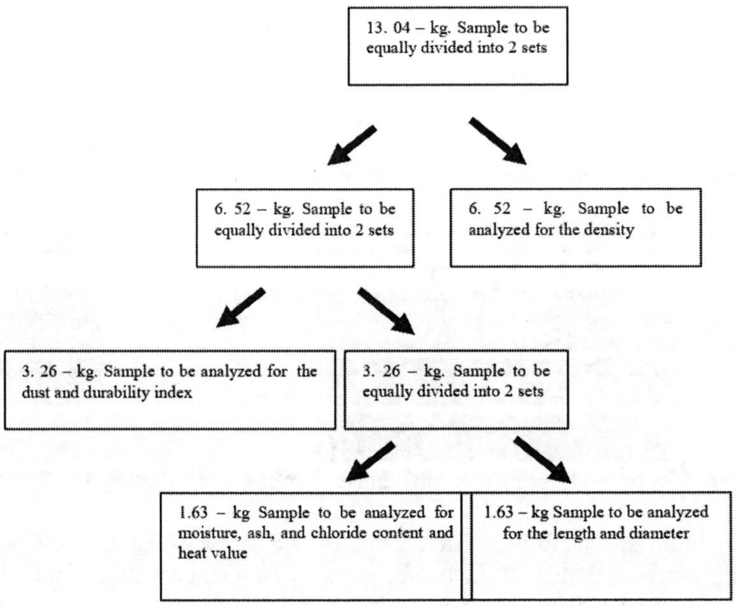

Figure 4. Divided Tests of the Biomass Fuel according to the ASTM C 702-98

IOP Conf. Series: Materials Science and Engineering **886** (2020) 012002 doi:10.1088/1757-899X/886/1/012002

The analysis of the fuel pellet properties are as follows:
1. Bulk density according to the ASTM C 702-98 [13]
2. Length and diameter according to the PFI Standard [10]
3. Durability according to the PFI Standard [10]
4. Dust percentage according to the PFI Standard [10]
5. Ash content according to the ASTM D 3174 [14]
6. Moisture values according to the ASTM D 3173 [15]
7. Chloride content according to the PFI Standard [10]
8. Heating values according to the ASTM D 5865 [16]

2.4 Analysis of Economic Value of the Fuel Pellet Production from the Coffee Grounds, Acacia Wood and Sawdust-mixed Corncobs

Apart from its fuel properties, analysis of the economic value of the fuel pellet production from the coffee grounds, acacia wood and sawdust-mixed corncobs were carried out to determine the practical and sustainable feasibility for a community production project. Thus, the rates of return from the production project were assessed whether the benefits obtained from the production are sufficient/worth the investment and management cost or not; whether the production project can be effectively and sustainably implemented in the community or not. In addition, the fuel pellet production was also analyzed to find out the risk factors.

2.4.1 The rates of return from the production project consist of indicators: Net Present Value, NPV, Reduction Rate, RR and Payback Period, PB, as the following details:

(1)The Net Present Value, NPV refers to the cash flow projection throughout the production project at the rate of return as required, which can be found out through the following eq.1 [18]:

$$NPV = \sum_{t=1}^{n} \frac{CF_t}{(1+i)^t} \quad (1)$$

Where n is the production period (year); t is the respective number of the production period (year) $t = 1, 2, 3, \ldots, n$, ; CF_t is the net cash(in baht) flow at year t = received cash flow at year t minus spent cash flow at year t ; and i is the reduction rate or the rate of return (%) as required. Criteria in judgment are: The investment in the production yields a good value when NPV ≥ 0 and it is not worth when NPV <0.

(2)The Reduction Rate, RR refers to the rate which makes the net cash flow throughout the production project to be zero and it can be found out through the following eq.2 [18]:

$$0 = \sum_{t=1}^{n} \frac{CF_t}{(1+IRR)^t} \quad (2)$$

Where n is the production period (year); t is the respective number of the production period (year) $t = 1, 2, 3, \ldots, n$,; CF_t is the net cash(in baht) flow at year t = received cash flow at year t minus spent cash flow at year t ; and RR is the Reduction Rate (%). Criteria in judgment are: The investment in the production yields a good value when RR $\geq$ cost of the production project and it is not worth when RR $<$ cost of the production project. In this research, the cost of the production project is the current interest rate of general commercial banks.

(3)The Payback Period, PB refers to the period where the cumulative received cash flow is equal to the spent cost of the production project. It is an indicator to tell the risk status of a project. Projects that have short PBs will have lower risks than those with longer PBs. Criteria in judgment are: The project will be accepted when PB $\leq$ target PB and not accepted when the PB> the target PB. This shows that the investment of the project is not worth.

2.4.1 The Sensitivity Analysis (SA) was used to analyse the risk in the production project. This method is particularly useful when a project does not meet the expectations. And, it is to ensure that there are some factors affect the benefits and expenses of the production project.

Besides, it was done to assess whether it will still be worthwhile to proceed with the project or not. The SA is used to analyze the changes in net cash flow due to various factors, as the following main steps.

(1) All variables are to be well defined.

(2) The scope of possible data for each variable is also to be well defined.

(3) NPV, RR and PB are to be calculated one by one on condition that variable value is changed one by one while all other variables are constant. Then, the relationship of NPV, RR and PB to each variable will appear.

In this research, the SA was carried out through five factors including fuel price, amount of fuel produced per day, machinery price, labor cost and the number of production days per year.

3. Results and Discussion

3.1. Results from the Fuel Properties

To study properties of the compressed pellet biomass from coffee grounds, acacia wood sawdust-mixed corncobs were used. The coffee grounds, acacia wood and sawdust-mixed corncobs were compressed and formed into biomass fuel sticks, which were used to test according to the international PFI Standards. It is apparent from the test that the moisture and ash contents and length did not exceed 1 per cent and met all three standard criteria. The durability achieved the standard whereas the bulk density was below 615.59kg/m^3, and is considered as fell short of the standard. Further analysis showed that the dust value, which was greater than 1 per cent, was higher than the standard value. The diameter met the standards, ranging from 5.84 to 7.25 mm. The chloride values of all three types of the biomass that were below 300 ppm; also reached the standards. The heating values of the coffee grounds, acacia wood and sawdust-mixed corncobs were found as 23.56 MJ/kg, 20.43 MJ/kg and 19.46 MJ/kg respectively (see Figure 5 and 6).

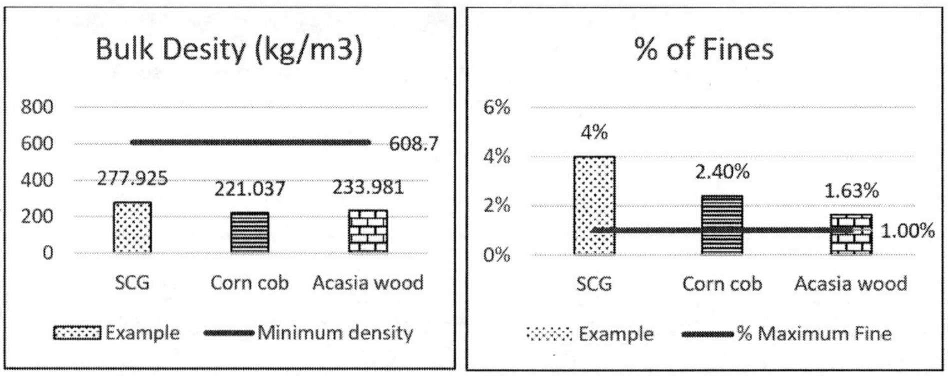

Figure 5. The bulk density and fine properties of the fuel pellets biomass from coffee grounds, acacia wood and corncobs

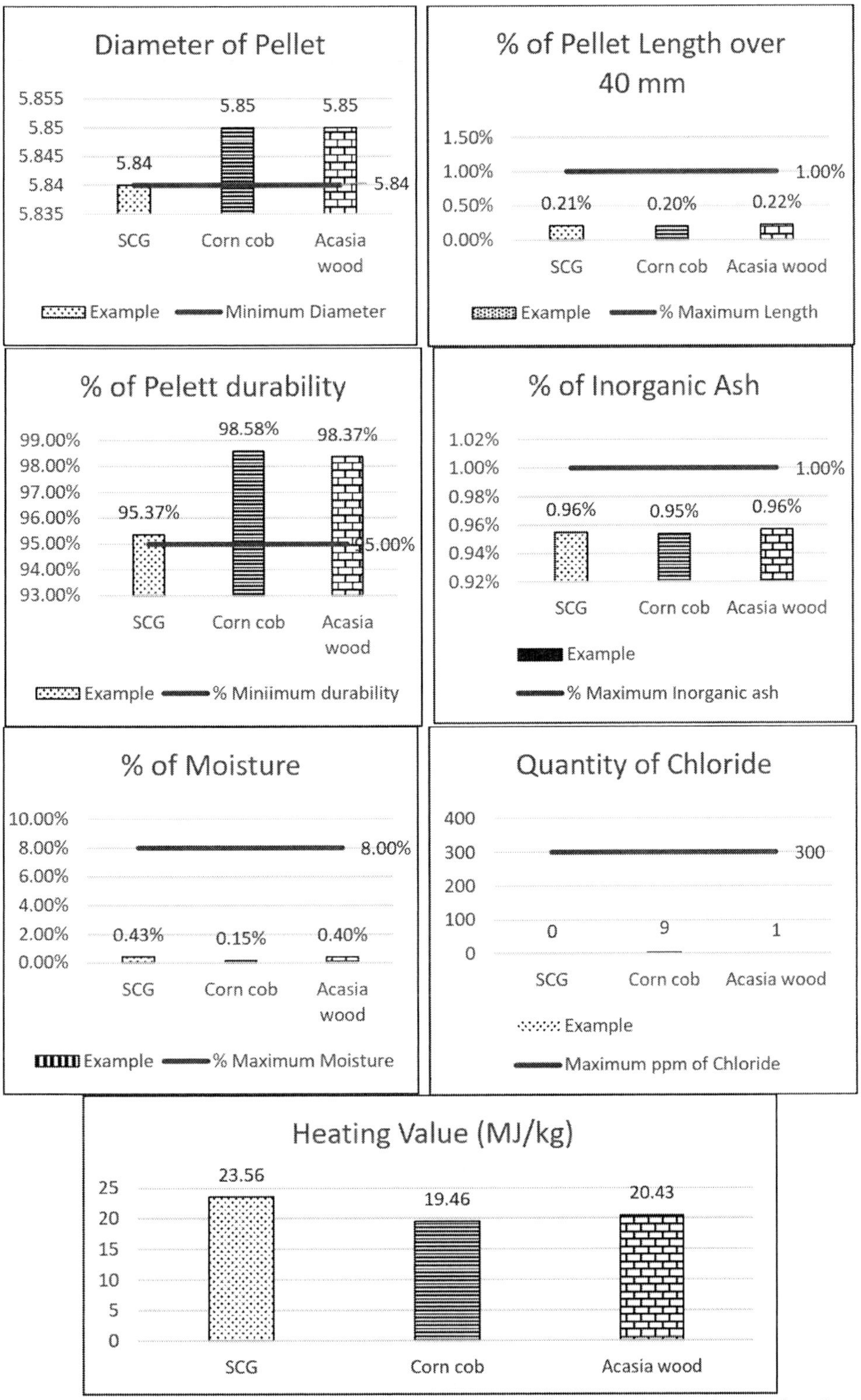

Figure 6. The properties of the fuel pellets biomass from coffee grounds, acacia wood and corncobs

IOP Conf. Series: Materials Science and Engineering **886** (2020) 012002 doi:10.1088/1757-899X/886/1/012002

As mentioned above, the bulk density and dust value were not up to the PFI Standards. This may be because of the improper condition of materials before pelletization. The mixture ratio that the pellet biomass can be successfully made is 1 kilogram biomass per 0.2 liter tapioca starch. The heating values of the pellet fuel were ranged from 19.46 to 23.56 MJ / kg, and appeared to be higher than that of firewood, which is about 15.98 MJ / kg [1]. Regarding the economic worthwhile analyses, the focus is on the cost-effectiveness to replace the fuel from eucalyptus wood and the non-commercial purpose.

3.2 Economic Value Analyses

The economic value analyses of the fuel pellet production from the coffee grounds, acacia wood and sawdust-mixed corncobs were conducted based on the effective pellet compression ratio of the coffee grounds, acacia wood and sawdust-mixed corncobs with a maximum of 0.2 liter/kg of the tapioca starch. The key indicators including the Net Present Value (NPV), the Reduction Rate, the Payback Period (PB) were analyzed for the returns of the production project. Then, the risk analysis of the project investment was carried out by the SA of the returns for the various factors.

Table 1, provides the base data for the economic value analyses. As shown in Table 2 and Table 4, the base data are used to find income, investment cost, expenses and net cash flows throughout the production project. In addition, through the net cash flows throughout the production project, the NPV, RR and PB can be calculated and can be seen from Table 6.

Table 1. Base data for the economic value analyses

Consideration Topics	**Base data for economic value analyses**
Benefits from the production project	-Using the biomass pellet fuel produced from coffee grounds, acacia wood and sawdust-mixed corncobs can replace the normal household fuel and the biomass pellet fuel production is non-commercial. -Referring to Eucalyptus charcoal, it costs 7 baht/kilogram and its heating value is 29.29 MJ/kg. - so biomass pellet fuel from coffee grounds with heating value 23.56 MJ / kg in this research will be able to replace the equivalent cost = 7x23.56/29.29 = 5.63 baht/kg. - so biomass pellet fuel from acacia wood with heating value 20.43 MJ / kg in this research will be able to replace the equivalent cost = 7x20.43/29.29 = 4.88 baht/kg. - so biomass pellet fuel from sawdust-mixed corncobs with heating value 19.46 MJ/kg in this research will be able to replace the equivalent cost = 7x19.46/29.29 = 4.65 baht/kg.
Investment cost	-A grinding machine costs 35,000 baht and screw compressor costs 120,000baht with manufacturing power of 40 kilograms/hour.
Expenses	-Labor cost 300 bah /day/person =600 baht/day -Electric power cost = 0.16 baht/kilogram of fuel pellets -Water supply cost = 0.0013 baht/kilogram of fuel pellets -Tapioca starch cost = 0.78 baht/kilogram of fuel pellets
Amount of fuel pellets(produced / day)	-300 kilogram / day
Number of Production Days (day / year)	-52 days/year(collecting raw materials in the community and producing once a week)
Inflation rate	-3% (average inflation rate for the past 10 years) [17]
Reduction Rate required (discounted rate)	Using interest rate = 7%(referring to current MLR (Minimum Loan Rate as of June 2019)
Production project lifetime	- 10 years

Table 2. The cash flow projection from the coffee grounds

			Year of	1	2	3	4	5	6	7	8	9	10
Benefit	5.63	฿/kg		90,418.00	93,130.54	95,924.46	98,802.19	101,766.26	104,819.24	107,963.82	111,202.74	114,538.82	117,974.98
	300	kg/วัน											
Total income				90,418.00	93,130.54	95,924.46	98,802.19	101,766.26	104,819.24	107,963.82	111,202.74	114,538.82	117,974.98
Labor cost	2	฿/kg		32,136.00	33,100.08	34,093.08	35,115.87	36,169.35	37,254.43	38,372.06	39,523.23	40,708.92	41,930.19
Electricity cost	0.16	฿/kg		2,570.88	2,648.01	2,727.45	2,809.27	2,893.55	2,980.35	3,069.77	3,161.86	3,256.71	3,354.42
Material cost	0	฿/kg		0	0	0	0	0	0	0	0	0	0
Starch cost	0.78	฿/kg		12,533.04	12,909.03	13,296.30	13,695.19	14,106.05	14,529.23	14,965.11	15,414.06	15,876.48	16,352.77
Water cost	0.0013	฿/kg		20.8884	21.52	22.16	22.83	23.51	24.22	24.94	25.69	26.46	27.25
Total expense	2.9413	฿/kg		47,260.81	48,678.63	50,138.99	51,643.16	53,192.46	54,788.23	56,431.88	58,124.83	59,868.58	61,664.64
Investment													
Grinding machine	35000	฿											
Compression machine	120000	฿											
Net cash flow			-155,000.00	43,157.19	44,451.91	45,785.47	47,159.03	48,573.80	50,031.01	51,531.94	53,077.90	54,670.24	56,310.35
NPV				-114,666.18	-75,840.16	-38,465.58	-2,488.18	32,144.27	65,482.04	97,573.55	128,465.37	158,202.36	186,827.68

Table 3. The cash flow projection from the acacia wood

			Year of	1	2	3	4	5	6	7	8	9	10
Benefit	4.88	฿/kg		78,405.76	80,757.93	83,180.67	85,676.09	88,246.38	90,893.77	93,620.58	96,429.20	99,322.07	102,301.74
	300	kg/วัน											
Total income				78,405.76	80,757.93	83,180.67	85,676.09	88,246.38	90,893.77	93,620.58	96,429.20	99,322.07	102,301.74
Labor cost	2	฿/kg		32,136.00	33,100.08	34,093.08	35,115.87	36,169.35	37,254.43	38,372.06	39,523.23	40,708.92	41,930.19
Electricity cost	0.16	฿/kg		2,570.88	2,648.01	2,727.45	2,809.27	2,893.55	2,980.35	3,069.77	3,161.86	3,256.71	3,354.42
Material cost	0	฿/kg		0	0	0	0	0	0	0	0	0	0
starch cost	0.78	฿/kg		12,533.04	12,909.03	13,296.30	13,695.19	14,106.05	14,529.23	14,965.11	15,414.06	15,876.48	16,352.77
Water cost	0.0013	฿/kg		20.8884	21.52	22.16	22.83	23.51	24.22	24.94	25.69	26.46	27.25
Total expense	2.9413	฿/kg		47,260.81	48,678.63	50,138.99	51,643.16	53,192.46	54,788.23	56,431.88	58,124.83	59,868.58	61,664.64
Investment													
Grinding machine	35000	฿											
Compression machine	120000	฿											
Net cash flow			-155,000.00	31,144.95	32,079.30	33,041.68	34,032.93	35,053.92	36,105.54	37,188.70	38,304.36	39,453.50	40,637.10
NPV				-125,892.57	-97,873.26	-70,901.41	-44,937.85	-19,944.89	4,113.76	27,273.01	49,566.50	71,026.59	91,684.43

Table 4. The cash flow projection from the sawdust-mixed corncobs

			Year of	1	2	3	4	5	6	7	8	9	10
Benefit	4.65	฿/kg		74,683.12	76,923.61	79,231.32	81,608.26	84,056.51	86,578.20	89,175.55	91,850.82	94,606.34	97,444.53
	300	kg/วัน											
Total income				74,683.12	76,923.61	79,231.32	81,608.26	84,056.51	86,578.20	89,175.55	91,850.82	94,606.34	97,444.53
Labor cost	2	฿/kg		32,136.00	33,100.08	34,093.08	35,115.87	36,169.35	37,254.43	38,372.06	39,523.23	40,708.92	41,930.19
Electricity cost	0.16	฿/kg		2,570.88	2,648.01	2,727.45	2,809.27	2,893.55	2,980.35	3,069.77	3,161.86	3,256.71	3,354.42
Material cost	0	฿/kg		0	0	0	0	0	0	0	0	0	0
starch cost	0.78	฿/kg		12,533.04	12,909.03	13,296.30	13,695.19	14,106.05	14,529.23	14,965.11	15,414.06	15,876.48	16,352.77
Water cost	0.0013	฿/kg		20.8884	21.52	22.16	22.83	23.51	24.22	24.94	25.69	26.46	27.25
Total expense	2.9413	฿/kg		47,260.81	48,678.63	50,138.99	51,643.16	53,192.46	54,788.23	56,431.88	58,124.83	59,868.58	61,664.64
Investment													
Grinding machine	35000	฿											
Compression machine	120000	฿											
Net cash flow			-155,000.00	27,422.31	28,244.98	29,092.33	29,965.10	30,864.05	31,789.97	32,743.67	33,725.98	34,737.76	35,779.90
NPV				-129,371.67	-104,701.41	-80,953.41	-58,093.17	-36,087.53	-14,904.53	5,486.58	25,115.41	44,010.46	62,199.14

Table 5. Results from the economic value analyses

Method	Type of fuel pallet	Value	Note
NPV	coffee grounds	186,837 ฿	NPV > 0 shows that the project investment is economically worthwhile.
	acacia wood	91,684 ฿	
	sawdust-mixed corncobs	62,119 ฿	
RR	coffee grounds	27.56%	RR > 7% interest rate shows that the project investment is economically worthwhile.
	acacia wood	17.88%	
	sawdust-mixed corncobs	14.61%	
PB	coffee grounds	4 years and 7 month	PB < Machine lifetime indicates that the production project is at risk at an acceptable level.
	acacia wood	5 years and 4 month	
	sawdust-mixed corncobs	6 years and 8 month	

As can be seen from Table 5 (above), it can be concluded that the production of the biomass pellet fuel is economically worthwhile because all three indicators have achieved the standards.

Results are shown in Tables 6, -10. The results are obtained from risk analysis by the sensitivity analysis of returns from biomass pellet fuel production; from all three types of raw materials, when various factors have changed from the expected results.

From Tables 6, –10, it can be seen that the return appears when the fuel price, amount of production per day (day), labor cost, number of production days (day/year), and machinery price have changed, in comparison with the returns in base cases.

Table 6. Results from the sensitivity analysis: fuel price

Fuel price (baht / kg.)		RR	NPV
coffee grounds	5.63 (base price)	27.56%	186,828
	4.78 (-15%)	16.54%	79,404
	4.16 (-26%)	7%	0
acacia wood	4.88 (base price)	17.88%	91,684
	4.39 (-10%)	10.77%	29,583
	4.16 (-15%)	7%	0
corncobs mixing sawdust	4.65 (base price)	14.61%	62,199
	4.42 (-5%)	11.14%	32,623
	4.16 (-10%)	7%	0

Table 6, above illustrates if the fuel prices decrease, the RR and NPV decrease. Therefore, the fuel prices decrease by 26% (coffee grounds), 15% (acacia wood) and 10% (sawdust-mixed corncobs), respectively. It was observed that the lowest levels of the return for each type of biomass are acceptable. However, if the prices are lower than these, the production project will not be worth to proceed.

Table 7. Results from the sensitivity analysis: amount of fuel produced per day

The amount of fuel produced per day (day)		**RR**	**NPV**
coffee grounds	300 (base price)	27.56%	186,828
	285 (-15%)	16.54%	79,404
	222 (-26%)	7%	0
acacia wood	300 (base price)	17.88%	91,684
	270 (-10%)	10.77%	29,583
	355 (-15%)	7%	0
corncob s mixing sawdust	300 (base price)	14.61%	62,199
	285 (-5%)	11.14%	32,623
	270 (-10%)	7%	0

Table 7 shows if the amounts of fuel produced per day decrease, the RR and NPV also decrease. Thus, if the amounts of fuel produced per day decrease by 26% (coffee grounds), 15% (acacia wood) and 10% (sawdust-mixed corncobs), respectively. We can see that the lowest levels that the return for each type of biomass is acceptable. If the amounts of fuel produced per day are lower than these, the production project will not be worth to proceed.

Table 8. Results from the sensitivity analysis: machinery price

machinery price(baht)		**RR**	**NPV**
coffee grounds	155,000 (base price)	27.56%	186,828
	119,231 (+30%)	19.62	140328
	341,000 (+120%)	7%	0
acacia wood	155,000 (base price)	17.88%	91,684
	119,231(+30%)	11.39%	45,184
	246,450 (+59%)	7%	0
corncobs mixing sawdust	155,000 (base price)	14.61%	62,199
	186,000 (+20%)	10.33%	31,199
	217,000 (+40%)	7%	0

From Table 8, if the machinery prices increase, the RR and NPV decrease. In general, therefore, it seems that the machinery prices increase by 120% (coffee grounds), 59% (acacia wood) and 40% (sawdust-mixed corncobs), respectively. These findings suggest that the highest levels of the return for each type of biomass are acceptable. If the machinery prices are higher than the specified price, the production project will not be worth to proceed.

Table 9. Results from the sensitivity analysis: labor cost

labor cost (baht)		**RR**	**NPV**
coffee grounds	300 (base price)	27.56%	186,828
	390 (+30%)	19.88%	110,467
	519 (+73%)	7%	0
acacia wood	300 (base price)	17.88%	91,684
	360 (+20%)	12.12/5	40,778
	408 (+36%)	7%	0
corncobs mixing sawdust	300 (base price)	14.61%	62,199
	330 (+10%)	11.64%	36,746
	372 (+24%)	7%	0

As Table 9 shows, if the labor costs increase, the RR and NPV decrease. Thus, the labor costs increase by 73% (coffee grounds), 36% (acacia wood) and 34% (sawdust-mixed corncobs), respectively. The results of this study indicate that the highest levels that the return for each type of biomass are acceptable. If the labor costs are higher than this specified price, the production project will not be worth to proceed.

Table 10. Results from the sensitivity analysis: number of production days (day / year)

Number of Production Days (day / year)		**RR**	**NPV**
coffee grounds	52 (base price)	27.56%	186,828
	42 (-20%)	12.45%	43,596
	38 (-26%)	7%	0
acacia wood	52 (base price)	17.88%	91,684
	47 (-10%)	10.77%	29,583
	43 (-15%)	7%	0
corncobs mixing sawdust	52 (base price)	14.61%	62,199
	49 (-5%)	11.14%	32623
	47 (-10%)	7.4%	3,046

From Table 10, if the numbers of production days (day/year) decrease, the RR and NPV also decrease. Thus, the numbers of production days (day/year) decreased by 26% (coffee grounds), 15% (acacia wood) and 10% (sawdust-mixed corncobs), respectively. This result may be explained by the fact that the lowest levels that the return for each type of biomass are acceptable. If the numbers of production days (day/year) are fewer than specified, the production project will not be worth to proceed.

According to the risk analysis of the production of the biomass fuel pellets from coffee grounds acacia wood and sawdust-mixed corncobs, this research found that the returns of the production project are the most sensitive to the fuel prices, the amounts of fuel produced and the numbers of production days. Thus, the fuel prices, amounts of fuel produced and numbers of production days increase by 10% (sawdust-mixed corncobs), 15% (acacia wood) and 26% (coffee grounds) for the biomass fuel pellets. Besides, the production project's rates of return cannot be economically worthwhile (RR < 7 %, NPV < 0). Another important finding was that some factors can also have an impact on the production project i.e. the labor costs and machinery prices. Therefore, the management of fuel production for renewable energy should be done in communities especially with sufficient amounts of these products: coffee, acacia wood and corncobs grounds (approximately 300 kg) before compressing the fuel pellets. These results are likely to be related to suitability of the compress fuel pellets and adequate amounts of the raw materials with appropriate frequency rather than doing it frequently in small amounts.

4. Conclusion

This study has identified the use of coffee grounds acacia wood and sawdust-mixed corncobs to produce biomass pellet fuel with extrusion method using tapioca starch (e.g. binder). The research has also shown that fuel pellets can be extruded well. This study has found that generally, the ratio with the best fuel properties is 1 kg of acacia wood and sawdust-mixed corncobs per 0.2 liters of tapioca starch. These experiments confirmed that the coffee grounds, acacia wood and sawdust-mixed corncobs have the heating values at 23.56 MJ/kg, 20.43 MJ/kg and 19.46 MJ/kg, respectively. Besides, their moisture content and ash content was at 0.15-0.43 at 0.954-.0957%. While the durability index was at 95.37-98.52%. The investigation of the properties has shown that they are sufficient to be used as a renewable fuel and can replace the firewood in the community. However, based on the PFI's standard diameters ranging from 5.84 to 7.25 mm., the biomass pellet fuel's diameter, therefore, meets the standard. The chloride contents of all 3-type biomass which are less than 300 ppm; also meet the standards. The values of the bulk density range from 221.04 to 277.93 kg/m3 which less than 615.59 kg/m3, below the standard and the dust value which is greater than 1%. This may due to the improper condition of materials before pelletization. The findings of this study suggest that a more suitable binder should be acquired to increase the values of the bulk density and can be able to reduce the dust value to meet the standards. The economic feasibility of pellet production was analyzed using the Net Present Value (NPV), Reduction Rate (RR) and Payback Period (PB). The analysis results show that the production of biomass pellets from coffee, acacia wood and corncob Residues are economically feasible with RR = 27.56, 17.88, 14.61%, NPV = 186,837฿, 91,684฿, 62,119 ฿. The findings reported here shed new light on the returns which are worth the investment and can be implemented in practice. Meanwhile, the Payback Period of the fuel pellets from coffee, acacia wood and corncob residues are 4 years and 7 months, 5 years and 4 months, and 6 years and 8 months respectively. These periods are considered not too long as compared to the normal lifetime of the machines resulting in a low risk for loss for the production project. Based on the Sensitivity Analysis, it was found that the returns of the production project are most sensitive to the fuel prices, amounts of fuel produced and numbers of production days followed by the labor costs and machinery prices, respectively. Therefore, to manage the production of this fuel as renewable energy in the community or selling in the industrial sector, various raw materials should be collected adequately before the pellet fuel is processed. In general, therefore, it seems that this method is more suitable than compressing fuel pellets in small amounts frequently. Processing coffee grounds acacia wood and sawdust-mixed corncobs into biomass pellet fuel is taking agricultural waste materials to produce as alternative energy for communities or for further selling in the industrial sector. The general properties of the fuel can be used as an effective fuel. This is an effective way to use resources that can help solve the problem of eliminating agricultural waste and help reduce the impact on the environment at the same time.

Acknowledgements
The authors would like to thank Faculty of Engineering, Kampheang Saen Campus for financial support, and we would like to thank Mr Natthawut N. and Mr Nopparat M., for their helpful advice on machine fabrication and performing the test.

References

[1] Department of Alternative Energy and Energy Conservation 2017 *Energy Situation of Thailand January - December 2017*

[2] Sueb Nakhasathien Foundation 2014 *Thailand Forestry Report 2014* http://www.seub.or.th/index.php?option=com_content&view=article&id=1287:seubnews&catid=112:clipnews&Itemid=141, accessed 15 March 2014

[3] Department of Renewable Energy and Energy Conservation 2012 *25% alternative energy development plan and alternative energy in 10 years (2012-2021)*

[4] Intergovernmental Panel on Climate Change 2006 *2006 Guidelines for National Greenhouse Gas Inventories Volume 4: Agriculture, Forestry and Other Land Use* http://www.ipcc-nggip.iges.or.jp/public/2006gl/vol4.html

[5] Darongthai P (2015) *Green fuel bars to replace wood and charcoal* http://www. Dnp.go.th/research/Knowledge/green_fuel.htm, accessed 25 January 2015

[6] Kamoldara R Natthawud D Wera P Nigran H and Chawaroj J 2014 Development of an integrated waste management model in a village level: The study of Wangpong village, Mueangkaeosub districts, Mearim district. Chiangmai province. Chiangmai *Energy research center, Maejo University* (Thailand)

[7] Kittikorn S Warapong S Nattapong W and Natthawud D 2015 *Thaksin University Journal* **18**(1) 5-14

[8] Naruephat T and Patcharee P 2015 *Srinakharinwirot University Journal of Science and Technology* **7**(13) 15-26
[9] Ogwu I Y Tembe E T and Shomkegh S A 2014 *Journal of Research in Forestry, Wildlife and Environmental* **l6**(2) 13-21
[10] Pellet Fuels Institute 2011 *Standard Specification for Residential/Commercial Densified Fuel*
[11] Sawadkij A 2008 *Master of Engineering Thesis (Bangkok:Sripratthum University)* (Thailand)
[12] Petchhuailuek S 2006 *Proc. 2nd Thailand Energy Network Symposium (Nakhon Ratchasima)* (Thailand)
[13] American Society for Testing and Materials 1998 *ASTM C 702* (USA)
[14] American Society for Testing and Materials 1997 *ASTM C 3174* (USA)
[15] American Society for Testing and Materials 1997 *ASTM C 3173* (USA)
[16] American Society for Testing and Materials 1997 *ASTM C 5865* (USA)
[17] Bank of Thailand, http://www.indexpr.moc.go.th/price_present/cpi/stat/others/report_core1.asp?tb=cpig_index_country&code=93&c_index=a.change_year, accessed 19 June 2019
[18] Kangyi M 2006 *Engineering Economics* https://ocw.mit.edu/courses/chemical-engineering/10-490-integrated-chemical-engineering-i-fall-2006/projects/eng_econ_lecture.pdf

IOP Conf. Series: Materials Science and Engineering **886** (2020) 012003 doi:10.1088/1757-899X/886/1/012003

Influence of Soot Contamination in API CI-4 Engine Oil on Four-Ball Metallic Wear Using Electron Microscopy Image Analysis

Pitchaporn Oungpakornkaew[1,*], Panyakorn Rungsritanapaisan[1], Preechar Karin[1], Ruangdaj Tongsri[2], Dhritti Tanprayoon[2] and Katsunori Hanamura[3]

[1]Faculty of Engineering, King Mongkut's Institute of Technology Ladkrabang, Bangkok 10520, Thailand
[2]National Metal and Materials Technology Center, Pathumthani 12120, Thailand
[3]Departments of Mechanical Engineering, Tokyo institute of technology 179-0085, Japan

* Corresponding Author: pitchaporn.oung@gmail.com

Abstract
Engine lubricant has an important role in decreasing friction and the wearing of moving parts in the engine. However, soot contaminated in engine oil could change chemical and physical properties that affect the lubricant oil operation. Soot particles were simulated using carbon black in order to eliminate the effect of metallic ash and unburned hydrocarbon, which contained in particulate matter. The carbon black was blended in CI-4 lubricating oil. The investigation of metallic wear was associated with the use of a four-ball wear test. The amount of wear was compared by measuring the wear scar diameter of the worn surface on the steel balls. Scanning electron microscope (SEM) and confocal laser scanning microscopy were used for characterizing wear scar and surface roughness, respectively. In addition, the evidence of lubricant additive elements was detected on the wear surface by Energy Dispersive X-ray analysis (EDX). This research found that engine oil with soot contamination could lead to increasing amounts of abrasive wear by 83% and has approximately 1.1% larger wear scar diameter.

1. Introduction

An internal combustion diesel engine has been widely used for many decades due to its high torque and power. Contact of moving parts in the engine generates friction force, which decrease the engine efficiency. The lubricant serves several functions in tribological contact. It is crucial to separate the moving part for reducing the friction and avoiding wear and the engine lubricant also removes the frictional heat. [1] The fuel is combusted under high pressure and temperature. In order to reduce the peak combustion temperature for decreasing the oxides of nitrogen (NOx) emission, exhaust gas recirculation (EGR) has been used. Therefore, incomplete combustion occurs more easily and causes the particulate matter as a byproduct remaining in the exhaust gas. The combustion products have been recirculated rather than pass out through the exhaust pipe by the recirculation system. The carbon content in the cylinder indicates that soot contamination in the oil effect on increasing piston ring and liner wear and on oil quality. [2] To understand the used lubricant oil contamination P. Karin kept oil samples from 2,000 cc and 2,500 cc diesel engine cars and 7,684 cc trucks. The amount of soot contamination through the end of oil life was less than 1% by weight. The nanostructure of soot was investigated by TEM. The primary particle size was found in the range of 20-80 nm. and the mean size was 30 nm. [3] The effect of soot contamination on engine wear was studied from different kinds of tests. S. George uses a ball-on-flat-disk wear test for testing various soot content in oil. The result showed that wear scar diameter (WSD) due to soot contamination was approximately double the WSD of oil without soot. Meanwhile, at higher soot levels, the oil performance was worsened. [4] Green et al. (2006) suggested that wear mechanisms in soot contamination occur because of the reaction between soot and anti-wear additive producing deteriorated lubricant. Another point of view is the starvation of lubricant in the contact, where the soot agglomeration is greater than oil film thickness and blocks the lubricant enters the contact. The last suggestion wear of surface occurs by three-body abrasion since the soot acts as a third body. [5] Li et al. (2006) have found that hardness of primary soot particles can determine from the correlation between the plasmon energy of a soot particle measured by electron energy-loss spectroscopy (EELS) and soot hardness. The research shows that the primary soot particles are sufficiently hard to abrade diesel engine metal parts. The engine parts wear was observed by SEM and the size of grooves was associated with dimensions of primary particles of diesel engines. The soot particles were a response for abrasive wear in the engine. [6] The soot particles were simulated using carbon black. Supanamok et al. (2016) use an electron microscope to examine the primary soot particles (30-50 nm). It shows small differences between soot and carbon black at the nanoscale. Nevertheless, carbon black creates larger agglomerate diameter than extracted engine soot, greater by approximately 50 nm because of oxygen and hydrogen content on its surface creating a relatively polar surface that interacts with another polar carbon black. Carbon black particles presented higher carbon contents and lower ash and volatile. [7] The wear mechanism of contaminated lubricant is not clearly understood. Hence, carbon black was used for simulating soot contaminating in the CI-4 engine lubricant and a four-ball wear tester was used in this study.

2. Experimental Setup

2.1 Materials and sample preparation

The formulated engine oil, which has been used in this study is CI-4 following API standard. CI-4 is group III base oil adding calcium (Ca), zinc (Zn) and phosphorus (P) as an additive. Calcium contains in detergent or dispersant additive. Zinc and phosphorus are used for anti-wear corrosion inhibitors. The concentration of metallic additives was measured by x-ray fluorescence analyser (XRF) is shown in Table 1 and the oil conditions were tested according to ASTM standard.

Table 1. Engine oil properties.

		Base Oil	Group III SAE15-W40
		API Standard	CI-4
Calcium (Ca)	%wt	ASTM D6481	0.3205
Zinc (Zn)	%wt	ASTM D6481	0.1349

Phosphorus (P)	%wt	ASTM D6481	0.1254
Total Base Number	mgKOH/g	ASTM D2896(B)-11	10.5
Viscosity @40°C	cSt	ASTM D445-15a	112.6
Viscosity @100°C	cSt	ASTM D445-15a	15.18
Viscosity Index	-	ASTM D2270	141
Calculated mean oil film thickness at 75°C	Nm		55.6
Calculated minimum oil film thickness at 75°C	Nm		34.9

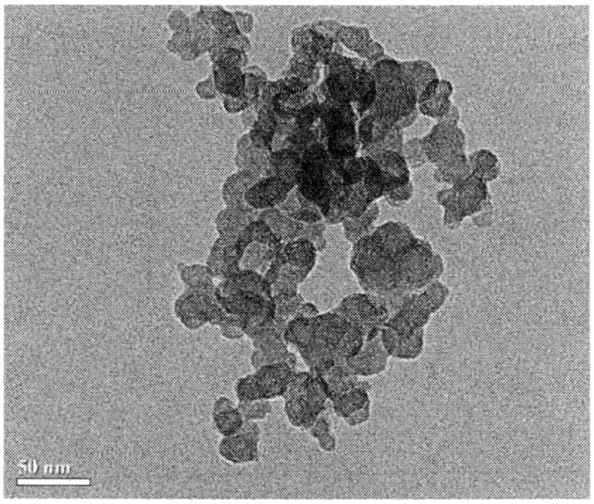

(a) Diesel engine soot

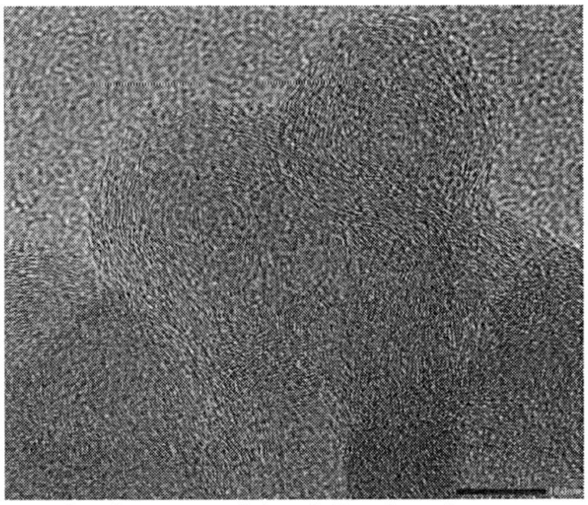
(d) Primary particle of diesel engine soot

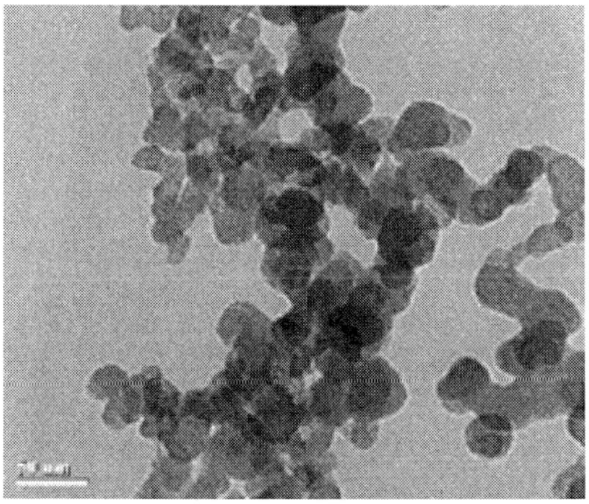
(b) Biodiesel engine soot

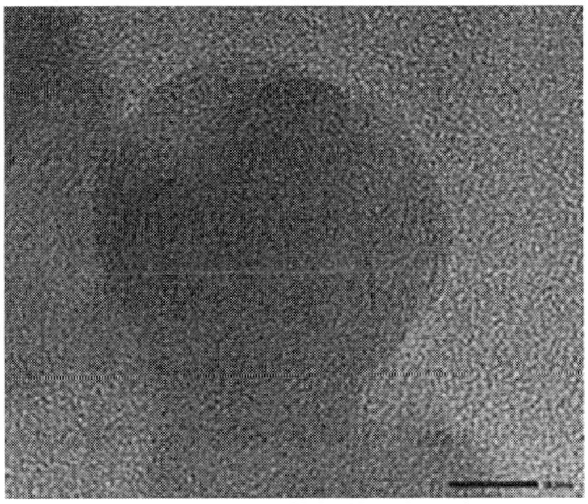
(e) Primary particle of biodiesel engine soot

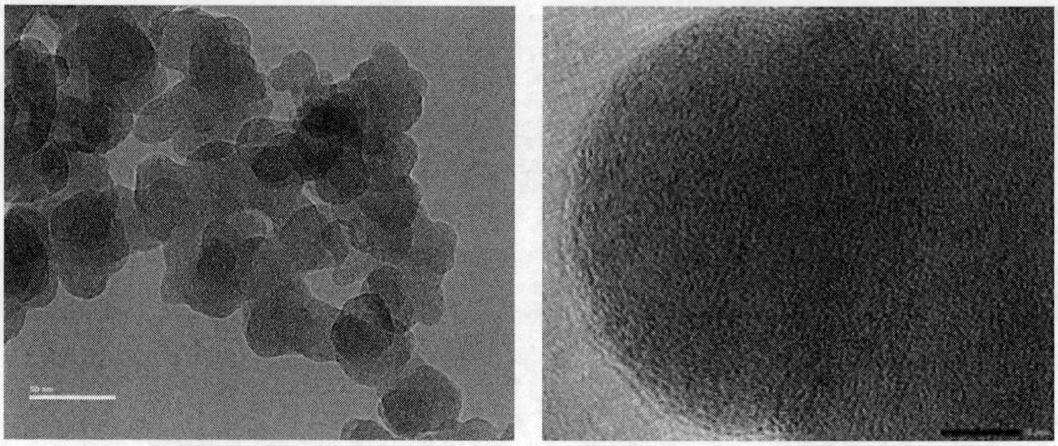

(c) N330 carbon black (f) Primary particle of N330 carbon black

Figure 1. TEM micrographs of (a) diesel engine soot ultrafine particle, (b) biodiesel engine soot ultrafine particle (c) commercial N330 carbon black at 100000x magnification (d) diesel engine primary particle (e) biodiesel primary particle and (f) N330 carbon black primary particle.

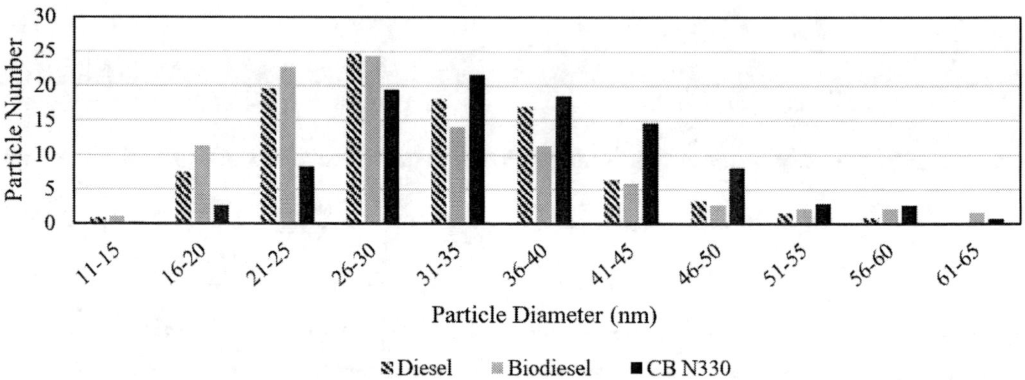

Figure 2. Primary particle size distribution of diesel, biodiesel soot and carbon black (CB N330)

Table 2. Carbon density and hardness comparison of diesel, biodiesel soot and N330 carbon black. [8]

		Diesel	Biodiesel	N330 CB
Carbon density	g/cm^3	2.01	1.82	1.89
Hardness	kg/mm^2	2,056	1,334	1,460

Commercial carbon black was used instead of a real engine soot. The chemical composition of carbon black mainly contains carbon without ash and unburned hydrocarbon component like the engine soot. Therefore, the carbon black and engine soot have similar particle size and crystallinity at the primary particle size scale as seen in Figure 1. The average primary particle size of carbon black N330, diesel and biodiesel engine soot were measured from TEM images were around 30, 28 and 30 nm, respectively. The distribution of primary particle size is revealed in Figure 2. It also can be seen from the TEM images that carbon black, diesel and biodiesel soot have similar morphology and physical property at the primary particle level. The TEM image of primary particle diesel, biodiesel soot and N330 carbon black were used for carbon atom density calculation by using image-processing software to convert TEM image to skeletonize image. The lines in skeletonized image represent the crystallite of carbon in soot and carbon black. The hardness of soot and carbon black can be calculated by carbon density and hardness relationship using carbon atom density and hardness data of diamond and diamond like-carbon, which are glassy carbon, amorphous carbon and diamond. The calculated result of carbon atom density and hardness are shown in Table 2. Diesel, biodiesel soot and N330 carbon black were calculated carbon atom density value is nearly the same and calculated hardness is higher than the steel balls of four-ball wear tester, which their harness is 867 kg/mm^2. The influence of soot contaminating in lubricant on wear was investigated by comparison of wear scar of lubricant without carbon black (CI-4) and with carbon black (CI-4 with CB). The CI-4 engine oil was mixing with carbon black N330 1% by weight to simulate the soot entrainment in the lubricant in a diesel engine.

2.2 Experimental Method

Wear quantity and mechanism are difficult to identify in the engine testing. Therefore, Four-ball test is one of the tribology test methods used to determine the relative wear preventive properties of lubricating fluid in sliding contact under ASTM D4172 standard. The test used four steel balls, three balls were fixed in the oil and another ball on the top rotated at 1,200 rpm and pressed with 392 N. The tested lubricant temperature was 75°C and the test duration was 60 minutes. The chromium alloy steel test balls were extra polished which their surface roughness is 0.005 microns following the AISI B3.12 standard. The hardness of test balls is in range 64-66 HRC. After the test, three balls at the bottom were measured the wear scar diameter using a high-resolution optical microscope to calculate average wear scar diameter of each condition. The surface roughness was measured by confocal laser scanning microscopy, capturing multiple two-dimensional images at different depths in a sample enables the reconstruction of three-dimensional structures of wear scar. The average surface roughness is determined using Ra value. Each ball was measured five times and calculated the average Ra value. In addition, the wear scar surface was characterized by Scanning Electron Microscopy (SEM) and Energy Dispersive X-ray analysis (EDX).

3. Result and discussion

The worn surface of steel balls (Figure 3) was measured average wear scar diameter is shown in Table 3. The result shows that wear scar diameter gets larger when adding the 1% wt of carbon black into CI-4 oil; it increases nearly 1.1% from new oil. As a result of carbon black contaminates in the lubricant, it causes more wear in the Four-ball wear test.

The surface roughness measured by confocal microscopy as seen in Figure 4. CI-4 oil with carbon black exhibits lower Ra value than new oil (CI-4) as shown in Table 3. The average primary particle size of carbon black N330 is 30 nm, which smaller than calculated minimum oil film thickness (Table 1). As a result, carbon black can enter into the sliding contact of the tested ball. Moreover, carbon black, which researched to be harder than steel ball, could act as the third body in the wear surfaces, and abrades the ball surface indicated by smoother surface. Thus, the Ra value of contaminated oil declines by 17.2%.

IOP Conf. Series: Materials Science and Engineering **886** (2020) 012003 doi:10.1088/1757-899X/886/1/012003

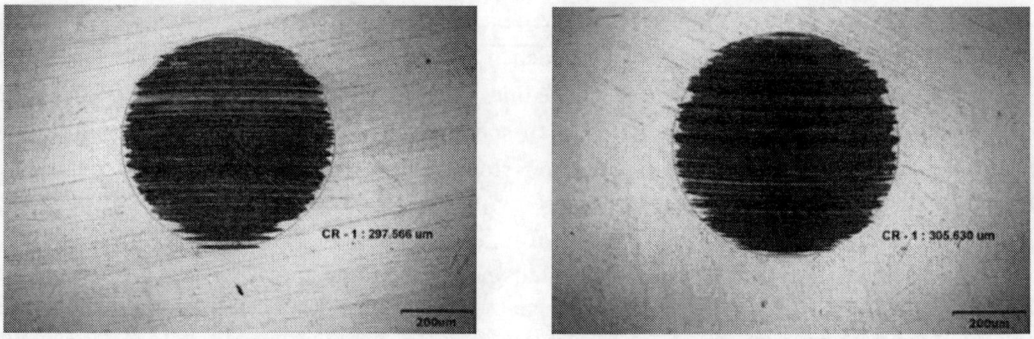

(a) CI-4 (b) CI-4 with CB

Figure 3. Wear scar diameter images of new oil and oil with carbon black.

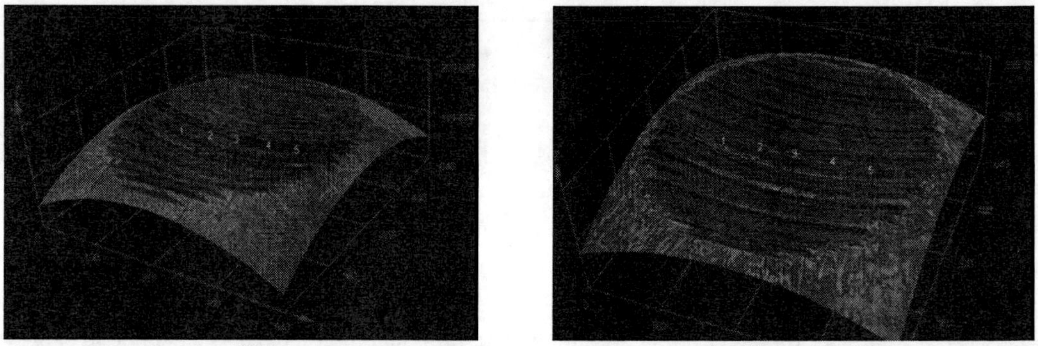

(a) CI-4 (b) CI-4 with CB

Figure 4. Surface roughness measurement of (a) CI-4 and (b) CI-4 with CB.

Table 3. Wear scar diameter and surface roughness.

	Average wear scar diameter (μm)	Ra value (μm)
CI-4	603.31	3.98
CI-4 with CB	609.71	3.30

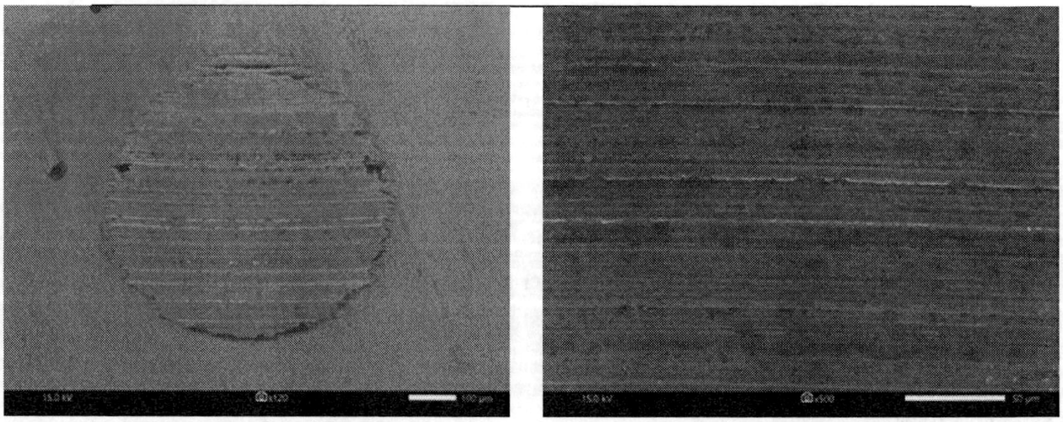

(a) CI-4 120x magnification (c) CI-4 500x magnification

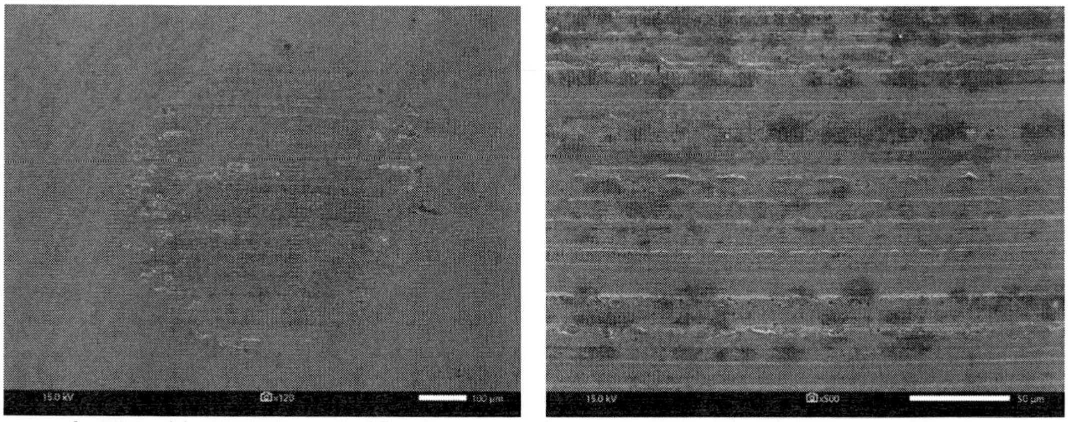

(b) CI-4 with CB 120x magnification (d) CI-4 with CB 500x magnification

Figure 5. SEM micrographs of (a) CI-4, (b) CI-4 with CB at 120x magnification, (c) CI-4 and (d) CI-4 with CB at 500x magnification wear scar.

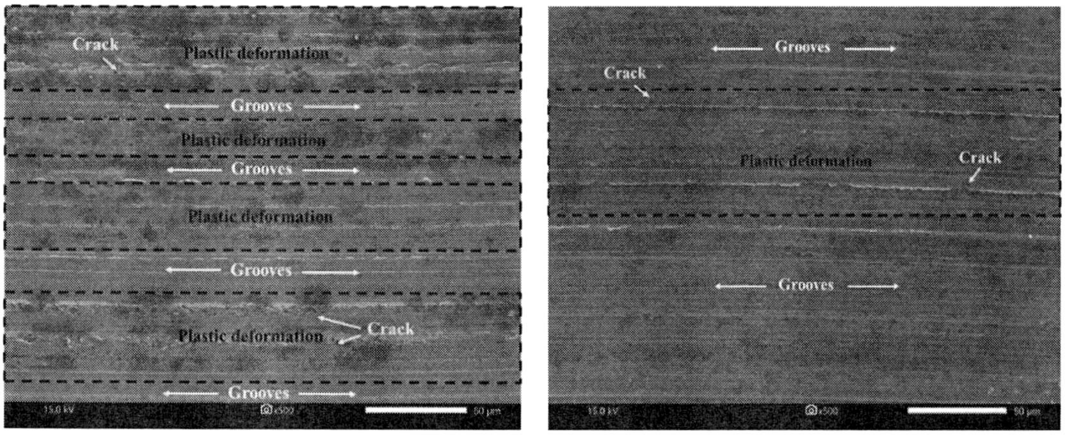

(a) CI-4 (b) CI-4 with CB

Figure 6. SEM micrographs of (a) CI-4 and (b) CI-4 with CB at 500x magnification with wear analysis criteria.

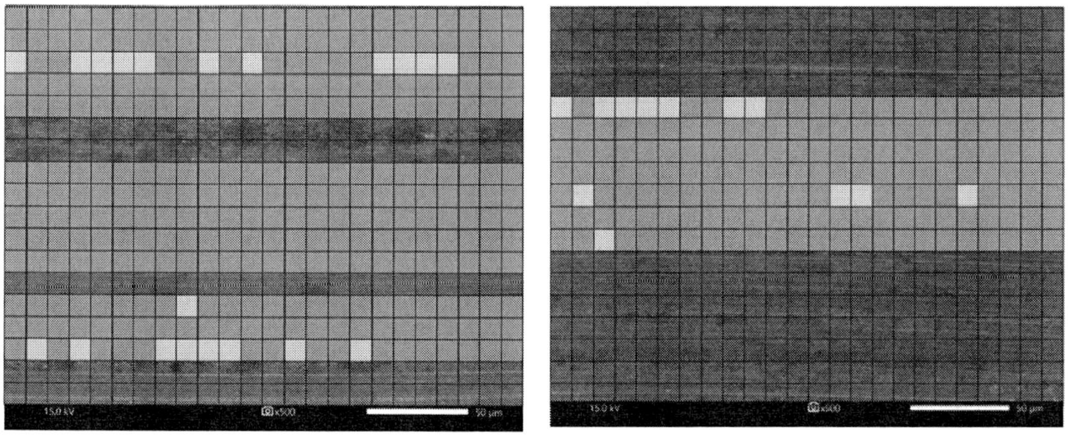

(a) CI-4 (b) CI-4 with CB

Figure 7. Colored wear analysis of (a) CI-4 and (b) CI-4 with CB at high magnification SEM micrographs.

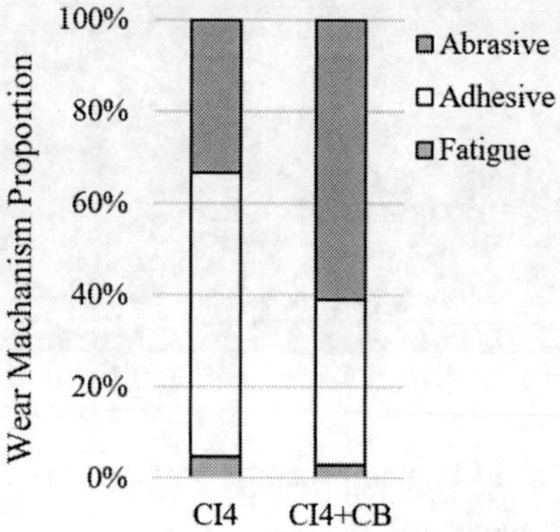

Figure 8. Quantitative wear mechanism analysis.

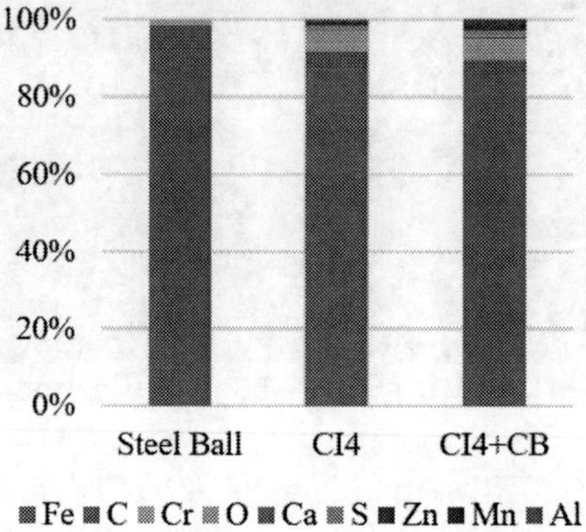

Figure 9. EDX analysis on the ball surface.

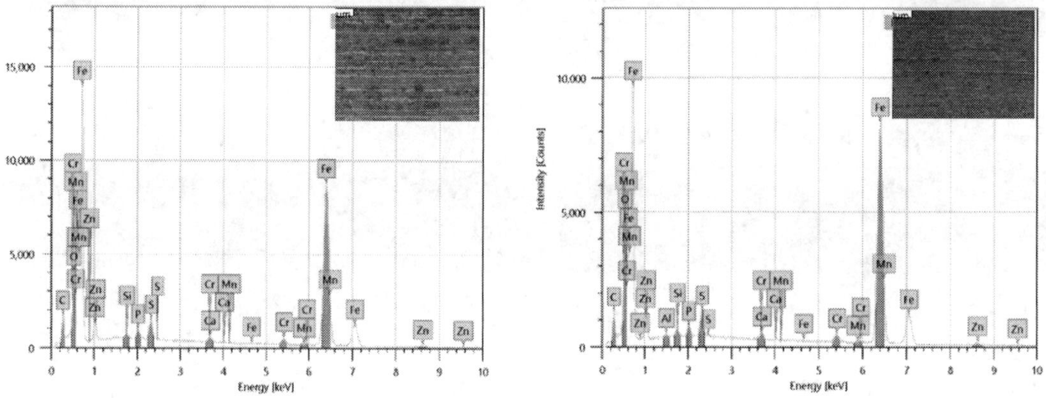

(a) (b)

Figure 10. EDX elemental spectrum analysis (a) CI-4 and (b) CI-4 with CB

The wear scars were observed by SEM as well. Figure 5 shows SEM micrographs of wear scar on the ball surface after the four-ball test at 120x and 500x magnification. At low magnification, the wear scar looks similarly between new oil (Figure 5 a) and oil with carbon black (Figure 5 b). However, the wear scar shows different surface characteristics at high magnification. Both cases exhibit grooves along the sliding direction of the ball where the abrasive wear occurs. The adhesive wear also can be observed in the SEM image as a plastic deformation on the ball surface. The result in the removal of wear debris larger than 20 μm indicate the adhesive wear. [9] Lastly, fatigue wear appears as a surface crack on the plastically deformed surface from the cyclic loading in a small section in two cases as shown in Figure 6. In order to analyze the wear mechanism, 24x18 squared blocks were created on the 500x SEM image and distinguish each squared block with different wear mechanism as seen in Figure 7. The blue colored blocks are determined as abrasive wear. Abrasive wear was pointed out by the clear block. Meanwhile, the delamination of the ball surface is defined by the yellow-colored block to specify the fatigue wear. The result of quantitative wear analysis showed the percentage of adhesive wear is highest in CI-4 sample. It shows 62% of the area in SEM image. On the other hand, CI-4 with carbon black has abrasive wear as a majority wear mechanism for 61% of the area. It demonstrates that the carbon black has a role in roughly increasing abrasive wear in the four-ball wear test by 83% in this research as can be seen in Figure 8.

EDX result of steel ball surface shows iron (Fe), carbon (C) and chromium (Cr) which are the composition of chrome steel ball as was seen in the bar chart in Figure 9. The additive elements are found on the wear surface of both CI-4 and CI-4 with carbon black as can be observed in Figure 9 and 10, which are calcium, zinc, and phosphorus similar to the XRF result in Table 1. It indicates that the additive elements are remaining on the wear surface.

4. Conclusion

This research is aimed to investigate the effect of soot contamination in lubricant on diesel engine wear. New oil and contaminated oil were tested by a four-ball wear test. The N330 carbon black is used as simulated soot contaminate in CI-4 oil. The result of wear scar diameter can represent the quantity of wear which occurs on the ball surface. The measurement from the optical microscope of soot-contaminated oil has a higher amount of wear with approximately 1.1% larger wear scar diameter as a result of soot existence in the lubricant, while the roughness of lubricating oil with soot is 17.2% lower. The carbon black has been investigated that it is harder than the steel ball, therefore carbon black and also small particles that entered into the sliding contact between the wear surface may polish the ball surface. Nevertheless, only the particle size that smaller than oil film thickness can involve in this mechanism. SEM micrographs also illustrate the small groove along the sliding direction, it indicates that the carbon black is abrade the ball surface. Consequently, analyzing the high magnification SEM image of tested oil with carbon black shows a higher percentage of abrasive wear than adhesive wear and fatigue. The adhesive area is increased 83% from the wear scar area of non-contaminating oil which is indicated by the sliding groove along the worn surface. Furthermore, the trace of additive containing in CI-4 oil is presented on the wear surface.

Acknowledgements

This research was supported by Thailand Research Fund (TRF) and Research and Researchers for Industries (RRI) for the research funding (MSD 60I0142) and National Metal and Materials Technology Center (MTEC) for laboratory equipment. In addition, the authors are also grateful to Bangchak Corporation (BCP) in supporting the lubricant oil and giving suggestions for this research.

References

[1] Torbacke M 2014 *Lubricants* (West Sussex: John Wiley & Sons Ltd)
[2] Dennis A J Garner C P and Taylor D H C 2010 *SAE Technical Paper Series* **1**
[3] George S Balla S and Gautam M 2007 *Wear* **262** 1113–1122
[4] Green D A Lewis R and Dwyer-Joyce R S 2006 *Proc. the Institution of Mechanical Engineers, Part J: Journal of Engineering Tribology* **220** 159–169
[5] Jao T C Li S Yatsunami K Chen S J Csontos A A and Howe J M 2006 *Lubrication Science* **16** 111– 126
[6] Karin P Supanamok C and Hanamura K 2016 *Journal of Research and Applications in Mechanical Engineering* **4** 126–134

[7] Supanamok C Karin P Benyajati C and Hanamura K 2015 *Proc. Int. Conf. on 2015 JSAE Annual Congress Proceedings (Yogohama)* 1673-1679

[8] Rungsritanapaisan P Karin P Amornprapa W Tanprayoon D Tongsri R and Hanamura K 2019 *Proc. Int. Conf. on Small Engine Technology Conference 2019 (Hiroshima).*

[9] Fazal M Haseeb A and Masjuki H 2013 *Energy Conversion and Management* **67** 251-256

IOP Conf. Series: Materials Science and Engineering **886** (2020) 012004 doi:10.1088/1757-899X/886/1/012004

Biodiesel Fuel (BDF) Synthesis by using Ion Exchange Resin Based on Ultrasonic Irradiation and Its Application

Ning. Zhu[1]*, Mingchi Kuo[2] and Chung Hsin Han[2]

[1]Shizuoka Institute of Science and Technology, 2200-2 Toyosawa, Fukuroi, 4378555, Japan
[2]Graduate School of Shizuoka Institute of Science and Technology, 2200-2 Toyosawa, Fukuroi, 4378555, Japan

* Corresponding Author: Ning Zhu: zhuing@me.sist.ac.jp, Tel:+81-538-450232, Fax: :+81-538-450120

Abstract
Ultrasound irradiation with NaOH proved an important method in catalyst to synthesize BDF. However, the synthesis of BDF remains a major challenge since a large quantity of water is needed to wash the BDF after it is produced. It is now well established from a variety of studies, that solid catalysis does not need a large quantity of water to wash the BDF after its production. The concept of zeolites is used as solid catalysts to synthesize BDF and BDF and its ratio yields up to 62%. This study uses ion exchange resin (henceforth, IER) as the catalysts for BDF production. Firstly, this research examines the emerging role of IER in the increase of its basic characteristics. Secondly, the prepared IER is characterized using SEM and X-ray diffractometer. Thirdly, the BDF synthesis experiment is carried out using the prepared IER based on ultrasonic irradiation. Finally, diesel engine performance test is carried out to investigate the thermal efficiency and exhaust gas concentration. The current study found that BDF yield ratio reached 44.4% when IER was prepared under the condition of 6mol NaOH preparation.

Keywords: BDF synthesis, Solid catalyst, Ultrasonic irradiation, Exhaust gas concentration

1. Introduction

Exposure to global warming and energy crisis has been shown to be related to adverse effects in drastic economic development and large quantity usage of fossil fuel. However, biodiesel fuel (henceforth, BDF) could be a contributing factor to alternative fuel for the automobile. There is a growing body of literature that recognises methods of synthesizing BDF and its applications [see 1-8].

There is some evidence to suggest that BDF comes from vegetable oil so that it is clean and sustainable. It can be also used for any automobiles with diesel engines. Due to the support and being promoted from the EU and the USA government, recent trends in the application of BDF have led to a proliferation of studies that BDF can be used as B5 or B10. However, a major problem with this kind of application in Japan is only 100% BDF is permitted to be used on diesel automobiles. While the production cost of the BDF is considerably higher than diesel fuel and it declines in the wide use of BDF. Studies over the past two decades have provided important information on employing ultrasound irradiation with NaOH as the catalyst to synthesize BDF. One major issue in early BDF research concerned a large quantity of water is needed to wash the BDF after it is produced. However, far too little attention has been paid to the development of a new method based on the solid catalysis because it can decrease water-washing in the production. Existing research recognises the critical role played by zeolites as solid catalysts to synthesize BDF and BDF and its yield ratio is up to 62%. Rather, this paper proposes a new methodology for using IER as the catalysts for BDF production. Firstly, ion exchange resin is prepared to increase its basic characteristics. Secondly, prepared IER is characterized by using SEM and X-ray diffractometer. Thirdly, the BDF synthesis experiment is carried out by using the prepared IER based on ultrasonic irradiation. Finally, diesel engine performance test is carried out to investigate the thermal efficiency and exhaust gas concentration. This study showed that the BDF yield ratio reached 44.4% when IER was prepared under the condition of 6mol NaOH preparation.

2. A principle of BDF synthesis

Figure 1 presents the BDF synthesis principle of using IER as the catalyst to produce. Vegetable oil (triglyceride) was then mixed with alcohol (methanol) at proper ratio under the condition of ultrasonic irradiation for some time. When a transesterification reaction was carried out, the BDF and the by-product of glycerin was yielded.

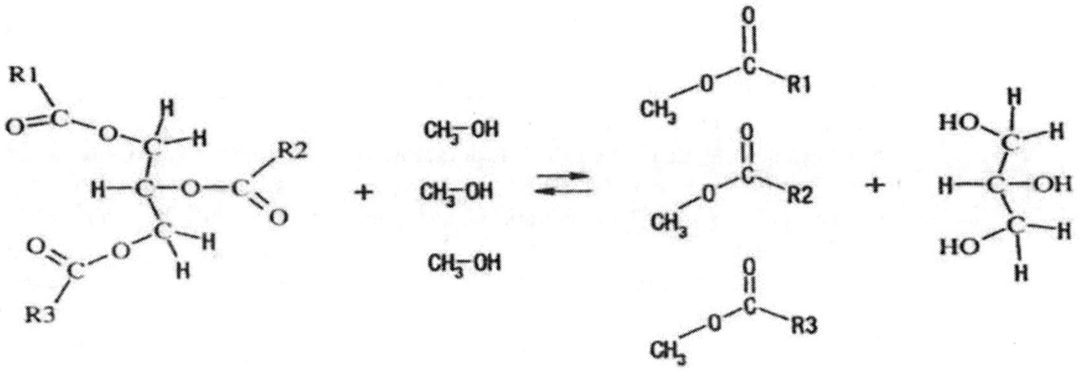

Vegetable
oil Alcohol(methanol) BDF Glycerin
(triglyceride)

Figure 1. BDF synthesis principle

3. An experiment of BDF synthesis

3.1 Catalyst preparation and evaluation

In this paper, the IER used for BDF synthesis was called ORGANO. IR120B NA and its specifications are set out in Table 1.

Table 1. IER model number

Construction	Classification	Ion form	Mean diameter(mm)	Operating temperature (°C)	Effective PH range
styrene	strong acidity	Na+ / H+	0.60 ~ 0.80	120	0 ~ 14

Figure 2 shows the size and shape of the IER and its SEM analysis before preparation. IER has the round shape whose average external diameter is around 0.6-0.8 mm, which are smaller than zeolites. According to the SEM analysis result, the surface area of IER is very large, which could absorb a large amount of liquid.

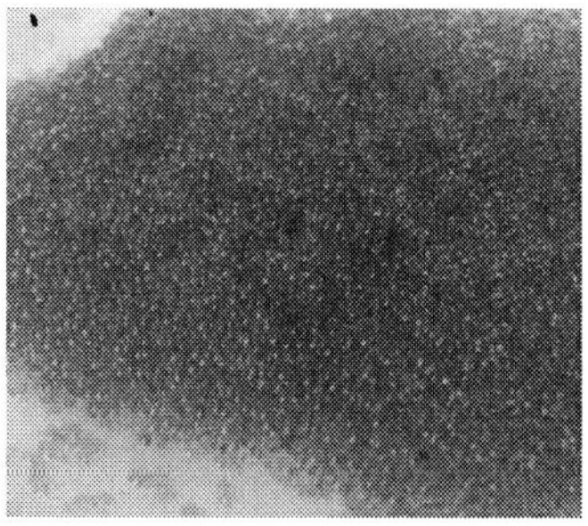

(a) IER

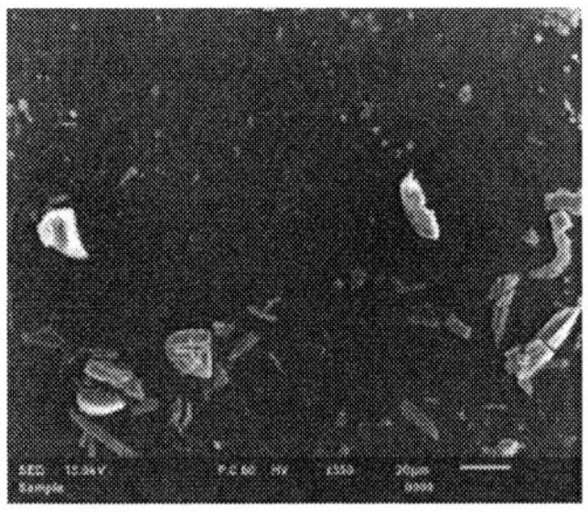

(b) IER before sintering by SEM

Figure 2. The shape of IER and its SEM analysis before preparation.

Since base strength could properly influence the BDF yield ratio, before BDF synthesis, IER was prepared as follows:

Step 1: Soak samples of IER with different NaOH or CaOH concentration (6 and 9 mol) for 24 hours
Step 2: Wash the samples by using methanol and then dry the samples
Step 3: Sinter the samples by an electric furnace at the temperature of 100°C for two hours

After IERs were prepared, SEM was used to investigate the changed quantity of $Na+$ that was added to the IER. An X-ray diffractometer is used to characterize their base strength characteristics.

3.2 A synthesis of BDF

Prior to batch type BDF synthesis, the mixture of the reactants of vegetable oil and methanol was prepared inside a glass tube, which is immersed into the ultrasonic reactor (called Hondex) with hot water (see Figure 3). Then, the power of the ultrasonic reactor was switched on at frequency of 28 kHz. From the data of BDF synthesis in Table 2, it is apparent that the IER was used as the solid catalyst while the Zeolite A5 is also used for BDF synthesis.

Table 2. BDF synthesis condition

Volume ratio of vegetable oil to methanol	5:1
Methanol purity	99.5%
Catalyst	IER & Zeolite A5
Quantity of Ion exchange resin	0.33wt%

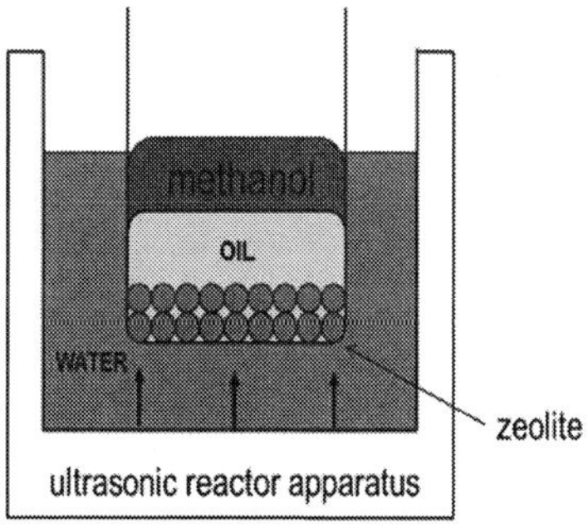

Figure 3. BDF synthesis by using a batch method

After BDF synthesis, BDF components are detected and BDF yield ratio is measured by GCMS (Shimadzu GCMS-QP2020).

4. Engine performance test and exhaust gas concentration

A small diesel engine (called KM170F) was used to obtain an engine performance test from the thermal efficiency and the exhaust gas concentration. The specification of the engine is listed in Table 3. The main advantage of the testing is that it is to confirm the possibility of using 100% BDF or the mixture of BDF with diesel fuel. The results obtained from the engine performance test experimental system are shown in Figure 4. This system consists of the diesel engine, fuel flow meter, fuel tank, electric load, revolution indicator and Watt indicator. After the engine is started, the first step in this process was to adjust the revolution number to 3300 rpm under an electric load of 500W, 1000W, 1500W and 2000W, respectively. Then, fuels of BDF with different mixture ratio (0%, 10%, 20%, 30%ıı 40%, 50%, 60%, 70%, 80%, 90% to 100%) was used. After the fuel consumption rate, output power, intake temperature, exhaust temperature, exhaust gas concentrations (e.g. smoke, CO and NOx) were measured, the engine thermal efficiency was calculated.

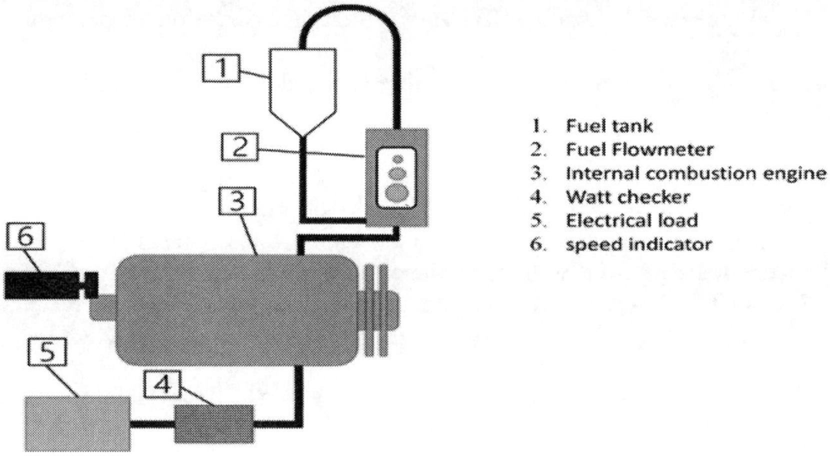

Figure 4. Experimental system for engine performance test

Table 3. Specification of the diesel engine

Engine type	type	Bore diameter x Stroke [mm]	Displacement [cc]	Rated output [kw / rpm]	Noise level (7m) [db]	Cooling system	Lubrication system	Rated speed [r / min]
KM170F	direct injection diesel engine	70×55	211	2.0/3600	77	Forced air cooling	splash lubrication	3600

5. Results and discussions

5.1 Catalyst evaluation

It was found that Na+ content change after the preparation of IER, as illustrated in Figure 5. Based on the SEM analysis, it is also found that the Na+ content increased when the preparation with higher NaOH concentration was used. In the case of IER with 6mol and 9 mol NaOH, the Na+ content was in a range of 16.8% and 22.2%.

IOP Conf. Series: Materials Science and Engineering **886** (2020) 012004 doi:10.1088/1757-899X/886/1/012004

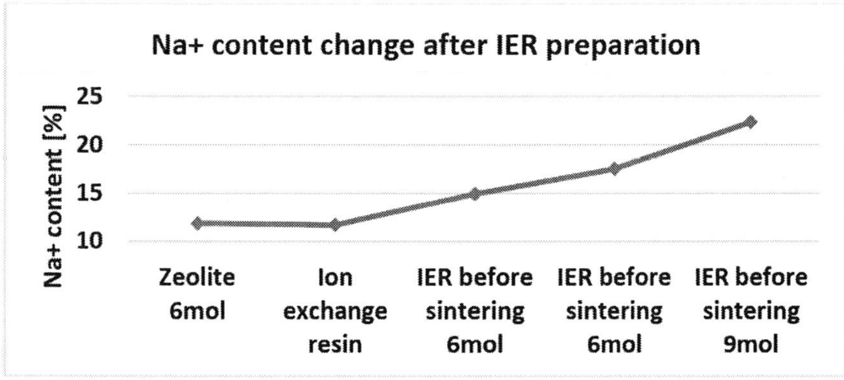

Figure 5. Na+ content change after IER preparation

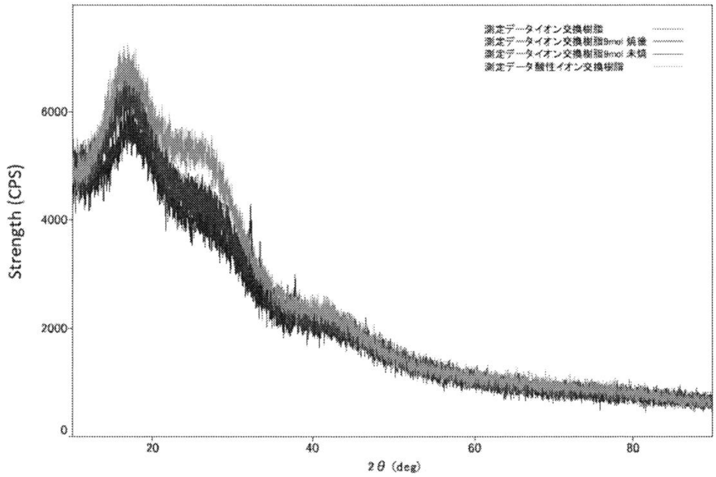

Figure 6. X-ray diffractometer result for A3

Before the preparation, the Na+ contend was only under 12%. It revealed that the preparation has increased the Na+ content and enhances the basic strength. Figure 6 presents an X-ray diffractometer. Further analysis showed that the forms are similar to general IER, prepared IER. However, there was a slight change in the strength. It is apparent that the structure of IER of before and after preparation did not show a significant difference. There was a slight change of the strength with the influence of the liquid NaOH.

5.2 BDF synthesis

BDF synthesis result is demonstrated in Figure 7. It is observed that BDF could be produced under the condition of using IER with 6mol NaOH and A5 Zeolite with 6mol NaOH as the catalyst, respectively. While less BDF was produced for IER with 6 mol CaOH preparation. To compare the photos of the BDF produced by using a different catalyst, BDF yield ratio was 41.4% for IER catalyst, and it appeared to be no higher than the zeolite. It was also found that that the broken zeolite under ultrasonic irradiation might strengthen the BDF synthesis reaction and increase the BDF yield ratio.

Figure 7. BDF synthesis result

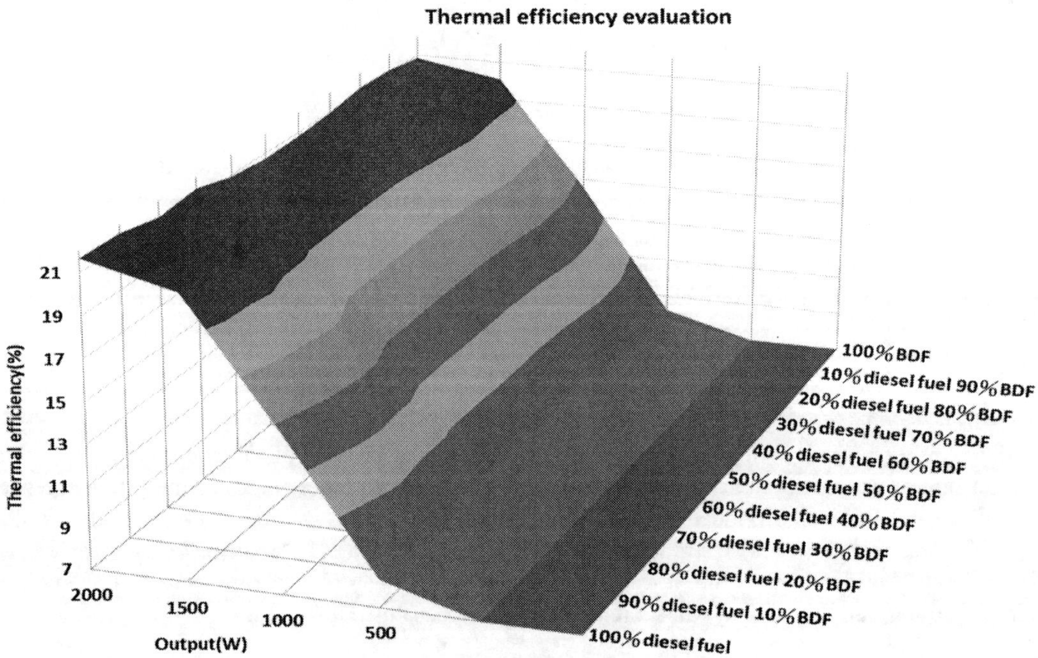

Figure 8. Thermal efficiency

5.3 Results of engine performance test

As can be seen from Figure 8, it reported the efficiency of the thermal under different electric load for a different mixture of BDF to diesel fuel. Further analysis showed that its efficiency was around 21% for all the experimental conditions. Thus, the lower the fuel mixture ratio of diesel fuel to BDF, the higher the thermal efficiency. In the case of 100% diesel fuel, the efficiency of the thermal reached 21.4% while in the case of 100% BDF, the thermal efficiency was by 17.4%.

Smoke concentration in the form of the optical absorption coefficient for the electric load of 0w and 2000W is shown in Figure 9. With the higher electric load, the Smoke concentration would increase rapidly. But in the case of 100% BDF, there was a slight decrease in the smoke concentration (0W electric load).

Figure 10 provides a CO concentration. With more percentage of BDF, CO concentration would decrease due to extra oxygen brought by the BDF contained the generation of CO. However, for the load of 1500W, CO concentration appeared to be the lowest for each experimental condition. It can thus be suggested that the best air-fuel ratio was observed for the electric load of 1500W.

NOx concentration is shown in Figure 11.With higher electric load, the NOx concentration would become higher because the higher electric load lead to the larger consumption of fuel. It is possible, therefore, that the increase in temperature of the combustion may be due to the increase in the generation of NOx. In case of higher ratio of BDF to diesel fule, it seems that the NOx concentration tended to increase to some extent. This may be explained by the influence of the extra oxygen coming from BDF, which increased the combustion temperature.

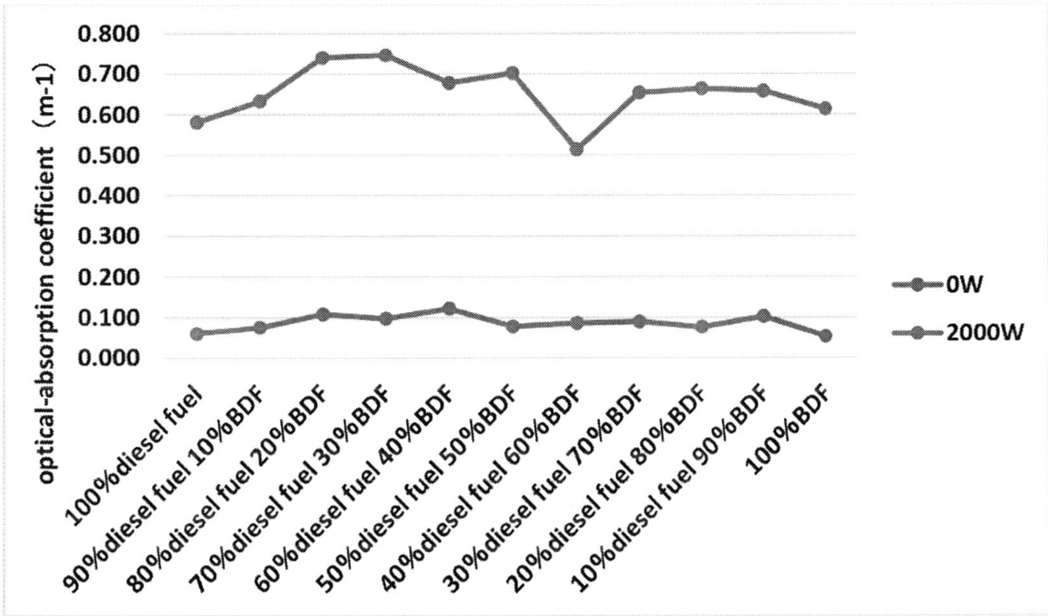

Figure 9. Smoke concentration for electric load of 0w and 2000W

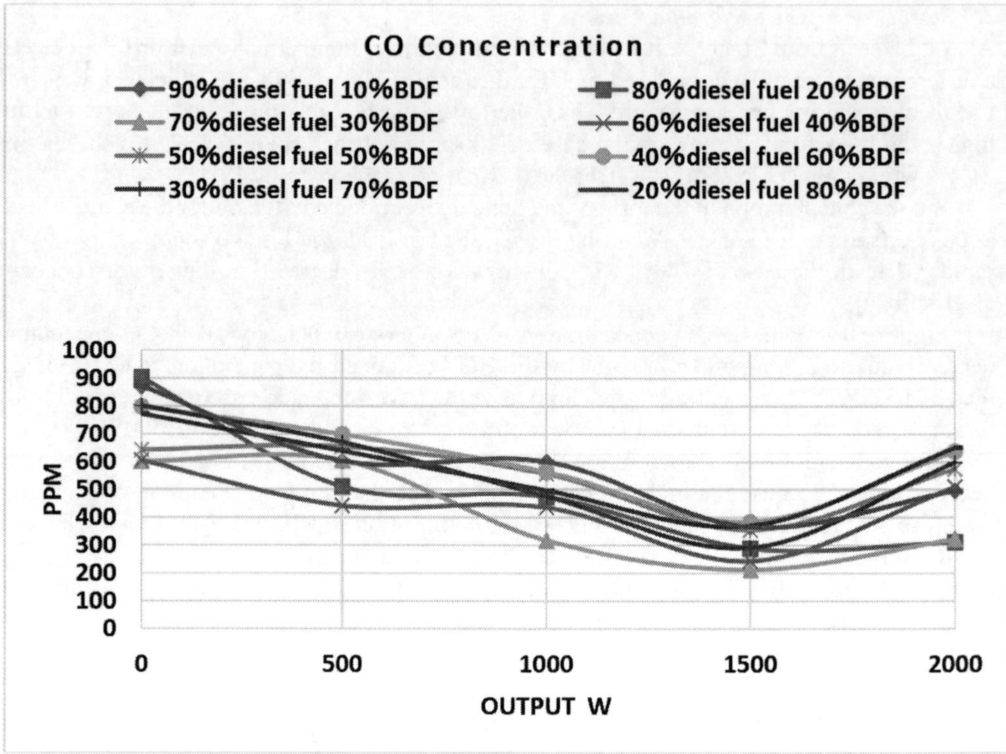

Figure 10. CO concentration

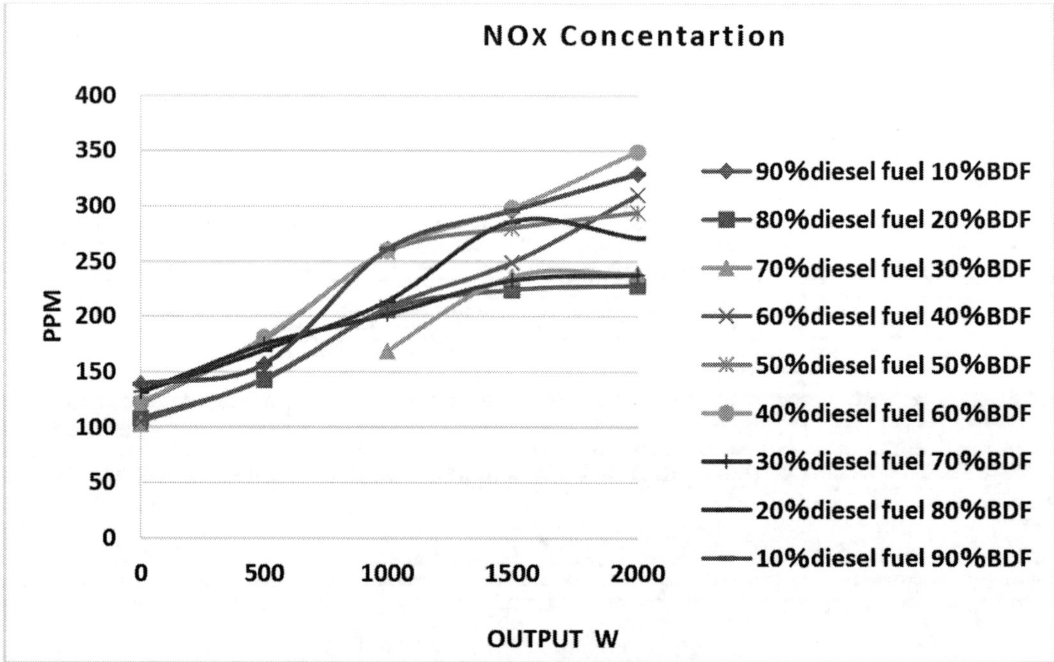

Figure 11. NOx concentration

6. Conclusion

This study has identified the method of using IER as the catalyst for BDF synthesis using IER preparation and BDF synthesis experiment. The mixture fuel of different ratio of BDF to diesel was also used for engine performance test and thermal efficiency, as well as the exhaust gas concentration was measured. This study has shown that the prepared IER of 6 mol NaOH, BDF yield ratio reached 41.4%. These experiments confirmed that the efficiency of the thermal was around 21% for each experimental condition. With a higher ratio of BDF to diesel fuel, the smoke concentration and CO concentration would become lower and NOx concentration became higher.

References

[1] Furugawa S Shimada M Okada M Hirobashi L Szuki Y 2004 *Proc. Japan Chemical Society* (Japan) **84**(2) 1295-1296
[2] Dszuki T Hanehara H Chita J 2005 *Proc. JSME of Kansai Branchy Conference* (Japan) **80** 1247-1248
[3] Yamane K Shimamoto J Ueda J 2001 *Transactions of JSA* **32**(2) 25-30
[4] Yamane K Shimamoto J Inoue H 2001 *Proc. JSME of Kansai Branchy Conference* (Japan) **76** 1249-1250
[5] Furuta S Matsuhashi H Arata K 2004 *Journal of Catalysis Communicationse* **5**(12) 721-723
[6] Zhu N Ttsuchiya T Ito N Kato N 2005 *Proc. 3rd International Energy Conversion Engineering Conference* 5543
[7] Zhu N Guo M C Zhang T C 2017 *Proc. the 8th Thailand Society of Mechanical Engineers International Conference on Mechanical Engineering* (Bangkok:Thailand)
[8] Zhu N Guo M C Zhang T C 2018 *Proc. the 9th Thailand Society of Mechanical Engineers International Conference on Mechanical Engineering* (Phuket:Thailand)

Acoustic Power Prediction of High BPR Turbofan Engine with Chevron Nozzle during Take-off and Landing

Rapee Ujjin[1,*], Choosak Ngaongam[1], Chatmongkol Phoomee[1] and Jirantanin Kanabkaew[1]

[1]Department of Aviation Maintenance Engineering, College of Engineering, Rangsit University, Pathumthani 12000, Thailand

* Corresponding Author: Rapee.u@rsu.ac.th

Abstract
Aircraft noise is a major cause of noise pollution around the vicinity of an airport. Jet noise from engines is a common condition, which has considerable impact on overall sound levels. Adding chevron pattern at the fan nozzle is one of the noise reduction enhancements found in current turbofan engine design features. This paper aims to study the power level of sound created in jet shear from the combination of ambient stream, fan and exhaust gas. Computational fluid dynamic modelling of a high bypass turbofan engine with a round shape and chevron nozzle were carried out during take-off and landing conditions. It was found that during take-off, the acoustic power level of the engine without a chevron nozzle was 105.95 dB, while the chevron nozzle was capable of reducing the sound level to 43.57 dB. During landing, the sound level of the round shape and chevron nozzle model were 145.77 dB and 47.51 dB, respectively. The turbulent intensity area at the fan cowling nozzle, where the ambient air stream and bypassed air from the fan mixed together, produced the highest sound power level. Therefore, slower aircraft speed during final approach produced a higher noise level than during take-off due to higher turbulent intensity and larger jet shear area around the high bypass fan cowling nozzle. The chevron pattern nozzle was capable of reducing turbulent intensity, which in turn reduced the acoustic power level.

Keywords: Acoustic power level, Turbofan, CFD, Aircraft noise, High bypass ratio

1. Introduction

Aircraft noise during take-off and landing is the leading cause of noise pollution at many airports around the world, especially in urban areas [1]. Noise from aircraft operations is a complex problem when compared to other noise sources as well as aircraft noise prediction method, which accounts for airframe, high lift systems, propeller and propulsive [2]. Recent evidence suggests that the aircraft noise is associated with gas turbine engines, particularly from high bypass ratio (BPR) turbofan engines [3]. The present high BPR of turbofan engines is between ranges of 10-12 BPR [4].

Since high bypass turbofan engines are designed for optimum fuel efficiency, many airlines utilized aircraft equipped with high BPR turbofan engines. High BPR turbofan engines have been subjected to continuous improvement predominantly to produce higher thrust with lower noise. Rolls-Royce is the only engine manufacturer to use a three-spool engine design allowing the power from the intermediate spool to deliver a lower thrust at idle during the flight until landing, which helps reduce fuel consumption [5].

It has previously been observed that noise from turbofan engines can be reduced by using jet noise reduction techniques, which involve the installation of acoustic liners such as zero-splice intake liner, bypass duct liner and core duct liner [6]. Studies of aeroacoustics research show that the addition of the chevron pattern at the nozzle can reduce sound pressure levels [7, 8].

The numerical prediction had been made to obtain turbulence quantities as well as acoustic quantities reasonably well for both round and chevron nozzles [9]. The mixing of air stream with air from the fan exit at different temperatures and velocity results in turbulence, which in turn increases noise. It has been found from the CFD analysis for acoustic performance evaluation of different shapes of chevron that managing the mixing area couple with a proper chevron pattern could reduce the sound pressure level [10].

Factors found to be influencing in patterning and geometry of chevron that can reduce noise and turbulence to a considerable degree have been explored in several studies [11]. Experimental testing of the chevron nozzle in subsonic flow was performed and found that each chevron reduced the shear velocity behind it, which results in noise reduction [12]. An investigation of different lengths and widths of chevron was extensively carried out at the supersonic flow. It was found that the shorter the length of the chevron, the less impact there was on noise reduction when compared with the longer one. Nevertheless, it is noteworthy that the noise reduction is significantly better than without a chevron pattern [13].

A high BPR dual-stream jet in a scaled model was also tested. It is now well established from a variety of studies, that a chevron nozzle could reduce the overall noise. The number of chevron teeth might also affect noise reduction capability [14]. The observation of flow characteristics to produce jet noise was done using a simulation of turbulent jet noise, resulting in the thickness nozzle boundary layer would affect noise. The chevron nozzle placed under a specific condition requires a particular arrangement and boundary layer thickness to reduce noise [15].

The aim of this paper is to predict acoustic power level by using CFD simulation of high bypass turbofan engine with round and chevron nozzle models with boundary conditions from numerical calculations and actual power setting parameters during take-off and landing.

2. Acoustic Power

2.1 Sound power level (SPL)

The sensitivity of a human's ear is expressed in logarithmic, which is measured using a decibel scale, as known as sound pressure level (SPL) [16]. This is given in terms of the root mean square of the fluctuating pressure-time history P_{rms} and reference pressure, P_{ref}, is defined by:

$$\mathrm{SPL} = 20\log_{10}\left(\frac{P_{rms}}{P_{ref}}\right) \tag{2.1}$$

Where: SPL = sound pressure level (dB),
P_{rms}= root mean square of pressure fluctuation (Pa),
P_{ref} = reference sound power (Pa)

2.2 Sound intensity level (IL)

Sound intensity level (IL) is a logarithmic measure of the power of a sound relative to a reference value [17, 18]. Sound intensity level, I_L also measured in dB, is defined by:

$$I_L = 10\log_{10}\left(\frac{I}{I_{ref}}\right) \tag{2.2}$$

IOP Conf. Series: Materials Science and Engineering **886** (2020) 012005 doi:10.1088/1757-899X/886/1/012005

Where: I_L = sound intensity level (dB),
I = sound intensity (w/m^2),
I_{ref} = reference sound intensity (w/m^2)

2.3 Effective perceived noise level (EPNL)

The most commonly used metric to obtain a measurement of the perceived loudness of a sound is the dB(A) level. The dB(A) level is subjected to the sensitivity of the human's ear and can be directly measured using a sound level meter. Other measures of annoyance include a perceived noise level (PNL) and an effective perceived noise level (EPNL). They are used in assessing aircraft noise [16]. Noise evaluation measure for the aircraft is the effective perceived noise level in EPNdB [19]. There are three measurement points: lateral full-power, fly over and approach reference noise measurement points. For an aircraft with two turbofan engines, it will use the following formulae to calculate maximum permissible noise levels and corresponding speed of the aircraft while being assessed. The equations are as follows:

For full-power noise reference,

$$EPNdB = 80.87 + 8.51 \log M \tag{2.3}$$

For approach noise reference,

$$EPNdB = 86.03 + 7.75 \log M \tag{2.4}$$

For take-off speed,

$$V_{T/O} = V_2 + 10\ kt \tag{2.5}$$

For approach speed,

$$V_{L/D} = V_{ref} + 10\ kt \tag{2.6}$$

Where: $EPNdB$ = effective perceived noise level (dB)
M = maximum take-off mass (Ton),
$V_{T/O}$ = take-off speed (kt),
V_2 = take-off safety speed (kt),
$V_{L/D}$ = landing speed (kt),
V_{ref} = landing reference speed (kt)

3. 3D Simulation

3.1 CFD modelling

A full-scaled turbofan engine model was constructed in SolidWorks simulation software. Geometries of the engine were taken from Trent 1000 high bypass turbofan aircraft engine specifications [20]. The fan air inlet diameter and length of the engine were 2.850 and 4.738 m, respectively. Figure 1(a) shows a side view of the turbofan engine with a round nozzle and (b) is an isometric view of the round nozzle model.

(a) (b)

Figure 1. Round nozzle model (a) side view (b) isometric view.

The number of chevrons and patterns were adopted from Boeing 787 Dreamliner aircraft as shown in Figure 2(a), side view of a turbofan engine with added chevron patterns at the bypass nozzle. Figure 2(b) shows an isometric view of the chevron nozzle model.

As can be seen in Table 1, the simulation was performed using external flow analysis with three streams at mean sea level conditions. Both models were simulated with take-off and landing conditions and complied with aircraft noise measurement procedures [19], which were calculated by using equations (2.5)-(2.6). Ambient airstream velocity for take-off and landing were determined to be 170 and 146 kt. respectively.

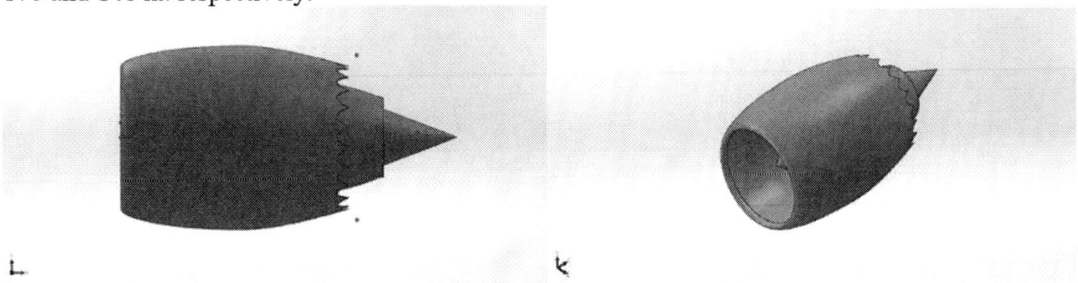

(a) (b)

Figure 2. Chevron nozzle model (a) side view and (b) isometric view.

Table 1. Boundary conditions for 3D simulation.

Boundary	Value	Unit
Air density (ρ)	1.225	kg/m^3
Air pressure (p)	1.01325x10^5	N/m^2
Air viscosity (μ)	1.7894x10^{-5}	kg/m·s
Air temperature (T)	288.16	K

3.2 Numerical calculation

Air velocity, temperature and pressure produced at the fan section were calculated from the following equations (3.1)-(3.4), which were based on a 3-spool turbofan engine with a bypass ratio of 10 [21]. The equations are as follows:

$$P_{02} = P_a(1 + \eta_d \frac{\gamma_c - 1}{2} M_a^2)^{\gamma_c/\gamma_c - 1} \tag{3.1}$$

$$T_{02} = T_a(1 + \frac{\gamma_c - 1}{2} M_a^2) \tag{3.2}$$

$$P_{03} = P_{02}\pi_f \tag{3.3}$$

$$T_{03} = T_{02}(1 + \frac{\pi_f{}^{\gamma_c - 1/\gamma_c} - 1}{\eta_f}) \tag{3.4}$$

Where: P_{02} = fan inlet pressure (N/m^2),
P_a = local air pressure (N/m^2),
η_d = diffuser efficiency (-),
γ_c = specific gas ratio (-),
M_a = Mach number (-),
T_{02} = fan inlet temperature (K),
T_a = local air temperature (K),
P_{03} = fan outlet pressure (N/m^2),
π_f = fan pressure ratio (-),
T_{03} = fan outlet temperature (K),
η_f = fan efficiency (-)

The exhaust gas velocity, pressure and temperature were validated with an engine operation data sheet [20] during full-power and idle operations before performing the simulation. It can be seen from the data in Table 2 that the boundary conditions to be used in the simulation during take-off and landing. Figure 3 represents the computational domain of the chevron nozzle model.

Table 2. Boundary conditions during take-off and landing.

Configuration	Boundary	Ambient stream	Fan	Exhaust gas
	Velocity (m/s)	87.45	284.53	632.55
Take off	Temperature (K)	288.16	297.85	1048.24
	Pressure (N/m^2)	1.01325x10^5	1.01325x10^5 [a]	1.97932x10^5
	Velocity (m/s)	75.0	280.87	255.8
Landing	Temperature (K)	288.16	297.66	473.98
	Pressure (N/m^2)	1.01325x10^5	1.01325x10^5 [a]	1.01325x10^5 [a]

[a]Use P_{MSL} since $P_{Critical} < P_{MSL}$

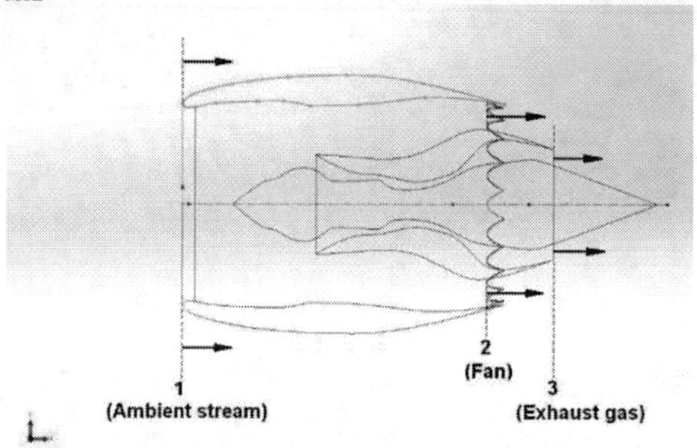

Figure 3. Chevron nozzle model computational domain.

4. Result and discussion

The table below illustrates peak velocity, acoustic power level and turbulent intensity, which were taken from the observation area around the fan nozzle where the highest noise was produced. From this data (see Table 3), it can be seen that the peak velocity at the nozzle of the round model was less than the chevron model for both take-off and landing. The chevron pattern increased velocity by 5.4% during take-off, and 21.5% during landing. However, the acoustic power level was reduced by 58.9% during take-off and 67.4% during landing. A possible explanation for this might be that chevrons were more effective in noise reduction at low power operation.

Table 3. Results of peak value at fan nozzle area.

	During take off		During landing	
	Round model	Chevron model	Round model	Chevron model
Velocity (m/s)	476.21	501.98	408.35	496.04
Acoustic power level (dB)	105.95	43.57	145.77	47.51
Turbulent intensity (%)	1.88	0.37	7.91	0.44

This study confirms that the landing phase with a low power setting produced a higher noise than take-off with full-power. The difference in velocity between take-off and landing was 14.2%. The round nozzle model increased noise levels from 105.95 dB to 145.77 dB, which was an increase of 37.6% while the chevron nozzle increased noise level by 9% from 43.57 dB to 47.51 dB. A possible explanation for these results may be the turbulent intensity increased from 1.88% to 7.91%, which was considered to be in the medium to high-turbulent flow. The chevron nozzle reduced turbulent intensity from 1.88% to 0.37% during take-off and from 7.91% to 0.44% during landing. It has been suggested that turbulent flow from the chevron nozzle was considered to be low-level turbulence.

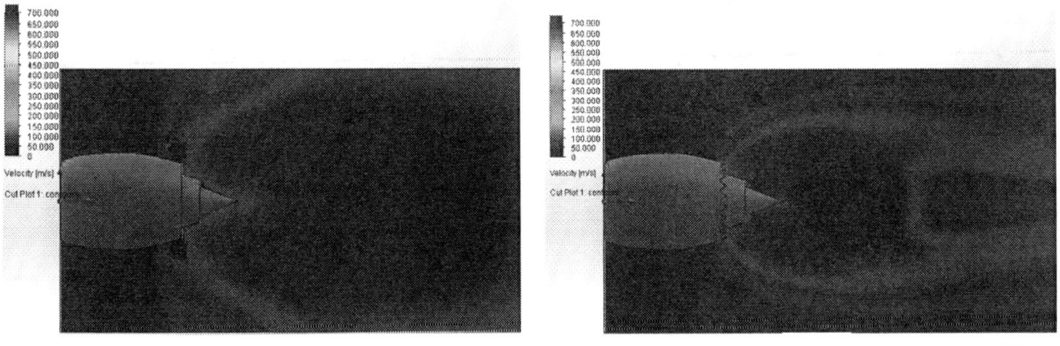

(a) (b)

Figure 4. Velocity plot during take-off (m/s) (a) Round nozzle (b) Chevron nozzle.

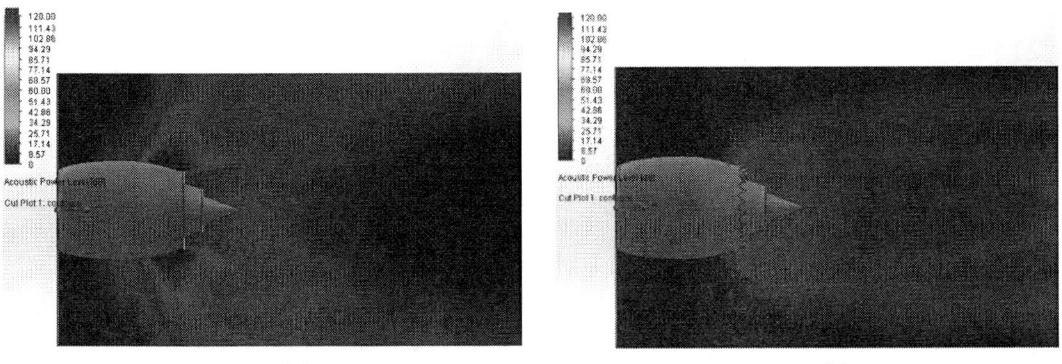

(a) (b)

Figure 5. Acoustic power level plot during take-off (dB) (a) Round nozzle (b) Chevron nozzle.

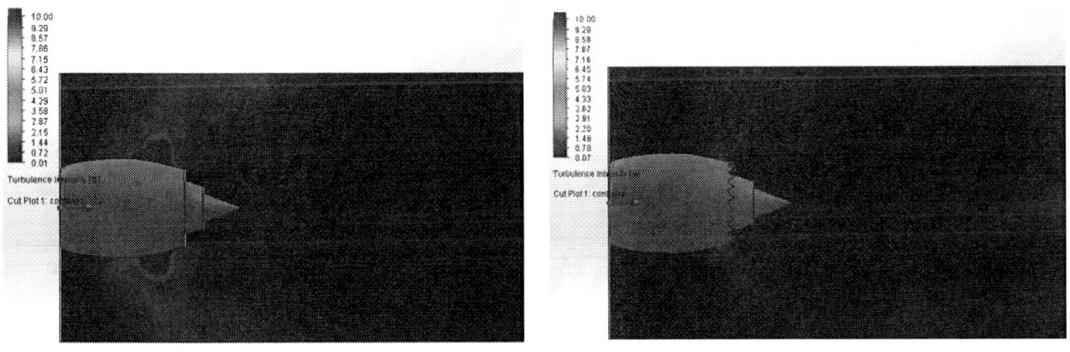

(a) (b)

Figure 6. Turbulent intensity plot during take-off (%) (a) Round nozzle (b) Chevron nozzle.

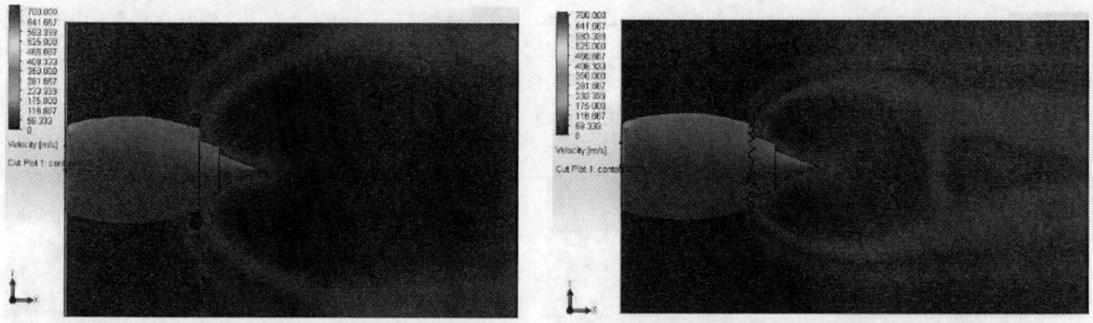

(a) (b)

Figure 7. Velocity plot during landing (m/s) (a) Round nozzle (b) Chevron nozzle.

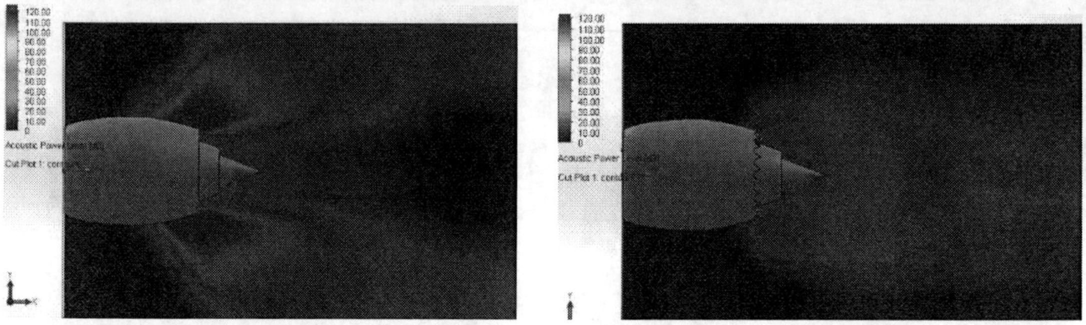

(a) (b)

Figure 8. Acoustic power level plot during landing (dB) (a) Round nozzle (b) Chevron nozzle.

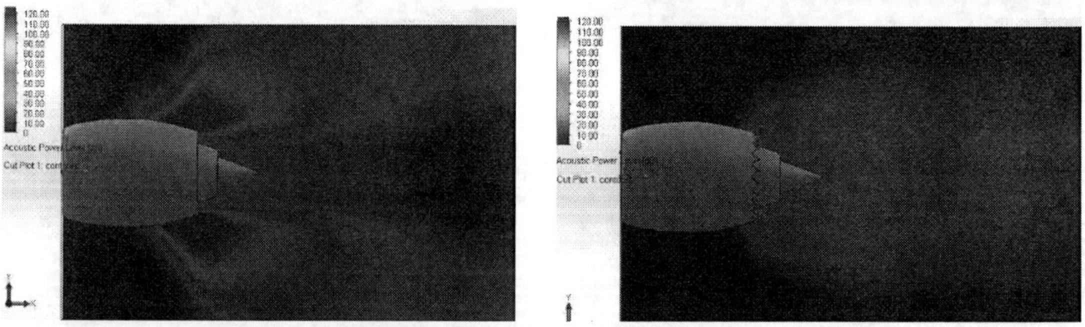

(a) (b)

Figure 8. Acoustic power level plot during landing (dB) (a) Round nozzle (b) Chevron nozzle.

From the data in Figures 4-9, it shows the velocity, acoustic power level and turbulent intensity plot for round and chevron nozzle models during take-off and landing.

As can be seen from the Figures (above), it is apparent that the chevron nozzle was capable of reducing noise by up to 98.26 dB during landing and 62.38 dB during take-off. A slower flight during landing produced a higher turbulent intensity, which caused higher noise regardless of the engine power setting. Maximum noises allowable in EPNdB from equations (2.3)-(2.4) were calculated for a 228-ton maximum take-off mass aircraft, which were 101 dB for take-off and 104.3 dB for landing. On the other hand, the round nozzle model produced a larger jet shear area and higher turbulent intensity, which exceeded the allowable noise level

5. Conclusion

This study has identified that turbofan models with round and chevron nozzles were simulated using take-off and landing conditions, which complied with the International Civil Aviation Organization (ICAO) measurements of aircraft noise [19]. The results of this investigation show that the prediction of acoustic power level for both take-off and landing were demonstrated successfully using actual engine operation parameters. The results of this study indicate that this simulation predicts the acoustic power level closest to the real flight operation. It was found that the chevron is capable of reducing noise level down to permissible levels for operations. The present study adds to the growing body of research that indicates turbulent intensity was the most crucial aspect of noise generation in a high bypass ratio turbofan engine.

References

[1] Durmaz V 2011 *EMAJ.* **1**(2) 13-20
[2] Torija A J and Self R H 2017 *JATM.* **67** 157-68
[3] Filippone A 2014 *Prog. in Aero. Sci.* **68** 27-63
[4] Epstein A H 2014 *AIAA J.* **52**(5) 901-911
[5] Howie D 2015 *The magazine Rolls-Royces.* **146** 12-15
[6] Kempton A 2011 *Acoustic Liner for Modern Aero-Engines* Rolls-Royce (UK: University of Southampton)
[7] Kanmaniraja R Freshipali R Abdullah J Niranjan K Balasubramani K and Kumar V 2014 *Int. J. Aero Mech. Eng.* **8**(9) 1530-1536
[8] Bridges J and Brown C A 2004 *AIAA. Paper.* 2004-2824
[9] Tide P S and Babu V 2009 Numerical predictions of noise due to subsonic jets from nozzles *Appl. Acoust.* **70** 321-332
[10] Sadanandan R Dheeraj R Aswin B Akshay K and Radhakrishnan A 2018 *Int. Res. J. of Eng. & Tech.* **5**(5) 3201-3204
[11] Tide P S and Srinivasan 2008 *Int. Mech. Eng.* **223** 51-67
[12] Rask O Harrison S Munday D Harris C Mihaescu M and Gutmark E 2007 *AIAA. paper.* 2007-5631
[13] Henderson B and Wernet M 2015 *JSV.* **351** 119-142
[14] Lee I Zhang Y and Lin D 2019 *Appl. Acoust.* **150** 246-267
[15] Fontaine R A 2014 *PhD Thesis* University of Illinois at Urbana-Champaign USA
[16] Glegg S and Devenport W 2017 *Aeroacoustics of low Mach Number Flows* (London: Elsevier) chapter 1 pp 3-8
[17] Garrett S 2017 *Understanding Acoustics* (Switzerland: Springer) chapter 10 pp 546-52
[18] Anselmet F and Mattei P 2016 *Acoustics, Aeroacoustics and Vibrations* (New Jersey: Wiley) chapter 5 pp 149-55
[19] ICAO 2011 *Annex 16: Environmental Protection* Volume I-Aircraft Noise (Canada: ICAO)
[20] EASA 2017 *Rolls-Royce Trent 1000 Type Certificate Data Sheet: E.036* (United Kingdom: EASA)
[21] Sforza P M 2017 Idealized cycle analysis of jet propulsion engine *Theory of Aerospace Propulsion* (London: Butterworth-Heinemann) chapter 3 pp 85-171

Automatic Balancing System in Quadcopter with Change in Center of Gravity

Kamolwat Chaisena[1], Benjamaporn Nenchoo[1] and Suradet Tantrairatn[1,*]

[1]Suranaree University of Technology, Nakhonratchasima, 30000, Thailand

* Corresponding Author: suradetj@sut.ac.th

Abstract
The purpose of this paper is to explore an automatic balancing system using the proportional integral derivative or PID feedback control algorithm in quad-rotor Helicopter, so-called quadcopter, with the center of gravity, moved on account of payload. Nowadays, the quadcopter widely used for a variety of applications due to high stability performance. In many applications in quadcopter usually carried a payload in the middle of its frame to maintain the balance during flight. However, the application of quadcopter could be a contributing factor to increase a payload on one side and can lead to an improper position of the center of gravity. This study set out to investigate the use of the mass of battery and landing gear to create a counterbalance to compensate for the payload. The stabilization of the balancing system was implemented by the PID control algorithm to control the counterbalancing of the quadcopter. The current study found that the PID control algorithm works well to maintain the balance in a quadcopter with the center of gravity moving during flight.

Keywords: Balancing System, PID control, Unbalance of Quadcopter

1. Introduction

Since it reported in the early 1900s, a multi-rotor helicopter has been attracting a lot of interest. In the 1920s, there has been an increasing interest in multi-rotor helicopter since it has been used in the military. One of the greatest challenges of the application of a multi-rotor helicopter is its huge, overweight, and instability in control. In a few decades, the technology in the field of Unmanned Aerial Vehicle or UAV has been developing continuously. This type of UAV aircraft was developed to a smaller size to suit the application. It also used in military and civil missions.

Nowadays, a quadcopter is considered as multi-rotor aircraft and has been widely used due to its lightweight and high-lift force, vertically take-off and landing. It has become a central issue for hovering in the air and remains stable for flight control. This type of UAV aircraft consists of four rotors at the end of the arm. Its movement is allowed an adjustment of the rotor speed independently so that it can control a pitch, roll and yaw. The quadcopter can play an important role in addressing the issue of carrying a payload. For instance, it can be used as a camera for aerial video and photography, gripper for pick and place the object, transportation of goods. Generally, the payload is usually attached in the center of mass of the aircraft body to maintain stability during flight. However, in some applications, the payload is mounted out in front of the body because there is not enough space in the center of the aircraft frame.

Studies of quadcopter show the importance of improving PID control so that it can increase the stability in aircraft control [1-4]. Moreover, several attempts have been made to investigate an effect of variations in center of gravity such as the response of rotating speed of the rotors [5], the PID control used to stabilize the quadrotor with dynamic change in center of gravity on account of payload [6]. In particular [7], this research has established the solution to maintain the stability of multi-rotor during flight using landing gear. And, it has been acting as a tail in a mechanical system to create a counterbalance to compensate the payload.

In previous research [8], we have presented an automatic balancing system in quadcopter using the aircraft component as lithium polymer battery, which energizes the whole system to create a counterbalance to compensate the payload attached on the front of the quadcopter frame. The method of the balancing system implemented by PID, which was used to control the stabilization. The battery had linearly driven a servomotor to create a counterbalance. As a result, the response of the previous system was quite slow because the movement was changed from rotation to translation. In the previous work, it had not been mentioned about the improvement of PID control.

This paper is to explore the relationship between a new model of a quadcopter and automatic balancing system using a combination of the mass of battery and landing gear. It is also to create a counterbalance to have a faster response and to present the process of improvements in PID control of stabilization.

2. Concept and Method

This section discusses the model and design of the quadcopter and concept of balancing system. The S-500 PCB quad-rotor frame was used. The quadcopter consists of four motors, there are two motors rotate clockwise (CW) direction and two motors rotate counterclockwise (CCW) direction as shown in Figure1. The moment force around the vertical axis of the aircraft body was compensated for each other.

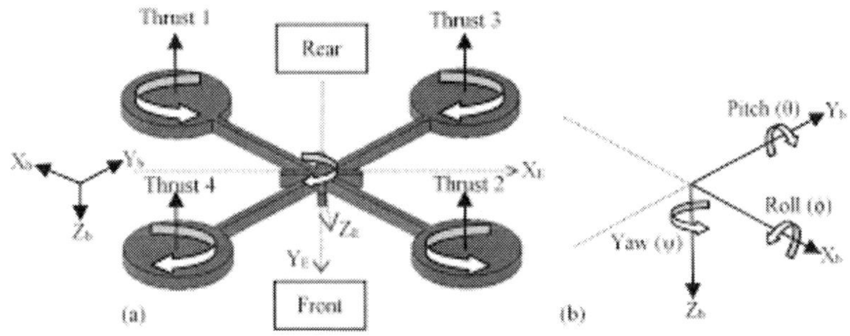

Figure 1. A free body diagram of quadcopter X [9]

In this study, the balance situation with the equally PWM signal value of four motors used as a condition. The unbalance situation simulated by installing a payload on the front of the aircraft frame. The PWM signal value of four motors was unable to have equal values. From this data, the PWM signal values of a pair of front motors showed greater value than a pair of rear motors. The lightweight aluminum bar is attached to the front of the aircraft frame so that it was able to adjust the position of aluminum block to simulate the movement of the center of gravity when carrying a payload (see Figure 2). Figure 3 provides a free body diagram of a quadcopter with three situations.

IOP Conf. Series: Materials Science and Engineering **886** (2020) 012006 doi:10.1088/1757-899X/886/1/012006

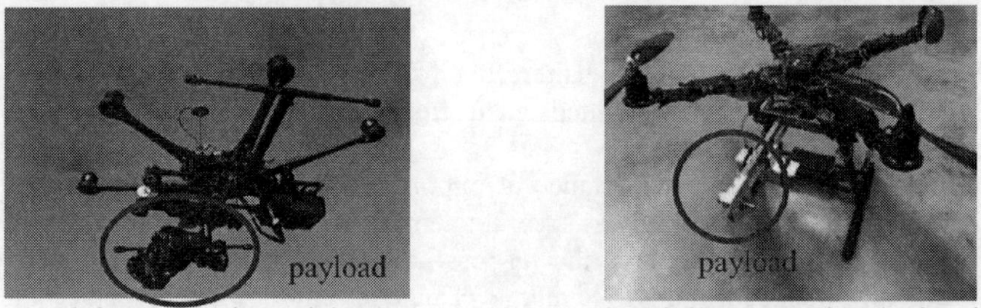

a. Unbalance system in a real situation b. Unbalance system in a simulate situation

Figure 2. Qu.adcopters with payload

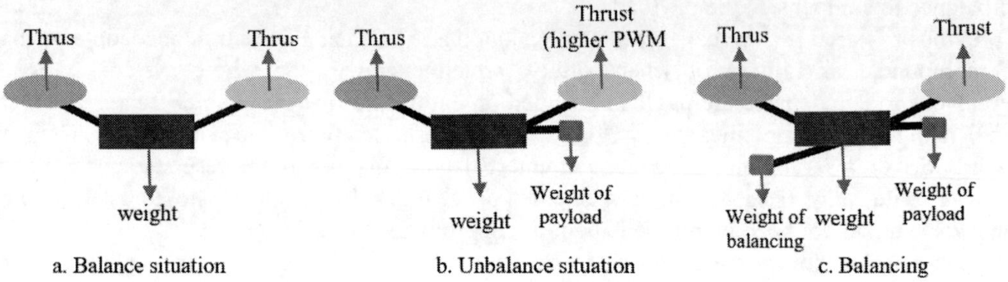

Figure 3. Free body diagram of a quadcopter with three situations

For the balancing mechanism, a couple of high - torque RC servo motor was used as an actuator at the landing gear joints to fold the rear of the aircraft frame and to compensate the payload. As can be seen from Figure 4 below, the battery was mounted at the bottom of the landing gear to increase a balancing capacity.

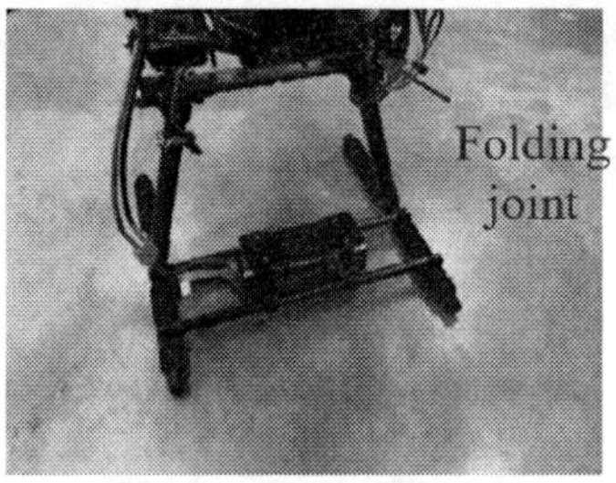

Figure 4. Battery attached with landing gear

The advantage of this balancing mechanism is a faster response than previous research [8] because the folding angle of landing gear depends on the angular angle of faster servomotor. Table 1 shows the aircraft specifications of some parameters.

Table 1. Quadcopter Specifications

Parameter	Size (grams)
Take-off Weight	1,695
Battery	235
Payload	330
Landing Gear	240

3. System Integration

The method of the study based on the main hardware of the quadcopter using the lithium battery as a payload and power source to create the balancing force. Pixhawk board for control stability and movement of quadcopter and receive PWM signal from the transmitter through the receiver.

Moreover, there is an Arduino board as a processing unit to control the servo motor for controlling and balancing the system (see Figure 5 below).

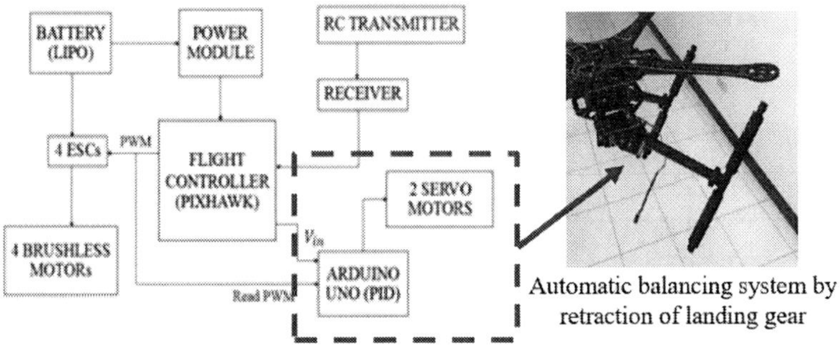

Figure 5. A diagram of a hardware system

By the way, it can be seen that complete systems need more than one hardware system, it also needs software and controller to control the systems. It is apparent from the diagram that the proportional – integral – derivative or PID control is applicable in many industrials and a wide range of research. This controller response is able to correct by tuning three values as Kp, Ki and Kd to create minimum errors.

It suggested that PWM values of a pair of front rotors have equal value and set as a set point. It also has equal value and set as a process variable. As shown in Figure 6a, the setpoint value as an input was sent to the PID controller to control servomotor until a process variable converges effectively to a set point. Besides, the process balancing system starts with tuning PID value with trial and error method in the Arduino board. Arduino board receives PWM values from Pixhawk and sets these values as setpoint and process variable. These values sent to the PID controlling process to control servomotor for balancing system. The schematic diagram provided in Figure 6b.

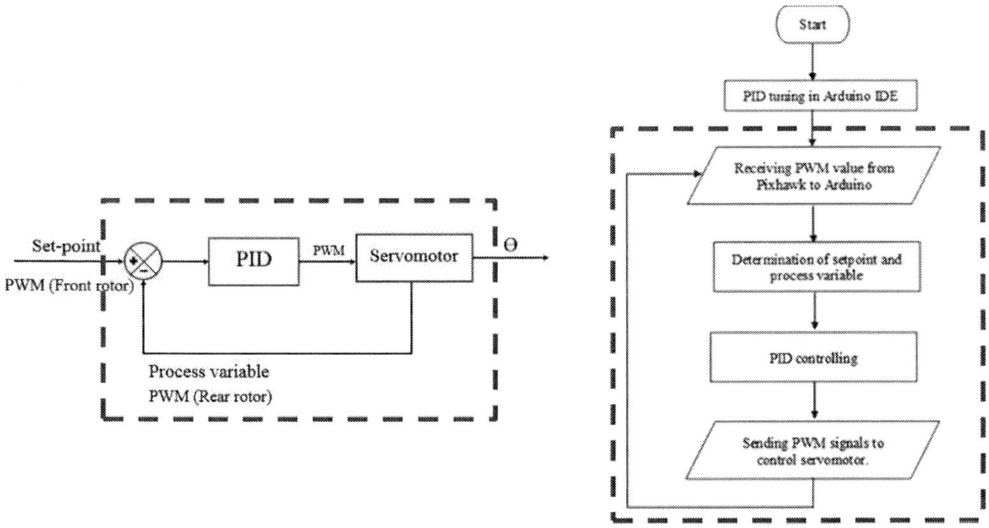

a. A PID block diagram

b. A schematic diagram of the balancing system

Figure 6. An automatic balancing system diagram

4. Experiments and Results analysis

This section shows the experiments and results of the study in four cases. It consists of a test that is with a payload and the test without a payload. It also comprises of a test that is with balancing and a test with a static balancing. To study the behavior of the PWM signal in various cases, a flight plotter program was used to analyze the data.

4.1 Testing without a payload

From the data in Figure 7, it is apparent that the stability of quadcopter is the most important issue since the position of any devices that set in quadcopter has an effect on PWM signals. We can see that the PWM signal values almost have equal values about 1700 (see Figure 7 below).

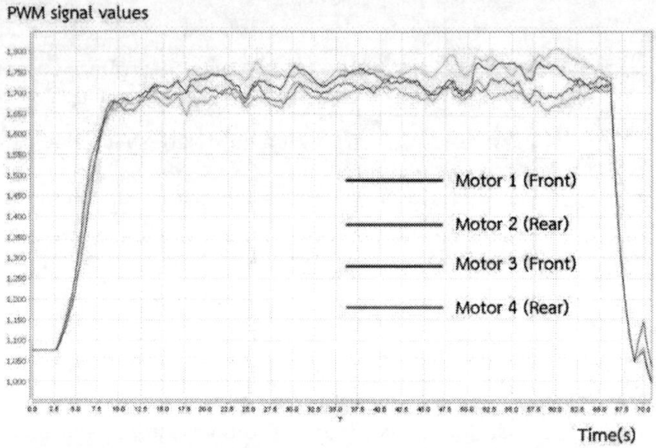

Figure 7. PWM values response in case with balance.

4.2 Testing with a payload

As Figure 8 shows, there is the 330 grams payload attached in front of the frame (15 cm), and it displaces between payload and center of gravity. The result shows that the PWM mean value of a pair of front motors (approx. 1700) was greater than a pair of rear motors (approx. 1550). Hence, the system of quadcopter has to maintain stability by increasing PWM values of the heavier side, as shown in Figure 8.

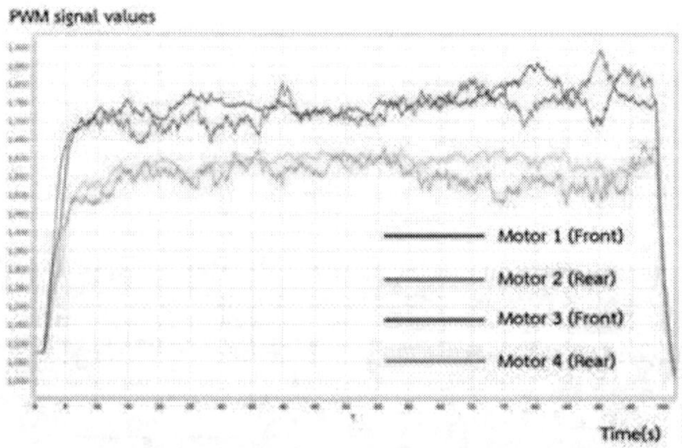

Figure 8. PWM values response in case with unbalance

4.3 Testing with balancing

As seen from the data in Figure 9, the folding of landing gear to the backside used as a counterbalance for the balancing system. The trial and error method used to find PID gain value (Kp, Ki, and Kd). It also used to control the rotation of the servo motor. After trial and error tuning, there are Kp, Ki, and Kd values that suit this system and they have chosen as 0.0375, 0.0175 and 0, respectively.

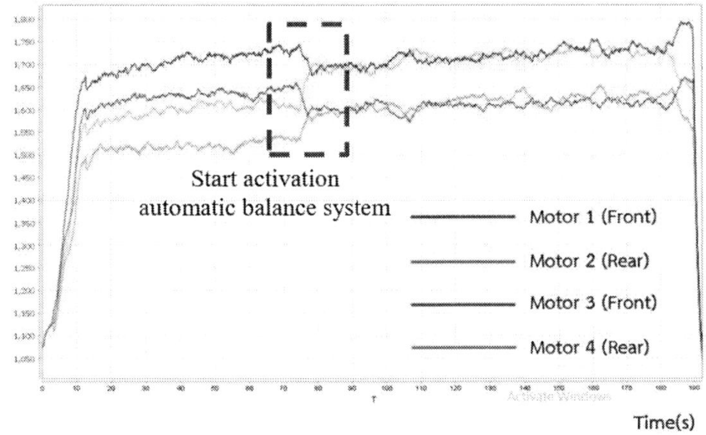

Figure 9. PWM values response in case with balancing.

As can be seen from Figure 8, the result demonstrates that the PWM value of a pair of the motor in the same diagonal line has converged to each other. The settling time is about three seconds and without overshoot of servomotor.

4.4 Static balancing test

After testing all cases, there is a static balancing test to make sure that these gain values are able to maintain the stability of the quadcopter. By testing the tilt angle of the quadcopter in three cases, the data shows balance, unbalance, and balancing. The 330 grams payload was attached at front of the frame in unbalance and balancing test with displacement 15, 17.5 and 20 cm, respectively. A payload at the center of gravity of quadcopter (Kp, Ki, and Kd values) was chosen in testing 4.3, and used in balancing test (see Table 2)

Table 2. Result of static balancing test

Displacement (cm.)	Angle (Deg.)				
	Balance	Unbalance	Balancing	error	Landing gear retraction
15	0.91	15.09	2.70	1.78	44.95
17.5	0.93	17.88	1.86	0.92	45.58
20	0.54	20.54	1.60	1.06	76.08

As shown in Table 2, it is apparent that the displacement of the payload has an effect on the angle of landing gear retraction by increasing displacement. Also, it is able to increase the angle of landing gear retraction. From the data, we can see that displacement of the payload has direct variation to an angle of landing gear retraction (see Figure 10 below).

Balance Unbalance Balancing

Figure 10. Tilt angle in any static testing case.

IOP Conf. Series: Materials Science and Engineering **886** (2020) 012006 doi:10.1088/1757-899X/886/1/012006

5. Conclusion

This study has found that generally, the balancing system by folding landing gear with 475 grams counterbalance payload due to the building of quadcopter with 330 grams of payload. The most obvious finding to emerge from this study is that the PID controller used in this balancing system for maintaining the stability of quadcopter with Kp, Ki and Kd values as 0.0375, 0.0175 and 0, respectively. The results can also be summarised as follow:

5.1 In balance case (without payload): PWM signal values of 4 motor have equal value about 1700.

5.2 In unbalance case (with payload): PWM signal values of a pair of the front motor is greater than a pair of the rear motor.

5.3 In balancing case: PID controller able to control a servo motor with three seconds of settling time, without overshoot and maintain the stability of a quadcopter.

References

[1] Li J and Li Y 2011 *Proc. 2011 IEEE Int. Conf. on Mechatronics and Automation (Beijing)* (China)

[2] Praveen V and Pillai A S 2016 *IJCTA.* **9**(15) 7151-7158

[3] Wang P Man Z Cao Z Zeng J and Zhao Y 2016 *Proc. 2016 Int. Conf. on Advanced Mechatronics Systems (Melbourne)* (Australia) pp 498-503

[4] Qasim M Susanto E and Wibowo A S 2017 *Proc. 5th Int. Conf. on Instrumentation, Control, and Automation* pp 109-114

[5] Xu X Liu C and Ye B 2016 *Proc. 8th Int. Conf. on Modelling, Identification and Control* pp 1348-353

[6] Ariyanto M Paryanto M and Naniwa T 2016 *Proc. 8th Int. Conf. on Information Technology and Electrical Engineering (Yogyakarta)* (Indonesia) pp 1-6

[7] Molina J and Hirai S 2017 *Proc. Int. Conf. on Unmanned Aircraft Systems (Miami)* (USA) pp1731-1736

[8] Chaisena K Chamniprasart K and Tantrairatn S 2018 *Proc. Int. Conf. on Engineering Science and Innovative Technology (Phang-Nga)* (Thailand) pp 1-5

[9] Kuantama E 2017 *Int. J. Computers Communication and Control* **12**(4) 519-532

IOP Conf. Series: Materials Science and Engineering **886** (2020) 012007 doi:10.1088/1757-899X/886/1/012007

Assessment on Effect of Load Frequency to Durability of Chassis of the Multi-Purpose Forest Fire Fighting Vehicle

Xa Hoi Nguyen[1],[*], Van Van Luong[2], Thanh Quang Nguyen[3], Tuong Tran Van[4]

[1]University of Fire, Hanoi, Vietnam
[2]Vinh Long University of Technology Education, Vietnam
[3]Hanoi University of Industry, Vietnam
[4]Vietnam National University of Forestry, Vietnam

* Corresponding Author: xahoinguyen@gmail.com, Telephone +84913377240

Abstract
Chassis is the main structural component of the automobile; it is subject to any damage due to fatigue or destruction during the operation of the automobile. Therefore, it is important to research and assess the durability of the chassis. In particular, studying the effect of load frequency on chassis durability is very important in the assessment of chassis durability. An objective of this study was to investigate the effect of load frequency, from the road surface and engine to chassis durability when the multi-purpose forest fire fighting vehicle operates in some cases when moving on the road. The assessment of the effect of load frequency on the chassis durability of the multi-purpose forest fire fighting vehicle contributes to input parameters for the determination and assessment of fatigue problem and breaking strength of the chassis.

Keywords: Chassis, load frequency, multi-purpose forest fire fighting vehicle, chassis durability

1. Introduction

The multi-purpose forest fire-fighting vehicle is manufactured by Vietnam and has been considered as one type of equipment in integrating many forest fire fighting functions. Their major functions include cutting trees, cleaning garbage grass, opening roads to create a fire isolation corridor; and using at firing sprinkler with wide spray area; creating high-pressure wind spray on the fire. Also, this vehicle uses sandy soil in place of fire extinguishing and is based on the basis of URAL 4320 vehicles. One of the main obstacles is vibrations can cause high-stress concentration resonance on the frame destruction structure or may cause fatigue on the frame during operation.

a) Base vehicle Ural 4320 b) Multifunction forest fire fighting was built

Figure 1. Multi-purpose forest fire fighting vehicle

The main equipment of multifunction forest fire fighting vehicle is shown in Figure 2 [11].

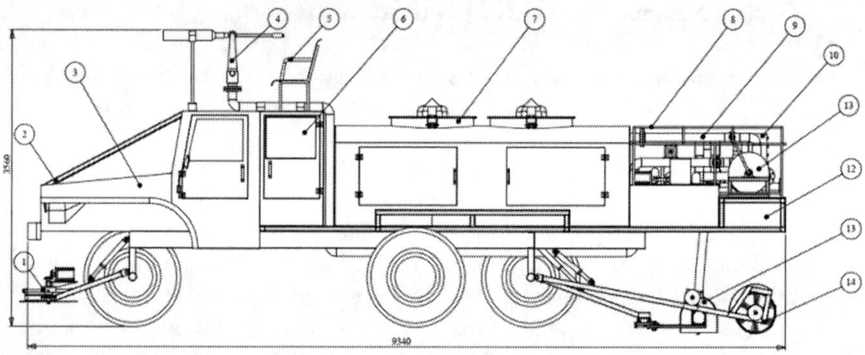

1) Tree cutting mechanism; 2) Tree cleaver; 3) Body; 4) Water sprayer; 5) Seat; 6) Cabin; 7) Water tank; 8) Back cover; 9) Pipe; 10) Table; 11) Ploting small nozzle; 12) Floorboard; 13) Dig system; 14) Lawn cutting mechanism.

Figure 2. General design multifunction forest fire fighting vehicle

Table 1. Specification of multifunction forest fire fighting vehicle

Order	Specification	Unit	Value
1	Power	Hp	180
2	Fire extinction speed	meter/minute	12
3	Flame height is extinguished	m	9
4	Translational speed	km/h	25
5	Longitudinal grade vehicle works	degree	$< 15^0$
6	Cross slope vehicle works	degree	$\leq 5^0$
7	Empty weight	kg	17.230
8	Full weight	kg	12.130
9	Container capacity water	m^3	5,1
10	Pump delivery	m^3/h	10
11	Length of intake water	m	300

Studies of the fluctuation of automobile chassis show the importance of the fluctuation of automobile chassis. For example, the durability of the chassis can contribute to ensuring passive safety and can improve the life of the chassis. The chassis is of interest because it is a system of beams that plays the role of receiving and transmitting reacting force during the operation of automobiles with different road conditions. In addition, the chassis is also influenced by vibrations from engines and power train. A primary concern of chassis is the vibration of car-mounted detail assemblies can cause fatigue at installation locations. A considerable amount of literature has been published on the durability of frames in static and dynamic problems. No previous study has investigated the effect of frequencies of external forces on the frame strength [3, 4, 9, and 10]. Currently, there are no data on the durability of the frame to prevent forest fire engines under the influence of frequency according to the method of theoretical testing by software. The importance and originality of this study are that it contributes to assessing the influence of frequency to the durability of light truck frames in particular and the scientific basis in the process of completing the design of forest fire engines.

2. Building models

2.1. Building a chassis model

Traditionally, chassis of multi-purpose forest fire fighting vehicle model has been built by Solidworks software according to the actual size of the chassis. The chassis is made of 40CR steel material with a specific weight 7850 kg/m^3, modulus 205 GPa, tensile strength 980 Mpa. The yield point is at 785 Mpa.

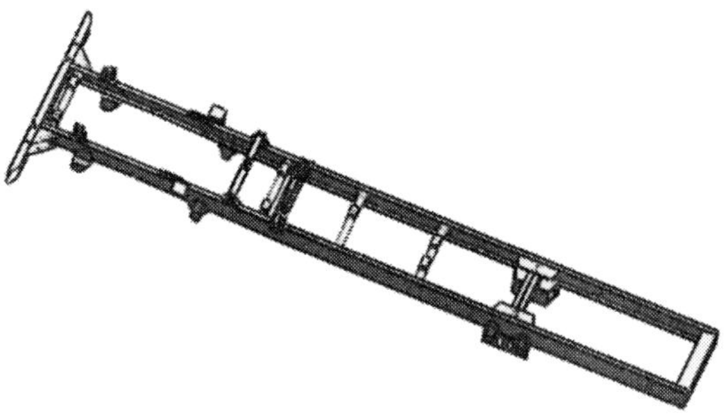

Figure 3. A 3D chassis model

Figure 4. A finite element model

In this study, the authors have simplified the actual geometric structure elements without affecting the results of the problem such as holes and removed studs. To identify longitudinal beams and cross beams with absolute rigidity, welds of cross beams with longitudinal beams were not included. For analysis of entire load of engine, cargo boxes are distributed on the surface of the chassis.

2.2. Building finite element model

The construction of finite element model was used in designing and analyzing problems. The meshing process of the chassis model used in this study is based on the structural characteristics of the frame and the lower model control criteria of ANSYS software [1, 2, and 8]. Using of this process is useful in determining the accuracy of finite element model used.

To reduce the problem size and to increase accuracy for finite element model while still ensuring computational time, hexahedral elements with 20 nodes (Hex 20) and tetrahedral elements with 10 nodes were used (see Tet 10). As can be seen from Figure 4, the finite element model of multi-purpose forest fire fighting vehicle chassis comprises of 174223 elements, 376982 nudes and average element quality that reaches 0,725.

3. Analyze results

3.1. Chassis's own vibrations

The chassis's own vibration analysis is a linear analysis method to determine the frame's own vibration frequency [4, 5]. Determine the specific vibration frequency of the chassis to avoid resonance occurs. Therefore, the determination of the frame's own vibration frequency is the basic math in the design process. In the separate oscillation analysis, when the mass matrix [M] and stiffness matrix [K] constant, there is no value in weighting on the chassis.

In this study, the linear analysis method was used to investigate the frame's vibration frequency [4,5]. To avoid the resonance, the specific vibration frequency of the chassis was used. For designing a basic math process, the determination of the frame's vibration frequency was used.

$$[\mathrm{M}]\{\ddot{u}\}+[k]\{u\}=0 \qquad (1)$$

Table 1 provides the results of the analysis of the mode of vibration and mode of the frame's vibration by Ansys software.

Table 2. The form and frequency of individual vibration of multi-purpose forest fire fighting vehicle chassis

Mode	Frequency (Hz)	Total Deformation (mm)	Mode	Frequency (Hz)	Total Deformation (mm)
1	17,812	3,4996	11	100,58	4,6058
2	31,436	2,4011	12	103,55	8,8978
3	54,127	4,1007	13	111,42	13,564
4	63,506	2,347	14	114,53	2,8216
5	64,404	4,5987	15	115,75	14,297
6	76,248	5,0434	16	124,61	6,1435
7	77,961	4,4542	17	129,88	6,6524
8	82,337	0,3608	18	133,73	6,9162
9	98,081	3,1975	19	138,64	15,304
10	100,21	3,2097	20	140,1	14,795

The result of the analysis of table 2 shows the chassis fluctuates with 20 different modes of vibration from 0 to 82 Hz. At the initial frequencies, the chassis is mainly bending in the Y direction. Strong vibrations occur only at node 3 onwards. The table above illustrates the large displacement value appears only in some modes. However, this displacement only occurs in horizontal bars.

IOP Conf. Series: Materials Science and Engineering **886** (2020) 012007 doi:10.1088/1757-899X/886/1/012007

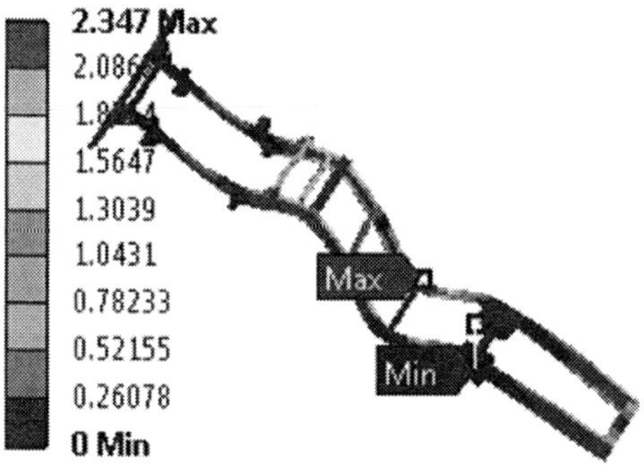

Figure 5. Natural oscillation of the chassis at frequency 63,506Hz

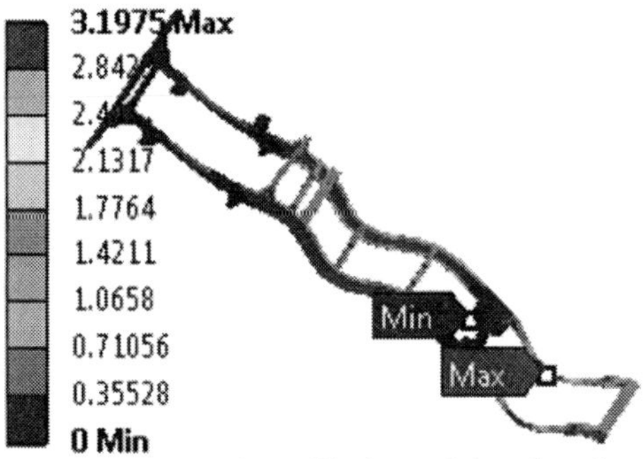

Figure 6. Natural oscillation of the chassis at frequency 98,081Hz

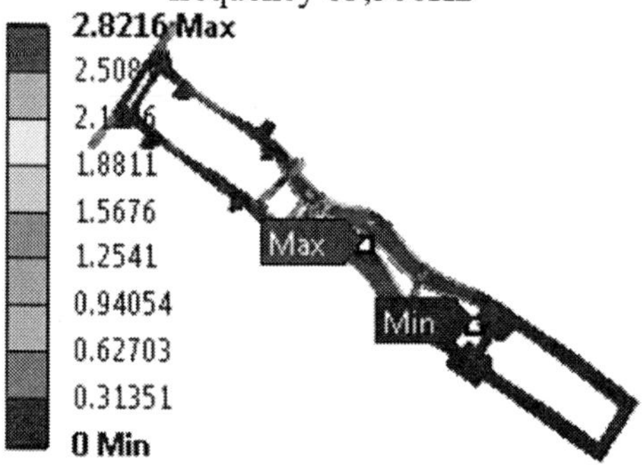

Figure 7. Natural oscillation of the chassis at frequency 114,53Hz

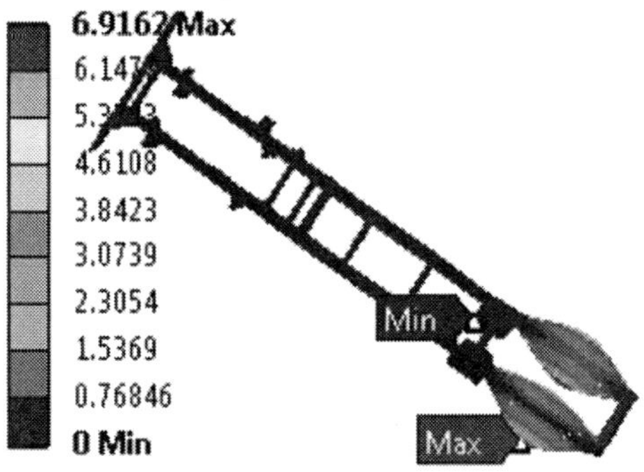

Figure 8. Natural oscillation of the chassis at frequency 133,73Hz

3.2. Influence of road surface bumps to vibrating chassis

The analysis of random vibration [6] was used to determine the parameters of the frame structure such as deviation, displacement, velocity, acceleration. This study also involves the response of the frame model to vibration and fatigue life structure.

Regarding random vibration analysis, the current study found that Ansys uses energy spectra or Power Spectral Density (henceforth PSD) to determine the reaction of the frame structure with the stimuli of the external force acting at random frequencies. Further analysis showed that the average value of the probability data can be distributed by Gaussian. Spatial frequency n is calculated by the number of cycles per meter of the road (cycle/m). The line profile is a variation of the bumping height along the longitudinal axis of the road. PSD is the energy spectral density function of the bumping height of the pavement calculated by frequency n or angular frequency Ω:

$$G_d(n) = G_d(n_0)\left(\frac{n}{n_0}\right)^{-w} \text{ or } G_d(\Omega) = G_d(\Omega_0)\left(\frac{\Omega}{\Omega_0}\right)^{-w} \tag{2}$$

The n is for spatial frequency (cycle/m). The symbol n_0 is the reference value of n (taken $n_0 = 0,1$ cycle/m). The $G_d\ (n_0)$ is the frequency spectrum at the frequency number n_0 (taken according to table 2). The symbol Ω is an angular frequency (rad/m), Ω_0 =1rad/m. To calculate the types of motor road, w index is usually chosen by 2.

ISO 8608: 1995 distinguishes the types of roads (streets, highways and non-road terrain) according to the energy spectrum density. It is also divided into eight categories with conventional symbols from A to H. The letter A is the best quality road surface type, the next letters describe the poor quality sugar and the last letter H is the worst road condition.

Table 3. ISO 8608: 1995 road classification [7]

Type of road	Parameters describing bumpy space frequency, n			
	k	$G_d(n_0)$ [10^{-6} m^3]		
		Min	Average	Max
A – B	3	--	24	25
B – C	4	25	26	27
C – D	5	27	28	29
D – E	6	29	210	211
E – F	7	211	212	213
F - G	8	213	214	215
G - H	9	215	216	217
H +	--	217	--	--
		$n_0 = 0,1$ cycle/m		

Further analysis revealed that the chassis is a structure subjected to complex forces with random fluctuations when traveling on uneven roads. In addition, the chassis is subjected to random vibrations of the engine during operation, powertrain and wheel axles. The chassis during operation is affected by many vibrations so that the random vibration analysis uses the power spectral density with the values determined according to the standard in ISO 8608: 1995 [7] with test speed at 20 km/h and type of D-E and E-F.

Table 4. Power spectral density in D-E road space

n_c/m^{-1}	0,125	0,25	0,5	1	2	4	8
$G_d(n_c)/10^{-6}$ m^3	134,4	33,6	8,4	2,1	0,525	0,13125	0

Table 5. Random load density according to D-E road frequency

f/Hz	0,6944	1,3888	2,7777	5,5555	11,111	22,222	44,444
$G_d(f)$ 10^{-6} (m^3 /Hz)	24,192	6,048	1,512	0,378	0,0945	0,023625	0

During the calculation process when the vehicle moves, the energy density value is changed according to the formula.

$$f = v.n_c \quad \text{and} \quad G_d(f) = \frac{G_d(n_c)}{v} \tag{3}$$

Consider the road E-F, power spectral density space and random load density according to frequency available:

Table 6. Power spectral density in E-F road space

n_c/m^{-1}	0,125	0,25	0,5	1	2	4	8
$G_d(n_c)/10^{-6}$ m^3	134,4	33,6	8,4	2,1	0,525	0,13125	0

Table 7. Random load density according to E-F road frequency

f/Hz	0,6944	1,3888	2,7777	5,5555	11,111	22,222	44,444
$G_d(f)$ 10^{-6} (m^3 /Hz)	24,192	6,048	1,512	0,378	0,0945	0,023625	0

The present results of random vibration calculations are analyzed on Ansys. The change in velocity and acceleration according to the frequency are shown in Figures 9 to 12.

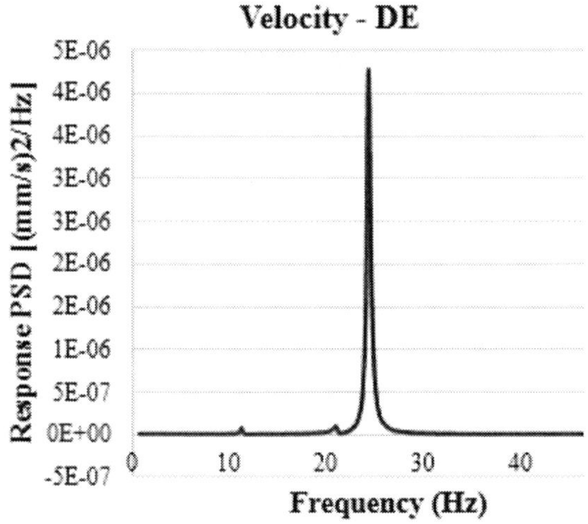

Figure 9. Speed change according to frequency in road D-E

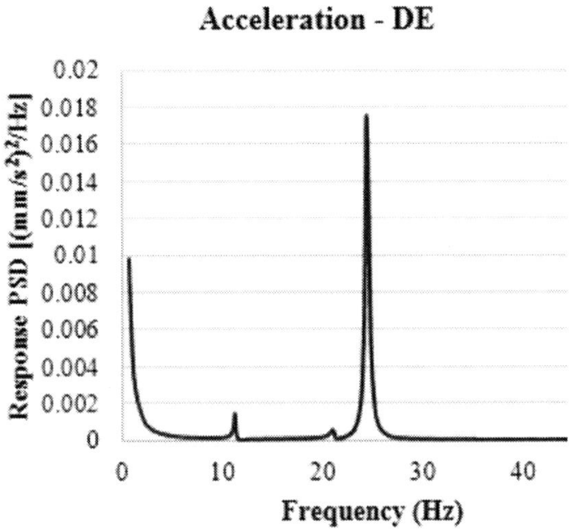

Figure 10. Acceleration change according to frequency in road D-E

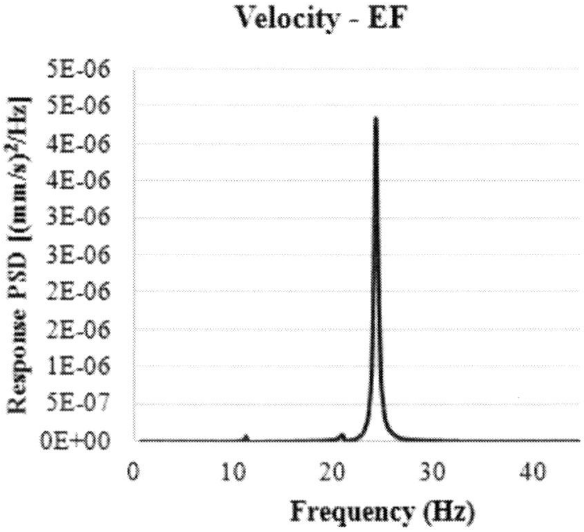

Figure 11. Speed change according to frequencyin road E-F

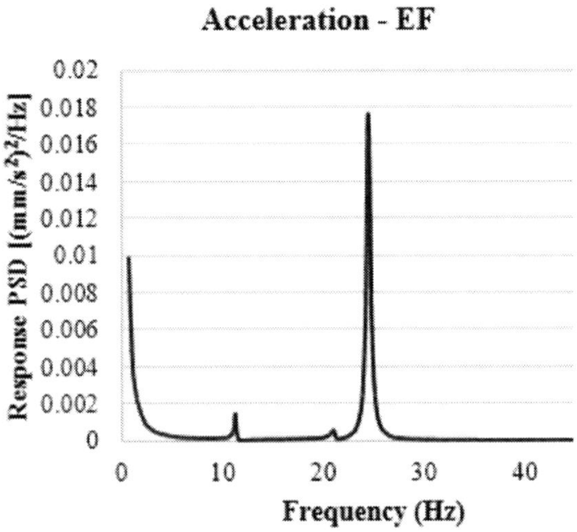

Figure 12. Acceleration change according to frequency in road E-F

The analysis results show that on D-E and E-F lines when the vehicle moves at 20 km/h, the change in speed and acceleration according to the largest frequency appears at 24, 42 Hz. It seems possible that these results are due to the chassis is not resonated when traveling on the D-E and E-F lines at a speed of 20 km/h. A possible explanation for these results may be the largest oscillator frequency value are outside the individual oscillator of the frame.

4. Conclusion

This study set out to present the influence of the frequency of some external forces to stimulate the durability of multi-purpose forest fire fighting frame in some practical working conditions. The present study was designed to study of durability and design of a part assembly. The empirical findings in this study provide a new understanding of the effect of other detail clusters on the chassis such as powertrain and engine assemblies, and to extend our knowledge of evaluation of the oscillation and frame durability.

References

[1] Bhise S Dabhade V Pagi S Veldandi A and Chodankar V 2017 *International Journal of Scientific & Engineering Research* 8(3) 158–166
[2] Sirisha K P Narayana, R L Gopichand A Srinivas C and Balaji G R 2013 *International Journal of Engineering Research and Applications* 3(4) 2318–2323
[3] Karaoglu C and Kuralay N S 2005 *Elsevier Science B.V. Finite Elements in Analysis and Design* 38 1115–1130
[4] Rao K S Musalaiah G Mohana K and Chowdary K 2016 *Indian Journal of Science and Technology* 9(2) 1–5
[5] Fui T H and Rahman R A 2007 *Jurnal Mekanikal* December 2007(24) 56–67
[6] Hai-fei W Kun-kun J and Zi-peng G 2014 *Journal of Chemical and Pharmaceutical Research* 6(3) 849–852
[7] The International Organization for Standardization 1995 *Road ISO 8608:1995*
[8] MoavenivS 2008 *Finite Element Analysis: Theory and Application with Ansys*, 3rd edition (Pearson)
[9] Patil H B Kachave S D and Deore E R 2013 *IOSR Journal of Mechanical and Civil Engineering* March – April 2013(1) 44–49
[10] Rahman R A Tamin M N and Kurdi O 2008 *Jurnal Mekanikal* December 2008(26) 76–85
[11] Luong V V Nguyen Q T and Tran T V 2018 *Proc. 9th TSME International Conference on Mechanical Engineering*(Phuket:Thailand)182–18

Determination of Dynamic Loads from the Road Surface Acting on the Chassis by Experimental Methods

Van Van Luong[1,*], Thanh Quang Nguyen[2], Hong Quan Le[3], Van Anh Le[4], Phuc Hoa Tran[5], Xa Hoi Nguyen[6], Tuong Tran Van[7]

[1]Vinh Long University of Technology Education, Vietnam
[2,3,4,5]Hanoi University of Industry, Vietnam
[6] University of Fire, Hanoi, Vietnam
[7]Vietnam National University of Forestry, Vietnam

* Corresponding Author: vanlv@vlute.edu.vn, Telephone +84902933271

Abstract
Chassis is the main structural component of an automobile, which is subjected to the full weight load being suspended. In the process of moving, the vehicle can both move and cut the tree while moving and sweeping the garbage as well as moving water. During the operation,the chassis is subject to dynamic loads depending on the driver and road conditions. The variation of dynamic loads causes fatigue failure to reduce the life of the chassis. Therefore,the determination of dynamic loads as input for chassis durability problem is necessary. This paper presents the research results on the determination of dynamic loads from the road surface acting on the chassis of multi-purpose forest fire fighting vehicles by experimental methods. Experiments are performed when vehicles are moving on forest roads, under the actual working conditions of the vehicle.

Keywords: Chassis, Dynamic loads, multi-purpose forest fire-fighting vehicle.

1. Introduction

In Vietnam, most durability tests are carried out on test bench with static load modes. In this way, they are limited in assessing the durability of vehicles in actual operating conditions. Therefore, they need to use dynamic loads as input parameters for durability determination models that have increasingly been developed in recent years. The chassis is a girder system bearing loads installed and loaded, receiving and transmitting jets during vehicle operation. Besides, it is influenced by vibrations from engines, powertrain, etc. A study of the load mode applied on the chassis is only within atheoretical model but there are no practical tests to assess the reliability of this model [1, 2, and 3]. Stemming from that reason, the author proposed a method to determine the dynamic load from the road surface to the cement frame by an experimental method with existing equipment in Vietnam. The purpose of the experiment is to measure the pavement reaction force acting on the sphere when the vehicle passes the standard scale. The reaction value from the road surface to the axle via suspension will affect the chassis. Experimental results are compared with the theoretical results to verify the simulation model to determine the dynamic load, as the reason for the calculation of durability of a multi-purpose forest fire truck.

Calculations that determine the impact on the chassis are carried out on multi-purpose forest fire fighting vehicle manufactured by Vietnam. This is a vehicle integrating many forest fire fighting functions including cutting trees, cleaning garbage grass, opening roads to create a fire isolation corridor; fire sprinkler with wide spray area; create high-pressure wind spray on the fire; using sandy soil on spot to extinguish the fire has been designed based on the vehicle URAL 4320, which has three active main axles used in the military field.

Studies to determine the force reaction from the road surface to the rear axle are limited, especially the method of experimentally determining when the vehicle is moving on the road. Many studies on the durability of rear axle cover under the effect of the effective loads [7, 8, 9, and 10]; these studies indicate the stress and deformation that appear on the active bridge cover. In their research, M H Trinh and colleagues [11] have determined the dynamic load from the road acting on the rear axle of the truck by experiment method, tested on real roads, vehicles passing through sinusoidal models when the vehicle is moving at different speeds.

Figure 1. Experimental multi-purpose forest fire fighting vehicle

2. Model of load determination on the bridge shell

2.1. Vertical load impacts on the rear axle of multi-purpose forest fire fighting vehicle

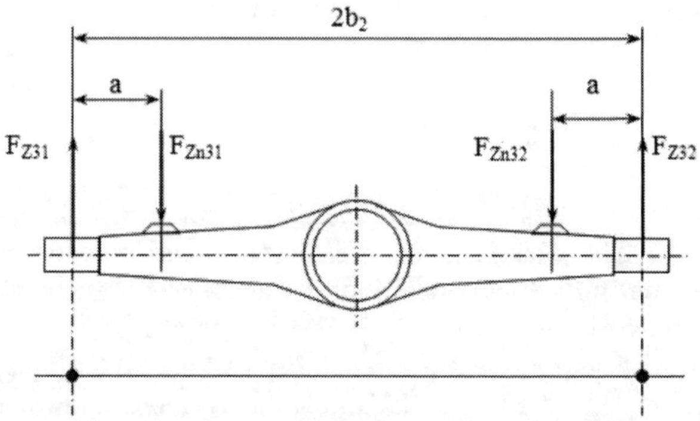

Figure 2. Diagram of forces acting on active rear axles of multi-purpose forest fire fighting vehicle

As can be seen in the diagram, the shell is subject to the reaction of the jet from the road surface Fz31, Fz32 (N) and force from leaf spring Fzn31 và Fzn32 (N); a shows the distance from the point of the jet from the road surface to the point of the force from leaf spring (m); b displays the distance from two points of the jet from the left and right pavement (m).

2.2. Determination of dynamic load impacting the bridge shell after active multi-purpose forest fire fighting vehicle

Generally, when the vehicle moves on the road, the forces acting on the axle are dynamic and they are determined by solving the differential equation system describing the general dynamic model of the vehicle. In this study, the author limited the testing of vertical Fz31 and Fz32 reaction tests from the road surface to the axles. To verify and compare with the theoretical model developed, the author uses the input agitation function, which is the sinusoidal scale of the road surface. The height of the sinusoidal model is determined by the following formula:

$$h(x) = \begin{cases} \frac{1}{2} H\left(1 - \cos\left(2\pi \frac{x}{L}\right)\right) & when \quad 0 < x < L \\ 0 \qquad when \quad x \le 0 \quad , \quad x \ge L \end{cases}$$

H - maximum surface roughness height, H= 0.1m; L - surface roughness length (L= 0.6m); v - velocity of vehicle movement.

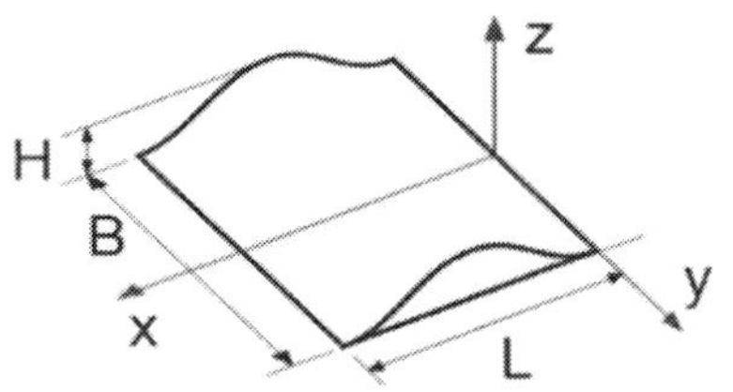

Figure 3. The surface roughness described in length

3. Experiment to determine dynamic load from road surface impacting rear axle of a multi-purpose forest fire fighting vehicle

3.1. Experimental purpose

The objective of the experiment is to measure the dynamic load from the road surface acting on the bridge in the process of passing the standard scale. Experimental results are compared with calculation results by dynamic models to verify theoretical models.

3.2. Measurement parameters

The parameter to be measured in the experiment is the force exerted from tires onto the rear axle in a vertical direction F_z (N)

3.3. Methods and measuring devices

3.3.1. Principle of measuring dynamic loads acting on rear axles

The method of measurement determines the forces applied from tires to the rear axle vertically with the application of the Tenzo switch with bending load on any beam. Figure 4 is a diagram showing the measurement structure system that determines the bending load on a beam. The measuring system includes sensors, which receive, amplify, convert, process, display and store measuring results. The electrical signal (current or voltage) from the sensors is amplified many times, then through the A/D converter into a digital signal transmitted to the software installed in the computer (PC).

IOP Conf. Series: Materials Science and Engineering **886** (2020) 012008 doi:10.1088/1757-899X/886/1/012008

3.3.2. Experimental equipment

To measure the vertical legal force, the research team used a four-resistor device. Four tenzo resistors are pasted in parallel and arranged symmetrically on the upper and lower sides of the bridge at 12 o'clock and 6 o'clock positions (the upper surface of 02 leaves, the lower surface of 02).

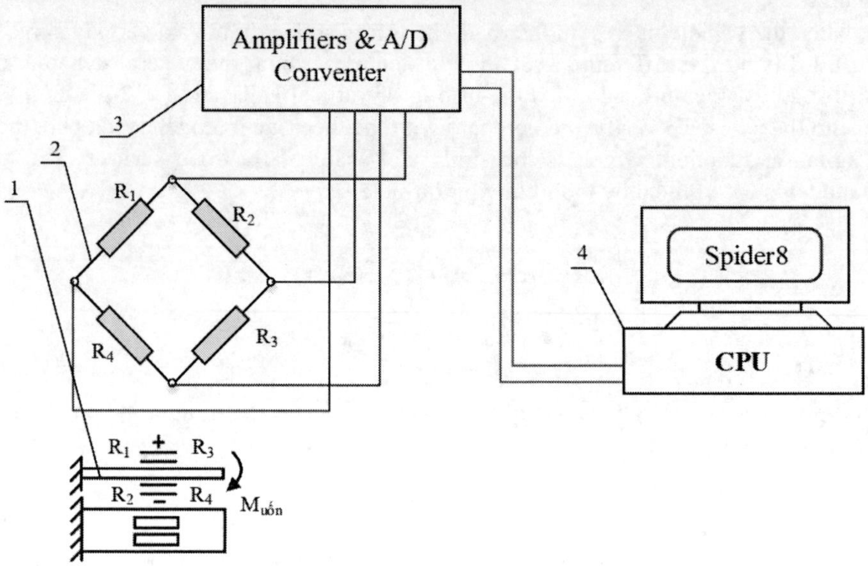

Figure 4. Diagram of the normal force

1) Bridge beam; 2) Tenzo; 3) Amplifiers and A/D Converter; 4) Computer.

Experimental devices include:

a. Resistor bridge

A diagram of the axle resistorand a measurement resistor bridge is shown in Figure 4 above. Tenzo resistors are applied directly to the outer surface of the axle. Before that, one must clean the shaft surface to achieve the required roughness, clean with gasoline and acetol.

To measure the vertical force acting on the rear axle proactively, four Tenzo resistive tenzo leaves (120V) are affixed to the upper and lower surfaces of the axle, facing on two leaves, and facing below two parallel leaves and opposite. Four Tenzo are arranged passes through the active axle axis (Figure 5). The resistors are connected into a bridge circuit as shown in Figure 6, called a resistor bridge (i.e. measurement bridge), the bridge circuit has 4 nodes, in which 2 nodes are connected to the feeding source (U) and 2 remaining nodes perform signal transmission function (U_d)

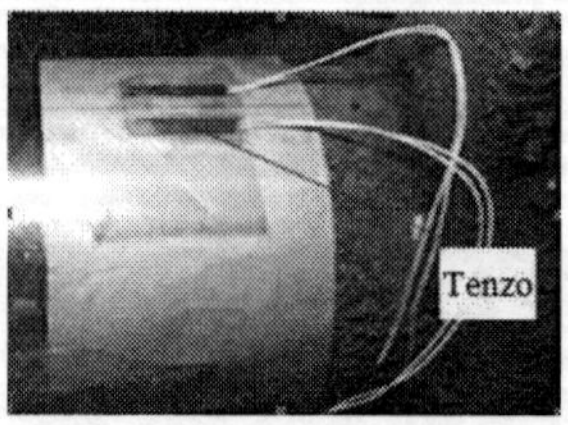

Figure 5. Tenzo affixed on the experimental shell

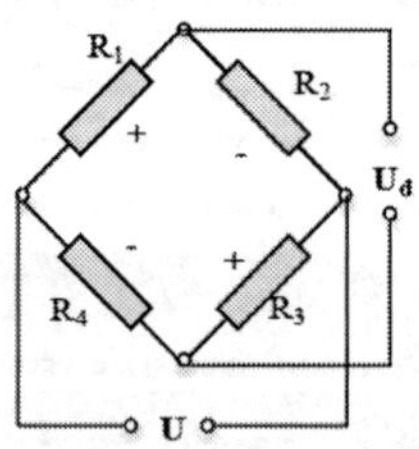

Figure 6. Tenzo bridge circuit

b. Experimental equipment

The force signal measured on the axle is received and amplified by the Spider8 device. This is a signal amplification device and an A/D converter is connected to a computer manufactured by the German HBM company.

c. Calibration of force measurement value

Calibrate the value of the measure of normal force to determine the relationship between the measured voltage in the bridge resistor and force from the road surface applied to the wheel in units of measurement, is N. Tenzo resistors are calibrated with the Spider8 measuring device and the Z4 load cell. To create a vertical force from the loads acting on the bridge, the team used hydraulic jacks in conjunction with the standard Z4 load cell to act on the bridge vertically.

To calibrate the measuring device, the research team progressively increased the load applied to the load cell Z4 by hydraulic jerks (lifting the axle), now under the effect of loads placed on the vehicle as an axle. deformed, change the resistor value and voltage sent to the computer. The measurement results are shown in Table 1 and the calibration graph in figure 8.

Figure 7. Calibration of force measurement value

Table 1. Results of calibration of the normal reaction

Force [kN]	0	5	10	15	20	25	30	35	40	45	50
Volt output [mV]	0	4,7	9,5	13,8	18,3	22,6	27,1	31,2	35,5	39,8	44,2

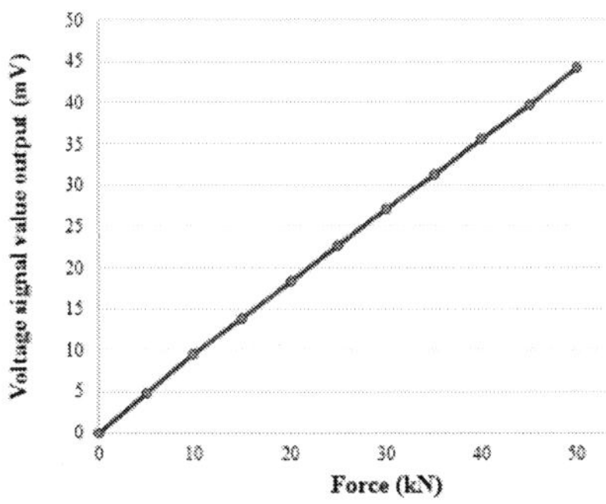

Figure 8. The calibration graph between the voltage signals measured at the resistor and the vertical force on the bridge

3.4. Experimental options

The experiment was conducted on a fine, flat, dry, bumpy asphalt road selected as random bumps on a road with a height of 0.1m, a length of 0.6 m. (Figure 9). In fully loaded vehicles, rear right wheels go up bumping at the speed of 10 km/h, 15km/h, and 20km/h.

Figure 9. An experimental stretch of road with wheels

3.5 Experimental results

The experimental results are compared with simulation results under the same conditions (when vehicles on the rear wheels have to be disguised with the format speed of 10, 15, 20 km/h.) to assess the accuracy of the model. Theoretical simulation results were simulated by Matlab Simulink on the dynamic model of the vehicle [6] when the vehicle passed the sinusoidal model (section 2.2), moving at a changing speed of 10, 15, 20 km/h. The final test results are the average results of 3 experiments, the author has evaluated and compared the experiments on the same road surface and the same speed of movement, the value is equivalent. In the comparison of this result, due to the simulated and experimental content is relatively similar, the researcher uses the method of graph overlap to evaluate the similarity of rules and frequencies between research and reason theory and experiment. Besides, to evaluate errors, compares the maximum force values at the vertices of the graph (to compare the maximum amplitude of dynamic loads). The comparison results in each case are shown in the graphs and tables below:

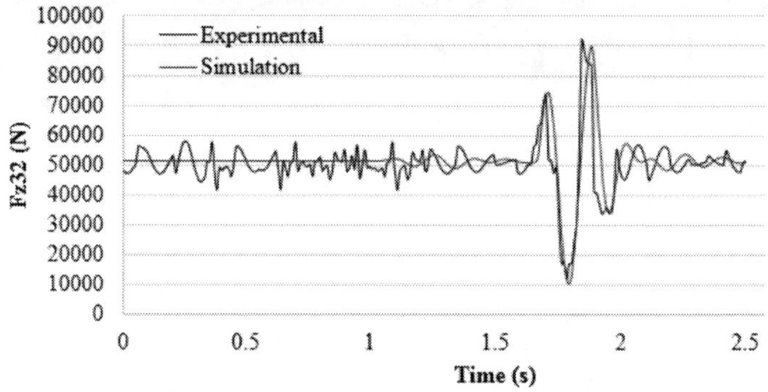

Figure 10. Force reaction when the right wheel bump into the rough (F_{z32}) at speed 10 km/h

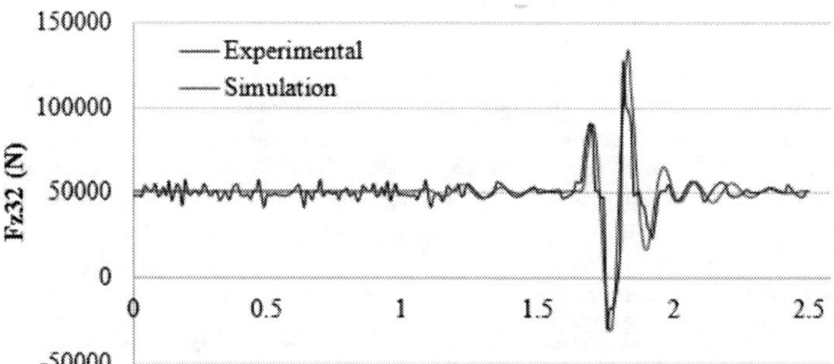

Figure 11. Force reaction when the right wheel bump into the rough (F_{z32}) at speed 15 km/h

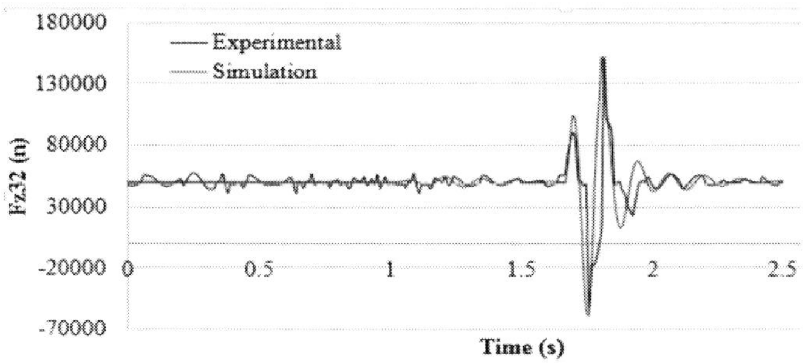

Figure 12. Force reaction when the right wheel bump into the rough (F_{z32}) at speed 20 km/h

The experimental results show the right wheel description of these physical phenomena, on the same type of road when the vehicle is moving at a higher speed, the force reaction will be bigger. When the rear wheel collides with a bump, the determined jet value corresponding to the speed of vehicle 10, 15, 20 km/h is 91, 78; 124, 8; 149, 8 kN. When the wheel comes in contact with the bump, the bridge deformation bends, and changes the voltage value of the Wheatstone bridge From the voltage value we can determine the value of the force reaction. At this point, the value of the force reaction increases (the first peak forces). When the wheel passes over the bump, because the bridge is more bent, the force reaction is higher than the value when the wheel has just come into contact with the bump (the second peak forces). When passing over the bump, under the influence of the weight of vehicle and the suspension system, the vehicle returns to its original state

Table 2. Comparisons between the force reaction simulations (Sim) and experiments (Exp) results

	V = 10 km/h			**V = 15 km/h**			**V = 20 km/h**		
	Sim	Exp	Error (%)	Sim	Exp	Error (%)	Sim	Exp	Error (%)
F_{z32max} (kN)	89,73	91,78	2,23	134,3	124,8	7,02	152,7	149,8	1,86
F_{z32min} (kN)	9,98	12,01	16,8	-30,8	-28,7	6,49	-50,8	-57,6	11,83

A comparison between simulation and experiment results show that the law of transforming vertical forces according to theoretical and measured calculations is the same. The maximum difference between simulation results and experimental results is 12.01%. With accepted assumptions when building simulation models, such deviation is acceptable. The difference between calculation results by the model and experimental results is not too huge which shows that the dynamic model of forest fire engines correctly describes the physical processes occurring in the system, with acceptable accuracy and reliability.

4. Conclusion

This paper presented the methods and results of testing the forces applied to the rear axle of a multi-purpose fire truck to verify the theoretical model used in calculating dynamic loads. The authors have selected experimental methods and equipment suitable to conditions in Vietnam In order to compare calculation results with experiments, this experiment is performed under the same conditions as theoretical simulations. The experiments were performed when the rear wheels had to pass through the bumps at speeds of 10, 15 and 20 km/h. These experimental results show that the rules are similar to the ones obtained from survey calculations by theoretical models. The maximum deviation between room tissue results and the test results is determined by maximum points at acceptable levels.

References

[1] Mitra A Benerjee N Khalane H A Sonawane M A Joshi D R and Bagul GR 2013 *IOSR J. Mech. Civ. Eng.* 22-33

[2] Sulaiman S Samin P M Jamaluddin H Rahman R A and Burhaumudin M S 2012 *Proc. Mechanical and Materials Engineering (Penang)* (Malaysia) 108-112

[3] Hassaan G A and Mohammed NA 2015*Int. J. Computer Techniques* 2(1) 56-62.

[4] Grishkevich A I 1984 *Proyektirovaniye transmissiy avtomobiley. Spravochnik.* Mashinostroyeniye (Moscow:High school publisher)

[5] Nunney M J 2007 *Light and Heavy Vehicle Technology* (UK:Butterworth-Heinemann)

[6] Luong V V Nguyen Q T and Tran T V 2018 *Proc. 9th TSME Int. Conf. on Mechanical Engineering (Phuket)* (Thailand)182-189

[7] Monica P and Babu K J M 2018 *International Research Journal of Engineering and Technology* **5**(7) 1270-1277

[8] Firat M 2011 *Engineering Failure Analysis* **18** 2137–2148

[9] Ambiger H and Reddy B 2016 *International journal for scientific research and development* **4**(5) 633-636

[10] Narayana S S 2012 *Indian journal of applied research* **1**(10) 88-90

[11] Trinh M H Tran P H and Tran T T 2017 *Proc. 10th National Mechanics Conference (Ha Noi)* (Vietnam)

Conceptual Design and Optimization of 2-seater Seaplanes

Cheewan Aroonsaengmanee[1,*], Boonchai Watjatrakul[1], Udomkiat Nontakaew[1]

[1]Department of Mechanical and Aerospace Engineering KMUTNB, 1518 Pracharat Soi 1, BKK, 10800, Thailand

* Corresponding Author: cheewanaroon@gmail.com

Abstract
This paper presents the conceptual design and optimization of 2-seater seaplane. A design code was developed from MATLAB with optimization toolbox. Design synthesis contains aircraft weight, performance, aerodynamics, aircraft stability, engine sizing and ship hull design according to user requirements and Standard Specification for Design and Performance of a Light-Sport Airplane. The seaplane take-off weight is used as the objective function to minimize subjected to design constraints. The optimization process makes use of the MATLAB GA-hybrid scheme (Genetic algorithm and Fmincon) which is faster and more efficient than the genetic algorithm or gradient-based approach. The optimum designs were obtained. The influences of design requirements and constraints were investigated. The outcome of the program is an optimum design of a single-engine 2-seater seaplane.

Keywords: Conceptual design, Seaplane, Optimization, GA, LSA

1. Introduction

The seaplane has been widely used in sea travel because of advantages that makes it faster than a ship and can manage to take-off and land on both ground and water, making it superior to normal aircrafts that need an airport to operate. The seaplane was invented in 1910 but after 1950 their performance has not been improved that much and the idea of designing seaplane stopped [1]. Seaplanes can be divided into two categories: a flying boat and a floatplane. In this study, we focus on the flying boat. Although there are some researches/studies on seaplanes, very few of them have found a conceptual design. Most of them, however, focused on improving specific areas such as surface effect [2] and water landing control [3]. Aircraft design is a multidisciplinary problem. The multivariate optimization method (MVO) [4] has been used popularly in conceptual and preliminary design stage [5-8] because of capability in searching an optimum design from a large number of design variables and non-biased design decision making. The optimization method used in this work is MATLAB GA-hybrid.

IOP Conf. Series: Materials Science and Engineering **886** (2020) 012009 doi:10.1088/1757-899X/886/1/012009

The objective of this work was to obtain an optimum design of single-piston engine of a-2-seater seaplane according to standard specification for design and performance of Light-Sport Airplane requirements [9] and study the influence of design variables and the constraints of the design.

2. Design Algorithm and Problem Formulation

The design algorithm of design synthesis analysis and optimization [10] and Design synthesis modules are shown in simplified block-diagram in figure 1. The input data is the value of design variables. These variables are the ones that affect the design and can be changed by the optimizer and fixed by variables such as range, payload, load factor, etc. After all the data was fed into the design synthesis, all values then were calculated before sending them to the optimizer.

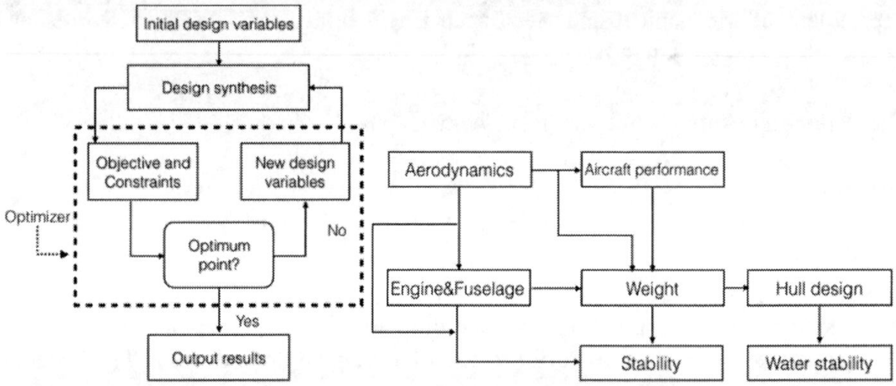

Figure 1. Design and optimization algorithm (left) and Design synthesis modules (right)

Aerodynamics module contains aerodynamic parameters calculation such as lift coefficient, drag coefficient and air density. Aircraft performance module [11] contain cruise, ground take-off, water take-off and rate of climb. The Weight module follows Raymor's weight equations [11]. Engine & fuselage module calculates engine and fuselage size. Stability module [12] calculates CG, static margin and trim parameters. Hull design [1] calculate the shape of the hull that is capable of taxiing and take-off from water. The water stability calculates parameter such as metacentric height in both directions.

2.1 Objective function

In this work problem, take-off weight ($\square_{\square\square}$) need to be minimized and was chosen to be an objective function. $\square_{\square\square}$ is a combination of all aircraft components weight as follows:

$$\square_{\square\square} = (\square_{\square\square\square} + \square_{\square\square\square\square})_{\square\square\square\square\square} + (\square_{\square\square\square\square} + \square_{\square\square\square\square} + \square_{\square\square\square} + \square_{\square\square\square} + \square_{\square\square} + \square_{\square\square\square\square} + \square_{\square\square\square\square})_{\square\square\square\square\square\square\square\square} \quad (1)$$

where $\square_{\square\square\square}$ is the weight of the payload
$\square_{\square\square\square\square}$ is the weight of fuselage structure
$\square_{\square\square\square\square}$ is the weight of wings
$\square_{\square\square\square\square}$ is the weight of vertical and horizontal stabilizer
$\square_{\square\square\square}$ is the weight of the airframe systems, equipment, passenger furnishing, etc.
$\square_{\square\square\square}$ is the weight of the installed engine
$\square_{\square\square}$ is the weight of landing gears
$\square_{\square\square\square\square}$ is the weight of fuel
$\square_{\square\square\square\square}$ is the weight of hulls structure

$\square_{\square\square}$ was categorized under 2 groups which are fixed and variable in weight. The fixed weight group contains the payload and fuselage weight, calculated from design requirements which cannot be changed by the optimizer. The variable weight group contains wings, tail, system, fuel, landing gear and hull weight, which have been calculated from design variables that can be changed during the optimization process.

2.2. Design constraints

Design constraints are the criteria to assure that the design requirements are met. Constraints were set from designer requirements and standard specification for the design and performance of a Light-Sport Airplane. These constraints are listed as follows:

1. The estimated take-off weight must be equal to the calculated take-off weight.
2. The take-off length required for both land and water must not be longer than that specified in requirements.
3. Stall speed must not exceed the requirement.
4. The cruise lift coefficient must not be greater than the buffet limit.
5. The required power must not be greater than the available power in the cruise.
6. The seaplane static margin must be in range to be stable.
7. The seaplane must be trimmable in requirement conditions.
8. The seaplane must be stable in water.
9. All of the seaplane components must be in an allowable layout range.

2.3 Problem formulation

The design problem can be formulated as a general mathematical problem of optimization as follows:

Find design variables $\square = \{\square_{\square\square\square}, \square\square, \square_\square, \Lambda_{\square/4}, \square, \square_\square, \square, \square, \square_\square, \square_\square, \square_\square, \square\square\square_{\square\square}, \square\square\square_{\square\square}, \square, \square\square\square_\square, \square\square\square_\square\}$ (2)

Minimize: $\square_{\square\square} = \square(\square)$ (3)

Subject to: $\square_{\square\square} \leq \square_\square \leq \square_{\square\square}$ $i = 1, n$ (bound/side constraints)

$\square_\square(\square) \leq 0$ $j = 1, y$ (inequality, equality constraints)

where n is the number of design variables
y is the number of constraints

$\square_{\square\square\square}$	is the take-off weight	$\square_\square$	is the wings' CG position
$\square\square$	is aspect ratio	$\square_\square$	is the tail's CG position
$\square_\square$	is the wings area	$\square_\square$	is the engine CG position
$\Lambda_{\square/4}$	is sweep angle at quarter cord	$\square\square\square_{\square\square}$	is horizontal tail volume
$\square$	is taper ratio	$\square\square\square_{\square\square}$	is vertical tail volume
$\square_\square$	is cruise speed	$\square$	is spray coefficient
$\square$	is power available	$\square\square\square_\square$	is hull slenderness ratio
$\square$	is the cruise altitude	$\square\square\square_\square$	is float slenderness ratio

2.4 Optimization and programming

The optimization method used in this work is MATLAB GA-hybrid. The seaplane design program was coded in MATLAB included Objective function, Constraints and design synthesis modules can be linked to the Optimizer. MATLAB GA-hybrid is a scheme to optimize function by using Genetic

IOP Conf. Series: Materials Science and Engineering **886** (2020) 012009 doi:10.1088/1757-899X/886/1/012009

algorithm and another optimization method. GA can reach the optimum region quickly but slow to reach an optimum point. However, combining it with another method, which is better at finding a local optimum point, will make the optimizer faster than GA and more efficient than gradient-based optimizer. This gradient-based optimizer tends to have difficulty in finding global optimum point when the initial point is far from the optimum one. A common technique is running a few generations of GA to approach optimum area, then it uses the GA solution as an initial point for the gradient-based optimizer to search the optimum point. The Genetic algorithm [13] is an optimization method inspired by Charles Darwin's theory of natural evolution and a part of Evolutionary algorithms. The algorithm uses the concept of a natural selection process that selects the fittest individuals to reproduce next populations by using Selection, Crossover and Mutation as three main operators.

3. Results and Discussions

3.1 Optimum design results

Mission requirements, optimization value and constraints are shown in Table 1, Table 2, and Table 3 respectively.

Table 1. Mission requirements

	Requirements
No. of seats	2
Range (nmi)	354.2
Ground takeoff length (ft)	≤550
Water takeoff length (ft)	≤ 700
Stall speed (ktas)	≤61
Cruise speed (ktas)	98
Rate of climb (ft per min)	≥1000
Takeoff weight (lbf)	≤1430
Static margin	5 ≤ □□ ≤ 15
Propulsion	1 piston engine

Table 2. Optimized design for a full set of the design variables

Variables	Initial value	Lower bound	Upper bound	Optimized value
$□_{□□□}$(□□□)	1200	1	2500	1287.04
□□	12	5	20	5
□□($□□^2$)	157	100	300	179.94
$\Lambda_{□/4}$(□□□□□□)	50	0	100	0.0006
□	0.5	0.5	1	0.5
$□_□$(□□□□)	98	98	98	98
□(□□□)	100	50	150	85
□(□□)	8200	5000	20000	5000

$□_{□}$(□□)	1	0.1	11.5	8
$□_{□}$(□□)	6	11.5	23	23
$□_{□}$(□□)	1	0.1	12	10.86
$□□□_{□□}$	0.7	0.7	1	0.7
$□□□_{□□}$	0.06	0.05	1	0.05
□	0.07	0.0548	0.0975	0.0975
$□□□_{□}$	2.5	1	10	3.02
$□□□_{□}$	5	1	10	4.42

Significant parameters	value
$□_{□□□□}/□_{□□}$	0.116
$□/□_{□□}$	0.066
$□_{□□}$/□□	7.15
Static margin	7.22%

Table 3. Constraints

Design Constraints(≤ 0)	Optimized value
$□_{□□□} - □_{□□}$	-210.4245
$0 - □□□□_{□□□}$	-149.2602
$□_{□,□□} - □_{□,□□□}$	-0.3449
$□□□_{□□} - □□□_{□□□}$	-0.0000
$□_{□□□□□,□□□} - □_{□□□□□,□□}$	-26.7525
$□_{□□□□□□,□□} - □_{□□}$	-42.7222
$□□_{□□□} - □□_{□□□}$	-0.0222
$□□_{□□□} - □□_{□□□}$	-0.0778
$□_{□□□} - □_{□□□}$	-4.6264
$□_{□□□} - □_{□□□}$	-4.7309
$□□□_{□□□□,□□} - □□□_{□□□,□□}$	-0.2562
$□_{□□□□□,□□□} - □_{□□□□□,□□}$	-0.0000
□□□□ − □□□	-0.0000
□□□□□ − □□□□	-0.6410
$□_{□□□} - □_{□□□□□}$(=0)	0.0000

In the optimization process, GA was performed first and the optimized value was used as an initial condition for Fmincon. The design variables behaviour from GA to Fmincon is shown in Figure 2.

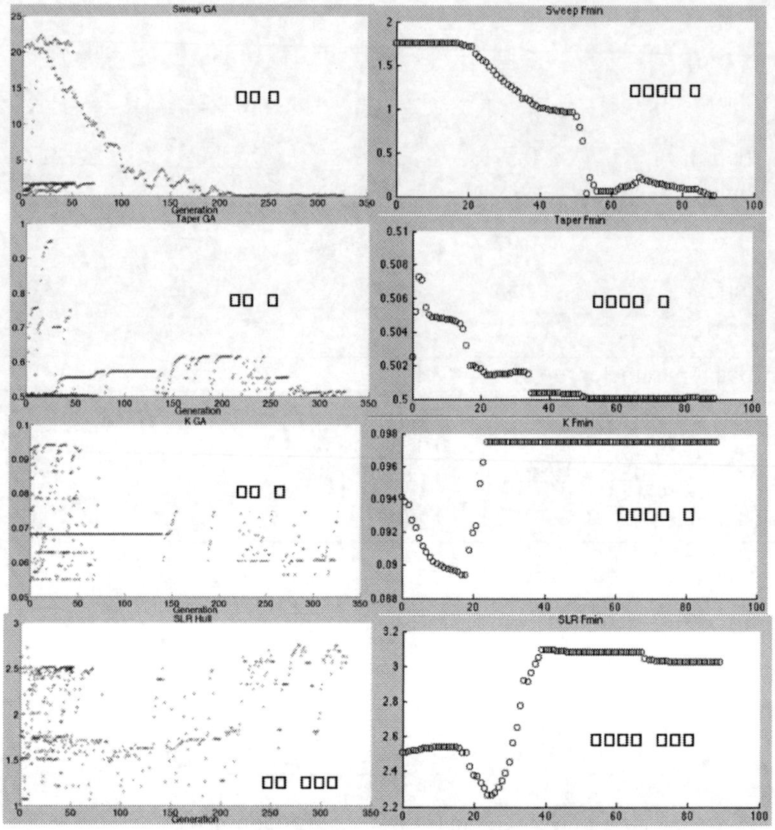

Figure 2. Optimization processing graph GA (left) and Fmincon (right)

As the results show, design-driving parameters are the rate of climb and water take-off distance because both are at the edge of the design constraints boundary (see Table 3) and affect aircraft sizing, directly. The optimizer attempts to reduce water-resistant as much as possible by reducing stall speed, increasing spray coefficient (K) and slenderness ratio (SLR) (see Figure 2). Both rate of climb and water take-off distances are constraints that determine engine size. The seaplane might be lighter if the designer can relieve the requirement from design drivers. The optimizer shows that sweep angle is not essential for a low-speed aircraft because sweep angle makes wings' structure heavier, but does not improve flight performance as shown in Table 2 and Figure 2 where the value was converged to zero. Taper ratio was converged to the minimum value to minimize wings' weight as well. The seaplane geometry base on the optimized value is shown in Figure 3 and component weight breakdown is shown in Table 4.

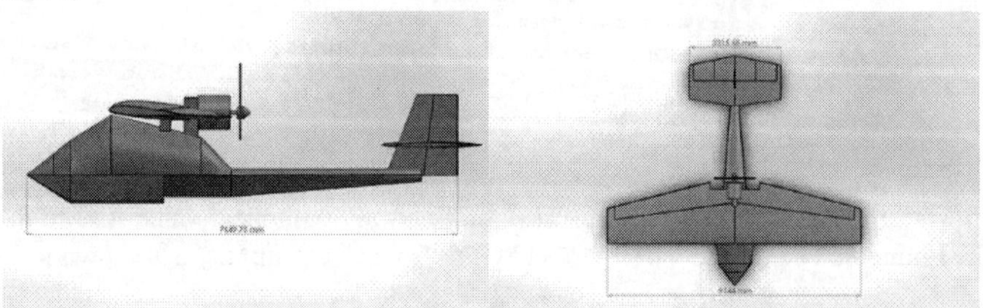

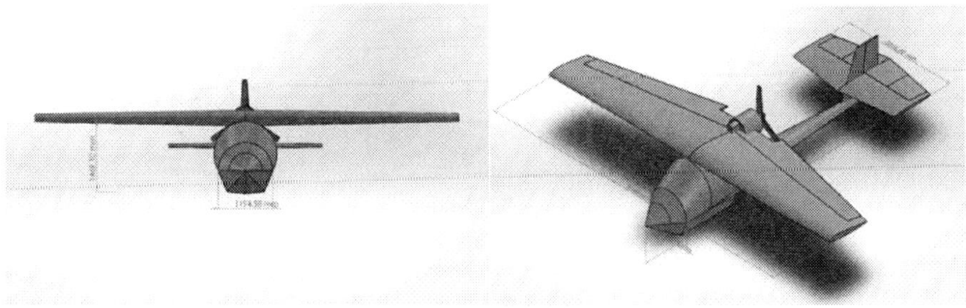

Figure 3. The baseline Seaplane geometry

Table 4: Components weight

Components(lbf)	value	Components(lbf)	value
□□□□□	49.4	□□□□	400
□□□□□	136	□ □□□□	29.5
□□□□	128.7	□□□□	195.4
□□□	28.9	□□□□□	149.3
□□□□□	169.9	□□□□□□□□	737.74

3.2 Trade-off studies

The cruise speed and water take-off distance were selected to be studied since they are parameters which important to set mission requirements. Results are shown in Tables 5 and 6, respectively.

As can be seen in Table 5, the increasing cruise speed leads to increased required power and reducing the wings' area, all of which prompt to higher weight and stall speed, even though the power available is higher than lower cruise speed cases but it did not affect water take-off distance, drastically.

To reduce water take-off distance effectively, stall speed was decreased by reducing the wings' area. However, a higher wing area means a higher structure weight, which leads to/ result in a higher required power.

Table 5. Optimize design according to cruise speed

V(ktas)	SW(□□^{2})	TO weight(lbf)	P (bhp)	Ground TO(ft)	Water TO(ft)	V stall(ktas)
88	172.6	1222.85	80.5	487.17	700	34.08
93	176.1	1253.12	82.59	488.35	700	34.16
98	179.94	1287.04	84.95	489.58	700	34.25
103	184.1	1324.61	87.59	490.86	700	34.35
108	182.37	1366.28	93.97	482.81	700	35.05
113	179.61	1414.88	102.36	473.43	700	35.94
118	177.66	1468.69	111.57	465.90	700	36.82

Table 6. Optimize design according to water take-off distance

Water TO(ft)	SW(□□^{2})	TO weight(lbf)	P (bhp)	Ground TO(ft)	V(ktas)	V stall(ktas)
450.00	225.67	1487.96	108.5	397.2	98	32.88
500.00	208.83	1407.88	98.66	423.37	98	33.25
600.00	192.80	1325.90	88.12	467.82	98	33.58
700.00	179.94	1287.04	84.95	489.58	98	34.25
800.00	172.30	1267.15	83.76	501.44	98	34.73
900.00	167.75	1255.48	83.07	509.10	98	35.03
1000.00	164.78	1247.91	82.63	514.38	98	35.24

To verify the program, comparison to real seaplane is conducted. Progressive Aerodyne Sea Ray is a light sport amphibious aircraft, which was chosen for the comparison process. Design requirements such as cruise speed, take-off distance, load, ROC and range were determined in a range around Progressive Aerodyne Sea Ray specification. The seaplane that was obtained from the program has identical performance to Sea Ray but some characteristics were different because the design synthesis model may not be the same. These comparison results are shown in Table 7 below.

Table 7. Comparison result

	Weight	P(bhp)	SW(□□^{2})	Length	b	V stall	Ground TO	Water TO
Sea ray	1370 lbf	100	157	22ft5"	30ft10"	42 ktas	350 ft	472 ft
Case1	1331 lbf	105	165	23ft	35ft4"	36 ktas	400 ft	493 ft

4. Conclusion

This article presented a comparison and a trade-off study. The optimum design was obtained. The results have shown the design drivers such as rate of climb and water take-off distance that affect engine sizing and variables that were minimized to a reduced weight such as a sweep angle and taper ratio. This study will help a designer to know significant parameters and could set a proper design requirement, which lead to a lighter, and a better-optimized design.

References

[1] Leyva C and Leonel A 2012 Glasgow: Glasgow Theses Service
[2] Muskat R and Ericson J A 1973 *Computers&Structures* **3** 529
[3] Nebylov A V Nebylov V A Panferov A I and Shepeta A P 2007 *Proc. 17th IFAC Symposium on Automatic Control in Aerospace* **40** 669
[4] Chacksfield J E 1984 *Proc. 13th Congress of the international Council of the Aeronautical Science* **1** 954
[5] Fioritia M Boggeroa L Corpinoa S Prakashab P S Ciampab P D and Nagel B 2018 *Transportation Research Procedia* **29** 135
[6] Korsuna O N Sergeevb S A and Stulovskiia A V 2019 *Procedia Computer Science* **150** 361

[7] Panzeri M Savelyev A Anisimov K d'Ippolito R and Mirzoyan A 2018 *Transportation Research Procedia* **29** 289
[8] Benaouali A and Kachel S 2019 *Aerospace Science and Technology* **92** 766
[9] American Society for Testing and Materials 2017 Standard Specification for Design and Performance of a Light Sport Airplane *ASTM international*
[10] Watjatrakul B 2006 *The journal of KMITNB.* **16** 10
[11] Gudmunsson S 2014 *General aircraft design: applied methods and procedures.* Oxford: Elsevier
[12] Napolitano M R 2012 *Aircraft Dynamics: From Modeling to Simulation.* Danvers: John Wiley & Sons

IOP Conf. Series: Materials Science and Engineering **886** (2020) 012010 doi:10.1088/1757-899X/886/1/012010

Hybrid Turbo Compound Fan Engine an Eco-Efficient Propulsion System for Aviation

H P Berg[1*], A Himmelberg[1] and T Poojitganont[1]

[1]BTU Cottbus-Senftenberg, Chair of Combustion Engines and Flight Propulsion, Siemens-Halske-Ring 14, 03046 Cottbus, Germany

* Corresponding Author: peter.berg@b-tu.de

Abstract
The path to electric propulsion systems depends on the development of powerful, compact and very light energy storage system with a high storage density. In order to create an environment-friendly intermediate solution in the medium term, especially for aircraft engines of medium power classes, it makes sense to use an innovative, electrically parallel hybrid unit based on rotary engines and electric boosters for start and climb phases as a propulsion system. An aero-engine application, based on the HSD concept (HSD = hybrid super-drive – hybrid Wankel rotary engine) for different hybrid-parallel propulsion systems is presented. In this article, the introduced technology is based on a currently produced family of multi-component Wankel rotary engines. The proposed HSD concept uses a chamber volume of 650ccm per rotor. The projected power range (0.8 to 1.36 MW) is covered by an innovative hybrid electric parallel turbo compound concept together with 2x4 rotary engine units. A novel, air-bearing turbo engine with integrated electric drive is used as a turbocharger for charging the rotary engine. For the aero-engine technology, the counter-rotating fan (CRF) concept shown as an example for an aircraft technolo¬gy carrier (light jet) leads to a further increase in the efficiency of a complete system.

Keywords: rotary engine, hybrid combustion engine, aero-engine, air bearing, counter rotating fan, electric flight

1. Electric and hybrid-electric flight

The efficiency potential of current engine systems has been explored to extensive studies [33]. To meet these guidelines, many manufacturers from the aerospace industry, including the new developments for innovative engines (for example of coupled turbo- and piston engines [16]) want to introduce overall concepts for electrical flight [18]. Bauhaus Luftfahrt, as a research institute – cooperating inter alia with Airbus, Liebherr Aerospace and MTU, has developed in this context with the CE-liners a concept aircraft. With such an aeroplane, the electric fly with a commercial jet should be possible in 2035 [4]. All-electric aviation needs right overall concepts for new technologies. Thus, a huge developing expenditure will still be necessary for the battery technology, the fuel cell technology, the refueling, loading and in particular for the airport infrastructure [18]. Hybrid energy systems with a combustion engine, generator and electric motor can make a contribution to reducing pollution in a transition phase (motivating for CO2 reduction) using technologies from the automotive industry. The problem seems to be particularly the supply of the required energy performance for a corresponding flight, and the realization of the power to weight ratio and safety of the required drive.

2. Technologies for the hybrid aero engine

The proposed propulsion technology is a hybrid drive consisting of a high-performance turbo compound Rotary engine (HSD) together with an efficient and compact electric motor. For the propulsion, the adaptation of modern counter-rotating fan is being considered. The advantages of using a turbo rotary engine as a hybrid engine, in combination with a suitable electric motor, could be demonstrated with an example of a regional commuter plane with propeller engine in [1].

2.1 Technologies for the core engine

The basic concept for the selected internal combustion engine is essentially based on the architecture of the KKM5 0 0 family from the company Wankel Supertec GmbH of Cottbus (www.wankelsupertec.de, [5]). In recent years, they have successfully developed the rotary piston engine system for diesel fuel. The electric, air-bearing turbomachine, which is used for charging the engine, based on technology from the company Euro-K GmbH of Cottbus (MTiG, www.euro-k.de [6], figure 1) and on the fundamental work of the authors [7].

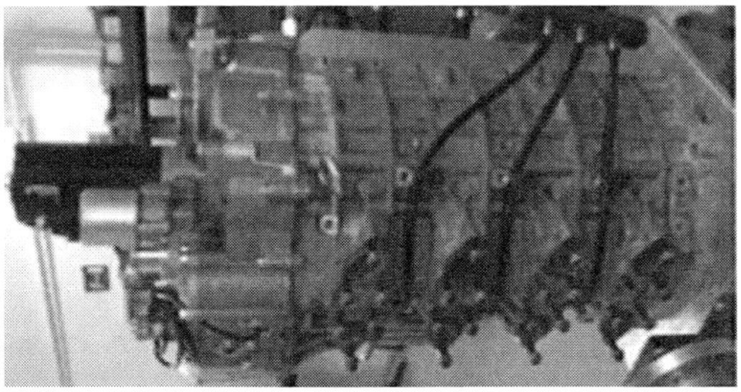

Figure 1. 4 Rotor Wankel Rotary engine bench (Industrialized technology, [5])

The Wankel engine technology with an air-bearing e-turbo engine (see figure 12) provides a high power density and low vibration. Additionally, the modular design is an advantage for the development of the new drive concepts for general aviation in the future. The TBO times required for the aeronautical potential of this system have already been proven in the past through the use of the individual components in various applications. The technological basic development of the engine concept in particular for diesel and kerosene application has been described in various publications ([14], [15], [19]).

The combustion process has been extensively investigated and developed in the past, which is reflected in today's standard applications of the rotary motor system [10]. The power to weight ratio, which is very important for aviation applications, can be archived by Wankel engines up to 3 kW/kg. This can be located in the area between the commercial gas turbine aero engines without afterburner (up to 6 kW/kg actual in service) and the in-service aircraft piston engines (1kW/kg). By supercharging the core engine with the so-called MTiG-system (microturbine with integrated generator) with air bearings, oil-free cabin air can be guaranteed.

The state of development on the Wankel-HSD system (figure 1) can be described as follows:

- The basic technology of the Wankel engine for multi-fuel combustion (gasoline, diesel, kerosene) was developed largely and validated [5], [12].

IOP Conf. Series: Materials Science and Engineering **886** (2020) 012010 doi:10.1088/1757-899X/886/1/012010

- The basic technology which has found its niche markets outside the aviation industry is used for special applications and is constantly being developed.

2.2 Electrical drives for the representation of the hybrid component

Siemens already has developed lightweight and powerful aircraft engines for aeronautical applications. With the electric drive SP260D, an output of 260kW, the power to weight ratio of 5.2kW / kg with a maximum continuous speed at 2500rpm and a maximum torque of 1000Nm, η=95%, a test aircraft could already start and be operated electrically [13], [19], [2]. Such an air drive can be used and adapted as a hybrid extension with an internal combustion engine. At the same time, in inflight operations, using a generator is required.

The research group "Bauhaus Luftfahrt" assumes in a draft of the CE-Liner, that with a storage capacity of 2000Wh/kg in light batteries, which have to be developed, also for larger aircraft the electric flight is possible [4], [21]. KIT/FG Energiewirtschaft in Karlsruhe, however, has already developed battery materials with a capacity of 450Wh/kg [19], [20]. With such a system, the battery volume and for a-15-minute take-off boost, the capacity can be reduced to the extent that an installation in a commuter aircraft without increasing the weight to volume ratio of the aircraft is possible. With this available battery system and power electronics a power to weight ratio of 2.9kW/kg for the hybrid HSD-System instead of 3.6kW/kg [26] for small turboprop systems (e.g. PW535B) can be realized. For power electronics, a compact system from the automotive industry can be used with products from the manufacturer Continental [22]

2.3 Hybrid systems in aeronautical propulsion

In [1], the parallel hybrid system has been discussed using the example of a Beechcraft 1900. This aircraft uses originally a serial drive system. In this paper, the focus will be in the field of parallel hybrid systems in the range to 1500kW. These drive systems could be implemented as a mild- or medium-hybrid or a full hybrid concept. Figure 2 shows the position of the engine benches in the fuselage. In this paper, the HSD concept is visualized using the example of a sheathed counter prop fan (CRF). This approach is chosenbecause for business jets the possible speed range during flight is reasonable in a flight Mach number range up to 0.75. Furthermore, a technology demonstrator with the selected hybrid drive technology could be implemented easier in this class of aircraft range than in the commercial aircraft sector. In the selected aircraft concept, a good integration of the HSD drive in the fuselage resulted at an angle of $\alpha_{HSD\,1-2} = 90°$ (see figure 2). Because the HSD is shorter than comparable shaft engines, therefore the torque transmission of the main engine to the CRF appears to be easier. Thus, the torque transmission to the angular gear (see Figure 3) at engine speed n_{Motor} can be carried with a direct shaft (L-configuration). When using a shaft engine, a planetary gear and, because of the longer machine length, a Z-arrangement of the powertrain is required. In such an angular gear, the required speed $n_{CFRa,b}$ can be transmitted.

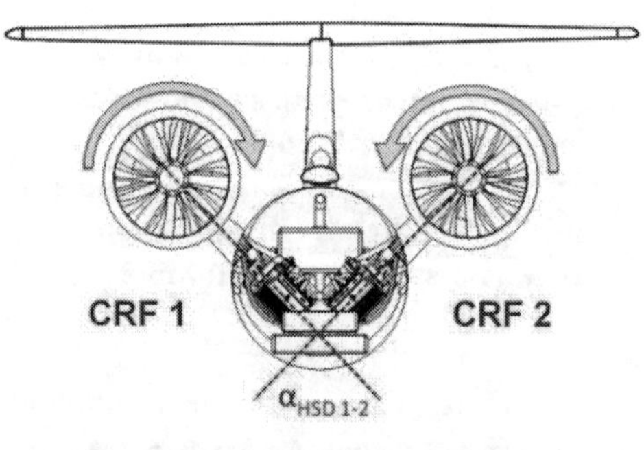

Figure 2. HSD-drive concept for the Business Jet MariE-1 with an engine angle of $\alpha_{HSD\,1-2} = 90°$

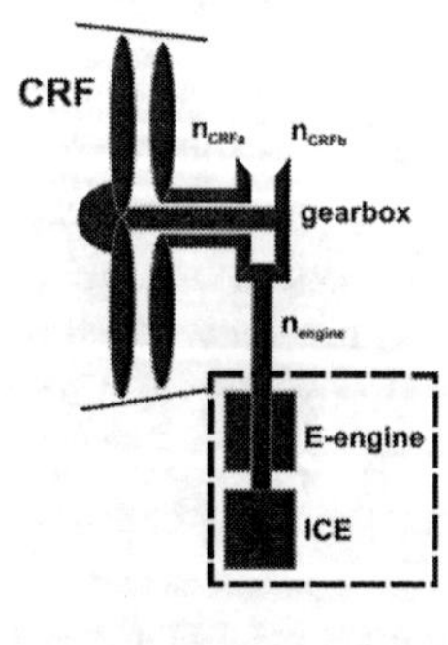

Figure 3. Transmissions-concept of the HSD-CRF-Power plant with $n_{CFRa}/n_{Motor} = -n_{CFRb}/n_{Motor} = 1{:}4\ to\ 1{:}2$ (L-Configuration)

The term "HYBRID" has a double meaning in the HSD engine. On the one hand, a hybrid combustion process with stratified charge takes place in the air-compressing Wankel engine. By a high-pressure direct injection, the fuel is supplied into the combustion chamber and ignited there [10]. Otherwise, in a hybrid concept, the Wankel engine is very well suited for a combination with electrical drive systems. The system can also be classified in the parallel hybrid category.

Until now, only a serial hybrid electric aircraft from Diamond Aircraft [31] has been realized. The advantages and disadvantages of the serial principle for aeronautical applications have been discussed in [1] and [24]. In a parallel hybrid aerospace propulsion system, electric motor and internal combustion engine can be operated simultaneously. With this addition of drive power, the individual drive machines can be designed to be smaller and more effective. This leads to a reduction in the space and weight installation. However, this reduction will be repaid partly through energy storage (battery) and partly via power electronics. This depends on the chosen level of hybridization H_{Hybrid} of the drive system [34]:

$$H_{Hybrid} = P_{Elektro}/(P_{eff/HSD} + P_{Elektro}) \quad (1)$$

This study shows that a mild hybrid system with $H_{Hybrid} < 0.23$ can certainly be a viable variant for the aviation in the future. Full-Hybrid Systems use the combustion engine only for range extending and are possibly an option for extending the range of an electrical flight system. Figure 4 shows the basic structure of the explored powertrain. A Wankel Turbo-Compound is connected with a Counter-Rotating Fan Propulsor by a shaft and gearbox-system. The electric boost system is mounted on the outgoing shaft from the Wankel system. Therefore, the combustion engine and booster can act in parallel with the drive train, which adds the torques of the individual drives. This allows a weaker design of the engines, saving costs, weight and space. Currently, hybridization levels of 25% to 40% appear to be appropriate.

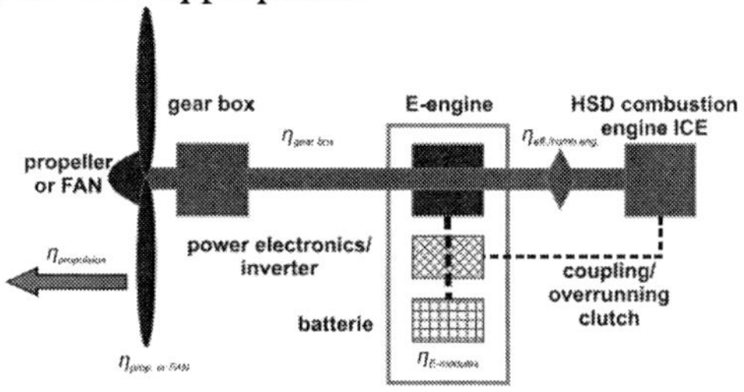

Figure 4. Basic structure of the powertrain in ICE-Modus

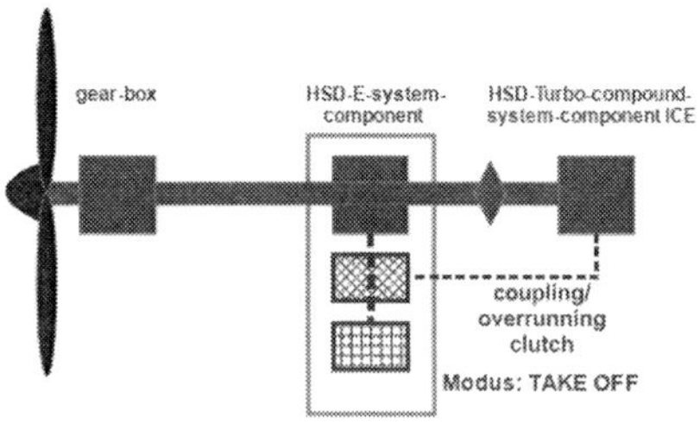

Figure 5. Boost function during takeoff. E-machine and ICE (red) made max. performance.

Using this degree of hybridization, the energy storage capacity is designed according to the "take-off mode". E-machine(s) and internal combustion engine(s) (ICE) are both working at maximum power (boosting, see Figure 5, red parts are working). In cruising mode, the ICE works and the electric motor can support it. Additionally, the E-Machine can increase the load of the ICE to higher torque and charges the energy storage. This is very helpful when engine damage occurs and also improves the entire energy management (particularly in the landing approach).

Figure 6. *MariE-1* as a technology platform for 2 HSD CRF engines

Parameter	Symbol	Value
Property		
Length	L	18,49m
Max. takeoff weight	MTOW	9.300kg
Wingspan	S	15,137m
Wing area	A	28,472m²
Passenger		14
Crew		2
Height	h	6,20m
Cell-diameter	d	2,10m
Performance		
Engine	P	2x 1,365MW
Type		HSD-rotary wankel-Hybrid
Range	R	3.300km
Max. cruise speed	Ma	Mach 0,75
Service Ceiling	h_{Cruise}	11.000m

Table 1. Design parameter for the business jet MariE-1

In the descent phase, mainly the electric machine works, the internal combustion engine only works as a support for the propulsion system. The electric motor can act as a generator to fill the battery at steep descent so that the drive can now regain potential energy for the use by taxiing on the ground. Figure 6 shows a technology platform for the use of 2 HSD CRF engines. On landing, the electric motor and the ICE both operate. This is necessary to make an effective thrust reversal and makes it possible to provide enough power for a go-around. During the subsequent, "taxi operation point" only the electric engine works (figure 7). To develop a combined hybrid drive, a power split system is realized, which represents a combination of serials and parallel concepts. At present, weight disadvantages at a power weight of 1.5kW/kg for conventional Hybrid Engines compared to 3.5kW/kg MTOW for small Turbine Engines [26] do not recommend a hybrid operation in an aircraft system. So, there still is no economical hybrid system in an aircraft. For these reasons, a single-shaft parallel hybrid rotary engine weight with a MTOW of 3kW/kg was chosen for the selected system (see also Figure 7) and optimized with near future technical capabilities for the flight systems. It should be determined if the economic use of such a system reduces emissions and increases the power to weight ratio of hybrid development.

Figure 7. "Aircraft Engine 2030" / HSD concept: hybrid turbo Wankel Twin-Pack engine [25]

3. Use of the HSD8650-40 in a business jet

To investigate the applicability of the hybrid engine application in a real-world system, a business jet for up to 14 passengers and a-2-man crew was designed (MariE-1, Figure 6). The established specifications and boundary conditions were based on the in-service aircraft. The use of a fan concept was realized, which allows reaching cruising speed at Ma = 0.75. The higher cruising speed compared with propeller aircraft makes it possible to provide a higher speed during takeoff and landing safely, so that the aircraft can be queued in the normal approach management on large airports [14].

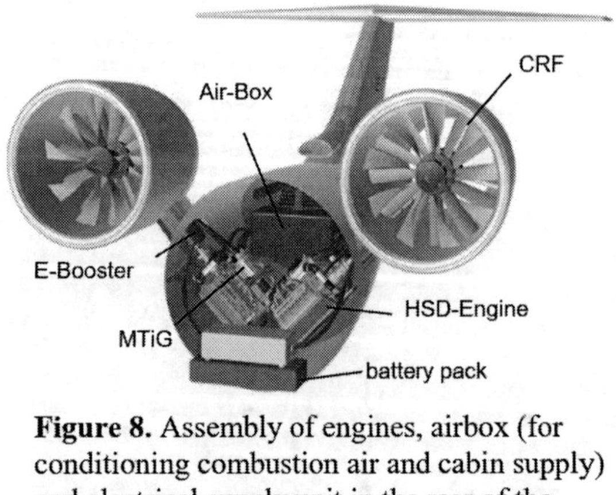

Figure 8. Assembly of engines, airbox (for conditioning combustion air and cabin supply) and electrical supply unit in the rear of the virtual technology demonstrator MARI-E

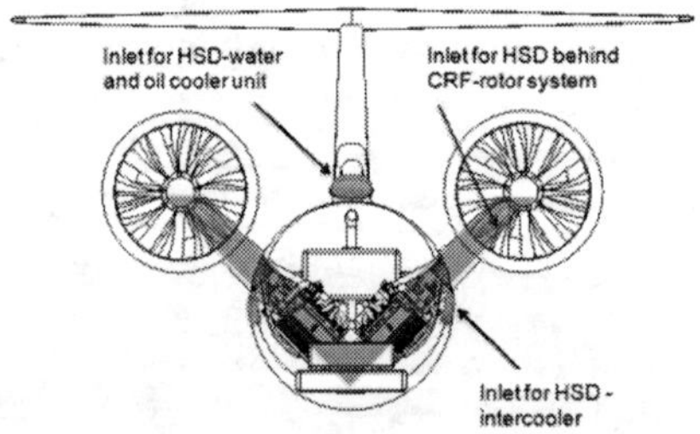

Figure 9. Compact V-type and air intakes

Advanced engine developments are adjusted to high bypass ratios. This allows for more economical flight operations by improving external efficiency. Currently, engines are under development, with a bypass ratio of up to 1: 24.5, such as at an ANEA configuration [23]. The main part of the generation of propulsion is taken from the fan with the bypass stream.

With the MTU CRISP concept of a counter-rotating fan (CRF) an effective drive concept was developed. With this concept, operating costs can be reduced and it may be achieved a significant reduction of aircraft noise [32]. Therefore, outer efficiency increases further and can reach up to 93% [27]. Now, if such a concept is not driven by a gas turbine, but with the above-HSD, the benefits of an effective turbo-compound concept are combined with a modern propulsion system. For these reasons, the technology platform will be equipped with hybrid drives of the type CRF-HSD-8650E-40.

3.1 A selected aircraft as a technology carrier

For an analysis of the applicability of the CRF-HSD engine system, an aircraft of the class "Light Jets" was modelled (Figure 6). It was ensured that advanced performance parameters were observed [17]. In addition, it should be possible to integrate the intended HSD engine with all the necessary auxiliary units in the fuselage twice (for redundancy).

3.2 Aircraft Engine System Architecture

Based on the experiences with the basic engine family (rotary piston engines of chamber volume class 500ccm) own preliminary investigations (as shown in [1]) were carried out for aircraft engines for the year 2030 with the project name HSD (Hybrid Super-Drive) (650ccm chamber volume class) were carried out. Deviating from the basic engine concept, this system has been extended with an enlarged rotor, a parallel-connected electric motor, an e-turbocharger (based on the MTiG) and a sufficient e-energy storage.

3.3 A Wankel rotary engine as a fan drive.

Based on the experiences with the basic engine family (rotary piston engines of chamber volume class 500ccm) own preliminary investigations (as shown in [1]) were carried out for aircraft engines for the year 2030 with the project name HSD (Hybrid Super-Drive) (650ccm chamber volume class) were carried out. Deviating from the basic engine concept, this system has been extended with an enlarged rotor, a parallel-connected electric motor, an e-turbocharger (based on the MTiG) and a sufficient e-energy storage.

Compared with/In comparison with the reciprocating engine, the Wankel engine with the same performance has a lower system weight. They may also be optimized for good specific fuel consumption, which is possible by an advanced stratified charge method [11]. This new aircraft engine concept covers the range up to 1500kW. For this purpose, it uses the advantages of the Wankel rotary engine technology combined with a highly charged supercharger (MTiG technology = microturbine with an integrated generator and air-bearings = E-turbocharger [24]).

In the presented technology platform, two Twin-Pack Systems of HSD8650E-40 with 1365 kW should drive two CRF-fans, so that the required performance parameters can be achieved. In this configuration, an electrical power proportion of 40% on the Wankel-HSD-power was found to be sufficient for the load requirement. For further investigation, a machine with a corresponding booster is examined. For this purpose, each has a twin pack formed of two motor benches, equipped with a four-rotor Wankel engine (see Figure 1, Figure 9). The connection is made via a gear assembly and the CRF-speed is set via the angular gear (see figure 3 for transmission concept). With a chamber volume of 650 cc per rotor and charging with a pressure ratio of $\pi_{turbo} = 3$ (reached 4.5 in height operation) can be achieved at a ground level output of up to 975kW for the part of the internal combustion engine.

To increase the engine power in the start and climbing phase a 390kW e-motor is coupled with the drive train.

IOP Conf. Series: Materials Science and Engineering **886** (2020) 012010 doi:10.1088/1757-899X/886/1/012010

3.3.1 The fan concept

Modern engines currently use the "Geared Turbofan" (GTF), in which speeds of the fan and low-pressure turbine are changed via a reduction gear. Both components operate at a reduction ratio of 3:1 in their optimum ranges [26]. To realize a larger fan diameter, the bypass ratio can be increased up to 1:12. Thus, the same propulsion can be realized with lower consumption in modern engines. Using a counter-rotating fan (CRF) this ratio can be further increased. After the development of CRISP by MTU [28], [27] further investigations at DLR were performed. A realization of such an engine on an aircraft was held by the Kuznetsov NK-93 [29]. This forms a by-pass ratio at 1:16.6. Further developments are currently being carried out at DLR [42], [27], [28]. The selected CRF here has a size of 50 inches and provides thrust performance of more than 17kN in Climb mode.

3.3.2 The operation of the hybrid engine

The hybrid engine is operating with a permanently connected electric motor at the output of the Wankel engine. The drive banks are coupled via the gear set in the HSD unit and used to drive the engine shaft (CRF). The electric motor is disposed on the shaft of the drive, to supply a power assist (see Figure 4, 6, 8). In the start and climb phase, the system together provides sufficient energy to reach the altitude and, ultimately, the cruising speed. With the energy stored in the battery, both engines of the technology carrier can be used up to 15 mins at full load. The electric motor can selectively supply power when the internal combustion engine due to the reduced air pressure at higher flight levels provides less energy. Another control option is the electric turbocharger whose compressor pressure ratio can be increased and adapted to the respective requirements.

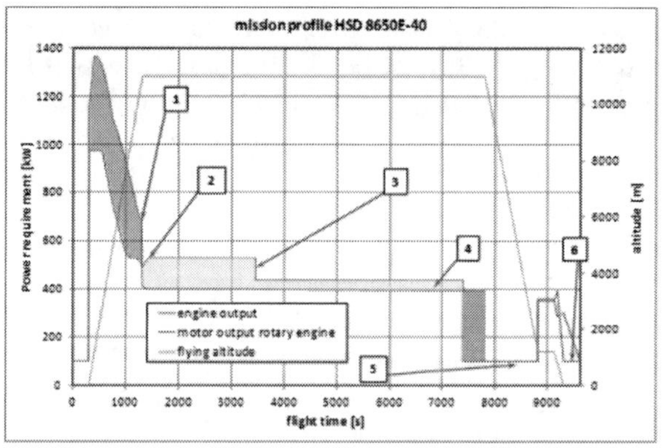

Figure 10. Mission Profile for a 2.5-hour cruise (1700km) Of the HSD engine during the complete flight

Description of the flight phases for the mission profile:

1. End of the climb, shutdown E-engine
2. Reach cruise speed, a plane in cruise mode, the battery will be charged, adjusting charge pressure in the internal combustion engine
3. Battery charged, reduction of the power of the internal combustion engine by reduction of the charge pressure
4. Take over flight by E-engine to adjust the battery charge for landing management
5. Aircraft in descent, Flight management by use/recuperation E-machine/battery with minimum power rotary engine
6. landing, Taxi possibly by E-engine

Additional requirement E-engine

Performance overheads Combustion engine for battery charging/cabin supply

In cruise mode, the E-machine operates as a generator and supplies the excess power of the combustion engine for charging the energy storage (battery). After reaching the maximum charge of the accumulator, the power of the engine can be reduced and the load can be controlled with optimal consumption. Here, also the charge pressure through the MTiG is adjusted to the required value. Thereby an eco-efficient operation is guaranteed. Based on a case study with an expected range of 1700 km, the driving of the system can be demonstrated. The simulations have shown that, depending on flight conditions, fuel consumption can be up to 50% lower than the consumption of a turbofan engine in such an aircraft.

4. Performance, efficiency and fuel consumption (rates CRF-HSD-engine / GTF Geared Turbo Fan-engine)

For a first comparison, a scenario with an approximately two-hour cruising flight was selected. Thermodynamic methods were used to investigate how the energy consumption of a GTF-system and a hybrid HSD system evolved during a flight task. To produce the required shaft power for the technology platform, hybrid HSD 8650 Wankel engines with a degree of hybridization of 28.6% is used (figure 9). These engines are connected through a transfer driveshaft to the transmission unit of the CRF. The available drive power is given by the degree of hybridization H_{hybrid} by [35]:

$$P_A = \frac{P_{eff/HSD} \cdot \eta_{gearbox} \cdot \eta_{CRF}}{(1-H_{hybrid})} \tag{2}$$

To evaluate the drive system, the comparison of the efficiencies of the different subsystems is useful. For the reciprocating engine of the HSD-engine, the efficiency can be written as follows [35]:

$$\eta_{eff/ICE} = \eta_{burn\,out} \cdot \eta_{thermal} \cdot \eta_{HP} \cdot \eta_{CC} \cdot \eta_{mech} \tag{3}$$

(HP = high-pressure loop, CC = charge cycle loop, mech = mechanically.)

For the entire hybrid HSD system, the efficiency of the electrical parts must be considered. The following equation results for the complete drive unit:

$$\eta_{eff} = \frac{W_e}{E_{zu}} \tag{4}$$

For this purpose, the supplied energy from battery and fuel for the internal combustion engine must, by introducing a dimensionless energy ratio $\varepsilon = E_{Fuel}/E_{Bat}$, be summarized as follows:

$$\eta_{eff} = \frac{\eta_{ICE} \cdot E_{Fuel} + \eta_{E-Masch.} \cdot E_{Bat}}{E_{Bat}+E_{Fuel}} \tag{5}$$

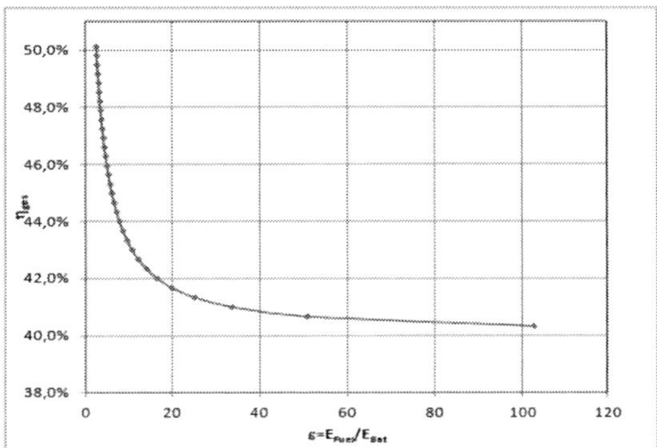

Figure 11. Comparison of the efficiency of the drive system as a function of the ratio of the energy contents of battery and fuel

Derived from this, the equation for the total efficiency can be set up:

$$\eta_{eff} = \frac{\eta_{ICE} \cdot \varepsilon + \eta_{E-Masch.}}{1+\varepsilon} \tag{6}$$

Figure 11 shows the efficiency of hybrid combination engines as a function of dimensionless energy distribution between battery capacity and fuel-energy content.

An evaluation of this analysis shows that with a high degree of electrification, the hybrid system has a higher efficiency than a single internal combustion engine can deliver. The mechanical efficiency of the internal combustion engine, therefore, was assumed to be 41%. When using modern battery systems, new electric motors and other electrical components, energy efficiency up to 75% can be achieved from the electrical system.

Particularly at high levels of hybridization, the propulsion of a 10-ton aircraft with consumption in the take-off phase of 0.01kWh/kg-aircraft-weight until reaching cruising altitude shows a strong increase in its efficiency. Similarly, a cruise fuel consumption of about 0.1kWh/kg/aircraft weight

shows an improvement. A high battery weight can be compensated by the efficiency of the drive. The efficiency of the overall machine is heavily dependent on the part of the power consumption of the hybrid shares in the respective flight phase. Weight-optimized, the efficiency is higher with a higher content of electrical Engine. Weight-optimized, the efficiency is higher at a higher ratio of the electric drive.
With the airspeed to the jet exit speed ratio $\zeta = C_{airplane}/W_{nozzle}$ follows with the quality criterion for the implementation of the jet power into propulsive power (external efficiency $\eta_{propulsion} = \left(\frac{2\cdot\zeta}{1+\zeta}\right)$), the total efficiency (with an adjusted nozzle) [30], [36]:

$$\eta_{total} = \frac{thrust\ performance}{\dot{E}_{fuel}^{rel}} = \eta_{internal} \cdot \eta_{propulsion} \quad (7)$$

$$\eta_{total} \approx \eta_{eff/Hybrid\ System} \cdot \eta_{gearbox} \cdot \eta_{CRF} \cdot \left(\frac{2\cdot\zeta}{1+\zeta}\right) \quad (8)$$

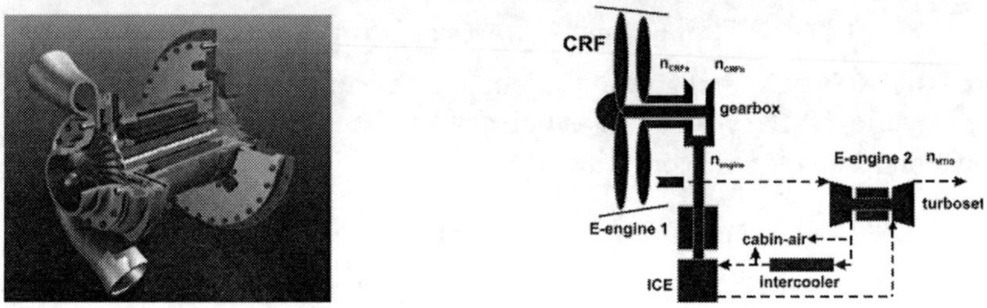

Figure 12. MTiG, electrical coupled MTiG in the complete turbo-compound-system as a central turbocharger (E-Engine 2: n_{MTiG}=70000 to 80000rpm), E-engine 1: HSD-Booster

With increasing altitude, the exhaust exergy losses are reduced because the described shaft engine concept is a highly efficient and turbo-compound engine with high efficiency $\eta_{eff/HSD}$. Because of the electrical supported E-central turbo, the system is also very flexible (MTiG, see Figure 12). As in the case of the described application, the MTiG gets a branched and charged partial airflow by the CRF with a pressure ratio of $\pi_{CRF} = 1.3$, the electric power output is increased in the MTiG against the basic version (see [1]) in the following ratio:

$$\frac{P_{MTiG,CRF}}{P_{MTiG,base}} \approx \frac{(\pi_{base}\cdot\pi_{CRF})_{CRF}^{\frac{\kappa-1}{\kappa}}-1}{(\pi_{base})_{base}^{\frac{\kappa-1}{\kappa}}-1} \quad (9)$$

The boost pressure can, as already described in the HSD concept, be adjusted in such a way that the effective efficiency (limiting pressures of the Wankel core engine: compression pressure 70bar, combustion maximum pressure 102 bar) in combination with the altitude-related decrease in temperature compared to the base value could be increased:

$$\eta_{eff/HSD\_altitude} > \eta_{eff/HSD\_ground\ level} \quad (10)$$

Further, in the described power class, the range of effective efficiency of the HSD drive compared with a gas turbine engine is better in the following estimated ratio:

$$\xi_{eff} = \frac{\eta_{eff/HSD}}{\eta_{eff/gasturbine}} \approx (1.0)1.15\ to\ 1.2 \quad (11)$$

On the propulsion efficiency side, if the CRF concept is compared to a geared fan (see Figure 13) with a modern bypass ratio of 10 to 14, then $\eta_{propulsion}$ results in an advantage in the selected design point (Ma = 0.75) from:

$$\xi_{propulsion} = \frac{\eta_{propulsion,CRF}}{\eta_{propulsion,geared\ fan}} \approx 1.06\ to\ 1.09 \quad (12)$$

$$\xi_{eff} = \frac{\eta_{gearbox/HSD}}{\eta_{gearbox/gasturbine}} \cdot \frac{\eta_{CRF/HSD}}{\eta_{FAN/gasturbine}} \cdot \delta_{external} \tag{13}$$

where $\delta_{external}$ represents a factor that evaluates the external efficiencies between a geared turbofan and an HSD engine. This factor shows a small advantage for the HSD-concept [25].

By parallel hybrid system technology, while weighting the different phases of flight (see Figure14), a further increase of the efficiency can be settled. This can be shown in terms of the fuel consumption ratio over the entire flight. The value must be within the following range:

$$0.78 = \frac{1}{1{,}27} > \frac{B_{HSD,CRF}}{B_{geared\,fan}} > \frac{1}{1{,}44} = 0.7 \tag{14}$$

This would mean that an HSD CRF only consumes about 60% of the fuel consumption of a comparable geared turbo fan. This value certainly represents a theoretical limitation. Three effects can explain it:

1. Although the average internal efficiency is slightly worse in cruise mode by the phase-wise use of the electrical component, the effect of the more fuel-efficient mode of the HSD-combustion engine predominates by a balanced load adaption.
2. In the climb phase, the energy of the charged battery significantly increases the efficiency of the system in the flight phase of the electrically assisted flight.
3. The third effect is defined by the approach phase. Here, the Wankel engine can be adjusted in a very low load point relative to a conventional engine, since the electric motor takes over the descent control and recuperation. This flight phase can be realized theoretically without consumption when the HSD-combustion engine component is switched off and only the electric motor works. For safety reasons, however, such a shutdown is omitted.

Comparing a GTF with a CRF-HSD reveals, in particular, the propulsion advantage of a Counter Rotating Fan and the advantage of an optimized Hybrid Wankel rotary engine compared to a small turbine engine. In particular, the improved efficiency of the internal combustion engine component is significant. This is made possible by the fact that the combustion engine part can be used in the partial load range of the entire system in the effective mode. Fully-loaded phases are supported by the more effective E-System. With this configuration, higher efficiency of the aircraft propulsion can be achieved.

In a simulation of a geared turbo engine with CRF, the hybrid HSD-System at least has already an efficient advantage of nearly 16%. Because modern propulsion systems show no great efficiency losses in part-load behaviour, an adjustment of this efficiency was omitted in the simulation.

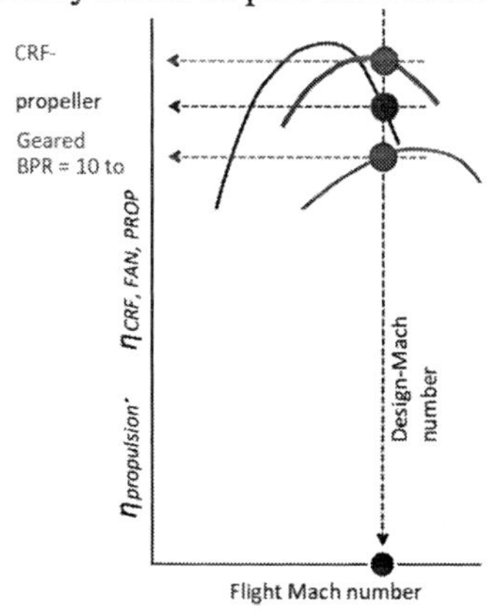

Figure 13. Propulsive efficiency as a function of the flight Mach number. Estimation: Ma = 0.75 (according to [36])

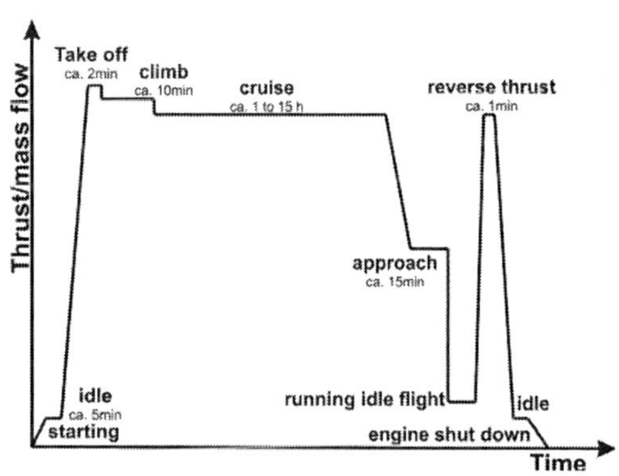

Figure 14. Flight phases with a turbofan engine [30]

IOP Conf. Series: Materials Science and Engineering **886** (2020) 012010 doi:10.1088/1757-899X/886/1/012010

A High overall appreciation in efficiency is obtained with short overall flight times because of the stronger effect of the evaluation of the relationship between an electric motor and combustion engine. For long flight times (3300 km maximum range), the efficiency of the hybrid system must be reduced due to the prolonged use of the pure combustion engine Thus, the following result is obtained for the fuel consumption ratio for the described exemplary engine:

$$\mathbf{0,6} > \left(\frac{B_{HSD,CRF}}{B_{geared\,fan}}\right)_{minimum} > \mathbf{0,44} \tag{15}$$

On average, therefore, the best possible reduction in consumption of the CRF-HSD engine compared to a fictitious geared engine in the lower thrust class (as is used in a comparable jet) of 30% is quite possible.

$$\left(\frac{B_{HSD,CRF}}{B_{geared\,fan}}\right)_{min.\ average\ value} \approx \mathbf{0,3} \tag{16}$$

5. Conclusion

By simulating the propulsion of a business Jet from the size class light jets, this study demonstrated that a parallel hybrid drive like the described HSD-engine can be used for eco-efficient flying. By coupling innovative components and technologies, a system has been created allowing a reduction in fuel consumption up to 30% compared to present engines, depending on the flight mission.

Such a system can support to initiate the target system in 2030 completely changing to a fully electric flying system as a technology bridge to meet the European Commission's emissions reduction targets in aviation.

References

[1] Berg H P Malenky U Himmelberg A and Mykhalyuk M 2015 *Proc. Deutscher Luft- und Raumfahrtkongress 2015 (Rostock)* **64** (Germany)

[2] Isikveren A T Pornet C Vratny P C and Schmidt M 2016 *Journal of Aircraft* 246-261

[3] Europäische Kommission 2011 Mitteilung der Kommission an das Europäische Parlament, den Rat, den Europäischen Wirtschafts- und Sozialausschuss und den Ausschuss der Region, Brüssel, 8.3.2011,DokumentCOM_2011_0112 https://eurlex.europa.eu/LexUriServ/LexUriServ.do?uri=COM:2011:0112:FIN:de:PDF*acces sed 12.*02.2020

[4] Isikveren A T Seitz A Vratny P C Pornet C Plötner K O and Hornung M 2012 *Proc. DGLR Conference (Munich)*

[5] Wankelsupertec GmbH 2016 Cottbus *www.wankelsupertec.de, accessed* 18.08.16

[6] EURO-K GmbH Cottbus 2016 *www.euro-k.de, accessed* 18.08.16

[7] Berg H P 2014 *Entwicklung einer wirkungsgradoptimierten Mikrogasturbine für eine KWK - Schlussbericht zum Zuwendungsnachweis F&E-große Richtlinie OP18 (Cottbus)* 80140661

[8] Izweik H T 2006 *Interner Bericht zur numerischen Berechnung der Gasströmung in einem rotierenden Modell mit AVL Fire* am LS VFA/BTU Cottbus-Senftenberg

[9] Poojitganont T Berg H P Izweik H T and Himmelberg A 2007 BTU Cottbus-Senftenberg, Germany - *Multi-fuel Wankel Engine for Small Aircrafts and UAVs Applications: Strategies of the Engine Development*, ISABE-2007-1128

[10] Berg H P and Antoshkiv O 2008 *Proc. ATZ / MTZ - Konferenz Virtual Powertrain Creation*

[11] Berg H P 2009 *Proc. ATZ/TÜV Süd Konferenz, Engine.Tech 2009 (Wolfsburg)* (Germany)

[12] Klotz R Eiermann D and Kolossow S 1999 *Abschlussbericht zum Projekt Untersuchung der Tauglichkeit des WANKEL-Kerosin-Kreiskolbenmotors für den Einsatz in der allgemeinen Luftfahrt* der Wankel-Rotary GmbH

[13] Siemens AG 2016 Erstflug am 24.06.16, http://www.siemens.com/presse/elektromotor-flugzeug, *accessed* 09.08.16

[14] 328 Support Services GmbH: 2013 *Dornier 328-300 (Light Jet) – data sheet* https://328.eu/wp-content/uploads/2013/06/328-300-lightjet.pdf, *accessed* 11.2.2020
[15] Vratny P C Gologan C Pornet C Isikveren A T and Hornung M 2013 *Proc. 4 CEAS Int. Conf. the European Aerospace Societies (Linköping)* (Germany)
[16] Kaiser S Seitz A Donnerhack S and Lundbladh S 2015 *Proc. 51st AIAA/SAE/ASEE Joint Propulsion Conference, Propulsion and Energy Forum (Orlando)* AIAA 2015-4028
[17] Filippone A 2000 *Progress in Aerospace Sciences* **36** 629f
[18] DLR *Emissionsarm elektrisch Fliegen: Was sind die Herausforderungen für elektrische Antriebe in der Luftfahrt?* – Interview mit Prof. J. Kallo, Institut für Thermodynamik, Universität Ulm, http://www.dlr.de, *accessed* 18.08.2016
[19] Landgraf M 2015 *Forscher steigern Energiedichte von Lithiumspeichermaterialien*, Pressemitteilung des KIT Karlsruhe, 16.03.2015
[20] Ren S Chen R Maawad E Dolotko O Guda A A Shapovalov V Wang D Hahn H and Fichtner M 2015 *Adv. Sci.* **1500128** 1-6
[21] Empl D Lorenz L Jeßberger C and Plötner KO 2014 *Proc. DLRK 2014 (Berlin)* 340117
[22] Pudenz K 2015 *Continental Leistungselektronik wird kompakter und leistungsfähiger*, Springer für Professionals, Nachrichten aus der Branche - Automobil- und Motorentechnik vom 27.04.2015
[23] Schwarze M C and Zold T 2013 *Proc. DGLR Conference (Stuttgart)* 301447
[24] Berg H P Malenky U Antoshkiv O and Mattke R 2015 *Proc. DLRK 2015* 370070
[25] Berg H P and Malenky U 2015 *Designstudie zum HSD-Flugmotorenkonzept*, Vorarbeiten zum Flugmotor 2030, LS VFA/ BTU CS
[26] Bräunling W J G 2009 *Flugzeugtriebwerke* (Springer-Verlag Berlin Heidelberg).
[27] Aulich A K Goerke D Blocher M Nicke E and Kocian F 2013 *ASME TURBO Expo 2013* GT2013-94259
[28] Kaplan B Nicke E and Voss C 2006 *ASME Turbo Expo 2006* GT2006-90363.
[29] Gruppe B 2010 *Triebwerke Aus Der Udssr Und Gus* (Books LLC) ISBN-13: 978-1158866908
[30] Hennecke D K, Wörrlein K *2000 Flugantriebe und Gasturbinen*, Technische Universität Darmstadt, 2. Aufl. WS2000/2001
[31] Diamond Aircraft 2011 *Diamond präsentiert das erste seriell-hybride Elektroflugzeug der Welt DA36 E-Star* Pressemitteilung vom 23.06.11
[32] Küffner G 1987 *Spitzentechnik in Deutschland, 10. Kapitel - Ein Triebwerk mit verbessertem Vortriebswirkungsgrad und hoher Wirtschaftlichkeit (MTU Motoren- und Turbinen-Union)*, Frankfurter Allgemeine; Wiesbaden: Gabler
[33] Gmehlin T C Hüttig G and Lehmann O 2008 *Untersuchungsbericht Im Auftrag des Bundesministerium für Umwelt, Naturschutz und Reaktorsicherheit IG I 5, 11055 (Berlin)* FKZ UM 07 06 602/01
[34] Hofmann P 2014 *Hybridfahrzeuge* (Springer-Verlag Wien)
[35] Golloch P 2005 *Downsizing bei Verbrennungsmotoren* (Springer-Verlag Berlin Heidelberg).
[36] El-Sayed A F 2016 *Performance Parameters of Jet Engines* (London:Springer) DOI 10.1007/978-1-4471-6796-9

Acknowledgements for photographs
Figures 2, 3, 4, 5, 6, 7, 8, 9, 10, 11 and 12 from Chair of Combustion Engines And Flight Propulsion, BTU Cottbus-Senftenberg, Germany, Figure 1 from [5], Figure 7 from [25], figure 13 from [36] and Figure 14 from [30].

Six-Degree of Freedom Mathematical Dynamic Model of a Light-Sport Aircraft

Sinchai Chinvorarat[1,*], Boonchai Watjatrakul[1], Pongsak Nimdum[1], Teerawat Sangpet[1], Tosaporn Soontornpasatch[1], and Pumyos Vallikul[1]

[1]Department of Mechanical and Aerospace Engineering, King Mongkut's University of Technology North Bangkok 1518 Pracharaj 1, Bangkok, 10800, Thailand

* Corresponding Author: sinchai.c@eng.kmutnb.ac.th

Abstract
The paper considers the applications of six-degree of freedom mathematical model of a light-sport aircraft and analytically reveals aircraft dynamic response both longitudinal and lateral-directional stability. The model composes of both the kinematics equations of motion and the kinetics equations of aerodynamic forces and moments, and it is known as the aerodynamic model equations. Simulation results show responses of the perturbed dynamic system at the trim condition, and indicate the dynamic stability of both short-period pitching oscillation and phugoid in longitudinal axes. The spiral, roll subsidence, and dutch roll modes in lateral-directional axes are dynamical stable as well. These are essential to understand and evaluate the dynamic behavior, stability, safety and other aspects of the designed aircraft through mathematical model before conducting operational flight test.

Keywords: mathematical model, dynamic stability, light-sport aircraft

1. Introduction

A two-seater light amphibious airplane special project supported by Thai Research Fund (henceforth, TRF) aims to build a light amphibious airplane called "NAX-5" in which categorized into a light-sport aircraft (henceforth, LSA). In short, LSA is defined as a small non-pressurized aircraft, fixed-wing, one crew and one passenger, 600 kg maximum take-off weight (650 kg for seaplane), and driven by a fixed pitch propeller connect to a single piston engine. A certificate of airworthiness can be issued to the LSA that comply with ASTM F2245 standard for the safety and performance. One of the compliance method through the certification process is analytically proven in both longitudinal and lateral-directional aircraft stability [1], [2]. Before the aircraft will be demonstrated the aerodynamic stability through a flight test program, an aircraft designer normally prefers to investigate the dynamic response of initial design by proper simulation [5], [6], [7].

A mathematical model of aircraft usually describes the aerodynamic forces and moments in terms of relevant measurable quantities such as control surface deflections, aircraft linear velocity, angular velocities, and the orientation of the aircraft respect to the relative wind [11]. When the mathematical model is parametric, aerodynamic parameters can be quantified as the functional dependence of the aerodynamic forces and moments on measureable quantities. In general, aircraft mathematical model is non-linear, comprised of complex aerodynamic function depends on aircraft flight configuration [9], [10]. To ensure the aircraft safety and performance design, dynamics simulation

IOP Conf. Series: Materials Science and Engineering **886** (2020) 012011 doi:10.1088/1757-899X/886/1/012011

through a mathematical model is required before conducting the first flight test so that suitable engineering correction can be made.

This paper presents a six-degree of freedom equation of motion of an aircraft model. With a small perturbation theory, one can analytically linearize the general equation of motion composed of kinematics equations of motion and kinetics equations of forces and moments at trim. Dccouple technique is used to simplify the equation of motion into a set of longitudinal equations and lateral-directional equations [8]. Laplace transform technique is also used to achieve a set of the input-output transfer functions.

Abbreviations

U, V, W	linear velocity vector component about the axes ox, oy and oz
X, Y, Z	moment vector component about the axes ox, oy and oz
p, q, r	angular velocity vector component about the axes ox, oy and oz
L, M, N	moment vector component about the axes ox, oy and oz
I_x, I_y, I_z	moment of inertia about the axes ox, oy and oz
I_{xy}, I_{yz}, I_{xz}	product of inertia about the axes ox, oy and oz
u, v, w	linear disturbance velocities
$\mathring{X}_u, \mathring{X}_v, \mathring{X}_w$ and so on	aerodynamic stability derivatives
ξ, η, ζ	aileron deflection angle, elevator deflection angle, and rudder deflection angle
ϕ, θ, ψ	components of airplane attitude in roll, pitch, and yaw angle
α	incidence angle
β	side slip angle
m	mass
τ	thrust
g	gravity
V_o	velocity of the airplane

2. Light sport aircraft model

The "NAX" project initiated by the Royal Thai Navy research team has been started since 2002, aiming to build a light versatile amphibious airplane that can serve in military and civilian purposes. The initial version of airplane, NAX-3, has been designed from the concept of the EDRA Aeronautic Super Pétrel, an amphibious pusher configuration biplane. The NAX-3 has demonstrated its satisfaction flying qualities in both performance and safety, but still, the total weight of airplane is above 650 kilograms which are the number classified as a light sport airplane category stated by the Civil Aviation Authorities of Thailand (CAAT).

The new design of NAX-5 project has started with the concept of building an amphibious pusher high-wing configuration [12]. The airplane main structure is mostly made from composite, gains benefits in lightweight control and enhances the strength of the structure. With retractable landing gears, the airplane can take-off and landing on a short airfield or water runway. A more powerful 115 Hp, Rotax 914 piston engine with three blades constant pitch propeller is installed as "pusher" high-wing configuration. The airplane basic design performance is shown in Table 1.

Figure 1. A two-seater light amphibious airplane "NAX-5"

Specifications:		
Crew:	1	person
Capacity:	1	person
Length:	22.13	ft
Wingspan:	32.33	ft
Wing area:	112.89	ft^2
Cabin width:	4.33	ft
Empty weight:	880.00	lbs
Max fuel capacity:	39.06	gallon
Max takeoff weight:	1,430.00	lbs
Performance:		
Maximum speed:	179.81	mph
Cruise speed:	141.47	mph
Stall speed:	53.98	mph
Endurance:	5	Hrs
Rate of Climb:	863.75	fpm
Take off Distance:	975.00	ft
Landing Distance:	910.00	ft

Table 1. "NAX-5" specification and performance

The flying motion of the aircraft analyzing at the center of gravity *cg* coincident with the origin o in an orthogonal axis set $(oxyz)$, in the general body, as shown in Figure 2. Then, the force and moment equation comprised the generalized six-degree of freedom equations of motion. The uniform mass distribution of an airframe can be expressed as the following equations:

Figure 2. Motion referred to generalized body axes.

$$
\begin{aligned}
&m(\dot{U} - rV + qW) = X_a + X_g + X_c + X_p + X_d \\
&m(\dot{V} - pW + rU) = Y_a + Y_g + Y_c + Y_p + Y_d \\
&m(\dot{W} - qU + pV) = Z_a + Z_g + Z_c + Z_p + Z_d \\
&I_x\dot{p} - (I_y - I_z)qr - I_{xz}(pq + \dot{r}) = L_a + L_g + L_c + L_p + L_d \\
&I_y\dot{q} + (I_x - I_z)pr + I_{xz}(p^2 - r^2) = M_a + M_g + M_c + M_p + M_d \\
&I_z\dot{r} - (I_x - I_y)pq + I_{xz}(qr - \dot{p}) = N_a + N_g + N_c + N_p + N_d
\end{aligned} \qquad (1)
$$

Assuming that the disturbing forces and moments on the right side of the equation are the effects resulting from aerodynamic, gravitational, aerodynamic flight control moment, power from engine, and atmospheric disturbances. The equations are in general highly non-linear, and unpractical to achieve their solution by analytical mean. To continue solving the analytical solution, linearization of these equations can be accomplished by restraining the motion of the aircraft to small perturbation at the trim condition. If the aircraft at the undisturbed axis $(oxyz)$ subjects a small perturbation of linear disturbance velocities (u, v, w) and angular disturbance velocities (p, q, r) about the trim; therefore, in the disturbed motion, the overall linear velocity vector components of the aircraft at the center of gravity are given by

$$\begin{aligned} U &= U_e + u \\ V &= V_e + v = v \\ W &= W_e + w \end{aligned} \tag{2}$$

At the flying circumstance in regular trim (none of the roll, sideslip, and yaw angle), there is no effect of an undisturbed atmosphere $(X_d = Y_d = Z_d = L_d = M_d = N_d = 0)$. Weight and moment about any axis are considered $(L_g = M_g = N_g = 0)$. The gravitational force components are $X_g = -mgsin\theta_e - mg\theta cos\theta_e$, $Y_g = -mg\psi\ sin\theta_e + mg\phi cos\theta_e$, $Z_g = mgcos\theta_e - mg\theta sin\theta_e$.

The aerodynamic terms are expressed by Taylor's expansion and neglected all small terms (i.e. $X_a = X_{ae} + \frac{\partial X}{\partial u}u + \frac{\partial X}{\partial v}v + \frac{\partial X}{\partial w}w + \frac{\partial X}{\partial p}p + \frac{\partial X}{\partial q}q + \frac{\partial X}{\partial r}r + \frac{\partial X}{\partial \dot{w}}\dot{w}$, and $\frac{\partial X}{\partial u} = \mathring{X}_u$ for shorthand notation). Also in the steady trimmed, $X_{ae} = mg \sin\theta_e$, $Y_{ae} = 0$, $Z_{ae} = -mg\ cos\theta_e$, $L_{ae} = M_{ae} = N_{ae} = 0$. The aerodynamic control derivatives (i.e. pitching moment $M_c = \frac{\partial M}{\partial \xi}\xi + \frac{\partial M}{\partial \eta}\eta + \frac{\partial M}{\partial \zeta}\zeta$) can be used to describe the response of all primary aerodynamic control such as the elevator, ailerons, and rudder. The relationship between the trust from the engine and throttle level angle in power term is also expressed in shorthand notation such as, $Z_p = \mathring{Z}_\tau\tau$. The contributions of the final equation are also written comparably, therefore, the equations of motion for small perturbations can be shown as

$$\begin{aligned} m(\dot{u} - qW_e) &= \mathring{X}_u u + \mathring{X}_v v + \mathring{X}_{\dot{w}}\dot{w} + \mathring{X}_w w + \mathring{X}_p p + \mathring{X}_q q + \mathring{X}_r r - mg\theta cos\theta_e + \mathring{X}_\xi\xi + \mathring{X}_\eta\eta + \mathring{X}_\zeta\zeta + \mathring{X}_\tau\tau \\ m(\dot{v} - pW_e + rU_e) &= \mathring{Y}_u u + \mathring{Y}_v v + \mathring{Y}_w w + \mathring{Y}_{\dot{w}}\dot{w} + \mathring{Y}_p p + \mathring{Y}_q q + \mathring{Y}_r r + mg\phi cos\theta_e + mg\psi sin\theta_e + \mathring{Y}_\xi\xi + \mathring{Y}_\eta\eta + \mathring{Y}_\zeta\zeta + \mathring{Y}_\tau\tau \\ m(\dot{w} - qU_e) &= \mathring{Z}_u u + \mathring{Z}_v v + \mathring{Z}_w w + \mathring{Z}_{\dot{w}}\dot{w} + \mathring{Z}_p p + \mathring{Z}_q q + \mathring{Z}_r r - mg\theta sin\theta_e + \mathring{Z}_\xi\xi + \mathring{Z}_\eta\eta + \mathring{Z}_\zeta\zeta + \mathring{Z}_\tau\tau \\ I_x\dot{p} - I_{xz}\dot{r} &= \mathring{L}_u u + \mathring{L}_v v + \mathring{L}_w w + \mathring{L}_{\dot{w}}\dot{w} + \mathring{L}_p p + \mathring{L}_q q + \mathring{L}_r r + \mathring{L}_\xi\xi + \mathring{L}_\eta\eta + \mathring{L}_\zeta\zeta + \mathring{Z}_\tau\tau \\ I_y\dot{q} &= \mathring{M}_u u + \mathring{M}_v v + \mathring{M}_w w + \mathring{M}_{\dot{w}}\dot{w} + \mathring{M}_p p + \mathring{M}_q q + \mathring{M}_r r + \mathring{M}_\xi\xi + \mathring{M}_\eta\eta + \mathring{M}_\zeta\zeta + \mathring{M}_\tau\tau \\ I_z\dot{r} - I_{xz}\dot{p} &= \mathring{N}_u u + \mathring{N}_v v + \mathring{N}_w w + \mathring{N}_{\dot{w}}\dot{w} + \mathring{N}_p p + \mathring{N}_q q + \mathring{N}_r r + \mathring{N}_\xi\xi + \mathring{N}_\eta\eta + \mathring{N}_\zeta\zeta + \mathring{N}_\tau\tau \end{aligned} \tag{3}$$

Equation (3) are small perturbations equations of motion describing the transient response of the aircraft at trim. They compose of six simultaneous linear differential equation written conventionally with inputs. Since longitudinal and lateral dynamics are fully coupled in the equations of motion, but the longitudinal and lateral coupling can be negligible when considering small perturbation of motion. As a result, these equations can be simplified by decoupling the longitudinal and lateral motion.

Decoupled equations of motion

Longitudinal equations of motion

The motion is described by the longitudinal plane of symmetry (*oxz* plane), therefore only axial force, normal force, and pitching moment are considered in the equations of motion. By this assumption, the motion variables and their derivative in lateral along with aileron and rudder deflections are all zero, thus

$$\begin{aligned} \mathring{X}_v = \mathring{X}_p = \mathring{X}_r = \mathring{Z}_v = \mathring{Z}_p = \mathring{Z}_r = \mathring{M}_v = \mathring{M}_p = \mathring{M}_r = 0 \\ \mathring{X}_\xi = \mathring{X}_\zeta = \mathring{Z}_\xi = \mathring{Z}_\zeta = \mathring{M}_\xi = \mathring{M}_\zeta = 0 \end{aligned}$$

Also, with the assumption that the aircraft is in level flight

$$\theta_e = W_e = 0$$

The simplified longitudinal equations of motion are rearranged as

$$\begin{aligned} m\dot{u} - \mathring{X}_u u - \mathring{X}_{\dot{w}}\dot{w} - \mathring{X}_w w - \mathring{X}_q q + mg\theta &= \mathring{X}_\eta\eta + \mathring{X}_\tau\tau \\ -\mathring{Z}_u u + (m - \mathring{Z}_{\dot{w}})\dot{w} - \mathring{Z}_w w - \left(\mathring{Z}_q + mU_e\right)q &= \mathring{Z}_\eta\eta + \mathring{Z}_\tau\tau \\ -\mathring{M}_u u + \mathring{M}_{\dot{w}}\dot{w} - \mathring{M}_w w + I_y\dot{q} - \mathring{M}_q q &= \mathring{M}_\eta\eta + \mathring{M}_\tau\tau \end{aligned} \tag{4}$$

IOP Conf. Series: Materials Science and Engineering **886** (2020) 012011 doi:10.1088/1757-899X/886/1/012011

Lateral-directional equations of motion

The motion contains yaw, roll, and sideslip; hence, yawing moment, rolling moment, and side force are considered in the equations of motion. The elevator deflection, thrust variation, all motion variables and their derivative in longitudinal do not influent lateral-directional motion and do not take into account, thus

$$\mathring{Y}_u = \mathring{Y}_{\dot{w}} = \mathring{Y}_w = \mathring{Y}_q = \mathring{L}_u = \mathring{L}_{\dot{w}} = \mathring{L}_w = \mathring{L}_q = \mathring{N}_u = \mathring{N}_{\dot{w}} = \mathring{N}_w = \mathring{N}_q = 0$$

$$\mathring{Y}_\eta = \mathring{Y}_\tau = \mathring{L}_\eta = \mathring{L}_\tau = \mathring{N}_\eta = \mathring{N}_\tau = 0$$

Also, with the assumption that the aircraft is in level flight

$$\theta_e = W_e = 0$$

The lateral-directional equations of motion are simplified and rearranged in the form

$$\begin{aligned} m\dot{v} - \mathring{Y}_v v - p\mathring{Y}_p - \left(\mathring{Y}_r - mU_e\right) r - mg\phi &= \mathring{Y}_\xi \xi + \mathring{Y}_\zeta \zeta \\ -\mathring{L}_v v + I_x \dot{p} - \mathring{L}_p p - I_{xz}\dot{r} - \mathring{L}_r r &= \mathring{L}_\xi \xi + \mathring{L}_\zeta \zeta \\ -\mathring{N}_v v - I_{xz}\dot{p} - \mathring{N}_p p + I_z \dot{r} - \mathring{N}_r r &= \mathring{N}_\xi \xi + \mathring{N}_\zeta \zeta \end{aligned} \tag{5}$$

3. The longitudinal and lateral-directional response transfer functions

When small perturbation motion is taken into consideration and thrust remains constant in trim, the Laplace transform with the assumption of zero initial conditions of longitudinal equation (4) can be written in the matrix form

$$\begin{bmatrix} \left(ms - \mathring{X}_u\right) & -\left(\mathring{X}_{\dot{w}} s - \mathring{X}_w\right) & -\left(\left(\mathring{X}_q\right) s - mg\right) \\ -\mathring{Z}_u & -\left(\left(\mathring{Z}_{\dot{w}} - m\right) s + \mathring{Z}_w\right) & \left(\mathring{Z}_q + mU_e\right) s \\ -\mathring{M}_u & -\left(\mathring{M}_{\dot{w}} s + \mathring{M}_w\right) & \left(I_y s^2 - \mathring{M}_q s\right) \end{bmatrix} \begin{bmatrix} u(s) \\ w(s) \\ \theta(s) \end{bmatrix} = \begin{bmatrix} \mathring{X}_\eta \\ \mathring{Z}_\eta \\ \mathring{M}_\eta \end{bmatrix} [\eta(s)] \tag{6}$$

By Cramer's rule, one can write the longitudinal transfer functions as

$$\frac{u(s)}{\eta(s)} = \frac{N_\eta^u(s)}{\Delta(s)}, \quad \frac{w(s)}{\eta(s)} = \frac{N_\eta^w(s)}{\Delta(s)}, \quad \frac{\theta(s)}{\eta(s)} = \frac{N_\eta^\theta(s)}{\Delta(s)}$$

$$\frac{N_\eta^q(s)}{\Delta(s)} = \frac{q(s)}{\eta(s)} = \frac{s\theta(s)}{\eta(s)} = \frac{sN_\eta^\theta(s)}{\Delta(s)}$$

Similarly, we can rewrite the lateral-directional in the matrix form, as shown in equation (5).

$$\begin{bmatrix} \left(ms - \mathring{Y}_v\right) & -\left(\left(\mathring{Y}_p\right) s + mg\right) & -\left(\left(\mathring{Y}_r - mU_e\right) s\right) \\ -\mathring{L}_v & \left(I_x s^2 - \mathring{L}_p s\right) & -\left(I_{xz} s^2 + \mathring{L}_r s\right) \\ -\mathring{N}_v & -\left(I_{xz} s^2 + \mathring{N}_p s\right) & \left(I_x s^2 - \mathring{N}_r s\right) \end{bmatrix} \begin{bmatrix} v(s) \\ \phi(s) \\ \psi(s) \end{bmatrix} = \begin{bmatrix} \mathring{Y}_\xi & \mathring{Y}_\eta \\ \mathring{L}_\xi & \mathring{L}_\zeta \\ \mathring{N}_\xi & \mathring{N}_\zeta \end{bmatrix} \begin{bmatrix} \xi(s) \\ \zeta(s) \end{bmatrix} \tag{7}$$

By Cramer's rule, one can write the lateral-directional transfer functions as

$$\frac{v(s)}{\xi(s)} = \frac{N_\xi^v(s)}{\Delta(s)}, \quad \frac{N_\xi^p(s)}{\Delta(s)} = \frac{p(s)}{\xi(s)} = \frac{s\phi(s)}{\xi(s)} = \frac{sN_\xi^\phi(s)}{\Delta(s)}, \quad \frac{N_\xi^r(s)}{\Delta(s)} = \frac{r(s)}{\xi(s)} = \frac{s\psi(s)}{\xi(s)} = \frac{sN_\xi^\psi(s)}{\Delta(s)}$$

$$\frac{v(s)}{\zeta(s)} = \frac{N_\zeta^v(s)}{\Delta(s)}, \quad \frac{N_\zeta^p(s)}{\Delta(s)} = \frac{p(s)}{\zeta(s)} = \frac{s\phi(s)}{\zeta(s)} = \frac{sN_\zeta^\phi(s)}{\Delta(s)}, \quad \frac{N_\zeta^r(s)}{\Delta(s)} = \frac{r(s)}{\zeta(s)} = \frac{s\psi(s)}{\zeta(s)} = \frac{sN_\zeta^\psi(s)}{\Delta(s)}$$

Incidence and sideslip response transfer functions

For other related variables, one can easily compose incidence angle in the longitudinal decouple equation by the assumption for small perturbation motion incidence as

$$\alpha \cong \tan \alpha = \frac{w}{V_o} \tag{8}$$

since $U_e \to V_o$ implying the perturbation has tendency going to zero, then incidence α can be written as aircraft vertical velocity over the steady air stream velocity.

Similarly, the sideslip angle (β) can be acquired from the lateral-directional equation by the relationship

$$\beta \cong \tan \beta = \frac{v}{V_o} \tag{9}$$

Aerodynamic stability derivative

All aerodynamic derivative appeared in the linearized equation of motion play significant roles in determining the accuracy of the mathematical models. Analytically, the aerodynamic derivative describing a dynamic characteristic of aircraft can be determined from the relationship of various aerodynamic coefficients and parameters which are linearized about the operating point of interest. However, since the flow conditions across the airframe are complicated, the explanation for the mathematical aerodynamic phenomena must result in compromise. Wind tunnel measurement and flight test measurement are considered to be most reliable methods to gain the correct aerodynamic coefficients and parameters of the specific aircraft model, but a number of equipment and expenses must be taken into account.

In this paper, all longitudinal stability derivative and lateral-directional stability derivative are obtained from NASA CR-1975 Handling Quality Light Aircraft report [3], [4] referring to the NAX-5 design dimension and configurations.

4. Simulation results

To simulate and verify aircraft stability dynamic through the equations of motion, the initial input data of the aircraft characteristic is determined and inserted into the simulation program. The flight condition corresponding to level cruising flight is the altitude at 5,000 ft. and aircraft's speed at Mach 0.2.

By using MATLAB, the transfer functions of both longitudinal and lateral-directional aircraft motions are shown in equations (10) and (11)

Longitudinal

$$\begin{aligned}
\frac{u(s)}{\eta(s)} &= \frac{-1.18(s-10.76)(s^2+3.72s+7.662)}{(s^2-0.0306s+0.0465)(s^2-3.46s+5.1833)} \\
\frac{w(s)}{\eta(s)} &= \frac{-9.075(s+28.59)(s^2+0.044s+0.085)}{(s^2-0.0306s+0.0465)(s^2-3.46s+5.1833)} \\
\frac{q(s)}{\eta(s)} &= \frac{-5.364s(s+0.0708)(s+1.82)}{(s^2-0.0306s+0.0465)(s^2-3.46s+5.1833)} \\
\frac{\theta(s)}{\eta(s)} &= \frac{-5.364s(s+0.0708)(s+1.82)}{(s^2-0.0306s+0.0465)(s^2-3.46s+5.1833)} \\
\frac{\alpha(s)}{\eta(s)} &= \frac{-0.1881(s+28.59)(s^2+0.044s+0.085)}{(s^2-0.0306s+0.0465)(s^2-3.46s+5.1833)} \\
\frac{\gamma(s)}{\eta(s)} &= \frac{0.1881(s+0.026)(s+7.299)(s-7.21)}{(s^2-0.0306s+0.0465)(s^2-3.46s+5.1833)}
\end{aligned} \tag{10}$$

The phugoid stability mode is determined by the first pair of complex poles with the value of damping ratio $\zeta_p = 0.0712$ and undamped natural frequency $\omega_p = 0.215$ rad/s. The short-period pitching oscillation mode is also determined by the second pair of complex poles with the value of damping ratio $\zeta_p = 0.761$ and undamped natural frequency $\omega_p = 2.28$ rad/s. The characteristic of these modes indicates the degree of stability of the aircraft.

The dynamic responses of the aircraft from a pulse elevator input (5 degrees pulse, held for 5 seconds and return to zero and another -5 degrees pulse, held for 5 seconds and return to zero) are shown in Figure 4. The results of this study clearly show dynamic stability both short-period pitching oscillation and phugoid modes. It should be noted that each response variable shows a different scale in term of each stability mode.

Lateral-directional

$$\frac{\beta(s)}{\xi(s)}=\frac{1.411(s+0.124)(s+9.22)}{(s+0.061)(s+8.81)(s^2+0.46s+1.91)}$$
$$\frac{p(s)}{\xi(s)}=\frac{15.41(s-0.012)(s^2+0.702s+1.53)}{(s+0.061)(s+8.81)(s^2+0.46s+1.91)}$$
$$\frac{r(s)}{\xi(s)}=\frac{-0.684(s-0.65)(s+0.715)(s+14.5)}{(s+0.061)(s+8.81)(s^2+0.46s+1.91)}$$
$$\frac{\phi(s)}{\xi(s)}=\frac{15.37(s^2+0.659s+1.52)}{(s+0.061)(s+8.81)(s^2+0.46s+1.91)}$$

$$\frac{\beta(s)}{\zeta(s)}=\frac{0.0963(s+0.0445)(s+8.81)(s+18.1)}{(s+0.061)(s+8.81)(s^2+0.46s+1.91)}$$
$$\frac{p(s)}{\zeta(s)}=\frac{0.5963(s-0.011)(s-1.45)(s+7.85)}{(s+0.061)(s+8.81)(s^2+0.46s+1.91)}$$
$$\frac{r(s)}{\zeta(s)}=\frac{-1.719(s+8.823)(s^2+0.055s+0.0938)}{(s+0.061)(s+8.81)(s^2+0.46s+1.91)}$$
$$\frac{\phi(s)}{\zeta(s)}=\frac{0.5061(s-1.77)(s+7.71)}{(s+0.061)(s+8.81)(s^2+0.46s+1.91)} \quad (11)$$

The spiral mode is determined by the first real of complex poles with the value of the time constant $T_s=\frac{1}{0.061}\cong 16.39\ sec$. The roll subsidence mode is determined by the second real pole with the value of the time constant $T_s=\frac{1}{8.81}\cong 0.1135\ sec$. The oscillatory dutch roll mode is also determined by the pair of complex poles with the value of damping ratio $\zeta_d=0.166$ and undamped natural frequency $\omega_d=1.38$ rad/sec.

Since all poles of the system transfer function have negative real parts, the mode characteristics indicate that the aircraft is aerodynamically stable. The dynamic responses from an aileron impulse input and a rudder impulse input of the aircraft are shown in Figures 5 and 6, respectively.

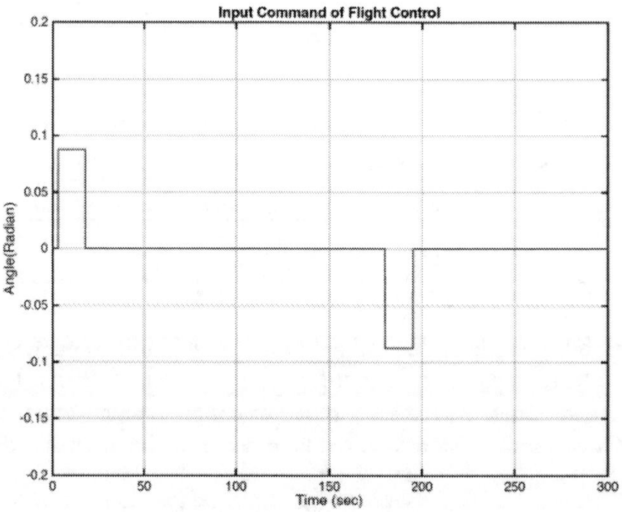

Figure 3. Impulse input

IOP Conf. Series: Materials Science and Engineering **886** (2020) 012011 doi:10.1088/1757-899X/886/1/012011

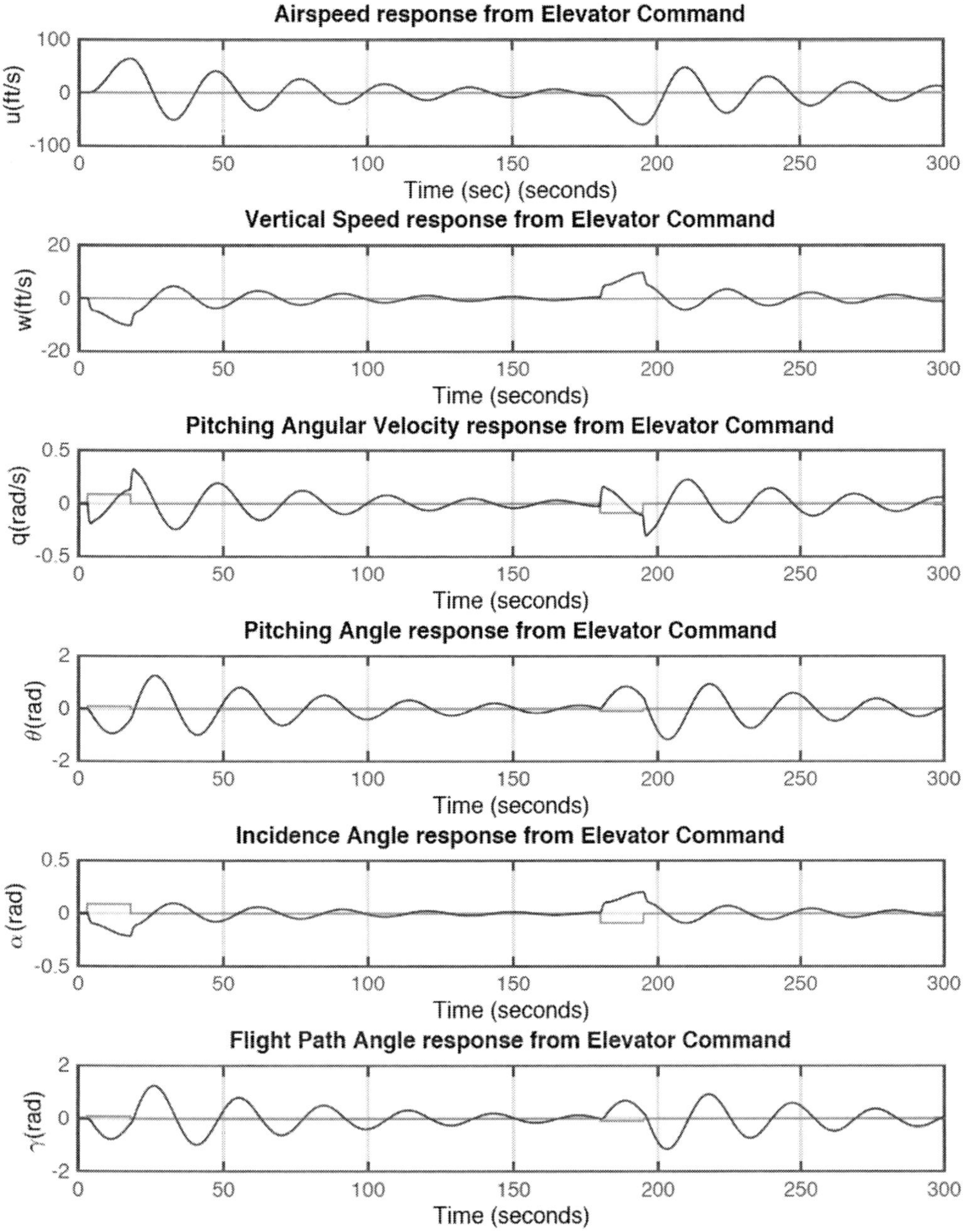

Figure 4. Aircraft dynamic response from elevator command

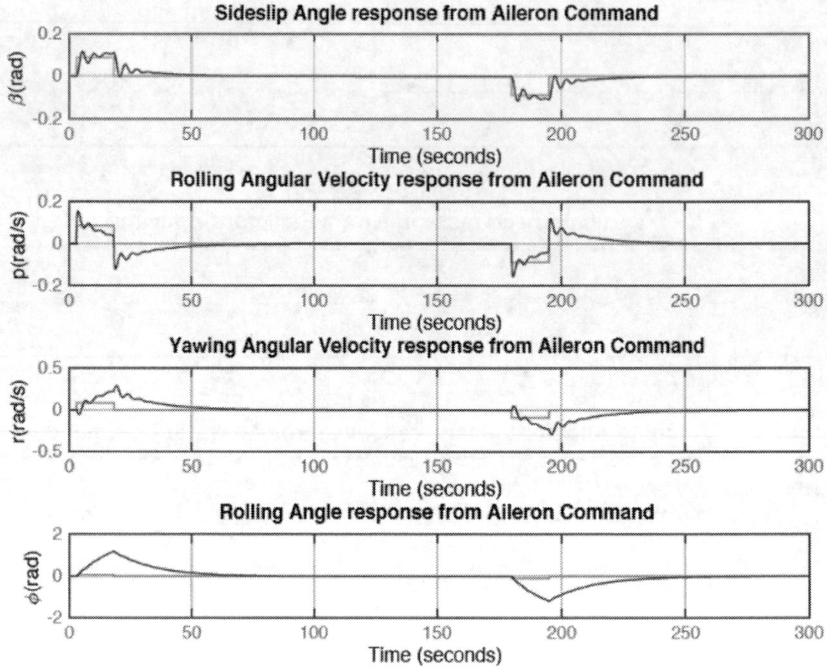

Figure 5. Aircraft dynamic response from aileron command

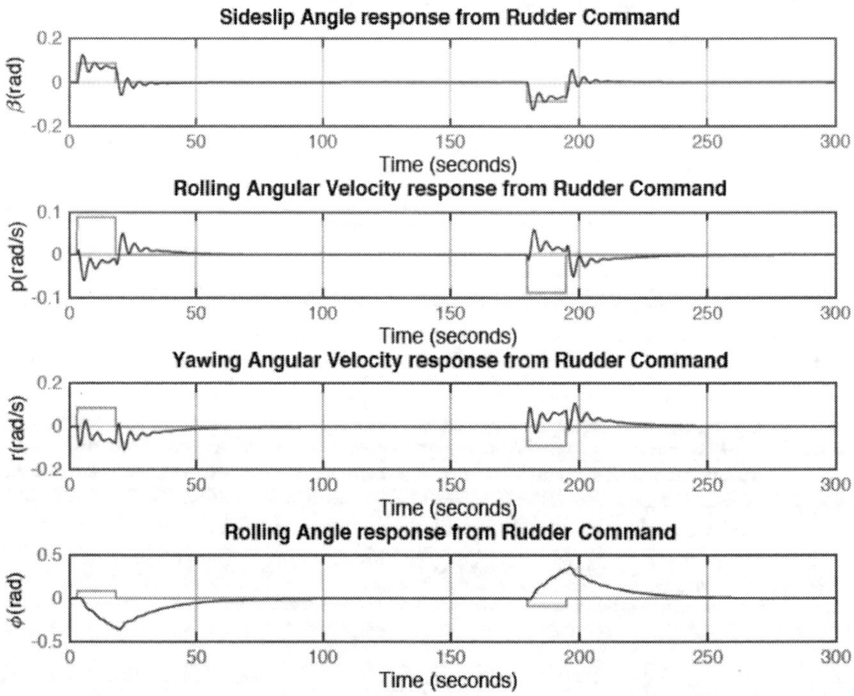

Figure 6. Aircraft dynamic response from rudder command

5. Conclusions

This study set out to investigate the mathematical model of a light-sport at the trim condition. The model composed of kinematics equations of motion and kinetics equations of forces and moments. This study has identified the system transfer function in both longitudinal and lateral-directional axes. It also shows a stable aerodynamic response with a suitable degree of stability. The investigation of the assumption of level flight and stable thrust has shown that the longitudinal equations are disturbed with five degrees of elevator deflection in both directions. The results revealed that it does not only make airspeed, vertical speed, pitch rate, pitch angle, incidence angle and flight path angle deviated, but it also turns to steady value in such a period. One of the more significant findings to emerge from this study is that when five degrees of aileron and rudder deflection is made, the roll rate, yaw rate and sideslip angle turn to steady-state values in a short time. However, the roll angle keeps slowly changing because a small degree of stability allows some changes to the aircraft attitude. These findings suggest that in general the adverse roll can be noticed from the graph since a non-minimum phase numerator term appeared in some system transfer function. Overall, this study strengthens the idea that the stability of aircraft dynamic and ensure safety before conducting actual test flight should be encouraged.

Acknowledgements

We would like to thank the Thailand Research Fund (TRF) and Department of Mechanical and Aerospace Engineering, KMUTNB for the research cooperation and financial support throughout the project. A special acknowledgement goes to the Royal Thai Navy (RTN) for invaluable information toward the end of this project.

References

[1] Teper G L 1969 *STI TechnicalReport176-1* (Washington D.C.:USA) 20546
[2] Hopkin H R 1970 *Aeronautical Research Council, Reports and Memoranda No.3562* (London:UK)
[3] Smetana F O Summey D C Johnson D W 1972 *NASA contractor report NASA CR-1975* (Washington D.C.:USA)
[4] Heffley R K Jewell W F 1972 *NASA Contractor Report NASA CR-2144* (Washington D.C.:USA)
[5] Barnett S 1975 *Introduction to mathematical control theory* (Oxford:UK)
[6] Stribling C B 1984 *Basic aerodynamics* (London:Butterworth & Co.)
[7] Shevell R S 1989 *Fundamentals of flight, second edition* (New Jersey:PrenticeHall)
[8] Cook M V 1994 *Aeronautical Journal* **98** 978
[9] Hancock G J 1995 *An introduction to the flight dynamics of rigid aeroplanes* (Hempstead:Ellis Horwood Ltd.)
[10] Fossen T 2011 *Mathematical models for control of aircraft and satellites second edition* (Department of Engineering Cybernetics)
[11] Angi N Huminic A 2015 Incas Bulletin **7**(4) 33–39
[12] Chinvorarat S Watjatrakul B Nimdum P Sangpet T Soontornpasatch T Vallikul P 2019 *IOP Conf. Series: Materials Science and Engineering* **501** 012026
[13] Dean J Morton S McDaniel D Clifton J Bodkin D 2008 *AIAA Atmospheric Flight Mechanics Conference and Exhibit* 6378

IOP Conf. Series: Materials Science and Engineering **886** (2020) 012012 doi:10.1088/1757-899X/886/1/012012

Simulation of Light-Weight Truck LF3070G1's Tire Dynamics

Thanh Quang Nguyen[1,*], Xa Hoi Nguyen[2], Hong Quan Le[3], Van Anh Le[4], Phuc Hoa Tran[5], Tien Tan Nguyen[6]

[1,3,4,5,6] Hanoi University of Industry, Vietnam
[2] University of Fire, Hanoi, Vietnam

* Corresponding Author: nguyenthanhquang@haui.edu.vn, Telephone +84903404601

Abstract
Being accurately analyses of the dynamics of automotive tires is essential for a wide range of technologies. There are two primary aims of this study: 1. To investigate the use of a Finite Elements Method (henceforth, FEM) in the light-weight truck' tire (LF3070G1) 2. To implement in Ansys Workbench software in the simulation. Two analyses of the tire dynamics were used. The first analysis was used when the tire is statically balanced between supporting the vehicle weight and the internal air pressure. The second analysis includes the dynamical balance while the vehicle is operating at different weight loads. Each analysis includes two parts. The first part was the mode shapes and the natural frequencies of the tire since it has an influence on the tire's stiffness and damping constant. The second analysis employed numerical simulations and it was carried out to determine the tire's time-dependent maximum deformation.

Keywords: Truck tire, tire model, air pressure, mode shape, deformation

1. Introduction

The wheel of Light-weight truck generally comprises a rubber tire, rim and wheel disc. The disc is where the wheel is mounted (to the wheel hub). Parameters of the wheel are primarily determined by the speed and load-bearing capacity of the truck. The main requirements of light truck wheels are high fatigue strength and long service life and high load capacity [1]. Extensive research has shown that tire modeling and its analysis are used for a variety of purposes. For instance, an analysis of vehicle collision accidents is used for modelling a tire force. Data for this study were collected using the estimation of the physical parameters, and it involved a statistical method of the tire forces experiment. This study also provided results for the dynamics of vehicles with tire broken and loads the limits corresponding [2].

Previous research findings have also revealed the impact of the road micro-profile on the duration and the tire of the vehicle wheel contact with the road surface driving at a different speed. Frequency characteristics of suspension motion and regularities of vertical movement of the wheel were identified after dividing the investigated road section according to driving modes. The study has attempted to analyze the wheel contact with the road surface and identified correlations. It was found that it enables engineers to determine vehicle stability on selected quality roads [3]. The study indicated longitudinal and transverse forces distributed over the tire-road contact area. The analysis of the use of the lumped parameters dynamic friction model for traction-braking control purposes was explored. The distribution of normal forces in the tire-road contact area at different vehicle speeds was determined [4].

IOP Conf. Series: Materials Science and Engineering **886** (2020) 012012 doi:10.1088/1757-899X/886/1/012012

Regarding the Ansys Workbench 14.0 software, it is used to conduct the static analysis of a tire model developed from natural rubber which has a hard texture and to identify factors regarding the linear isotropic elastic behavior of the specific rubber. The maximum inflation pressure of 0.2206 MPa and maximum load capacity of 515 kg is obtained from the tire size specification load index. The von mises stress the value of 0.9448 MPa obtained on application of a vertical load of 5.15 kN is lower than the specified pressure on vertical loading [5].

2. Tire models

For the simple vehicle model, a tire may be represented by an elastic and viscous resisting system of rubber components. With the model complexity increases with higher frequencies are analyzed, tire needs to be represented in greater detail. The tires model are characterized by pressure, operating conditions, loads and its nonlinearities nature.

The fiala tire model

In this model, lateral and longitudinal forces are computed by the model using as input parameters. The maximum forces are of the model used to take into account frictions coefficient μ_{max} and μ_{min} at full adhesion and sliding of normal conditions, not including slip conditions in the braking or driving while turning [6].

The magic Formula tire model

The magic formula tire model is used by curve fitting an analytical expression of the generic empirical equation in different driving situations. The model is able to describe tire behavior the characteristics for either side force, torque or longitudinal force [7].

The FTire (flexible ring tire) model

The wheel is modeled as a rigid body masses interconnected by elastic elements. The tire vibration behavior at the interface of interaction between the tire and the uneven road is modeled in high precision [8].

The swift model

The swift model considers the tire dynamics in rotational and forward directions. It is an empirical model that is originally based on the Magic Formula combined with the rigid body model. One disadvantage of the swift model is the variety of wheel parameters used as well as simulation results in low accuracy and may not be well suited to bigger tires [9].

The hohenheim model

The hohenheim tire model has been validated for large agricultural tires. The model is a combination between an empirical and a physical model. It uses a rigid body comprises of non-linear spring and damper elements to represent the forces and torques in different directions [10].

Other models

In dynamic analysis, tires are generally presented by empirical models. There are three principal tire models used: the elastic model, the string model, and beam model. In the elastic model, each small element of the contact patch surface is considered to act independently. The string model, lateral displacement of each element is resisted by the tension between the elements. The beam model does not allow discontinuous of either. The string and beam models are sometimes combined [11].

3. Simulation of the light truck tire

Tire structure and meshing

Tire's specifications of light truck depend on the respective requirements of various vehicle's types and sizes, and total weight, and speed, and operating conditions. The dimensions, load ratings, specified inflation pressures and limited speeds are the main parameters of tire.- DIN 7804 is the German standard for Light duty commercial vehicles [1, p. 576].

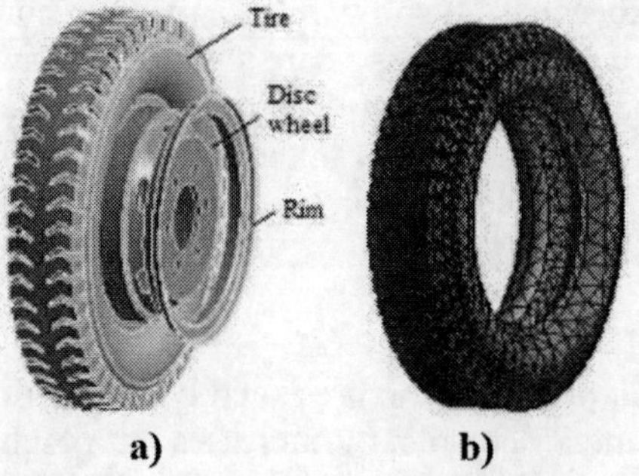

Figure 1. Wheel and tire of light truck
a) Wheel construction, b) Tire meshing (nodes: 77862, elements: 40722)

The light truck LF3070G1 has a payload of 3000 kG, and weigh 7150 kg. The number of wheels has six units in size, ranging from 8.25-16. The tire of the light truck LF3070G1 is made from natural rubber, as shown in Table 1.

Table 1. Properties of the light truck with rubber tire

No	Property	Value	Unit
1	Density	1000	Kgm^-3
2	Modulus	0.25	
3	Material Constant MU1	6.1803E+05	Pa
4	Material Constant A1	1.3	
5	Material Constant MU2	1180	Pa
6	Material Constant A2	5	
7	Material Constant MU3	-9810	Pa
8	Material Constant A3	-2	

Load on tire

The tire is used to the static nominal load since the wheel rolls slowly on a flat road surface. The tire is also subjected to the dynamic forces as the vehicle moves straight ahead over an uneven road surface or due to vehicle maneuvers such as cornering, and braking and accelerating process. As can be seen from Figure 2, other forces can also be caused by the air compressed in the tire, and errors in the assembling the rubber to the rim and wheel disc [1, p. 574].

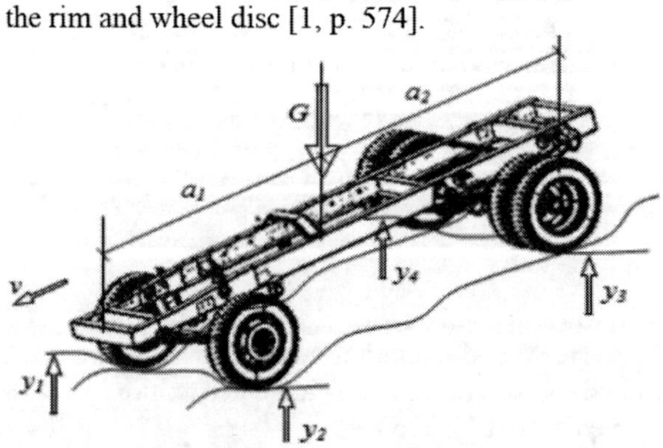

Figure 2. Load acting on the wheel of light truck

The static loading and dynamic loading are calculated as in equation (1).

$$y_i = Y_i \sin\left(\frac{2\pi v}{\lambda_0}t + \frac{2\pi}{\lambda_0}(a_1 + a_2)\right) \tag{1}$$

Where Y_i is vertical displacements of the front and rear tires, a_1,a_2 is the distance from the center of mass of the hung masses to front and rear axles, v is the speed of vehicles [12].

4. Results Analysis

4.1 Vibration modes of the tire

To simulate the response of tire to road roughness, the vibration modes can be represented by a parallel set of spring, mass, damper systems set. Thus, one of the masses resonates at each of the simulation modes. Some mode shapes are provided in Figure 3.

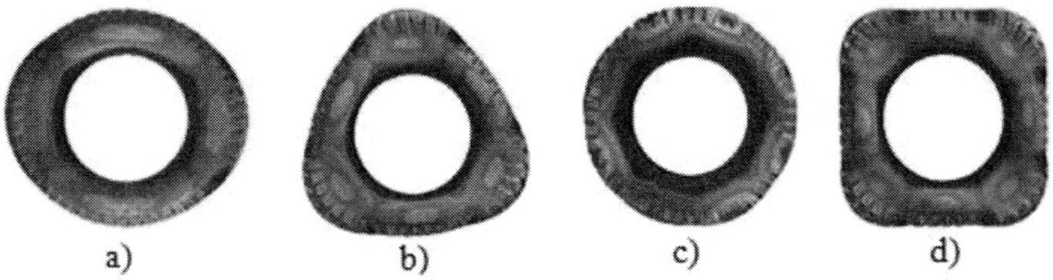

Figure 3. Vibration frequencies (f) and deformation (Df) modes of the light truck's tire
a) Mode 12: f = 52.546 Hz, Dfmax = 17.585 mm
b) Mode 15: f = 61.726 Hz, Df max = 15.341mm
c) Mode 17: f = 68.099 Hz, Dfmax = 13.457 mm
d) Mode 20: f = 74.28 Hz, Dfmax = 15.597 mm

When the parallel second-order systems are subjected to random input, each will provide input to the wheel hub in each particular vibration frequency. These inputs will be summed in amplitude and phase at the axle of the light truck.

4.2 Influence of tire air pressure and the vertical load

The tire is used with the prescribed air pressure. It has a good impact on moving vehicle safety, such as durability at high loads and speeds, ability to transmit high braking and forces to the road surface. It also indicated aquaplaning behavior on the wet surface. The impact of vertical load and the tire air pressure on a tire stiffness is given in equation (2).

$$k_r = \frac{G_R}{\Delta Z_{max}} \tag{2}$$

k_r represents a stiffness of the tire depending on the air pressure, the properties of the rubber, and the structure of the tire. G_R is a normal load of the tire, while ΔZ_{max} is a vertical load.

The air pressure and normal load may be used as calculation, as shown in equations (3) and (4).

$$p_m = \frac{G_R}{A_{total}} \tag{3}$$

$$G_R = G_S \cdot \frac{l_s}{0.315} + 490{,}223 \quad [N] \tag{4}$$

IOP Conf. Series: Materials Science and Engineering **886** (2020) 012012 doi:10.1088/1757-899X/886/1/012012

While p_m is the average air pressure in the tire. The A_{total} is a total of the contact patch area. G_S is a calibrated mass weight of 50 kgs used in the experimental. l_s [m] is used to measure the distance from the center mass G_S to the axes rolling of the arm.

By using the simulation in the Ansys Workbench software, the results of the study indicated the impact of vertical load and the tire air pressure on the tire endurance. It was also found that there is a local singular deformation on the outer surface of the tire at the pavement contact area (see Figure 4). The next images are enlarged to better describe the depth and width of the deformation mark. This deformation will change in depth when changing the tire pressure inside the tire. The tire will be damaged and causing unsafety of vehicle in the deeper part over the specified limit.

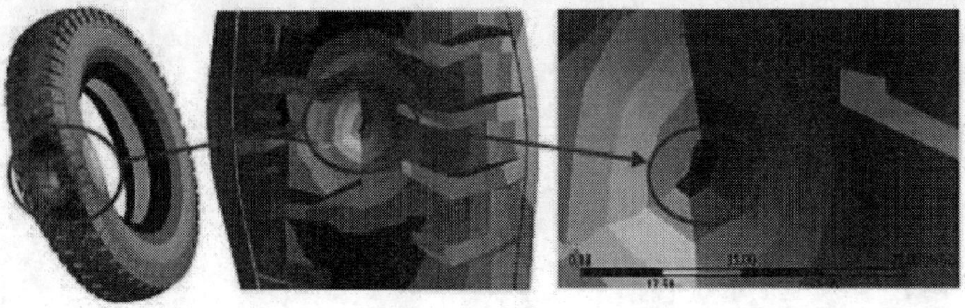

Figure 4. Influence of tire air pressure and the vertical load

Figure 5 presents the results from the data of depth of deformation in time in four different pneumatic pressure levels of the tires (0.3 Mpa, 0.4 Mpa, 0.5 Mpa and 0.6 Mpa). The results of this study indicate that at the lowest air pressure (0.3 Mpa), the depth of deformation will be greatest, and at the high air pressure (0.6 Mpa), the depth of deformation will be minimal. This study found that it is better to choose the air pressure level of tires ranging from 4-5 Mpa.

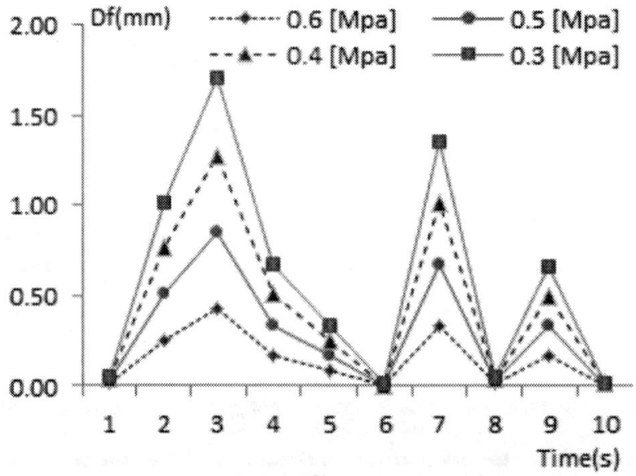

Figure 5. Transient of the tire of light truck

The present study was designed to determine the effect of dynamic analysis of light truck tires with differences parameters such as rubber material properties, tire structure, loading on the tire. The current study found that there is a tire pressure occurs when the parameter changed using simulation. The remaining parameters will be the given variable. The cases of tire pressure and deflection corresponding results are presented in Table 2. Thus, on the same characteristics of tires, through simulation of dynamic analysis, we can choose the optimal air pressure to ensure the tire durability and vehicle safety.

Table 2. Total deflection of the tire

Air pressure	0.3 [Mpa]	0.4 [Mpa]	0.5 [Mpa]	0.6 [Mpa]
Static [mm]	9.6414e-003	8.6072e-003	7.5753e-003	7.4662e-003
Harmonic [mm]	1.2725e-003	1.6967e-003	2.1209e-003	2.545e-003
Transient	7.03E-03	5.86E-03	4.69E-03	3.52E-03

5. Conclusion

This study set out with the aim of assessing the importance of the dynamic simulation conducted in Ansys Workbench software. It was designed to determine the effect of the air pressure on the deformation of rubber of the tire. This study confirms that the lower pressure, the more deformed of the tire in the high pressure. When the tire is highly deformed, it will increase the road contact, the rolling resistance, It was also found that it can reduce the vehicle's movement speed, and increase fuel consumption and causing the damage of the tire.

Regarding the depth of the singular deformation, the extreme limits are measured and ranged between 9.6414e-003 (mm) and 7.4662e-003 (mm) at center of the contact point. Values were obtained and the loads are beyond the prescribed limits. These findings may help us to understand it is the best to choose the air pressure level in tires, ranging from 4-5 Mpa.

References

[1] Adler U 2018 *Automotive Handbook, tenth edition* (Society of Automotive Engineers)
[2] Inhwan H 2016 *Automobile Engineering* **231**(1) 16–26
[3] Vidas Ž Loreta L Edgar S 2014 *Transport* **29**(4) 430–438
[4] Aguilar M J Alvarez I L 2015 *Journal of Applied Research and Technology* **13** 461–471
[5] Chinedum O M Chika E M Festus I A Ravindranatha M 2016 *Proc. the World Congress on Engineering* IIWCE (London:UK)
[6] Blundell M Harty D 2004 *The multibody systems approach to vehicle dynamics* (eBook:Elsevier)
[7] Andrius R Henrikas S 2017 *Proc. 10th International Scientific Conference Transbaltica* (Elsevier) 335–341
[8] Bin L Xiaobo Y Ankang J Yunqing Z James Y 2015 *Proc. the ASME 2015 International Design Engineering Technical Conferences & Computers and Information* (Massachusetts:USA)
[9] Besselink I J M Schmeitz A J C Pacejka H B 2010 *Proc. the 21st symposium of the International Association for Vehicle System Dynamics* (Stockholm:Sweden)
[10] Paul W 2018 *Journal of Terramechanics* **75** 3–14
[11] John C D 1996 *Tire, Suspension and Handling, second edition* (Pennsylvania:SAE International)
[12] Nguyen T Q 2019 *Applied Mechanics and Materials* **889** 461–468
[13] Cristian M 2015 *Review of the Air Force Academy* **1**(28) 149–154.

Analysis of Torsional Vibration Reduction on Automobile Cardan Shaft by using Composite Materials

VanAnh Le[1,*], Thanh Quang Nguyen[2], Hong Quan Le[3], PhucHoa Tran[4], TienTan Nguyen[5], Xa Hoi Nguyen[6]

[1,2,3,4,5] Hanoi University of Industry, Vietnam
[6] University of Fire, Hanoi, Vietnam

* Corresponding Author: anhlevan@haui.edu.vn, Telephone +84966399188

Abstract
The Cardan shaft is of interest because it transfers torsional loads from the gearbox to the rear axle for vehicle motion. Central to the entire discipline of the Candan shaft is the concept of the dynamics of the Cardan shaft. This concept can be very complex because one head of the shaft is fixed to the gearbox, while the other end can move in the vertical plane when the vehicle is running on an uneven road. The Cardan shaft is twisted and bent at different rotational speeds. When the torsional oscillation frequency of the Cardan shaft coincides with one of its natural frequencies, it can induce resonance and causing structural damage. The paper presents the analysis of the Cardan shaft's natural frequencies with two different materials: conventional sheet metal and composite material. The calculations can be used as the basis for designing the Cardan shaft in the vehicle powertrain system of the vehicles. These results further support the idea of using composite materials for the Cardan shafts in the automotive powertrain system, and composite materials for some other shaft parts in the vehicle.

Keywords: Cardan shaft, torsion vibration, composite material

1. Introduction

A driveline on automotive usually consists of a Cardan shaft, half shafts, transmission gears systems and bearings, as well as other components. For all-wheel-drive (henceforth, AWD) vehicles, the driveline includes a front Cardan shaft for the front axle, a rear Cardan shaft for rear axle (or more than one Cardan shaft is used). For front-wheel drive (henceforth, FWD) vehicles, the most usually is the Cardan shaft. In the Cardan shaft design, the bending and torsional natural frequencies should be higher than the specific frequencies corresponding to the maximum rotation speed of the shafts, to prevent the resonance caused by the rotating unbalance force. The natural mode frequencies of parts should be designed or tuned to avoid resonance [1]. Fuel efficiency and the weight of the vehicle are two of the most important parameters to be considered in automobile design. To reduce power loss in the powertrain, composite materials can be used to design Cardan shafts with less weight while maintaining the required strength and stiffness [2]. A researched by performing FEA analysis on two different composite materials are E-GLASS and E-CARBON reducing Von-Mises stress and weight compared to structural steel. This can be concluded that the composite material can be used in the Cardan shaft [3].

2. Composite materials selection

Figure 1 shows the Cardan shaft for the automobile. The Cardan shaft is used to connect the gearbox to the different rear axle, and to transfer power from the engine to differential automobile through universal joints assembly. Cardan shaft has to withstand high rotational speeds and has to change its length while transmitting the torque required by the vehicle. The second method used to identify the motion of the universal joints which is proportional to the angle between the driving and driven shafts. It also affects system vibrations. Constant velocity operation of Cardan shaft obtained if the axes of the input and output shafts are truly parallel. There are, however, torque and velocity variations in the elements of conventional Hookes joints which give rise to secondary rocking couples; these affect the support bearing of the input-output shaft. Joint angles should thus be as small as possible to minimize these effects. Due to the limited installation location of the driveline system, the use of composite materials is an effective solution in this case.

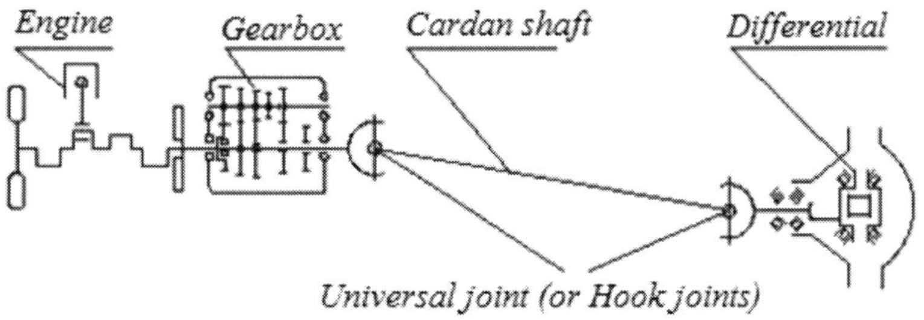

Figure 1. Cardan shaft of the driveline for Automobile

This research studies replacing the conventional steel materials with composite materials for Cardan shaft. This method is particularly useful in studying reduction in the allowable working stress on the Cardan shaft, obtaining a given life expectancy, and analysing different composite materials for best-used selection.

The objective of this research is to replace C45 carbon steel material of the Cardan shaft assembled two Hook joints with composite material for a single shaft. The material mechanical properties of C45 carbon steel and E-glass epoxy, kevlar epoxy composite material are shown in Table 1. E-glass epoxy is a composite material that has been used extensively in engineering.

Table 1. Properties of materials

Mechanical Properties	Steel C45 carbon	E-glass Composite	Kevlar Epoxy Composite	E-carbon
Density (kg/m^3)	7850	1580	1402	1451
Young's Modulus (Pa)	2.0e+11	7.23e+10	9.571E+10	5.92E+10
Tensile Yield Strength, (Pa)	2.5E+08	7.8E+08	2.5E+08	5.13E+08
Tensile Ultimate Strength, (Pa)	4.6E+08	3.1E+07	4.6E+08	5E+07
Poisson's ratio	0.3	0.28	0.34	0.3
Shear Modulus (Pa)	7.679e+10	1.37E+10	3.57E+10	2.28E+10
Bulk Modulus (Pa)	1.67E+11	2.65E+10	9.97E+10	4.93E+10

Figure 2 presents the parameters of the Cardan shaft. The base length is 1450 mm, while the outer diameter is 90 mm, and the inner diameter 78 mm.

IOP Conf. Series: Materials Science and Engineering **886** (2020) 012013 doi:10.1088/1757-899X/886/1/012013

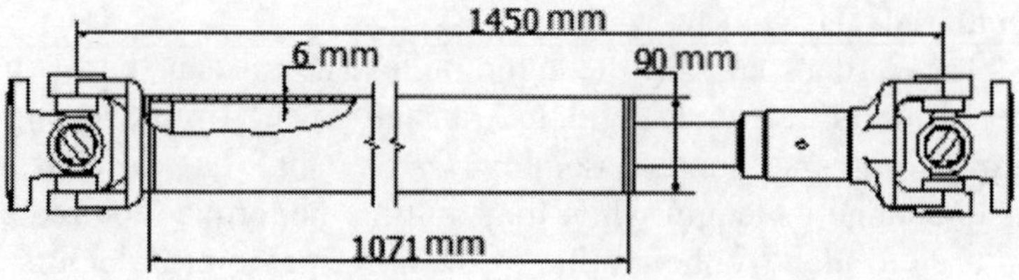

Figure 2. Geometric properties of the Cardan shaft

3. Simulation results and discussion

3.1 Modal analysis

The finite element analysis used in this study was Ansys Workbench software. This method is particularly useful in studying the mode shape, deformation, stresses and strain value in the Cardan shaft [4,5]. To evaluate the modal frequency and mode shape to prevent resonance, a modal analysis of other materials of Cardan shaft. The modal analysis of the Cardan shaft used in this study includes the modes that must be verified to avoid mode coincidence and to have sufficient separation from vehicle body modes. In the operation, the Cardan shaft undergoes the excitations of both torque modulation and unbalance. Thus, both torsional and bending vibrations play a vital role in the study. For torsional vibrations, the analysis procedure is similar to the engine crankshaft system. When the rotational speed of a shaft is sufficiently high, the unbalance excitation frequency could be identical to the bending natural frequency of the shaft. Under this situation, resonance occurs, and the corresponding specific speed is referred to as the critical speed. Operation at critical speed is likely to lead to failures of the shaft and its system. Regarding an elastic shaft with two ends connecting to universal joints, it can be modeled as a simply-supported beam on the two ends. Its critical speed is as the following equation (1):

$$f_n = 3.56\sqrt{\frac{EIg}{WL^4}} \tag{1}$$

in which E is the modulus of elasticity, I is the moment of inertia, g is the acceleration due to gravity, L is the length, and W is the weight of the Cardan shaft [1].

The boundary conditions including frictionless support, the fixed support, rotational velocity and moment were used to simulate the same Cardan shaft. The torsional moment is 320 Nm and is applied to all the body with the rotational velocity of 2200 rpm (230.308 rad/sec). The torsional moment and boundary conditions on the Cardan shaft are illustrated in Figure 3.

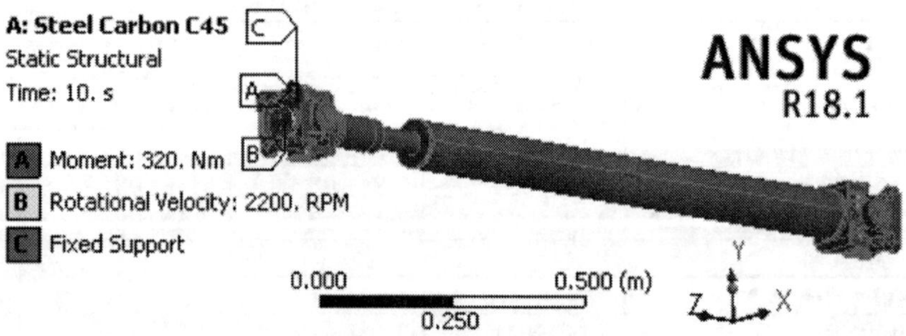

Figure 3. The torsional moment and working boundary conditions of the Cardan shaft

As shown in Figure 4, the mesh used in this study were 381312 nodes and 215054 elements.

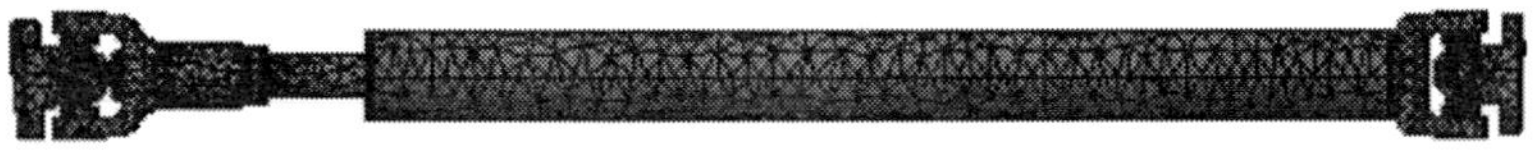

Figure 4. Meshing of the Cardan shaft

The results of mode shape and maximum total deformation in four types of materials of Cardan shaft indicate that the highest natural oscillation frequency is in the 10th mode where the Kevlar Cardan shaft has the highest nature oscillation frequency at 1156.7 Hz. The remaining axes are the steel Cardan shaft (707.3), the E-glass Cardan shaft (609.87 Hz), the E-carbon Cardan shaft (894.72 Hz). It was also found that the deformation is equivalent to each mode shape. The deformation of E-carbon and Kevlar Cardan shaft is equivalent coincidentally.

3.2 Static analysis

From this data, we can see that the static analysis resulted in four types of materials of the Cardan shaft: maximum total deformation, maximum equivalent elastic strain, and minimum, maximum equivalent von-misses stress. It can be seen from the data in Figure 5 that the deformation of the E-glass Cardan shaft reached the highest point (2.9509m) and was at the smallest point (0.82602m) of the Kevlar Cardan shaft. The equivalent elastic strain maximum has the highest by 4.86 (m/m) of E-glass material, as shown in Figure 6.

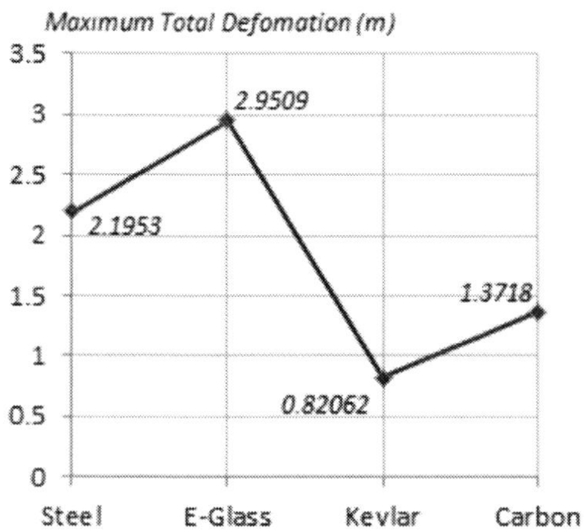

Figure 5. Total deformation (m)

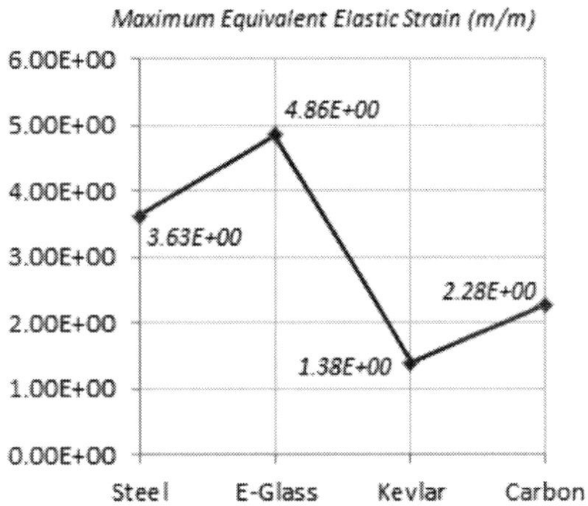

Figure 6. Equivalent elastic strain (m/m)

As can be seen from Table 2, it reported the statistical comparison of the four types of materials, the steels Cardan shaft specifications showed 100% reduction in weight. The proportion of other materials slightly decreased; E-glass (76.4%), Kevlar (82.1%) and Carbon (81.5%).

Table 2. Statical analysis results

Materials	Steel	E-glass	Kevlar	Carbon
Weight (kg)	29.042	6.8444	5.1869	5.3682
Reduction in Weight (%)	100	76.4	82.1	81.5
Max. total deformation (m)	2.1953	2.9509	0.82062	1.3718
Max. equivalent elastic strain (m/m)	3.6275	4.8607	1.3834	2.2779
Max. equivalent (Von-Mises) stress (pa)	6.7057e+11	1.5803e+11	1.1978e+11	1.2398e+11

The minimum and maximum of equivalent (Von-Mises) stress reached the highest (6.71E+11 Pa) of steel material of the Cardan shaft, as shown in Figure.7.

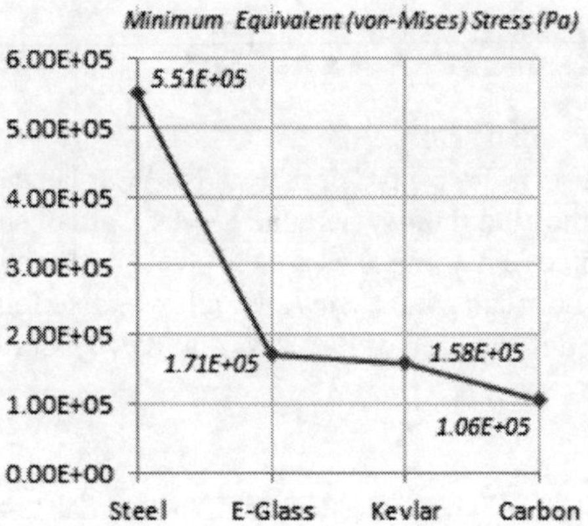

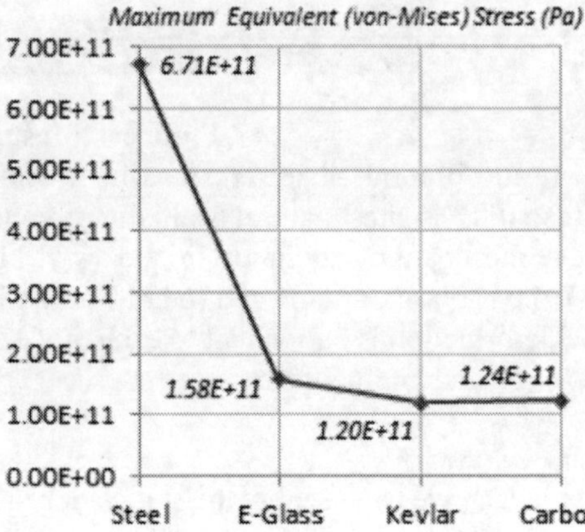

Figure 7. Equivalent (Von-Mises) Stress (Pa)

Figure 8 compares the weight of C45 steel and composite E-glass, Kevlar, E-carbon materials of the Cardan shaft.

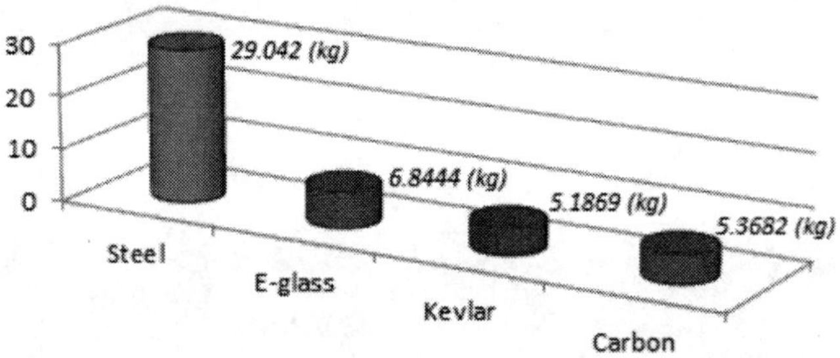

Figure 8. Graphical comparison of weights in four types of materials of Cardan shaft

3.3 Dynamic analysis

The analysis of the dynamic was shown to have the Cardan shaft working conditions (see Table 3). The moments included the active torsional moment in the whole body of the Cardan shaft (320 Nm), resistance moment (125 Nm) which contains a phase angle of 30 degrees (0.5236 rad). The rotational motion velocity is 2200 rpm (230.308 rad/sec), and the acceleration inertial is 1.41 m/s^2.

Table 3. Results of dynamic analysis

Materials	C45 steel	E-glass	Kevlar	Carbon
Harnomic Maximum Total Deformation (m)	2.35E-03	2.08E-04	1.22E-04	2.1922E-04
Harnomic Maximum Equivalent Elastic Strain (m/m)	6.88E-02	8.03E-03	2.89E-03	5.40E-03
Harnomic Maximum Equivalent (von-Mises) Stress (Pa)	1.36E+10	1.09E+09	2.74E+08	3.1674e+08

In this study, the Harmonic in Ansys Workbench software was used to analyze the deformation, stresses, and strain dynamics value in the Cardan shaft. The results, as shown in Figure 9, indicated that deformation of E-carbon of the Cardan shaft showed the highest, and so does the equivalent (Von-Mises) steel of the Cardan shaft (see Figure 10).

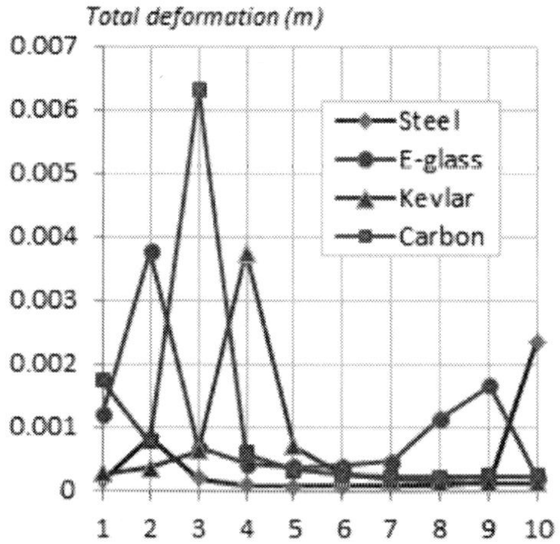

Figure 9. Dynamic deformation

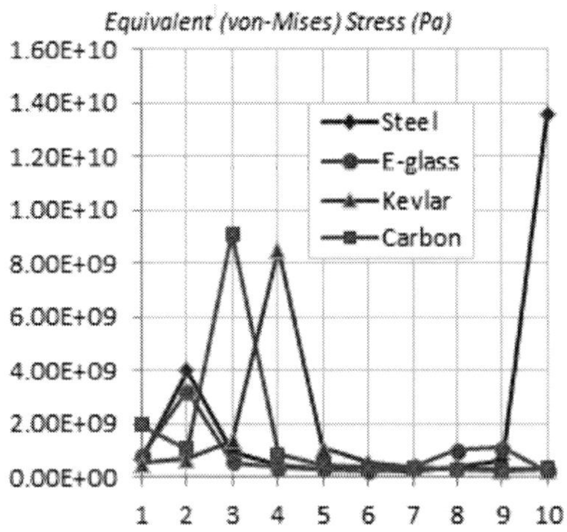

Figure 10. Dynamic equivalent stress

4. Conclusion

This project was undertaken to evaluate the use of composite materials. To analyze the Cardan shaft's natural frequencies with four different materials, the conventional sheet metal, the E-glass composite, Kevlar, and E-Carbon material were adopted. The investigation of the effectiveness of Kevlar composite materials for the Cardan shafts in the automotive powertrain system have shown that the Kevlar composite materials are the best specifications. The second major finding was that the static analysis was reduced in weight (82 %). Also, the total deformation was at 0.82062 m, and the equivalent elastic strain was at 1.3834 m/m. While the equivalent (Von-Mises) stress showed 1.1978e+11 (Pa) and was the smallest (see Table 2). Concerning the dynamic analysis, the results of the total deformation were at 1.22E-04 m, and the equivalent elastic strain was 2.89E-03 m/m. These experiments confirmed that the equivalent (Von-Mises) stress was at 2.74E+08 Pa, and is the smallest (see Table 3). These findings suggest that, in general, the composite is gradually deformed, if one want to replace steel by the composite in the automobiles, one should consult engineers. Because they have been improved with a solid surface that meets the demanding requirements of the body, chassis. These findings have significant implications for the understanding of how to reduce costs in the process as it has many options which it should help to give more style, size and color of the parts or components of the automobiles.

References

[1] Chen G S 2012 *Vehicle Noise, Vibration, and Sound Quality* (Pennsylvania:SAE International)
[2] Kushwaha D Chauhan G S Rai V S 2016 *International Journal of Current Trends in Engineering & Research* **2**(6) 595–603
[3] Vinodh K S Sampath V Baskar P 2015 *Research Journal of Recent Sciences* **4**(9), 9–15
[4] Dill E H 2011 *The Finite Element Method for Mechanics of Solids with ANSYS Applications* (Florida:CRC Press)
[5] Chen X Liu Y 2014 *Finite Element Modeling and Simulation with ANSYS Workbench* (Florida:CRC Press)

The Effect of Road Surface Roughness to Recommended Speed of Vehicles

XaHoi Nguyen[1,*], ThanhQuang Nguyen[2], Phuc Hoa Tran[3]

[1] University of Fire, Hanoi, Vietnam
[2,3] Hanoi University of Industry, Vietnam

* Corresponding Author: xahoinguyen@gmail.com, Telephone +84913377240

Abstract
Road safety is one of the major concerns of the transportation system planners and managers in any nation. The principle goal of road safety is to reduce the frequency and severity of traffic accidents. The design and manufacturing of high-quality road directly play an important role in achieving better road safety. This research presents the model describing the relationship between road surface roughness with the vehicle's recommended safe velocity. The result shows that the road surface roughness measured by the International Roughness Index (henceforth IRI) is indirectly proportional to the vehicle's recommended safe velocity. In particular, if the recommended safe velocity is 100 km/h, the surface roughness IRI should not exceed 1.9 m/km. An implication of this is the possibility that using the simulation on MATLAB Simulink software should have the appropriate IRI concerning the designed speed.

Keywords: Road roughness, road safety, traffic speed

1. Introduction

Road safety is the combination of multiple parameters including geometry, vision distance, width, over operating conditions, pavement quality, road surface conditions, vehicle and driving behavior.

Of particular concern are deterioration of the road surface and the reduction of road safety. Road surface roughness is a parameter quantifying of the degradation of the road surface. Recent evidence suggests that the relationship between roughness with pavement life is being widely used in road surface management. Questions have been raised about the change in the estimated road life. It is now well established from a variety of studies, that the model also uses climatic conditions, initial structures, loads of vehicles involved in traffic and surface maintenance variables, [1].

To determine road surface roughness according to longitudinal sections, an existing standard ISO 8608 is used [2].

It is only since the work of Múčka (2018) that the study of road surface roughness has gained momentum. Analysis of 27 scientific papers simulated the road surface according to ISO 8608 such as that conducted by Múčka (2018) have shown that road layers were determined mainly for reference corner space-frequency $\square_0$ = 1 rad/m (L_0 = 6.28 m) and for angular frequency, $n_0 = 1/2\pi$ cycle/m (L_0 = 6.28 m) or n_0 = 0.1 cycle/m (L_0 = 10 m). The results showed a significant difference between the simulated lines and the spectral properties of the actual lines. Thus, a type-A road is proposed for good or medium quality roads (highways). While types B and C are suitable for low-quality surfaces (grade 2, and grade 3 or local), [3].

To analyze the impact of the surface roughness of the vehicle's motion to the movement of the vehicle at different speeds, a simulation based on the automobile model was used to describe the contact of the tire with the road surface. The IRI and abnormal changes in road surface configuration are determined by dedicated road test equipment. The vehicle speed is closely related to road surface roughness and affects the dynamics of the suspension system and the stability of the movement, [4].

According to the Vietnam transportation standard, evenness of the road is checked and evaluated through the IRI. The experiment of the study is carried out using collecting, surveying designed enhancement, and constructing of automobile road data bank since it can forecast the investment in repairing and upgrading transportation roads [5]. The image of road surface roughness and pavement degradation is provided in Figure 1.

Figure 1. Road surface roughness and pavement degradation

The more the roughness index values increases, the more accident rates increases. Also, the number of car accidents is larger than that of cars accidents, but on the highway, the accident rate does not increase with roughness.

To date, the driver's driving behavior is governed by road rules in developed countries. Vietnam is one of the countries that gradually follow the driving rules. To overcome this problem, the rule is applied to limit the speed of the vehicles. Studying the effect of road safety, the driver's behaviour should be taken into account, especially limiting speed and being aware of the safety factors. Figure 2 presents the driver's behavior on IRI during the control of a series of routes to comment on driving comfort. It has been suggested that the lower IRI, the better the ride comfort [6].

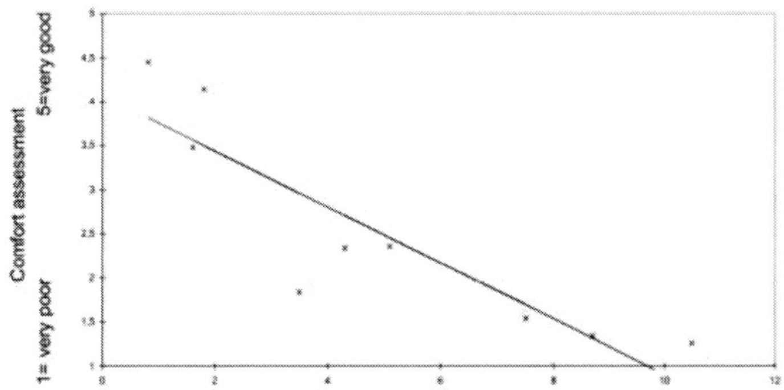

Figure 2. Results of investigation of the driver's behavior with road surface roughness

IOP Conf. Series: Materials Science and Engineering **886** (2020) 012014 doi:10.1088/1757-899X/886/1/012014

As can be seen from Figure 2 above, the roughness of the road surface has a direct impact on vehicles collision. The braking distance increases with the specified speed of vehicles running from up to 50 km/h. Besides, longer wavelengths in vertical configuration have an impact on the ride quality of trucks compared to cars, [7].

2. Relation of the speed of movement with the coefficient of drag

The rolling resistance coefficient of the road surface depends on the movement speed of the car. When designing highways, rolling resistance coefficient is used as shown in formula (1).

$$f_1 = f_o\left[1+0.01(V-50)\right] \quad (1)$$

f_o represents the rolling resistance coefficient at vehicle speed under 50 km/h, while it depends on the type of road surface. The fo is valued at 0.01 for concrete road, and the value of 0.3 is used for discrete sandy road. V is used to refer to the vehicle speed, km/h.

When designing natural soil roads (i.e. rural roads, forestry roads), rolling resistance coefficients are used as illustrated in the following formula (2).

$$f_2 = \xi G\sqrt{\frac{H}{D}} \quad (2)$$

Where ◘ is the coefficient that varies from 0.6 to 1, depending on the state of the soil. While H is the depth of the wheel track, and D is the wheel diameter.

3. The roughness of the road surface

Traditionally, it has been argued that the road surface roughness is widely accepted and used by road management agencies globally. It has been also acknowledged the roughness as deviation or irregularity of the intended longitudinal profile (called the real plane) of the road surface. The indicators of the road surface roughness is determined as current use rate (present serviceability rating as PSR), current usability (present serviceability index as PSI), number of turns (ride number as RN), and riding comfort index (RCI).

The IRI index is defined as the surface roughness with a wavelength of 0.5 m to 50 m in the longitudinal section. It is also measured by recording the movement of the axle and suspension during the vehicle moves at a constant speed. In the industry-standard (22TCN-277-01), the IRI is defined more specifically since the surface roughness reflects the longitudinal roughness of the pavement with a wavelength of specific amplitude. Also, the IRI appears to affect the dynamic characteristics of the vehicle, driving quality, drainage conditions and causes of dynamic loads impact on the road.

The roughness is first calculated with an IRIqc wheel on a track using a single-track model and is then determined by averaging the two IRIqc values of each wheel on the rear axle with two dual wheels.

In the national standard TCVN 8865: 2011, the method of measuring IRI by the indirect method is given in formula (3) with a vehicle speed of 30 km/h.

IRI = 0.1268. (Shockiness) - 0.3905; (R2 = 0.9396) (3)

IRI is calculated in formula (4) with vehicle speed of 45 km/h

IRI = 0.1302. (Shockiness) -1.1044; (R2 = 0.9625) (4)

The relationship between the roughness of the road and the risk of collision is given in the equation (5) below [see 8].

Crash rate = 0.0049 (Ro) 2 - 0.4948 (5)

It should be noted that the Crash rate (CR) is the collision rate, and the Roughness (Ro) is the average value of roughness in length (kilometre).

4. Relationship of the roughness with the speed of the vehicle

Figure 3 provides an experimental study that causes accident rates to increase at 50 km/h and 70 km/h [8].

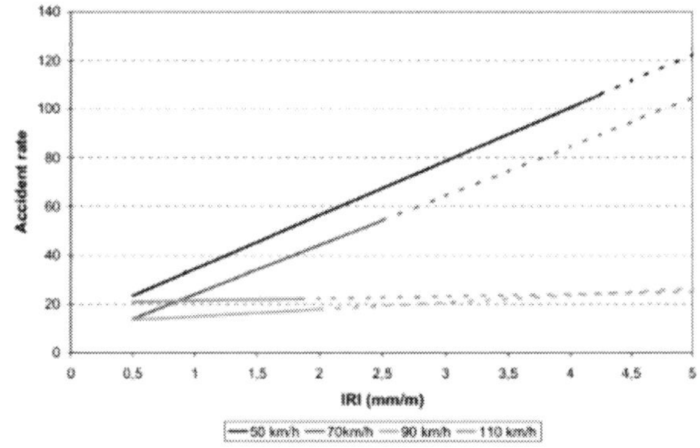

Figure 3. The relationship of roughness with the speed of movement of the vehicle through the number of accidents

As shown in Figure 4, the relationship with speed depends on the type of vehicle, the difference between a car and heavy truck (HV) when running on the same surface roughness [9].

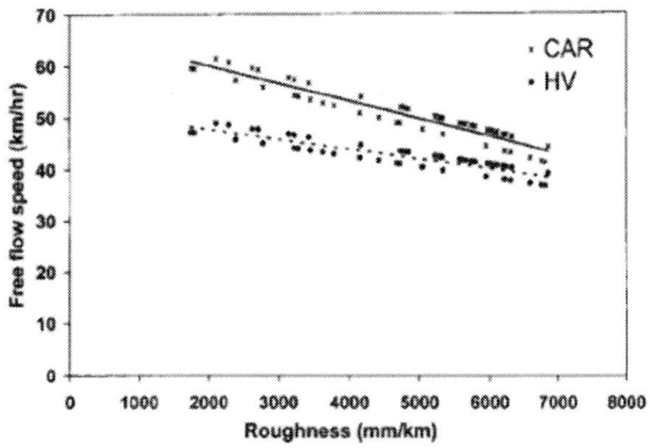

Figure 4. Relation of vehicle speed and roughness according to vehicle type

Figure 5 illustrates the limit of the roughness given for each speed. It indicated that the lower rate of IRI, the speed is being decreased. Therefore, for high speeds (100 km/h) the road surface must have a higher quality level than the lower speed zone (50 km/h).

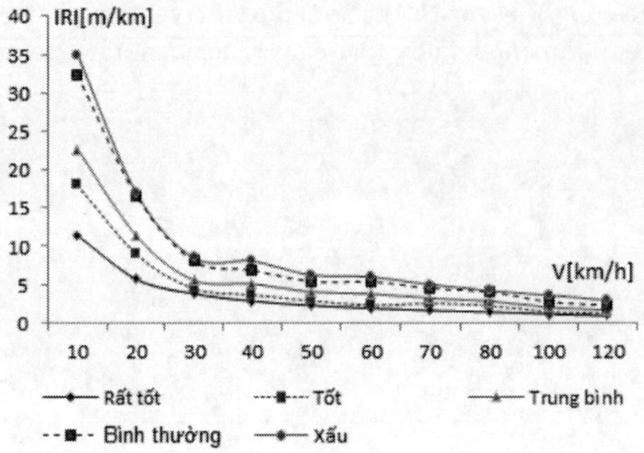

Figure 5. Speed limited according to road roughness

5. Examine the surface movement [9]

In the previous section, the measurement results were adopted to determine the exact surface roughness. The two methods used in this study involved direct and indirect measurement. Measurements were selected for their specialized agency about public transport.

In most recent studies, the application of software has been developed and can simulate virtual alternative to the experiment, which still gives reliable results. This study used Matlab Simulink tool to simulate in the module called sldemo_VariableTransportDelay as an assisted application.

The model of the vehicle and the road is provided in Figure 6. Hi (t) is the bumping at the front wheel over time (Front wheel vertical displacement). And, Ho (t) is the bumping at the rear wheel or if determined by displacement which is rear wheel vertical displacement.

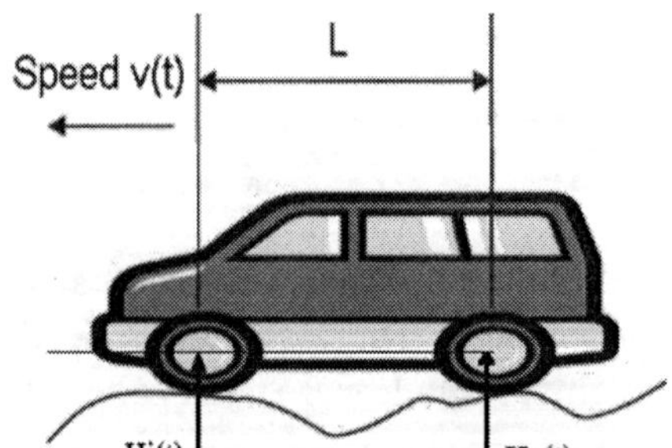

Figure 6. Vehicle-road model and survey parameters

The block diagram of the simulation module is shown in Figure 7. The purpose of the simulation is to determine the "bumping" of the wheel in the formula when moving at different speeds 15, 30, 50, 70, 80, 90, 100 and 120 km/h with different IRI indicators on the road.

IOP Conf. Series: Materials Science and Engineering **886** (2020) 012014 doi:10.1088/1757-899X/886/1/012014

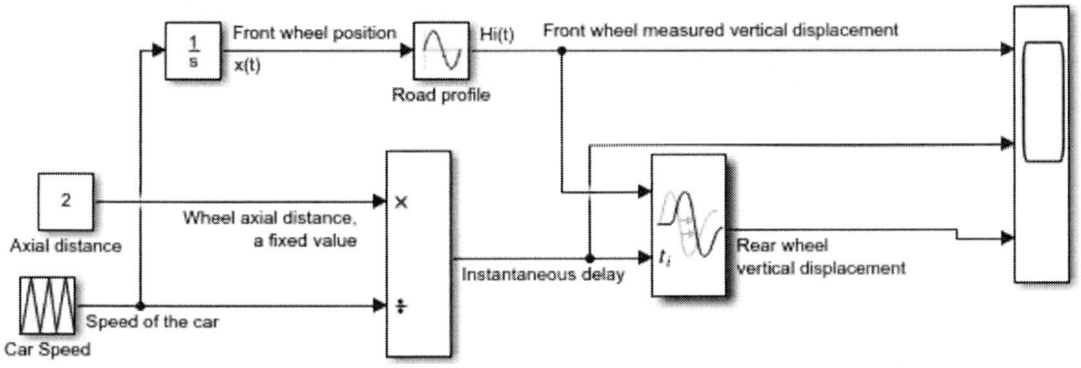

Figure 7. Block of kh□i Simulink model "sldemo_VariableTransportDelay"

In particular, the vehicle speed and the IRI are the parameters set, moving vertically of the front or rear wheel is the parameter definition in the simulation. As can be seen in Figure 8, the frequency of "bumping" increases when the vehicle speed increases. The more increase of the simulation time with the longer the actual distance, the more increase in the number of measurement points. In this stage, we cannot count manually, since we can also use the meter to record the results and counting the analysis of the number of "bumping" of the wheels.

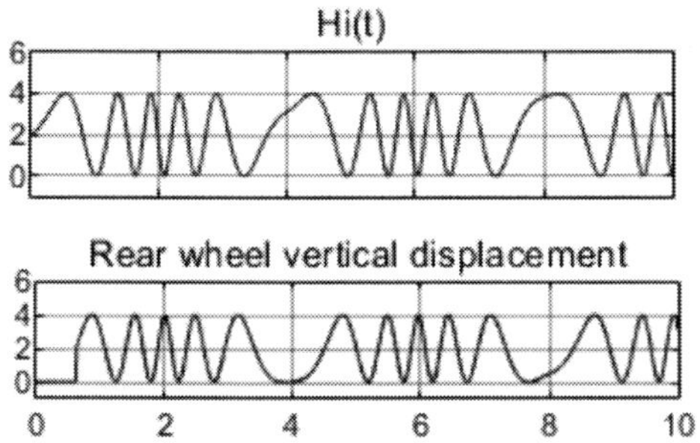

a) At speed 15 km/h

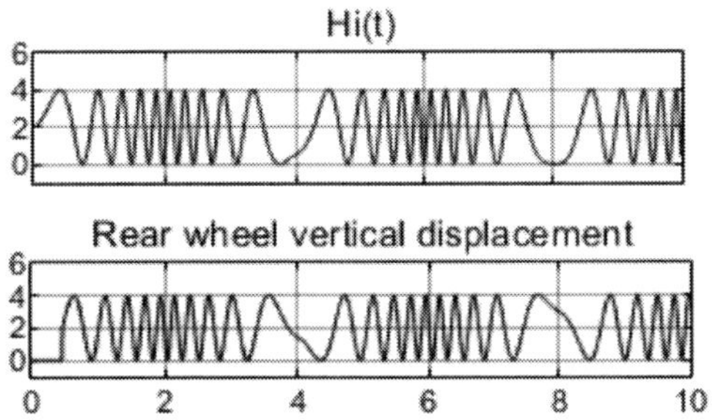

b) At speed 30 km/h

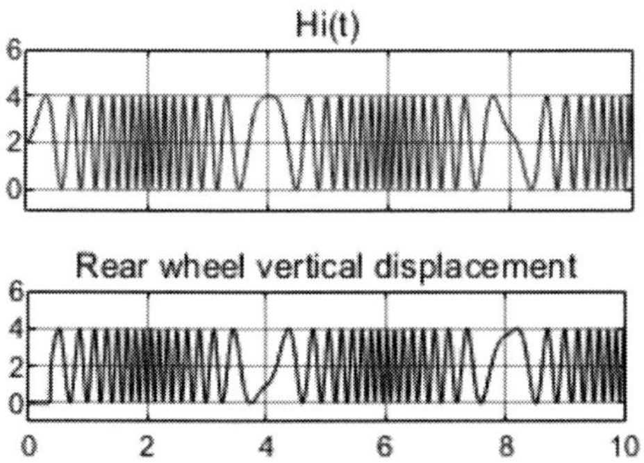

c) At speed 50 km/h

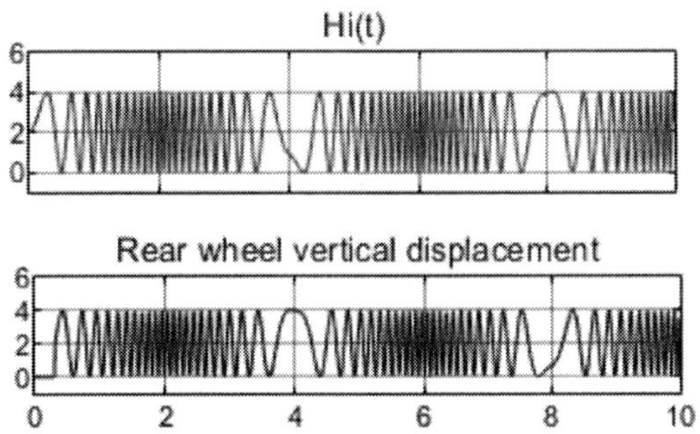

d) At speed 70 km/h

IOP Conf. Series: Materials Science and Engineering **886** (2020) 012014 doi:10.1088/1757-899X/886/1/012014

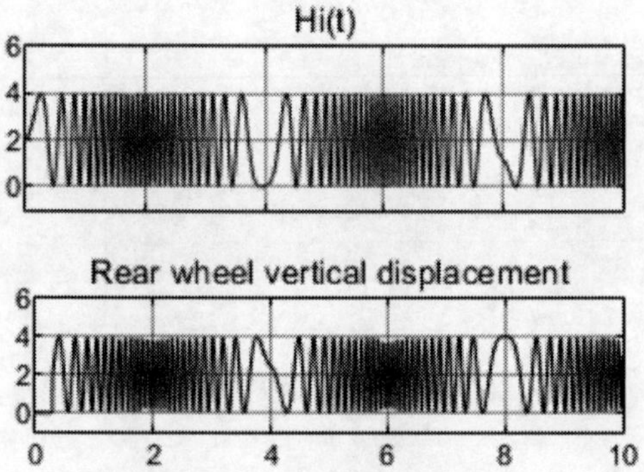

e) At speed 80 km/h

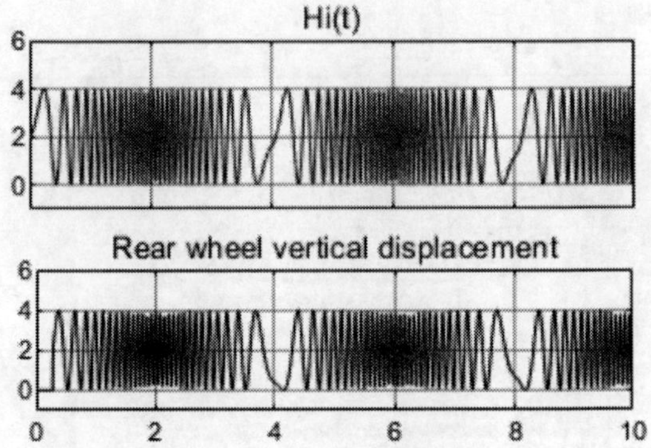

f) At speed 90 km/h

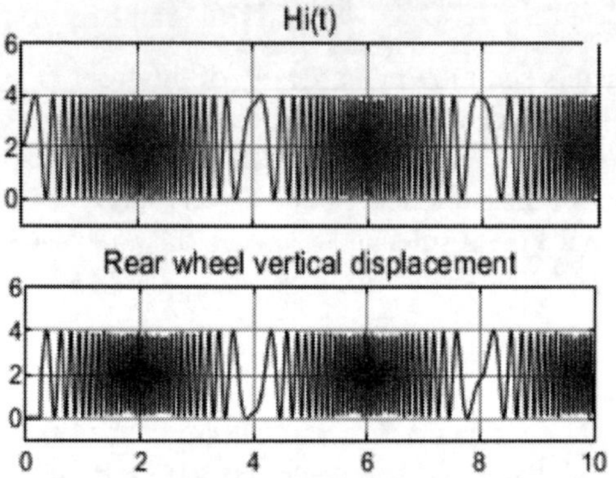

g) At speed 100 km/h

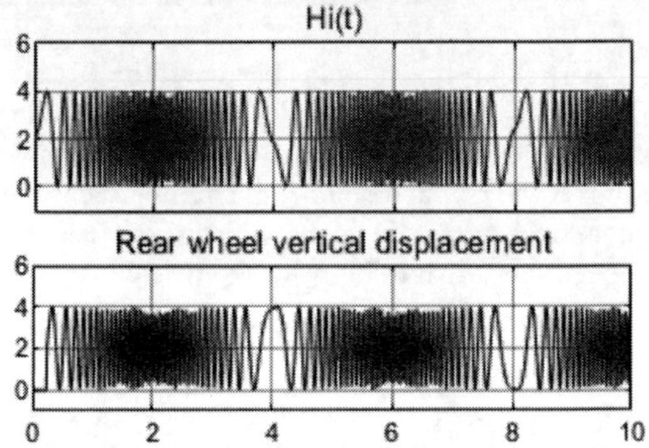

h) At speed 120 km/h

Figure 8. Simulation results determine the "shockiness" of the wheel

6. Conclusion

This study has shown that the vehicle's speed is the main parameter used in conjunct with the change of the roughness of the road surface. It has argued that these two parameters will indicate the driving behavior. The analysis of two parameters in the "Driver-Vehicle-Road" relationship supported by the current findings. These experiments confirmed that road surface roughness may benefit to road safety. One of the more significant findings to emerge from this study is that driving to the environment that has the roughness changes, the driving speed and driving concentration might have been reduced. Hence, the roughness will have been challenging in the operation. In general, therefore, it seems that the roughness could be one of the factors that can adjust the speed limit in-vehicle control.

The results of this investigation show that the roughness value is in the range of 0 to positive numbers but there is no upper limit. However, it will be limited according to road use requirements. In a highway, the study revealed that the IRI is 1.9 m/km maximum. It has been suggested from the study that the sections should be longer than 500 m with the results of the IRI between 4.2 and 5.3 m/km to create the feel for drivers.

The findings will be of interest to use the existing modules of Matlab Simulink software to simulate the determination of the "bumping" of the wheels and it will make the study reliable and time-saving. More research using applications regarding these similar problems and real experiments should be carried out, in particular of public transport research.

References

[1] Hunt P D Bunker M J 2004 *Proc. 6th International Conference on Managing Pavements* (Brisbane:Australia)
[2] International Standardization Organization 1995 *ISO 8608: Mechanical Vibration–Road Surface Profiles–Reporting of Measured Data* (Switzerland: International Standardization Organization)
[3] Peter M 2018 *Journal of Testing and Evaluation* 46(1) 405–418
[4] Vidas Ž Loreta L Edgar S 2018 *Transport* **29**(4) 430–438
[5] The Vietnam National Standard 2011 *Vehicle road surface - measuring and evaluating method of determining the degree of flatness according to the international roughness* (Vietnam Ministry of Science and Technology) TCVN 8865
[6] Ihs A 2004 *Proc. 6th International Conference on Managing Pavements* (Brisbane:Australia)
[7] Peter C Robert D Neil J 2012 *Proc. Australasian Road Safety Research, Policing and Education Conference* (Wellington:New Zealand) 1–12
[8] Cairney P Bennett P 2008 *Proc. 23rd ARRB Conference – Research Partnering with Practitioners* (Adelaide:Australia)
[9] MatWorks 2019 https://www.mathworks.com/help/simulink/slref/simulating-systems-with-variable-transport-delay-phenomena.html, accessed 15 November 2019.

Reliability-based Design Optimization of Classical Wing Aeroelasticity

S Sleesongsom[1*], S Yooyen[1], P Prapamonthon[1] and S Bureerat[2]
[1]Department of Aeronautical Engineering, International Academy of Aviation Industry, King Mongkut's Institute of Technology Ladkrabang, Bangkok 10520, Thailand
[2]Sustainable and Infrastructure Development Center, Department of Mechanical Engineering, Faculty of Engineering, KhonKaen University, KhonKaen City 40002, Thailand

* Corresponding Author: suwin.se@kmitl.ac.th

Abstract
Flutter speed of aircraft is very important and needs to be firstly specified before a certification applied for a new aircraft by airworthiness regulator to make sure that the aircraft is free from flutter in its flight envelope. By assuming geometrical and physical parameters known, the speed is usually estimated from deterministic analyses in a design stage. In practice, some parameters are finitely measured by observing, especially for the geometrical parameters, material properties and so on due to the random in nature, which causes uncertainty of information often called uncertainties. The purpose of this paper is to combine reliability analysis and optimum design of aeroelastic aircraft wing. The classical two-dimensional wing with a typical airfoil section is used as an example in this study. To quantify uncertainty in the design of flutter speed, the discrete-time aero-elastic model and worst-case scenario are applied. Furthermore, the comparison between optimum design with/without reliability is provided in this study. The results show the proposed technique leads to the flutter speed being more conservative and realizable compared with the traditional technique.

Keywords: Distributed combustion, Thermal efficiency, Experiment

1. Introduction

According to FAA and EASA regulations for airworthiness, a new aircraft can be allowed for flight if a company can define its actual aero-elastic characteristics. The technique of simulation is selected to study the aero-elastic characteristics in a design process to ensure that an aircraft developed passes the regulation in the final stage. The aero-elastic phenomena can be separated into two groups as static and dynamic aeroelasticity. The most important static aeroelastic phenomena are divergence speed, control reversal and lift effectiveness, while the dynamic aeroelastic characteristics include flutter speed. The flutter and divergence speeds are called critical speed when considering the lowest of them as a design criterion. The flutter analysis programs for low-speed air-vehicles were studied by Nantasenee et al. [1]. The aero-elastic phenomena have been considered in aircraft and adaptive or morphing aircraft wing design [2-6]. A new reduced-order model for aeroelastic analysis was studied in Sleesongsom and

Bureerat [7] and Sleesongsom [8]. The aeroelastic phenomena have been studied for a long time, but they are random in nature. Material properties are evaluated by performing a finite number of experiments and by observing. We use the properties to design in engineering with developing the innovation to improve our human life. Obviously, changes in the material properties may cause damage to the engineering equipment and may be dangerous to human. In engineering design, the safety factor is used for increasing design reliability, but the high value of the safety factor causes a high price. The exactly unknown material property is called uncertainty. If we can define it, we can increase the reliability of a designed object. Moreover, it is known that dealing with uncertainty by using safety factors or worst-case scenarios in design causes inefficient design and overly conservative. The theory of uncertainty is important to increase the structural reliability analysis in designing aeroelastic optimization to offer the correct aeroelastic evaluation for designing purposes.

The techniques for handling uncertainty have been studied by many researchers especially in aircraft design, in the last decades. The most well-known and straight forward technique to deal with uncertainties is called a Monte Carlo simulation (MCS) [9]. This technique is in the group of probability analysis that has been used in studying the configuration variation of a Goland wing [10]. The model of aero-elastic analysis for finding flutter speed is analysed based on linear aerodynamics. A further technique in this group is polynomial chaos expansion (PCE), which is a well-known technique to decompose a random process in terms of deterministic functions of space and time, and orthogonal functions of random variables. This technique has been used to quantify the uncertainty of flutter speed due to composite plate wings layup with uncertain ply orientations [11]. Later, this technique has been extended to reduce the number of random variables [12]. In recent years, this technique has been extended to study the uncertainty qualification of a generic UAV wing by the same author [13].

Composites material is considerably popular for use in various structural applications due to its high mechanical strength-to-weight ratio, stiffness, resistance to corrosion, thermal insulation properties and the low volume-to-weight ratio [14]. Furthermore, combining structural reliability analysis with aeroelastic simulation in a simple form that has been studied and given a correct flutter speed evaluation to design results. Two types of commonly used material modelling are isotropic and composite materials. All of them have been employed with MCS, a first-order reliability method (FORM), a second-order reliability method (SORM), and including a response surface model coupled with FORM and MCS [15]. However, the anisotropic material behaviour of laminated composite structures leads to complicated analysis and design approaches. It implies that high fidelity computational simulation using Finite Element Analysis (FEA) can be more time-consuming. The MCS in combination with FEA is a direct approach for quantifying uncertainty, but a million of finite element runs is required. To reduce such a time consuming and impossible-to-do task, the use of surrogate models is needed as showed by the work [11, 12].

Another technique to quantify the uncertainty into a model excepts as mentioned above, it handles the uncertainties as an objective or constraint functions is called Reliability-Based Design Optimization (RBDO). This method aims to ensure that the constraints are acceptable by enforcing the threshold according to the probability failure [14].

Reliability has been used to increase the realization of structural topology, in which the uncertainty factors are varied and make the optimum design results impractical or unrealizable [16]. To address the problem, two strategies are robust topology optimization (RTO) [17] and reliability-based topology optimization (RBTO) [18] have been used to collect uncertainties into a topology optimization problem. The first technique is performed optimization design at the same time with considered variability of the system with respect to uncertainties, while the last method is performed optimization design with failure probability constraints to achieve the reliable design. Both methods can be based on probabilistic [19] or non-probabilistic [20-21] models. The first model is the most popular one due to its progress, but this technique requires a very good knowledge of uncertainties distribution. The knowledge of uncertainties distribution requires a large amount of objective data, which needs a huge computation time. The second model is non-probabilistic where the well-known techniques are a convex model [20] and a fuzzy set method [22]. The convex model required only bounds of uncertainties from a small number of samples to reduce time-consuming tasks, whereas the solution is too conservative. The fuzzy set model is an alternative technique to collect the uncertainties into RBTO by using a level set to softly separate between the members and non-members of the set. It makes the model to get an acceptable solution better than the convex method [16]. Another well-known technique in the group of non-probability approach is the worst-case reliability, which needs the interval of uncertainties without accurate probability distributions. This technique is called anti-optimization, which has been proposed by Elishakoff [23].

One strategy is to quantify the uncertainties in the group of probability techniques, although very time consuming, it uses approximation techniques to the model of real behaviours. In this strategy, a surrogate model is used to approximate the finite element model leading to significantly faster computing compared to actual function evaluations. The various surrogate models have been used for structural reliability. The advanced Kriging model (AKM) proposed by Zhang et al. (Date) is applied to approximate the mechanical model of a structure [24]. Examples of the Kriging models have been used with the surrogated model-based uncertainty are as Ordinary Kriging, Blind Kriging, and Co-Kriging [25]. Later, several Kriging-based strategies have been proposed for structural reliability analysis [26-27] including an adaptive sampling method to construct a surrogate model [28-29].

This work aims to study the reliability effect in an optimization problem of aeroelastic behaviour of classical aircraft wing model without statistical information of model-parameters and mechanical properties. The aircraft wing modelled as a two-dimensional classical wing model is used for demonstration. The computation cost is low for this model, so the model can combine with non-probabilistic and optimization technique without using an approximation approach.

2. Aeroelastic Model and Numerical Validate

Nowadays, reduced-order modelling (ROM) is used to study the design of aeroelastic structures. The traditional ROM is the ROM with a static correction technique. It has been used in analyzing unsteady incompressible aerodynamic flows with using only few unsteady flow eigenmodes to constructed the reduced-order unsteady flow model. Later this technique is adjusted to alternative ROMs, which is expected to increase its performance. They are constructed only based on the wake or body vortices without using a static correction technique [7, 8]. Aeroelastic analysis in this research applied ROM based on a vortex lattice method (VLM) in which the eigenvalue problem is defined only based on unknown wake vortices for flutter analysis of a two-dimensional airfoil for studying reliability design optimization.

2.2.1 Aeroelastic validation

In this section, we propose the flutter analysis validation of a classical aeroelastic model, which is used in reliability design optimization in this research. The model is considered as the aircraft wing, which has a typical airfoil section as shown in Fig.1. The model has two degrees of freedom, which are plunged and pitched [8]. The linear and torsion springs are attached at the elastic axis to restrain motions in plunge and twist. The validation model parameters include mass ratio μ is 20, the static imbalance x_θ= 0.2, the radius of gyration r_θ is 0.5, and the location of elastic axis stay after the mid chord at-0.1. The frequency ratio R of uncoupled modes is 0.3. The chord is modelled with 10 vortex elements, while the wake is modelled using 100 vortex elements. The length of wake vortices is 10 times of the chord length. The result is shown in Fig. 2 where the reduced flutter speed is 1.95, which is in good agreement with the previous result [8]. This model used ROM based on wake vortex, where the numbers of body and wake vortices are 10 and 100, respectively. Furthermore, ROM exploits only the first 10 aerodynamic modes. Our aeroelastic result is in agreement with the previous result [8].

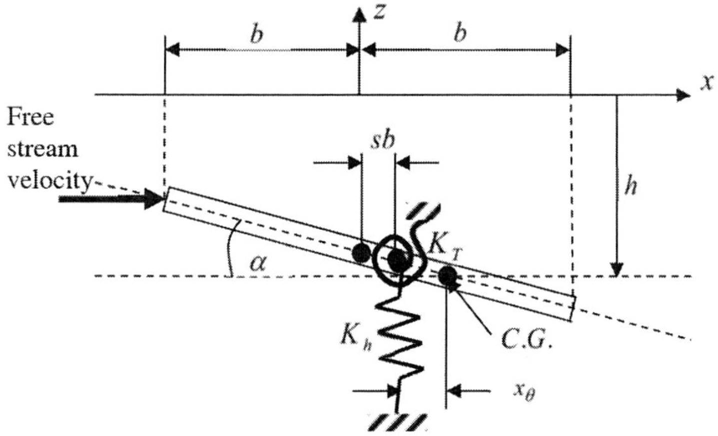

Fig.1 Typical model with plunging and pitching model

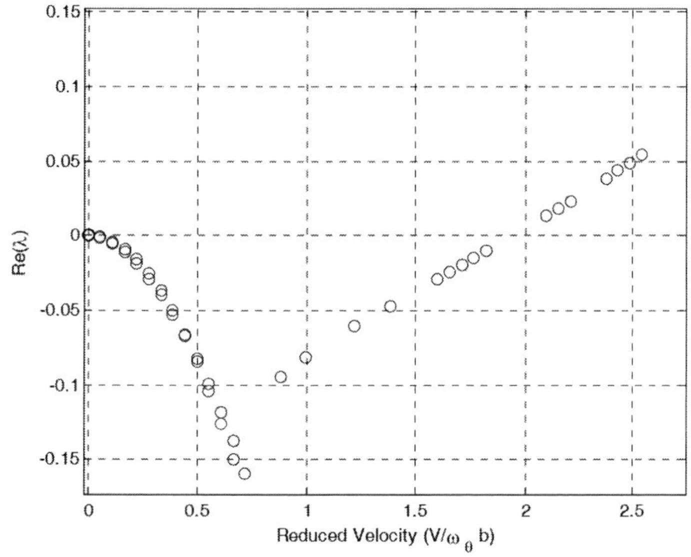

Fig. 2 Flutter analysis of typical section model using the discrete-time vortex lattice method

3. Design example

In this section, the reliability design optimization of aeroelasticity and design demonstration refers to a maximization of reduced flutter speed ($v_f(\boldsymbol{x})$) and maximizing the worst-case scenario ($\max(\boldsymbol{g}(\boldsymbol{a}))$), when considering the non-dimensional lift. The reduced flutter speed and non-dimensional lift are not over than the prescribed value. Design variables $\boldsymbol{x}$ are the static imbalance (x_θ), the ratio of natural frequency (R) and the location of the elastic axis after the mid chord (s). Uncertain parameters ($\boldsymbol{a}$) are the radius of gyration $r_\theta \in (0.45, 0.5)$, and the mass density of air $\rho_{air} \in (1.15, 1.2)$. The anti-optimization problem is formulated as

$$\min\{-v_f(x), -\max(g_i(x,a))\} \tag{1}$$

Subject to

$$\frac{g_i(x,a)}{\bar{g}_i} - 1 \le 0,\ i = 1,2,\ldots,N$$

$$x_l \le x_i \le x_u,\ i = 1,2,\ldots,n$$

$$a_l \le a_i \le a_u,\ i = 1,2,\ldots,m$$

where the number of constraints, design variables and uncertainties are N=2, n =3, and m =2, respectively. $\bar{g}_i$ is the allowable value of flutter speed and non-dimensional lift. The limit of the reduced flutter speed is 11 and the non-dimensional lift is 1.0. For the non-dimensional lift for the two-dimensional airfoil, it should not exceed 1.0. The objectives are two conflicting criteria in nature since it is impossible to increase both the flutter and the worst-case scenario. The optimizer is an adaptive multi-objective real-code population-based incremental learning and differential evolution (aRPBIL-DE), which has been presented in our very recent work [30]. The number of population size for aRPBIL-DE is 50 performs in 100 iterations with the external Pareto archive size is 100. Other parameters are initially assigned similar to [30].

4. Design results

From the design example, having solved the optimization problem, the obtained result is shown in Fig. 3. The technique used for clustering selection is from our previous work [31]. The selected solutions are obtained from the front, which is shown in Fig. 4. The minimum reduced flutter speed obtained is 7.740, while the maximum worst-case scenario is 1.704 at the parameter bound. The maximum reduced flutter speed obtained is 11, while the maximum worst-case scenario is 2.0 at the parameter bound. The lowest value for reduced flutter speed is 7.740, while the highest is 11. The highest reduced flutter represents the deterministic optimum solution, which gives the highest worst case too. Fig. 5 shows some selected solution and the corresponding flutter speed. The designer can choose the conservative solution which compromises between the flutter speed and uncertainty from the Pareto front. The box-plot in Fig. 6 shows distributions of the design variables of all solutions in the Pareto front in which R and x_θ are more variation dimensions.

5. Conclusions

This paper focused on a combination of reliability analysis and optimum aeroelastic design of an aircraft wing to find a correct flutter speed. The classical two-dimensional aeroelastic wing model is used as an example in this study. The discrete-time aeroelastic model and worst-case scenario are applied to quantify an uncertainty into the design of flutter speed of a classical wing and provide a comparison between solutions with and without reliability analysis. The results show the proposed technique leads the flutter speed of classical typical section of an aircraft wing in design optimization to be more conservative, and realizable when compared with the deterministic design optimization technique. The designer can add the experience to select a more conservative solution from the solution set.

The extension of the proposed technique is to design a real aircraft wing, which is expected to make it more efficient and realizable.

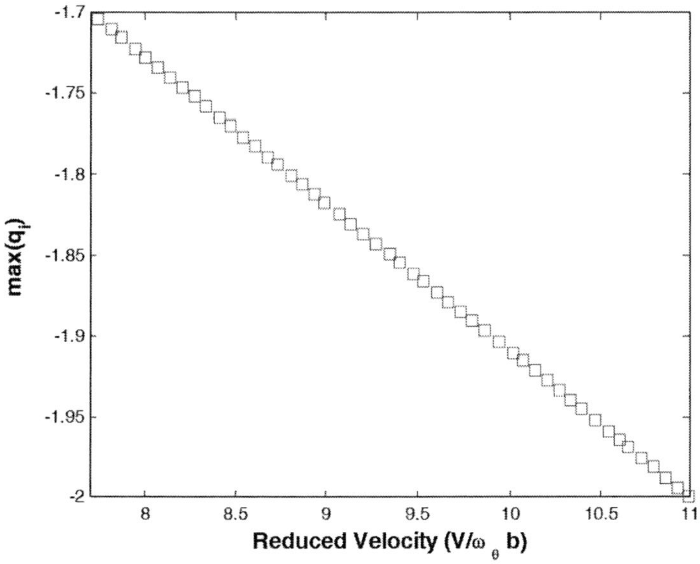

Figure 3. Approximate Pareto front

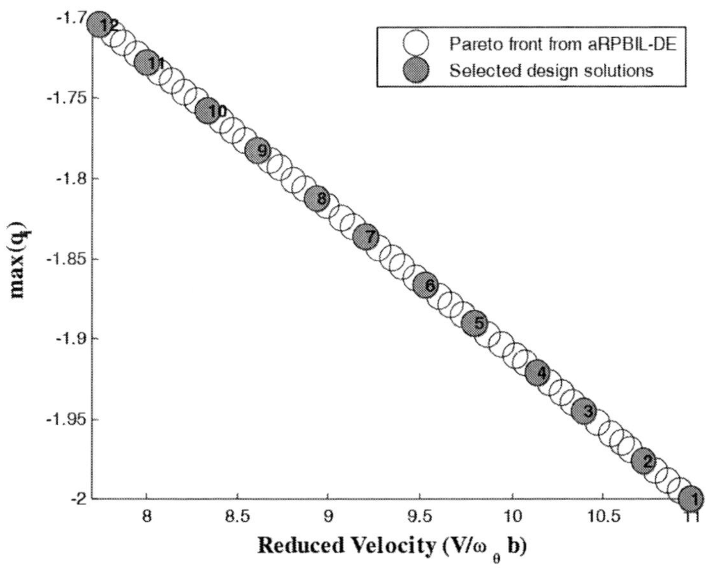

Figure 4. Pareto front and selected solutions

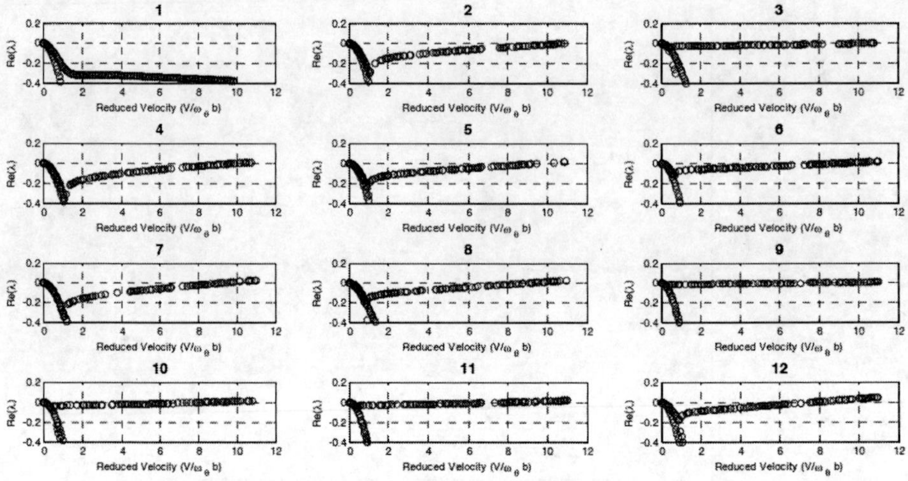

Figure 5. Reduced flutter speeds from selected solutions in figure 4

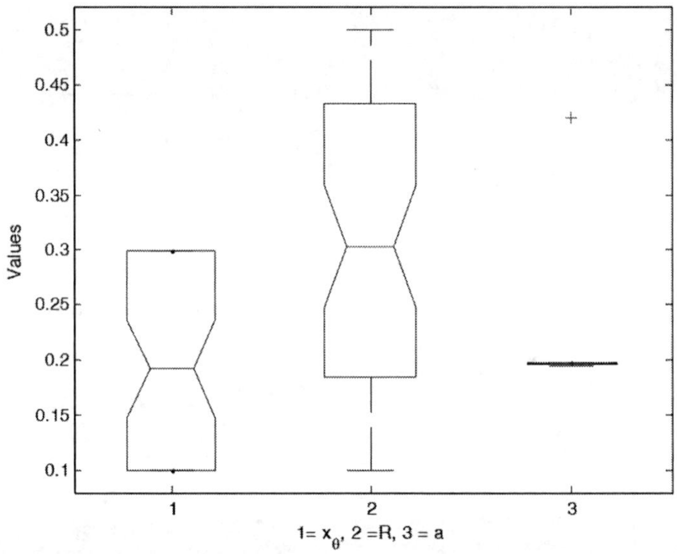

Figure 6. Distribution of design variables

Acknowledgments
The authors are grateful for the financial support provided by King Mongkut's Institute of Technology Ladkrabang, Thailand Research Fund (RTA6180010).

References

[1] Nantasenee S Sleesongsom S Bureerat S 2009 *Proc. 23rd Conference of Mechanical Engineering Network of Thailand* (Chiang Mai:Thailand)
[2] Sleesongsom S Nuntasenee S Benjapiyaporn J Bureerat S 2010 *Proc. 1st TSME International Conference on Mechanical Engineering* (Ubon Ratchathani:Thailand)
[3] Sleesongsom S Bureerat S 2011 *Applied Mechanics and Materials* **52-54** 308–317
[4] Sleesongsom S Bureerat S 2013 *Engineering Optimization* **45**(1) 107–122
[5] Sleesongsom S Bureerat S Tai K 2013 *Structural and Multidisciplinary Optimization* **48**(6) 1109–1128
[6] Sleesongsom S Bureerat S 2015 *Mathematical Problems in Engineering* **2015** 730626
[7] Sleesongsom S Bureerat S 2013 *Journal of Engineering* **2013** 326496
[8] Sleesongsom S Winyangkul S Bureerat S 2015 *Proc. International Conference on Power Electronics and Energy Engineering* 98–101
[9] Beran P Stanford B 2013*Lecture Notes in Computational Science and Engineering* **92** 59–103
[10] Lindsley N Beran P 2007 *AIAA Atmospheric Flight Mechanics Conference and Exhibition* (Hilton Head:SC)
[11] Manan A Cooper J 2009 *Journal of Aircraft* **46**(2) 601–607
[12] Scarth C Cooper J E Weaver P M Silva G H C 2014 *Computers and Structure* **116**(1) 84–93
[13] Cook R G Wales J A Gaitonde A L Jones D P Cooper J E 2018 *AIAA Aviation Forum* (Atlanta: Georgia)
[14] Scarth C Cooper J E 2018 *Structural and Multidisciplinary Optimization* **57** 1695–1709
[15] Borello F Cestino E Frulla G J 2010 *Aerosp. Eng.* **23**(4) 327–338
[16] Yin H Yu D Xia R 2018 *Comput. Methods Appl. Mech. Engrg* **333** 197–217
[17] Dunning P Kim H 2013 *AIAA J.* **51**(11) 2656–2664
[18] Kharmanda G Olhoff N Mohamed A Lemaire M 2004 *Struct. Multidiscip. Optim.* **26** 295–307
[19] Kalman R E 2002 *Ann. Statist.* **30** 1292–1294
[20] Moens D Hanss M 2011 *Anal. Des.* **47**(1) 4–16
[21] Wang L Wang X Xia Y 2014 *Acta Mech.* **225** 413–430
[22] Moller B Graf W Beer M 2000 *Comput. Mech.* **26** 547–565
[23] Elishakoff I Colombi P 1993 *Comput Methods appl Mech Eng.* **104**(2) 187–209
[24] Haeri A Fadaee M J 2016 *Composite structures* **149** 26–32
[25] Mukhopadhyay T Chakraborty S Dey S Adhikari S Chowdhury R 2017 *Arch Computat Methods Eng* **24** 495–518
[26] Wang J Sun Z Yang Q Li R 2017 *Reliability Engineering and System Safety* **167** 494–505
[27] Paiva R M Crawford C Suleman A 2014 *AIAA Journal* **52**(4) 711–724
[28] Wen Z Pei H Liu H Yue Z 2016 *Reliability Engineering & System Safety* **153** 170–179
[29] Yun W Lu Z Jiang X 2018 *Structural and multidisciplinary design optimization* **58**(4) 1383–1393
[30] Sleesongsom S Bureerat S 2019 *Lect. Notes Comput.* **11656** 47–58
[31] Sleesongsom S 2008 *Proc. of Natural Computation* (Jinan:China) 92–96

IOP Conf. Series: Materials Science and Engineering **886** (2020) 012016 doi:10.1088/1757-899X/886/1/012016

Aerodynamics, Stability and Control Analysis of Tactical Solar Power UAV

J Kampoon[1,*], W Jiajan[2], and J Klongtrujrok[3]
[1,2,3]Department of Aeronautical Engineering Navaminda Kasatriyadhiraj Royal Air Force Academy, 171/1 Phaholyothin Rd, Saimai District, 10220, Bangkok, Thailand

* Corresponding Author: janewit_kamp@rtaf.mi.th, +662-534-8057, +662-534-8057

Abstract
This paper addresses the fundamental aerodynamics, stability, and control analysis of a solar power Unmanned Aerial Vehicle (UAV) as a part of the preliminary design stage. The tactical solar-powered UAV addressed in this paper is primarily developed for intelligence, surveillance, and reconnaissance (ISR) missions at low altitude. Moreover, such design also has a dual-use capability, which could be utilized in both military and civilian domains. The aerodynamic characteristics of the UAV are obtained from both computational fluid dynamics analysis technique and the wind-tunnel testing at the design operating Reynolds number ranging from $1x10^5$ – $4.5x10^5$. The fundamental aerodynamics coefficient consists of lift, drag, and pitching moment variations versus the angle of attack. The stability and control analysis was carried out based on the small disturbance theory in correlation with the XFLR5 software. The results show that the tactical solar power UAV design could achieve high aerodynamic efficiency at a 4-degree angle of attack which is corresponding to the lift-to-drag ratio of 20.05. Also, the analysis results confirm that the design possesses positive static and dynamic stability at the design cruise flight condition.

Keywords: Aerodynamics, Stability and Control, Solar Power UAV

1. Introduction

Due to the ability of harvesting additional energy from the solar ray, the unmanned aircraft equipped with solar-powered technology could achieve flight endurance further than that of the conventional platform which is solely relying on the battery cells. Hence the popularity of the solar-powered UAV research has increased over the years [1-5]. Based on the current technology the solar-powered UAV could be developed to deliver high flight endurance from 5 hours up to the complete perpetual flight, provided that the degradation of the subsystem components is neglected. Moreover, the solar-powered UAV could be employed for both military and civilian applications (dual-use). The Royal Thai Air Force (RTAF)'s solar power UAV concept presented in this paper is designed primarily for the military's Intelligence, Surveillance, and reconnaissance (ISR) operation at low altitude with enhancing flight endurance [6]. This paper will discuss the key aerodynamic characteristics of the aforementioned design concept as well as the stability and control analysis conducted as part of the preliminary design phase.

The first part of this paper will discuss the design overview of the solar-powered UAV concept followed by the discussion on the design's aerodynamic characteristics obtained from both computational and experimental approaches. Next, the stability and control analysis will be discussed. Finally, the conclusions from the analysis will be addressed along with suggestions for future work.

1.1. The tactical solar power UAV Design Concept

Following the discussion during the design review process, it is decided that the empennage design configuration is to be changed from a T-tail into a V-tail arrangement for the lower part counts. The wing planform geometry is also changed from pure rectangular into the prismatic shape with a constant chord at the inboard section for better aerodynamic characteristics while providing a structural and manufacturing advantage. The revised concept design illustrated in figure 1 has a total wing area of 1.231 m^2, in which 69% of the total planform area is occupied by the solar cells array. The revised V-tail design layout is conducted to satisfy the required 'projected' horizontal and vertical planform area. The overall dimension and mass properties of the revised design concept are illustrated in table 1.

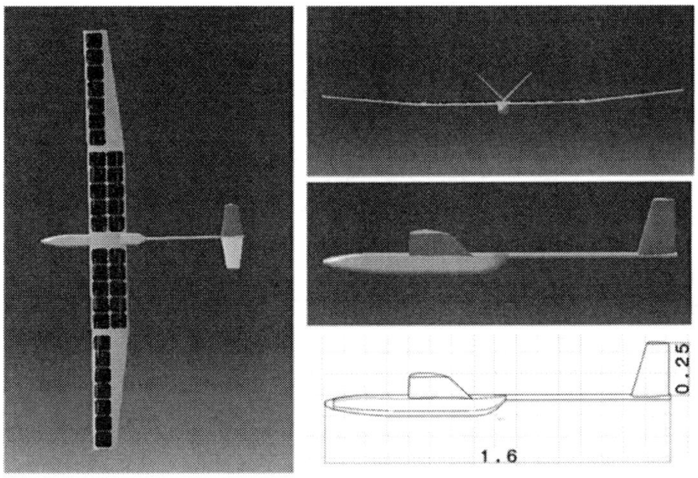

Figure 1. The tactical Solar Power UAV's layout

Once the configuration of the design concept is defined and frozen, the preliminary design phase could now be executed. In this paper, the aerodynamic and stability and control analysis is the main discussion.

Table 1. The tactical solar power UAV's geometric and mass properties.

Wing		Tail		MTOW	3.88 kg
airfoil	SG6043	airfoil	NACA0012	Ixx	0.6449 $kg.m^2$
wingspan	3.37 m	Type	V-tail	Iyy	0.3422 $kg.m^2$
aspect ratio	13	V-area	0.124 m^2	Izz	0.9807 $kg.m^2$
wing area	1.23 m^2	H-area	0.0715 m^2	Ixz	-
incidence angle	2 deg			CG	@37.88% MAC

2. Aerodynamics Characteristics

The aerodynamic characteristics of the design concept are observed using both computational and experimental approaches. Two computational aerodynamic tools including the Vortex Lattice Method (VLM) and the commercial Computational Fluid Dynamics (CFD) software are implemented. Also, an aerodynamic experiment using a sub-scale model in a subsonic wind tunnel is performed. Both computational and experimental studies are conducted at the Reynolds number ranging from Re $1x10^5$ - $4.5x10^5$ based on the design operating speed. The lift, drag, and pitching moment coefficient (C_L, C_D, and C_m) of the air vehicle are observed at various angle-of-attack. The stability and control analysis are conducted based on the small disturbance assumption implemented in the XFLR5 software tool.

2.1 Aerodynamics calculation by XFLR5 and CFD

The XFLR5 is an open-source design tool that could be implemented for an initial aerodynamic study as well as stability and control analysis in the early stage of the aircraft design studies [7-9]. The solar-powered UAV model implemented in the XFLR5 consists of wings and V-tail components, which will be used for further stability and control analysis. The 3-D wing and plane analysis are performed in XFLR5 based on the Vortex Lattice Method (VLM).

As the aerodynamic result obtained in XFLR5 are reasonable only in the linear region. The CFD analysis is implemented to observe the aerodynamic characteristics of the non-linear region such as stall. In this study, the CFD simulations are performed using the Spalart-Allmaras turbulence model through the ANSYS FLUENT V16. Figure 2 illustrates the air vehicle model implemented in the XFLR5 and CFD environment

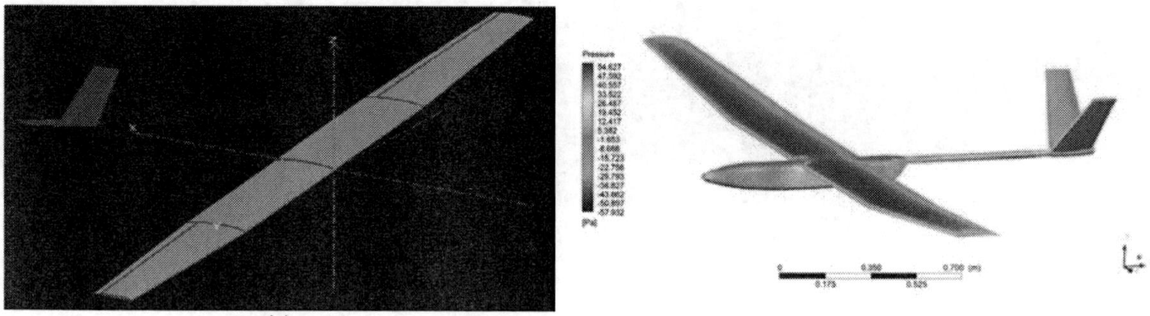

(a) (b)

Figure 2. The tactical Solar Power UAV Model in (a) XFLR5, (b) CFD

2.2 Aerodynamics by wind-tunnel Testing

The aerodynamic experiment of the air vehicle is performed in the Subsonic Research Wind Tunnel (SRWT) facility (figure 3(b)) located at the Aeronautical Engineering and Aviation Division at the Navaminda Kasatriyadhiraj Royal Air Force Academy (NKRAFA). The wind tunnel has a closed-circuit arrangement with the 1.5 m x 2 m x 5 m (H, W, L) test section size capable of delivering the maximum airspeed of 80 m/s. Due to the test section size restriction that limits the model width to be under 1.5 meters, the experiment is conducted using the sub-scale force model with the 1:2.67 scaling factor. The wind tunnel model is fabricated in-house using the additive manufacturing machine followed by the hand surface finishing process. The structural integrity of the finished model is confirmed by the sand-bag loading method which simulates the maximum lift condition. The aerodynamic forces and moments are measured by the six-component sting-type force balance which capable of 2669 N (normal force), 1334 N (axial force), 712 N (side force), 113 N-m (pitching moment), 56.5 N-m (yawing moment), and 56.5 N-m (rolling moment) maximum load limit. The model is attached to the balance unit using an offset type mounting to provide adequate clearance between the balance mounting system and the test model. The lift, drag, and pitching moment coefficients are at the angle of attack ranging from -5.0 to 15.0 degrees.

(a) (b)

Figure 3. (a) The tactical Solar Power UAV Model (b) The SRWT at NKRAFA

2.3 Aerodynamic characteristics Comparison

The comparison of the air vehicle's aerodynamic characteristics is illustrated in figure 4. The lift coefficient variation with angle-of-attack from all three approaches shown in figure 4 (a) is correlated well within the linear region ranging from -1.0 - 6.0 degree, in which the lift curve slope of 0.3 per degree is observed. The lift coefficient comparison between CFD and wind tunnel experiment shows a good correlation in which a slightly lower lift curve slope is observed from the CFD computational result.

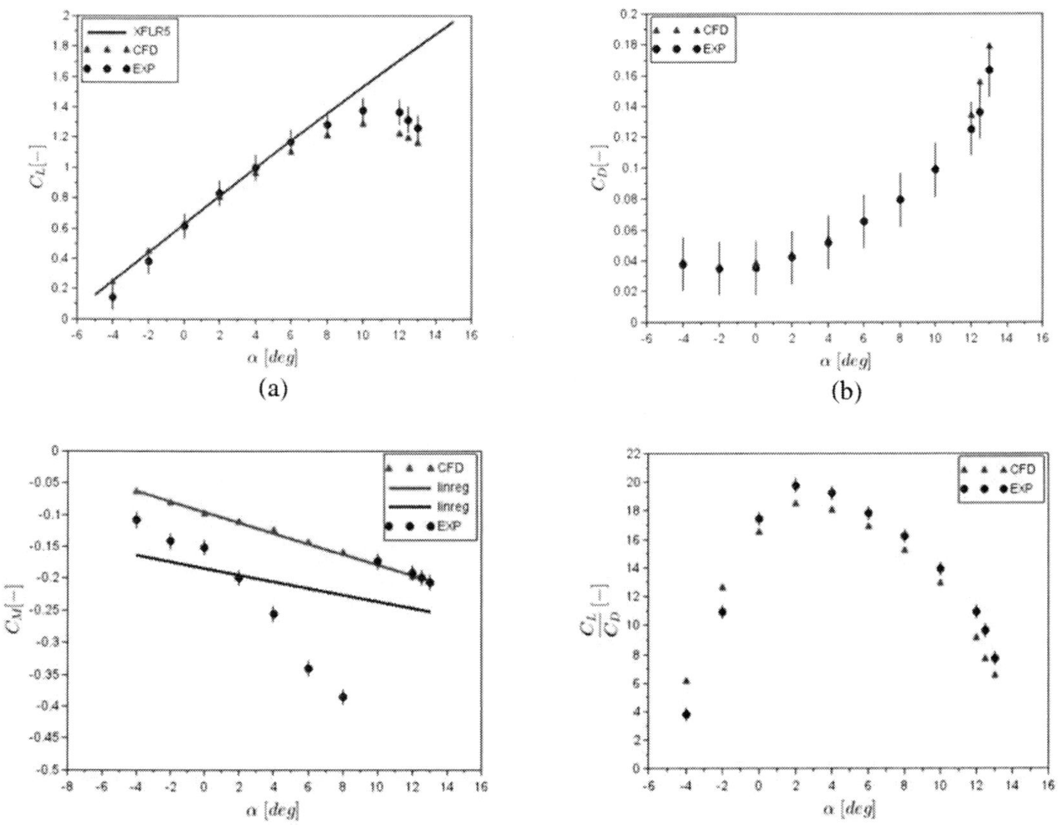

(a) (b)

(c) (d)

Figure 4. The Fundamental aerodynamics characteristics comparison

IOP Conf. Series: Materials Science and Engineering **886** (2020) 012016 doi:10.1088/1757-899X/886/1/012016

It is also observed that the critical angle-of-attack is similar between CFD and wind tunnel experiments. According to the data from the wind tunnel experiment, the air vehicle could achieve the maximum lift coefficient of 1.495 at 10 degrees angle of attack.

The drag coefficient variations versus angle-of-attack obtained from CFD and wind tunnel experiments, shown in figure 4(b), are well correlated with the minimum drag which is equal to 0.037 at -2 degrees angle-of-attack. The pitching moment coefficient variations with angle-of-attack shown in figure 4(c) show that, despite the negative pitching moment tendencies which indicated that the design has favorable 'positive' static longitudinal stability, there is no correlation between CFD and wind tunnel experimental results. This is due to the effect of the model installation mounting arrangement which gives a large vertical separation between the model's center of gravity and the balance's electrical measurement center. Figure 4 (d) illustrates the lift-to-drag ratio variation with angle-of-attack. It is observed that the air vehicle could achieve maximum aerodynamic efficiency at 2.0-degree angle-of-attack with the maximum lift to drag ratio of 20.05.

3. Stability and control

The stability and control characteristics of the air vehicle could be determined by various approaches such as (i) empirical calculation and simulation [7-9], (ii) wind-tunnel testing [10], and (iii) flight testing [10]. In the early design stage, an empirical calculation and simulation are commonly employed as it could provide some advantages in terms of engineering effort and resource requirements while delivering reasonable results [11-12]. In this design study, the XFLR5 software tool is implemented to determine the static and dynamic stability and control of the solar-UAV design. The data from table 1 is used as input parameters for XFLR5 calculations based on the design flight conditions. The mass properties of the airframe and subsystems arrangement are illustrated in figure 5. The calculation output expressed as longitudinal and lateral-directional stability derivatives are illustrated in table 2 and table 3, respectively. Note that all values are expressed in per-radian.

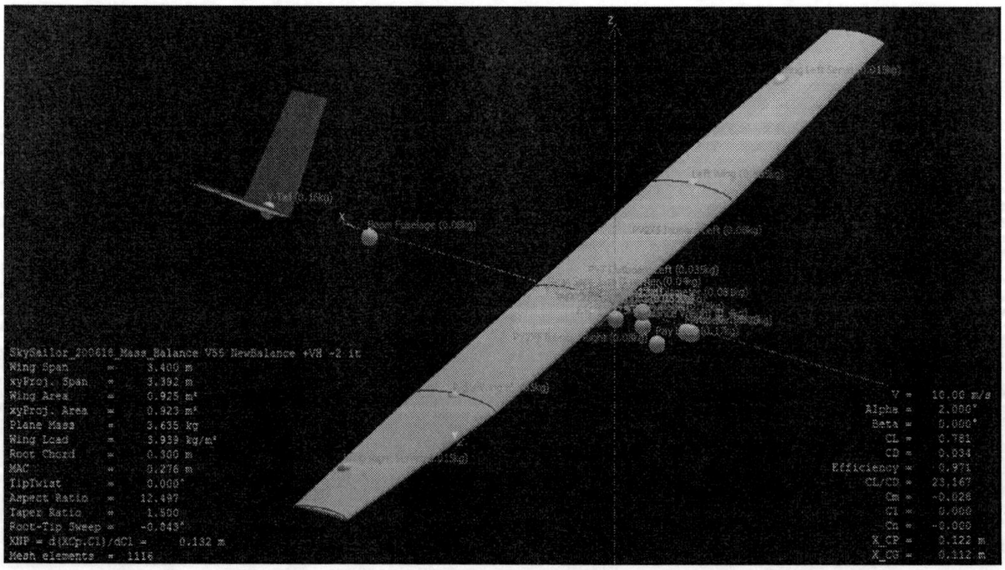

Figure 5. Weight and Balance model in XFLR5 for stability and control derivatives calculation

Table 2. Longitudinal stability and control derivatives

$C_{X_u} = -0.0469$	$C_{X_\alpha} = 0.53020$		
$C_{m_u} = -0.0043$	$C_{m_\alpha} = -0.3423$	$C_{m_q} = -13.937$	$C_{m_{\delta_E}} = -1.0304$
$C_{Z_u} = 0.00610$	$C_{L_\alpha} = 5.50370$	$C_{L_q} = 6.08290$	$C_{L_{\delta_E}} = 0.00150$

Table 3. Lateral-Directional stability and control derivatives

$C_{Y_\beta} = -0.13613$	$C_{Y_p} = -0.15746$	$C_{Y_r} = 0.125480$		
$C_{l_\beta} = -0.06683$	$C_{l_p} = -0.62025$	$C_{l_r} = 0.198220$	$C_{l_{\delta_r}} = 0.013120$	$C_{l_{\delta_a}} = 0.4102$
$C_{n_\beta} = 0.029971$	$C_{n_p} = -0.10014$	$C_{n_r} = -0.02526$	$C_{n_{\delta_r}} = -0.04225$	$C_{n_{\delta_a}} = 0.0015$

3.1. Static stability

As stated in the UAV airworthiness certification guidelines [13-14], the air vehicle should be demonstrated that the stability characteristics in all axes are positive. The static stability is crucial as the aircraft cannot be dynamically stable without being statically stable. Using the XFLR5 as an analysis tool it is confirmed that the solar-powered UAV design concept possesses positive static stability in longitudinal, lateral and directional axes as it gives negative pitching moment curve slope, positive yawing moment curve slope, and negative rolling moment curve slope. This would ensure that the aircraft's initial tendency to respond back to the equilibrium state once being disturbed by vertical and lateral gusts.

3.1.1. CG location variation

The location of the center of mass is an important factor affecting the aircraft's stability and control characteristics. Based on the airframe mass estimation and the expected arrangement of subsystem components defined at the conceptual design phase, the center of mass of the aircraft is located at the distance of 37.88% of the wing mean aerodynamic chord measured longitudinally from the leading edge of the wing at constant chord section. As the aircraft's static longitudinal stability characteristic is determined from the position of an aircraft's center of mass and its aerodynamic center along the longitudinal axis, it is important to determine the neutral point which considered as a center of mass location where the aircraft possess neutral static longitudinal stability. A sensitivity study to observe the pitching moment coefficient curve slope ZX variations due to the changes in the center of mass position is performed. Based on the sensitivity data shown in figure 6, the neutral point of the solar-powered UAV is determined at the distance of 12.2 cm measured longitudinally from the wing leading edge at the constant chord section. Finally, the aircraft's static margin could be calculated based on the wing mean aerodynamic chord and the longitudinal position of the aircraft's center of mass and aerodynamic center.

$$K_n = \frac{X_{NP}-X_{CG}}{\acute{c}} \tag{1}$$

IOP Conf. Series: Materials Science and Engineering **886** (2020) 012016 doi:10.1088/1757-899X/886/1/012016

Where X_{NP} is the position of the neutral point, X_{CG} is the position of the CG, $\acute{c}$ is the mean aerodynamic chord, and K_n is the static margin. It is observed that the solar-powered UAV design concept has a static margin value of 7.24%, which is considered moderate.

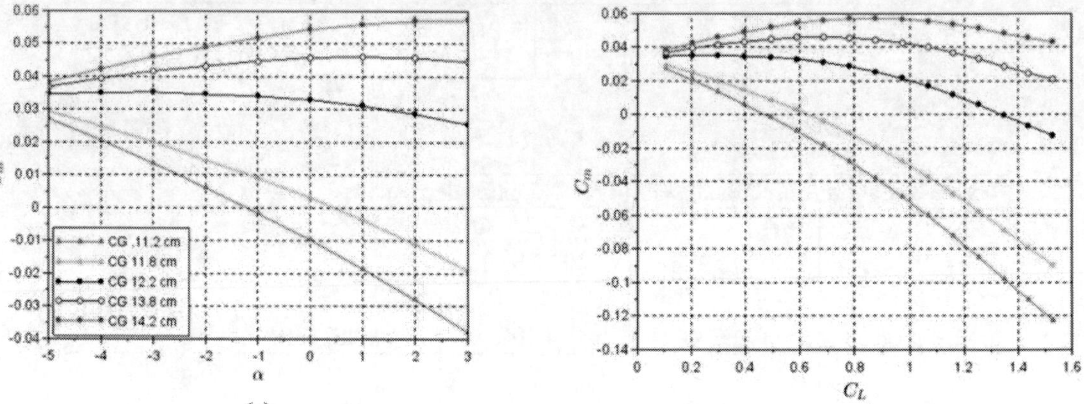

(a) (b)

Figure 6. Sensitivity Analysis of CG variation and Neutral Point Determination

3.2. Dynamic stability

In addition to the static stability, the aircraft is also required to be dynamically stable to achieve favorable flying quality. To observe the dynamic stability characteristics, the equation of motion in six degrees of freedom (6DOF) of the solar-powered UAV is linearized using the small disturbance theory around the trim flight condition. The linearized equation of motion for the longitudinal axis is expressed as

$$\begin{bmatrix} \Delta\dot{u} \\ \Delta\dot{w} \\ \Delta\dot{q} \\ \Delta\dot{\theta} \end{bmatrix} = \begin{bmatrix} X_u & X_w & 0 & -g \\ Z_u & Z_w & u_0 & 0 \\ M_u + M_{\dot{w}} Z_u M_u + M_{\dot{w}} Z_w M_q + M_w u_0 & & & 0 \\ 0 & 0 & 1 & 0 \end{bmatrix} \begin{bmatrix} \Delta u \\ \Delta w \\ \Delta q \\ \Delta\theta \end{bmatrix} + \begin{bmatrix} X_{\delta_e} & X_{\delta_T} \\ Z_{\delta_e} & Z_{\delta_T} \\ M_{\delta_e} + M_{\dot{w}} Z_{\delta_e} M_{\delta_T} + M_{\dot{w}} Z_{\delta_T} & \\ 0 & 0 \end{bmatrix} \begin{bmatrix} \Delta\delta_e \\ \Delta\delta_T \end{bmatrix} \quad (2)$$

and for the lateral-directional axis

$$\begin{bmatrix} \Delta\dot{v} \\ \Delta\dot{p} \\ \Delta\dot{r} \\ \Delta\dot{\phi} \end{bmatrix} = \begin{bmatrix} Y_v\, Y_p & -(u_0 - Y_r) & g\cos\theta_0 \\ L_v\, L_p & L_r & 0 \\ N_v N_p & N_r & 0 \\ 0\ 1 & 0 & 0 \end{bmatrix} \begin{bmatrix} \Delta v \\ \Delta p \\ \Delta r \\ \Delta\phi \end{bmatrix} + \begin{bmatrix} 0 & Y_{\delta_r} \\ L_{\delta_a} & L_{\delta_r} \\ N_{\delta_a} & N_{\delta_r} \\ 0 & 0 \end{bmatrix} \begin{bmatrix} \Delta\delta_a \\ \Delta\delta_r \end{bmatrix} \quad (3)$$

where $x_{Long} = [\Delta u \Delta w \Delta q \Delta\theta]^T$ is the longitudinal state vector and $x_{Lat} = [\Delta v \Delta p \Delta r \Delta\phi]^T$. is the lateral-directional state vector. The input vectors in the longitudinal and lateral-directional axis are expressed as $u_{Long} = [\Delta\delta_e \Delta\delta_T]^T$ and $u_{Lat} = [\Delta\delta_a \Delta\delta_r]^T$. respectively. The $\Delta u, \Delta v, \Delta w$ represent linear velocities along x-axis, y-axis, and z-axis, and $\Delta p, \Delta q, \Delta r$ represent angular velocities about the x-axis, y-axis, and z-axis. The pitch and bank attitude are represented by $\Delta\theta$ and $\Delta\phi$, respectively. Other parameters in the square brackets are the dimensional stability and control derivatives which are derived from table 2 and table 3. The mathematical model is implemented in the Scilab/Xcos dynamic system modelling framework.

3.2.1 Longitudinal dynamic stability

The open-loop dynamic longitudinal stability characteristics of the solar-powered UAV is obtained by solving equation (2) subjected to the fixed control inputs assumption. By establishing the initial steady-state level-flight condition at 9.45 m/s airspeed and 500 m flight altitude, the aircraft is shown to has stable phugoid as well as short period motion after being disturbed by an external gust representing as a singlet (non-cyclic) external pulse. Table 4 illustrated the longitudinal motion characteristics of the solar-powered UAV design. The open-loop dynamic response of the aircraft along the longitudinal axis is illustrated in figure 8.

Table 4. The UAV longitudinal motion characteristics

	Phugoid	Short period
Characteristic roots	$-0.08154 \pm 0.5954i$	$-8.134 \pm 3.546i$
ω_n (rad/s)	0.108	0.916
ζ	0.096	8.816

The responses of the phugoid motion expressed by variables $\Delta u, \Delta\theta, \Delta\alpha$. is plotted as illustrated in figure 8 (a)-(c) in which $\Delta\alpha$ is derived from Δw, indicating the aircraft's Phugoid motion is lightly damped in which the oscillation would converge and fully settled within 90 seconds. The short-period longitudinal responses are plotted as illustrated in figure 8 (d)-(f), which shows the highly-damped oscillatory motion with settling time equal to 1.0 second. This confirms the aircraft's stable longitudinal dynamic stability under the specified flight speed and altitude with the stick-fixed condition.

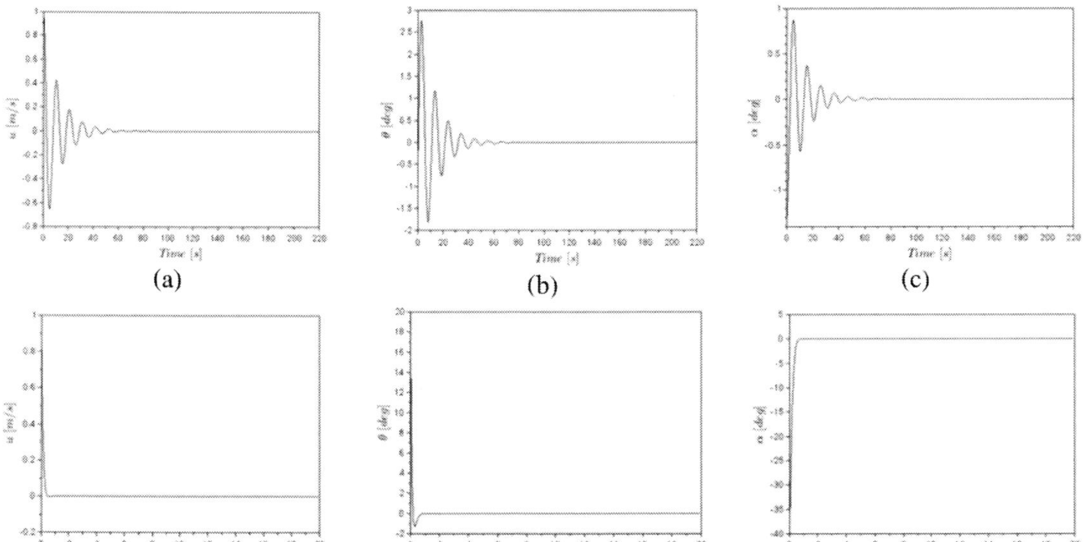

Figure 8. The Open-loop longitudinal dynamic response, Phugoid (a) – (c) and Short Period (d) – (e)

3.2.2. Lateral- Directional dynamic stability

Similar to the longitudinal stability analysis, the open-loop lateral-directional dynamic stability characteristics of the solar-powered UAV is obtained by solving the equation (3). Subjected to the same steady-state flight condition used in a longitudinal case, a singlet external disturbance is generated to introduce a sideslip resulting in a small bank angle $\Delta\phi$. The lateral-directional dynamic response of the aircraft in pure rolling and roll-yaw coupling (dutch-roll and spiral) motions are observed. Figure 5 Illustrates the characteristic roots, natural frequencies, damping ratio, and time constant of all three motions are illustrated in Table 5.

Table 5. The UAV Lateral-directional motion characteristics

	Dutch roll	Roll	Spiral
Characteristic roots	$-1.044 \pm 2.721i$	-28.67	0.1505
ω_n (rad/s)	0.389	-	-
ζ	0.472	-	-
Time constant (sec)	-	0.348	4.607

As illustrated in figure 9, the time history of the aircraft response in dutch-roll motion, considering motion variables $\Delta p, \Delta r, \Delta\phi$, indicates the convergent oscillation behaviour in which the aircraft returns to its trim bank angle after 10 seconds. It is also observed that for pure rolling motion, the aircraft response with a stable roll motion with time constant equal to 0.348 seconds and returns to its trim bank angle within 1.74 seconds.

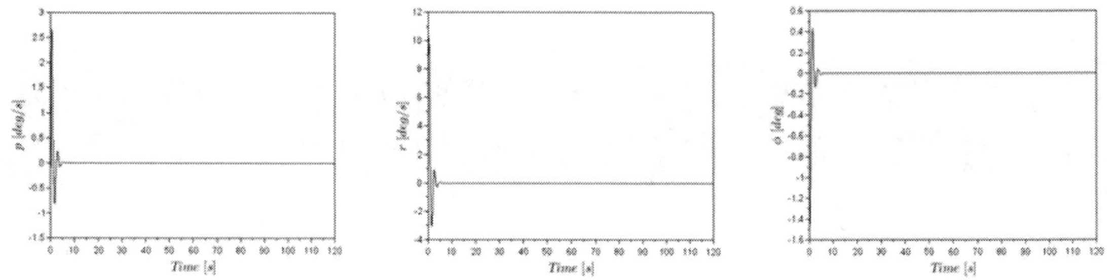

(a) (b) (c)

Figure 9. Open-loop Lateral dynamic response, Dutch roll motion

The aircraft response in a spiral motion, however, is observed to be unstable in which the amplitude of the aircraft bank angle is doubled in each cycle with a time constant of 4.607 seconds.

3.3. Control of UAV

The UAV has three aerodynamic control surfaces consist of ailerons and ruddervator (the combination of the solar-powered UAV concept employing ailerons for roll control). As the empennage design of the aircraft is of the V-tail type, the ruddervator coupling control surface system is implemented for both roll and yaw control of the aircraft. An overview of the control surface arrangement is illustrated in figure 10. The aileron is designed with a constant-chord distribution along with the outboard wing panel. Table 6 shows the geometric parameters of the aileron including chord ratio, span-wise location of the aileron's inboard and outboard end, and the design deflection angles. The rate of change of pitching moment due to elevator deflection (pitch control power) and the rate of change of yawing moment due to rudder deflection (yaw control power) are estimated at -1.030/rad and 0.04224/rad, respectively.

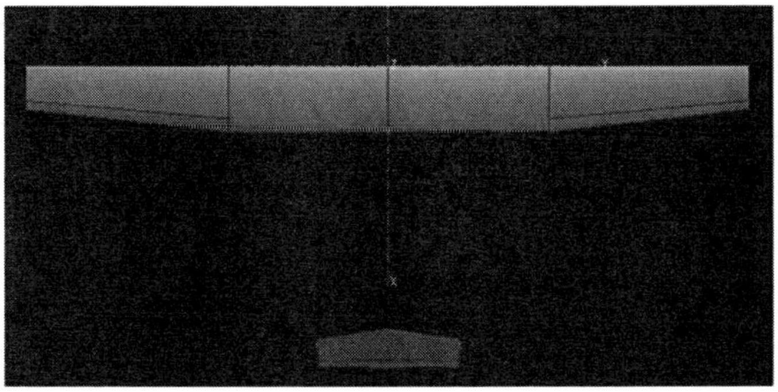

Figure 10. The UAV primary control surfaces, ailerons and ruddervator

Table 6. The ailerons geometric data

$\frac{c_a}{c}$	$b_i\frac{b}{2}$	$b_o\frac{b}{2}$	$\delta_{a_{max}}$(degree)	
			up	down
0.3	0.44	1.0	20	20

4. Conclusion

In this paper, the fundamental aerodynamic characteristics, as well as the stability and control of the solar-powered UAV, are investigated at the design flight conditions using various methods. The fundamental aerodynamic characteristics obtained from CFD and wind tunnel testing are compared for validation. The stability and control characteristics obtained from XFLR5 software shows that solar power UAV is flyable. However, to complete the preliminary design phase the performance and structural analyses need to be performed. Moreover, after the preliminary design of the solar-powered UAV is completed the detailed design needs to be conducted before the prototype could be fabricated. Then the ground and flight-testing program could be established to verify the aircraft's flight performance as well as its flying qualities.

Acknowledgments

The Authors would like to acknowledge the assistance from all of the staffs from the aeronautical engineering department and aviation technology division, NKRAFA. This research is possible by the financial support granted by the NKRAFA's Research and Development Promotion Fund (RDPF-NKRAFA), under the contact number 7/2561.

References

[1] Hosseini S Mesbahi M 2016 *J. of G.C. and D.* **39**(9) 1980–1993
[2] Guo Z Chen X Hou Z Guo J 2011 *Proc. 11th AIAA Aviation Technology, Integration, and Operations Conference* (Verginia Beach:Virginia)
[3] Maleki M 2011 *Proc. 49th AIAA Aerospace Sciences Meeting including the New Horizons Forum and Aerospace Exposition* (Orlando:Florida)

[4] Stern T Steele K Yang L Kruse M 2018 *Proc. AIAA Propulsion and Energy Forum, Joint Propulsion Conference* (Cincinnati:Ohio)
[5] Hartney C 2012 *Proc. 50th AIAA Aerospace Science Meeting Including the New Horizons Forum and Aerospace Exposition* (Nashville:Tennessee)
[6] Jiajan W Kampoon J Klongtrujrok J Yuthayanon I 2019 *IOP Conference Series Materials Science and Engineering* **501**(1) 012011
[7] Joshua J Azhaguraja P Banumathi P Abirami P 2015 *Int. J. ME. and IE Techno* **2**(2) 117–121
[8] Kampoon J 2017 *NKRAFA J. Sci. and Techno.* **13**(1) 25–33 (in Thai)
[9] Kampoon J 2018 *NKRAFA J. Sci. and Techno.* **14**(1) 86–97 (in Thai)
[10] Fabrizio N Agostino M Piealuigi V 2010 *Proc. AIAA GNC Conference* (Canada)
[11] Zhian Y Hudan S 2017 *Auto., Control and Intelligent Systems* **5**(2) 29–32
[12] Lui C 2012 *Proc. 2nd International Conference on Com. Appl. System Modeling*
[13] The North Atlantic Treaty Organization 2007 *NATO Standardized Agreement 4671, 1st edition*
[14] The North Atlantic Treaty Organization 2014 *NATO Standardized Agreement 4703, 1st edition*

IOP Conf. Series: Materials Science and Engineering **886** (2020) 012017 doi:10.1088/1757-899X/886/1/012017

Simulation and validation of Airflow Past Hand-Shape Mold Using OpenFOAM

K Loksupapaiboon[1,2] and C Suvanjumrat[1,2,*]

[1]Department of Mechanical Engineering, Faculty of Engineering, Mahidol University, Nakhon Pathom, 73170, Thailand
[2]Laboratory of Computer Mechanics for Design (LCMD), Department of Mechanical Engineering, Faculty of Engineering, Mahidol University, Nakhon Pathom, 73170, Thailand

* Corresponding Author: chakrit.suv@mahidol.ac.th

Abstract
This paper aims to present a wind tunnel, aiming to examine the effect of airflow past the hand-shape mold especially airflow around the gap between the fingers of the mold. Part of the aim of this project is to investigate the behavior of airflow past the mold on the different angles, which was valuable to vulcanize rubber glove in the oven. The wind tunnel experiments were performed on the various angle of attack of hand-shape mold from 0 to 180 degrees under the airflow velocity of 5 to 20 m/s. The 3D simulations were carried out by using open source code software, OpenFOAM. The k-ε model was used to simulate the turbulent flow past hand-shape mold. The pressure-velocity coupling problem and the convection-diffusion term were solved by using a SIMPLE algorithm and upwind differencing scheme, respectively. The drag force by airflow on the rubber glove mold which obtained by the computational fluid dynamics (henceforth CFD) method was compared with the experimental data. The comparison between CFD simulation and the experimental data showed a fairly close agreement and the average error was less than 13.96%. Further research in this field would be of great help in developing a model to optimize the mold installation inside the rubber glove oven.

1. Introduction

Drying process is widely used to release moisture of food products, color or lacquer coating and rubber product. In the production of rubber glove, the baking is an important process for drying the liquid that coated on the hand-shape mold surface. This liquid consists of cleaning water, coagulant and compound rubber. The convection oven is the most favorite equipment used for drying liquid. One major issue found about the internal structure of the convection oven and the shape of the hand-shape mold is the wetness of the coating liquid on the mold surface remains. The wetness is associated with increased risk of a non-uniformity of air velocity around the gap between the fingers of hand-shape mold. This could be a contributing factor to prevent waste from the drying process. Scholars have long debated the impact of the uniformity of velocity and temperature of the air inside the oven, which could affect the dryness of the products. Smolka et al. [1] studied the uniformity of airflow and thermal parameters inside the drying oven. These parameters could indicate the efficiency and homogeneity of the product being dried. The controlling of these parameters was very difficult in experimentation. The simulation methods were used to study and analyze the airflow instead. The CFD was employed to predict the airflow and thermal distribution that can reduce time and cost from trial and error. The simulation results also suggest that the uniformity around the products could increase the drying quality. Most of CFD software is commercial such as ANSYS-FLUENT, FLOW-3D, PHEONICS and COMSOL. This software has an expensive license cost. OpenFOAM (henceforth Open Source Field Operation and Manipulation) is one of open-source CFD software, which can be applied and developed codes for solving a fluid problem by using C++ language. It is used under GNU GPL (General Public License) without license cost [2]. The fluid flow problem together with heat transfer and turbulence flow behavior could be solved by OpenFOAM software [3-6].

The Reynolds-Averaged Navier-Stokes (henceforth RANS) equations are the set of average equations of motion over the time for the fluid flow that is primarily equations to describe the turbulent flow. RANS equations are so-called Reynolds stress, which is used to describe eddy effects, which is difficult to calculate. Many models were developed to calculate the Reynolds stress term and apply for the bluff body flow. The lift and drag force are important parameters from the CFD simulation for validating the turbulence model [7]. Kurec et al. [8] studied the airflow past a car with a rear wing, causing the down force on the car body. Four turbulence models were used to simulate the airflow. The force components that obtained from simulation were compared with experimental data. The results showed that the r-k-ε and k-k_l-ω turbulence models gave the force component and were closer to the experiment. The k-ε turbulence model was chosen to evaluate the effectiveness of airflow past complex structure i.e. the cyclist following by motorcycle.

The CFD results showed that the motorcycle following the cyclist decreased the drag force which was unwanted for aerodynamics benefits [9]. Suvanjumrat [10] has demonstrated that the lift and drag coefficient can be used to compare turbulence models for the flow past NACA0015. It appears that the Menter SST k-ω model with the SIMPLE algorithm and LUD scheme was the suitable model. This study also highlighted that k-ε turbulence model can be applied to predict the airflow past hand-shape mold by using OpenFOAM software. The viscous flow was solved by using Semi-Implicit Method for Pressure Linked Equation (SIMPLE) algorithm and upwind differencing scheme. To validate the simulation model, the wind tunnel experiment is carried out. The results also showed that optimized angle can cause the good airflow past gab between fingers of hand-shape mold. There was an increase in the drying efficiency of the rubber glove. It is suggested that the CFD technique will be used to analyze the arrangement of hand-shape mold inside the batch type oven to avoid the blockage between molds.

2. Simulation Methods

2.1 Governing Equations

The governing equations of fluid dynamics consist of three fundamental physical principles: conservative of mass, momentum and energy. These equations are solved together to describe fluid flow behaviour. The generic form of the conservative equation for turbulent flow and incompressible fluid is as follows:

$$\frac{\partial \bar{\phi}}{\partial t}+div\left(\bar{\phi}\bar{U}\right)=\frac{1}{\rho}div\left(\Gamma grad\left(\bar{\phi}\right)\right)-\frac{1}{\rho}div\left(\rho\overline{u'\phi'}\right)+\bar{S}_{\phi} \quad (1)$$

where $\bar{\phi}$ is the mean general variable of flow properties, ρ is the fluid density, Γ is the diffusion coefficient, $\rho\overline{u'\phi'}$ is the Reynolds stress term and $\bar{S}_{\phi}$ is the source term.

The Reynolds stress term was based on the fluctuation of turbulent flow. The calculation of Reynolds stress term requires extra equations or turbulence models which are derived from the mathematic model.

2.2 Turbulence Model

2.3 To identify the concept of the turbulence model, it is assumed that the Reynolds stress related to the eddy viscosity and the mean velocity gradient is known as eddy viscosity model, as can be seen from the following equation (2).

$$\tau_{ij} = -\rho\overline{u_i'u_j'} = \mu_t\left(\frac{\partial \overline{u_i}}{\partial x_j} + \frac{\partial \overline{u_j}}{\partial x_i}\right) - \frac{2}{3}\delta_{ij}\rho k \tag{2}$$

where μ_t is the turbulent viscosity, δ_{ij} is the Kronecker delta and k is the turbulence kinetic energy. The turbulent viscosity is written by:

$$\mu_t = \rho C_\mu \frac{k^2}{\varepsilon} \tag{3}$$

where C_μ is the dimensionless constant and ε is the rate of dissipation of turbulence energy.

The standard k-ε turbulence model is widely used to predict turbulent flow conditions. This model is the two equations model which consist of two extra transport equations, k and ε equations [11].

$$\frac{\partial \rho k}{\partial t} + div\left(\rho k \overline{\boldsymbol{U}}\right) = div\left(\left(\frac{\mu_t}{\sigma_k}\right) grad\left(k\right)\right) + 2\mu_t S_{ij}\cdot S_{ij} - \rho\varepsilon \tag{4}$$

$$\frac{\partial \rho \varepsilon}{\partial t} + div\left(\rho \varepsilon \overline{\boldsymbol{U}}\right) = div\left(\left(\frac{\mu_t}{\sigma_\varepsilon}\right) grad\left(\varepsilon\right)\right) + C_1\frac{\varepsilon}{k}2\mu_t S_{ij}\cdot S_{ij} - C_2\rho\frac{\varepsilon^2}{k} \tag{5}$$

where S_{ij} is the velocity gradient yield which can be shown as follows:

$$S_{ij} = \left(\frac{\partial U_i}{\partial x_j} + \frac{\partial U_j}{\partial x_i}\right) \tag{6}$$

The constant variables for the standard k-ε turbulence model comprise C_μ=0.09, σ_k=1.00, σ_ε=1.30, C_1=1.44, and C_2=1.92.

3. Experiment

The wind tunnel experiment, model WT300, was used to study the airflow past hand-shape mold The structure and the equipment of the wind tunnel experiment are set out in Figure 1. The wind tunnel was installed the 2-axis load cell under the testing section. The hand-shape mold was connected to the 2-axis load cell inside the testing section. It was designed to adjust the angle of attack from 0 to 360 degree. The first angle of attack or 0 degree starts from the flow direction parallel to the hand-shape mold. The thumb side of mold was assigned to be attacked by the inlet flow. The angle of attack for the experiment consisted of 0, 45, 90, 135 and 180 degrees. Regarding the suction side of the wind tunnel, the grille was installed to decrease the turbulent behavior. The axial fan and the inverter were used to control airflow velocity. The airflow velocity of the experiment including 5, 10, 15 and 20 m/s was prepared. The Reynolds numbers 38,265, 76,530, 114,795 and 153,061 were also chosen in the experiment. The airflow velocity was measured by pitot tube anemometer at the front of the mold. The drag force due to the airflow past hand-shape mold was measured by the 2-axis load cell under the steady-state condition. The sampling rate to record the drag force was 1 Hz along with the time of five minutes by using the data acquisition, MicroStrain SG-Link.

4. CFD model

The simulation of the airflow past hand-shape was considered to be a steady flow; therefore, the transient term was eliminated. The open-source code CFD software, OpenFOAM, was used to simulate the airflow.

Figure 1. The wind tunnel experiment model WT300 with hand-shape mold and

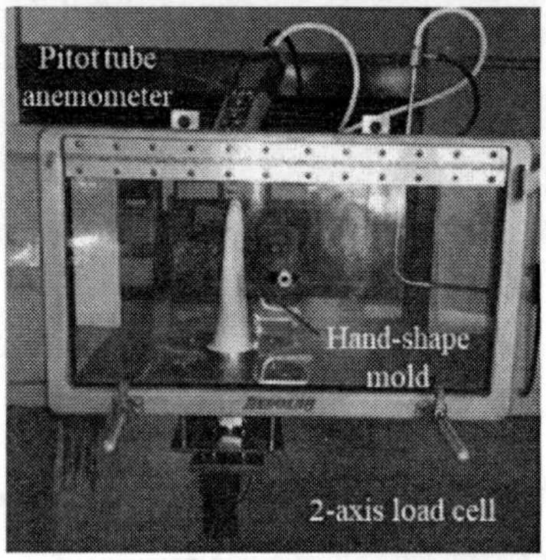

Figure 2. 2-axis load cell and pitot tube anemometer.

The computational domain was created by the 3D model. It has the total length of 13H, the width of 10W and the height of 5W, where H and W are height and width of the hand-shape mold, respectively (Figure 3). In this study, the height and width of the hand-shape mold is equal to 250 and 120 mm, respectively. The 3D unstructured cell was used to divide the computational domain for the discretization. The finest cells are created on the mold surface, as shown in Figure 4 . The best finest cell size which affected the number of total cells was carried out by a grid independence test. The drag force by airflow was used to arbitrate the convergence result. The range of the finest cell size varies from 1.0 to 1.5 mm with the increments of 0.1 mm. No significant differences were found between the grids independence test and the convergence. Furthermore, the computer that used in this research was out of memory when the finest cell size was less than 1.0 mm. Interestingly, the computer used in this research was out of memory when the finest cell size was less than 1.0 mm. However, the y+ of 2.31 was in the viscous sub-layer. These results suggest that it can be used to simulate the airflow past hand-shape mold in all cases with wall functions.

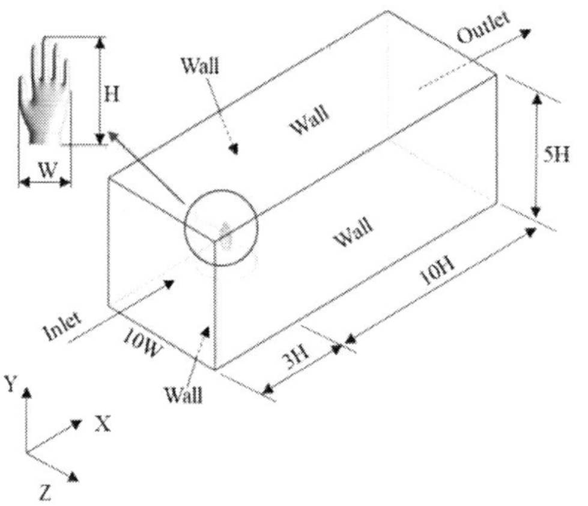

Figure 3. Domain for CFD simulation of the airflow past hand-shape mold and

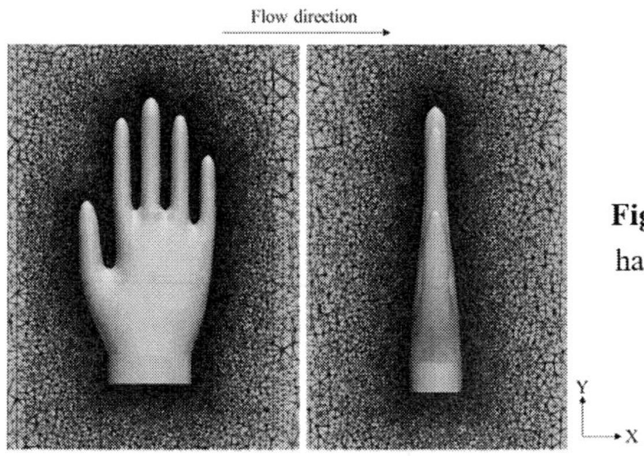

Figure 4. Finest cells structure around the hand-shape mold.

0 degree 90 degree

The boundary conditions were assigned to the domain including inlet, outlet and wall. The inlet was assigned to be the uniform flow according to the experiment while the outlet was assigned to be an ambient pressure. The outer surface of the domain and the surface of the hand-shape mold were assigned by no-slip wall condition. The angle between the hand-shape mold and X-Y plane (flow direction) could be adjusted according to the experiment. These simulation cases conducted a uniform inlet velocity according to the physical experiment composing 5, 10, 15 and 20 m/s. The angle of attack was also conducted on five angles including 0, 45, 90, 135 and 180 degrees. Further analysis showed that the drag force was investigated by 20 simulating cases. All CFD cases, the standard k-ε model governed the turbulent flow. However, the flow simulation with turbulence model was valid only in a fully developed turbulent flow which was far from the wall. In the near-wall region, the turbulence model could not be calculated. Therefore, two methods could be used to resolve this problem. The first method was created the first cell center keeping on the viscous sublayer (y≤1). While the second one was wall functions which modeled the turbulent flow on the near-wall region. In this study, the wall functions were used to calculate the effect on the near-wall instead to generate the first cell center for making the y+ equal or less than 1.The pressure-velocity couple problem and convective term of the governing equations were solved by the SIMPLE algorithm and upwind differential scheme, respectively [12]. The air which flows through the domain was considered to be an incompressible fluid with the density and dynamics viscosity of 1.225 kg/m^3 and 1.838x10^{-5} kg/(m·s), respectively.

5. Results and Discussion

The purpose of the study is to investigate the simulation of the airflow past hand-shape mold which is based on the CFD model. The results gained can be compared with the experimental data. The drag force countered the airflow was calculated by the pressure on the hand-shape mold. At a various angle of attack, it was compared with the measuring data by the 2-axis load cell. As can be seen in Figure 5, it shows the comparison of the graph of drag force on the various angle of attack when the airflow velocity increases. These results suggest that the more the drag force increases, the wider the angle of attack can be extended from 0 degrees to 90 degrees. Then, the drag force can be decreased when the angle of attack is increased to 180 degrees. The results of this study indicate that the CFD show significant difference than the experiment at the angle of attack of 0 and 180 degrees. It is exceptional for other angles since they show a significant difference than the CFD results. This behavior was caused by the structure of the cells of hand-shape mold on the trump side and little finger side were quite complex. These areas past by airflow at the angle of attack of 0 and 180 degrees. This problem can be resolved by decreasing y+ on these areas. The comparison of drag force between the CFD and experiment had an average error of 10.73, 15.29, 14.54 and 15.29% along with the airflow velocity of 5, 10, 15 and 20 m/s, respectively. The results were satisfied with the accuracy of the CFD model.

At the angle of attack of 0 degrees, the inlet flow of 20 m/s past the hand-shape mold can be illustrated by color contour (see Figure 6 below). The airflow was not past through the finger gap of hand-shape mold. The blue color contour was appeared. Figure 7 shows the velocity contour of airflow past the finger gap of hand-shape mold at the angle of attack of 90 degrees. This study indicate that the airflow past the finger gap of hand-shape mold very well. In this regards, the airflow at a different angle of attack could be different from the airflow behavior around the gap between the finger. All in all, the suitable alignment of hand-shape mold inside the batch type drying oven should equal to 90 degrees for the most airflow past the gaps between the finger of the hand-shape mold.

6. Conclusion

In this investigation, the aim was to study the steady flow simulation of the airflow past a complex hand-shape mold. The study was carried out by using the open-source CFD software, OpenFOAM. The simulation results were compared with experimental data. These findings suggest that in general, the CFD model had good agreement with experiment. The average error of the CFD model was less than 13.96%. These findings have significant implications for the understanding of how the airflow behavior between the gap of the finger of hand-shape mold was different in the different angle of attack. The present study provides that the suitable angle of attacks was 90 degrees so that the open source software can be used to analyze the suitable alignment of hand-shape mold inside the batch type oven. This would be a fruitful area for further work.

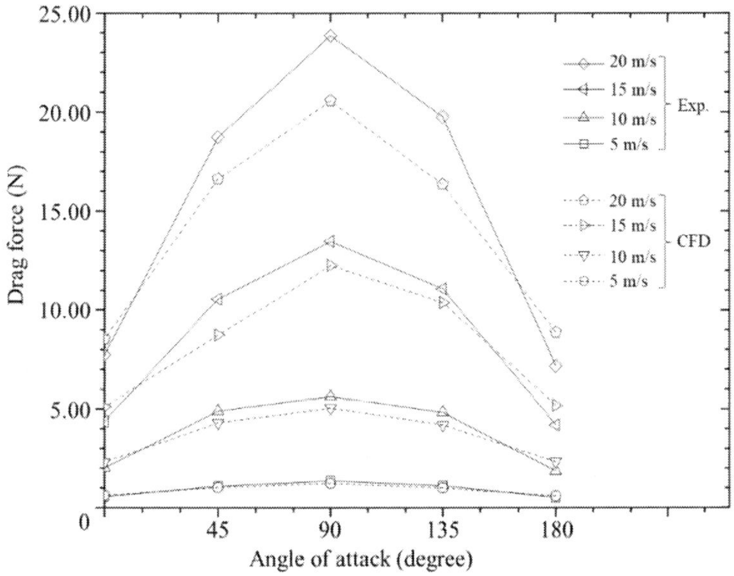

Figure 5. The drag force comparison between CFD and experiment at various angle of attack and velocity.

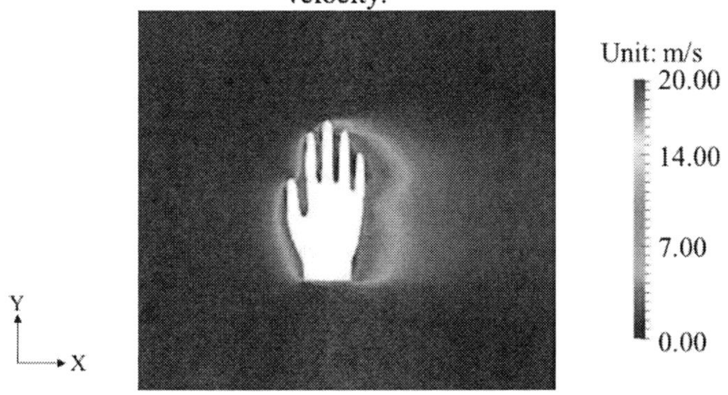

Figure 6. The velocity contour of airflow past hand-shape mold at the angle of attack of 0 degrees by the inlet velocity of 20 m/s.

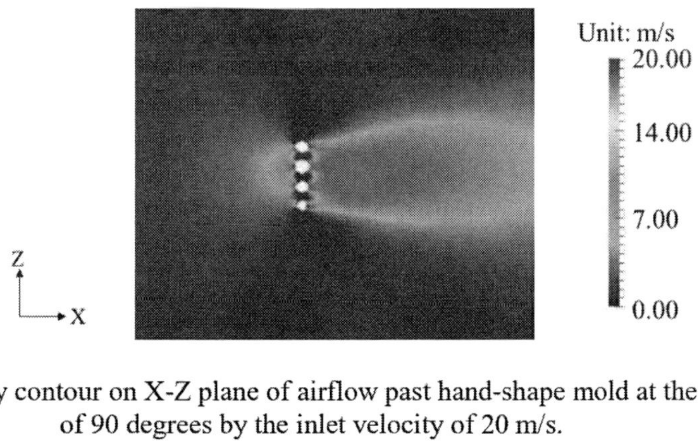

Figure 7. The velocity contour on X-Z plane of airflow past hand-shape mold at the angle of attack of 90 degrees by the inlet velocity of 20 m/s.

Acknowledgments
This research was granted by The Thailand Research Fund (TRF) and W.A. Rubbermate Co., Ltd under Research and Researchers for Industry Program, Grant No. PHD61I0003.

References
[1] Smolka J Nowak A J and Rybarz D 2010 *J. Food Eng.* **97**(3) 373
[2] Greensheilds C J 2019 *OpenFOAM The OpenFOAM Foundation User Guide version 7* (UK: OpenFOAM Foundation Ltd.,)
[3] Flores F Garreaud R and Munoz R 2013 *Comput. Fluids* **82** 1
[4] Prasad B Hino T and Suzuki K 2015 *Ocean Eng.* **102** 87
[5] Suvanjumrat C 2017 *Engineering Journal* **21**(5) 225
[6] Zang B Vevek U S Lim H D and New T H 2018 J. Comput. Sci. **28** 18
[7] Roberson E Choudhury V Bhushan S and Walters D K 2015 Comput. *Fluids* ***123*** *122*
[8] Kurec K Remer M Mayer T Tudruj S and Piechna J 2019 *Int. J. Mech. Sci.* **152** 384
[9] Blocken B Toparlar Y and Andrianne T 2016 *J. Wind Eng. Ind. Aerod.* **155** 1
[10] Suvanjumrat C 2017 Engineering Journal **21**(3) 207
[11] Kaunder B E and Spalding D B 1973 *Comput. Method Appl. M.* **3** 269
[12] Versteeg H K and Malalasekera 2007 *An Introduction to Computational Fluid Dynamics The finite Volume Method 2nd edition* (London: Pearson Education Ltd.,)

IOP Conf. Series: Materials Science and Engineering **886** (2020) 012018 doi:10.1088/1757-899X/886/1/012018

Nonlinear Responses of An Unbalanced Overhung Rotor-Short Journal Bearing System with Some Bifurcation Results

P Atayagul[1], T Kamnerdthong[1], and S Songschon[1,*]

[1]Department of Mechanical Engineering, King Mongkut's University of Technology Thonburi, Bang Mot, Bangkok, 10140, Thailand

* Corresponding Author: szathys.son@mail.kmutt.ac.th

Abstract
Both autonomous and non-autonomous nonlinear systems display complex responses including the transient and long run responses. An overhung rotor-short journal bearing system (ORSJB) is the excellent paradigm exhibiting nonlinear responses and of practical importance. The unbalanced overhung rotor-journal bearing system is formulated by using modified Laval-Jeffcott rotor model. The nonlinear fluid forces model of the short journal bearing is adopted and rearranged in a suitable form for numerical analysis. This paper investigates numerically a bifurcation point, transient and long run responses of the ORSJB system. The rotational speed is preferable bifurcation parameter herein. The ORSJB system without unbalanced eccentricity for the selected parameters yields the bifurcation value of $\Omega = 3.0091$. Limits cycles in the long-run responses with $e_U = 0.002\ m$ and without unbalanced eccentricity at long run are presented in the bifurcation regime at the speed of $\Omega = 6.6869$ and $\Omega = 0.01671$ respectively. The transient response with $e_U = 0.002\ m$ at the speed of $\Omega = 4.7142$ is elucidated. The evident results in this paper give only a partial view of bifurcation behaviours. Bifurcation analysis is a useful tool for design and operation overhung rotor dynamic systems.

1. Introduction

There is a growing body of literature that recognizes some advantages of plain journal bearings, for instance, the ease of manufacturing and lubricant feeding. A considerable amount of literature has been published on nonlinear dynamics of rotor-journal bearing system [1]. Previous research investigated various topics and developed theoretical results in the domain of a plain journal bearing [2, 3, 8]. Short plain journal bearing is the prominent bearing type commonly selected for theoretical study. The chaos of symmetrical in-board rigid rotor interacted with fluid forces in laminar regime from supporting short journal bearings was investigated analytically and experimentally in [2] while the turbulent regime of fluid force dynamics was studied theoretically in [3]. Surveys such as that conducted by Wang et.al [4] explored the analytical model of fluid film force of finite length journal bearing for laminar regime. In-board Laval-Jeffcott rotor with unsymmetrical clearance was analyzed at a fractional speed of whirling as in [5]. Performance of plain journal bearing with effects of viscosity and clearance was analyzed in [6]. Most models used for nonlinear analysis for Laval-Jeffcott rotor journal systems are of an inboard type. To date, most studies have carried out an overhung rotor for nonlinear motion with the effects of clearance and rubbing [7]. Whilst there have been few empirical investigations into the nonlinear effect of plain journal bearing forces. Overhung rotor dynamic systems with journal bearings have been used wildly in engineering applications such as gas turbine engines, marine propeller shaft systems, to name but a few. This article will formulate governing equations for the proposed overhung Laval-Jeffcott rotor journal system and investigate numerically on effects of nonlinear fluid force with various rotational speeds.

2. An Overhung Rotor-Journal Model and Equations of Motion

Figure 1 presents the 4-DOF model of an overhung rotor journal bearing system.

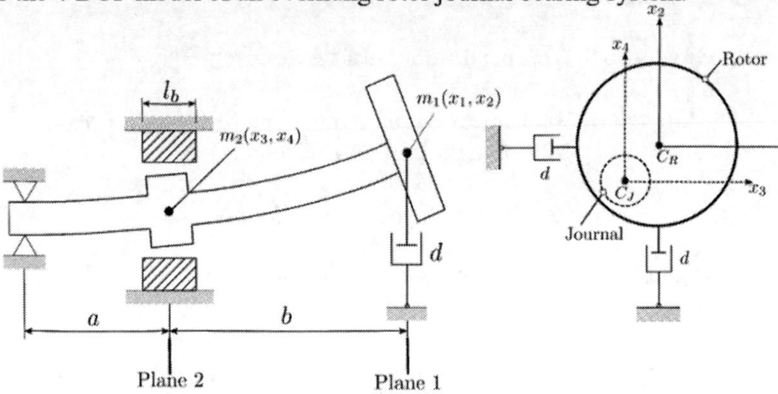

Figure 1. An Overhung Rotor-Journal Model

As can be seen from Figure 1 above, the overhung Jeffcott rotor with a journal bearing is proposed. From Figure 1 above we can see that a simple supported bearing is assumed at the left end. It is apparent from this figure that the assumption results has no transverse motion on the left-most plane. While a short plain journal, bearing with clearance is placed at plane 2. The r_B and r_J are radii of bearing and journal respectively. Jeffcott rotor is the massless shaft with point masses at journal and rotor. The relation of length portions can be written as $l = a + b$. The stiffness of the shaft for transverse deflections is $k = \frac{3E_Y I}{lb^2}$ where I the area moment of inertia of the shaft is. The viscous damping with equally damping coeffcients (d) is assumed at the rotor plane 1.

As shown in Figure 1 above, the kinetic energy (T) and potential energy (U) of the system in the can be written in (1) and (2).

$$T = \frac{1}{2}m_1(\dot{x}_1^2 + \dot{x}_2^2) + \frac{1}{2}m_2(\dot{x}_3^2 + \dot{x}_4^2) \tag{1}$$

$$U = \frac{1}{2}k(x_1 - K_a x_3)^2 + \frac{1}{2}k(x_2 - K_a x_4)^2 + m_1 g x_2 + m_2 g x_4 \tag{2}$$

where $K_a = l/a$. The Lagrangian is defined as $\mathcal{L} = T - U$. By using $D_t(\partial_{\dot{x}_i}\mathcal{L}) - \partial_{x_i}\mathcal{L} = Q_{x_i}, i = 1,2,3\ and\ 4$, one can write the governing equations of the above system as in (3)-(5).

$$m_1\ddot{x}_1 + k(x_1 - K_a x_3) = -d\dot{x}_1 + m_1 e_U \omega^2 \cos(\omega t) \tag{3}$$

$$m_1\ddot{x}_2 + k(x_2 - K_a x_4) = -d\dot{x}_2 + m_1 e_U\, \omega^2 \sin(\omega t) - m_1 g \tag{4}$$

$$m_2\ddot{x}_3 + kK_a(K_a x_3 - x_1) = f_3 \tag{5}$$

$$m_2\ddot{x}_4 + kK_a(K_a x_4 - x_2) = f_4 - m_2 g \tag{6}$$

where the "˙" represents the "$\frac{d}{dt}$". Q_{x_i}, $i = 1,2,3,4$ are generalized forces in the directions of x_i appearing on the right-hand side of (3)-(6) and e_U is the eccentricity of unbalance mass at the rotor. The f_3 and f_4 are the fluid forces from bearing acting on the journal. These are sources of nonlinear terms dominating behavior of the systems summarizing in the next section.

IOP Conf. Series: Materials Science and Engineering **886** (2020) 012018 doi:10.1088/1757-899X/886/1/012018

3. Fluid Film Force of Short Journal Bearing

It can be seen from the data in Figure 1 that a plain short journal bearing was selected. Figure 2 presents description of the quantities used in this study. Regarding a circular journal bearing, $h(\theta,t) = c - e(t)\cos(\theta - \gamma(t))$ is where the bearing clearance is. The eccentricity is $c = r_B - r_J$ and e, respectively. We write $ecos(\gamma) = x_3$ and $esin(\gamma) = x_4$, and thus $h(\theta,t) = c - x_3 cos\theta - x_4 sin\theta$ is provided.

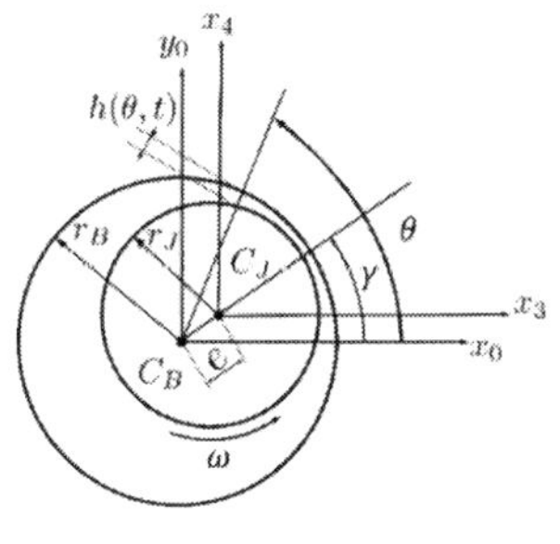

Figure 2. Journal Bearing Parameters

$$\partial_z \left(\frac{h^3}{12\mu}\partial_z p\right) = \frac{1}{2}\omega\partial_\theta h + \partial_t h \tag{7}$$

In terms of the Reynolds equation [1], it is specifically shown for a short bearing imposing ∂_θ p=0 with p(θ,z). It is facing dimensional pressure from the fluid film, and is called a dynamic viscosity of the fluid (μ). The Reynolds equation (7) for a short journal bearing can be rewritten in a non-dimensional form with zero boundary conditions at both ends of the bearing along the axial axis. We use some capital Latin and lowercase Greek letters for non-dimensional quantities, such as $X_i = \frac{x_i}{c}$, $\zeta = z/r_B$, $H = \frac{h}{c}$, $\tau = \omega t$ and $P = \frac{p}{6\mu\omega\left(\frac{r_B}{c}\right)^2}$. After some manipulations, we can obtain the fluid forces in the equation (8)

$$\begin{bmatrix} f_3 \\ f_4 \end{bmatrix} = \Gamma\Omega \begin{bmatrix} F_3 \\ F_4 \end{bmatrix} \tag{8}$$

where F_3 and F_4 are non-dimensional fluid film forces from the bearing acting on the journal. $\Gamma = \frac{\mu r_B l_B^3}{2g^{0.5}c^{2.5}m_q}$ is bearing's characteristic number where $m_q = m_2 + m_1 K_a$. $\Omega = \omega\sqrt{c/g}$ is the non-dimensional rotational speed. We have verified and adopted the adorable results from [2] to compute the F_3 and F_4. We define " $' = \frac{d}{d\tau}$ " , $Q = (1 - X_3^2 - X_4^2)^{1/2}$, $V = X_4 cos\alpha - X_3 sin\alpha$ and use $G(X_3, X_4, \alpha) = \frac{2}{Q}\left[\frac{\pi}{2} + \text{atan}\left(\frac{V}{Q}\right)\right]$ with $\alpha = atan\frac{X_4 + 2X_3'}{X_3 - 2X_4'} - \frac{\pi}{2}sign\left(\frac{X_4 + 2X_3'}{X_3 - 2X_4'}\right) - \frac{\pi}{2}sign(X_4 + 2X_3')$. It should be noted that the findings from the experiment provide some support for the $\pi - film$ model of fluid film pressure that is used in this formulation. We then obtain the non-dimensional fluid film forces as:

$$F_3 = \frac{C_F}{2}(3X_3V_V - \sin(\alpha)\,G - 2\cos(\alpha)\,S) \tag{9}$$

$$F_4 = \frac{C_F}{2}(3X_4V_V - \cos(\alpha)\,G - 2\sin(\alpha)\,S) \tag{10}$$

where $C_F = -\frac{\sqrt{(X_3-2X_4')^2+(X_4+2X_3')^2}}{Q^2}$, $V_V = \frac{2+VG}{Q^2}$ and $S = \frac{X_3 cosa+X_4 sina}{1-(X_4 cosa+X_4 sin\alpha)^2}$

The fluid film forces in the equations (9) and (10) are sources of nonlinear influencing in the considering rotor-bearing system. The F_3 and F_4 represent in dimensionless forms, however, the system equations from (3) to (6) are all in dimensional quantities. Use of the groups of non-dimensional quantities gives us more portrayal of the system behavior. The next section of the study was concerned with the non-dimensional system of (3) to (6).

4. Non-dimensional Equations and State Space Representation

As previously mentioned in Section 3, the non-dimensional fluid film forces are obtained and they are also coupled with the variables $\boldsymbol{X_i}$ and $\boldsymbol{X_i'}$. The transformation of all quantities in non-dimensional forms was used to compute the responses of this system. To get all the equations in non-dimensional forms, the systematic method is used to represent in state-space form for the ease of computing the responses.

4.1 Non-dimensional Equations

Starting with the equations (3) to (6), we can normalize the equations and make use of some more dimensionless quantities. The appropriated non-dimensional quantities to be mentioned are D=d/m_1 √(c/g) , K_1=kc/(m_1 g), K_2=kc/(m_2 g), and E=e_U/c. The equations (3) to (6) can be written, in non-dimensional forms, as in (11) to (14).

$$X_1'' + \frac{D}{\Omega}X_1' + \frac{K_1}{\Omega^2}(X_1 - K_a X_3) = E cos\tau \qquad (11)$$

$$X_2'' + \frac{D}{\Omega}X_2' + \frac{K_1}{\Omega^2}(X_2 - K_a X_4) = E sin\tau - \frac{1}{\Omega^2} \qquad (12)$$

$$X_3'' + \frac{K_a K_2}{\Omega^2}(K_a X_3 - X_1) = \frac{m_q \Gamma}{m_2 \Omega} F_3 \qquad (13)$$

$$X_4'' + \frac{K_a K_2}{\Omega^2}(K_a X_4 - X_2) = \frac{m_q \Gamma}{m_2 \Omega} F_4 - \frac{1}{\Omega^2} \qquad (14)$$

4.2 State Space Representation

Representation of the system of equations (11) to (14) in state-space form is preferred for writing computing programs. The following equations (15) to (24) are in the form of ready to use for coding in the Julia Programing Language

$$X_1' = X_5 \qquad (15)$$

$$X_2' = X_6 \qquad (16)$$

$$X_3' = X_7 \qquad (17)$$

$$X_4' = X_8 \qquad (18)$$

$$X_5' = -\frac{D}{\Omega}X_5 - \frac{K_1}{\Omega^2}(X_1 - K_a X_3) + EX_9 \qquad (19)$$

$$X_6' = -\frac{D}{\Omega}X_6 - \frac{K_1}{\Omega^2}(X_2 - K_a X_4) + EX_{10} - \frac{1}{\Omega^2} \qquad (20)$$

$$X_7' = -\frac{K_a K_2}{\Omega^2}(K_a X_3 - X_1) + \frac{m_q \Gamma}{\Omega m_2} F_3 \qquad (21)$$

$$X_8' = -\frac{K_a K_2}{\Omega^2}(K_a X_4 - X_2) + \frac{m_q \Gamma}{\Omega m_2} F_4 - \frac{1}{\Omega^2} \tag{22}$$

$$X_9' = X_9 - X_{10} - (X_9^2 + X_{10}^2)X_9 \tag{23}$$

$$X_{10}' = X_9 + X_{10} - (X_9^2 + X_{10}^2)X_{10} \tag{24}$$

5. Results and Discussion

The results obtained from the preliminary analysis of numerical bifurcation of the unbalanced rotor-journal bearing system are shown. In general, non-autonomous nonlinear systems can be represented as $\boldsymbol{X}' = \boldsymbol{F}(\tau, X_i; \Omega)$ where Ω is the bifurcation parameter. It is possible for some non-autonomous systems to express as $\boldsymbol{X}' = \boldsymbol{F}(X_i; \Omega)$ and our system depicted in the equations (15)-(24) is one of the cases. We have chosen the non-dimensional rotational speed Ω as the bifurcation parameter in this investigation. In the process of bifurcation investigation, all X_i are treated as coordinates and the others considered as parameters changing one at a time. The (X_1, X_2) and (X_3, X_4) are of interest because they illustrate the orbits of the rotor and the journal centers. The physical values of the system to yield the parameters are $a = b = 0.4\,m$, bearing length $(l_B) = 0.01\,m$, shaft diameter $(d_s) = 0.03\,m$, $E_Y = 200 \times 10^9 N/m$, $r_B = 0.02\,m$, $c = 1.0 \times 10^{-4}\,m$, $\mu = 50 \times 10^{-3} Pa \cdot s$, $m_1 = 5\,kg$, $m_2 = 1\,kg$, $g = 9.81\,m/s^2$ and $d = 50\,Ns/m$. These values yield the parameters of $D = 0.03192754$, $K_1 = 0.37997689$, $K_2 = 1.89988446$, and $\Gamma = 0.1451$. In practice, the bearing's characteristic number can be changed independently since it depends upon the chosen oil viscosity. All numerical results of the orbits are coded in the Julia Programing Language.

We firstly consider the absence of the unbalanced eccentricity $e_U = 0 = E$. As shown in Figure 3, the equilibrium positions depend on the rotational speeds. It is apparent from this figure that the solid line means stable equilibrium points while the dashed line represents unstable equilibrium points. From this data, we can see that if one starts at some initial conditions at the speeds lower than the critical speed, then the orbits will converse to the equilibrium points at that speed. We now search for the critical speed or the bifurcation point by analyzing the Jacobian matrix of the equations (15) to (22) by setting $X_9 = X_{10} = 0$. They have been implemented on MATLAB and the bifurcation value is $\Omega = 3.0091$ $(\omega = 9{,}100\,rpm)$. The selected orbits at the speed of $\Omega = 6.6869$ $(\omega = 20{,}000\,rpm)$ being in bifurcation regime are shown in Figure 4. The initial conditions used to generate results in this study start with equilibrium points. One should keep running the computation until the orbits reach the limit cycles.

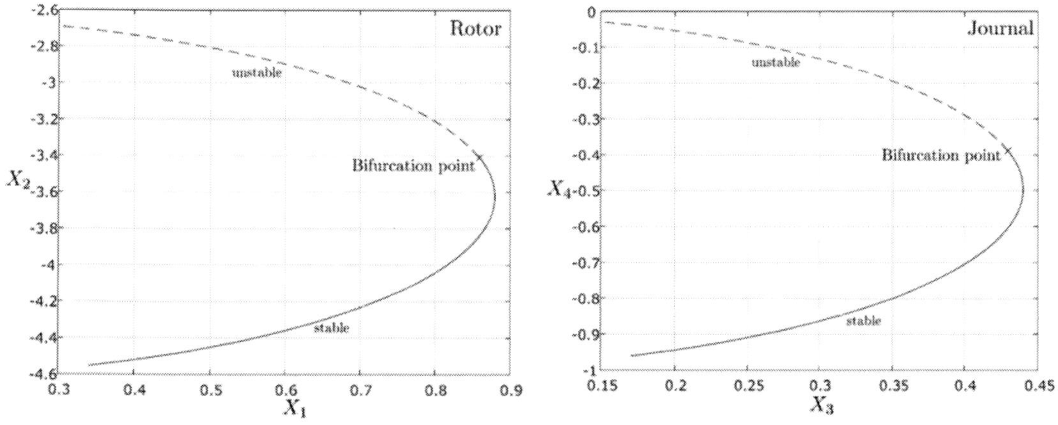

Figure 3. Equilibrium positions of centers of the rotor and journal at various speeds

IOP Conf. Series: Materials Science and Engineering **886** (2020) 012018 doi:10.1088/1757-899X/886/1/012018

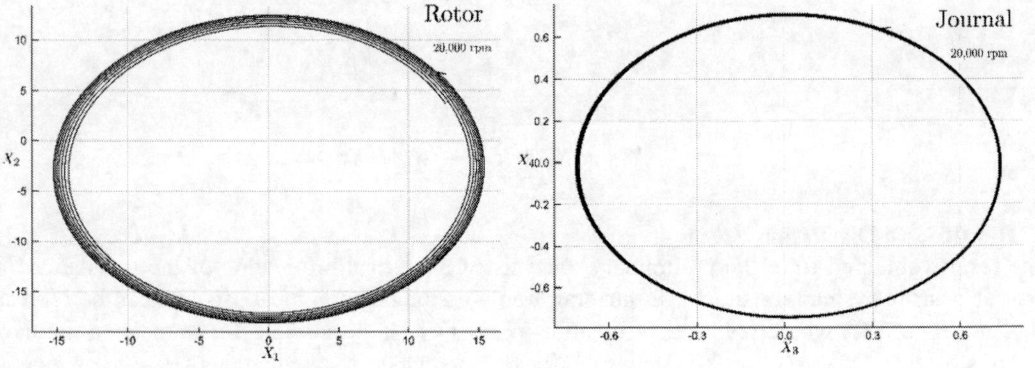

Figure 4. The bifurcations of the rotor and journal at rotational speed of $20{,}000\ rpm$ without unbalanced eccentricity

In the presence of unbalanced mass at the rotor with the unbalanced eccentricity $e_U = 0.002\ m$, the selected results at the rotor were illustrated in Figure 5 and Figure 6. All the selected results herein start to compute from equilibrium points at the corresponding rotational speeds. We have picked, for instance, the orbits at the rotor center to illustrate. In Figure 5, rotational speed is of $\Omega = 0.0167\ (50\ rpm)$ below the bifurcation value from the previous case but it reaches into the limit cycle at a very early time ($\tau = 100$). Figure 6 compares the rotational speed is higher than the bifurcation point of the absence of the unbalanced case at the speed of $\Omega = 4.7142\ (\omega = 14{,}100\ rpm)$. The latter case in Figure 6 shows more complex with higher amplitude of the orbit during the transient response. It eventually arrives the limit cycle with bigger amplitudes than the unbalanced case and it is not shown here.

.

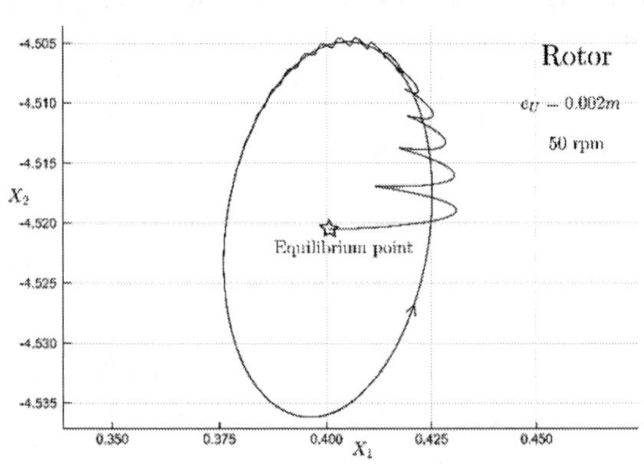

Figure 5. Limit cycle of the rotor with e_U=0.002 m at the speed of 50 rpm

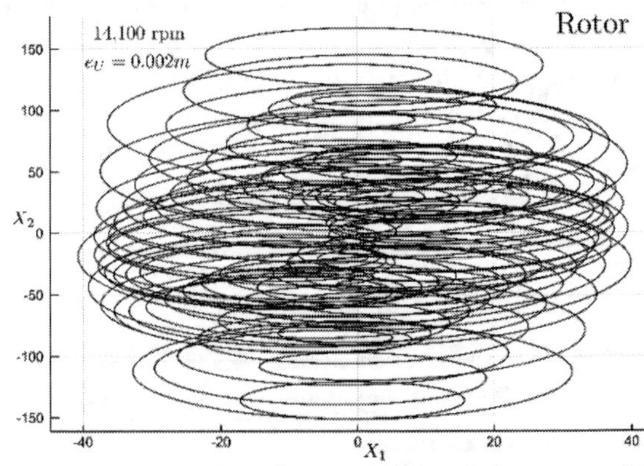

Figure 6. Transient response of the rotor with e_U=0.002 m at the speed of 14,100 rpm

6. Conclusion

The aim of the present research was to examine an overhung rotor-journal model with nonlinear fluid film forces. his project was undertaken to formulate the governing equation and arrange a non-dimensional form that suited for computation. In this study, the results of the analysis of the numerical bifurcation are also represented. Regarding the absence of an unbalanced eccentricity ($e_U = 0$), it revealed that the bifurcation value is found at $\omega = 9{,}100\ rpm$. The limit cycle after the bifurcation point at $\omega = 20{,}000\ rpm$ is investigated. In presence of unbalance, we demonstrate both results of higher and lower speeds of bifurcation point. This study has identified the occurrence of the limit cycle at a very low speed of ω=50 rpm well below the bifurcation point of the previous case. At high speed of $\omega = 14{,}100\ rpm$ starting from the equilibrium point, it encounters more complex orbits with large amplitude. The contribution of this study has been to confirm numerical bifurcation simulation is a valuable tool to design and predict a behavior of rotor dynamics. Further research should be carried out to establish the behavior of the overhung journal-rotor system.

References

[1] Kramer E 1993 *Dynamics of rotors and foundations* (Berlin : Springer-Verlag)
[2] Adiletta G Guido A R and Rossi C 1996 *Nonlinear Dynam.* **10**(3) 251-269 DOI: 10.1007/BF0004
[3] Hashimoto H Wada S and Ito J 1987 *J. Tribol.-T ASME* **109**(2) 307-314 DOI: 10.1115/1.3261357
[4] Wang Y L Liu Z S Kang W J and Yan J J 2011 *J. Mech. Eng. Sci.* **226**(5) 1345-1355 DOI: 10.1177/0954406211418302
[5] Childs D W 1982 *J. Eng. Power-T ASME* **104**(3) 533-541 DOI: 10.1115/1.3227
[6] Prashad H 1988 *Tribol. T,* **31**(2) 303-309 DOI: 10.1080/10402008808981827
[7] Zilli A Williams R J Ewins D J 2015 *J. Eng. for Gas Turb. Power* **137**(6) 0605001(1-10) DOI: 10.1115/1.4028844
[8] Miura T Inoue T and Kano H 2015 *J. Vib. Acoust.* **139**(3) 031012(12) DOI:10.1115/1.4036098

IOP Conf. Series: Materials Science and Engineering **886** (2020) 012019 doi:10.1088/1757-899X/886/1/012019

Numerical Simulation of Skin Formation: Modeling of Scale Peeling

Shu Oba[1] and Katsuya Nagayama[1,*]

[1]Kyushu Institute of Technology Graduate School, 680-4, Kawazu, Iizuka, Fukuoka, 820-8502, Japan

* Corresponding Author: nagayama@mse.kyutech.ac.jp, Tel&Fax 81-948-29-7778

Abstract
Skin is formed by cell division, and the cells eventually mature into corneums and drop off. This process is known as turnover. With normal turnover, skin disease rarely occurs. However, the disturbance of turnover leads to skin inflammation and associated effects such as scale formation. The mechanisms that underpin scale formation are not fully understood. This study analyses the impact of scale formation in atopic dermatitis using a numerical simulation. The numerical simulation model of the skin was constructed by theoretically setting the main parameters i.e. transepidermal water loss (henceforth *TEWL*) of atopic dermatitis, thereby allowing the elucidation of the mechanism of scale formation. As a result, thick corneum and scale of atopic dermatitis were successfully modeled. In addition, the variation of desmosome distribution in stratum corneum was shown to be associated with scale formation.

1. Introduction

There is growing interest in the beauty field and skin disease regardless of gender. The prevalence of skin disease is rising and is associated with an increase in active research projects [1]. For example, atopic dermatitis is a disease whose cause and mechanism have not been completely elucidated. Besides, an established treatment that leads to the cure of the patient is unavailable; most strategies focus on coping therapies. There is an urgent need to address the clarifying of the mechanism.

Skin is formed by the proliferation and the differentiation of cells, which mature into corneum and drop off. The elucidation of the formation process would provide insights into numerous other phenomena related to skin. Despite the importance of observing the formation process inside the skin, there remains a paucity of evidence on skin formation.

The main aim of this study is to investigate the numerical simulation model of skin formation using a particle method. The research data in this thesis is drawn from the analysis method under a range of experimental conditions. In 2018, the analysis method made it possible to change corneum thickness according to skin environment (e.g., *TEWL*) [2]. In our study, the developed model (verified in the past [3], [4]) was used to elucidate the mechanism of scale formation in cases of atopic dermatitis.

IOP Conf. Series: Materials Science and Engineering **886** (2020) 012019 doi:10.1088/1757-899X/886/1/012019

2. Analysis target

Skin can be roughly divided into three layers: subcutaneous tissue, dermis, and epidermis. Figure 1 shows a cross-sectional view of the skin. The epidermis is used in the analysis and is composed of the basal layer, stratum spinosum, granular layer, and stratum corneum. Skin cell proliferation occurs daily in the basal layer, followed by migration through the stratum spinosum, granular layer, and stratum corneum. The cells eventually mature into corneum and drop off. This whole process is known as turnover.

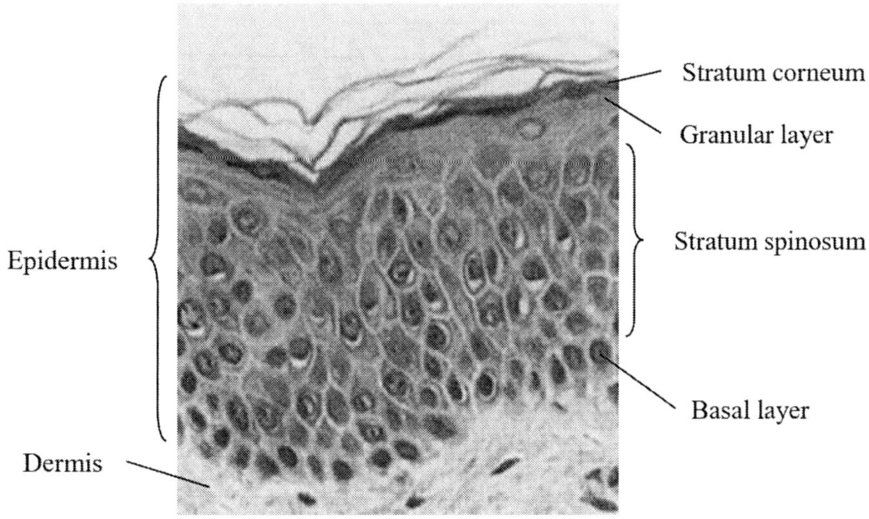

Figure 1. A sectional view of the skin [5].

3. Methods

3.1 Calculation model

3.1.1 Particle method [6]

Particle method was used in this model. This method does not depend on mesh because it does not use mesh and treats calculation point as a moving particle. Each particle was regarded as a single cell.

Body and spring forces work between particles. The balance of these forces contributes to particle movement.

3.1.2 Free water and combined water

The skin is combined with two types of water: free water and combined water. The free water can move free in the skin. In contrast, the combined water cannot move freely in the skin because of the combination with corneum cells.

3.1.3 Water concentration and peeling calculations [2]

Water concentration and peeling calculations were introduced to each particle.

Skin water (distinguished from sweat) evaporates through stratum corneum; this phenomenon is called the transepidermal water loss (*TEWL*). It depends on skin condition (i.e., dry or moist). *TEWL* is given as follows:

$$TEWL = \frac{c_0}{\dfrac{l}{D} + \dfrac{1}{h(1-\mu)}} \tag{1}$$

c_0 is the free water concentration at the deepest level of the stratum corneum. l is the corneum thickness. D is the water diffusion coefficient. h is the evaporation coefficient and μ is the humidity. Equation (1) was derived from the relationship between *TEWL* and corneum thickness by our lab in 2018, referring to the problem of one-dimensional steady heat transfer. The skin free water concentration is given as follows (see also equation 1).

$$\frac{dc}{dx} = -\frac{TEWL}{D} \tag{2}$$

In equation 2, c is represented the free water concentration and x is the position from the deepest level of the stratum corneum to the stratum corneum surface.

Corneum cells are attached by proteins called desmosome, which is generated simultaneously with cornification and must be degraded for peeling to occur. The activity of the enzyme that degrades desmosome are water-dependent [7]. As can be seen from the equation 2 above, the desmosome concentration can be given as follows.

$$\frac{dDesmo}{dt} = -\frac{2.04}{7}\left\{r_1\left(\frac{c}{c_0}\right) + r_2\right\} \tag{3}$$

In equation 3, *Desmo* presents the desmosome concentration. While t is the time, and r_1 $(= c_0 / (1 \times 10^{-12}))$ is the free water concentration at the deepest level of the stratum corneum. And, r_2 is the combined water concentration.

3.1.4 Scale model [8], [9]

The mechanism of scale formation has not been clarified. The solution was then assayed for the mechanism of scale formation using the scale model for calculating scale formation method. As shown in Figure 2, desmosome concentration and variation were chosen in this study. In the case of atopic skin, corneum particles attached on the corneum particle that desmosome is completely degraded are peeled together because desmosomes are more likely to be distributed (see Figure 3 below).

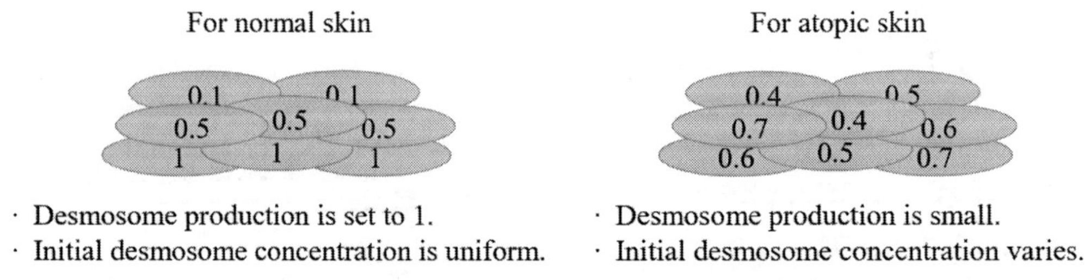

Figure 2. Comparation of normal skin and atopic skin.

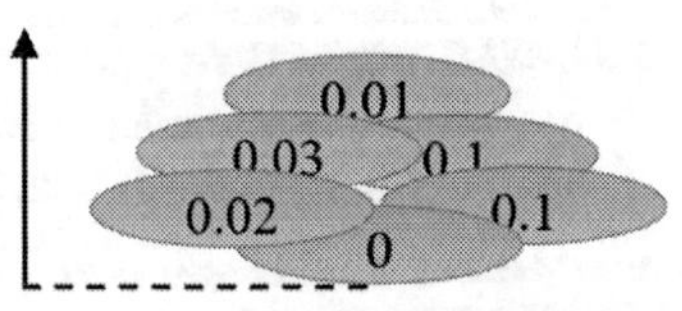

Figure 3. Scale formation mehod.

4. Analysis conditions

4.1 Initial setting

It can be seen from the data in Figure 4 that the analysis area is 300 × 300 × 300 μm^3. The initial placement consists of blood vessels; dermis and basal layer from the bottom (see Figure 4). The boundary condition is a periodic boundary condition. Each particle is spherical, with a diameter of 10 μm. The shape of the particles changes in the layer above stratum spinosum; their thickness decreases, c_0, r_1, r_2 and μ are set to 0.4 × 10^{-12} g/μm^3, 0.4, 0.2 and 0.6, respectively.

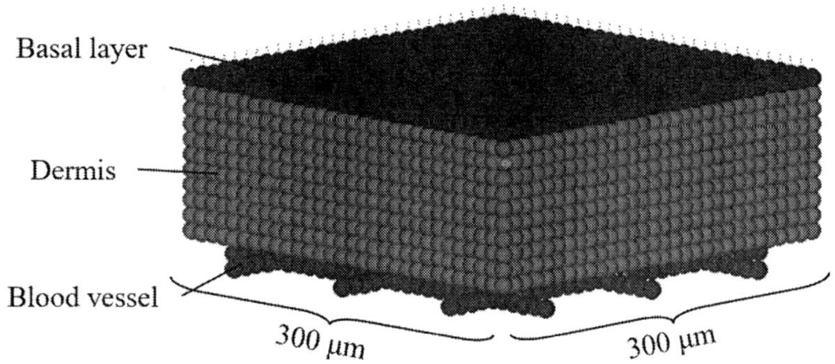

Figure 4. The initial placement of the analysis model.

4.2 Parameters

In this analysis, the patterns of both atopic skin and normal skin were modeled and compared. Table 1 summarizes the parameters of the two skin conditions. The basal cells of atopic skin divide at twice the rate of those of normal skin [10]. Water diffusion coefficient D in atopic skin is 10 times larger than that in normal skin because of the defect in skin barrier function. Evaporation coefficient h in atopic skin is larger than that in normal skin because of dry skin. For these reasons, we have set out D and h properly. Initial desmosome concentration [8] and variation [9] in atopic skin are less and larger than those in normal skin respectively. The results of the analysis are summarised based on the following conditions in Table 1.

Table 1. Setting of parameters.

	Normal skin	Atopic skin
Division frequency [/day]	0.2	0.4
Water diffusion coefficient D [μm^2/s]	0.1	1.0
Evaporation coefficient h ×10^{-4}[μm/s]	0.268	3.95
Initial desmosome concentration	1	0.5 ~ 0.8

5. Results of analysis, and discussion

Figure 5 confirms that a thick stratum corneum is formed and sections of corneums are scattered (referred to as scale).

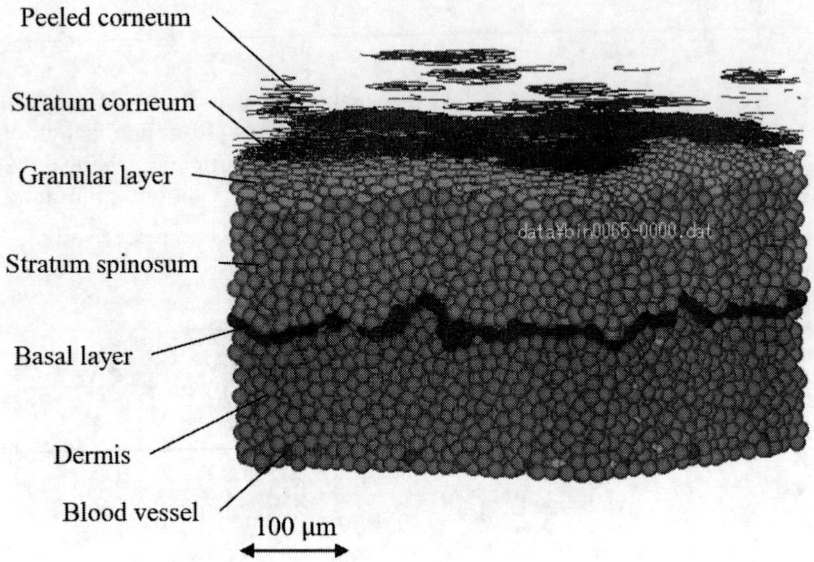

Figure 5. Analysis image of atopic skin.

Figure 6 confirms that more water is lost from atopic skin than normal skin due to the former's drier property. Also, the results obtained from the analysis revealed that stratum corneum is thicker in atopic skin than normal skin. Atopic skin is lack of ceramide, which is an inter-keratinocyte lipid with a role in water retention. It is apparent from the figure that low levels of this lipid interfere with skin barrier function, leading to water loss and dry skin in individuals with atopic skin. This phenomenon corresponds to the results shown in Figure 6 below.

As can be seen in Figure 6, the plot of water loss versus time is undulating because the skin is repeatedly drying and wetting. It also shows an inverse relationship to the thickness of the stratum corneum, as shown in Figure 7. There is an interaction between the skin condition (dry or wet) and the corneum thickness and is well-documented. The balance of the skin can be maintained by adjusting the barrier function according to the external environment. As such, the results of this analysis are consistent with the real skin phenomenon.

The results, as shown in Figure 8, indicate that there is a relationship between the variation of desmosome distribution in stratum corneum and scale formation. It also demonstrates that the greater the variation of desmosome distribution in stratum corneum is, the higher the appearance rate of thick scale increases. From the obtained results, we can see that the skin is peeled easily. It is interesting to note that the variation of desmosome distribution is related to scale formation.

IOP Conf. Series: Materials Science and Engineering **886** (2020) 012019 doi:10.1088/1757-899X/886/1/012019

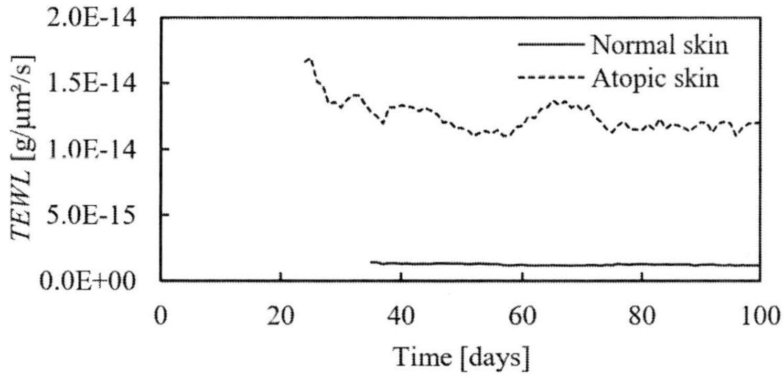

Figure 6. Transepidermal water loss (*TEWL*).

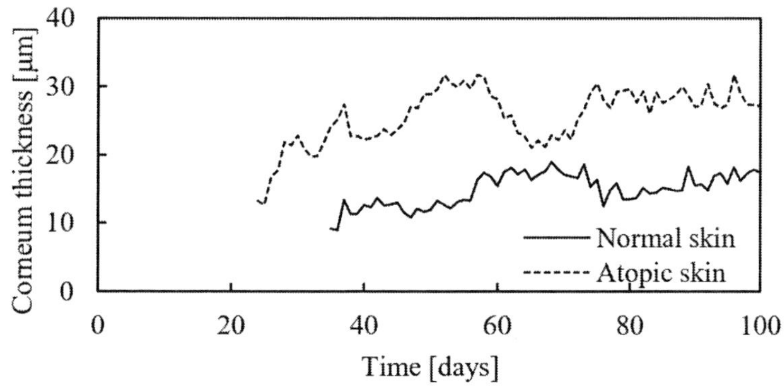

Figure 7. Corneum thickness.

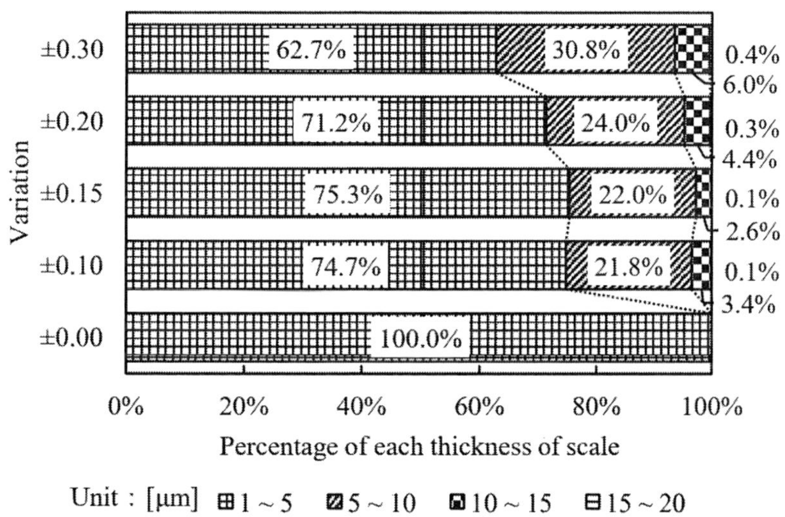

Figure 8. Relationship between the variation of desmosome distribution and scale formation.

6. Conclusion

In this investigation, the aim was to model atopic dermatitis with a particle method. Regarding the analysis of the skin, the free water, combined water, water concentration calculation and peeling calculation were introduced in this study. The scale model (constructed by the relationship between desmosome concentration and variation) were introduced in the analysis model of atopic dermatitis. Suitable parameters for normal skin and atopic skin analyses were theoretically set.

The evidence from this study suggests that the introduction of free water and combined water can be contributed to the analysis of real skin which is closely related to real skin. Water concentration and peeling calculations brought the real skin turnover phenomenon. These data suggest that it was possible to reproduce the formation of thick stratum corneum and scale. The findings of this investigation on comparison of atopic skin and normal skin (e.g., *TEWL* or corneum thickness) show the degree of dryness of atopic skin. These findings have significant implications for the understanding of how the variation of desmosome distribution in stratum and corneum scale formation are associated.

References

[1] Nagayama M A 2004 *Physica D Nonlinear Phenomena* **194**(3-4) 151-165
[2] Shobuda K and Nagayama K 2018 *Proc. 9th TSME Int. Conf. on Mechanical Engineering (Phuket)*
[3] Nagayama K Uehara T Amano Y and Tanahashi M 2015 *J. Biosciences and Medicines* **3** 45
[4] Nagayama K Matsuoka S Morisaki N and Taguchi H 2015 *J. Biosciences and Medicines* **4** 1
[5] Saita T 2009 *Introduction to dermatopathology diagnosis*, 2nd edition (Japan:Nankodo) 2
[6] Kurihara T and Nagayama K 2016 *Proc. 7th TSME Int. Conf. on Mechanical Engineering (Chiang Mai)*
[7] Koyama J 1999 *J. SCCJ.* **33** 16
[8] Lee U H Kim B E Kim D J Cho Y G Ye Y M and Leung D Y M 2017 *Br. J. Dermatol.* **176** 537
[9] Igawa S Kishibe M Honma M Murakami M Mizuno Y Suga Y Seishima M Ohguchi Y Akiyama M Hirose K Yamamoto A I and Iizuka H 2013 *J Dermatol Sci.* **72** 54
[10] Watanabe M Tagami H Horii I Takahashi M and Kligman A M 1999 *Arch. Dermatol.* **127** 1689

IOP Conf. Series: Materials Science and Engineering **886** (2020) 012020 doi:10.1088/1757-899X/886/1/012020

Influence of the Wavelength of Cut-In Sinusoidal Trailing Edge Shape to the Aerodynamics Characteristic of the Airfoil

P Rattanasiri[1,*] and R Perrin[2,3]

[1]Department of Mechanical Engineering, Faculty of Engineering, Burapha University, Chon Buri, 20130, Thailand
[2]Department of Aeronautical Engineering, International Academy of Aviation Industry, King Mongkut's Institute of Technology Ladkrabang, Bangkok 10520, Thailand
[3]Department of Mechanical Engineering, Faculty of Engineering at Sriracha, Kasetsart University Sriracha Campus, Sriracha, Chonburi 20230, Thailand

* Corresponding Author: pareecha@eng.buu.ac.th

Abstract
The influence of the cut-in sinusoidal trailing edge shape with different wavelengths on the aerodynamics characteristic has been parametrically investigated by numerical unsteady RANS simulation, open-source code; Code_Saturne. The results were compared with the benchmark baseline and blunt profile trailing edge shape. The geometry of NACA0012, NACA4412 and NACA4415 airfoil with a small modification to obtain a zero thickness trailing edge are selected as a baseline profile. The blunt trailing edge is a cut-offs at the trailing edge for 10% of the chord. Three wavelengths of sinusoidal trailing edge shape at 0.25c, 0.50c and 0.75c with 0.05c amplitude are selected, where *c* is the airfoil chord length. The flow is studied at high Reynolds numbers (*Re*) 10^6 for the angle of attack 5 degrees. The results show the change in lift and drag characteristics with changing of NACA profiles and the modified trailing edges.

1. Introduction

There is evidence that the turbulent boundary layer (henceforth TBL) trailing edge noise is one of many causes of an airfoil self-noise [1-2]. It is now well established that a modification of the trailing edge geometry could reduce this noise [2-6]. The main challenge faced by many researchers is the NACA0012 with cut-in sinusoidal trailing edge shape at "Re" =5000 and the angle of attack 5 degrees provides 10% more of the lift force compare with the baseline trailing edge shape. However, it may cause the drag penalty of 2.5% [7].

The specific objective of this study was to extend the investigation of an airfoil with cut-in sinusoidal trailing edge shape in ref [7] at the high *Re*. Cross-sectional profiles were NACA0012, NACA4412 and NACA4415. The trailing edge shape was a cut-in sinusoidal shape. The first aim is to understand the aerodynamics characteristics due to the changes of wavelengths of the sinusoidal shape. The second one is to investigate the influence of camber on the flow. Finally, this study set out to assess the effect of thickness of asymmetric airfoil's profile to the flow.

2. Theoretical approach

2.1 Total drag prediction

Physically, the aerodynamic forces are calculated by the pressure (P) and wall shear (τ_w). If the total lift force and the total drag force on the airfoil are predicted, the lift and drag coefficient can then be estimated by C_L = (Total lift) / (0.5 $\rho V^2 sc$) and C_D = (Total drag)/(0.5 $\rho V^2 sc$), where ρ is the fluid density, c is the aerofoil's chord, s is the span.

The airspeed (V) is considered in term of Reynolds number, $Re = (\rho VL)/\mu$, which based on airfoil characteristic length (L) and where μ is the fluid viscosity.

To predict the accurate aerodynamic forces of the flow pass fully immersed body, Reynolds Averaged Navier Stokes (RANS) simulation has proved to provide a reasonably accurate result.

2.2 RANS

By assuming the flow is incompressible, the continuity equation and the momentum equation can be written as:

$$\frac{\partial \overline{U_i}}{\partial x_i} = 0 \tag{1}$$

$$\rho\left(\frac{\partial \overline{U_i}}{\partial t} + \frac{\partial \overline{U_i U_j}}{\partial x_j}\right) = -\frac{\partial \bar{P}}{\partial x_i} + \frac{\partial}{\partial x_j}\left\{\mu\left(\frac{\partial \overline{U_i}}{\partial x_j} + \frac{\partial \overline{U_j}}{\partial x_i}\right)\right\} - \rho\frac{\partial \overline{u_i' u_j'}}{\partial x_j} \tag{2}$$

where i is Cartesian coordinates in X, Y and Z and U_i are the Cartesian mean velocity components (U_x, U_y, U_z). The Reynolds stress tensor ($\overline{\partial u_i' u_j'}$) is represented in the turbulence closure.

2.3 Code Saturne

The Code Saturne has been proposed which can solve the Navier-Stokes equations for various types of flow. This open-source software is based on a co-located Finite Volume approach and SIMPLEC type algorithm that accepts three-dimensional meshes. It can be built on any type of cell (tetrahedral, hexahedral, prismatic, etc.). The ready-to-used graphic user interface and capability for parallel code coupling make it simple for any user. The code is selected and validated by many references [8-10]. The unsteady solver is selected in this work and a RANS-Shear Stress Transport (SST) turbulence model with wall function, implemented in the Code_Saturne is employed. Second, the order linear upwind (SOLU) scheme is selected to discrete convection terms. The time step is 0.01 s with at least 20000 iterations. The selected criterion for convergence is RMS residual, which is less than 10^{-8}. The run type used in this study is Intel CORE i7, 4 logical cores, with 2GB RAM.

3. Numerical modelling

3.1 Airfoil modelling

As shown in Figure 1, the profile of NACA0012, NACA4412, NACA4415 is modelled for 1 m chord length (c). To obtain a zero thickness trailing edge (TE), the geometry with small modification is chosen as a baseline profile, as shown in Figure 2 (a). As can be seen from Figure 2 (b), the blunt TE is selected by cutting the baseline profile at 0.9c from the leading edge. Figure 2 (c) provides the sinusoidal equation that is implemented into the trailing edge to build the cut-in sinusoidal TE.

$$\bar{x}(z) = c - h - h\cos\left(\frac{2\pi}{\lambda}z\right) \tag{2}$$

As shown in the equation illustrated above, the symbolic variables can be defined as: c is the chord length, h is the waviness amplitude and λ is its wavelength, $\bar{x}(z)$ is the physical coordinate in the chord direction (x-axis), and z is the spanwise coordinate in the computational domain. While s is the span-wise length of the computational domain. Table 1 provides a list of TE's parameter used in this study.

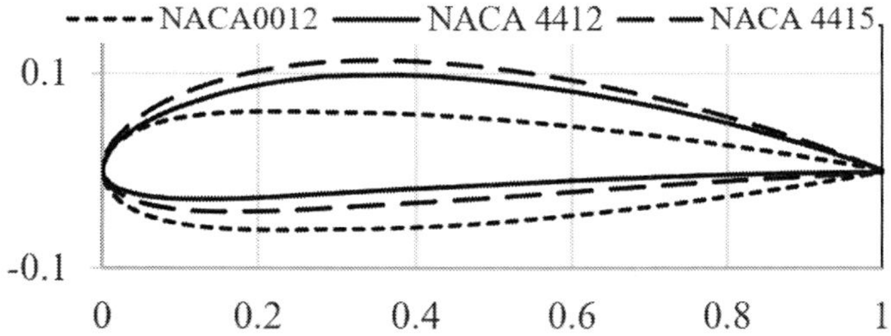

Figure 1. NACA0012, NACA4412 and NACA4415.

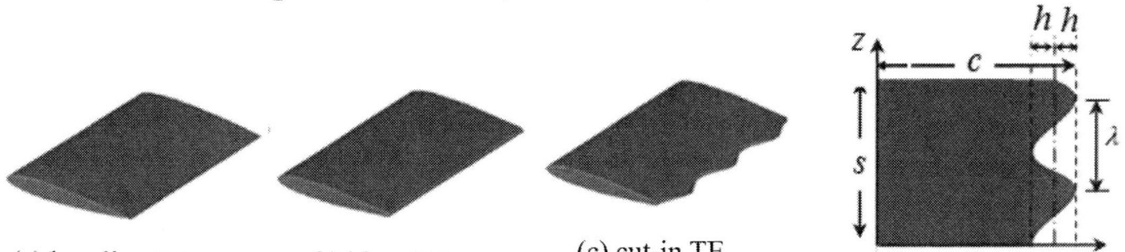

(a) baseline TE (b) blunt TE (c) cut-in TE (d) Sinusoidal parameters

Figure 2. Airfoil with baseline, blunt and cut-in sinusoidal trailing edge shape.

Table 1. Parameters of the sinusoidal trailing edge.

Case	baseline	blunt	L25	L50	L75
s/c	0.125	0.125	0.125	0.250	0.375
λ/c	0	0	0.25	0.50	0.75
h/c	0	0	0.05	0.05	0.5

3.2 Model domain and boundary condition

The fluid domain extends up to $20c$ and $20c$ in the upstream and downstream direction (x), respectively. And it can be extended up to s/c in the span-wise direction (z) (see Table 1). Symmetry conditions are imposed at both side boundaries. The flow inlet velocity (V) used in this experiment is 1 m/s with 1% turbulent intensity. A fluid density (ρ) is a 1 kg/m^3, and dynamics viscosity (μ) is 10^{-6} Pa.s. It consists of the chord $Re = 10^6$. An aerofoil surface is implemented with smooth wall condition.

An appropriate mesh strategy and mesh resolution are set to capture the effect of the boundary layer and wake flow. Consequently, the result obtained showed high fidelity. The sample of domain mesh is presented in Figure 3. As Figure 4 shows, there is an isometric view of the structure mesh of airfoil and side meshes. A structured mesh with local refinement is built around aerofoil and in the wake regions. The study included the refinement at the walls ($y^+ \approx 30$) in accordance with the selected model so-called *2 scales (log law) wall function* that is also implemented in the Code Saturne.

4. Result

4.1 Mesh convergence

To assess the accuracy of the numerical simulation, the mesh convergence was used. Figure 5 shows the result of mesh convergence. Due to the simplicity of structured mesh, the convergence of meshing from coarse (50k), medium (100k), medium-fine (150k) to the fine mesh (200k) can be obtained for

IOP Conf. Series: Materials Science and Engineering **886** (2020) 012020 doi:10.1088/1757-899X/886/1/012020

C_L and C_D. Regarding the accuracy and time consumption prospect, the medium mesh was selected in this study.

Assessment of the results for the baseline TE was carried out using a comparison of XFoil's estimation and experimental data from the references [11] and [12].

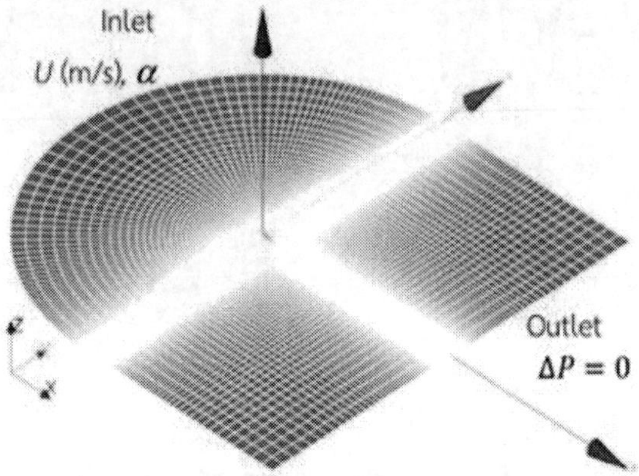

Figure 3. Fluid domain and boundary conditions

(a) airfoil and the side view of peak wall

(b) airfoil and the side view of trough wall

Figure 4. Isometric view of structure mesh in simulation domain

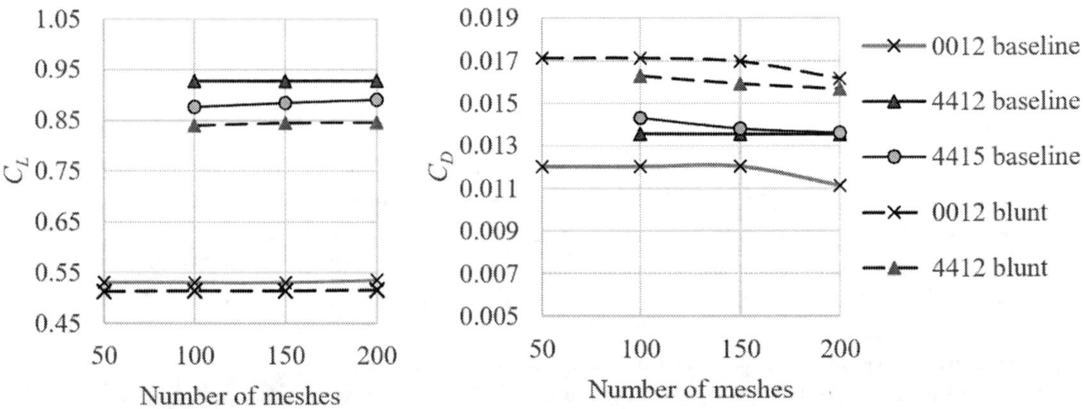

Figure 5. Mesh convergence C_L and C_D of airfoil with baseline and blunt TE.

4.2 Interpretation of forces results

The chord length (c) is an important parameter for estimating of C_L and C_D. C_L and C_D. It can be scaled in two ways. The first one is to scale by using the physical chord length c = 1 m, 0.95 m and 0.9 m for baseline, cut-in and blunt TE, respectively. Second, it can be carried out by using the baseline chord length c = 1 m.

These results were found to be different. This inconsistency may be different scaling. This result may be explained by the fact that the ratio of lift and drag coefficient (C_L/C_D) for both cases are the same (see Figure 6). For example, the NACA 4412 was found to be better performance than 4415 and 0012. The results obtained from C_L/C_D show a similar trend for modified TE. The finding showed that the lowest and the highest of C_L/C_D are NACA with blunt TE and baseline TE. It is possible; therefore, that the wavelength has no effect on C_L/C_D of NACA with cut-in sinusoidal TE.

Table 2 presents the percentage differences of lift coefficient ($\%C_L$) and drag coefficient ($\%C_D$) of modified TE airfoils compared to baseline TE airfoil using both scalings. The equations are as follows:

$$\%C_L = \frac{C_{L,modified\,TE} - C_{L,baseline}}{C_{L,baseline}} \times 100\%, \quad \%C_D = \frac{C_{D,modified\,TE} - C_{D,baseline}}{C_{D,baseline}} \times 100\%$$

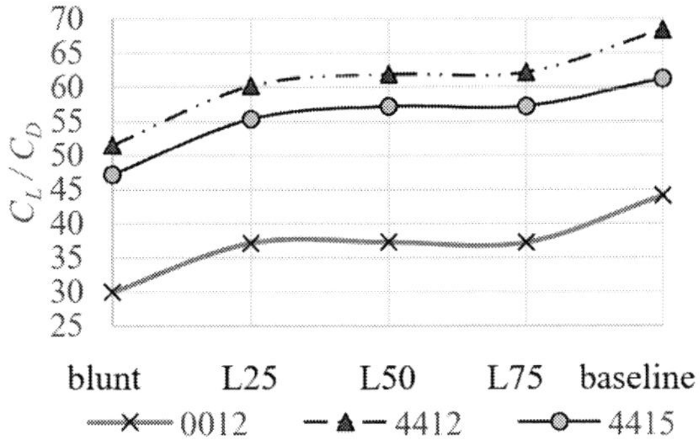

Figure 6. C_L/C_D

Table 2. $\%C_L$ and $\%C_D$ for modified TE compared with baseline.

	$\%C_L$ *(using physical chord lengths)*			$\%C_D$ *(using physical chord lengths)*			$\%C_L$ *(using baseline chord length)*			$\%C_D$ *(using baseline chord length)*		
	0012	4412	4415	0012	4412	4415	0012	4412	4415	0012	4412	4415
blunt	7.8	0.6	8.1	58	33	40	-3.0	-9.5	-2.7	42	20	26
L25	4.5	-0.4	5.3	24	13	16	-0.7	-5.4	0.0	18	7	11
L50	4.4	-0.1	4.1	23	10	11	-0.8	-5.1	-1.1	17	5	6
L75	4.6	-0.1	4.4	24	10	12	-0.6	-5.1	-0.8	18	4	6

Note: 'Minus' sign of $\%C_L$ means the decreasing of lift when compared with baseline TE.

4.3 Scaling with physical chord lengths

In this section, C_L and C_D of airfoils have been scaled by the physical chord lengths, c = 1 m, 0.95 m and 0.9 m for baseline, cut-in and blunt TE, respectively (see Figure 7). Table 2 shows $\%C_L$ and $\%C_D$ for modified TE compared with baseline (see column 2 -7 above).

4.3.1. Blunt and baseline TE

By comparing the blunt TE with baseline, the results of 0012 revealed that lift increases by 7.8% and drag penalty up to 58%. The effect of the trailing edge shows a significant difference between the symmetric airfoil (0012) and the airfoil with camber (4412). In contrast, it shows less drag penalty (25%) for 4412 than that of 0012.

The results of the baseline TE suggest that the airfoil 4412 provides more drag than 0012 (with the bunt TE). On the other hand, the airfoil 4412 provides less drag than 0012.

The 8.1% increasing of lift and 44% of drag penalty for 4415 shows that the airfoil with camber (44xx) and the thicker airfoil 4415 provide higher lift than that of 4412 except for higher drag penalty.

4.3.2. Influence of cut-in sinusoidal trailing edge shapes

To compare the cut-in sinusoidal TE with a baseline for 0012, all wavelengths were found to be a similar trend of increasing of lift about 4.4% to 4.6% with the drag penalty 23% to 24%. An implication of this is the possibility that the wavelength does not effect for 0012.

Regarding 4412, all wavelengths showed a similar trend of lift decreasing 0.1% to 0.4% with the drag penalty 10% to 13%. This result may be explained by the fact that the influence of cut-in sinusoidal TE can be observed from comparing 0012 and 4412. We can see that the effect of camber is almost vanished for drag. Both NACAs has similar drag.

For 4415, the results show the influence of wavelength. While there is 5.3% lift increment, and 16% drag penalty for L25. The data showed similar results for L50 and L75, increasing of lift by 4.1% to 4.4% and increasing of drag by 11% to 12%. To compare 4412 and 4415, the effect of the cut-in sinusoidal trailing edge is used for the thicker airfoil, with a small increase of drag penalty.

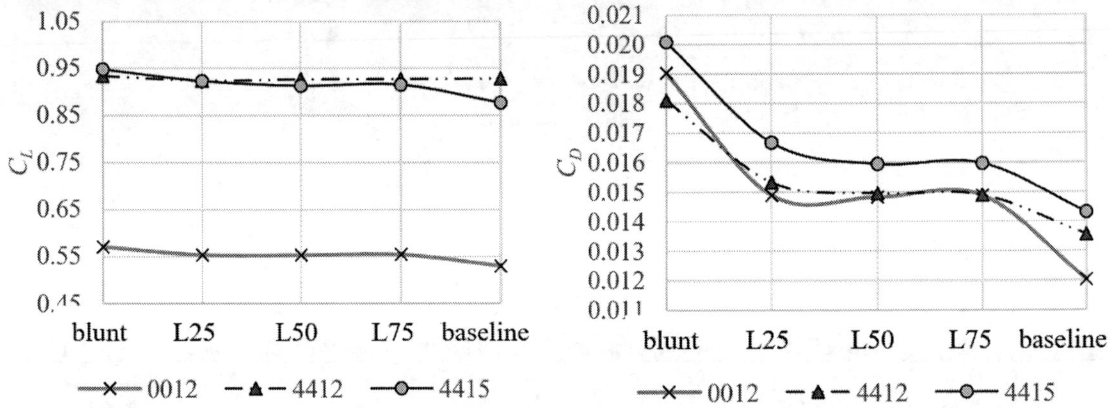

Figure 7. C_L and C_D scaled by c = 1, 0.95 and 0.9 m for baseline,

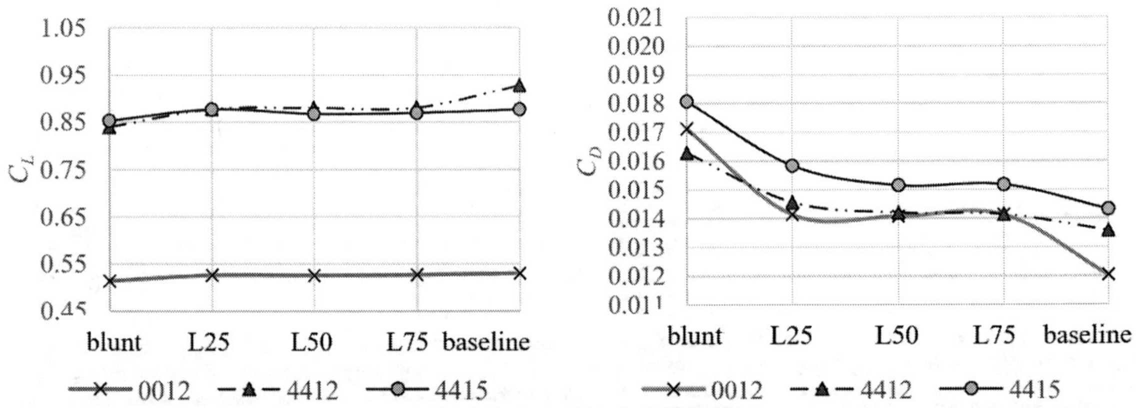

cut-in and blunt TE, respectively

Figure 8. C_L and C_D scaled by baseline chord length c = 1 m.

4.4 Scaling with baseline chord length c = 1 m

As can be seen from Figure 8, it provides the results obtained from C_L and C_D of the airfoil which scaled by c = 1 m. The $\%C_L$ and $\%C_D$ is also represented in Table 3 (Columns 8-13).

As shown in Table 3, the lift coefficients of the modified TEs are higher than the baseline case when the physical length (c = 1, 0.95 and 0.9 m. for baseline, cut-in and blunt TE, respectively) is used (see columns 2-4). It can thus be suggested that it is lower when c = 1 m is used (columns 8-10).

Regarding the comparison of the blunt TE and the baseline TE, the lift is reduced more than 2.7% for all profiles. While comparing cut-in sinusoidal TE with baseline, all wavelengths showed a similar trend of decreasing of lift and was less than 1.1% for 0012 and 4415 and it can decrease from 5.1% to 5.4% for 4412.

The results of the comparison of the influence of airfoil's characteristics were camber, thickness, and the lift and drag. The results indicated the same trend as in the previous discussion (see Section 4.3).

5. Conclusion

This study set out to present the cut-in sinusoidal trailing edge shape with three wavelengths (0.25c, 0.50c and 0.75c) with 0.05c amplitude is parametrically investigated by numerical unsteady RANS simulation, open-source code; Code_Saturne. The results obtained were used to compare with the benchmark baseline and blunt profile trailing edge shape. This study has identified the geometry of NACA0012, NACA4412 and NACA4415 airfoil with a small modification to obtain a zero thickness trailing edge, and to be used as a baseline profile. This study has found that generally the cut-off baseline TE at 10% of chord from the trailing is modelled as the blunt trailing profiles. The flow is simulated at high Re of 10^6 for the angle of attack 5 degrees.

The influence of wavelength on aerodynamics characteristic has been observed. One of the more significant findings to emerge from this study is that the wavelength does not have an effect on C_L/C_D of NACA with cut-in sinusoidal TE. It also suggests that modification of TE can increase the lift coefficient with a drag penalty (when aerodynamics coefficients are calculated using physical chords of the airfoils).

The research has also shown that the drag penalty of an airfoil with modified TE occurs from 20% and up to 50%, and it depended on the TE shape. The results of this research support the idea that the desired modified modelling recommended from this study is the magnitude of $C_D \approx 0.1C_L$, the 8% increase of lift with 40% of drag penalty in case of blunt TE or the 5.3% of lift increment with 16% of drag penalty in case of cut-in sinusoidal TE of NACA 4415. These data suggest that the modified TE with the same airfoil thickness can be achieved through using a cambered profile rather than the symmetric profile.

References

[1] Howe MS 1978 *J. Sound Vib.* **61** 437–465
[2] Howe M S 1991 *J. Acouts. Soc. Am.* **90**(1) 482–487
[3] Hussai U Malook S U Shabir B Ali O and Ahmad S M 2017 *Proc. AIAA AVIATION Forum (AIAA 2017-4470) 35th AIAA Applied Aerodynamics Conf.* (Denver) 1–10
[4] Jantapremjit P Rattanasiri P Tipsiri A Arkadumnuay T Truekhakit W and Wilson P A 2017 *Proc. 8th TSME Int. Conf. on Mech. Eng.* (Bangkok) 59–67
[5] Rattanasiri P Jantapremjit P Tipsiri A Arkadumnuay T Truekhakit W and Wilson P A 2017 *Proc. 8th TSME Int. Conf. on Mech. Eng.* (Bangkok) 784–791
[6] Prigent S L Buxton O R H and Bruce P J K 2017 *AIAA J.* **55**(5) 1590–1601
[7] Rattanasiri P and Perrin R 2018 *Proc. 9th TSME Int. Conf. on Mech. Eng.* (Phuket) 712–719
[8] Archambeau F Méchitoua N and Sakiz M 2004 *Int J Finite Volumes* 1
[9] Fournier Y Bonelle J Moulinec C Shang Z Sunderland A G and Uribe J C 2011 *Comput. Fluids* **45**(1) 103–108
[10] Vogeltanz T 2015 *Proc. IEEE/AIAA 34th Digital Avionics Systems Conference* (Prague) 5E3-1–5E3-15
[11] Munson B Young D and Okiishi T 1998 *Fundamentals of Fluid Mechanics* 3rd Edition (New York:John Wiley & Sons Inc.)
[12] Harris C D 1981 *NASA Technical Memorandum* **81927**

IOP Conf. Series: Materials Science and Engineering **886** (2020) 012021 doi:10.1088/1757-899X/886/1/012021

Influence of the trailing edge shape on the aerodynamic characteristics of an airfoil at low Re number using RANS

R. Perrin[1,2,*], P. Rattanasiri[3], E. Lamballais[4] and S. Yooyen[2]

[1]Department of Mechanical Engineering, Faculty of Engineering at Sriracha, Kasetsart University Sriracha Campus, Sriracha, Chonburi 20230, Thailand
[2]Department of Aeronautical Engineering, International Academy of Aviation Industry, King Mongkut's Institute of Technology Ladkrabang, Bangkok 10520, Thailand
[3]Department of Mechanical Engineering, Faculty of Engineering, Burapha University, Chonburi, 20131, Thailand
[4]Institute PPRIME, Department of Fluid Flow, Heat Transfer and Combustion, CNRS Université de Poitiers ENSMA, Téléport 2, Futuroscope Chasseneuil Cedex, France

* Corresponding Author: rodolphe.perrin@univ-poitiers.fr

Abstract
Recent studies have demonstrated that periodic spanwise modifications of the Trailing Edge of an airfoil can significantly reduce the noise produced and can also increase its aerodynamic performances. This study aims to analyse the effects of such modifications on the aerodynamic performances of a profile by numerical simulation, with a particular emphasis on low Reynolds numbers typical of Unmanned Aerial Vehicles, Micro Aerial Vehicle, and/or small wind turbines. In the range of Re numbers considered here, the flow presents laminar separation eventually followed by transition and reattachment (depending on Reynolds number, Angle of Attack and Free Stream Turbulence).
As the standard for industrial applications is still the RANS approach, thanks to its moderate computational cost, this approach is considered here and a first step consists in evaluating the ability of recently proposed transition models (Menter, 2015; Ge, 2014; Kubacki, 2016) to predict the characteristics of the flow for such geometries. A baseline NACA0012 profile is considered, together with modified Trailing Edges (blunt and serrated). Simulations are performed using Code_Saturne (2nd order finite volume), and the influence of the Free Stream Turbulence, Angle of Attack, and Reynolds number on the aerodynamics performances is examined. The results are compared with data from the literature.

1. Introduction

This paper is a first step of a study on the effect of Trailing Edge (henceforth TE) modifications on the aerodynamic performances of wings at low Reynolds number. The specific objective is to assess the performances of recently proposed RANS transition models ([1-3]) by making a comparison with results from the literature. NACA0012 profiles, with baseline and modified TE, are considered and the influence of Angle of Attack (henceforth AoA), Re number and Free Stream Turbulence (henceforth

FST) are examined. The configuration and numerical set-up are described in the following section. Section 3 presents the main results and conclusions and perspectives are given in Section 4.

2. Configuration and numerical set-up

The three airfoils analysed in the study of [4] using Direct Numerical Simulation (henceforth DNS), are considered: a standard NACA0012 with straight TE (denoted 'baseline TE' in the following), a blunt TE airfoil which is obtained by truncating the NACA0012 profile at a distance $0.133c$ from the TE of the baseline airfoil (where c is the chord of the baseline airfoil) and a serrated cut-in TE with a triangular shape and spanwise wavelength of $0.266c$, as shown in Figure 1. At a trough of the serrated TE airfoil, the cross-section is the same as the blunt TE one, and at a peak, the cross-section is the same as the standard NACA0012 one.

Simulations are carried out using the open-source CFD software called *Code_Saturne,* developed by EdF (see http://www.code-saturne.org). The software is based on a 2nd order accurate finite volume method using a collocated arrangement on unstructured grids and velocity-pressure coupling is performed using a predictor-corrector algorithm of SIMPLEC type. In the present study, a 2nd order upwind scheme is selected for the convection terms of each transport equation.

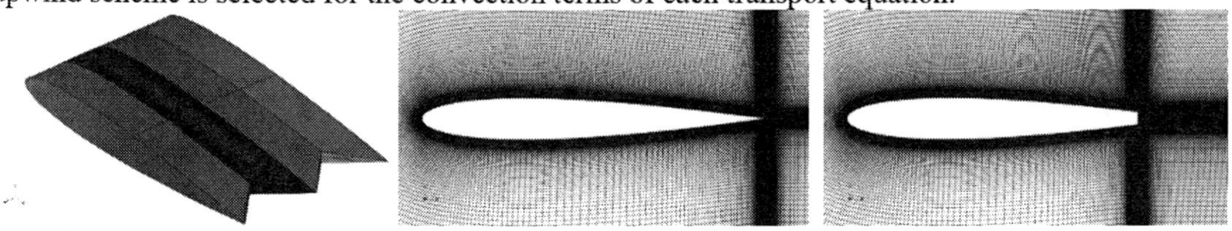

Figure 1. Geometry of the serrated TE (left) and mesh at a peak (middle) and at a trough (right).

Meshes have been generated using the open-source platform *Salome*. A C-type grid topology is employed. The computational domain extends a radius $20c$ around the leading edge and the outlet boundary is located $20c$ downstream the trailing edge. 2D simulations are carried out for the baseline TE and blunt TE cases, and grids of 140000 to 160000 cells have been found necessary for these cases. Grids are refined in the direction normal to the walls so that $y^+ \leq 1$ at the walls and the number of points around the surface of the airfoil is 438, with a higher density on the upper surface to resolve the separation, transition and reattachment. For the 3D simulations (serrated TE), the domain extends half a wavelength in the spanwise direction (symmetry conditions are used on the span boundaries) and the mesh contains 11 million cells.

RANS transition models from [1], [2] and [3] have been implemented in *Code_Saturne* and compared in this paper. Each model is based on a $k-\omega$ model (SST in the case of [1]). Models [1] and [3] uses a transport equation for the intermittency γ, designed so that $\gamma = 0$ in the laminar regions of the boundary layer and increases to 1 in the turbulent regions after the transition. The variable γ mainly acts on the original turbulence model by weighting the production term in the equation for k. In the model by [2], γ is given by an algebraic relation and other modifications are made in the original model. A detailed description of the equations of each model can be found in the original papers.

The angle of attack is introduced through the free-stream velocity vector imposed on the external boundary. Concerning the turbulent variables k and ω, inlet values at a distance of $20c$ upstream of the leading edge are first selected, and the analytical solution of the $k-\omega$ model corresponding to the case of decaying isotropic turbulence is imposed on the external boundary. The inlet values are chosen to impose a FST intensity $Tu=u'/U$ at the leading edge location.

As explained in the following sections, simulations with different Re, Tu and TE shape are carried out. For each case, AoA is varied from 0° to 12° by steps of 1°. Although not discussed in the paper, some of the cases considered are unsteady mainly because of the presence of a vortex shedding in the wake. For these cases the simulations are carried out until a statistically steady regime is reached and averages are computed. In the following, all quantities are made dimensionless using the inflow velocity and the chord of the baseline airfoil.

3. Results

3.1 Comparison with reference data for the baseline case at Re=50000

The models by Menter et al [1], Kubacki et al [2] and Ge et al [3] are first compared with reference data for the baseline TE NACA0012 at Re=50000. As noted by [5], a lot of disagreements is found between published experimental results in the literature, mainly because at this low Reynolds number, the flow is very sensitive to small variations of external conditions. Recently, Direct Numerical Simulations and Large Eddy Simulations could be performed for this case and the DNS results of [6] (at AoA=5° and 8°) will be used here as a reference for comparison between the models. The experimental results of [5] are also selected for comparison because of their reasonable agreement with the DNS results.

Figure 2 presents the iso-contours of the U-component of the velocity and streamlines obtained with the model by Kubacki et al., [2] at AoA=5°. As can be seen, the presence of a separation bubble on the suction side can be identified. Although not shown in the paper, visualisation of the turbulent kinetic energy and intermittency, for instance, clearly shows a laminar separation, followed by transition and turbulent reattachment, in agreement with the literature.

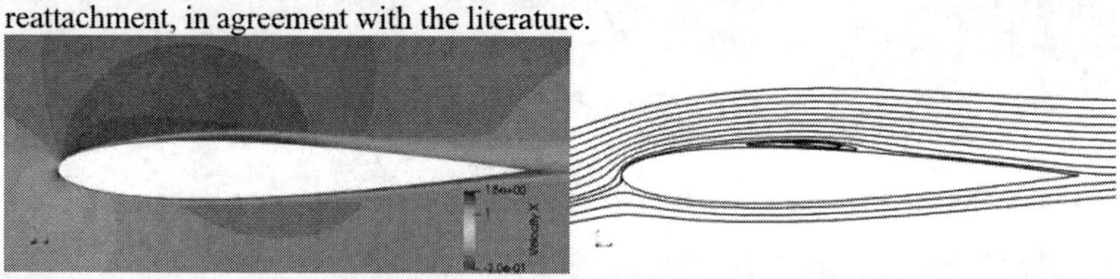

Figure 2. Iso-contours of U (left) and streamlines (right) at Re=50000; AoA=5° predicted by the model of Kubacki et al (2016).

Figure 3 (on the left side) shows the mean lift and drag coefficients for different AoA, which are predicted by the different models. One aspect consistently reported in the literature is the nonlinearity of the curve *Cl(AoA)* at low Re, which contrasts with results from the thin airfoil theory. It appears that this aspect is better predicted by the models by [1] and [2]. In particular, the model by [1] predicts the negative slope of *Cl(AoA)* at very low AoA reported by the experiment (smaller steps in AoA should be considered to assess if this behaviour is also predicted by the model by [2]). Except at very low AoA, the model by [2] achieves the best agreement with experimental values on a large range of AoA. The nonlinear character of the lift curve is poorly predicted by the model given in [3]. Furthermore, a very long time of convergence was needed in comparison with the computations using the other models. Thus, only the models provided in [1] and [2] have been considered in the following.

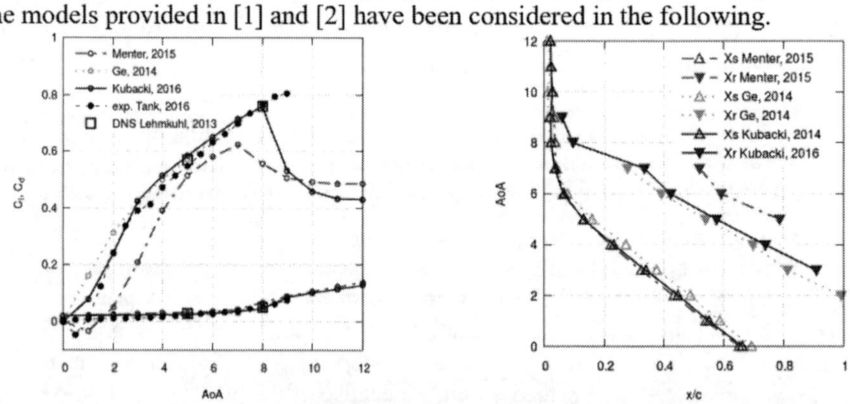

Figure 3. Comparison of models; left: lift and drag coefficients *Cl* and *Cd*; right: locations of separation *Xs* and reattachment *Xr*

Figure 3 (on the right side), shows the position of the separation and reattachment (identified using the mean friction coefficient) for different AoA. In agreement with the literature, e.g. [7], it is first observed that no reattachment occurs at low AoA, and that for AoA>3°, both separation and reattachment moves towards the leading edge when AoA increases. It is also seen that the decrease of the length of the separation bubble with AoA, reported in the literature, is well predicted. Regarding the comparison of the models, it is clearly observed that the reattachment predicted using the model in [1] is strongly delayed comparing to the one predicted by the model [2].

A closer analysis can be done from Figure 4 and Figure 5, which represent the pressure and friction coefficients *Cp* and *Cf* from both models and compared to DNS results from [6] at AoA=5° and AoA=8°.The shape of the profiles of *Cp* and *Cf* are typical of laminar separation bubbles with transition followed by reattachment. The start of the constant pressure region ("plateau") is located close to the separation, the end of the plateau is close to the transition, and the point at which the rate of the pressure recovery changes is found close to the reattachment. Concerning the *Cf* profile, low negative values are found in the laminar region of the bubble upstream of transition and an increase of |*Cf*| values is observed after the transition, before the mean reattachment. At AoA=5°, an excellent agreement is achieved with the DNS results using the model from [2]. The model by [1] appears to predict correctly the location of separation and transition, but the rate of the pressure recovery appears too slow after the transition and the position of the reattachment is predicted too much downstream. Thus, it appears that while the start of the transition is well predicted by the model by [1], this transition is too slow and delays the reattachment. This also explains why no reattachment is observed (on figure 2 right) using this model for AoA<5°, contrary to the results of the other models. At AoA=8°, the same default is observed with the model by [1] concerning the region after transition and the best agreement with DNS is also achieved by the model [2] concerning the *Cp* profile. However, it can be noticed on the *Cf* profiles that the transition appears too early with the model by [2], while the location of the start of transition is well predicted by the model by [1].

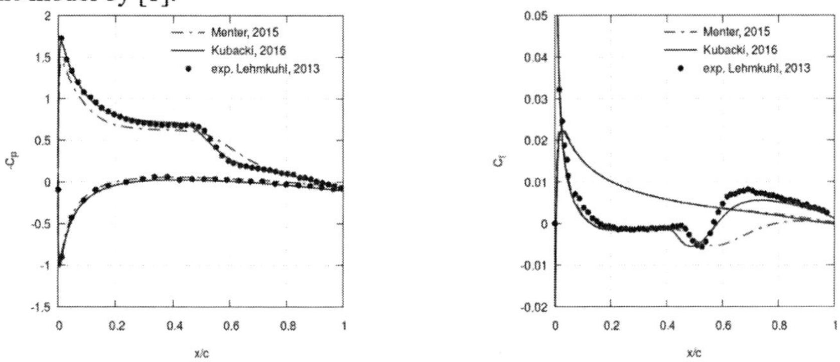

Figure 4. Pressure coefficient *Cp* (left) and friction coefficient *Cf* (right) at Re=50000; AoA=5°

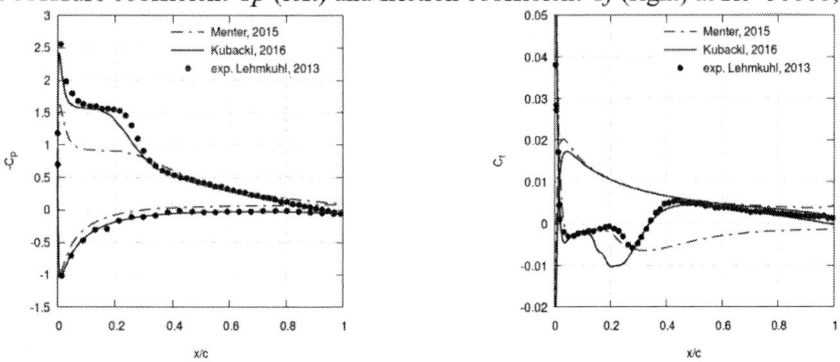

Figure 5. Pressure coefficient *Cp* (left) and friction coefficient *Cf* (right) at Re=50000; AoA=8°

3.2 Reynolds number influence

In this section, the effect of a variation of the Reynolds number in this regime is examined by comparing the precedent results (at Re=50000) with simulations at Re=100000, and for the same Free Stream Turbulence characteristics. According to the literature (e. g. [7]), an increase in Re number from 50000 to 100000 results in a slight shift of the separation in the downstream direction (for AoA at which the bubble is present) and a significant shift of the reattachment toward the upstream direction, which can be explained by an earlier transition. This results in a shorter separation bubble length. Concerning the aerodynamic performances, this increase of Re number results in a higher maximum lift coefficient, which happens at a higher stall AoA.

Figure 6 (on the left side) shows the lift and drag coefficients whilst Figure 6 (on the right side) reports the positions of separation and reattachment predicted by both models at different AoA. It can be seen that the trends reported in the literature are well predicted by both models, except the slight shift of the separation predicted by the model from [1]. It is also noticeable that the two models are in much better agreement at Re=100000 than at lower Re. Particularly, it is observed that the reattachment occurs at much lower AoA, and that the nonlinearity of the *Cl* curve is strongly reduced with the increase of Re number.

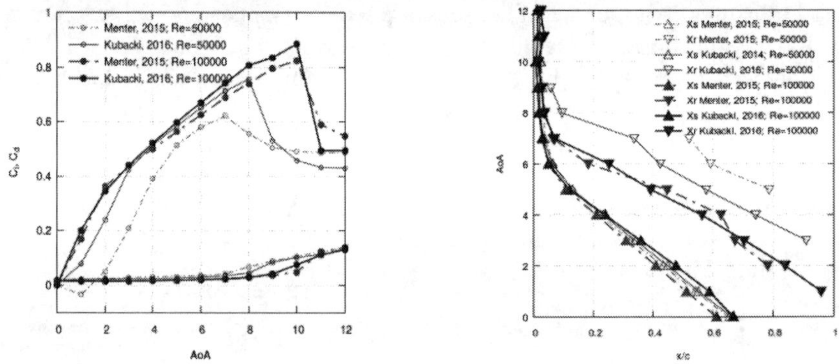

Figure 6. Influence of Reynolds number: *Cl* and *Cd* (left) and positions *Xs* and *Xr* (right).

3.3 Free Stream Turbulence intensity influence

According to [7] and other references, it is generally accepted that increasing the Free Stream Turbulence intensity results in effects similar to that obtained by increasing the Re number. While an increase in FST intensity delays separation, the transition and reattachment occur earlier, and the resulting separation bubble is significantly shorter. An increase of the maximum lift coefficient and stall angle of attack is also reported in the literature. Figure 7 presents aerodynamic coefficients, as well as separation and reattachment positions predicted by both models for different turbulent intensities *Tu=u'/U* (measured at the leading edge). For both models, no difference is obtained by lowering the turbulence intensity compared to that imposed in the previous sections. For increased *Tu*, no increase in the maximum lift coefficient and stall angle is observed for both models. The model by [1] predicts a slightly delayed separation, as well as an earlier reattachment, in agreement with the literature. In contrast, very minor changes are observed using the model by [2].

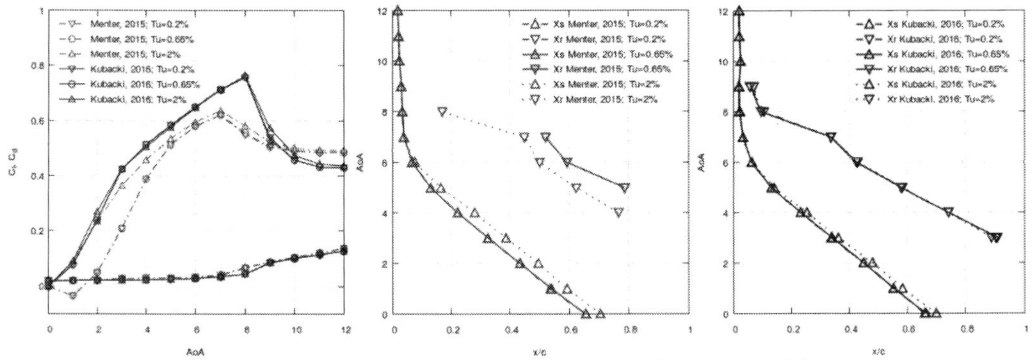

Figure 7. Influence of Free Stream Turbulence: *Cl* and *Cd* (left) and positions *Xs* and *Xr* using the model of Menter et al (2015) (middle) and the model of Kubacki et al (2016) (right)

3.4 Trailing Edge shape influence

In this section, the prediction of the flow around wings with a modified TE is analysed. The same geometries as in [4], which uses DNS, are selected for comparison and assessment of the models. The results for the blunt TE are presented in Figure 8 and Figure 9 (on the left side). It is reminded that the lift and the drag from all cases are made dimensionless using the chord of the baseline airfoil. It can be seen from the figures that the position of separation and reattachment are only slightly modified by the TE modification both being slightly shifted toward the downstream area, in agreement with [4]. Concerning the lift coefficient, both models are in better agreement at low AoA, for which no reattachment occurs. It is also noticeable that a small increase of the stall angle is predicted by the model in [2], in line with results reported in the literature.

Concerning the pressure coefficient profile at AoA=5°, the most noticeable feature is a reduction of the minimum *Cp* for the blunt TE, compared to the baseline TE (see Figure 9 on the left side). This effect, well predicted by both models, is in good agreement with the results of [4]. As [4] explains that the origin of this behaviour is inviscid, it is not surprising that simulations with both models predict well this effect. Although not shown in the paper because of space limitation, it is also noticeable that a vortex shedding is predicted by the model by [2] in the wake, which has to be associated with the bluntness of the TE, also in good agreement with the results of [4]. This shedding is not predicted by the model of [1] (the predicted flow is steady), probably because of the different wake velocity profile resulting from the too long transition and delayed reattachment discussed in section 3.1.

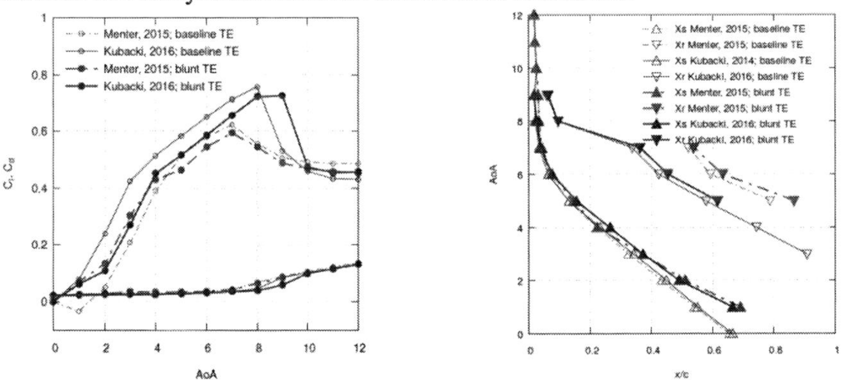

Figure 8. Influence of Trailing shape: lift and drag coefficient (left) and position of separation and reattachment (right) for the baseline and blunt trailing edge.

The pressure coefficient obtained for the serrated TE (3D) case is represented at the spanwise locations of a peak and a trough in Figure 9 (on the right side). It is reminded that the section of the wing at a trough corresponds to the blunt TE geometry. In agreement with the results of [4], the pressure distribution is almost identical at peak and trough spanwise locations in the upstream part of the wing.

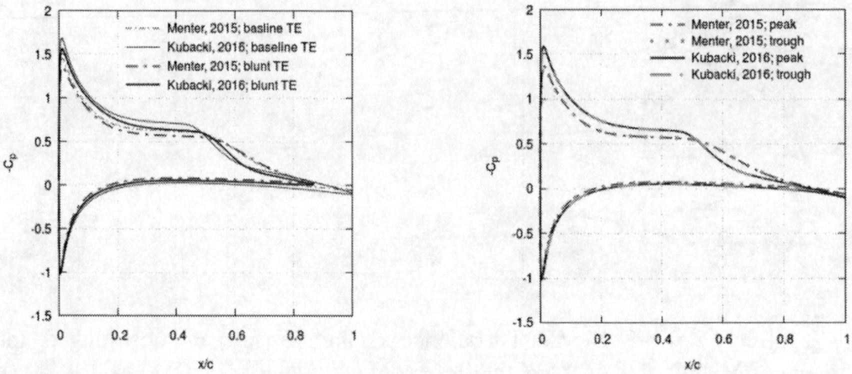

Figure 9. Influence of Trailing Edge shape: pressure coefficient *Cp* (Re=50000; AoA=5°); left: baseline and blunt trailing edge; right: serrated trailing edge (*Cp* at peak and trough).

4. Conclusion and perspectives

As a first step of a study on the effects of TE modification of airfoil at low Re number using RANS, the models by Menter et al [1], Kubacki et al [2] and Ge et al [3] were implemented in Code_Saturne and assessed for the flow around a NACA0012 profile, for different Re, different Tu and different TE. For the cases considered here at Re=50000, the model by [2] leads to the closest agreement with results from the literature (e.g. [6]), while the model by [1] predicts a too slow transition. However, the effects of the variations of parameters such AoA and Tu seems to be better predicted by the model of [1], especially considering the start of the transition. Although further investigation would be necessary, it is possible that a recalibration of some terms in the model of [1] for low Re case would allow better predictions. Concerning the TE modification effect, a close agreement has been achieved with the results of [4] and it is therefore expected that the used model will be suitable for a more detailed study on the effects of spanwise wavelength and amplitude of the serrations.

Acknowledgments

Computations have been performed on the supercomputer facilities of the "Mésocentre de calcul de Poitou Charentes".

References

[1] Menter F R Smirnov P E Liu T and Avancha R 2015 *Flow, Turbulence and Combustion* **95**(4) 583-619
[2] Kubacki S and Dick E 2016 *International Journal of Heat and Fluid Flow* **62** 344-361
[3] Ge X Arolla S and Durbin P 2014 *Flow, turbulence and combustion* **93**(1) 37-61
[4] Thomareis N and Papadakis G 2017 *Physics of Fluids* **29**(1) 014101
[5] Tank J Smith L and Spedding G R 2017 *Interface focus* **7**(1) 20160076
[6] Lehmkuhl O Rodríguez I Baez A Oliva A and Pérez-Segarra C D 2013 *Computers & Fluids* **84** 176-189
[7] Huang R F and Lee H W 1999 *Journal of aircraft* **36**(6) 965-972

IOP Conf. Series: Materials Science and Engineering **886** (2020) 012022 doi:10.1088/1757-899X/886/1/012022

A Two-Dimensional Computational Study of a Single Right Trapezoidal Cylinder Subjected to Unsteady Laminar Flow Regime at Low Reynolds Number

Sodsai Lamtharn[1] and Monsak Pimsarn[2]*
[1, 2]Mechanical Engineering Department, Faculty of Engineering, King Mongkut's Institute of Technology Ladkrabang, Chalongkrung Rd. Ladkrabang, Bangkok, 10520, Thailand

* Corresponding Author: monsak.pi@kmitl.ac.th

Abstract
A right trapezoidal cross-section cylinder is used as guided bars for a barred tee in the long pipeline and piping system to protect the moving of pigs across the branched pipe. The cylinder must have the integrity to withstand to fluid flows-induced vibration which concerns the synchronization between vortex shedding frequency of fluid flows with a natural frequency of the cylinder. Therefore, this article aims to numerically investigate the vortex shedding frequency responses which are in terms of Strouhal number, flow force coefficients and in order to provide additional data for use in the absence of more applicable data for engineering design of the single right trapezoidal cylinder in structural integrity point of view. In this study, the simulation of the single right trapezoidal cylinder immersed in the uniform oncoming laminar flow regime with low Reynolds number; $Re = 100$, was performed. There are various sharpening angles in this study; defined as angle which is measured from virtual line parallel with height base side to slant side or (90°- *acute angle°*) of the cylinder, in the range of 0°, 15°, 22.5°, 40° and 60°, accompanying with the cylinder side ratio (B/A) which is the ratio of the longest base side-to-height base side, in the range of 1, 2, 3, 4, 5, 6 and 7. In the simulation, the Direct Numerical Simulation (DNS) with a finite volume method, an incompressible flow and constant fluid properties were employed. The computational results were validated against the published data. The effect of the sharpening angle and side ratio on the response of the single right trapezoidal cylinder were presented and discussed. The Strouhal number and flow force coefficients at fully ten cycle of saturated flow have been calculated. The simulated results show that increasing of sharpening angle, the Strouhal number slightly increases whilst the root mean square lift coefficients significantly increases. Additionally, as the side ratio increases, the Strouhal number and the root mean square lift coefficients decreases.

Keywords: single right trapezoidal cylinder, unsteady flow, incompressible flow, laminar vortex shedding

IOP Conf. Series: Materials Science and Engineering **886** (2020) 012022 doi:10.1088/1757-899X/886/1/012022

1. Introduction

A right trapezoidal cross-section cylinder is used as guided cylinder bars for a barred tee in the long pipeline and piping system to control propel of pigs in desired partway line and to protect the moving of pigs across the branched pipe for various internal pipe activities, such as batching fluid product, cleaning, separating fluids and inspecting these systems. The pigs are a spherical or circular cylindrical shaped device, which can be pushed through the main pipe from the pigs' launcher station by pressurized fluid flows through the desired main pipeline, branched pipe and together with all piping component, pipe supports, valves and inline instruments to the pigs' receiver station. The main pipe is a major portion of the long pipeline and piping system that composes of many branch connections between the pig traps station. When the system is filled with pressurized liquid to start-up operation, the fluid flows past the main pipe and flowing into a branched pipe together with past over the barred tee. To control propel of pigs in the desired way and protect the moving of pigs across the branched pipe, the barred tees are provided as shown on figure 1 (left) and (right), the guided cylinder bars attached cross inside the branched pipe with fillet-welded joints at the connection point. Normally, the barred tee is installed on all branch pipes, with a diameter larger than half or one-quarter of the main pipe diameter.

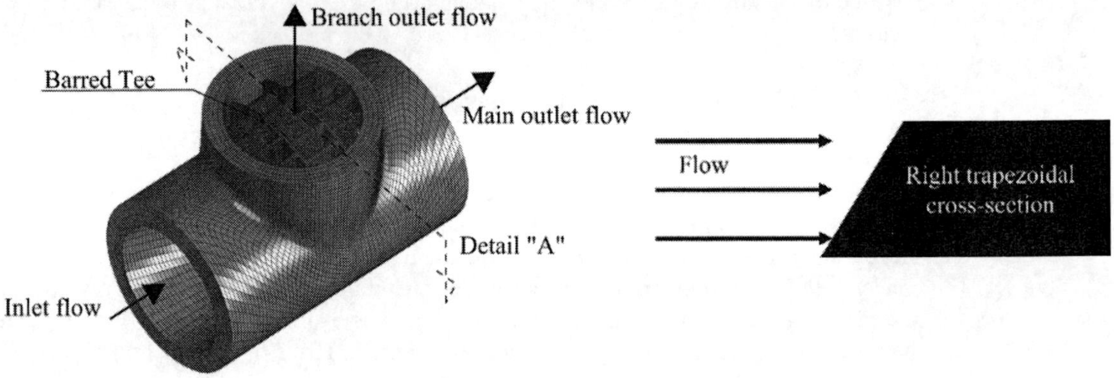

Figure 1. (Left) Schematic of flow passing the main pipe, branched pipe and barred tee, (right) Details cross-section of the cylinder bar

As the flow passes over a barred tee which composes of a square, a rectangular and a right trapezoidal cylinder bar cross-section, the cylinder bar must have the integrity to withstand to fluids flow-induced vibration which concerns the synchronization between vortex shedding frequency with the natural frequency of the cylinder bar, in the aspect of engineering design. The synchronization vibration of them will cause the cylinder bar subjected to strong vibration eventually leading to fatigue damage. Then, the bars will detach from the branched pipe wall and flow into the main pipe. This causes the pigs stuck, valves face scraper or crash and the bars flow into other inline pieces of equipment, such as a pump impeller, a flow control valve, etc. Additionally, the strong cylinder bar vibration can result in pipe wall local stress amplification, especially, at the fillet-welded region.

Most of the research on flow past over a cylindrical blu□ bodies in the past has been intensively performed for circular, square, rectangular, equilateral triangle and equilateral trapezoidal cylinders rather than a right trapezoidal cylinder. For a comprehensive review on circular cross-section, the details are referred to the articles by Williamson [1] and Zdravkovich [2]. Mittal et al. [3] who carried out simulation study for flow past circular cylinder to investigate the effect of the blockage ratio at $Re \leq 150$, the blockage ratio $\gamma = 1\%$ and 5% with a stabilized space-time finite element formulation. Sohankar et al. [4, 5, 6] performed 2-D simulations for a square and a rectangular (B/A = 1-4) cylinder subjected to unsteady flow at approaching angle $\alpha = 0°$-$90°$, $Re \leq 200$ to study the effect of modeling parameters including the outlet of boundary condition. They employed the time step $\Delta t = 0.025$ and mesh size next to cylinder $\zeta = 0.004$ for accurate simulation. The proper values of $X_d \geq 25$ and $X_u = 11.1$ were utilized, where X_u and X_d are defined as dimensionless distance measured from both the inlet and outlet

IOP Conf. Series: Materials Science and Engineering **886** (2020) 012022 doi:10.1088/1757-899X/886/1/012022

boundaries to the most upstream corner and the most downstream corner of the cylinder, respectively. At B/A = 1, the root mean square (RMS) lift coefficients, *CL RMS* and Strouhal number, *St*, increase with the increase of *Re* up to the angle of incidence of 45° after their decrease. A convective Sommerfeld boundary condition (CBC) is better than a traditional Neumann boundary condition (NBC) for flow around a square cylinder at non-zero an angle of approaching and the CBC more effective in reducing the CPU time, reducing the upstream influence from the outlet and reducing the necessary downstream extent. Sohankar et al. [6] performed numerical study and recommended for $Re < 100$ use proper value $X_d > 15$ units and for $Re = 100\text{-}200$ use $X_d > 10$ units. For CBC and the upstream influence of the outlet is effectively damped out on NBC at approximated $X_d > 25$ units. At Re = 100 - 200 with $\alpha = 0°$, the blockage ratio γ reduced from 5% to 2.5% and slightly decrease on the RMS lift coefficients and Strouhal number. Zhao et al. [7] provided the RMS lift coefficients and Strouhal number from the studied/measured influence of an angle of incidence for flow over a square cylinder subjected to VIV at $Re = 100$ with $\alpha = 0°$, 45°, and 22.5° and $\gamma = 2.5\%$. More additional studied for numerical simulations of 2-D unsteady flow past over a square cylinder at low and moderate Reynolds numbers are Sharma and Eswaran [8], Singh et al. [9], Sahu et al. [10] and Sen et al. [11]. Sohankar et al. [12], Robichaux et al. [13], Darekar and Sherwin [14] and Saha et al. [15] have been carried out for the simulation study of 3-D flow around a square cylinder at low and moderate Reynolds numbers, the reader is referred to the review these works. Okajima [16] conducted experimental on flow past over various rectangular cylinder in both a water tank and a wind tunnel to determine Strouhal number as a function of Reynolds number and the width-to-height ratio. The results show that the Strouhal number has a minimum value at the width-to-height ratio of 1:2 and a maximum value at 1:3. Okajima [17] and Okajima et al. [18] carried out simulation for flow past cylindrical rectangular cross-section and investigated the e□ect of blockage ratio on both oscillating and stationary rectangular cylinder. Arnab and Dalal [19] carried out 2-D numerical simulations of a triangular cylinder which placed in free-stream at $10 \leq Re \leq 250$ subjected to laminar flow with finite volume and more study on the RMS lift coefficients and Strouhal number. Alawadhi [20] numerically studied the 2-D incompressible laminar flow past over a triangular cylinder in a horizontal channel for a stationary and undergoing vertical oscillating motion with a fixed $Re = 100$. Jiang and Cheng [21] present a DNS study of the 2-D, 3-D on a square cylinder subjected to flow around in order to investigate the characteristics of wake flow and the force due to hydrodynamic at Re = 10-400. They found that a local maximum has been occurred at Re =168, for the 2-D *St-Re* relationship. Chung and Kang [22] studied effect height ratio of trapezoidal with laminar vortex shedding for Re = 100, 150 and 200. The result has shown that the Strouhal number for trapezoidal cylinders depends on the height ratio and *Re*.

The results indicate that when the cylinder's natural frequency is equal to the Strouhal number of the stationary cylinder, the RMS lift coefficients reaches its maximum and at a high oscillating frequency that does not affect to the oscillating amplitude. As mentioned above from literature and in the past, the most applicable engineering data for flow past circular, elliptic, square, rectangular, triangle and trapezoidal cylinder are emphasis intense investigated but more absence data for the right trapezoidal cylindrical cross-section to use in barred tee engineering design. Therefore, this article aims to numerically investigate the vortex shedding frequency responses which are in term of Strouhal number, flow force coefficients and in order to provide additional these data for use in the absence of more applicable data for engineering design of the single right trapezoidal cylinder in structural integrity point of view.

2. Numerical model

The numerical details, numerical methods, governing equations, boundary conditions and grid structure for the flow past over the single right trapezoidal cylinder are discussed below.

2.1 Numerical details and grid structure

The geometry for this computational study of flow is shown in figure 2 (left), a stationary 2-D single right trapezoidal cylinder with projected width d of the cylinder in the streamwise direction subjected to

IOP Conf. Series: Materials Science and Engineering **886** (2020) 012022 doi:10.1088/1757-899X/886/1/012022

unsteady laminar flow inlet with constant free-stream velocity U_∞ flowing into side *A* and *B* at some flow approaching angle, *α*. Here *B* is the longest base side, *A* is height base side, where sharpening angles *β* is defined as angle which is measured from a virtual line parallel with height base side to slant side or (90°- *acute angle*°) of the cylinder. As depicted in figure 2 (left) the center of cylinder and force coordinates are placed on origin, the *L* axis is the transverse direction and the lift force is positive, the *D* axis is a streamwise direction and the drag force is positive. The projected width *d* is used scale for length dimensional to dimensionless, to all force coefficients and Strouhal number, $St = f_S d/U_\infty$, by f_S is

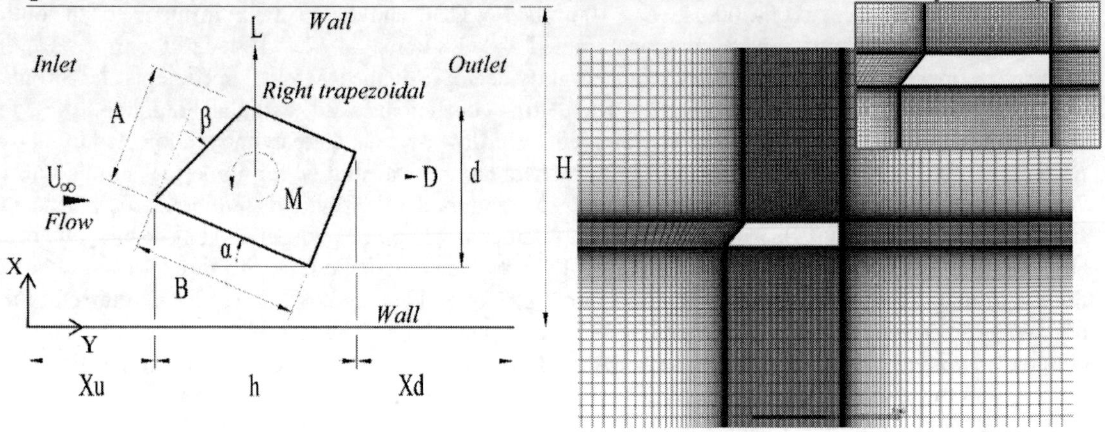

Figure 2. (Left) Schematic of the computational geometry, (right) mesh, the right top is a zoomed view

the frequency of vortex shedding. Scaling with U_∞ for all velocities and physical times are scaled with d/U_∞. The blockage ratio which is confined boundary parameter is defined as $\gamma = d/H$, where *H* is a vertical measurement from the upper wall to lower wall boundary of the computational domain.

In this study given *β* varied in the range of 0°, 15°, 22.5°, 40° and 60° accompany with the cylinder side ratio, *B*/*A* varied in the range of 1, 2, 3, 4, 5, 6 and 7. Some remaining key parameters came from the resultant studied work of Lamtharn and Pimsarn [23]. As re-written following, $Re = 100$, $d = 1$, $\alpha = 0°$, $\gamma = 0.05$, $H = 20$, $X_u = 10$ and $X_d = 15$, $\zeta = 0.004$. The non-uniform mesh distribution was used vary mesh size of 0.004-0.5 in an area that extends of 5 units all side from the nearest corner of the cylinder. The constant mesh size $\Delta = 0.5$, uniform distribution in the outside that area is employed, on each side of the cylinder is assigned 20 elements per unit length as mesh shown in figure 2 (right).

2.2 Boundary conditions

The inlet and outlet boundary are located at $X_u = 10$ units upstream and $X_d = 15$ units downstream from the most upstream and the most downstream corner of the cylinder, respectively. The upper and lower boundary is located at *H*/2 of cylinder walls. In table 1, the boundary conditions for this computational study is shown, where $U = U_\infty$ and *V* are a uniform flow velocity in the streamwise and transverse direction, respectively. The gage pressure is defined as *P*.

Table 1. Boundary conditions.

Boundary	Velocity	Pressure
Inlet	$U = 1, V = 0$	-
Outlet	$\frac{\partial U}{\partial X} = 0, \frac{\partial V}{\partial X} = 0$	$P = 0$
Upper and Lower	$Frictionless, \frac{\partial U}{\partial Y} = 0, V = 0$	-

IOP Conf. Series: Materials Science and Engineering **886** (2020) 012022 doi:10.1088/1757-899X/886/1/012022

Cylinder surface	*No-Slip*, $U = 0$, $V = 0$	-

The calculations initiated with the fluid flow velocity $V = 0$, $U = 0$, at rest, then in few time step (constant $\Delta t = 0.05$) the flow velocity is increased smoothly to $V = 0$, $U = 1$, and used 40 iteration per time step. The monitor convergence criteria in continuity and velocity are set for B/A = 1-4 of 0.001 and 0.00001 for B/A = 5-7 to monitor convergent against residual at the end of each iteration during iterative sequence same as Lamtharn and Pimsarn [23] and Sohankar et al. [4, 5, 6].

2.3 Numerical method

For this study in solving the unsteady, conservative, continuity and dimensionless form of the 2-D incompressible flow Navier–Stokes equations with a constant fluid property are given as follows.

Continuity:

$$\frac{\partial U}{\partial X}+\frac{\partial V}{\partial Y}=0 \tag{1}$$

Momentum-x:

$$\frac{\partial U}{\partial T}+\frac{\partial (UU)}{\partial X}+\frac{\partial (UV)}{\partial Y}=-\frac{\partial P}{\partial X}+\frac{1}{Re}\left(\frac{\partial^2 U}{\partial X^2}+\frac{\partial^2 U}{\partial Y^2}\right) \tag{2}$$

Momentum-y:

$$\frac{\partial V}{\partial T}+\frac{\partial (UV)}{\partial X}+\frac{\partial (VV)}{\partial Y}=-\frac{\partial P}{\partial Y}+\frac{1}{Re}\left(\frac{\partial^2 V}{\partial X^2}+\frac{\partial^2 V}{\partial Y^2}\right) \tag{3}$$

With

$$U=\frac{u_x}{U_\infty}, V=\frac{u_y}{U_\infty}, T=\frac{tU_\infty}{d}, X=\frac{x}{d}, Y=\frac{Y}{d}, P=\frac{p}{\rho U_\infty{}^2} \tag{4}$$

The Reynolds number is defined as $Re = U_\infty d/v$, where v is kinematic viscosity, $d = B\sin\alpha$, $(90\text{-}\beta)^\circ < \alpha \leq 90^\circ$ and $0^\circ \leq \beta \leq 60^\circ$ and $d = A\cos(\alpha+\beta)/\cos\beta + B\sin\alpha$, $0^\circ \leq \alpha \leq (90\text{-}\beta)^\circ$.

The Direct Numerical Simulation (DNS) with a finite volume method, non-uniform meshing with second-order implicit time discretization into eight-node quadratic quadrilateral elements are employed. An incompressible flow SIMPLEC code with constant fluid properties such as kinematic viscosity and density are used. The convective terms using a third-order QUICK scheme same as [4, 5, 6, 23]. The same numerical approach as mentioned above has been successfully adopted in the work of [23] for the simulations of flow past over a single square and right trapezoidal cylinder is used in the present study.

3. Results and discussion

Benchmark numerical model

Table 2. Comparison of results at Re = 100 for a fixed square cylinder subjected flow past around.

			Percentage Difference (%)	
Works by	*CL RMS*	*St*	*CL RMS*	*St*
Darekar and Sherwin [14] (3D, simulation)	0.1860	0.1460	1.1	0.7
Zhao et al. [7] (2D, simulation)	0.1908	0.1447	1.5	0.2
Sahu et al. [10] (2D, simulation)	0.1880	0.1486	0.0	2.5
Sen et al. [11] (2D, simulation)	0.1928	0.1452	2.5	0.2
Sharma and Eswaran [8] (2D, simulation)	0.1922	0.1488	2.2	2.7
Singh et al. [9] (2D, simulation)	0.1600	0.1470	14.9	1.4
Sohankar et al. [6] (2D, simulation)	0.1560	0.1460	17.0	0.7
Okajima [16] (experimental)	-	0.1400	-	3.4
Present	0.1880	0.1449	-	-

IOP Conf. Series: Materials Science and Engineering **886** (2020) 012022 doi:10.1088/1757-899X/886/1/012022

In order to avoid inaccuracy from a numerical simulation. The benchmark numerical model for a fixed square cylinder subjected flow past around ($B/A = 1, \beta = 0°, \alpha = 0°$) at $Re = 100$ is performed and the computational results were validated against the published data in literature as presented in Table 2. By using the RMS lift coefficients and the Strouhal number as an indicator the strong sensitivity [6, 7, 23] represent to another sensitivity numerical parameters from fully at least ten cycles saturated flow were calculated. These results indicated that the results of the RMS lift coefficients values predicted from in this study very good agreement with the published numerical results in Zhao et al. [7], Sharma and Eswaran [8], Sahu et al. [10], Sen et al. [11] and Darekar and Sherwin [14]. The percentage difference between published data works with the present study of these RMS lift coefficients is less than 3%. The numerical results of Strouhal number, *St* from different models are very close to each other. The percentage difference between numerically published works with the present study of these Strouhal number is less than 3% and about 3.4% with experimental works of Okajima [16].

3.1 Effect of sharpening angle and side ratio on RMS lift coefficients

A comprehensive numerically investigate of the influence of the sharpening angle from 0°, 15°, 22.5°, 40° and 60° with side ratio B/A = 1, 2, 3, 4, 5, 6 and 7 were studied for various quantities parameters such as flow force coefficients and vortex shedding frequency responses. In figure 3 (left top) shown the variations of RMS lift coefficients with sharpening angle for side ratio B/A = 1, 2, 3, 4, 5, 6 and 7 and (left bottom) shown that with side ratio for sharpening angle β = 0°, 15°, 22.5°, 40° and 60°. The RMS lift coefficients abruptly at around $B/A \leq 3$ monotonic increase as sharpening angle increases, in contrast, the RMS lift coefficients abruptly at around $\beta \geq 40°$ monotonic decrease as side ratio increases. For instance, the RMS lift coefficients at $B/A = 4$ is lower than about 4.4 times at $B/A = 1$ for $\beta = 0°$ and lower than about 2 times at same side ratio for β = 15°, 22.5°, 40°. For $\beta = 0°$ that is a square cylinder for $B/A = 1$ and rectangular cylinder for B/A = 2, 3, 4, 5, 6 and 7, the RMS lift coefficients slightly monotonic decrease from the square section to longer rectangular section.

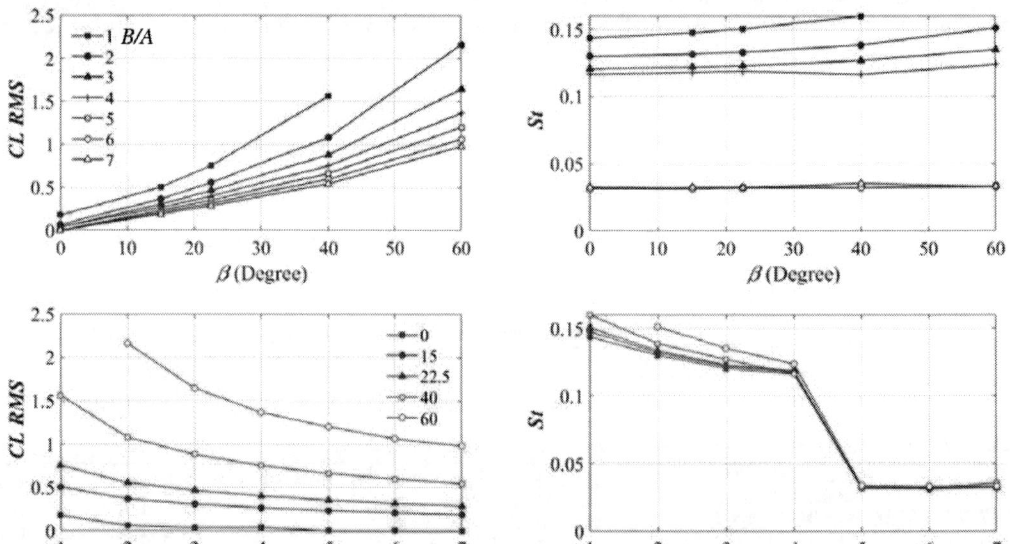

Figure 3. (left top) The *CL RMS* as a function of sharpening angle, (left bottom) the *CL RMS* as a function of side ratio, (right-top) The Strouhal number against the sharpening angle , (right bottom) the Strouhal number against the side ratio.

3.2 Effect of sharpening angle and side ratio on Strouhal number

In figure 3 (right-top) shown the variations of Strouhal number with sharpening angle for side ratio B/A = 1, 2, 3, 4, 5, 6 and 7 and (right bottom) shown that with side ratio for sharpening angle β = 0°, 15°, 22.5°, 40° and 60°. The Strouhal number slightly change with smoothly monotonic increase as sharpening angle increases, in contrast, the Strouhal number slightly change with smoothly monotonic decrease as side ratio increases until at B/A = 4 to 5, the Strouhal number abruptly decrease. It then Strouhal number slightly decrease again for larger side ratio as shown in figure 3 (right bottom). For instance, the Strouhal number at each B/A = 2, 3, 4, 5, 6 and 7 is lower than at B/A = 1 with some constant value for same sharpening angle. For β = 0° that is square cylinder for B/A = 1 and rectangular cylinder for B/A = 2, 3, 4, 5, 6 and 7, the Strouhal number slightly monotonic decrease from the square section to longer rectangular section.

4. Conclusion

The simulation of a single right trapezoidal cylinder immersed in a uniform oncoming unsteady laminar flow regime with low Reynolds number, Re = 100, and with various sharpening angles in the range of 0°, 15°, 22.5°, 40° and 60° accompany with the cylinder side ratio in the range of 1, 2, 3, 4, 5, 6 and 7 have been performed.

The increasing of sharpening angle with constant side ratio, the Strouhal number slightly increases whilst the root mean square lift coefficients significantly increase. As the side ratio increases with constant sharpening angle, the Strouhal number slightly decreases and the RMS lift coefficients evidently decrease. The larger side ratio is lesser flow force than smaller side ratio for the single right trapezoidal cylinder. The Strouhal number and the RMS lift coefficients in this study are provided to additional data for use in the absence of more applicable data for engineering design of the single right trapezoidal cylinder in structural integrity point of view further.

References

[1] Williamson C H K 1996 *Annu. Rev. Fluid Mech.* **28** 477–539
[2] Zdravkovich M M 1997 *Flow around Circular Cylinders* (New York: Oxford University Press)
[3] Prasanth T Behara S Singh S Kumar R Mittal S 2006 *J. Fluids and Structures* **22** 865–876
[4] Sohankar A Davison L Norberg C 1995 *Proc. Twelfth Australasian Fluid Mech. Conf.* 517-520
[5] Sohankar A Norberg C Davison L 1997 *J. Wind Eng. Ind. Aerodyn* **69-71** 189–201
[6] Sohankar A Norberg C Davison L 1998 *Int. J. Numer. Meth. Fluids* **26** 39–56
[7] Zhao M Cheng L Zhou T 2013 *Phys. Fluids* **25** 023603-25
[8] Sharma A Eswaran V 2004 *Numer. Heat Transfer, Part A* **45** 247–269
[9] Singh A P De A K Carpenter V K Eswaran V Muralidhar K 2009 *Int. J. Numer. Methods Fluids* **61** 658–682
[10] Sahu A K Chhabra R P Eswaran V 2009 *J. Non-Newtonian Fluid Mech.* **160** 157–167
[11] Sen S Mittal S BiswasG 2011 *Int. J. Numer. Methods Fluids* **67** 1160–1174
[12] Sohankar A Norberg C Davidson L 1999 *Phys. Fluids* **11** 288–306
[13] Robichaux J Balachandar S Vanka S P 1999 *Phys. Fluids* **11** 560–578
[14] Darekar R M Sherwin S J 2001 *J. Fluid Mech.* **426** 263–295
[15] Saha A K Biswas G Muralidhar K 2003 *Int. J. Heat Fluid Flow* **24** 54–66
[16] Okajima A 1982 *J. Fluid Mech.* **123** 379–398
[17] Okajima A 1990 *J. Wind Eng.Ind. Aerodyn.* **33** 171–180
[18] Okajima A Yi D Sakuda A Nakano T 1997 *J. Wind Eng. Ind. Aerodyn.* **67–68** 91–102
[19] Arnab A K Dalal A 2006 *Int. J. Numer. Meth. Fluids* **52** 801–821
[20] Alawadhi E M 2013 *J. Fluids Eng.* **135** 041202–10
[21] Jiang H Cheng L 2018 *Phys. Fluids* **30** 104107–13
[22] Chung Y J Kang S H 2000 *Phys. Fluids* **12** 1251–1254
[23] Lamtharn S Pimsarn M 2018 *MATEC Web of Conferences ICEAST 2018* **192** 02030

IOP Conf. Series: Materials Science and Engineering **886** (2020) 012023 doi:10.1088/1757-899X/886/1/012023

Flow Simulations of Modified Diffuser Augmented Wind Turbine

Saravana Kannan Thangavelu[1,*], Timothy Goh Leong Wan[1], and C Piraiarasi[2]
[1]Faculty of Engineering, Computing and Science, Swinburne University of Technology, 93300 Kuching, Sarawak, Malaysia
[2]Department of Architecture, Thiagarajar College of Engineering, Madurai 625 015, Tamilnadu, India

* Corresponding Author: sthangavelu@swinburne.edu.my

Abstract
In order for wind technology to compete with conventional sources of energy, in terms of energy production costs, researchers are working on different ways to increase the energy density in wind. Development of wind power augmentation system is one of the most promising concepts in this field. The main objective of this research is to propose a new diffuser design for Diffuser Augmented Wind Turbine (DAWT) by introducing air vents and vortex generators. This study also investigates the effect of wind velocity and air pressure on the proposed modified diffuser to develop a suitable diffuser for DAWT. The purpose of this modified diffuser is to create a low-pressure region in the centre of the diffuser to increase the inlet mass flow rate. Three modified designs were developed and were simulated using CFD techniques. Results showed an outstanding increment in wind speed, which is 8.56 m/s (2.14 times increment), over the 4 m/s initial wind speed in a modified diffuser (Model 3 Air vents with splitter). This modified diffuser is based on the parameters: rotor diameter (D) = 500 mm, diffuser length (L) = 1000 mm, diffuser open-angle (θ) = 12°, flange size = 221.73 mm and inner splitter with 14° open-angle (α).

Keywords: DAWT, Diffuser with splitter, Air vents, Wind Velocity, CFD

1. Introduction

Wind energy is the second largest source of energy and it is the fastest-growing renewable energy resource around the world. The main problem of using wind energy is the uncertainty of wind pattern. Low wind speeds result in less energy per unit volume of air passing through a turbine. This leads to higher electricity production cost from wind energy than from fossil fuels. In order for wind technology to compete with conventional sources of energy in terms of energy production costs, researchers are working on different ways to increase the energy density in wind [1]. One of the most promising concepts in this field is the growth of wind power augmentation systems. By the use of diffuser in the wind turbine, wind power augmentation system is also called a Diffuser Augmented Wind Turbine (DAWT). In DAWT, the diffuser or flanged diffuser generates separation regions behind it, where low-pressure regions appear to draw more wind through the rotors compared to a 'bare wind turbine'. DAWTs can provide an opportunity to extract power from unstable low wind speeds [2].

The concept of 'diffuser' is to increase the power output of a wind turbine by accelerating the wind velocity that approaches the wind turbine. To increase the wind velocity, low pressure would appear at the back of the wind turbine to act as a vacuum to suck the wind and accelerate it towards the blades. The region of the area surrounding or near the vortex would have typically lower pressure where the suction force will be formed by the vortex. Therefore, it will act as an accelerator to accelerate the wind velocity approaching the wind turbine. The vortex is used to generate a low-pressure region in the wake of the turbine. Vortex should appear at the wake of the diffuser but as little as possible inside the diffuser wall. If the vortices formed inside the diffuser, the pressure inside the diffuser might be lower than the pressure at the back of the wind turbine [3].

The performance increases of DAWT is proportional to the mass flow through the duct and that larger performances are possible by lowering back pressure levels at the exit. Many numbers of experimental and numerical studies reported in DAWT since 1950 presented the effect of diffuser length and open-angle of diffuser on the performance of the DAWT. In addition, some studies presented the effect of brim (flange) heights on the DAWT performance [4]. From the previous studies, it was concluded that DAWT exhibits the best advantages compared to other augmentation solutions in the production of power at high performance exceeding the Betz limit. The configurations of the shroud of DAWT included the simple diffuser, diffuser with an inlet shroud, converged - diverged diffuser, curved diffuser, flanged diffuser, and multilayer diffuser [3-9]. The optimum design for shroud will generally increase power obtained from any turbine used, but in order to reach optimum power from this turbine, a combination with a modified design for the turbine blade is necessary.

Kannan et al. [3] developed five different DAWT models using CFD and found that the flange and inner splitter are the effective features to increase the inlet air velocity of the diffuser. This research is the extension of this previous study having the objective to propose a new diffuser design for DAWT by introducing air vents and vortex generators. This study also investigates the effect of wind velocity and air pressure on the proposed modified diffuser to develop a suitable diffuser for DAWT. The purpose of this modified diffuser is to create a low-pressure region in the center of the diffuser to increase the inlet mass flow rate.

2. Methodology

Three different modified DAWT models were developed as shown in Figure 1 using SolidWorks and start from sketch the section cut on the front plane then revolve. All models are fixed to the inlet diameter (D) = 500 mm; diffuser length (L) = 1000 mm; thickness (t) =10 mm; open -angle (θ) = 12° and flange length (h) = 221.73 mm and the length diameter ratio (L/D) is 2. All models have a converging duct with the length l = 102.23 mm. However, model 3 has an extra small splitter inside it with the dimensions, splitter L_2 = 300 mm, splitter diameter D_2 = 150 mm and splitter angle α = 14°. All models have a converging duct with a radius of 106 mm and length of 102.23 mm.

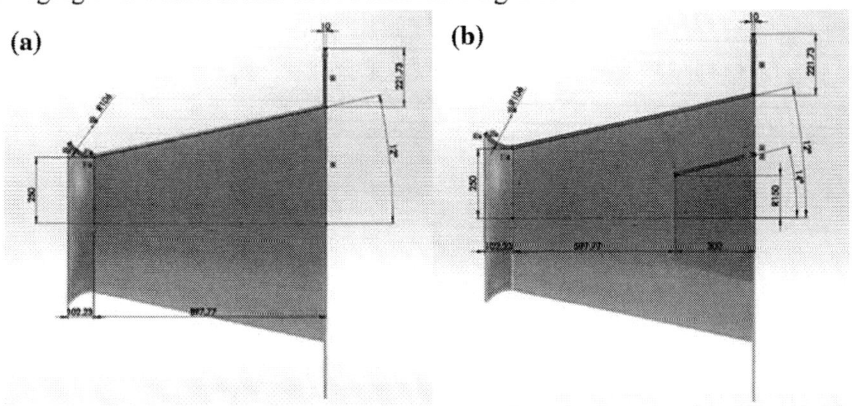

Figure 1. Front plane sketch of modified DAWT Models (a) Model 1 & 2 (b) Model 3

Most of the design are referred to and cited to the previous research by Kannan et al. [3]. From the previous research, it was found that the flange is the most efficient feature to improve the diffuser performance. Therefore, the flange feature is taken into the new design with other features to test the performance. As shown in Figure 2, Model 1 and Model 3 have eight air vents surrounding the outlet of the diffuser but Model 3 has a splitter inside it. The dimension for the air vents is 250 mm in length along the diffuser wall, 50 mm height and 202 mm in hypotenuse. Model 2 has eight pairs of vortex generators inside the outlet. The vortex generators are 5 mm in thickness, 350 mm in height and 100 mm in length with the open-angle of 35°. The air vents idea come from tornado and fire tornado generator; vortex generators are taken from the same application on a racing car; flange and splitter come from previous research [3].

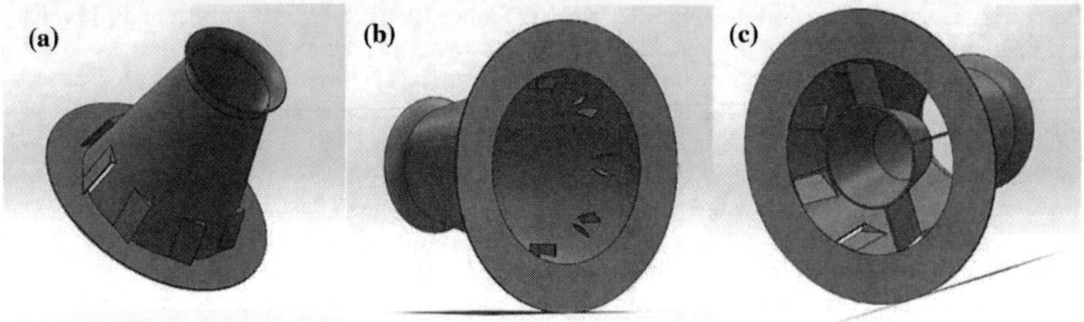

Figure 2. (a) Model 1 Air vents (b) Model 2 Vortex generator (c) Model 3 splitter with air vents

The geometry model is saved and uploaded as IGS file in ANSYS Fluent. As shown in Figure 3, a tunnel is created by using Envelope in *DesignModeler* with 10000 mm height, 10000 mm width and 20000 mm long. Then the model is subtracted from the tunnel by using subtract Boolean in *DesignModeler*. The fluid flow inside the tunnel is set as air and the model is made by aluminium. The mesh is set as hexahedron mesh, initial wind speed is as 4 m/s, initialization is hybrid, the solution method is SIMPLE and 500 interactions are set for calculation. The air velocity and pressure data of X-axis collected and plotted into graphs. Realizable k-epsilon model is used for the simulation with Single Reference Frames (SRF) model [1, 9]. Density and viscosity based flow solver is used. Grid density validation was done as similar to our previous study [9].

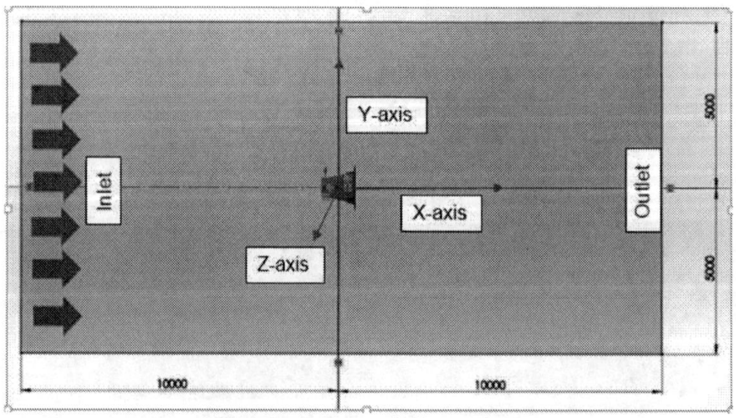

Figure 3. Tunnel dimension

3. Results and Discussions

The inlet air velocity and static gauge pressure at the diffuser inlet (X= 0 m) are taken for all models and reported in Table 1. For Model 1, the inlet air velocity increases up to 8.3079 m/s, which is 207.70%

increment (over 4 m/s initial wind velocity). Model 2 increases the inlet air velocity up to 6.1260 m/s, which is 153.15% increment, and Model 3 increases the inlet air velocity up to 8.5594 m/s, which is 213.985% increment. Model 3 is the best model to improve the inlet air velocity and 52.7% more than the previous model, 6.45 m/s as reported in Kannan et al. [3]. The suction power of the diffuser also can be determined by analysing the static gauge pressure at the inlet centre. The static gauge pressure from lowest to highest was as follows: Model 2 (-6.5172 Pa), Model 1 (-26.3601 Pa), and Model 3, -29.7099 Pa. A negative pressure region formed at the back of the diffuser help to draw more air out of the diffuser and cause the diffuser inlet mass flow rate increase [8-9].

Table 1. Air velocity and static pressure at the inlet (initial wind velocity 4 m/s)

Model	Inlet Air Velocity (m/s)	Static Gauge Pressure (Pa)
1	8.3079	-26.3601
2	6.1260	-6.5172
3	8.5594	-29.7099

Figure 4 shows the change of air velocity in the centre of the diffuser. The diffuser located at X= -0.102 m to 0.898 m and the rotor located at X= 0 m. The graph shows that the air velocity increases from initial air velocity, 4 m/s to highest peak when near the rotor position then decreasing. Model 2 has the lowest peak and decreases rapidly after the inlet. While Model 1 & 3 have a similar characteristic at the beginning due to similar geometry. However, Model 3 shows a slightly high peak with the splitter inside it. The splitter helps to increase the air velocity at the end of the diffuser or after X= 0.5m. The results showed that the splitter successfully was drawn more air out of the diffuser and increases the inlet air velocity.

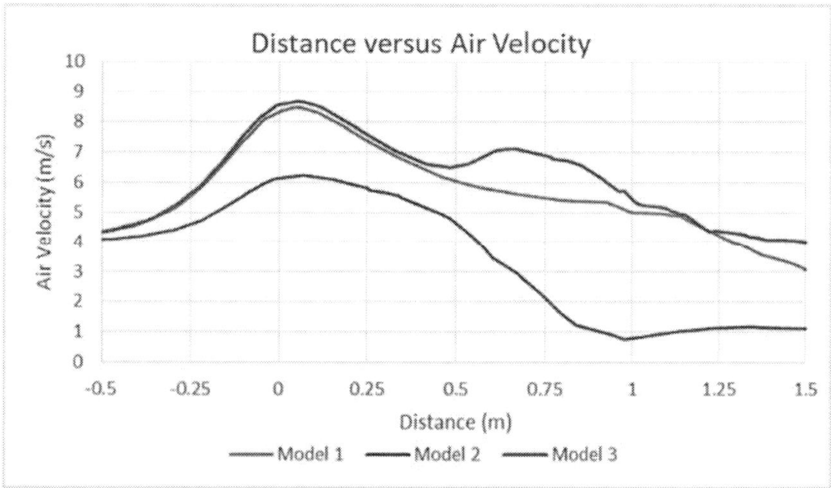

Figure 4. Air velocity across X-axis

Figure 5 shows the change of air static gauge pressure in the centre of the diffuser. The diffuser located at X= -0.102 m to 0.898 m and the rotor located at X= 0 m. All models have the lowest pressure at the inlet of the diffuser. Model 2 has the overall higher pressure compared with Model 1 & 3. Model 1 & 3 has a similar trend line but Model 3 has a slightly lower pressure than Model 1. The splitter at the end of the Model 3 also causes the pressure at the end decrease for the second time. A pressure lower than atmospheric pressure increases the suction power of the diffuser and draws more air.

All models can generate a low-pressure region behind the diffuser outlet due to the flange and larger outlet area. Model 3 has the lowest backpressure according to the analysis on the pressure contour as shown in Figure 6, which is around -11.587 Pa to -2.422 Pa. While Model 1 has a slightly higher backpressure than Model 3, from -6.523 Pa to -2.629 Pa. The backpressure of Model 2 is the highest and varies, approximately about -0.767 Pa. The lowest pressure region occurred at the rotor position where X = 0 m.

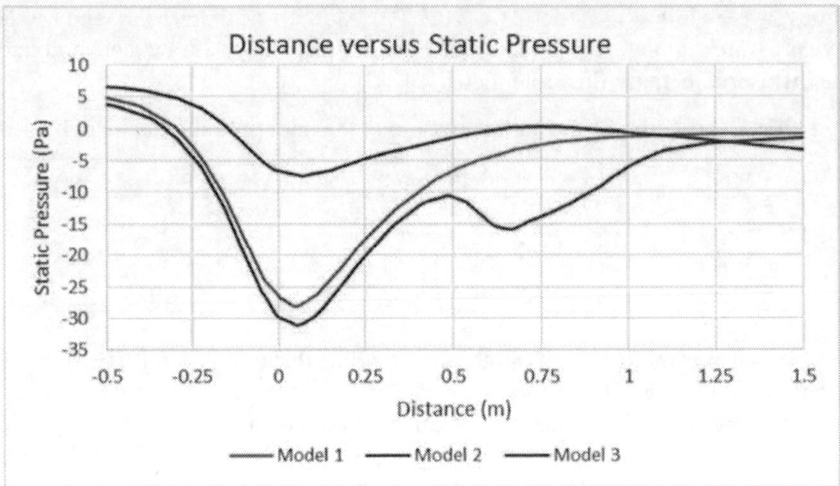

Figure 5. Static pressure across X-axis

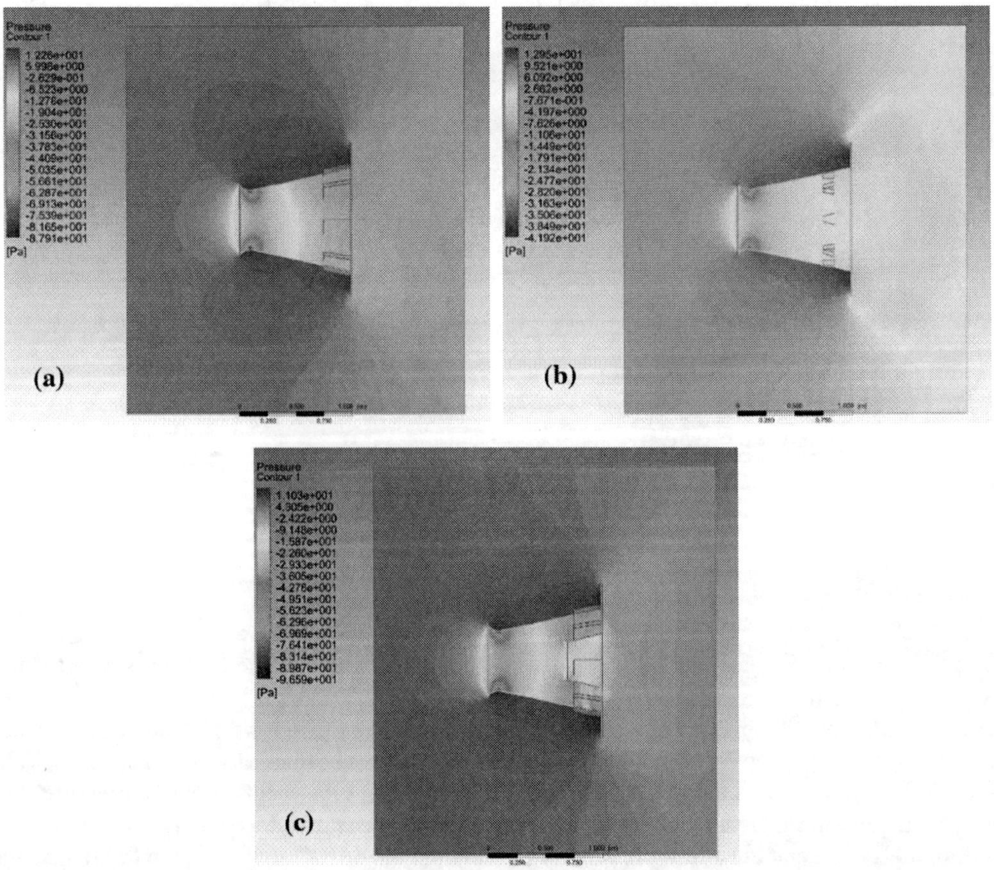

Figure 6. Static pressure contour across X-axis (a) Model 1 (b) Model 2 (c) Model 3

IOP Conf. Series: Materials Science and Engineering **886** (2020) 012023 doi:10.1088/1757-899X/886/1/012023

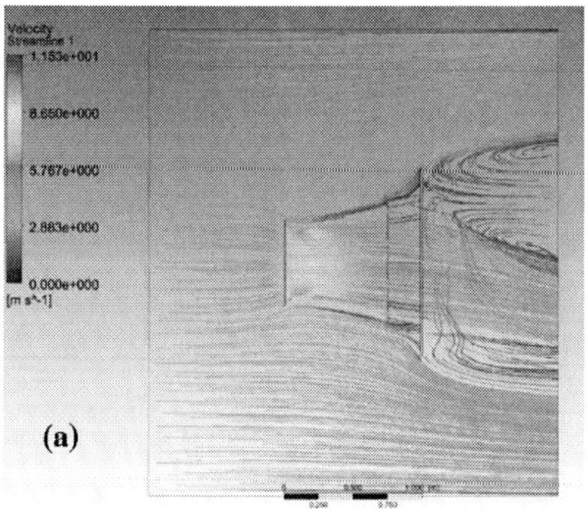

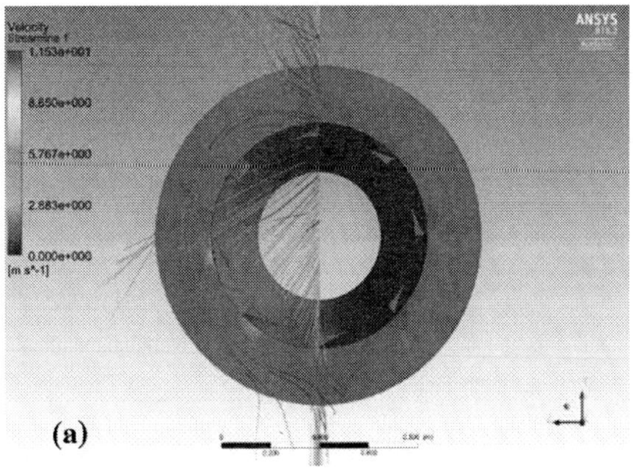

(a) (a)

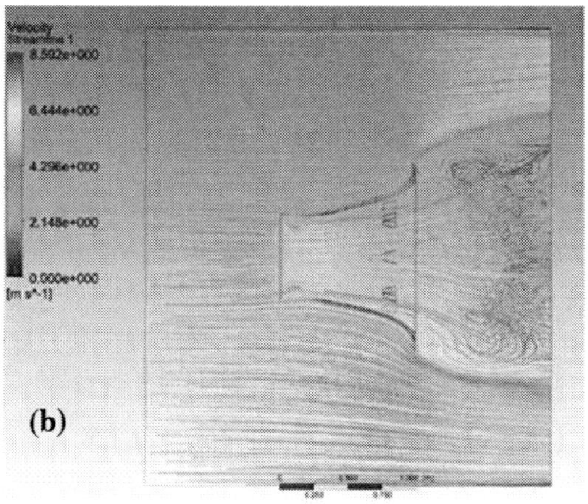

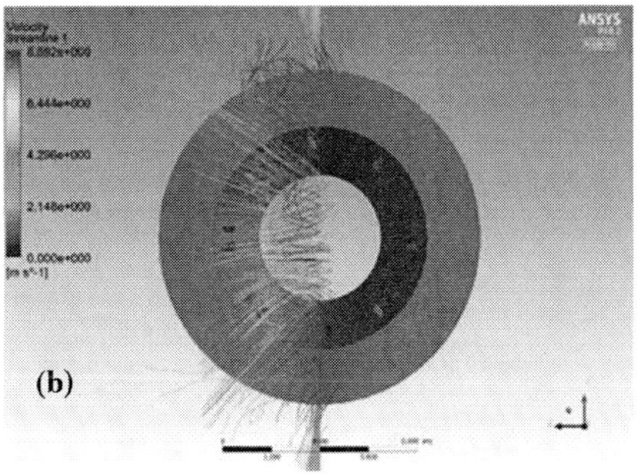

(b) (b)

(c) (c)

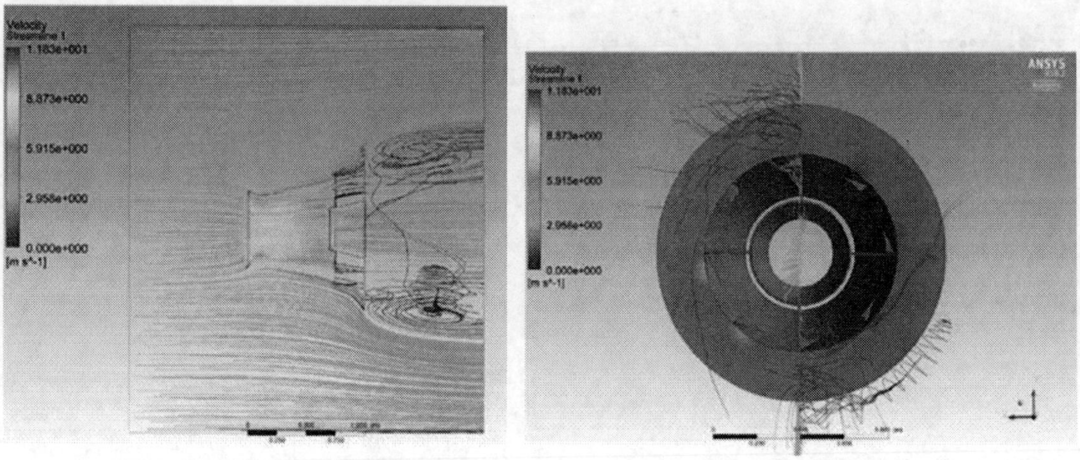

Figure 7. Velocity streamlines and end airflow (a) Model 1 (b) Model 2 (c) Model 3

Model 1 (Air vent features) and Model 3 (Splitter with air vents features) succeed to create a tornado at the end of the diffuser. Model 3 generates a more obvious tornado and the air velocity in the centre is faster. This is due to the splitter, which separated the outer airflow with inner airflow. Model 1, which does not have a splitter, had caused the airflow on both sides imbalanced. Model 2 (Vortex generator features) has a lower air velocity but fairly distributed. The vortices generated at the back of Model 1 and Model 3 are much larger and concentrated. Meanwhile, Model 2 has many small vortices at the back due to the vortex generators.

Figure 7 shows the streamlines on the Z-plane. Figure 7 explains the airflow pattern at the end of diffuser, in addition, how the modified features improve the performance of the diffuser. The airflow of the Model 1 starts to spin slightly after the end while Model 3 has a more obvious spinning of the airflow near a tornado. The airflow in the centre of Model 3 is more stable due to the splitter. The airflow of Model 2 bends towards one side. The higher air velocity flow occurs mostly in the centre of the diffusers.

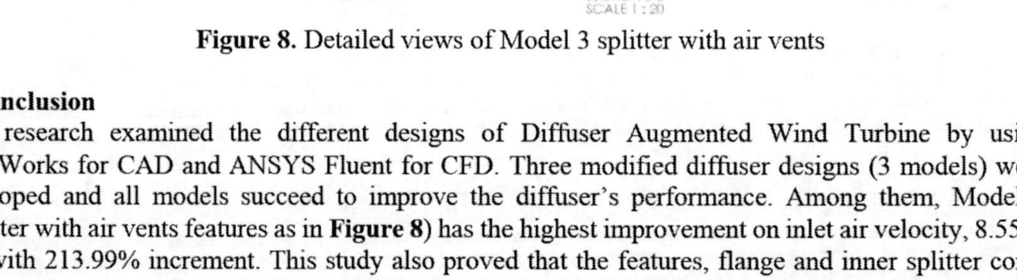

Figure 8. Detailed views of Model 3 splitter with air vents

4. Conclusion

This research examined the different designs of Diffuser Augmented Wind Turbine by using SolidWorks for CAD and ANSYS Fluent for CFD. Three modified diffuser designs (3 models) were developed and all models succeed to improve the diffuser's performance. Among them, Model 3 (Splitter with air vents features as in **Figure 8**) has the highest improvement on inlet air velocity, 8.5594 m/s with 213.99% increment. This study also proved that the features, flange and inner splitter come from previous research are feasible and effective to improve the diffuser's performance. The new idea, air vents and vortex generators also show their strength. The performance of the air vents is more significant. Air vents can provide horizontal airflows and cause the airflow at the end to behave like a tornado. The suction power of the tornado increases the mass flow rate and air velocity at the inlet.

References

[1] Thangavelu S K Goh C Y Sia C V 2019 *IOP Conf. Series: Mater. Sci. Eng.* **501** 012041
[2] Ohya Y Karasudani T Sakurai A Abe K I Inoue M 2008 *Wind Engineering and Industrial Aerodynamics* **96**(5) 524-539.
[3] Kannan T S Mutasher S A Lau Y H K 2013 *Journal of Engineering Science and Technology* **8**(4) 372-384.
[4] Al-Quraishi B A J Asmuin N Z B Mohd S B Abd al-wahid W A Mohammed A N Didane D H 2019 *The International Journal of Integrated Engineering* **11**(1) 178-206.
[5] Roshan S Z Alimirzazadeh S Rad M 2015 *J. Wind. Eng. Ind. Aerodyn.* **145** 270-279.
[6] Sorribes-Palmer F Sanz-Andres A Ayuso L Sant R Franchini S 2017 *Renew. Energ.* **105** 386-399.
[7] El-Zahaby A M Kabeel A E Elsayed S S Obiaa M F 2017 *Alexandria Engineering Journal* **56** 171-179.
[8] Jadallah A A Farag S R Hamdi J D 2018 *Journal of Engineering Science and Technology* **13**(7) 1891-1904.
[9] Hooi L B Thangavelu S K 2018 *IOP Conf. Series: Mater. Sci. Eng.* **297** 012057

A study of Rate of Temperature Change during Tire Curing Using Computational Fluid Dynamics

Watthanai Thamsamerdee[1], Kunthakorn Khaothong[2] and Jetsadaporn Priyadumkol[1,*]

[1]Department of Mechanical Engineering, Faculty of Engineering, Mahidol University, Nakhon Pathom, 73170, Thailand

[2]Department of Mechanical Engineering, Faculty of Engineering at Kamphaeng Saen, Kasetsart University, Nakhon Pathom, 73140, Thailand

* Corresponding Author: jetsadaporn.pri@mahidol.ac.th

Abstract

The most important procedure in tire manufacturing is curing. The purpose of the curing is to apply high pressure and temperature over a period of time to unvulcanised rubber so that the rubber is cured to the point of best material properties. In this research, fluid flow analysis was considered to understand airflow pattern in an autoclave for a better quality of the product. We present preliminary results of the temperature distribution of heat transfer in an autoclave using Computational Fluid Dynamics package, OpenFOAM. The turbulent model was verified by comparison with the experimental results to obtain an accurate heat transfer coefficient. It was found that the error was 12%. The results from the simulation and experiments are in good agreement. Consequently, this turbulent model can be used to determine the heat transfer coefficient to achieve a better and more efficient curing process.

Keywords: Computational Fluid Dynamics,Curing, Heat transfer coefficient, Simulation

1. Introduction

The curing method is the longest in tire manufacturing. The tires are cured for a certain condition in order to get the properties. The curing process is accomplished under pressure and temperature provided by an autoclave. Autoclaves are cylindrical pressure chambers with doors to manufacture the rubber parts. The curing method is energy-consuming to minimize the cycle of time and understanding the behaviour of fluid flow in the curing method, the simulations were performed using CFD software.

Finite volume and finite difference methods have been widely used to solve CFD problems [1]. These methods require mesh generation in the simulation. It is very complicated for simulations, especially when the mesh has a complex and large shape. Flow simulation based on time and mesh during the simulation [2-4]. Computational fluid dynamics has been used by many researchers to analyse the problems of heat transfers and fluid flow. In 2013, Hamlin [5] presented the distribution of steam in an autoclave using CFD analysis through the ANSYS program. The SST k-ω model was used as turbulence modelling. The SIMPLE scheme was used to solve the pressure-velocity coupling calculations. The results showed that the model can be used to predict the general trends of steam distribution in the chamber. Ismail et al. [6] studied an optimization of the sterilization process. The experiment carried out temperature and time for two different types of retorts or autoclaves which were water immersion and water spray. The temperature distribution was considered to optimize performance. In 2015, Mosna and Vignail [7] used a k-ω SST turbulence model to predict the fluid flow and energy equation within the baking autoclave for sterilizing the packaging. The results showed that non-tray was heater than tray type, which was caused by the tray blocking the air flow. Chen et al. [8] used ANSYS to simulate temperature distributions of the frame mould for the autoclave forming process. The k- ε turbulence model was selected. The result showed that the maximum temperature appears at the corner of mould and the lowest is at the centre of the back. Their simulation reported accuracy of 9% for the temperature distribution. In 2017, Yongqian et al. [9] studied autoclave variables including pressure, velocity and heating rate. They proposed a 3D symmetries model with Boussinesq equation which use a constant density fluid model. It was found that increasing the velocity improves the temperature uniformity, while raising the pressure and the heating rate have a negative impact on the temperature uniformity. Gao et al. [10] compared the RNG k-ε model and standard k-ε model on simulating the unsteady flow. The results showed that the RNG k- ε was the high precision model compared with benchmark data. Bohne et al. [11] presented airflow and heat transfer inside the autoclave. The heat transfer was measured by lumped mass calorimeters. The effect of calorimeter measurements had been compared with the simulation results. The experiment and simulation results were 74 $Wm^{-2}K^{-1}$and 60.5 $Wm^{-2}K^{-1}$, respectively. The simulation results showed a complicated turbulent flow pattern in the autoclave using the commercial software package ABAQUS CFD. Zhang et al. [12] compared various turbulent models for prediction of the airflow pattern and turbulent using FLUENT (version 6.2). These models were conducted to cover a wide range of CFD approaches which are Reynolds averaged Navier-Stokes (RANS), hybrid RANS, and detached eddy simulation (DES) and large eddy simulation (LES). This study used the relative error between numerical results and experimental results at measured significant criterion. The results show that if the error is less than 10%, the model accuracy is rated as A. Rating B is given to predictions with relative error of less than 20-30%. While rating C is given to the error about 30-50%. And if the error is larger than 50%, the model accuracy is rated as D.

In this study, the CFD was used to simulate using OpenFOAM to study the airflow pattern during curing processing. The heat transfer coefficient (HTC) results were validated against the value obtained from experiments [11]. To understand the behaviours of fluid flow, this method was also used to inspect the distribution of temperature inside the autoclave to develop the performance and decreasing the manufacturing times.

2. HTC measurement

The gas flow patterns in an autoclave are quite complex, for both empty and loaded autoclaves.
Arafath et al. [13] proposed experiments to measure the heat transfer coefficient (or HTC) distribution at calorimeter inside the autoclave. The results showed that calorimeter was promising options to measure HTC values. The heat transfer coefficient (HTC) in an autoclave can be measured by [11]:

$$h = \frac{mc_p\frac{dT}{dt}}{(T_\infty - T)A} \quad (1)$$

where m is the mass of the aluminum plate, T refers to the autoclave fluid temperature, $T_{\propto}$ the temperature within the calorimeter and A refers to the lower and upper free surfaces of the aluminum plate. The specific heat capacity Cp of the aluminum plate was determined using DSC-measurements and was modeled temperature-dependent

$$c_p = (-0.0101T^2 + 7.8934T - 657.8658) \quad (2)$$

3. Numerical model

Temperature masurement was carried out in an autoclave, based on the experiment [11], to be used for validation with the computational result. The dimensions of the autoclave using SolidWorks 2010 program are shown in Figure 1.

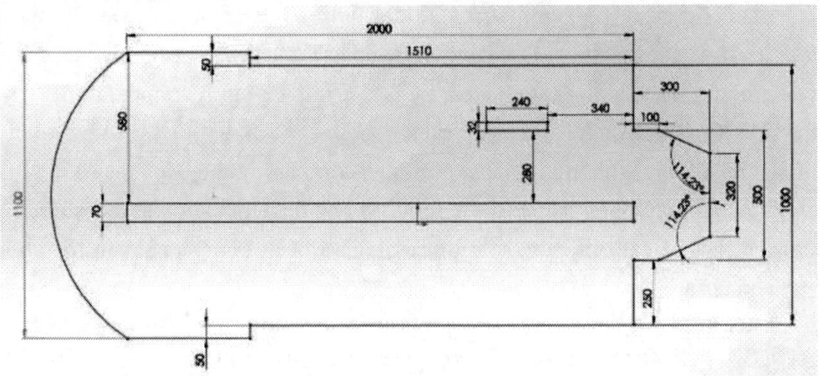

Figure 1. Dimensions and calorimeter position inside autoclave, in mm.

Computational fluid dynamics model of the autoclave was generated using OpenFOAM software. The transient analysis of the two-dimensional model was solved by the k-ω SST turbulence model for the analysis of turbulent flow. The domain is a quadrilateral of 50,600 meshes. The non-dimensional near- wall grid distance (y^+) of the surface of the calorimeter is between 3.5-6 [11]. The lower the value, the more effective the skin is. The boundary condition is set to have 2 ways velocity inlet 5.058 m/s with temperature 333.15 K and outlet condition is set to be pressure outlet 700,000 Pa as shown in Table 1.

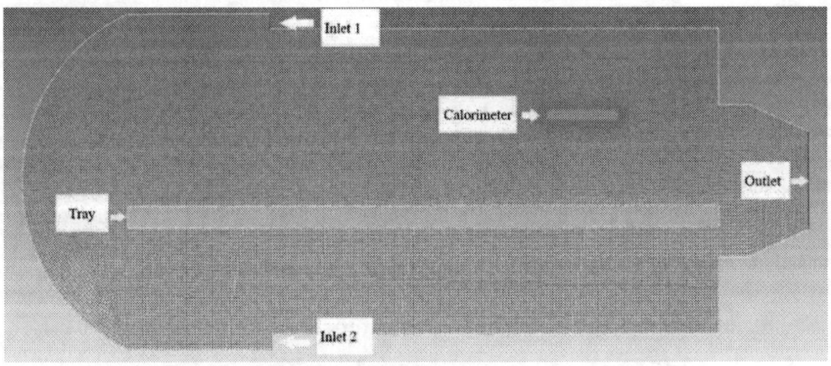

Figure 2. Two-dimensional computational domain.

IOP Conf. Series: Materials Science and Engineering **886** (2020) 012024 doi:10.1088/1757-899X/886/1/012024

Table 1. Boundary conditions in an autoclave

Boundary	Setting
Inlet 1 and 2	Velocity inlet 5.058 m/s temperature 333.15 K
	Kinetic turbulent viscosity 2.7e-5 m^2/s
Outlet	Pressure outlet 700,000 Pa
Calorimeter	Temperature 323.15 K
Other	Wall

4. Results and discussion

The results of the CFD simulation at various times in autoclave were reported. The post-processed result was compared with those obtained from the experiment at the height of 5 cm below the calorimeter for validation. The heat transfer coefficient (HTC) difference between those simulated and measured in this autoclave was obtained. The HTC values of the numerical and experimental were 74 $Wm^{-2}K^{-1}$and 82.9 $Wm^{-2}K^{-1}$, respectively. It was found that the data error between the simulation and experimental data was 12%. Therefore, using available CFD software OpenFOAM can predict a phenomenon of turbulence flows in autoclave correctly.

Figure 3 shows the temperature contour at various times which are 20, 40, 60, 80, 100 and 106 seconds, respectively. In 20 seconds, the airflow inlets entered to the autoclave and clashed, causing the turbulence of the air (fig.3a). As can be seen in Figure 3b-3e, the air began to split into two ways: the bottom and the top, after colliding with the object. The temperature of the bottom was lower than the top. The air above had a higher temperature below, but while the bottom was getting higher. Finally, the temperature reached the steady state condition at 106 seconds, as shown in fig. 3f.

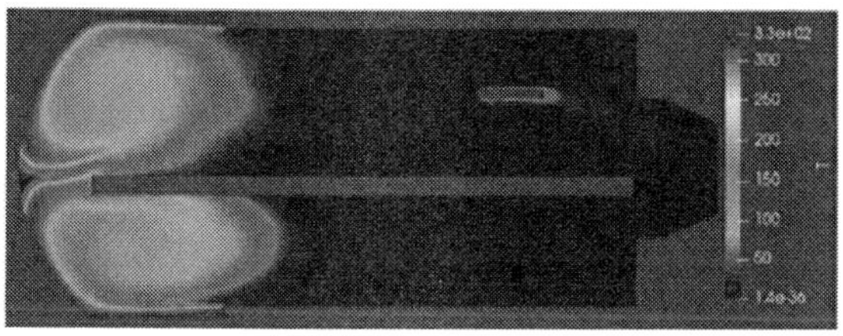

(a)

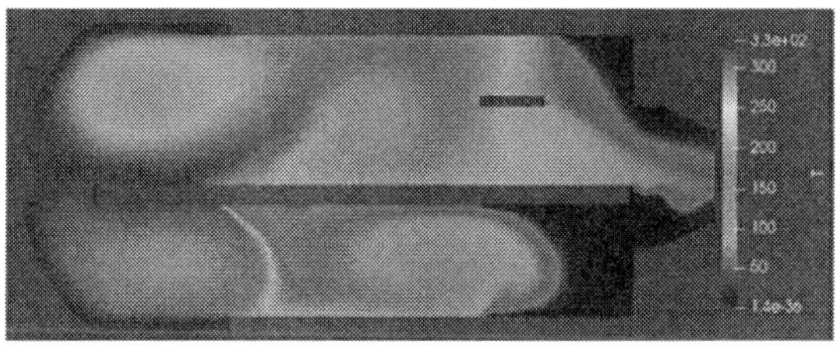

(b)

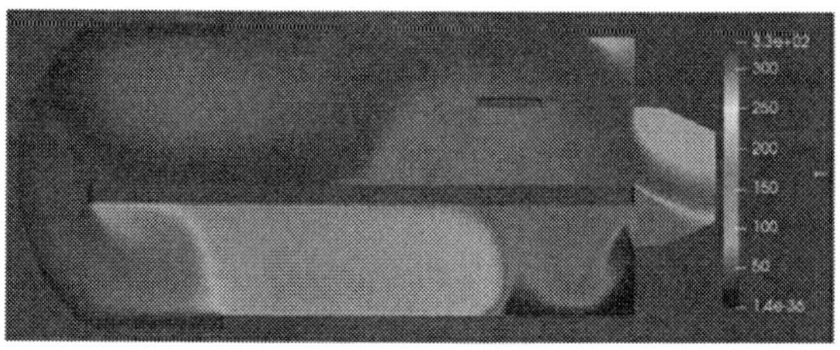

(c)

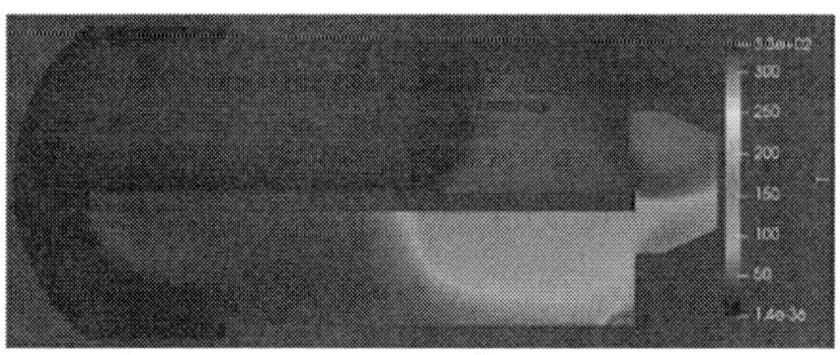

(d)

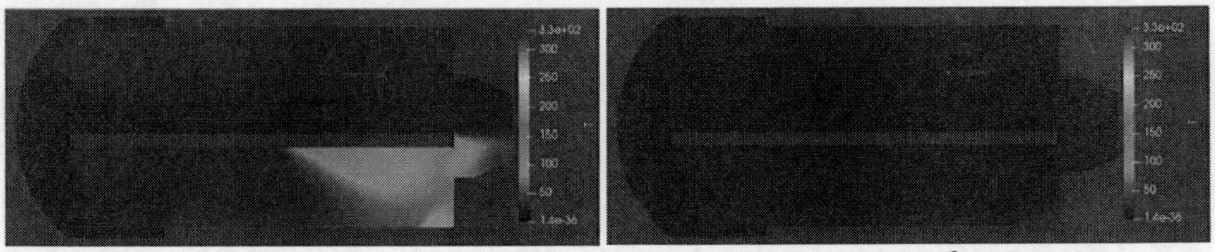
(e) (f)

Figure 3. Temperature contour at various times (a) 20 s (b) 40 s (c) 60 s (d) 80 s (e) 100 s (f) 106 s

5. Conclusion

We present preliminary results of a major systematic study of heat transfer in autoclaves using simple lumped mass calorimeters. A 2-D numerical simulation has been conducted to study the flow pattern. The result of CFD analysis with k-ω SST model had converged in terms of HTC. The k-ω SST model is an enhancement of the original k-omega model. It has the advantage that applied the viscous-affected region without further adjustment [14]. The result has been compared to experiment and appear to be in a good agreement with the experimental data [11]. It has been confirmed that the proposed model can be used for prediction of temperature distribution. The results show that the error was 12%, which suggests that the simulation using the OpenFOAM program of this research is more accurate than the simulation [11]. In the future, a 3-D calculation should be made and the results will be compared to experiment. Then, it will be possible to minimize the cycle time in the curing process.

Acknowledgments
The authors are grateful to the Laboratory of Computer Mechanics for Design (LCMD), Department of Mechanical Engineering, Mahidol University for valuable technical support.

References

[1] Pletcher R Tannehill J Anderson D 2013 *Computational Fluid Mechanics and Heat Transfer* (CRC Press)
[2] Smirnov P Zaitsev D 2004 *St. Petersburg Polytec. Un. J. Eng Sc. Techn.* **36** 70–81
[3] Zaitsev D Schur N 2006 *St. Petersburg Polytec Un. J. Eng Sc. Techn.* **47** 15–22
[4] Zaitsev D Smirnov P Yakubov S Balashov M 2012 *Software and Systems* **2** 32–35
[5] Hamlin P 2012 *Master of Science Thesis, Chalmer's University of Technology*
[6] Ismail I Fahmy A Azab A Abadir M Fateen S-E 2013 *J. Ag and Vet Sc.* **6**(4) 26–33
[7] Mosna D Vignali G 2015 *J. Food Eng.* **11**(6) 715–729
[8] Chen F Zhan L Xu Y 2015 *Iran Polymer J.* **24** 927–934
[9] Yongqian X Lihua Z, Minghui H Rulin S 2017 *Procedia Eng.* **207** 269–274
[10] Gao F Wang H, Wang H 2017 *Procedia Eng.* **205** 3970–3977
[11] Bohne T Frerich T Jendrny J Jürgens J P Ploshikhin V 2018 *J. Composite Materials* **52**(12) 1677– 1687
[12] Zhao Z Wei Z Zhiqiang Z Chen Q 2007 *HVAC&R Research* **13** 14–15
[13] Slesinger N Shimizu T Arafath A R A Poursartip A 2009 *Proc. the 17th International Conference of Composite Materials* 1–10
[14] El-Behery S Hamed M 2009 *Engineering and Technology* **53** 10–11

Efficiency and Emission Simulations of Hydrogen-Fuel City Buses

T Poojitganont[1,*], O Antoshkiv[1] , B Watjatrakul[2] and H P Berg[1]

[1]Chair of Combustion Engines and Flight Propulsion, Brandenburg University of Technology Cottbus - Senftenberg, Cottbus, 03046, Germany
[2]Department of Mechanical and Aerospace Engineering, King Mongkut's University of Technology North Bangkok, Bangkok, 10800, Thailand

* Corresponding Author: poojitth@b-tu.de

Abstract
In this study, the AVL Cruise is implemented to simulate driving parameters including the driving path, the power loss on engine accessories, the gear-shifting scheme and the engine's fuel cut-off strategy influencing the bus fuel efficiency and emission. The Mercedes Conecto LF city bus using a diesel engine with Standardised On-Road Test Cycles was firstly simulated and compared with available literature. The application of hydrogen fuel was then examined and its result was compared with the diesel fuel on the specific fuel consumption. In addition, the driving cycles in Bangkok were simulated. The results show the pattern on both driving cycles with different fuel are similar, however, the fuel consumption of H_2 is significantly less than in the case of Diesel. Moreover, it is also evident that the SORT driving cycle cannot represent the heavy traffic of Bangkok.

Keywords: City Bus, Modelling and Simulation, Performance Simulation, Hydrogen, Efficiency and Emission

1. Introduction

Regarding the pollution emission especially in big cities caused by the transport sector, governments worldwide generate a lot of policies targeting to enhance the efficiency of energy use, support independency from fossil fuels, and also boost alternative fuels [1]. Road transport in Europe has a significant share of total carbon dioxide emissions [2]. The transport segment is the second largest source of pollutant emissions in the EU and despite the measures taken, the volume of emissions is still growing [3]. An expansion of energy application including impact on the greenhouse gas and pollutant emissions are also related to this sector. Based on these statements the reduction of CO_2 has become an important task. In addition, since the fuel consumption and CO_2 emission are proportional, together with other emissions also sensitive to fuel consumption rate, it is clear to mention that the lowering of the fuel consumption induces to emissions decreasing effect as well.

The experimental methods are costly and time-consuming. Therefore, it is not proper to conduct many investigations with changing vehicle specifications and driving cycles. The simulation methodology, on the other hand, has advantages to get results in a short time along with many applications which are also available. Thus, extensive studies can be enabled with varying vehicle factors and driving conditions [4]. In the present works, a commercial program, i.e. AVL CRUISE, has been implemented to simulate the engine's and vehicle's characteristics based on the vehicle dynamics. Engine diagrams, gear shifting scheme, the gear ratio and other input parameters were applied to model the fuel consumption and other significant outputs in the simulation. The fuel consumption of city buses counts on many factors that characterize the technical parameters of the bus and the behavior of the driver, as well as traffic conditions [5, 6].

The alteration of the driving procedures is also an important factor, which influences fuel consumption [7]. Each of the base cycles composes of three sections. In each section, three phases were distinguished: acceleration, constant speed and braking. In this study, the basis cycle called SORT2 has been applied in the early steps/stages. Later the represented Bangkok driving cycle, named BKK2, was implemented.

The uneconomical driving style of the driver can cause an increase in fuel consumption by 5 - 10% compared to economic driving [1]. Secondly, the electric energy on board demands constantly increasing, resulting in a reduction in the overall efficiency of the bus and increased fuel consumption. Even though the drivers were on a similar road, the behavior of them influences the high deviation of fuel consumption quantity [8]. However, in this investigation, the driver was set as a constant factor, whereas the standard driver has been applied.

Based on the properties of the hydrogen-fuelled in the internal combustion engine, the advantage is including wide flammability range, high flame speed and propagation, however, its drawbacks are low ignition energy and small quenching length causing the premature ignition, which is the main problem especially in these kinds of engines [9].

2. Simulation Modelling

2.1. AVL CRUISE

AVL CRUISE is a system for modeling vehicles at the level of drive systems. It supports ongoing tasks during the analysis of vehicle and drive systems in all development phases, from conceptual planning, commissioning and actual testing. Its application covers the demand in the field of conventional car engines, through advanced hybrid systems and electric vehicles. The vehicle model in the AVL CRUISE consists of assemblies and subassemblies. This allows the user to model various vehicles, such as motorcycles, passenger cars, trucks and buses. In addition, it is possible to model all configurations of drive systems, such as hybrid systems, two-motor concepts or advanced transmission systems [10].

2.2. Bus Modelling

The 12-meter-long urban bus MAXI Mercedes Conecto 12LF has been chosen for investigation. The basic technical data of the bus including the drive units are shown in Table 1. These basic variables have been implemented into the simulation program. The power train unit is a supercharged self-ignition Diesel engine powered by a Common Rail system and meeting the Euro V purity standard.

Table 1. Mercedes Conecto 12 LF parameters [11].

Parameter	Valve
Dimensions	11950 x 2550 x 3076 mm
Wheelbase	5845 mm
Own weight	10860 kg
Engine	OM 926 LA

Type	4-stroke, in-line, 6 cyl.
Displacement	6370 cm^3
Power	205 kW
Torque	1120 Nm @ 1300 rpm
Fuel consumption	39 dm^3 / 100 km
Transmission	Voith 854.3 - 4 speed
Gear ratio	1-5; 2-1.43; 3-1.0; 4-0.7
Rear axle system	ZF AV-132 Gear ratio: 7.38

The model has been divided into three parts. The first one concerns the vehicle. It includes object properties such as mass, dimensions, location of the center of gravity. The second part involves the powertrain including engine, clutch and gearbox, whereas the third one relates to the driveline, for example, rear differential, brake system and tires. The schematic diagram of the bus model is illustrated in figure 1.

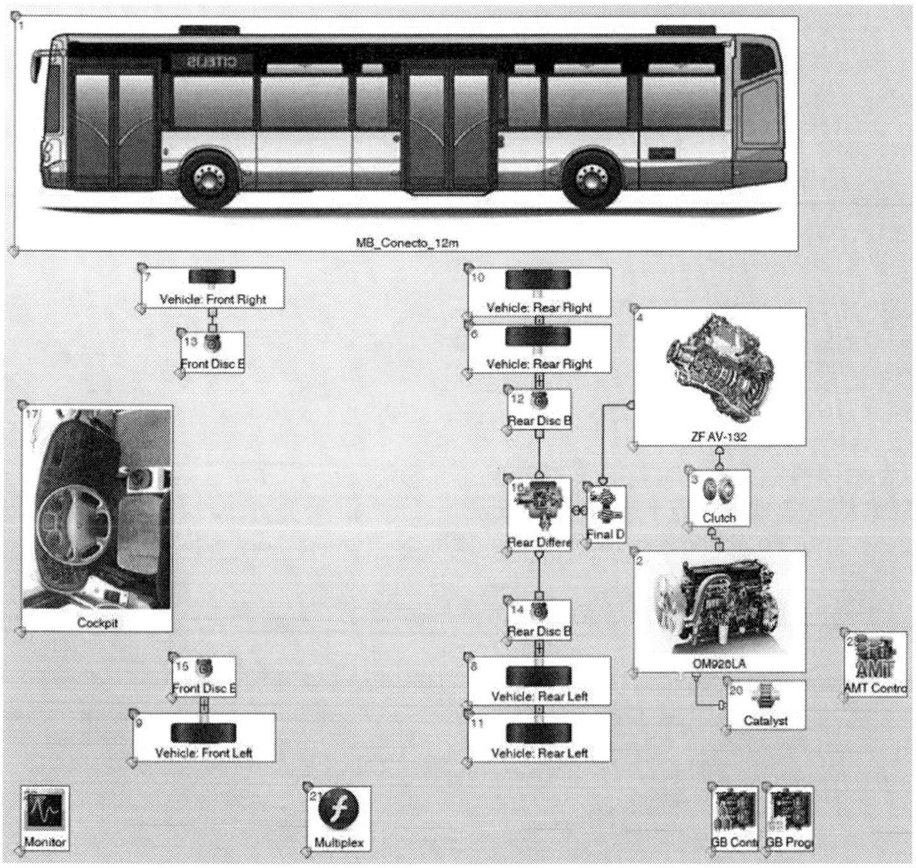

Figure 1. Schematic presentation of the bus model in AVL Cruise

2.3. *Driving Cycles*

As already mentioned the driving cycle is one important input parameter for the vehicle system simulation. It can be apprehended that the vehicle would undergo this driving path repeatedly thoroughly its route. By statistical it stands for an essential driving path for the vehicle fleet of a city. Moreover, this driving path is also supposed to be repeated uninterruptedly [12].

In this study, two driving cycles have been involved, i.e. SORT2 and BKK2. The explanation of them is introduced in the next section.

2.3.1. Standardised On-Road Test Cycles - SORT

SORT is a driving cycle for measuring the busses' fuel consumption under the aim of consumer protection and information [13]. It is created by the UITP - International Association of Public Transport in cooperation with the bus producers and public transportation providers mainly in Europe. SORT is employed to correlate between buses regarding their fuel consumption in true working conditions. There are three different driving cycles available, including SORT1, SORT2 and SORT3. They represent also the different traffic conditions, which are "Heavy Urban", "Easy Urban" and "Suburban", respectively.

In this study, the SORT driving cycle has been used in the early step to validate the simulation program with the selected publication [2]. Based on this reason, SORT2 has been applied. The characteristic of this driving cycle on a velocity versus distance and time is shown in figure 2.

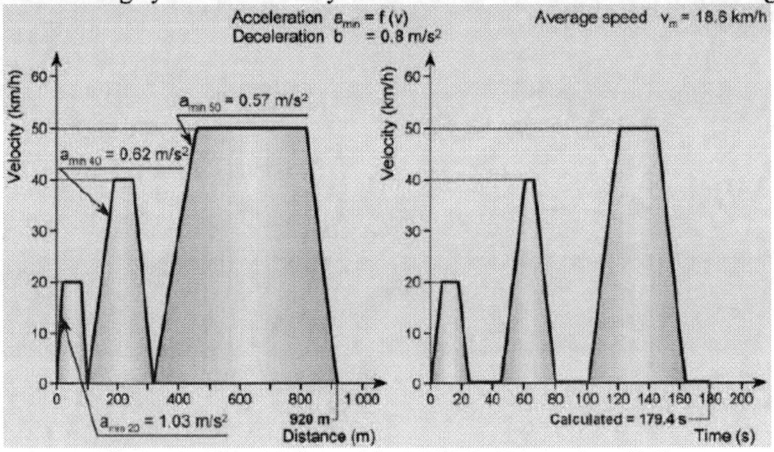

Figure 2. The driving cycle SORT2: Easy urban [13].

2.3.2. Bangkok Driving Cycle

In order to assess exhaust emissions and fuel consumption of newly registered automobiles in Thailand, the standard driving cycle of the European Community is ground on [14]. However, regarding the traffic condition especially in Bangkok, the European driving cycle cannot represent the exact situation.

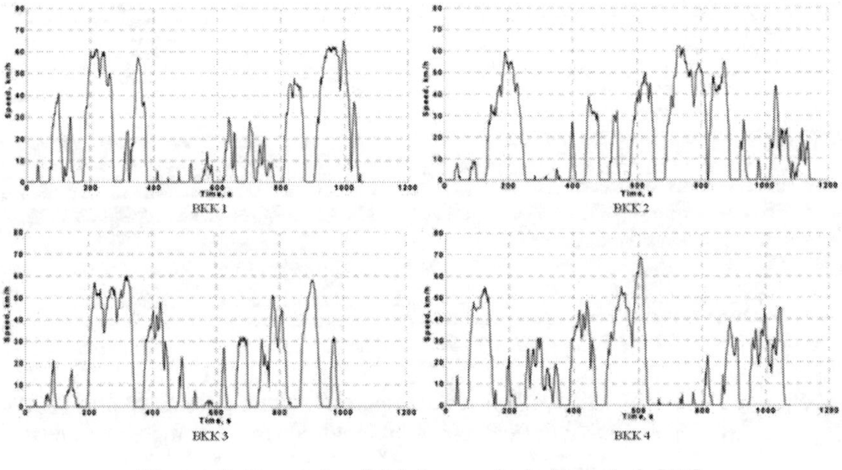

Figure 3. Example of driving cycle in Bangkok [12].

Tamsanya et al. [12] have proposed the driving cycle for Bangkok generated by means of numerical simulation based on the traffic flow data input. In their study, the traffic stream information at each selected junction in the city center of Bangkok observed. The highest traffic quantity takes place in the morning period from 7 to 9 O'clock. The physical speed and time outlines are contained small parts of driving information, named micro-trips, separated by idling time. It can be summarized that the main driving paths in Bangkok are short driving trips with low speeds and often stops. The usual stops and high-velocity fluctuation are the causes of extended emissions and fuel consumption, as pointed by Ericsson [15], Haan and Keller [16]. The generated driving cycles from the experimental input data are illustrated as an example in figure 3.

In order to select the most substitute for the true driving circumstances, the driving variables of the created driving cycles must be nearest to the pointed statistics. Their research work concluded that the most representative of Bangkok traffic is the BKK2 driving cycle.

3. Validation

The validation on AVL Cruise has been done earlier step of this study. The vehicle and engine characteristics following the selected publication [2] have been set up. The conducted simulation tests to confirm the possibility of using the AVL Cruise software to investigate the vehicle system. The 12-meter-long MAXI Mercedes Conecto 12LF equipped with the Mercedes engine OM926LA has been examined. The SORT2 has been applied for the driving cycle as well.

The comparing results are shown in this section. In figure 4 the comparing on engine power and acceleration along the driving cycle SORT2 is illustrated. The compatibility between the investigation in this work and the publication is significantly acceptable.

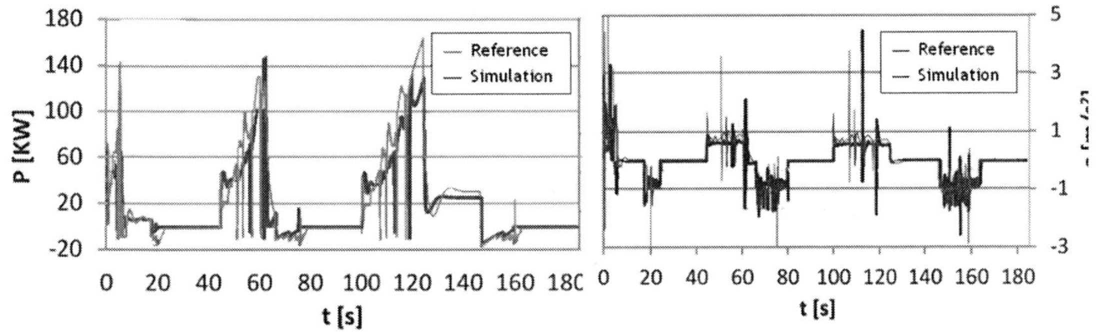

Figure 4. Comparison of engine power and vehicle acceleration with the reference data.

Further on emission CO_2 and fuel consumption are also presented in figure 5. The results are agreed in the right manner.

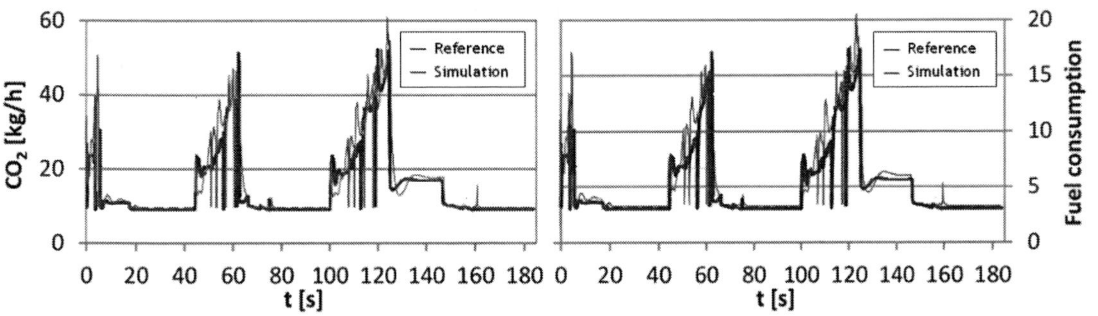

Figure 5. Comparison of CO_2 and fuel consumption with the reference data

4. Simulation Results and Discussion

As the aims to implement the complicated driving condition such as inside the city of Bangkok and also to propose the alternative way for public transportation using H_2 fueled engine. The simulation on the BKK2 driving cycle, which represent more closely to the real condition of traffic in Bangkok, has been accomplished. In addition, the investigation on changing Diesel to H_2 has also been achieved.

4.1. Variation of Driving Cycles

The results of acceleration on SORT2 comparison with the BKK2 driving cycle are illustrated in figure 6. On SORT2, the maximum acceleration of the vehicle is not more than 2.39 m/s^2, while the deceleration during the braking time does not exceed 2.48 m/s^2. Conversely, on BKK2 both maximum acceleration and deceleration are higher than the SORT2, their values are 3.47 m/s^2 and 4.36 m/s^2 as shown in the graph, respectively.

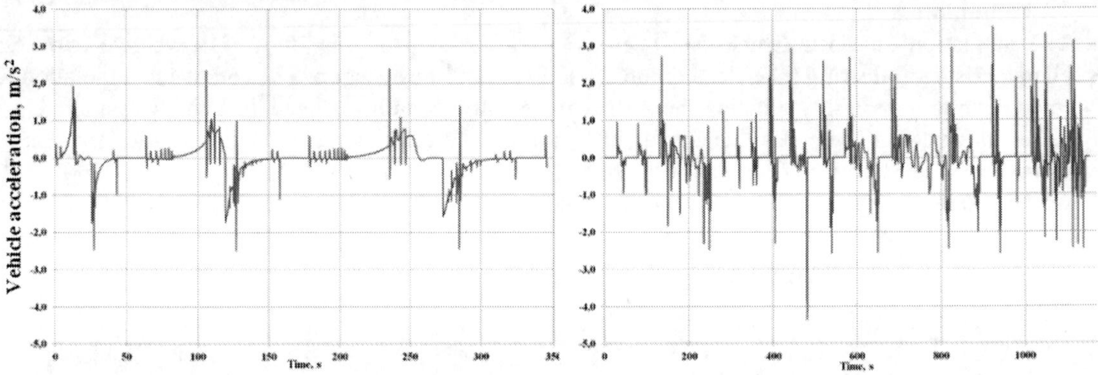

Figure 6. Diesel bus acceleration on SORT2 (left) and BKK2 (right).

Further comparison of engine power is exhibited in figure 7. It is evident that on BKK2 driving cycle the bus engine needs c.a. 33.3% more maximum power than on the SORT2. Its maximum value is reached 194.77 kW in 727 s of the test.

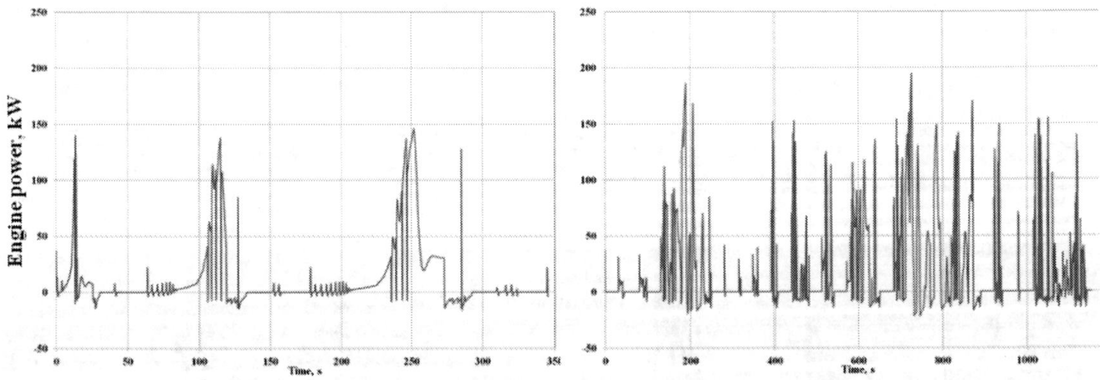

Figure 7. Diesel bus engine power on SORT2 (left) and BKK2 (right).

The emission of carbon dioxide depends on the fuel consumed by the drive unit. In the case of Diesel, the fuel consumption in SORT2 is approximately 69.50% of fuel consumption in BKK2. The results are presented in figure 8.

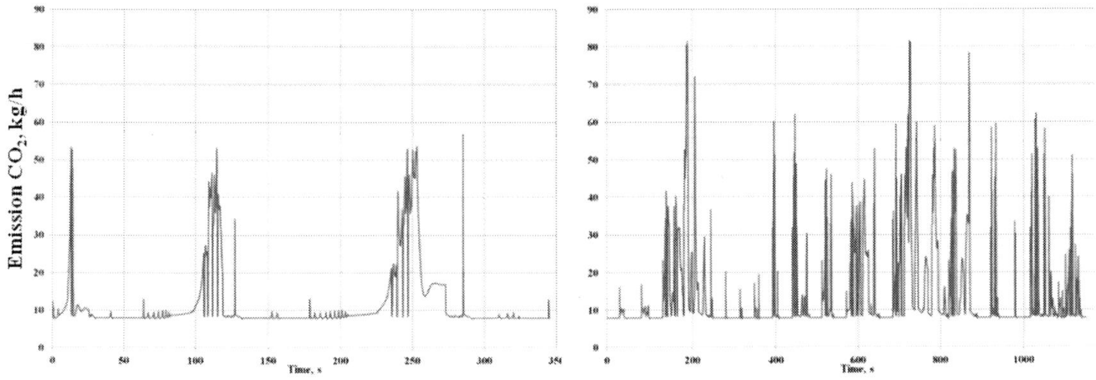

Figure 8. CO_2 from Diesel bus on SORT2 (left) and BKK2 (right).

4.2. Applying Hydrogen

In this phase of work, the hydrogen fuel has been applied based on the Diesel engine model configuration. The fuel properties including engine maps have been implemented. In addition, the weight-increase due to the larger hydrogen tank has been ignored.

The diagram on dimensionless specific fuel consumption between Diesel bus and hydrogen bus running on the BKK2 driving cycle is shown in figure 9. It is evident that the specific fuel consumption in different fuel on the same bus configuration has also the similar characteristic, however, the value between these two types of fuel are significantly different.

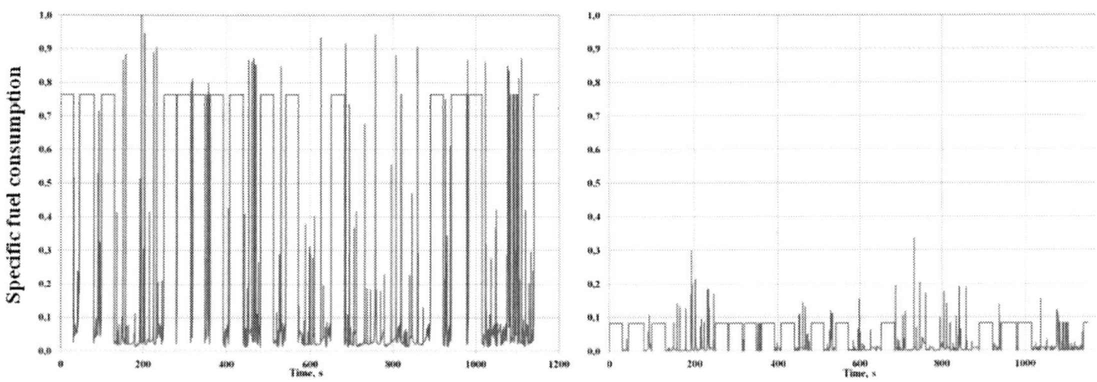

Figure 9. Specific fuel consumption on BKK2 driving cycle of Diesel bus (left) and H_2 bus (right)

Conclusion

In order to design, evaluate or develop the new vehicle and drive systems, the simulation method would be a bright solution because the consumption of costs and time can be economized. The vehicle specification and driving mode can be varied during the simulation. In addition, the components, i.e. gross weight, engine configuration, gear-shifting strategy, gear ratio, tire radius, etc., can be also modeled, tested and optimized in a short period of time.

The results for different driving cycles clearly confirm that the SORT – Standardized On-Road Test Cycles could not well represent the complicated traffic condition in Bangkok. Regarding the acceleration together with the CO_2 results, it can be concluded that the steady or cruising speed-driving pattern are cleaner than acceleration and deceleration pattern. It also corresponds to the former study that the Bangkok driving cycle causes a bigger quantity on both exhaust emissions and fuel consumption.

The combustion between hydrogen and air can also produce oxides of nitrogen (NOx) as a product. However, with its lower final combustion temperature, its amount emitted in the exhaust is significantly reduced. Moreover, the produced oxides of carbon, i.e. CO, CO_2, are not available, since carbon is not a component as in conventional fossil fuel. Based on the emission aspect, hydrogen is considered as the forthcoming fuel especially for the public transportation in cities, however, the pre-mature ignition and its storage are the big issues, which have to be solved in the near future as well.

Acknowledgments
Authors would like to thank Mr. Jedchan Velasquez for the simulation support. This research is accomplished under collaboration of the DAAD "Praxispartnerschaften zwischen Hochschulen und Unternehmen in Deutschland und in Entwicklungsländern" program between the BTU C-S and KMUTNB. The AVL-CRUISE software is also supported by the AVL List GmbH under the "AVL AST UNIVERSITY PARTNERSHIP PROGRAM". Several results are financially supported in the frame of the project by AiF Projekt GmbH.

References
[1] Ribau J Silva C 2017 *Proc. Int. Conf. on Vehicle Tech. and Intelligent Transport Systems* (Porto) 392
[2] Grabowski Ł Gęca M Biały M 2015 *Combustion Engines* **162** 475
[3] Wang B Luo Y Zhang J 2008 *Int. J. of Automotive Technology* **9** 501
[4] Oh Y Park J Lee J Eom M Park S 2014 *Transportation Research Part D* **32** 287
[5] Chłopek Z 2006 *J. of KONES Powertrain and Transport* **1** 237
[6] Antoshkiv O Berg H P Bondarenko E 2014 *Wärmemanagement des Kraftfahrzeugs IX; Haus der Technik Fachbuch*
[7] Biggs D Akcelik R 1987 *J. Transport. Eng.* **113** 101
[8] Vieira da Rocha T Can A Parzani C Jeanneret B Trigui R Leclercq L 2013 *Transp. Res. Part D: Transp. Environ.* **24** 17
[9] Lanz A Heffel J Messer C 2001 *Hydrogen Fuel Cell Engines and Related Technologies* (Palm Desert: College of the Desert) chapter 3
[10] AVL CRUISE, AVL Deutschland GmbH, accessed 13 June 2019 <https://www.avl.com/cruise>
[11] Der Conecto. Technische Information, A&O GmbH, accessed 13 June 2019 < https://daimlerbuses-printshop.com/en/mercedes-benz/technical-brochures.html?p=5 >
[12] Tamsanya S Chungpaibulpatana S Limmeechokchai B 2009 *Int. J. of Automotive Technology* **10** 251
[13] UITP Internatinal Association of Public Transport 2009 *Project SORT Standardised On-Road Tests Cycles*
[14] Thai Industrial Standards Institute (TISI) 2003 Thai Industrial Standard (TIS 2160-2003) Gasoline Engined Vehicles: Safety Requirements; Emission from Engine, Level 7. TISI. Bangkok
[15] Ericsson E 2001 *Transportation Research Part D* **6** 325
[16] Haan P Keller M 2000 *Atmospheric Environment* **34** 4629

Surrogate-assisted Meta-Heuristic method for Aerodynamic Design of an Aircraft Wing

Cho Mar Aye, Nantiwat Pholdee* and Sujin Bureerat

Sustainable and Infrastructure Research and Development Centre, Department of Mechanical Engineering, Faculty of Engineering, Khon Kaen University, 40002, Thailand.

* Corresponding Author: nantiwat@kku.ac.th

Abstract

This paper presents the aerodynamic design of an aircraft wing using a surrogate-assisted meta-heuristic (MH) method. The optimization problem is posed to find the wing shape in order to maximize its lift-to-drag ratio. Two surrogate-assisted MHs are presented which are the use of a surrogate model to predict the objective function directly and using a surrogate model to predict lift coefficient and drag coefficient separately before calculating the objective function. Computational fluid dynamic analysis is used for calculating the values of lift and drag coefficients while a Differential Evolution (DE) algorithm and a radial basis function are respectively used as MH and a surrogate model. From this study, the performance of the surrogate-assisted design approaches for aerodynamic optimization of an aircraft wing is obtained. The attained results are said to be the baseline for future study of surrogate-assisted aerodynamic optimization.

Keywords: Evolutionary algorithms, Metamodel, Aerodynamic of aircraft wing, Computational Fluid Dynamics

1. Introduction

In an aircraft design, aerodynamic shape optimization is necessary while Computational Fluid Dynamics (CFD) and optimization method involve important parts in the design activity. The optimization method that is commonly used for real engineering design activity and is divided into two groups as gradient-based and non-gradient-based methods. The gradient-based method is efficient with high convergence speed and searches consistency. However, they require function derivatives and usually can be trapped at a local optimum. For the gradient-free methods, the best-known method is meta-heuristics (MHs), which is said to be a global optimization method (they can avoid a local optimum trap) while function derivatives are not required. As a result, they can deal with any kind of objective function and design variables. However, most of MHs have a lack of search consistency and a large number of function evaluations required. It is difficult to use MHs for a problem with computational expense such as CFD based aerodynamic shape design [1][2] Therefore, the development of an optimization technique for a particularly computational expensive problem is an interesting research direction.

Nowadays, the use of surrogate-assisted MHs is presented for optimization of many computational expensive problems. The main idea is to use a surrogate model for estimating the real computational expensive function values and perform optimization by MHs [3][4][5]. The surrogate model is built based on many techniques such as polynomial regression, radial basis function, Kriging, support vector regression [3] and neural network. For the aircraft wing design for aerodynamic optimization, various researchers have studied surrogate assisted MHs and applied the approaches to various aerodynamic design optimization problems [6][7][8][9][3]. However, most of the work usually use the surrogate model for predicting the objective and constraint function values directly. The use of surrogate models for predicting a part of the parameter in the objective and constraint is rarely studied in the field of aerodynamic design.

In this study, a surrogate assisted differential evolution algorithm for 3D shape aerodynamic optimization of an aircraft wing is presented. The objective function is posed to maximize the lift to drag ratio while the design variables determine wing shape. Two surrogate-assisted DE are presented which are the use of a surrogate model to predict the objective function directly and using a surrogate model to predict lift coefficient and drag coefficient separately before calculating the objective function. Section 2 shows a short explanation of the procedures applied in this paper. Next section presents numerical simulations of CFD flow solver and optimization process. In Section 4, the results are expressed. Finally, conclusions are shown in Section 5.

2. Problem Statement /Statement of Problem

In general, aerodynamic performance optimization is one of the main challenges in aircraft design. It can be taken into consideration in all design stages i.e. conceptual, preliminary, and detail design. The most commonly used design merit function for aircraft aerodynamics is the lift-to-drag ratio since it determines aircraft range and endurance. Such a design objective can be predicted using a panel method for medium fidelity computation; however, it has been shown in the literature that the use of CFD simulation spectacularly agrees with wind tunnel test. Thus, aircraft aerodynamic design using CFD is one of the research issues in aerospace engineering. The optimization problem of a wing can be expressed in the following mathematical model

$$\text{Maximize } f(\mathbf{x}) = C_l/C_d, \qquad (1)$$

$$s.t.\ \mathbf{x}_l \le \mathbf{x} \le \mathbf{x}_u$$

where $f(\mathbf{x})$ is an objective function of the problem, $\mathbf{x}_l$ and $\mathbf{x}_u$ are lower bound and upper bound of the design variables vector $\mathbf{x}$. In this work, the objective function is to maximize the lift-to-drag coefficient (C_l/C_d) while the design variables are the Lancer wing airfoil sections. In this case, the Lancer wing is shaped using five air foil sections as shown in Figure 1. Therefore, there are five design variables ($\mathbf{x}=[x_1, x_3, x_3, x_4, x_5]^T$) in this study, where x_i represent the 4 digit NACA (00**) airfoils and $x_i \in$[Naca0010, Naca0011, Naca0012, Naca0013, Naca0014, Naca0015, Naca0016, Naca0017, Naca0018, Naca0019]. In the terminology of the four-digit NACA airfoil, the first digit is the value of the maximum camber (stated in percent of the chord) while the second digit is the maximum camber (measured from the leading edge in tenths of the chord). The last two digits represent the maximum thickness of the airfoil in percents of the chord [7]. The Lancer 200 wings (downloaded from openvsp.org) are used for the design example in this work.

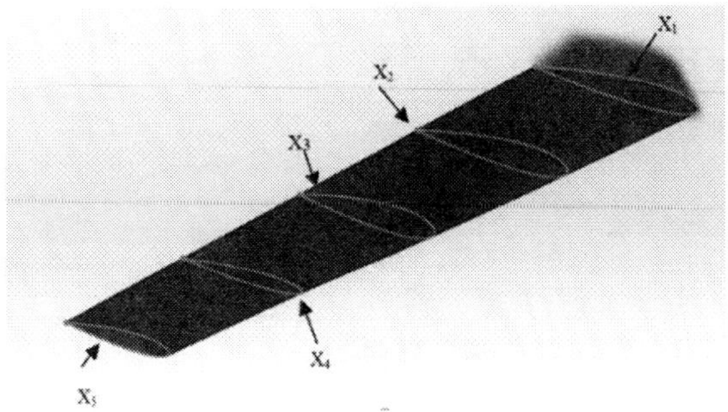

Figure 1. Lancer 200 aircraft wing based on five design variables of airfoil sections

3. Numerical simulation and optimization process

In this section, detail of CFD-based aerodynamic analysis and surrogate assisted MHs used in this study are given. CFD simulation is reliable and accurate for aerodynamic analysis, however, it is time-consuming to use. As a result, we have to exploit a surrogate model for an optimization process.

3.1. CFD Flow Solver and Mathematical Model

Simulation of fluid dynamics of engineering systems is achieved by numerical methods, in which a studied domain is discretized into a finite number of simple shape elements. The most popular technique for fluid dynamic analysis is a finite volume method often referred to as computational fluid dynamics (CFD) [10]. In this paper, the value of lift and drag coefficients are estimated by using CFD simulation. The details of a CFD model are as follows.

3.1.1. Aircraft Wing Geometries and Arrangement

Aircraft wings are constructed with various shapes and sizes for preferred flight characteristics of an airplane to obtain greater lift, balance or stability in flight. In this paper, we used the Lancer 200 aircraft wing [11] to demonstrate the optimal aerodynamic design.

3.1.2. Boundary Conditions

The working fluid of wing is used as air at atmospheric pressure. The inlet velocity, angle of attack and the temperature are set at 90 m/s, 5 degrees and 293.2K, respectively. Types of wall conditions were applied real wall condition to all surfaces. During the simulation process, the value of temperature is kept constant while all the walls of the inlet section, the outlet section and the wing are fixed to have an adiabatic condition.

3.1.3. Mathematical Model

Mathematical equations and physical properties can be used to solve the CFD flow solver problem. Three dimensional steady Navier-Stokes equations and governing equations are used to find the satisfied solution in this work. The governing equations of the Navier stroke equations can define the conservation of mass, momentum and energy of the incompressible fluid flow. Where the fluid properties were constant and three-dimensional laminar and turbulent flows are used in this simulation. These equations can be expressed as follows:
For laminar flow,

$$\text{Continuity equation:} \quad \frac{\partial(\rho\, u_i)}{\partial x_i} = 0 \tag{2}$$

$$\text{Momentum Equation:} \frac{\partial(\rho\, u_i u_j)}{\partial x_j} = -\frac{\partial \rho}{\partial x_i} + \frac{\partial}{\partial x_j}[\mu(\frac{\partial u_i}{\partial x_j} + \frac{\partial u_j}{\partial x_i})] \tag{3}$$

Energy Equation : $$\frac{\partial}{\partial x_i}(\rho u_j T) = \frac{\partial}{\partial x_j} \tag{4}$$

For turbulent flow

Continuity equation: $$\frac{\partial u}{\partial x} + \frac{\partial \upsilon}{\partial y} + \frac{\partial w}{\partial z} = 0 \tag{5}$$

Momentum Equation: $$\rho[u\frac{\partial u}{\partial x} + \bar{v}\frac{\partial u}{\partial y}] = -\frac{\partial \bar{p}}{\partial x} + \mu[\frac{\partial^2 \bar{u}}{\partial x^2} + \frac{\partial^2 u}{\partial y^2}] - \rho\frac{\partial(\overline{u'})^2}{\partial x} - \frac{\partial(\overline{u'v'})}{\partial y} \tag{6}$$

Energy Equation: $$\rho c_p[\overline{u\frac{\partial \bar{T}}{\partial x}} + \bar{v}\frac{\partial \bar{T}}{\partial y}] = k[\frac{\partial^2 \bar{T}}{\partial x^2} + \frac{\partial^2 \bar{T}}{\partial y^2}] - \rho c_p\frac{\partial(\overline{u'T'})}{\partial x} - \rho c_p\frac{\partial(\overline{v'T'})}{\partial y} \tag{7}$$

For the numerical discretization of the governing equations, the second-order upwind scheme was used to the advection term in these problems. It has been found that the mesh generated contains 200,255 nodes between the computational time and the accuracy of the calculation for optimization purpose [12]. In the steady-state flow, lift coefficient and drag coefficient equations can be written as follows

Coefficient of Lift,

$$C_l = \frac{2 \times L}{\rho \times v^2 \times A} \tag{8}$$

Coefficient of Drag,

$$C_d = \frac{2 \times D}{\rho \times v^2 \times A} \tag{9}$$

Where, ρ= density of air (kg/m^3)
D= Drag Force (N)
L=Lift Force (N)
A= Area of Wing (m^2)

3.1.4. CFD Analysis of Wing

Before solving the flow analysis, we control the mesh of the wing to get the accurate and efficient multi-physics solutions. The simulation is started by using with above equation to solve the problem iteratively as a steady-state condition. Finally, a postprocessor is applied for the analysis and visualization of the resulting solution. Inflow simulation, the solve time is extremely dependent on the number of cells in the computational domain. Figure 2 shown a three-dimensional CFD model of the Lancer wing used in this study. Constant speed of 95m/s air-fluid is set to flow to the wing in x-direction while the *K-epsilon* turbulent model is used for the CFD simulation.

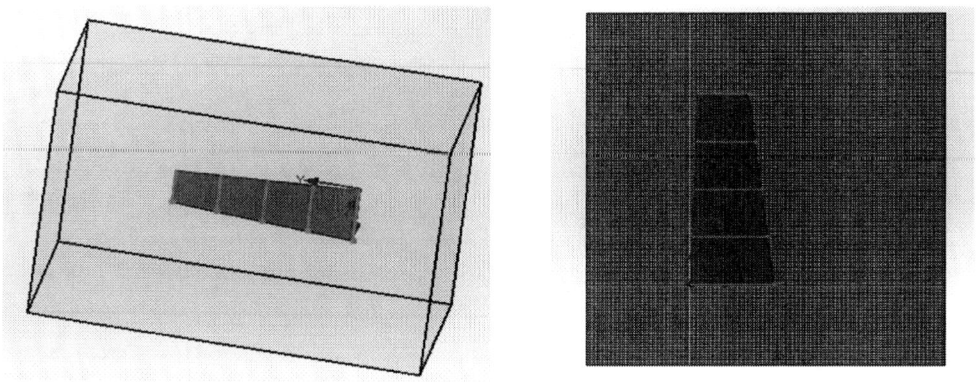

(a) (b)

Figure 2. (a) Enclosure of Lancer wing and (b) Meshing of wing model for one sampling

3.2. Design of Experiment

In building a surrogate model, design of experiment (DoE) is a procedure which is frequently employed to produce places of sample points within the design space [13]. A surrogate model is generally used to decide the locations of the simple points in the design space. There are two types of DoE) methods: classic DoE methods and modern DoE method. The modern DoE method include Latin Hypercube Sampling (LHS), Optimum Latin Hypercube Sampling(OLHS), Orthogonal Array design (OA) and uniform design (UD) [14]. In this paper, the LHS method is used to create the number of initial sampling points in the design space. Each sampling point represents a candidate aerodynamic shape. These sample points are made for the computational grid and analysed to find the corresponding aerodynamic data needed to estimate the objective function and constraints. Then, the data were saved to be database [10].

3.3. Surrogate Modelling

Surrogate modeling is an approximation of functions of design objective and constraints. The main purpose is to reduce computational time in cases that the functions are computationally too costly to evaluate [15]. The model will be constructed based on a set of sampling data which are calculated from the real expensive functions. We consider an m dimensional problem and n design variables to predict and evaluate aerodynamic function y: $R^m \rightarrow R$. Assume that the function y is sampled at sites[14].

$$S=[\mathbf{x}^{(l)},\dots,\mathbf{x}^{(n)}]^T \in R^{n\times m} \tag{10}$$

With the corresponding responses[13],

$$f_s = [f^{(1)}, \dots\dots, f^{(n)}] \tag{11}$$

Here n is the number of sampling points [14]. The pair (S, y_s) represents the sample data sets in the vector space[14]. There are many types of surrogate models such as Kriging, Polynomial response surface model (RSM), radial basis function (RBF) and support regression. A polynomial surface method has some disadvantages which need statistical treatment for getting the suitable design of experiment and degree of the polynomial based on the available data. RBF, on the other hand, can be best for consideration of acceptable accuracy, very fast and high robustness with small sample size. By comparing several surrogate model methods, the radial basis function network can be used easily which having good generalization, being flexible to input noise, behaving well even with unknown data and having online learning ability. It is a real value and an alternative interpolation method for a meta-model which is a distributed data and contains a very great application because it can interpolate all places in the space. It can be specified that the RBF method can fit distributed exactly multivariate data, thus preserving the CFD sample points. This method usually applies linear combinations of a radially symmetrical function based on Euclidean distance to estimate the response functions. It has been found that it is very accurate for interpolation in high dimensions and is ideal for interpolation of distributed data. [16].

3.3.1. Radial Basic Function Predictor and approximation

Radial basis functions are used as interpolation models or regression models. In this study, we typically build up the function in approximations of the form. The structure of RBF-based interpolation is completed as reported in [10]. Given a set of sampling design variables, $\mathbf{x}=\{\mathbf{x}^1,\ldots,\mathbf{x}^n\}$, which are the input vector corresponding to the function values, $f=(f^1,\ldots,f^n)^T$, an approximate function $\tilde{f}$ of any $\mathbf{x}$ can be expressed by RBF models have shown to provide a higher modeling accuracy and higher robustness (3)

$$\tilde{f}(\mathbf{x}) = \sum_{i=1}^{n} c_i \phi\left(\left\|\mathbf{x}-\mathbf{x}^i\right\|\right) \tag{12}$$

where $\left\|\mathbf{x}-\mathbf{x}^i\right\|$ distance between $\mathbf{x}$ and $\mathbf{x}^i$ and c_i is interpolation coefficients which can be calculated by solving the linear equation of

$$\sum_{i=1}^{n} c_i \phi\left(\left\|\mathbf{x}_i-\mathbf{x}_j\right\|\right) = f\left(\mathbf{x}_j\right) \text{ ; for } j=1,\ldots,n \text{ and} \tag{13}$$

where ϕ is a kernel function that determines the smoothness properties of the estimation scheme. Examples of radial basis kernel functions are

$$\text{Linear: } \phi(\mathrm{r}) = r, \tag{14}$$

$$\text{Cubic: } \phi(r) = r^3, \tag{15}$$

$$\text{Gaussian: } \phi(r) = e^{-\theta_i r^2}, \; 0 \le \theta_i \le 1 \tag{16}$$

$$\text{Quadratic: } \phi(r) = \sqrt{r^2+\theta_i^2}, \; 0 \le \theta_i \le 1, \tag{17}$$

where θ_i represent the shape parameters and $r=((x\text{-}c_i)^T+(x\text{-}c_i))^{1/2}$ is the radial distance. In this paper, the linear kernel function method is used.

3.4. Optimization search with Differential Evolution

Surrogate-assisted differential evolution was generally used to reduce computational time in evolutionary optimization problems, such as aerodynamic optimization [17]. Figure 3 shows the optimization procedure which involves overall steps in this work. The process starts with the generation of a set of sampling points using LHS. The CFD is performed for real C_l and C_d calculation. After that, the optimization search is performed using a Differential Evolution (DE) algorithm and the optimum results obtained which are used to calculate real function values based on CFD again. The number of sampling points used in this study is 20 while population size and number of iterations for DE are set to be 100 and 200, respectively.

IOP Conf. Series: Materials Science and Engineering **886** (2020) 012026 doi:10.1088/1757-899X/886/1/012026

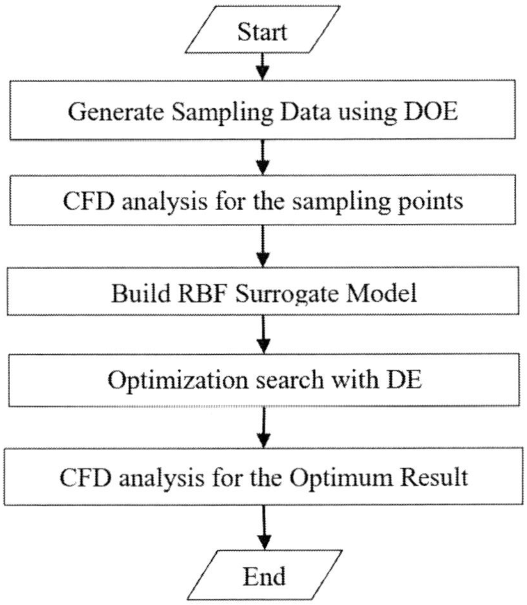

Figure 3. Flow chart of the surrogate-assisted DE optimization

4. Results

Figure 4 shows the 20 sampling wings from LHS, which are used for constructing 2 surrogate models after performing CFD analyses. Table 1 shows the values of C_l/C_d from 20 sampling points before performing optimization based on CFD. After performing optimization based on surrogate-assisted DE, the optimum results are reported in Table 2 and Figure 5 (a) and (b). Optimum C_l/C_d of Case A is based on the surrogate model that is used to directly predict C_l/C_d, while optimum C_l/C_d of Case B is based on the surrogate model that is used to separately predict C_l and C_d. For the Case-A, the optimum C_l/C_d obtained based on the surrogate model and CFD are 22.02 and 22.51 where the percentage of error of C_l/C_d between surrogate model and CFD is 2.23%. For the Case-B, the optimum C_l/C_d obtained based on the surrogate model and CFD are 20.382 and 22.07 where the percentage of error of C_l/C_d between surrogate model and CFD is 8.2%. It is found that the optimum C_l/C_d obtained from both Case-A and Case-B is better than all initial sampling points. The Case-A which performs optimization based on using a surrogate model to predict the objective function (C_l/C_d) directly obtain significantly better optimum results than the Case B which performs optimization based on using the surrogate model to predict C_l and C_d separately. Figure 5 (a) and (b) shows the optimum wing obtained for Case-A and Case-B in this paper.

Based on this study, for optimization of the aerodynamic shape of an aircraft wing with an objective to maximize Cl/C_d using surrogate-assisted DE, it was found that using the surrogate model to predict the objective function (C_l/C_d) directly is the more accurate and obtains the better optimum results than using the surrogate model to predict C_l and C_d separately.

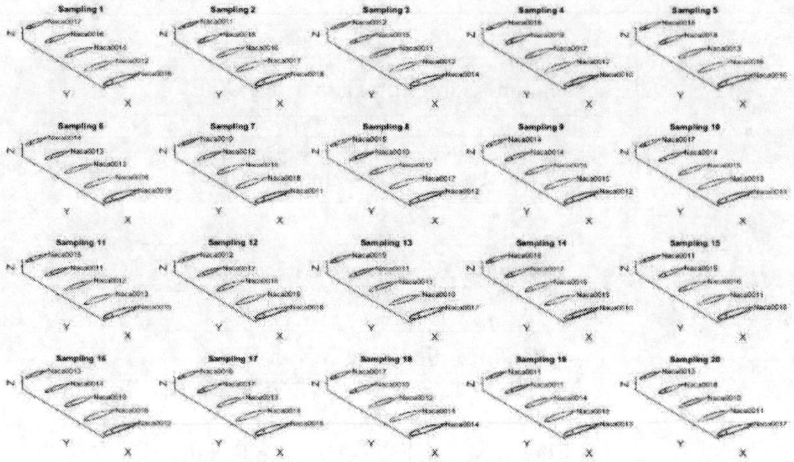

Figure 4. Twenty initial sampling point of the wing by LHS method

Table 1. The value of C_d /C_l of 20 sampling points from CFD

Sampling No.	1	2	3	4	5	6	7	8	9	10
C_l/C_d	11.19	11.52	10.17	12.81	9.548	14.96	7.68	8.14	11.16	11.15
Sampling No.	11	12	13	14	15	16	17	18	19	20
C_l/C_d	16.63	13.73	8.311	11.65	19.84	16.84	13.19	11.21	9.723	11.57

Table 2. Comparison of the predicted and the validated C_d /C_l at the obtained optimal point

	By Surrogate Model			By CFD			Error		
	C_l	C_d	C_l/C_d	C_l	C_d	C_l/C_d	C_l	C_d	C_l/C_d
Case-A	-	-	22.02	0.07337	0.00326	22.51	-	-	2.23%
Case-B	0.07584	0.00372	20.382	0.0724	0.00328	22.07	4.8%	13.4%	8.2%

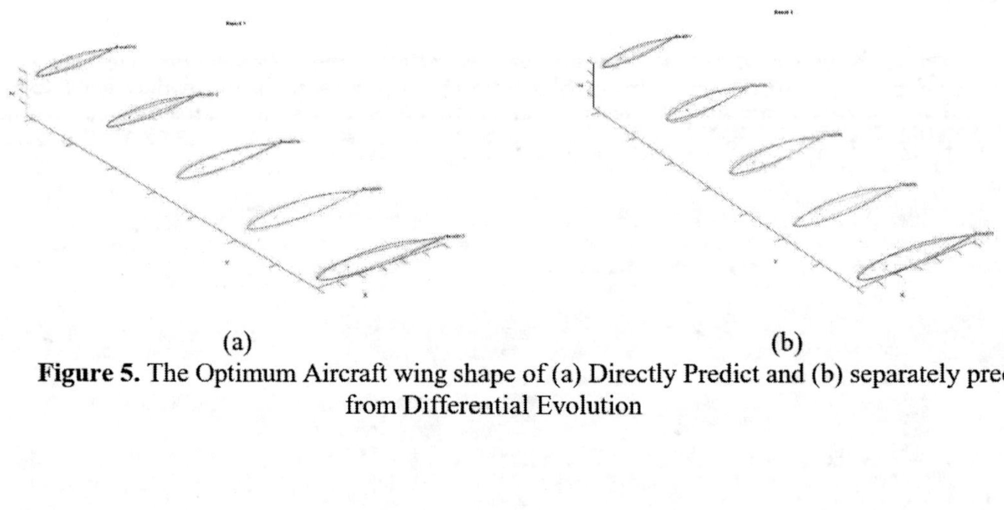

(a) (b)

Figure 5. The Optimum Aircraft wing shape of (a) Directly Predict and (b) separately predict from Differential Evolution

5. Conclusions

The aerodynamic shape optimization of an aircraft wing using surrogate-assisted DE was successfully conducted. The optimization problem is posed to find wing shape in order to maximize its lift-to-drag ratio. Two surrogate-assisted DE are presented which are the use of a surrogate model to predict the objective function directly and using a surrogate model to predict lift coefficient and drag coefficient separately before calculating the objective function. The results obtained found that using the surrogate model to predict the objective function (C_l/C_D) directly is the more accurate and obtains the better optimum results than using the surrogate model to predict C_l and C_D separately. For further work, the development of the more effective and efficient surrogate-assisted MHs for such a problem in order to get better results are the main focus.

Acknowledgement

The authors are grateful for financial support from KKU Scholarship for ASEAN and GMS Countries and Thailand Research Funds (TRF)

Reference

[1] Jameson A Fatica M 2002 *Icas* 1–24
[2] Li J Cai J Qu K 2019 *Struct. Multidiscip. Optim.* **59**(2) 403–419
[3] Skinner S N Zare-Behtash H 2018 *Appl. Soft Comput. J.* **62** 933–962
[4] Acerbi L Ma W J 2017 *Advances in Neural Information Processing Systems* **30** 1834–1844
[5] He X Li J Mader C A Yildirim A Martins J R R A2019 *Aerosp. Sci. Technol.* **87** 48–61
[6] Koreanschi A Gabor O S Acotto J Brianchon G Portier G Botez R M Mamou M Mebarki Y 2017 *Chinese J. Aeronaut.* **30**(1) 164–174
[7] Jouhaud J C Sagaut P Montagnac M Laurenceau J 2007 *Comput. Fluids* **36**(3) 520–529
[8] Tesfahunegn Y A Koziel S Leifsson L Bekasiewicz A 2015 *Procedia Comput. Sci.* **51**(1) 795–804
[9] Maruyama D Liu D Görtz S 2016 *Proc. VII European Congress on Computational Methods in Applied Sciences and Engineering M Papadrakakis V Papadopoulos G Stefanou V Plevris (eds.)* (Greece) 8787–8800
[10] Jakobsson S Patriksson M Rudholm J Wojciechowski A 2010 *Optim. Eng.* **11**(4) 501–532
[11] All-aero 2019 http://all-aero.com/index.php/53-planes-l-m-n-o/7237-neico-lancair-200--320--360
[12] Wang H Zhu X Du Z 2010 *Int. Commun. Heat Mass Transf.* **37**(8) 998–1003
[13] Han Z H Zhang K S Song W P Qiao Z D 2010 *J. Aircr.* **47**(2) 603–612
[14] Han Z H Abu-Zurayk M Görtz S Ilic C 2018 *Notes Numer. Fluid Mech. Multidiscip. Des.* **138** 257–282
[15] Koziel S Yang X S Zhang Q J Koziel S Leifsson L Yang X S 2013 *Simulation-Driven Des. Optim. Model. Microw. Eng.* (Imperial College Press:London) 41–79
[16] Mukesh R Lingadurai K Selvakumar U 2014 *J. King Saud Univ. - Eng. Sci.* **26**(2) 191–197
[17] Jin Y 2011 *Swarm Evol. Comput.* **1**(2) 61–70

IOP Conf. Series: Materials Science and Engineering **886** (2020) 012027 doi:10.1088/1757-899X/886/1/012027

Three-Dimensional Finite Element Analysis of Stress Distribution and Displacement to Design of Patient Lifting Equipment

M. Sakunsongbunsiri and P. Keangin*

Department of Mechanical Engineering, Faculty of Engineering, Mahidol University, 25/25 Phutthamonthon 4 Road, Salaya, Nakhon Pathom 73170, Thailand.

* Corresponding Author: pornthip.kea@mahidol.ac.th

Abstract
The patients with stroke, especially hemiplegia and paraplegia cases, need to be transferred whether from bed to wheelchair or wheelchair to bed and be transferred whether from wheelchair or bed to car or car to a wheelchair or bed while the patient is admitted to the hospital or stay at home. The accuracy of movement handling is required to be in accordance with the principles of physiology in order to increase the safety of patient moving. The utilization of patient lifting equipment will improve patient handling more efficiently for both patients and staff during health care. Patient-lifting equipment is absolutely helpful in lifting patient movement. The study on the analysis of stress distribution as well as the displacement of the device in the analysis of strength in the structure design is rarely found because of computer program limitations. For appropriate usage in each situation, it is important to apply a comprehensive understanding of principles of strength analysis to design patient-lifting equipment. The objective of the proposed research is to design a spreader bar of the patient lift equipment to support use for transferring from a wheelchair or bed to car or car to a wheelchair or bed. The details regarding concept design and analysing the strength of the structure is solved by finite element method (FEM) with a three-dimensional model. The simulated results obtained from the computational solution are investigated and validated against those obtained from the analytical solution. The study focuses attention on the effects of different angles of the spreader bar on the stress distribution and displacement at various points. It is found that the change in the angle adjustment of the spreader bar directly affects the change in the stress distribution, as well as displacement. The maximum stress distribution, as well as maximum displacement, are reduced when the degree increases. The obtained results can be a helpful information to use as the guideline to design a spreader bar of the patient- lifting equipment to transfer patients suitably under each situation.

Keywords: Finite element method (FEM), Patient lifting equipment, Stress distribution, Displacement

1. Introduction

Transferring patients who move themselves with difficulty including those with paralysis, whether lifting, carrying, supporting or repositioning them will affect the relatives, especially, the hospital's health service staff who will have the opportunity to recognize injuries such as musculoskeletal injuries, back pain and joint disease [1]. Work-related Musculoskeletal Disorder (WRMD) is a collective and descriptive term for symptoms, i.e. injuries of muscle, tendon or nerves, resulted or aggravated by work and characterized by discomfort impairment, disability or persistent pain in joints which of these problems are common among all healthcare workers [2]. According to the study [3], it has been found that the use of lifting equipment to move including the proper lifting can help prevent injuries to the musculoskeletal disorder and can reduce unexpected accidents while transferring the patients. Besides, it has been found in many studies that when the mechanical patient lifting equipment is used, the risk of musculoskeletal injury caused by moving and handling patients is reduced. Some research shows that the implementation of mechanical patient lift equipment helps to significantly reduce the psychophysical stress, injury rate, discomfort and workers compensation costs in healthcare workers [4,5].

The mechanical transfer lift equipment can be divided into three major categories, according to researches studied on type of mechanical transfer lift equipment, namely, the standing hoist, the mobile hoist, and the gantry/ceiling hoist [6]. There are many types of hoisting devices in the world market which differ in the function of use, types of spreader bar attached to the slings, height, length and number of suspension points, those are two-, four-, six-, and eight-point spreader bars [7], and varying degrees of spreader bars. The above device is suitable for transferring patients from wheelchair to bed or from bed to wheelchair, however, it is not suitable for transferring patients from inside the car to wheelchair or bed or transfer back from wheelchair or bed to the car. Therefore, the design of a new patient lifting equipment that can help reduce these limitations will help to develop health services efficiently and completely.

The main components of mechanical transfer lift equipment can be divided into 4 parts which are spreader bar, lift arm, mast and base. Many previous studies on the design of transfer lift equipment components such as Hakamiun et al. [8] designed a patient lift equipment for transporting a person. The equipment consists of a base, a frame, a lifting arm, an actuator as well as an attachment bar. From the base, the frame is extended upwardly and the lifting arm is coupled with an upper end of the frame. To move and lift the arm between the lower and raised position, the actuator is then paired with the lifting arm, which is connected to the attachment bar as well. In addition, a scale, used for weighing a person being lifted, is included in the apparatus. A portable patient transfer, in which a small wheelbase is included, is improved by Brandorff and Campbell [9]. From the wheelbase, a vertically disposed telescoping lifting column extending upwardly. For convenient transfer with a patient, the lift can be closed and temporarily attached to a wheelchair. Huang [10] designed a transfer apparatus for a patient using an electrical system with a linear actuator for moving the lifting arm between the lowered position and a raised position. The difference in the designed equipment was base of equipment has short plates in front and rear, and L invert shape long plate in left and right, which each invert L shape has a guiding slot for bolt for folding the device and length of the apparatus can be folded down. Lear [11] designed a patient lifting apparatus that includes a front and rear independently controlled cable lowering which a patient sitting in a sling suspended from the cables can be assisted out of the sling. It was found that the new design equipment was easy to use. The lifting patient device using the control panel having a micro-controller for remotely controlling the patient lift was presented by Tu [12]. The masts of the device can adjust the lifting angle, a pair of extendable legs extended or folded from the frontal side of the base used to maintain the balance of the device. Wang [13] designed a device for turning over and transferring the patient. A spreader bar of the device can be a turn driving device to control rotation and can be control rotation of the cantilever. A portable assistive lift was designed by Fakhrizadeh [14], the device comprised a vertical member comprising a handlebar and a central wheel controlled by the handlebar disposed at the top and bottom extremities. The highlight of this device was all parts can be disassembled. Runnels [15] presented an apparatus, systems, and methods for providing a mobility device lifting and positioning. A lifting and positioning apparatus included one support structure, a control unit configured to cause the apparatus to perform operation and lifting mechanism. There is much equipment designed for assisting patient handling, as mentioned earlier; however, the design of the patient lifting is still not comprehensive, especially the equipment for transferring the patient in the limited area such as transfer from wheelchair or bed to inside the car or from car to wheelchair or bed. The main limitation is designing a spreader bar component that is suitable for use in limited sizes. For utilizing patient lifting equipment appropriately, a comprehensive understanding of the influences of key parameters, for example, angle adjustment of spreader bar which affects the strength of the structure, will help the design of patient lifting equipment as a recommendation.

This study focuses on the design of a spreader bar that is the main component of patient lifting equipment. The effects of angle adjustment of the spreader bar on stress distribution and displacement are investigated. The angle adjustment of spreader bar on both sides of 0 degree, 30 degrees, 45 degrees, 60 degrees and 90 degrees are considered. The structural analysis of spreader bar design is carried out computationally by using SolidWorks simulation program. A Three-dimensional model of the spreader bar component of patient lifting equipment is constructed and simulated. In the validation of this simulated model, the accuracy of the proposed computational model is verified by validating the simulated results against the analytical solution, which uses a two-dimensional model for calculating. The simulated results will be used as an information for designing the patient lifting equipment which is used for lifting patients as patient care technologies.

2. Model and Material

The objective of the study is to design the spreader bar components of patient lifting equipment. The structural design and analysis are carried out to study on the stress distribution as well as displacement of spreader bar using Solidworks program. A three-dimensional model of spreader bar component of the patient lifting equipment is designed and constructed, the design corresponds to the sizes of the patient lifting equipment. The overall procedure of the study starts with the literature review. Then, the related data and knowledge on general methods concerning transferring patients are summarized. After that, the conceptual and preliminary designs are conducted, respectively. To study the strength of the structure, the structural analysis is carried out by computer simulation. The computer simulation method is selected in this study due to cost, size factor and convenience. Thereafter, Von-Mises stress distribution from the simulated results is validated against the results from the analytical solution which uses a two-dimensional model with the same configuration to verify the model's accuracy. The overall flowchart of the research methodology is shown in Figure 1. Figure 2 shows the spreader bar design for analysis of stress distribution and displacement at critical points of 1, 2 and 3, respectively. The angle adjustments of spreader bar on both sides of 0 degrees, 30 degrees, 45 degrees, 60 degrees and 90 degrees on the stress distribution and displacement are studied. The angle adjustments of the spreader bar are indicated in Figure 3. Figures 3(a)-(e) display the angle adjustment of spreader bar on both sides of 0 degrees, 30 degrees, 45 degrees, 60 degrees and 90 degrees, respectively.

For structural strength of the spreader bar, the alloy steel is used as a test material. This material is selected due to its a good combination of all general characteristics of steel strength resulting in high forming ability, high strength and elongation, corrosion resistance and suitable for use on cleanliness and hygiene. The mechanical properties of the material are presented in Table 1.

In the numerical analysis, the finite element method (FEM) is utilized/employed for solving the mathematical models in order to demonstrate the stress distribution and displacement. Triangular elements are used to discrete the three-dimensional model of the spreader bar and approximately calculate the stress distribution and displacement variations across each element. Appropriate required number of elements is identified by performing the grid independence test. Furthermore, a fine mesh is defined in the sensitive area for good calculation.

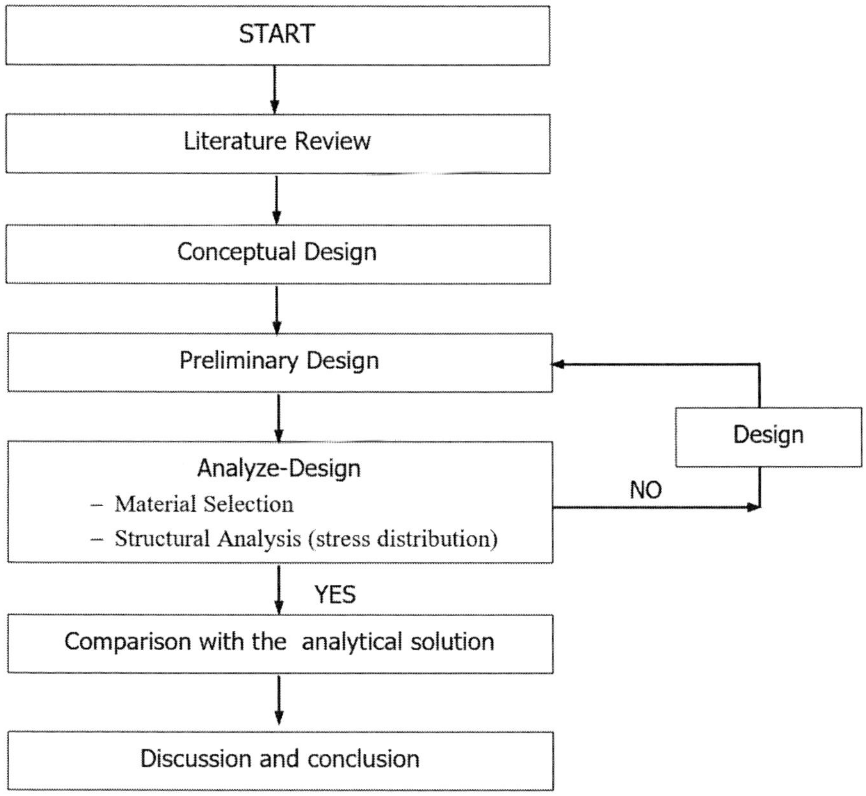

Figure 1. The flowchart methodology of this research

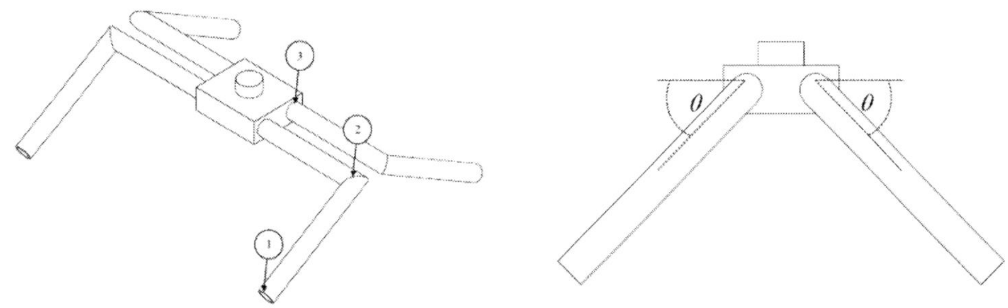

Figure 2. A Three-dimensional model of spreader bar designed by SolidWorks software

Table 1. The mechanical properties of material applied in the model

Property	Value	Unit
Modulus of Elasticity	210,000	MPa
Shear Modulus	78,990	MPa
Poisson Ratio	0.28	-
Yield Strength	620.42	MPa
Tensile Strength	723.83	MPa

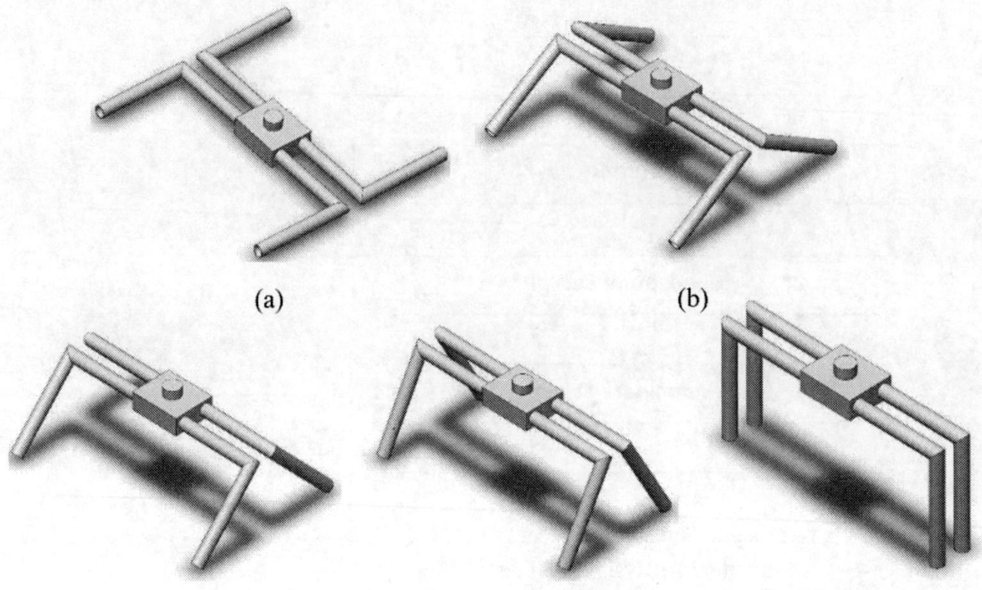

(a) (b) (c) (d) (e)

Figure 3. The angle adjustment of spreader bar on both sides of (a) 0 degrees, (b) 30 degrees, (c) 45 degrees, (d) 60 degrees and (e) 90 degrees

3. Methodology

Structural strength analysis using FEM is an effective and popular method. SolidWorks simulation software is applied to analyze the strength of the spreader bar to determine the appropriate stress and displacement for the safety of users. There is a total of 5 cases in which the data are collected. Von-Mises stress distribution and displacement at critical points of 1, 2 and 3 include the maximum stress and maximum displacement on the spreader bar in each case is considered.

3.1 Governing Equation

The basic strength equations to calculate the balance of a flexible solid in three-dimensional can be written in the form of partial differential equations as follows [16]:

$$\begin{aligned}
&\frac{\partial \sigma_x}{\partial x}+\frac{\partial \tau_{xy}}{\partial y}+\frac{\partial \tau_{xz}}{\partial z}+F_x=0\\
&\frac{\partial \tau_{xy}}{\partial x}+\frac{\partial \sigma_y}{\partial y}+\frac{\partial \tau_{yz}}{\partial z}+F_y=0\\
&\frac{\partial \tau_{xz}}{\partial x}+\frac{\partial \tau_{yz}}{\partial y}+\frac{\partial \sigma_z}{\partial z}+F_z=0
\end{aligned} \tag{1}$$

Where $\sigma_x, \sigma_y, \sigma_z$ are normal stress components in x, y, z direction (MPa), respectively

$\tau_{xy}, \tau_{yz}, \tau_{xz}$ are shearing stress components (MPa)

F_x, F_y, F_z are body force in x, y, z direction (N), respectively

3.2 Boundary Conditions

The characteristics of the spreader bar will be attached to the lifting arm of the patient lift equipment as shown in Figure 4. This study assumes that the weight in each point on spreader bar is constant and the same at the connection point. Based on general requirements for basic security and important competencies from Ministry of industry, casting materials are not damaged due to wear when stretching less-than 5 %. In addition, the structural strength analysis will specify the tensile factor of safety equal to 4. The patient lifting device is assumed that can lift the patient with the maximum weight equal to

135 kg, therefore the load used in the analysis is equal to 540 kg. The load is distributed evenly at the 4 ends of the spreader bar with equal to 135 N is assumed. Figure 4(a) shows the characteristics of the spreader bar will be attached to the lifting arm of the patient lift equipment and Figure 4(b) displays the mesh independence test. In stress distribution as well as displacement analysis, the boundary condition assigned at various points are illustrated in Figure 5. For evaluating a suitable number of elements which leads to the accuracy of the numerical results, the mesh independent solution is carried out. It is found that the number of elements in case the angle of 0 degrees approximately 625,014 elements, in case the angle of 30 degrees approximately 583,000 elements, in case the angle of 45 degrees approximately 573,023 elements, in case the angle of 60 degrees approximately 581,589 elements, and in case the angle of 90 degrees approximately 575,401 elements, as shown in Figure 5(a)-(e), respectively.

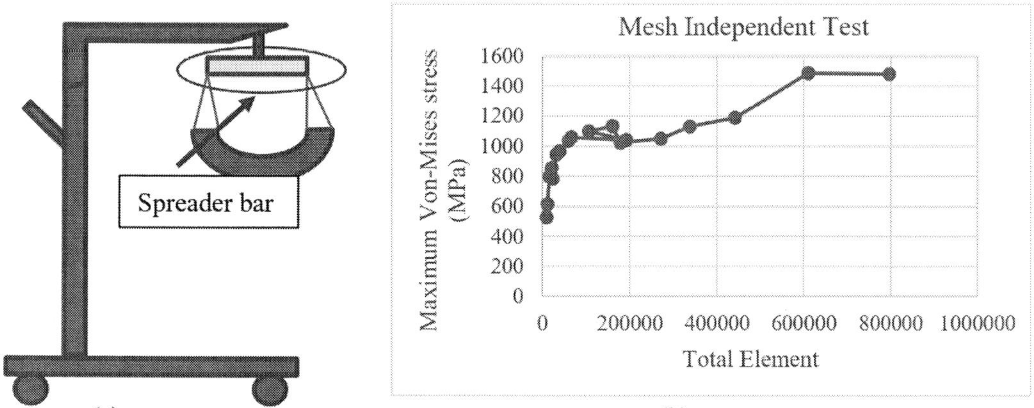

(a) (b)

Figure 4. The characteristics of the spreader bar in the calculation (a) The characteristics of the spreader bar will be attached to the lifting arm of the patient lift equipment and (b) Relationship between the total element and maximum Von-Mises stress

IOP Conf. Series: Materials Science and Engineering **886** (2020) 012027 doi:10.1088/1757-899X/886/1/012027

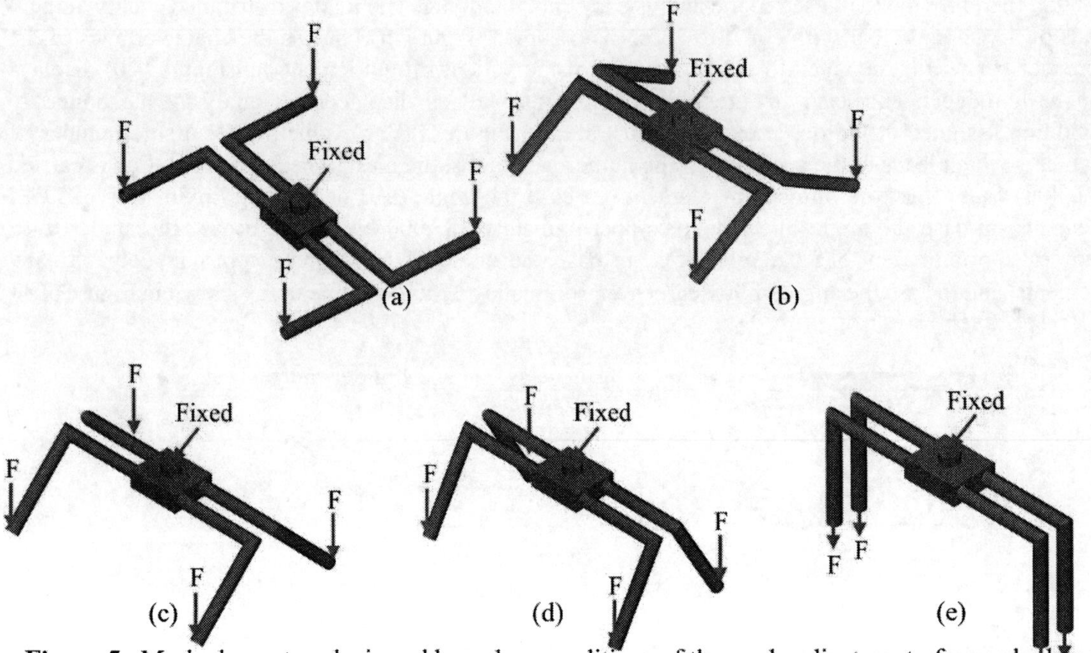

Figure 5. Mesh element analysis and boundary conditions of the angle adjustment of spreader bar on both sides of (a) 0 degrees, (b) 30 degrees, (c) 45 degrees, (d) 60 degrees and (e) 90 degrees.

4. Results and Discussion

4.1 Verifications of the model

To evaluate the accuracy of the present numerical models, Von-Mises stress distribution of the simulated result is validated against the analytical result in the same condition. Due to the spreader bar model studied is complex geometry, it is difficult to calculate by an analytical solution. Therefore, a simple spreader bar model of 0 degree is used for verifications of the model. The computational domain for validation in case the angle of 0 degree is shown in Figure 6. The two-dimensional strength equations for calculating the analytical solution is described by the following equation:

$$\sigma_1, \sigma_2 = \frac{\sigma_x - \sigma_y}{2} \pm \sqrt{\left(\frac{\sigma_x - \sigma_y}{2}\right)^2 + \tau_{xy}^2} \quad (2)$$

$$\sigma_v = \sqrt{\sigma_1^2 - \sigma_1\sigma_2 + \sigma_2^2} \quad (3)$$

Where σ_1, σ_2 are principal stresses (MPa)

σ_x, σ_y are normal stress components in x and y direction (MPa)

τ_{xy} is shearing stress component (MPa)

σ_v is Von-Mises stress (MPa)

For verifications of the model, the two-dimensional model is assigned with triangular elements. The number of mesh element which is selected to use in verifications of the model is 625,014 elements. The Von-Mises stress distribution result from the numerical solution of the present study is indicated in Figure 7. The comparison of the Von-Mises stress distribution of a spreader bar between the numerical solution and analytical solution is illustrated in Table 2. The results show that the maximum value of Von-Mises stress from numerical solution and one from the analytical solution are 287.7 MPa and 278.2

MPa, respectively. The error between the two solutions is equal to 3.415 %. This favorable comparison presents the accuracy of the numerical model confidently. However, some errors may occur in simulations by the numerical scheme.

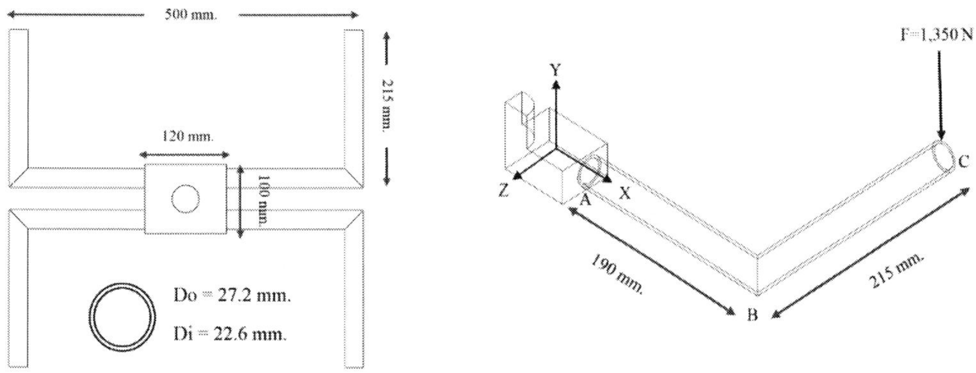

Figure 6. The computational domain for validation in case the angle of 0 degree

Table 2. Comparing solution methods between the analytical solution and numerical solution

	Analytical Solution	Numerical Solution	Error
Von-Mises stress	278.2 MPa	287.7 MPa	3.415 %

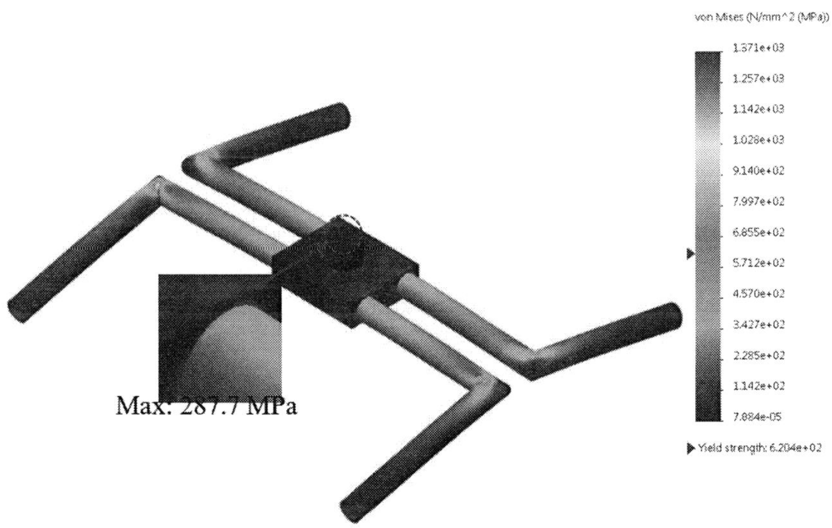

Figure 7. Von-Mises stress distribution result from the numerical solution of the present study

IOP Conf. Series: Materials Science and Engineering **886** (2020) 012027 doi:10.1088/1757-899X/886/1/012027

4.2 Effects of angle adjustment on the Von-Mises stress distribution

The Von-Mises stress distribution of spreader bar influenced by various angle adjustments is exhibited in Figure 8. Figure 8(a)-(e) shows the Von-Mises stress distribution in the case of the angle of 0 degree, 30 degrees, 45 degrees, 60 degrees and 90 degrees, respectively. The results showed the corresponding results in all cases, where the maximum Von-Mises stress distribution of them take place at the connection point. The comparison of the Von-Mises stress value by varying the angle adjustment of the spreader bar is presented in Figure 9. Figure 9(a) displays the maximum Von-Mises stress distribution influenced by various angle adjustments of spreader bar. The results show that in the case of 0 degree, the highest maximum Von-Mises stress distribution occurs and is followed by the case of 30 degrees, 45 degrees, 60 degrees and 90 degrees, respectively. Figure 9(b) displays the Von-Mises stress distribution at the critical points influenced by various angle adjustments. The three critical points of 1, 2 and 3 are exhibited in Figure 2. The results show that at the critical point 1, in the case of 0 degree causes the highest Von-Mises stress value that equal to 46.21 MPa while in the case of 90 degrees causes the lowest Von-Mises stress value that equal to 7.49 MPa. At the critical point 2, in case of 0 degrees causes the highest Von-Mises stress value that equal to 1,379 MPa while in the case of 90 degrees causes the lowest Von-Mises stress value that equal to 23.73 MPa. At the critical point 3, in the case of 0 degree causes the highest Von-Mises stress value that equal to 379.8 MPa while in case of 90 degrees causes the lowest Von-Mises stress value that equal to 269.7 MPa. Considering the Von-Mises stress value, it is seen that in the case of 90 degrees is most suitable for design specification because it has the lowest Von-Mises stress value. Furthermore, considering the maximum Von-Mises stress value, it is found that in the case of 60 degrees and in case of 90 degrees not exceed the yield strength of the material at 620.42 MPa, referred to the Table 1.

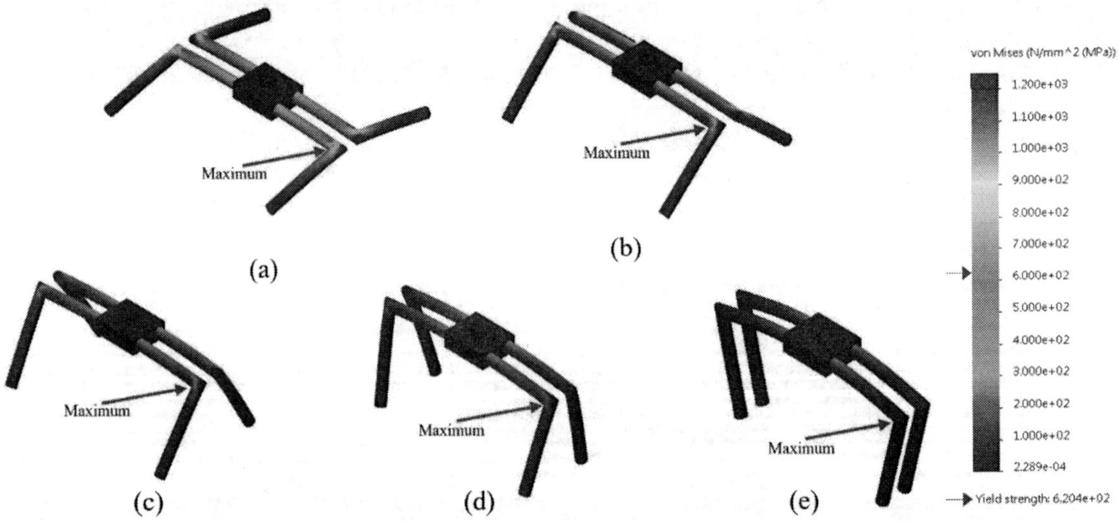

Figure 8. The effects of angle adjustment on the Von-mises stress distribution of spreader bar in case the angle of (a) 0 degree, (b) 30 degrees, (c) 45 degrees, (d) 60 degrees and (e) 90 degrees

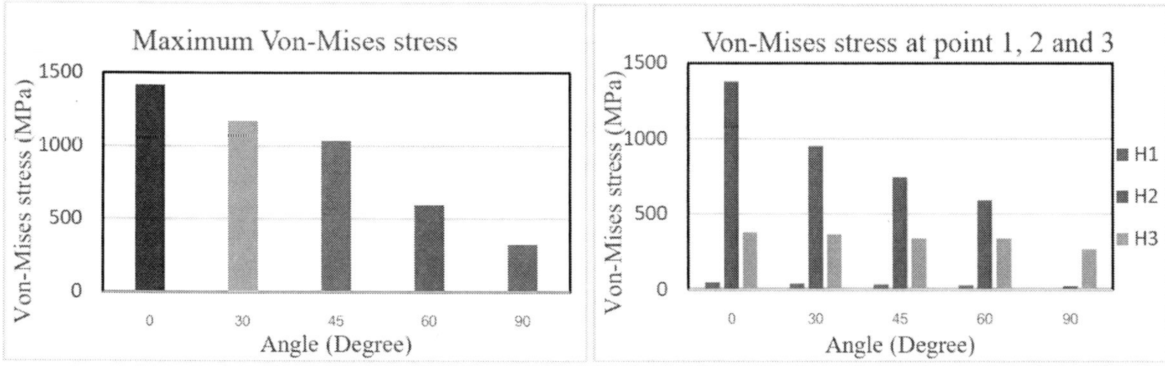

(a) (b)

Figure 9. Comparison of Von-Mises stress distribution by varying the angle adjustment of spreader bar: (a) maximum Von-Mises stress distribution and (b) Von-Mises stress distribution at critical points 1, 2 and 3.

4.3 Effects of angle adjustment on the displacement

Figure 10 shows the displacement of the spreader bar influenced by various angle adjustments. The difference in displacement by varying the angle adjustment of spreader bar in case the angle of 0 degrees, 30 degrees, 45 degrees, 60 degrees and 90 degrees are shown in Figure 10(a)-(e), respectively. The results showed the corresponding results in all cases, where at the position at the ends of the spreader bar, the maximum displacement take place. This is because of the distance between load and fulcrum point that is the point that must support the weight. The comparison of the displacement value by varying the angle adjustment of the spreader bar is displayed in Figure 11. Figure 11(a) demonstrates the effects of angle adjustment of the spreader bar on the maximum displacement in each case. As can be seen, the figure shows that in case of 0 degrees has the highest maximum displacement, followed by the in case of 30 degrees, in case of 45 degrees, in case of 60 degrees and in case of 90 degrees has the lowest value of maximum displacement, respectively. The results have shown that the maximum displacement value in case of 0 degrees is 9.017 mm, in case of 30 degrees is 7.896 mm, in case of 45 degrees is 6.588 mm, in case of 60 degrees is 4.944 mm and in case of 90 degrees is 2.336 mm, respectively. The effects of angle adjustment and position of the spreader bar on the displacement in each point are illustrated in Figure 11(b). In accordance with the above, the three critical points of 1, 2 and 3 are exhibited in Figure 2. The obtained results displayed that at critical points of 1 has the highest displacement value in case of 0 degrees that equal to 8.851 mm while the lowest displacement value in case of 90 degrees that equal to 2.269 mm, at critical points of 2 has the highest displacement value in case of 45 degrees that equal to 1.404 mm while the lowest displacement value in case of 90 degrees that equal to 1.181 mm and at critical points of 3 has the highest displacement value in case of 90 degrees that equal to 0.019 mm while the lowest displacement value in case of 30 degrees that equal to 0.015 mm. Considering the displacement value, it is found that in the case of 90 degrees is most suitable for design specification because it has the lowest displacement value which of the following corresponds to Von-Mises stress distribution value.

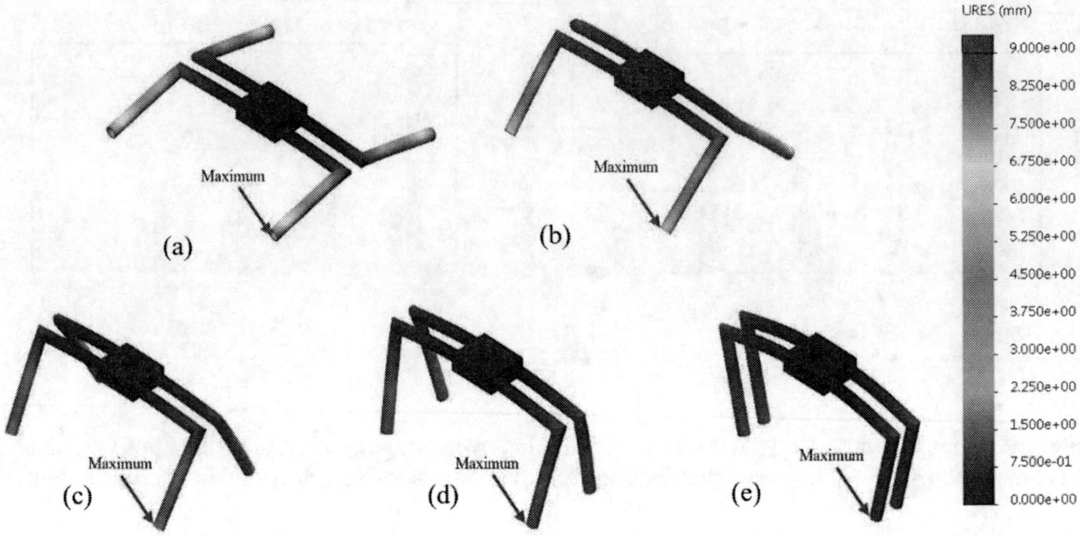

Figure 10. The effects of angle adjustment on the displacement of spreader bar in case the angle of (a) 0 degrees, (b) 30 degrees, (c) 45 degrees, (d) 60 degrees and (e) 90 degrees.

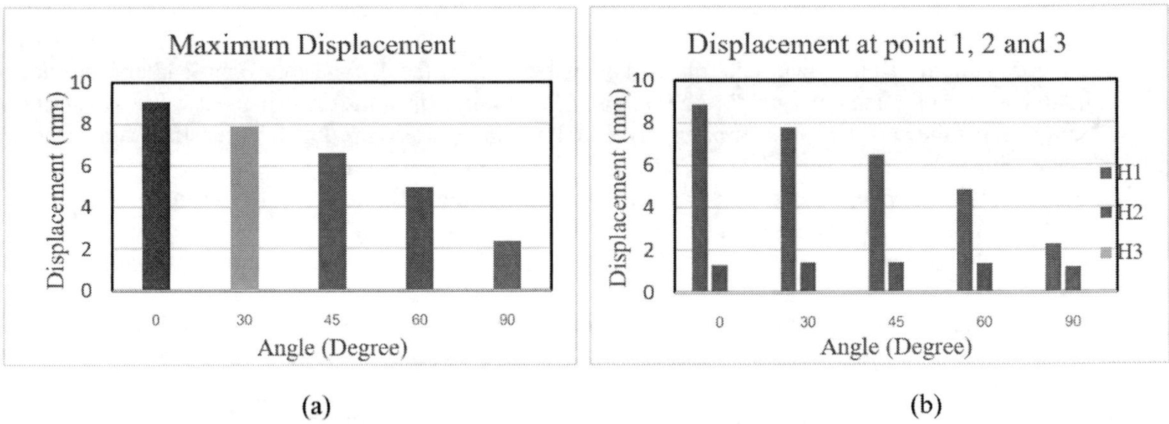

Figure 11. Comparison of displacement by varying the angle adjustment of spreader bar: (a) maximum displacement and (b) displacement at critical points 1, 2 and 3.

5. Conclusion

This study focused on the use of computer modeling and simulation of stress distribution and displacement to prepare the preliminary design of the spreader bar of patient lifting equipment. The patient lifting equipment is assumed that can lift the patient with the maximum weight equal to 135 kg, therefore the maximum weight used in the analysis is equal to 540 kg for considering the tensile factor of safety equal to 4. SolidWorks software is used in the design including analysis of the strength of spreader bar by FEM via three-dimensional model. The various angle adjustment on the stress distribution and displacement by varying the angle adjustment of the spreader bar is investigated. The simulated results obtained by implement of computer programs are validated against the calculated results obtained by analytical solution with the two-dimensional model. The results clearly show a good agreement of the analytical results and present simulated stress distributions. Considering of model-based testing for designing of 5 cases with different angle adjustment namely angle of 0 degree, 30 degrees, 45 degrees, 60 degrees and 90 degrees, it is found that in case of 0 degrees has the highest maximum stress distribution, followed by the in case of 30 degrees, in case of 45 degrees, in the case of 60 degrees and in the case of 90 degrees, respectively. The change in the displacement from the angle adjustment change gives the result that corresponds to the change of stress distribution. The simulation results show that in the case of 0 degree has the highest maximum displacement, followed by the in case of 30 degrees, in case of 45 degrees, in case of 60 degrees and in case of 90 degrees, respectively. The maximum displacement will be reduced when the degree increases. Considering the Von-Mises stress value and displacement value, it can be seen that in case of 90 degrees is most suitable for design specification because it has the lowest Von-Mises stress value and lowest displacement values. In addition, considering the highest maximum Von-Mises stress value, it is found that in case of 60 degrees and in the case of 90 degrees not exceed the yield strength of the material. Consequently, this favourable result lends confidence that the spreader bar will not be damaged when really used. The basic knowledge obtained from computer analysis can be used to improve the structure in order to build a prototype of patient lifting equipment. This research, studied the effects of angle adjustment of spreader bar on stress distribution and displacement to ensure that the spreading bar will not damage when actually used. The obtained results will be compared with the results of actual/real testing in the future.

Reference

[1] Risør B W Casper S D Andersen L L Sørensen J 2017 *Applied ergonomics* **60** 74-82
[2] Kee D Seo S R 2007 *International Journal of Industrial Ergonomics* **37** 207-212
[3] Zhao G Yee R B Mun K R Yu H 2017 *Applied ergonomics* **65** 389-397
[4] Ziqing Z Stobbe T J Collins J W Hsiao H Hobbs G R 2000 *Applied ergonomics* **31** 35-44
[5] Li J Wolf L Evanoff B 2004 *Injury Prevention* **10** 212-216
[6] Jusoh M A B M Ismail M T B Rashid H B Lokman A M Makhtar A K B 2018 *Proc. the 7th International Conference on Kansei Engineering and Emotion Research* 2018 229-238
[7] Baptiste A Cleerey M M Matz M Evitt C P 2008 *Rehabilitation Nursing* **33** 22-32
[8] Hakamiun R Genereux D P Falin M D Mutka M J Brooke J C 2001 *Patient Lift* US6289534B1
[9] Brandorff A Campbell G M 2002 *Compact portable patient lift* US6430761B1
[10] Huang C T 2005 *Foldable lift and transfer apparatus for patient* US6857144B1
[11] Lear A T 2010 Improved *Patient Lifting Apparatus* EP2197404A4
[12] Tu C T D 2015 *Method and apparatus for an automatic patient lift* US8935814B1
[13] Wang Y L 2016 *Device for turning over and transferring the patient* US6321398B1
[14] Fakhrizadeh M 2019 *Portable assistive lift* US10238564
[15] Runnels J 2019 *Apparatus and method for mobility device lifting and positioning* US10406047B1
[16] Singiresu S R 2005 *The Finite Element Method in Engineering, 4th edition, Basic Equations and Solution Procedure* chapter 3 279-308

Effects of Waveguide Position on Electric Field and Temperature Profile in Natural Rubber Gloves during Vulcanization Process Using Microwave Energy

P. Keangin[1,*], U. Narumitbowonkul[1], and P. Rattanadecho[2]
[1]Department of Mechanical Engineering, Faculty of Engineering, Mahidol University, 25/25 Phutthamonthon 4 Road, Salaya, Nakhon Pathom 73170, Thailand.
[2]Department of Mechanical Engineering, Faculty of Engineering, Thammasat University (Rangsit Campus), 99 mu 18, Paholyothin Road, Klong Nueng, Klong, Pathumthani, 12120, Thailand.

* Corresponding Author: pornthip.kea@mahidol.ac.th

Abstract
Natural rubber gloves (NRG) become a broader industry with fierce competition and demands. The vulcanization process is the most important and takes the longest time in the cycle period of the NRG manufacturing process, so it becomes of paramount importance to improve this process. The purpose of this research is to experimentally and computationally study the characteristics of the electric field as well as temperature profile in NRG during the microwave vulcanization process. The effects of position of waveguide placement are investigated for nine different positions. The experiment of NRG microwave vulcanization is performed for temperature distribution measurement. Three-dimensional models of NRG and microwave oven are considered. Finite element method (FEM) is utilized to solve a mathematic model of the transient Maxwell's equation coupled with the transient heat transfer equation in order to determine the electric field and temperature profile via computer simulation. The computer simulation results are validated against the experimental results. The validation confirms that the computational model is able to represent the practical transport phenomena in NRG with high accuracy. The outcomes clearly reveal that the position of waveguide effects on the electric field as well as temperature profile in NRG during the microwave vulcanization process. The positioning of waveguides on both the front and backsides on the opposite side of the microwave oven provides higher intensity electric field and higher temperature profile than the positioning of waveguides in the middle of both the front and the backsides of the microwave oven. In addition, the middle backside is the appropriate position for waveguide in which NRG is uniformly dry and not burnt or overheated. However, it is found that there is no direct relationship between the electric field and temperature changes. This research guides to essential aspects of the design of NRG vulcanization to improve the efficiency of microwave and heat delivering along with other heating systems to heat any sorts of material.

Keywords: Natural rubber gloves; Microwave heating; Electric filed; Temperature profile

1. Introduction

Nowadays, rubber is broadly used to be the raw material of various products such as vehicle tires, elastics, helmet, tires or tubes for motorcycle, parts of automotive and rubber bands, especially rubber gloves. Natural rubber gloves (NRG) is one of natural rubber (NR) products that becomes a broader industry. The fierce competition and high demands have forced the organization in this industry to continuously research and develop to improve the manufacturing process and the quality of its products [1-2]. The vulcanization of NRG is the most vital process that improves the strength and elasticity of NR by the chemical-thermal process. The heating process also takes the longest time in the cycle time of the NRG manufacturing process [3].

The conventional NR heating method used hot air as the heat source [4]. In the hot air heating process, the convection and conduction play the essential roles as heat transfer mechanisms [5]. The conductive heating by the hot air causes the very high initial rate of heat transfer on the surface of products because of the high difference of temperature between the hot air and the surface of products. Therefore, the surface temperature is numerously higher than the inner temperature and thus the nonuniformity in temperature distribution occurs and damage on the surfaces in the NR products during hot air heating process. Therefore, the combined heating of convection and radiation heat transfer were developed to improve the weakness of hot air heating about the temperature distribution as well as to optimize the process parameters of the vulcanization. Khamdaeng et al. [6] presented material parameters of NRG using the combined infrared and hot air vulcanization. However, the heat transfer in the infrared heating is subjected to the emission. In addition, the limitation of the combined system is the complex phenomena shown in the above-mentioned study [6]. The exploration of the effective heating method in the NRG manufacturing process becomes of paramount importance for the researches to improve the manufacturing process and the quality of the products.

Microwave energy is a famous alternative of heat source in this day and age. When the material undergoes microwave heating process, microwave impinges on the material and part of the energy is absorbed and converted uniformly into thermal energy by electromagnetic wave effects and delivering the energy to evenly heat the entire volume of the product called volumetric heating. Several advantages of microwave heating are quicker start-up, selective heating, higher uniformity, more precise regulation, better preventing overheating, preferable quality and salability of products, higher energy efficiency as well as environmentally friendly and safety. [7-8]

There are previous studies on the process using microwave energy to heat NR [9-14]. Bovtun et al. [9] studied the dielectric properties of rubber compounds during microwave vulcanization to investigate the effects of carbon black. The vulcanizations of carbon black, reclaimed rubber blends and NR were conducted by Sombatsompop and Kumnuantip [10] to monitor and compare the physical and mechanical properties during the microwave irradiation to convention thermal methods. Makul and Rattanadecho [11] computationally and experimentally analyzed the pre-curing process of NR-compounding using microwave energy through a rectangular waveguide. The results showed that partial cross-linking was produced by microwave energy at a temperature below the actual vulcanization. Aiyarak and Sunheem[12] designed and implemented the experiment to measure microwave attenuation for dry rubber content estimation of natural rubber latex during the microwave vulcanization process. Chen et al. [13] experimentally studied the effects of microwave heating and vulcanization process on temperature distribution in sheet rubber.

The waveguide is an important component which delivers microwave energy to the load in a cavity. It governs the characteristics of the electric field as well as temperature profile and effects on the heating pattern of products [14]. However, there are few studies on the design, positioning and operation of the waveguide that will be able to improve the efficiency of NRG vulcanization.

In the present study, the experimental and numerical analysis of NRG undergoing the 2.45 GHz-microwave vulcanization process within a microwave oven is performed. The study aims at examining the influences of the position of the waveguide on the electric field as well as temperature profile when the position of the waveguide is changed to the variations in nine cases i.e. a couple of middle fronts and backsides, middle front side, middle backside, couple of front right side and back left side, front right side, back left side, couple of front left side and back right side, front left side and back right side. The microwave power input of 100 W is applied to NRG Size M (Medium) for the duration of heating of 300 s. The three-dimensional computational models of NRG and microwave oven are modeled and the finite element method (FEM) is considered to solve this problem. The experiment is performed to validate the numerical results. This work is a basic of the design of NRG vulcanization to improve the efficiency of the heating process of NRG in the industry.

2. Experimental study

The experiment of NRG microwave vulcanization is performed for distribution of temperature measurement and record. The experimental apparatus diagram is illustrated in Figure 1. The dimension of the microwave oven cavity is 310 x 280 x 190 mm^3 in x, y and z-axis, respectively. The microwave oven operates at a widely conventional microwave frequency of 2.45 GHz. The microwave power input of 100 W is set for a duration of heating of 300 s as used in the industrial microwave vulcanization process. The experimental setup starts with the preparation of the mold of NRG where a ceramic *hand-shaped former* is dipped into 0.5-mm thick natural rubber latex (NRL). Then, place it in the microwave cavity at the central position and vertically along the waveguide area. The initial temperature of NRG measured by the thermocouple is 24 °C. The thermocouple K-type is installed inside the NRG to measure the temperature varied in positions and time during microwave vulcanization process [15-16] as shown in Figure 2 and record temperature by a data logger (Wisco DL2200 Data Logger). The dimension of commonly used NRG size M is 110 mm wide by 200 mm long as shown in Figure 3. In addition, Figure 3 illustrates the positions of temperature measurement (position A1-A5) which are located at 45 mm inside the NRG. Microwave energy generated by the microwave generator is transported to NRG located at the center of microwave cavity via waveguide having a cross-sectional area of 79 x 43 mm^2. The microwave energy delivered at the microwave power input of 100 W for the period of microwave vulcanization of 300 s and the distribution of temperature varied in positions of temperature measurement (consider the position of A1-A5) at various durations of heating (60 s, 180 s and 300 s) is examined.

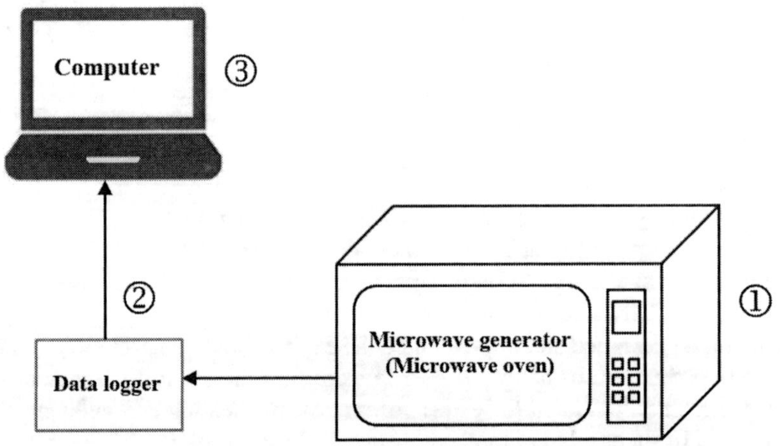

Figure 1. Experimental apparatus; ① Microwave generator ② Datalogger (Wisco DL2200) and ③ Thermocouple

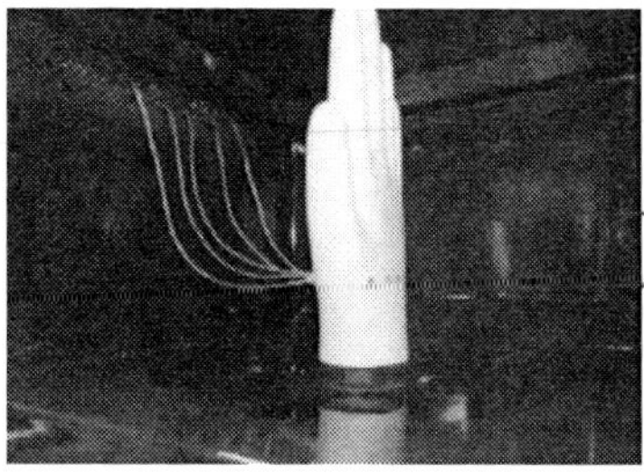

Figure 2. Temperature measurement within NRG placed in the cavity of a microwave oven

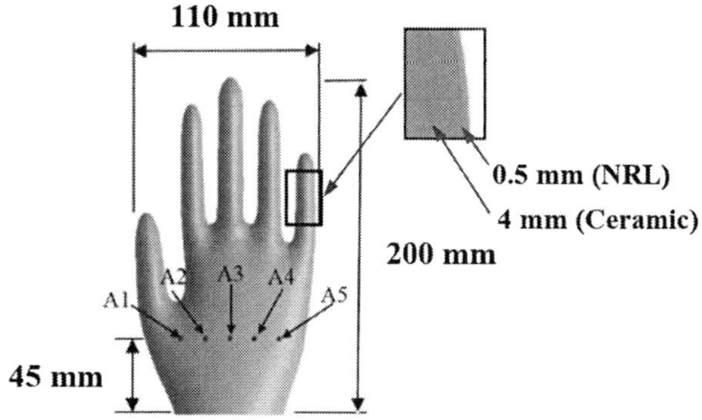

Figure 3. Dimensions of NRG Size M and positions for measuring temperature (A1-A5) within NRG.

3. Computer Simulation Study

3.1 Physics Model Formulation

Figure 4 shows the model of a microwave oven, rectangular waveguide and NRG used in the numerical study where their dimensions and those used in the experimental study are totally identical. The three-dimensional model of the microwave oven and NRG are presented to obtain the electric field as well as temperature profile results during microwave vulcanization process in a computer simulation. The NRL which is 0.5 mm thick and ceramic which is 4 mm thick are used in the computational analysis with the same condition to the experimental analysis. In a computer simulation, a uniform microwave propagates via a waveguide to the NRG, then penetrate the surface of NRG and resistive heating transforms the microwave into internal heat energy within the NRG.

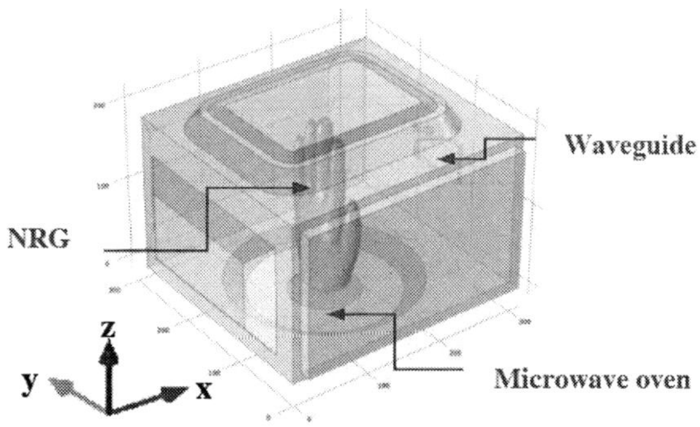

Figure 4. The model of a microwave oven and NRG used in the numerical study

3.2 Electromagnetic Wave Propagation Analysis

A mathematical model is formulated to determine the electric field as well as temperature profiles within the NRG during the microwave vulcanization process. The assumptions used in numerical analysis are as follows:

1) The multi-mode microwave used for vulcanizing NRG is purposed in a mathematical model.
2) The energy absorbed by air occupied the microwave oven can be ignored.
3) The model of NRG is considered stationary in simulation.
4) The waveguide is considered to be perfect conductors while the microwave oven's walls are considered to be impedance.
5) The NRG and ceramic have dielectric properties which are assumed to be constant in this model.

Maxwell's equations are mathematical models which describe the interdependence among the electric and magnetic field. In this numerical study, for analysing electromagnetic wave propagation in NRG and to predict the electric fields penetrated in NRG, the general form of Maxwell's equations is simplified as the following equation [17]:

$$\nabla\times\left(\frac{1}{\mu_r}\nabla\times\vec{E}\right)-k_0^2\left(\varepsilon_r'-\frac{j\sigma}{\omega\varepsilon_0}\right)\vec{E}=0 \qquad (1)$$

where μ_r is the relative magnetic permeability (H/m), $\vec{E}$ is the electric field intensity (V/m), ε_r' is the relative dielectric constant, σ is the electric conductivity (S/m) and ε_0 is the permittivity of free space (8.8542x10-12 F/m), $\varepsilon_r = n^2$ and n is the refractive index.

Figure 5 illustrates the boundary conditions for analyzing the electric field considered as follows:

In the numerical analysis, a microwave power transports into the microwave cavity through the waveguide. After that, it is absorbed by the NRG and so the NRG is heated as a result of increasing of thermal energy. To be consistent with the position of the waveguide of the microwave oven used in the experiment, it is assumed that the uniform microwave power irradiates to the left side of the NRG. That's why at this port of microwave oven, the port power level and port phase of wave excitation are specified. There is stored energy within electric and magnetic fields and thus it can be carried by the electromagnetic wave. The Poynting vector represents the energy flux transported by a propagating electromagnetic wave. Thus, the port boundary condition is using an electromagnetic wave propagation simulator specified to the boundary of the considered domain at the left as follows [18]:

$$S=\frac{\int\left(\vec{E}-\vec{E}_1\right)\cdot\vec{E}_1}{\int\vec{E}_1\cdot\vec{E}_1} \qquad (2)$$

In addition, the propagation constant is defined as:

$$\beta=\frac{2\pi}{c}\times\sqrt{f^2-\frac{c^2}{4\times a^2}} \qquad (3)$$

Where β is the propagation constant, c is the speed of light in a vacuum (3x10^8 m/s), f is the operating microwave frequency (2.45 GHz) and a is the depth dimension of waveguide (3.4 inches).

The waveguide is considered to be perfect conductors. The perfect conductor boundaries thus are applied to the walls of the waveguide:

$$\hat{n}\times\vec{E}=0 \qquad (4)$$

The wall surfaces of the microwave oven are considered to be impedance. The impedance boundary so applied to wall surfaces of the microwave oven:

$$\sqrt{\frac{\mu_0\mu_r}{\varepsilon_0\varepsilon_r - j\frac{\sigma}{\omega}}}\hat{n}\times\vec{H} + \vec{E} - (\hat{n}\cdot\vec{E})\hat{n} = (\hat{n}\cdot\vec{E}_s)\hat{n} - \vec{E}_s \tag{5}$$

Furthermore, at the internal boundary along with the interfaces between NRL and the ceramic, the continuity boundary conditions are applied as follows:

$$\hat{n}\times(E_{NRL} - E_{air}) = 0 \tag{6}$$

$$\hat{n}\times(H_{NRL} - H_{air}) = 0 \tag{7}$$

3.3 Heat Transfer Analysis

Coupling mathematical model of the propagation of electromagnetic wave with unsteady-state heat transfer along with the boundary conditions is solved for analyzing the thermal problem. Nevertheless, only within the NRG is considered in heat transfer analysis.

In microwave heating process, microwave propagates via a waveguide to the NRG and electric field is generated. Then, it is absorbed by NRG and converted into thermal energy. Increasing of thermal energy causes temperature within NRG to increase. In this research, the heat transfer analysis of microwave heating process is based upon the following assumptions:

1) The phase change of substance in the NRG does not occur.
2) The chemical reaction in the NRG does not occur.
3) The effect of shrinkage of NRG can be ignored.
4) The smooth contact surface of NRG is considered.
5) The NRG and ceramic have thermal properties which are assumed to be constant in this model.
6) The NRG is considered to be stationary in simulation, while the mold of NRG is rotated in a number of replications in the practical microwave vulcanization.

The unsteady-state heat transfer equation describes the transfer of heat in the internal energy within the NRG and is formulated to approximate the temperature profile within NRG as given by [19]:

$$\rho C_p \frac{\partial T}{\partial t} - k\nabla^2 T = Q \tag{8}$$

Where ρ is the density (kg/m^3), C_p is the heat capacity (J/kg·K), T is the temperature (°C), k is the thermal conductivity (W/m·K) and Q is the external heat source term (W).

Considering Eq. (8), the first term on the left-hand side of the equation indicates the transient term and the second one indicates the heat conduction term. While, considering the term Q on the opposite one, when microwave energy generates electric field and it is absorbed by NRG. The NRG heats up due to resistive heating that is an external heat source in the analysis of microwave heating process called microwave power absorbed as given in Eq. (9):

$$Q = 2\pi f\varepsilon_0\varepsilon_r'(\tan\delta)\vec{E}^2 \tag{9}$$

Where $\tan\delta$ is the dissipation factor. As the external heat source in this research is obtained from resistive heating generated by an electric field which is the result of the electromagnetic wave, the electric field in Eq. (9) is calculated via the electromagnetic wave propagation analysis mentioned before.

Figure 5 illustrates the boundary conditions for analyzing the electric field considered as follows:

As only within the NRG is considered in heat transfer analysis, the thermal insulation boundary condition is applied to the surface of NRL and ceramic:

$$-\hat{n}\cdot(-k\nabla T)=0 \tag{10}$$

The continuity boundary conditions applied at the interfaces between NRL and ceramic glove mold former in perfect contact that has the same temperature at the region of contact and does not store any energy are expressed as:

$$T_{NRL}(x_0,t)=T_{Ceramic}(x_0,t) \tag{11}$$

$$-k_{NRL}\frac{\partial T(x_0,t)}{\partial x}=-k_{Ceramic}\frac{\partial T(x_0,t)}{\partial x} \tag{12}$$

Where k_{NRL} and $k_{Ceramic}$ are the thermal conductivities of the NRL layer and ceramic layer, respectively.

Complex permittivity (ε), being a function of dielectric loss factor and dielectric constant, is a general form of the dielectric properties and can be expressed as [20]:

$$\varepsilon=\varepsilon'-j\varepsilon''=\varepsilon_0\left(\varepsilon_r'-j\varepsilon_r''\right) \tag{13}$$

where ε' is the dielectric loss factor, ε'' is the dielectric constant, ε_r' is the relative dielectric constant and ε_r'' is the relative dielectric loss factor.

The NRG is considered to be homogeneous and isotropic in electrical and thermal processes. The NRG and ceramic have dielectric and thermal properties in this computer simulation are assumed to be constant and presented in Table 1.

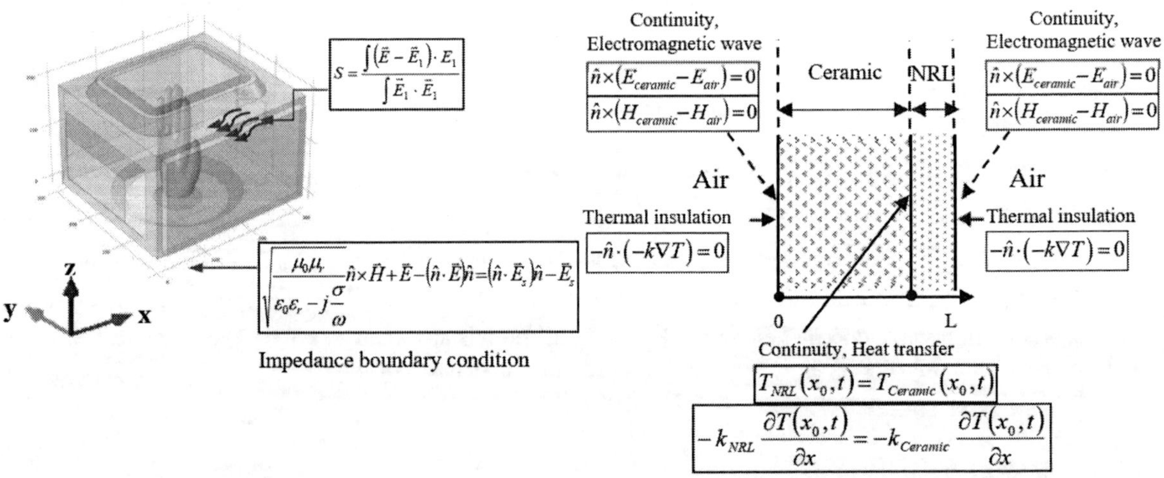

(a) (b)

Figure 5. The boundary condition for numerically analyzing the electromagnetic wave propagation as well as heat transfer within (a) microwave oven and (b) NRG during the microwave vulcanization process

Table 1. The dielectric and thermal properties of NRL and ceramic used for numerically analyzing the electromagnetic wave propagation as well as heat transfer during microwave vulcanization process [21-25]

Parameters	Emblem	Natural Rubber Latex	Ceramic
Electric conductivity (S/m)	σ	0	0
Relative dielectric constant	ε_r'	20	3.4
Relative dielectric loss factor	ε_r''	0.3	0
Thermal conductivity (W/m·K)	k	0.13	1.1
Density (kg/m^3)	ρ	975	2,200
Heat capacity (J/kg·K)	C_p	1894	480

4. Problem Statement

This study aims to examine the effects of various positions of the waveguide on the electric field as well as temperature profile in NRG during the microwave vulcanization process. The models of microwave oven with various positions of the waveguide and the condition studied are exhibited in Figure 6 and Table 2. Computer simulation analysis of the characteristics of the electric field as well as temperature profile within NRG and microwave oven undergoing microwave vulcanization process which operates at a microwave frequency of 2.45 GHz varies in the following parameters:

- NRG size M
- Microwave power input of 100 W
- Duration of heating of 300 s
- Positions of the waveguide of 9 conditions as shown in Figure 6 and Table 2

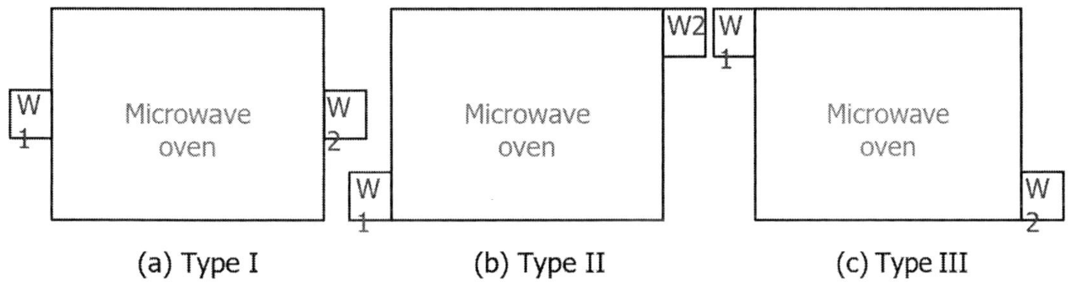

Figure 6. 2-D Physical models of microwave oven with various positions of the waveguide

Table 2. The conditions of positions of a waveguide for simulation analysis

Type	Models equipped with various waveguide	Waveguide 1	Waveguide 2
	Case 1	open	open
I	Case 2	open	close
	Case 3	close	open
	Case 4	open	open
II	Case 5	open	close
	Case 6	close	open

	Case 7	open	open
III	Case 8	open	close
	Case 9	close	open

In this study, the mathematic model coupling of electromagnetic wave propagation and heat transfer equations solved by FEM is carried over the entire domain. The coupled equations of transient Maxwell's with the transient heat transfer along with related boundary conditions via COMSOL™ Multiphysics software are solved. The heat is generated by electric filed as an external heat source of the heat transfer equation. The three-dimensional model of the microwave oven and NRG is assigned with triangular elements. The electric field and temperature variations across each element are then calculated approximately using the Lagrange quadratic elements. The approximately 431,110 elements of mesh are identified using a grid independence test as it is the appropriate number of elements required in the computer simulation study.

5. Results and Discussion

The present study is aimed at investigating the influences of the position of the waveguide on the electric field as well as temperature profile during NRG microwave vulcanization.

5.1 Validation of Computational Results Against Experimental Results

For verifying the accuracy of the computational model, the validation of the computational results is performed, in the identical conditions, against the experimental results. Figure 7 shows the comparison between distributions of temperature in NRG at various positions of temperature measurement (A1-A5), exhibited in Figure 3, using computer simulation and experiment based on microwave frequency of 2.45 GHz, microwave power input 100 W lasting for the duration of heating of 300 s and NRG size M. It depicts that the perfect agreement exists in the comparison between distributions of temperature in NRG from computer simulation and experiment with approximately 9.96 % difference in error values. The results also display that the distributions of temperature from computational and experimental study increase little by little at position A1 to the maximum at position A2, thereafter the distributions of temperature decrease gradually at position A3, A4 and A5. Therefore, this validation confirms that it is valid to represent the practical propagation of electromagnetic wave along with heat transfer within NRG undergoing microwave vulcanization process by the present computational model with high accuracy.

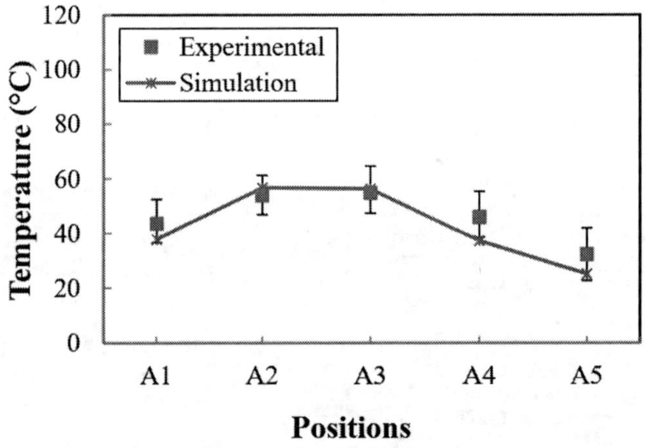

Figure 7. The comparison of distributions of temperature in NRG obtained by computer simulation results against the experimental results.

5.2 Computer simulation results from the effects of the position of the waveguide

Figure 8 exhibits the top view of the computer simulation results that is electric field as well as temperature profile, influenced by the variation of the position of the waveguide. For Cases 1-9, the electric field, as well as the temperature profile of NRG, is shown in Figure 8 (a)-(i), respectively. It does maintain that positioning of waveguides on both sides of the microwave oven results in good electric field distribution with high intensity. However, the overlap of their combined microwave in Case 1 (middle of both the front and the back sides of the microwave oven) results in the less absorbed microwave along with less hot spot zone. The highest intensity electric field occurs in cases which place the waveguide couple of front right side and back left side of the microwave oven in Case 7. However, it reveals that the change in the electric field is not consistent with the change in temperature. When the position of a waveguide in the microwave oven is the front left side in Case 8, it is found that electric field is better at distributing around and then is absorbed by NRG, converted to heat, consequently results in more a hot spot zone. An asymmetric model with a concave surface on the front side and the backside of microwave oven cause low-intensity electric field as well as low hot spot zone in cases which place the waveguide on the front right side in Case 5.

Figure 9 displays the temperature profile of NRG influenced by the variation of the position of waveguide during microwave vulcanization where the 2.45-GHz microwave is applied to heat NRG size M using microwave power input of 100 W for the duration of heating of 300 s. The results from Type I (Case 1-3) of positions of the waveguide are shown in Figure 9 (a) - (c). Figure 9 (a) exhibits that at the adjacent region of the middle of the palm of NRG, the influences of the position of waveguide during microwave vulcanization causes the hot spot zone with a maximum temperature of 77.13 °C in Case 1. Meanwhile, in Case 2, Figure 9 (b) illustrates that there is not much variation in temperature profile at the adjacent region of the middle finger with a maximum temperature of 57.84 °C. In Case 3, at the adjacent region of the middle finger, the little finger as well as the middle of the palm, microwave heating creates the hot spot zone with the maximum temperature of 135.58 °C as shown in Figure 9 (c). The results from Type II (Case 4-6) of positions of the waveguide are shown in Figure 9 (d) - (f). The temperature profile in Case 4 varies a bit at the region near the middle finger with the maximum temperature of 50.41 °C, likewise, it varies a bit at the palm in Case 5 with the maximum temperature of 25.13 °C as shown in Figure 9 (d) and 9 (e), respectively. In Case 6, the region in which the hot spot zone occurs with the maximum temperature of 211.38 °C includes the palm and the tip of the middle finger as shown in Figure 10 (f). The results from Type III (Case 7-9) of positions of the waveguide are shown in Figure 9 (g) - (i). Figure 9 (g) exhibits that the effects of position of the waveguide in Case 7 caused not much variation in temperature profile at the adjacent region of the middle of palm with the maximum temperature of 50.03 °C. However, the hot spot zone with a maximum temperature of 382.47 °C is reached in Case at the adjacent region of the tip of the middle finger and especially the middle of the palm as shown in Figure 9 (h). In Case 9, there is a bit of variation in temperature profile with the maximum temperature of 25.16 °C in the region of the middle fingertip as shown in Figure 9 (i). Because of the very high temperature shown in the simulation results, it has to be emphasized that the phase changes of substance in the NRG are ignored. It is discovered that the position in Case 3, Case 6 and Case 8 are good positions for the waveguide. Although the position in Case 8 results in the hottest spot zone with the highest temperature distribution in NRG, temperatures above 300 °C may cause a burning in NRG. Therefore, the positioning of the waveguide in the middle backside in Case 3 is the appropriate position for waveguide in which NRG is uniformly dry and not burnt or overheated.

IOP Conf. Series: Materials Science and Engineering **886** (2020) 012028 doi:10.1088/1757-899X/886/1/012028

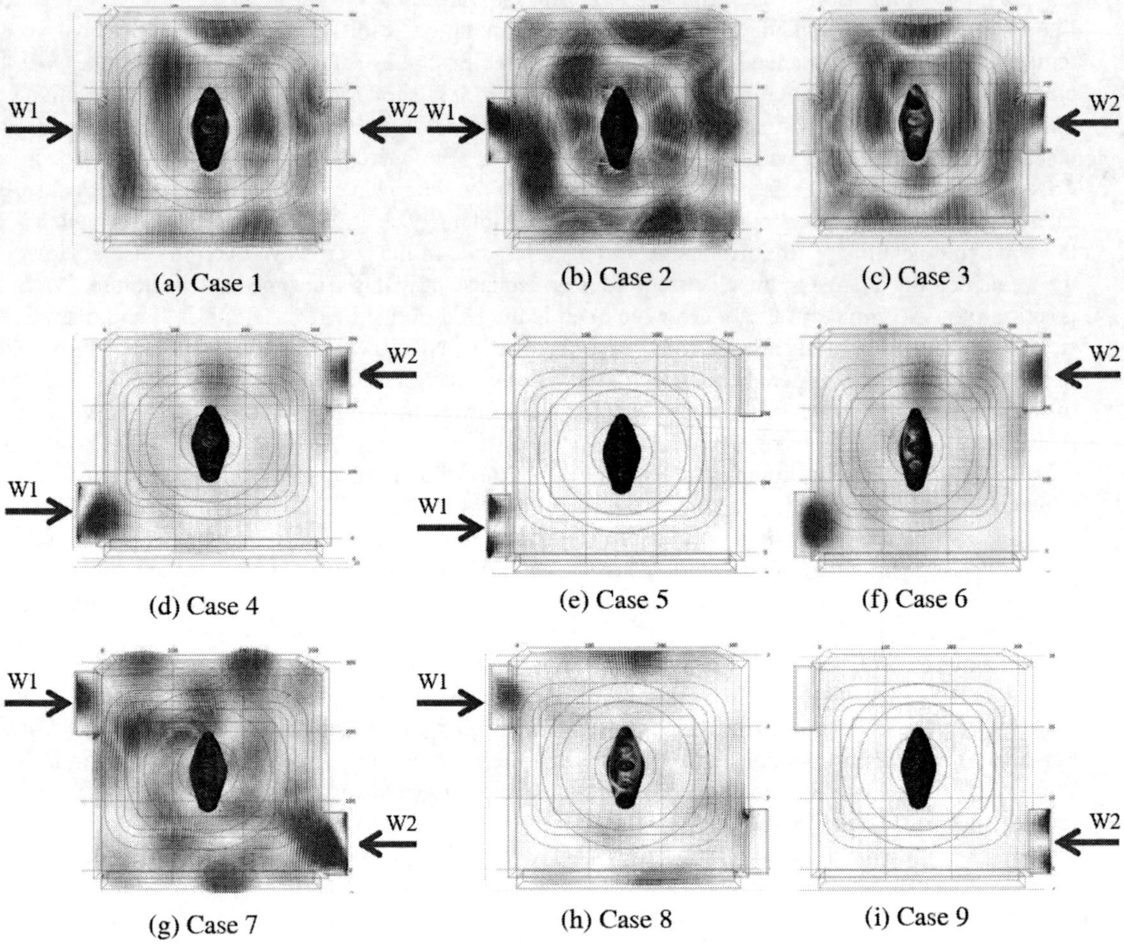

(a) Case 1 (b) Case 2 (c) Case 3

(d) Case 4 (e) Case 5 (f) Case 6

(g) Case 7 (h) Case 8 (i) Case 9

Figure 8. The top view of the electric field as well as temperature profile in NRG influenced by the variation of the position of the waveguide in each case
variation of the position of the waveguide in each case

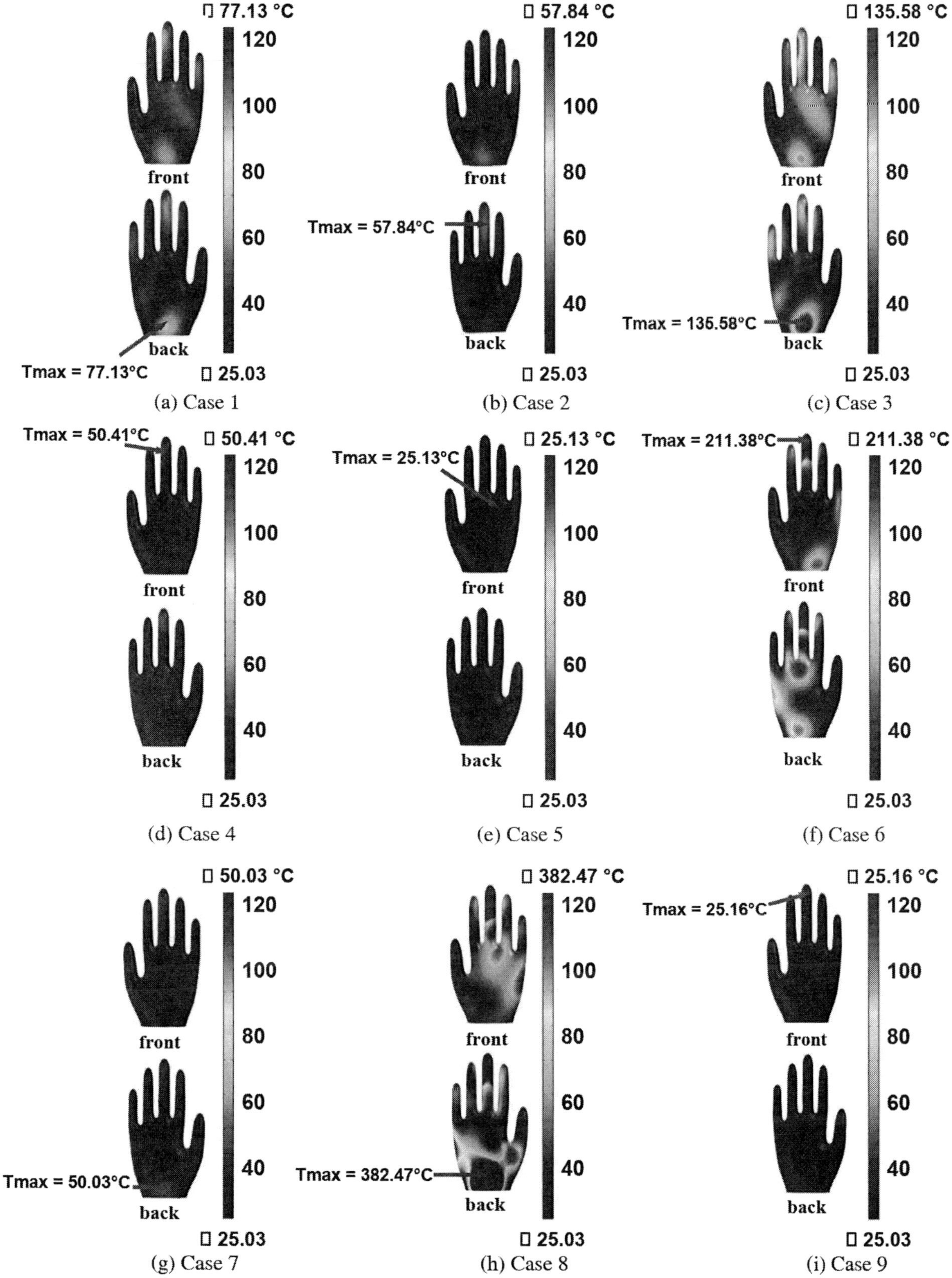

Figure 9. The effects of the position of the waveguide on the temperature profile in NRG in

6. Conclusions

In the present study, the computational and experimental study on the electric field as well as temperature profile in NRG during microwave vulcanization process is performed in order to investigate the effects of different position of the waveguide. The distribution of temperature from simulation results during microwave vulcanization is validated against experimental results. The obtained results showed that the middle backside (Case 3), the back left side (Case 6) and the front left side (Case 8) are good positions for the waveguide. However, positioning the waveguide at the front left side (Case 8) results in the hottest spot zone with the highest temperature distribution in NRG may cause burning in some areas. The positioning of the waveguide in the middle backside in Case 3 is the suitable position for waveguide in which NRG is homogeneously dry and not burnt or overheated. This investigation provides fundamental information and knowledge on heat transfer within NRG during microwave vulcanization process and can be applied to use in NRG industry as a guideline to improve the heating process of NRG and lead to improve the manufacturing process of NRG to gaining the competitive advantage.

Acknowledgment

The authors would like to gratefully acknowledge the Thailand Research Fund under the project no. RDG62T0013 for supporting this research.

References

[1] Yip E Cacioli P 2002 *J. Allergy Clin. Immunol.* 110(2 Suppl) S3-14
[2] Akabane T 2016 *Int. Polym. Sci. Technol.* 43(5) 45-50
[3] Jirasukprasert P Garza-Reyes J A Kumar V Lim M K 2014 *Int. J. Lean Six Sigma* 5(1) 2-21
[4] Schöppner V Brockhaus S Penner C Möckel J 2016 *Int. Polym. Sci. Technol.* 43(10) 1-6
[5] Hopmann C Lemke F Peter J P Gahlen P 2018 *Rubber World* 258(4) 22-26
[6] Khamdaeng T Panyoyai N Wongsiriamnuay T 2014 *Am. J. Appl. Sci.* 11(4) 648-655
[7] Keangin P Wessapan T Rattanadecho P 2011 *Appl. Therm. Eng.* 31(16) 3243-3254
[8] Jones D A Lelyveld T P Mavrofidis S D Kingman S W Miles N J 2002 *Conserv. Recycl.* 34(4) 75-90
[9] Bovtun V Stark W Kelm J Porokhonskyy V Yakimenko Y 2002 *Kautsch. Gummi Kunstst.* 55(1) 16-22.
[10] Sombatsompop N Kumnuantip C 2006 *J. Appl. Polym. Sci.* 100(6) 5039-5048
[11] Makul N Rattanadecho P 2010 *Int. Commun. Heat Mass Transfer* 37(7) 914-923
[12] Aiyarak P Sunheem P 2015 *Songklanakarin J. Sci. Technol.* 37(6) 713-718
[13] Chen H L Li T Liang Y Sun B Li Q L 2017 *Heat Mass Transfe*r 53(3) 1051-1060
[14] Gerling J F 2000 *Gerling Applied Engineering, Inc.* 1-8
[15] Potter D 1996 *National Instruments Corporation* 1-15
[16] Olmstead W E Brodwin M E 1997 *Int. J. Heat Mass Transfer* 40(7) 1559-1565
[17] Wessapan T Srisawatdhisukul S Rattanadecho P 2011 *Int. Commun. Heat Mass Transfer* 38 255-262
[18] COMSOL Multiphysics Reference Guide 1998–2012 COMSOL US7519518 US7596474 US7623991 (Patents pending)
[19] Narumitbowonkul U Keangin P Rattanadecho P 2014 *Proc. the 5th TSME International Conference on Mechanical Engineering 2014*
[20] Rattanadecho P Makul N 2016 *Drying Technol.* 34(1) 1-38
[21] Rands R D Ferguson W J Prather J L 1944 *J. Res. Natl. Bur. Stand.* 33 63-70
[22] Khalid K B 1992 *The Microwave Conference, APMC 92. Asia-Pacific* 2 611-616
[23] Hassan J Khalid K Mohammad W Yusoff D W 1997 *Pertanika J. Sci. & Technol.* 5(2) 179-190

[24] Sridee J 2006 Rheological Properties of Natural Rubber Latex Suranaree University of Technology Academic Year ISBN 9745335886
[25] Ortiz-Serna P Díaz-Calleja R Sanchis M J Floudas G Nunes R C Martins A F Visconte L L 2010 *Macromolecules* 43(11) 5094-5102

IOP Conf. Series: Materials Science and Engineering **886** (2020) 012029 doi:10.1088/1757-899X/886/1/012029

The Conception of an Autonomous Traveling Manual Wheelchair and a Study on its Localization

K Tobita[1,*], Y Shikanai[1], and T Murata[2]

[1]Shizuoka Institute of Science and Technology, 2200-2 Toyosawa, Fukuroi, Shizuoka, 437-8555, Japan
[2]Kanagawa Rehabilitation Center, 516 Nanasawa, Atsugi, Kanagawa, 243-0121, Japan

* Corresponding Author: tobita.kazuteru@sist.ac.jp

Abstract
In a care facility, it is one of the factors that the burden of a caregiver is increasing by guiding many cared persons in a wheelchair before and after a meal. Although many autonomous electric wheelchairs have been researched and developed, it is not reasonable to introduce them for the entire care receivers in a nursing home from the viewpoint of cost, time and trouble of maintenance. We are developing a detachable robot that makes a manual wheelchair run autonomously. This paper describes the basic concept, target specifications and design conditions of the experimental machines and feasibility study about the localization by using the optical flow sensor. In the future, we will proceed with the implementation of the autonomous driving function.

1. Introduction

In a nursing home, it is one of the factors that the burden of a caregiver is increasing by guiding many cared persons in a wheelchair before and after a meal. Besides, even in the rehabilitation center, when a patient moves to a training facility with a wheelchair, a trainee must accompany him and the burden on the trainee's work increases. It is thought that these can be eliminated by autonomously moving the wheelchair. A lot of research has been done on the autonomous movement of electric wheelchairs. An electric wheelchair that travels in a hallway using visual servo [1], a map creation using 3D SLAM, and an electric wheelchair that estimates its position [2]. In addition, Whill and Panasonic are carrying out demonstration experiments of autonomous movement with an electric wheelchair at the airport [3].

They are highly feasible for continuous use by individuals and in applications where operation with a high utilization rate is expected. However, the opportunity to use the autonomous function in a nursing home is limited only before and after events such as meals, and the operation rate is low. Also, it is not reasonable to introduce them for the entire care receivers in a nursing home from the viewpoint of cost, time and trouble of maintenance.

On the other hand, there are products such as SmartDrive [4], which is a drive unit to add motorization to a manual wheelchair, research examples in which the robot climbs the step in cooperation with the user [5], and cases of burden reduction for assistants by the addition of electrically assisted function [6]. However, no one adds an autonomous movement and guidance function to a manual wheelchair.

In this research, we propose a robot that drives and guides wheelchairs one by one by attaching an easily detachable autonomous drive unit to an already installed or inexpensive manual attendant-controlled wheelchair. This paper describes the basic concept of driving manual wheelchairs with a robot, the target specifications and design conditions of the experimental machines and feasibility study about the localization by using the optical flow sensor.

2. Basic concept and the design criteria

2.1 Operating environment and user
The use environment is indoor barrier-free. Corridor width of a nursing home is defined as more than 1,800mm in Japan [7]. The structure of the wheelchair should be able to be turned into a super soft turn so that direction changes can be made even if people are passing by in the 1,800 mm wide corridor. The users are an operator (caregiver) and a passenger (care receiver). There is no restriction on the caregiver but is a weight restriction on the care receiver. According to the age-specific physique data of the Ministry of Education, Culture, Sports, Science and Technology, the average weight of the elderly over 65 years old is less than 65 kg. [8]. Therefore, the weight of the care receiver was set to a maximum of 70 kg. with a margin. Besides, there is a possibility that a dangerous event may occur, such as the cared person lowering their feet below the foot support and coming into contact with the ground while traveling in a wheelchair.

2.2 Basic function
We plan to incorporate the following as basic functions:
(a) Simple coupler function
(b) Autonomous mobile function (self-position estimation, route generation)
(c) Obstacle detection, avoidance function
A simple connection device is incorporated so that the caregiver can easily attach and detach from the wheelchair.

The self-position estimation function recognizes its position on the map, generates a route to the destination, and incorporates an autonomous movement function that follows the route. At the time of connection, since many wheels come in contact with the ground, self-position estimation that is more robust to wheel slippage is required than in the case of an independent wheeled robot.

When there are obstacles such as persons and things on the way, the sensor detects this and incorporates the function to avoid it. It is necessary to detect the situation where the passenger's foot is likely to get to the ground and stop it.

2.3 Method of connecting a robot to a wheelchair
Methods for driving a manual wheelchair from the outside are roughly classified into two categories whether the robot pulls from the front of the wheelchair or pushes it out from the rear of the wheelchair. The pulling type is advantageous in that the sensor measurement range is wide because it does not obstruct the front of the robot, but it has the following disadvantages. When the wheelchair is turned on the spot, the turning radius increases and the movement increases, so the battery consumption is fast. The rigidity of the tip part, which is the connecting part with the wheelchair, is low. When reaching the dining room table, the robot needs to be removed from the wheelchair under the table and there is a risk that the caregiver's head may collide with the table.

On the other hand, the pushing type adopted in this study has the following advantages, although the detection range of the sensor is partially blocked and limited by the wheelchair itself and the passenger. Both the turning radius and the traveling distance can be reduced. It can be connected with a highly rigid tipping lever. Tipping levers are usually made rigid because they are used when an assistant steps on one foot and crosses a step. Desorption can be performed in a space where the caretaker does not have a head collision with the table.

2.4 Driving force

IOP Conf. Series: Materials Science and Engineering **886** (2020) 012029 doi:10.1088/1757-899X/886/1/012029

The autonomous mobile robot needs to be placed between the left and right rear wheels. Besides, the size of the robot is made as compact as possible to make it easy to secure the storage space in a nursing home. In order to drive the wheelchair on which the care receiver rides, it is necessary to take into consideration the mass of the wheelchair, the mass of the care receiver (weight), the mass of the robot, and the friction coefficient of the floor. The prototype is designed for MIKI BAL-01 (its weight is 13 kg). As described above, the weight of the care recipient is 70 kg.; the mass of the robot itself is set to 25 kg. by calculating the torque required for the motor, and the coefficient of friction of the floor is set to 0.4 assuming the floor of the nursing care facility as flooring. Regarding speed, there is a document that 3.7 km/h is suitable for wheelchair transfer by assistants [9]. However, the maximum speed was set at 2 km/h to reduce the anxiety of the cared person because of the machine as the driving source instead of the person.

3. Prototyped experimental machine

Figure 1 shows the appearance of the autonomous mobile robot prototyped based on the target specifications and design study, Figure 2 shows the appearance of the wheelchair when it is connected to a wheelchair and Table 1 shows the basic specifications

Although it became large from the standard of the target size, there is no trouble in connecting with the tipping lever. In order to give priority to the realization of the concept, this prototype does not incorporate a simple removal mechanism without tools.

Figure 1. Overview of the pushing robot.

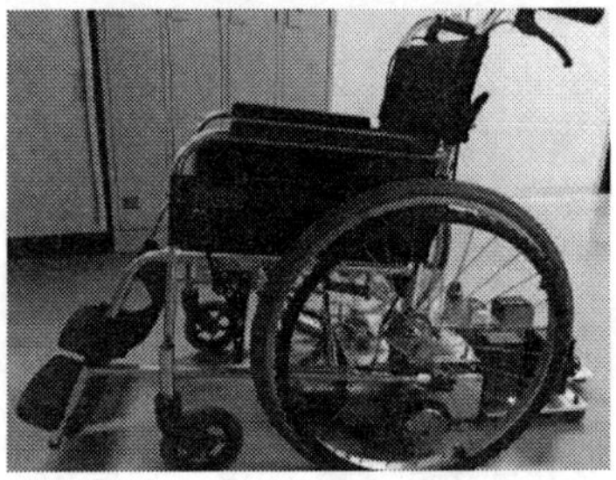

Figure 2. Connected condition.

Table 1. Specification of autonomous robot

Dimension	W458.0 × D718.7 × H284.5 [mm]
Motor	DC Brushless Motor: 180W Gear head: 1:26 Encoder: 1,024pulse
Velocity	2 km/h Max
Controller	Raspberry Pi 3B
Battery	24V 6Ah×4(Li-ion)

As the drive system, Maxon EC-i52 brushless motor was adopted in terms of the size, required torque and required rotation speed. By using Maxon's GP52C planetary gearhead (speed reduction ratio 1:26) and pulleys (speed reduction ratio 44:60), the required torque of the drive wheel is generated. The motor driver should be able to control the position, voltage, and rotational speed and be equipped with a heat sink. Likewise, ESCON 70/10 from Maxon was selected. The microcomputer used the Raspberry Pi 3 Model B, which has a USB port that can receive information from a two-dimensional range sensor and has Wi-Fi and Bluetooth communication functions. Two sets of two 14.4-V Li-ion batteries connected in series are used in parallel as the drive source. The microcomputer power supply supplies/provides 5 V, which is stepped down by the DC-DC converter.

Figure 3 shows a system-block diagram of the autonomous traveling robot. In order to detect obstacles, two-dimensional laser range sensor URG-04LX-UG01 (measurement range 240 deg, maximum measurement distance 5.6m) made by Hokuyo Electric Co., Ltd. is installed parallel to the floor surface at the lower front of the robot. The reason for installing it in the lower part is to detect as a dangerous state the above-mentioned cared person lowered his/her foot beneath the foot support. Because the robot is connected to the wheelchair and driven, many wheels touch the ground, and slippage of the drive wheels is expected. Therefore, the amount of movement is measured by an optical flow sensor directed to the ground without relying on odometry based on the rotational speed of the drive wheel. The optical flow sensor can measure the translation direction but cannot detect it in the turning direction, so it is fixed at a position away from the robot's turning center. If there is only one optical flow sensor, there are not enough degrees of freedom, so two sensors are attached to the front and back of the robot. The following equation is derived to perform relative self-position estimation.

$$\begin{pmatrix} x(k) \\ y(k) \\ \theta(k) \end{pmatrix} = \begin{pmatrix} x(k-1) \\ y(k-1) \\ \theta(k-1) \end{pmatrix} + \begin{pmatrix} \sqrt{\Delta O_{F,lo}(k)^2 + \Delta O_{F,la}(k)^2} \cos\theta(k) \\ \sqrt{\Delta O_{F,lo}(k)^2 + \Delta O_{F,la}(k)^2} \sin\theta(k) \\ \frac{\Delta O_{R,la}(k) - \Delta O_{F,la}(k)}{L} \end{pmatrix} \tag{1}$$

$x(k)$: x position at time step k
$y(k)$: y position at time step k
$\theta(k)$: Orientation at time step k
$\Delta O_{F,lo}(k)$: Relative longitudinal position by a front optical flow sensor
$\Delta O_{F,la}(k)$: Relative lateral position by a front optical flow sensor
$\Delta O_{R,lo}(k)$: Relative longitudinal position by a rear optical flow sensor
$\Delta O_{R,la}(k)$: Relative lateral position by a rear optical flow sensor
L: Interval between front optical flow sensor and rear optical flow sensor

In addition, the position and orientation are corrected based on the wall information measured by Laser Range Finder. A control system that follows the generated path is combined with the low-accuracy absolute self-location using a wireless ID tag. In this paper, self-position estimation is limited to verification of its feasibility with two optical flow sensors.

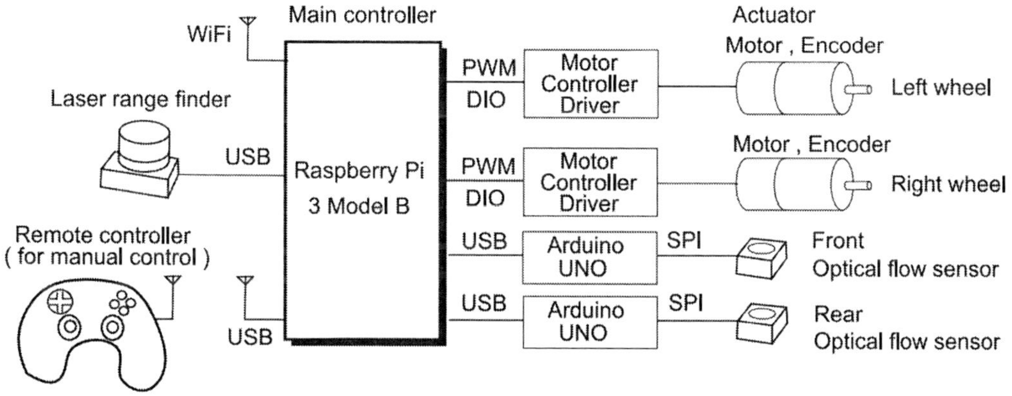

Figure 3. System block diagram

4. Basic Experiment

4.1. Detecting the environment with the Laser Range Finder

With the robot connected to the wheelchair on which a person is riding, the obstacle measurement situation by a two-dimensional range-finding sensor was investigated. In an indoor corridor with a width of 1,850 mm, which is about the same as a nursing home, sensor measurements were obtained when an obstacle was placed 1 m., 2 m., and 3 m. ahead.

Figure 4 shows the results. Figure 6 (a) shows the measurement results when the obstacle was placed 1 m ahead, (b) 2 m. ahead and (c) 3 m. ahead. In each case, the vertical axis is the X coordinate of the moving direction, and the horizontal axis is the Y coordinate perpendicular to the traveling direction.

In any case, obstacles installed in the riding direction are detected, and it is considered that there is no problem in obstacle detection during traveling. However, it can be seen that a part of the wall (about 1 m. to 3 m. away) is blocked by the front wheel of the wheelchair and cannot be detected. We confirmed that it is necessary to apply an algorithm that excludes this range for self-location estimation using wall information.

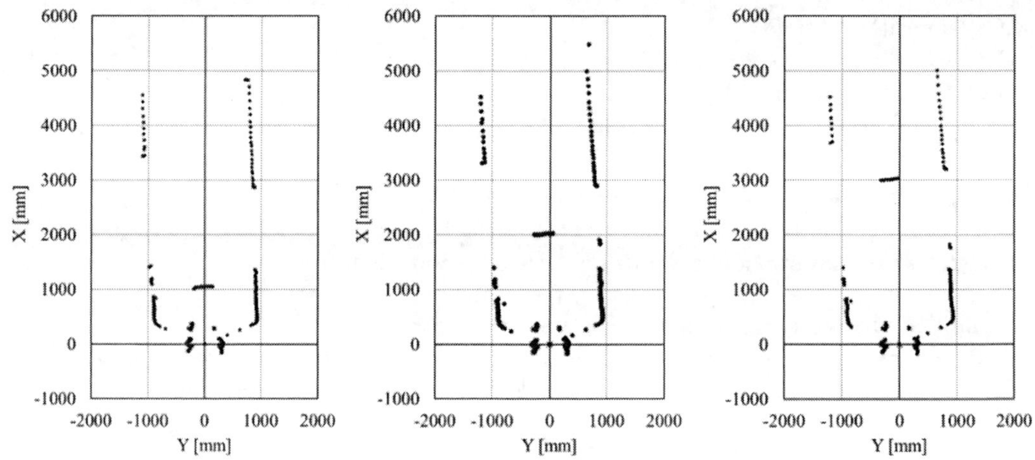

(a) 1 m. (b) 2 m. (c) 3 m.

Figure 4. Examples of measuring distance

4.2. Localization with the Optical flow sensor

Figure 5 shows two examples of the trajectory when the robot was traveling straight 1m. The X-axis is the direction of travel, and the Y-axis is the direction perpendicular to the direction of travel. The red dashed line means the trajectory drawn with a pen attached to the robot on graph paper laid on the ground, and the blue solid line means the trajectory calculated by Equation (1) from the values of the front and rear optical flow sensors. The tendency of the wobbling of the trajectory was similar between the two, but there was a difference in the accumulated error. Therefore, the value was calculated on the assumption that the error of the initial angle between the direction of the graph paper and the straight direction of the robot affected. In (a), the initial angle error was assumed to be 2 deg, and in (b) the calculation result when 0.5 deg was assumed is shown by the black dotted lines in the figure. The results showed that the error of the trajectory due to the error of the initial angle affected.

In this experiment, since two optical flow sensors were connected to different microcomputers, the two sensors were not synchronized, and it is considered that the timing difference affected them. Besides, since the current calculation algorithm was based on the front sensor, the error tended to increase if the front sensor missed the reading. However, the initial angle error and the error due to reading error can be solved by considering the correction by wall information detected by the Laser range finder.

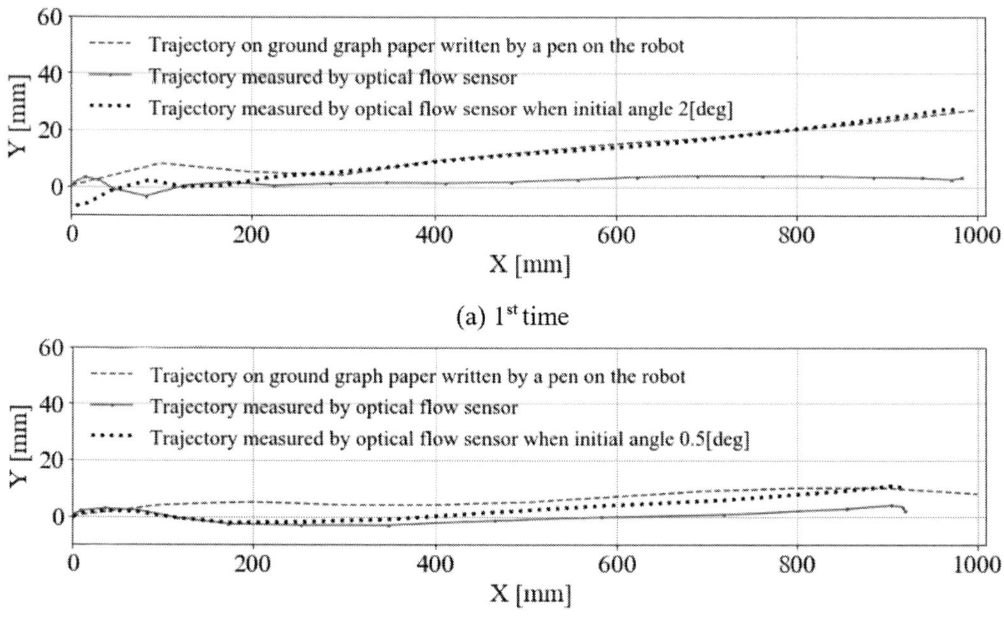

(a) 1st time

(b) 2nd time

Figure 5. Examples of the trajectory of the robot when forwarding 1 m.

5. Conclusion

In this report, to automate the mobility assistance of a wheelchair in a nursing home, we designed a removable autonomous traveling robot, set its concept, set target specifications, and made a trial design. Furthermore, the basic experiment of the prototyped robot was performed, the results and the consideration were reported, and the following was clarified.

- The concept of driving a manual wheelchair for people was shown to the board.
- It is necessary to estimate the position of the vehicle taking into account the effect of the partially-blocked area by the front wheel of the wheelchair.
- It was shown that a trajectory with both translational motion and swivel motion can be calculated using an optical flow sensor attached to the front and rear.

In the future, in addition to mounting linear tracking control, obstacle avoidance function using ranging sensors, navigation function, and so forth, a mechanism that can be connected easily is incorporated to promote demonstration experiments.

Acknowledgements

This research was funded by the NSK Mechatronics Technology Foundation grants. I would like to thank the NSK Mechatronics Technology Foundation, NSK Ltd. and its related companies.

References

[1] Pasteau F Narayanan V Babel M and Chaumette F 2016 *Robotics and Autonomous Systems* **75**(A) 28-40

[2] Yokozuka M Suzuki Y Takei T Hashimoto N and Matsumoto O 2010 *Journal of Robotics and Mechatronics* **22**(6) 758-766

[3] Uematsu H Imaoka N Nguyen D Sasai H Kitazawa K and Ando T 2018*Panasonic Technical Journal* **64**(1) 60-65

[4] SmartDrive 2019 *http://www.max-mobility.com/* accessed on 2 Jul 2019

[5] Ikeda H Katsumata Y Shoji M Takahashi T and Nakano E 2008 *Journal of the Robotics Society of Japan* **26**(2) 80-89

[6] Li P and Zhu C 2018 *Proc. the 2018 JSME Conference on Robotics and Mechatronics* 2A2-G04

[7] Health and Welfare Bureau for the Elderly, Ministry of Health, Labour and Welfare 2019 *About setting administration standard instruction guideline of pay nursing home, https://www.mhlw.go.jp/stf/seisakunitsuite/bunya/0000083170.html* accessed on 2 Jul 2019

[8] Ministry of Education, Culture, Sports, Science and Technology, *Age-based physical measurement results 2017, https://www.e-stat.go.jp/dbview?sid=0003289205* accessed on 2 Jul 2019

[9] Nagashima H Kato N Hosokawa M Tachiki F and Ito I 2003 *Annual meeting of the Japanese Society of National Medical Services* **58**(1) 168

Modelling Method Investigation of Drive and Motor for an Industrial Overhead Crane

Nattapong Suksabai, Jatuporn Waikoonvet, and Ittichote Chuckpaiwong*

Department of Mechanical Engineering, Mahidol University, 999 Phutthamonthon 4 Rd., Salaya, Nakhon Pathom, 73170 Thailand

* Corresponding Author: ittichote.chu@mahidol.ac.th

Abstract

Overhead cranes are commonly used to carry heavy objects from one place to another in a workspace. In common, 3-phase electric drive and induction motor are selected to drive a crane in three dimensions. Many mathematical models of a three- dimensional overhead crane were previously derived by energy equation to describe a relationship of applied force and crane movement. The problem was that the applied force could not be used for controlling most overhead crane systems whose motors were commonly controlled by velocity commands. Although, this problem was solved by creating an empirical model which contains the three functions: switch, rate limiter and linear second order, the further investigation and validation was still required. In this paper, the model of drives and motors of an overhead crane was investigated and validated by experiments using an industrial-grade 1-ton overhead crane. A unit step input of velocity command and actual velocity were used as data to create an empirical model. The model performance was experimentally evaluated by comparing with actual velocity and position of each time step. The discussion was also provided at the end.

Keywords: overhead crane, modelling, control process

1. Introduction

To transport a heavy object in a closed workspace, an overhead crane is a suitable choice. It operates on the rail system that is installed on top of the whole workspace, consists of the three components called a bridge, a trolley, and a hoist. Basically, an object is vertically lifted by a hoist which is mounted under the bridge and the trolley. The trolley can be moved along the bridge length which usually is defined as left and right direction. Eventually, the bridge moves in forward and backward direction. This simple but effective design makes overhead crane popular to manipulate the object in closed workspaces.

As global productions in many industrial cranes are increasing, overhead crane needs to be more automated. Using an operator to control crane may not suffice for today demands. A conventional overhead crane is controlled by an operator via push button to move an object in space. While an automated overhead crane promises to move an object to a target faster than a skilled operator, an automated crane relies on automatic control theory, which needs to be studied to archive a desirable control motion.

Mathematical model of an overhead crane is important for developing the control system. Previously, it was usually derived using Newton's Law [1], and Lagrange's equation [2]. Tuan et al. derived a mathematical model of an overhead crane and proposed the control technique for payload positioning. The control algorithm was implemented on lab-scale mini overhead crane where DC motors were used to actuate the crane movement. The result of payload positioning showed the successful control method based on the mathematical model of overhead crane [3]. Another related control algorithm based on a mathematical model was also published as well [4]. Lee Ho-Hoon implemented his control technique with AC servo motor, linear model of AC servo motor and nonlinear model overhead crane were together derived for the crane positioning and payload suppression of the control scheme. It was implemented on the crane prototype of 5.5x3.5 meter. A good result was also obtained for the prototype [5]. However, for an industrial crane, the mathematical model which describes the relation of force and acceleration may be useless. Unlike DC motor that is controlled by adjusting electrical voltage which implicitly rates to the force or even AC servo motor that can produce the linear force using its controller [5-6], an AC induction motor is controlled by an adjustable speed drive, so-called drive or inverter (alternatively called as VSD or VFD). The drive controls the AC motor using an adjustable electrical frequency that relates to angular velocity output. Since the model of the drive (inverter) and motor should base on the actual control parameter, the input should be a velocity command and the output should be an actual velocity or even distance. Sorensen and Singhose proposed the control technique for positioning of an industrial overhead crane and payload suppression using an ON-OFF relay. Regardless of the motor and drive model, the control technique results in the positioning error within 13 mm. and payload oscillation reducing it to approximately 20% [7]. Later, the positioning control and payload suppression were improved by developing of drive and motor model, Sorensen and Singhose proposed an empirical model of drive and motor where input was velocity command and output was actual velocity. It consists of three functions: switch, rate limiter and linear second order. The result showed that position error was less than 10 mm and payload oscillation was reduced to 5% of the maximum [8]. However, the validation of this model still requires further investigation in various aspects before using it in the automatic control design.

In this paper, section 2 will provide a detail of drive and motor modelling method proposed by Sorensen and Singhose [8], and the extended model for the validation process. Section 3 shows an experimental setup for the studying of drive and motor model. Experimental data, resulting system identification and performance of the model prediction will be expressed and discussed in section 4. Finally, section 5 is the conclusion of this paper.

2. Modelling of Drive and Motor

2.1 Proposed Model Development

By observing a crane movement during the experiment, Sorensen and Singhose [8] proposed an empirical model of the drive and motor for an industrial overhead crane. Initially, three velocity step commands were used as an input to the drive and motor then three output as actual velocity was collected to analyse the movement behaviour. Three velocity step commences are velocity step from rest to maximum (0 to 100%), from rest to medium (0 to 50%), and maximum to a maximum in another direction (100 to -100%). The corresponding three actual velocities were plotted separately; vertical and horizontal axis are actual velocity and time, respectively. In a case of rising velocity (0 to 100% and 0 to 50%), actual velocity raised from zero to 100% and 50% with infinite slope at the beginning and end and also minimum overshoot. The responding S-curve profile looked like a heavily damped second-order function. The difference of both settling times reminded that the system has some acceleration limit. In the case of transitional velocity when the motor is driven forward 100% and immediately commanded to 100% in the reverse direction, the system acted the same way before. That is, the S-curve profile decreased smoothly from 100 to 0% then increased smoothly from 0 to 100% in the reverse direction. As all mentioned an empirical model of drive and motor was proposed

IOP Conf. Series: Materials Science and Engineering **886** (2020) 012030 doi:10.1088/1757-899X/886/1/012030

as a combination of a second-order transfer function, a rate limiter, and a transitional velocity switch. When S-curve profile was represented by a second-order transfer function, a rate limiter described the difference of settling time of the two velocities commence, and transitional velocity switch was used to emulate an actual velocity profile at transitional velocity zone-for smoothing stop and smoothing start. Figure 1 shows the three elements of the empirical model.

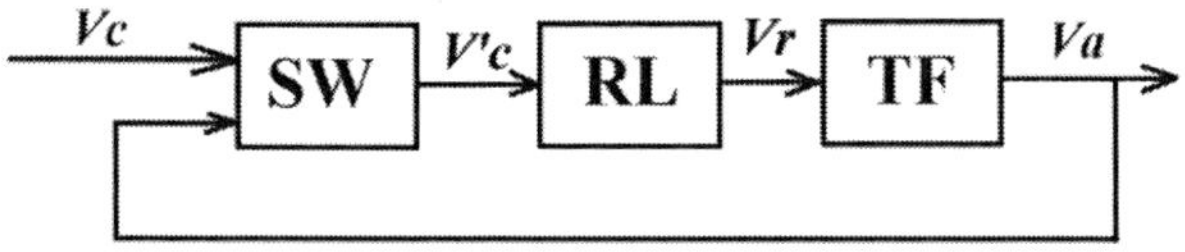

Figure 1. The representative of drive and motor model

Where V_c, V_r, and V_a are a velocity command, limited velocity command and actual velocity, respectively. The three blocks, *TF RL* and *SW*, are a second-order transfer function, a rate limiter, and a transitional velocity switch, respectively. *TF* is a standard form of a second-order transfer function, which is expressed in Equation (1).

$$\mathrm{TF} = \frac{V_a}{V_r} = \frac{\omega_n^2}{s^2 + 2\zeta\omega_n s + \omega_n^2} \tag{1}$$

ω_n and ζ are natural frequency and damping ratio of the system, which have to be identified. V_a is output and V_r is an input, which is filtered by the rate limiter before. Block *RL* is a rate limiter that limits the rate of change of velocity commence (V'_c). If the rate of change of input signal is greater than the rising limit, (*R*) and falling limit (*F*), block *RL* passes a signal not exceed *R* and *F* setting value. Otherwise, if the rate of change of signal input is less than raising limit (*R*) and falling limit (*F*) range, block *RL* passes an original input signal. Equations (2) and (3) express discrete function of the rate limiter. Subscribe *i* and *i*-1 are the time step at the current and one previous step, respectively.

$$\begin{aligned} V_r(i) &= \Delta t \cdot R + V'_c(i-1) && \text{rate} > R \\ V_r(i) &= \Delta t \cdot F + V'_c(i-1)\ ; && \text{rate} < F \\ V_r(i) &= V'_c(i) && F < \text{rate} < R \end{aligned} \tag{2}$$

Where

$$\text{rate} = \frac{V'_c(i) - V_r(i-1)}{\Delta t} \tag{3}$$

The transitional velocity switch is expressed in Equation (4), which is used to delay velocity command. When reversing velocity command is occurred and absolute actual velocity is greater than V_{switch}, zero will be sent to rate limiter instead of the velocity command. When absolute actual velocity is lower than V_{switch}, the velocity command will be passed transparently.

$$V'_c = \begin{cases} V_c, & \text{sign}(V_c) = \text{sign}(V_a) \\ V_c, & |V_a| \le V_{switch} \\ 0, & \text{otherwise} \end{cases} \tag{4}$$

As the proposed model, equation (1) – (4) are the representation of a drive and a motor of an industrial overhead crane.

IOP Conf. Series: Materials Science and Engineering **886** (2020) 012030 doi:10.1088/1757-899X/886/1/012030

2.2 Extended Model

According to the proposed modelling method, the three-step inputs as velocity command (V_c) should be sent to a motor drive, 0 to 100%, 0 to 50%, and 100% to -100% and actual velocity (V_a) must be measured. If the input and output were known, system parameters could be identified by some methods. However, hardware limitation exists in this study, actual velocity cannot be measured but positioned instead. Therefore, the proposed model is extended here using a satisfying assumption. That is, if there is no slip between wheel and rail, the crane position *P* can be calculated by integration of the actual velocity, which is shown in Figure 2.

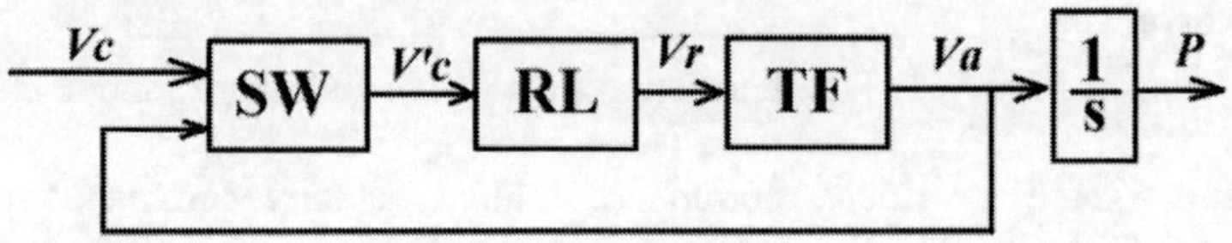

Figure 2. The representative of drive and motor model with an extended part

However, to relate the velocity command and the crane position, the velocity command (V_c) should also be identified, not in percentage but in the same unit (i.e. meter per second). Therefore, to model the drive and motor of the overhead crane according to the proposed model, all the mentioned parameters must be identified, including natural frequency and damping ratio for the *TF* block, raising and falling rate for *RT* block, V_{switch} for *SW* block, and velocity command in m/s.

3. Experimental Setup

A 1-ton industrial overhead crane was used to verify the proposed empirical model of a drive and a motor. To make the crane moved in a horizontal direction, 380 VAC motors and drives were used. These horizontal axes are normally called bridge and trolley, whose travelling spans were 23 and 9 meters, respectively. Figure 3 shows a hardware overview of the crane system. A wireless control pendant was used for sending a digital signal to PLC. PLC received that signal and sent a velocity step command (V_c) to motor drives or inverters. Two-position sensors measured bridge and trolley positions (*P*) and fed back to PLC for data recording. A PLC used in this work consisted of several units: CPU, option board, and analogue input units. The CPU unit is Omron NX1P2-1040DT, built-in with digital input ports for receiving signals from the wireless control pendant. NX1W-CIF11 is a serial communication option board, mounted on the CPU unit for sending velocity command to two YASKAWA AC Drives: V1000, called bridge inverter and trolley inverter. Modbus RTU protocol was selected for serial communication between PLC and inverters. NX-AD2203 is an analogue input unit, used for data recording from two DL50 P1123, range distance sensors, via 4-20 mA analogue output.

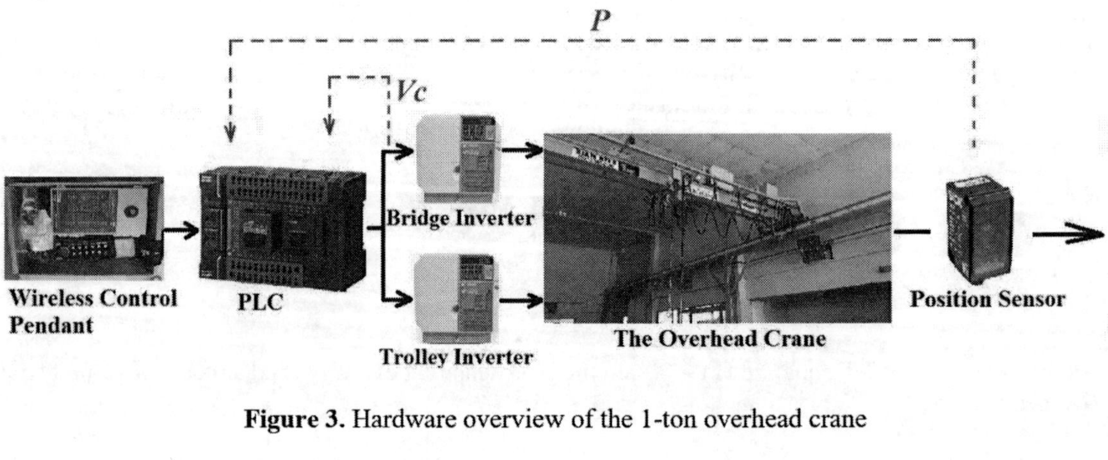

Figure 3. Hardware overview of the 1-ton overhead crane

4. Results and Discussion

4.1 Experimental Result

Due to the hardware setup as mention before, crane positions were used as outputs of the system, instead of actual velocity. Square-wave velocity commands were inputs of the system. Therefore, arbitrarily periods of two square-wave velocity commands (V_c) were sent to the bridge inverter and the trolley inverter with no payload in this testing. Figure 4 shows the plot of crane positions and velocity commands over time for bridge and trolley.

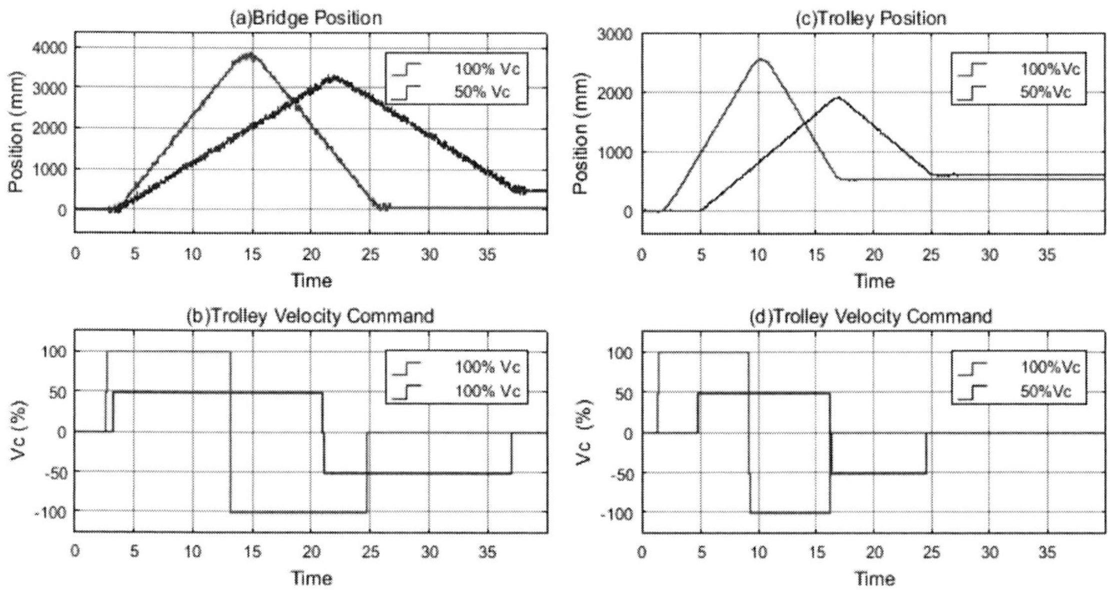

Figure 4. Crane positions and velocity commands over time

In figure 4, the red lines and the blue lines represent crane positions and velocity commands for 100% and 50%, respectively. Velocity commands and crane positions will be used for system identification in the next section.

4.2 Identification Result

To identify the proposed model parameters, a trial and error technique was used based on a rational principle. The proposed model consists of three elements that are second-order transfer function (*TF*), rate limiter (*RL*), and transitional switch (*SW*). For the *TF* block, it was originally defined as a heavily damped system, thus, given the system as critical damp at the first trial is reasonable. Then the natural frequency is easy to be tuned. For the *RL* block, the rising and falling rate is approximately defined by acceleration and deceleration times in the drive (inverter) parameter setting, which need to be fine-tuned later. For the *SW* block, the velocity switch is usually approximated to 10 to 20% of the maximum velocity. After a few trials, all above- mentioned parameters were tuned to fit the experimental data. For the bridge direction, natural frequency and damping ratio, rising rate, falling rate, V_{switch} , and maximum velocity are 4.8, 1, 1, -1, 12%, and 0.36 m/s, respectively. And for the trolley direction, natural frequency and damping ratio, rising rate, falling rate, V_{switch} , and maximum velocity are 10, 1, 1, -1, 15%, and 0.33 m/s, respectively. Figure 5 compares the data from the experiment and the proposed model of bridge and trolley, respectively.

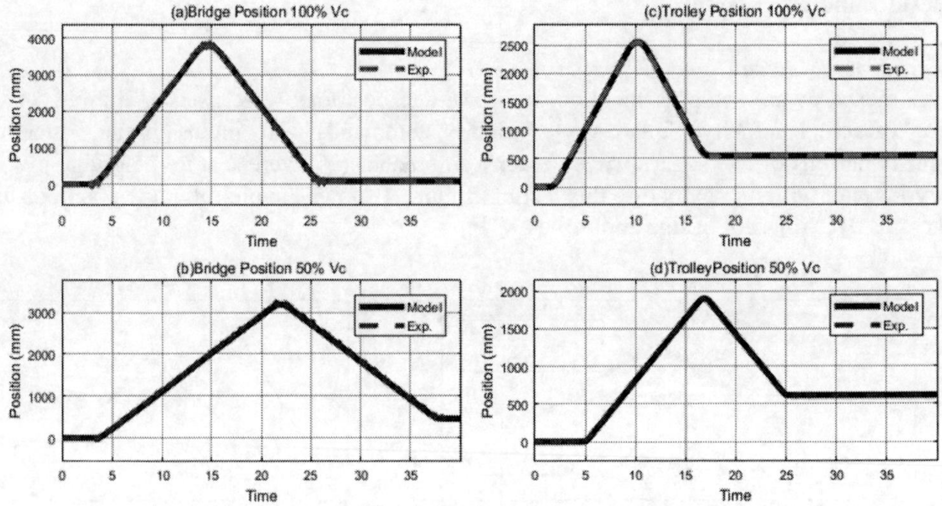

Figure 5. Comparison of experimental data and proposed model

Red lines and blue lines represent both bridge and trolley positions from experimental data of 100% and 50% velocity command. Black lines represent outputs of the proposed model, which are bridge and trolley positions. It can be seen that the outputs of the proposed model track the experimental data nicely

4.3 Discussion

Parameter identification result was done by trial and error technique. Figure 6 shows the performance of model prediction by plotting error over time. That said, a static and dynamic error is less than 20 and 40 mm, respectively. Although, using the trial and error method provides a good result of parameter identification, but a better result may be achieved by some powerful optimization techniques. For further investigation, parameter identification should be optimized by minimization of a cost function.

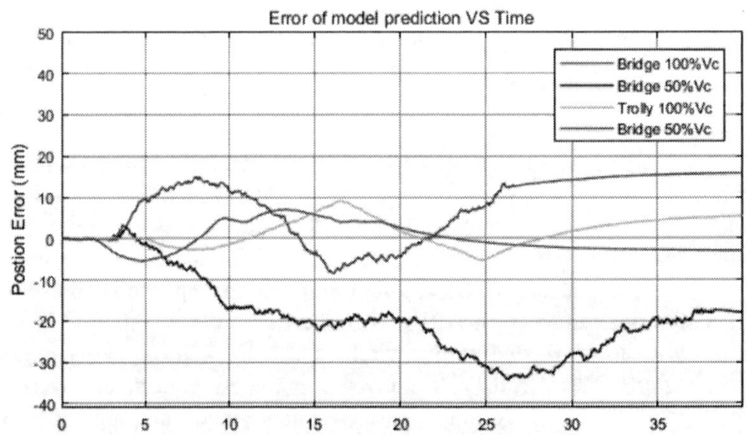

Figure 6. Showing of static and dynamic error

During the trial and error process, the natural frequency of the *TF* block and the velocity switch of the *SW* block depends on velocity command input. Therefore, the system parameters should be identified within a specific input range.

5. Conclusion

This paper investigated the model of drives and motors of an overhead crane, which was previously proposed by Sorensen and Singhose [8]. The modelling method was described to provide an understanding of the basic idea. The model was intently extended due to the hardware configuration, in the way that the crane position was measured instead of the crane velocity. The proposed model was verified by an industrial-grade 1-ton overhead crane. By using the trial and error method, the result showed that the proposed model provided a good prediction of the crane position. The static and dynamic errors were less than 20 and 40 mm., respectively. Inside the proposed model, it was found that natural frequency and velocity switch were input dependent, varying by velocity command inputs. To use this model, its parameters should be identified within a velocity command range.

Acknowledgment

This research was co-funded by National Science and Technology Development Agency (NSTDA) and Siam Kito Co., Ltd. We would like to thank Siam Kito Co., Ltd. for their technical supports and valuable advice from their expertise in the field.

References

[1] Kaur A Priyahansha A Kumari S and Singh T 2014 *IJAREEIE* **3**(5) 2278–8875
[2] Qian D Yi J and Zhao D 2011 *Proc. Int. Conf. the International Federation of Automatic Control*
[3] Tuan L and Moon S 2013 *Journal of Mechanical Science and Technology* **27**(3) 885–893
[4] Ngo Q and Hong K 2012 *IET Control Theory Appl.* **6**(5) 662–668
[5] Lee H 1998 *Journal of Dynamic Systems, Measurement, and Control* **120**(47) 471–476
[6] Padula F and Amamini R 2017 *IFAC Papers OnLine* **50**(1) 9126–9131
[7] Sorensen K and Singhose W 2005 *Proc. IFAC, 16th Triennial World Congress* (Prague)
[8] Sorensen K and Singhose W 2007 *Control Engineering Practice* **15** 825–837

The 10th TSME International Conference on Mechanical Engineering
10th – 13rd December 2019
Pattaya, Thailand

DRC0007

Simulated Braking Performance Comparison of an Electric Drum Brake and a Hydraulic Drum Brake Systems

Kittirattanachai A[1], and Watechagit S[1,*]

[1] Department of Mechanical Engineering, Faculty of Engineering, Mahidol University, Nakornpathom, 73170, Thailand

* Corresponding Author: sarawoot.wat@mahidol.ac.th

Abstract. The most important safety equipment for the operation of a vehicle is the brake equipment. This research attempts to develop an all-electric braking system that can be used in an electric vehicle. Since the drum brake requires lower actuation input than the disk brake in order to deliver the same amount of braking force. It seems logical to be used in the electric vehicle as far as minimizing energy usage is concerned. The proposed electric drum brake system conceptually is not new as it has been used as a trailer brake for a tractor-trailer vehicle. Adopting the trailer brake for a passenger vehicle requires proper downsizing and different braking control scheme. This paper presents the modelling and simulation results of the proposed electric drum brake. The main interest is its braking performance which can be investigated by the brake force profile or the brake force characteristic. The results are compared with the braking performance of the conventional hydraulic drum brake system which is using validated data from the literature. The simulation results show that the braking force of the electric drum brake system is two times higher than that is of the hydraulic drum brake system subjected to the same actuation force. The results from this paper will be valuable for the development of the controller in the future.

Keywords: Electromagnetic trailer brake, Electric drum brake, Brake performance and Mathematic models.

1. Introduction

The working principle of a drum brake has not been changed since its first commercial use in 1900. It has still been commonly used in commercial vehicles. One to of the main benefits of the drum brake is its high brake factor, which is defined by the ratio of the output braking force to the application force that activates the braking action. By comparing with other types of vehicle brake systems, e.g. a disk brake, subjected to the same application force, the drum brake generally produces higher braking force to the vehicle's wheel [1]. For the development of an electric braking system for an electric vehicle, and since the electric vehicle has limited resource for its operation, the drum brake seems to be a good reference system as it requires less actuation effort for the same braking force output. This research, therefore, investigates the development of the electric brake from the conventional drum brake system.

Reviewing pieces of literature related to the electric brake found that there have already been a few electric brake designs, including the magnetorheological brake [2-3], the electronic wedge brake [4-5] and the electromagnetic trailer brake. While the previous/former two are under their development phase, the latter has been used widely in the commercial trailer truck. Principally, the electromagnetic trailer brake is the drum brake with its application force or the brake actuation action which is produced by an electromagnetic mechanism. In other words, without the electromagnetic parts, this trailer brake would look similar to the general drum brake. It is used as an auxiliary brake rather than the main or the master brake. The adoption of the trailer brake to be used in a passenger vehicle is not as straight forward, especially regarding the controller design. Unfortunately, the detailed design and development of the electromagnetic trailer brake cannot be found in previous studies such as brake characteristic profile and brake mechanism theory.

In this paper, the mathematical models of both conventional hydraulic brake and the electromagnetic brake are hence developed. Only the conventional hydraulic drum brake model can be validated using data from the literature including brake characteristic and parameter of hydraulic drum brake for developing mathematic models. The resulted model will be used as a reference for the model of the electromagnetic trailer brake. To investigate the braking performance of the electromagnetic brake, brake force from both braking systems are simulated and compared with brake force profile. The results can help us to better understand the physics behind the operation of the electromagnetic brake. The developed model will also be used for controller design in the near future.

2. The mathematic model of the hydraulic drum brake and electromagnetic trailer brake systems

2.1 Hydraulic drum brake system

The hydraulic drum brake system works by the interaction of five parts. The pedal force from the driver actuates the braking action. The vacuum booster amplifies the pedal force. The master cylinder transmits force to the wheel cylinder via a hydraulic system. And the brake shoes which generates the braking force to the wheel. The schematic diagram of the hydraulic brake system and the flow of the braking power are shown in Figures (1) and (2).

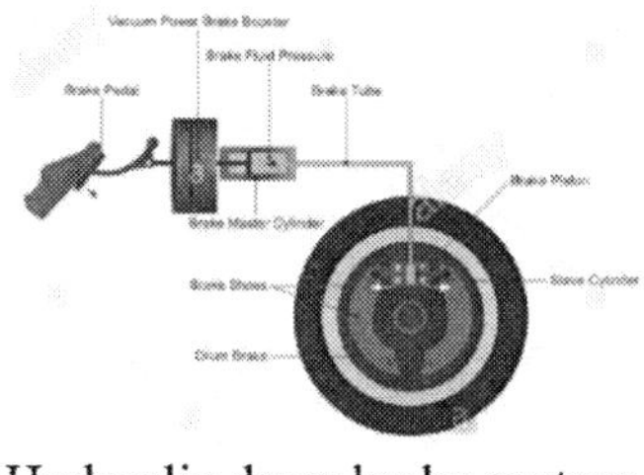

Figure 1. Hydraulic drum brake system [6].

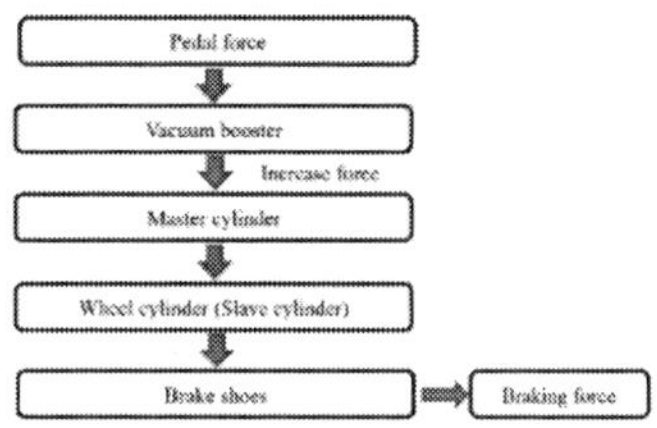

Figure 2. The transmissible force in the hydraulic drum brake system.

2.1.1. Model of the brake pedal

As can be seen in Figure 1, the pedal force is the transmitted force from the brake pedal, generated by the driver, to actuate the vacuum booster. The pedal force is calculated from taking moment at the clevis pin [7] as shown in equation (1).

$$F_n = F_d \frac{l_1}{l_2} \tag{1}$$

Here, F_n is the pedal force (N), F_d is the force input from the driver at the brake pedal (N), l_1 is the perpendicular distance from the pedal pivot to the center of input brake pedal (0.36 m) and l_2 is the perpendicular distance from the pedal to the clevis pin (0.04 m).

2.1.2. Model of the vacuum booster

In Figure 2, the pedal force (F_n) actuates the vacuum booster piston and vacuum diaphragm, which increases the force to push the master cylinder piston. The total force at the master cylinder piston, the

IOP Conf. Series: Materials Science and Engineering **886** (2020) 012031 doi:10.1088/1757-899X/886/1/012031

so called F_{pedal} which is the summation of the force from the pedal and the vacuum booster as shown in equation (2) [8].

$$F_n + \left(\left(\frac{\pi}{4}D^2 - \frac{\pi}{4}d^2\right) * (atm\ pressure - vacuum\ pressure)\right) = F_{pedal} \qquad (2)$$

Here, D is the diameter of the vacuum diaphragm (m), d is the diameter of vacuum booster piston, atm pressure is 101,315 ($\frac{\mathrm{N}}{\mathrm{m}^2}$) and the vacuum pressure is 55,000 ($\frac{\mathrm{N}}{\mathrm{m}^2}$). The vacuum booster is connected with the master cylinder.

2.1.3. Model of the master cylinder

The vacuum booster is connected to the master cylinder, which transmits the vacuum force to actuate the hydraulic pressure in the master cylinder. The pressure in the master cylinder (P_{mcp}) is calculated by considering all the forces acting across the master cylinder's piston as described by equation (3). This takes into account also the spring force and seals-friction force at the master cylinder's piston [9].

$$P_{mcp} = \frac{F_{pedal} - F_{spring0} - K_{spring}(x_1 - x_2) - F_{seal}}{A_{mc}} \qquad (3)$$

Here, x_1 is the piston displacement in the primary line (m), x_2 is the piston displacement in the second line. Assuming here $x_1 - x_2$ is 0.04 [1]. $F_{spring0}$ is the master cylinder spring pre-tension (138 N). F_{seal} is the seal friction force (80 N). A_{mc} is the master cylinder piston area (0.000491 m^2) and K_{spring} is the master cylinder spring constant (175 $\frac{\mathrm{N}}{m}$).

2.1.4. Model of the drum brake

The drum brake model takes the force from the brake shoe [10] and calculates brake torque and brake force. Forces and dimensions inside the drum brake are shown in Figure 3.

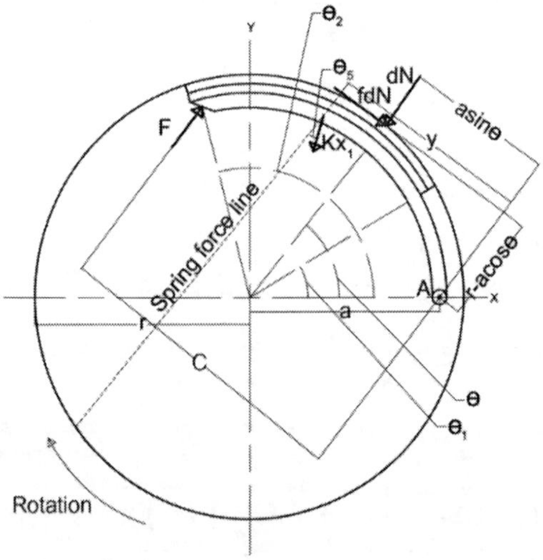

Figure 3. Drum brake schematic.

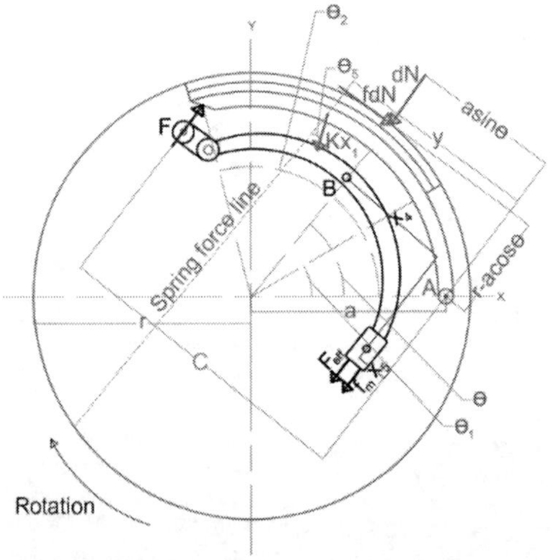

Figure 4. Electromagnetic trailer brake schematic.

Figure 3 illustrates the schematic diagram of the rear right wheel of a vehicle. The pressure from the master cylinder actuates the wheel cylinder, which then creates the actuation force (F). This is described by equation (4). This actuation force activates brake force at the leading and trailing shoes.

$$F = P_{mcp} * A_{wc} \qquad (4)$$

Here, A_{wc} is the wheel cylinder piston area (0.00045238 m^2). The normal force (dN) occurs at any angle θ from the hinge pin, which moves brake lining to touch midpoint of wheel hub. The proportional of pressure and maximum pressure are substituted in the magnitude of the pressure formula, which is described by equation (5).

$$dN = \frac{p_0 br \sin\theta \, d\theta}{\sin\theta_a} \tag{5}$$

p_0 is the maximum pressure at an angle θ_a (1.5708 rad) of the right-hand shoe, which can be found by equation (6). b is the width of the shoe (0.05 m) and r is the radius of the drum (0.160 m).

$$\theta_a = 1.5708\,(rad) = \frac{\theta_1 + \theta_2}{2} \tag{6}$$

By applying the condition of static equilibrium, the maximum pressure of the right-hand shoe (p_0) can be calculated by taking clockwise (M_f) and anticlockwise (M_N) force moment at point A. From Figure 3 one can see that, each force vector has components in the x-axis and the y-axis term, which includes force in wheel cylinder (*F*), spring force (Kx_1), friction force (fdN) and normal force (dN). These forces are used for the calculation of clockwise (M_f) and anticlockwise (M_N) moment. Hence, p_0 can then be written as shown in equation (8). Here it should be noted that the integral terms are the averaged friction areas.

$$F = \frac{M_N - M_f}{c} \tag{7}$$

$$p_0 = \frac{Fc + (Kx_1 \cos(\theta_5)\, y) - (Kx_1 \sin(\theta_5)\, x_3)}{br(a\int_{\theta_1}^{\theta_2} \sin^2\theta d\theta - f\int_{\theta_1}^{\theta_2} \sin\theta\,(r - a\cos\theta) d\theta)} \tag{8}$$

Based on the above equation, c is the distance between the apply force (*F*) and the hinge (0.243 m). K is the spring constant of the leading shoe and trailing shoe (3,394.85 $\frac{\text{N}}{m}$). x_1 is the right-hand shoe's spring extending range (0.01 m). θ_5 is the directional angle of the spring force (0.05236 rad). y is the length of the spring to the hinge (0.125 m). x_3 is the point of spring to the hinge (0.07 m). a is the distance between the drum brake center and the hinge (0.120 m). And f is the friction coefficient between the lining and the drum surface (0.48).

The brake torque of the right-hand shoe (τ_{right}) is represented by equation (9), which is the sum of the friction force fdN times the radius of the drum.

$$\tau_{right} = f p_0 b r^2 (\cos\theta_1 - \cos\theta_2) \tag{9}$$

The maximum pressure of the left-hand shoe (p_a) can be developed by similar fashion and can be described by a similar equation as (8). In this case, the brake torque of the left-hand shoe (τ_{left}) is then described by a similar equation as (9). Furthermore, the total braking force can be described by equation (10).

$$Total\ braking\ force = \frac{\tau_{right} + \tau_{left}}{r} \tag{10}$$

2.2 Electric drum brake system

The mechanism of the input force applying at the brake shoe is different from the case of the hydraulic drum brake as it is now generated by the electromagnetic subsystem. The magnet sticks to the drum surface by the application of electric current from the trailer brake control. As the drum moves, the lever hence moves which causes the shoes to push against the drum [12]. The schematic of the electric drum brake is shown in Figure 4.

IOP Conf. Series: Materials Science and Engineering **886** (2020) 012031 doi:10.1088/1757-899X/886/1/012031

2.2.1 Model of electric drum brake

As shown in Figure 4, the electromagnetic mechanism is the schematic with the dark-black line, overlaid onto the schematic of the hydraulic drum brake. The magnet system consists of the electric circuit subsystem and the electromagnetic subsystem. The latter one produces the actuation force to push the shoe. The electric circuit subsystem can be formulated by equation (11). This electric circuit generates a magnetic flux. The electromagnetic circuit subsystem generates a magnetic suction force as shown in equations (12) and (13)[13].

$$U = iR + N\frac{d\phi}{dt} \tag{11}$$

$$Ni = \phi\frac{l_m}{\mu_0 S_0} \tag{12}$$

$$F_m = \frac{\phi^2}{\mu_0 S_0} \tag{13}$$

U is voltage (Volt). i is the coil current (Ampere). R is the resistance in the coil (3 ohms). N is the coil number of turn. ϕ is magnetic flux (Wb). l_m is the effective length of the magnetic conductor (0.003 m). μ_0 is the permeability of vacuum ($4\pi * 10^{-7}$) and S_0 is the cross-section area of the air gap (0.0003 m^2). The friction force between the magnet and the wheel hub (f_m) is represented by equation (14) [13].

$$f_m = \frac{\mu_m \phi^2}{\mu_0 S_0} \tag{14}$$

Here, μ_m is the friction coefficient between the magnet and the wheel hub assuming value 0.48.

The force that generates the magnet movement is the effective driving force, F_{eff}, which takes into account the maximum driving force, rolling resistance force and aerodynamic force in the vehicle model. The actuation force that pushes the brake shoe (F) can be found by considering the moment by the effective force (F_{eff}), the friction force between the magnet and the drum surface (f_m), and force pushed brake shoe (F), around point B in Figure 4. The result shows in equation (15).

$$F = \left|\frac{-(F_{eff} + f_m) * x_4}{c - x_4 - x_5}\right| \tag{15}$$

x_4 is the distance from point of spring to the magnet (0.02648 m) and x_5 is the distance between the magnet and the hinge (0.005 m). F from equation (15) is used to calculate the brake force and the brake torque in a similar fashion as shown in section 2.1.4. In particular, F in equation (15) will be substituted in equation (7) and the term ($-F_{eff}x_5 - f_m x_5$) will be added in the numerator of equation (8).

3. The validation of the hydraulic drum brake

The hydraulic brake model is validated by comparing the contact force (N) of the leading shoe with the data from the literature [11]. The input in the model is the actuation force F. The model then generates the normal force F_d and hence the contact force. Table 1 shows the data from hydraulic drum brake testing obtained from the literature [11].

Table 1. The constraint parameter of validation.

Cases	F (N)	F_d (N)	f	Contact force (N)
1	4,119	365.30	0.48	32,656
2	4,217	377.10	0.48	32,754
3	4,315	388.90	0.50	32,852
4	4,413	400.72	0.50	32,950
5	4,511	412.60	0.50	33,048
6	4,609	424.36	0.50	33,146
7	4,707	436.18	0.51	33,244
8	4,805	448.00	0.50	33,342
9	4,903	459.81	0.48	33,440

3.1 Validation results of the hydraulic drum brake

Table 2 and Figure 5 show the simulation results as compared to the brake testing data from the literature.

Table 2. The comparison of simulation results.

Cases	Contact force (N) from literature	Contact force (N) from simulation	% error
1	32,656	32,658	0.006
2	32,754	32,755	0.003
3	32,852	36,457	9.888
4	32,950	36,467	9.644
5	33,048	36,488	9.427
6	33,146	36,509	9.211
7	33,244	38,420	13.472
8	33,342	36,581	8.854
9	33,440	33,522	0.244

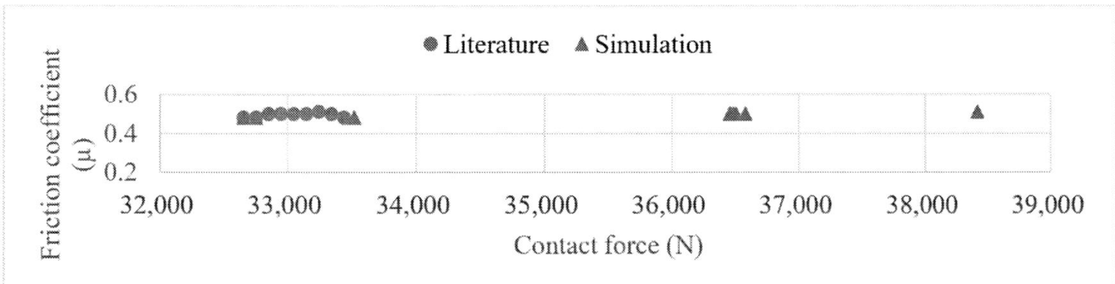

Figure 5. Contact force vs friction coefficient.

Based on the results shown, the contact force calculated by the hydraulic model deviates from the testing data by 6.75% on average. The error is shown here mainly due to the uncertainty of the shoe's geometry, i.e. θ_1 and θ_2, as well as the non-uniform contacting surface between the wheel hub and the shoe. The error though seems to be much, but, when considering the general magnitude of the brake force in the real application, 5%-10% brake force difference does not produce much difference in terms of braking performance, i.e. stopping distance. Therefore, the developed hydraulic model is then considered acceptable to be used as a reference model for the electric drum brake development here.

IOP Conf. Series: Materials Science and Engineering **886** (2020) 012031 doi:10.1088/1757-899X/886/1/012031

4. Simulation result and analysis of hydraulic drum brake and electric drum brake

The simulated actuation force that pushes the brake shoe from the hydraulic brake and the electric drum brake are compared in this section. The simulation is setup to brake the vehicle with its mass of 2,400 kg, and its initial speeds of 14, 25, and 36 m/s. Using equations (7), (8), (11), (12), (13), (14) and (15) the actuation force of the electric drum brake can be calculated.

4.1 The brake actuation force simulation

The simulation results showing the brake actuation force for both hydraulic drum brake and electric drum brake for each initial vehicle speeds are shown in Figure 6.

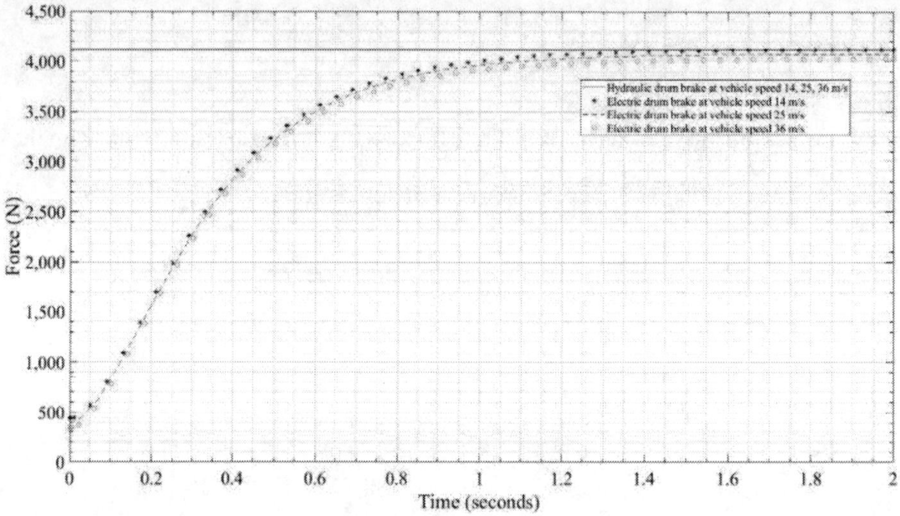

Figure 6. Force push shoe of hydraulic drum brake vs electric trailer brake

From figure 6, the response time of the hydraulic drum brake performs around 0.01 second. This is because the model does not take into account the dynamic of the hydraulic system during the filling phase. Therefore, the force quickly increases to its steady state. In other words, it is assumed that the hydraulic fluid is filled in the system. Hence the wheel cylinder can be actuated instantaneously.

The response times of the electric drum brake is around 0.9 second. The steady state actuation forces in this case are 4,118.08, 4,074.46, 4,024.68 N for the vehicle speed 14, 25, 36 m/s, respectively. Higher actuation force is produced at low vehicle speed because the dry road friction coefficient at low speed is higher than the one that is of the high vehicle speed. This affects the maximum driving force for effective force (F_{eff}) input in equation (15).

4.2 The braking force simulation

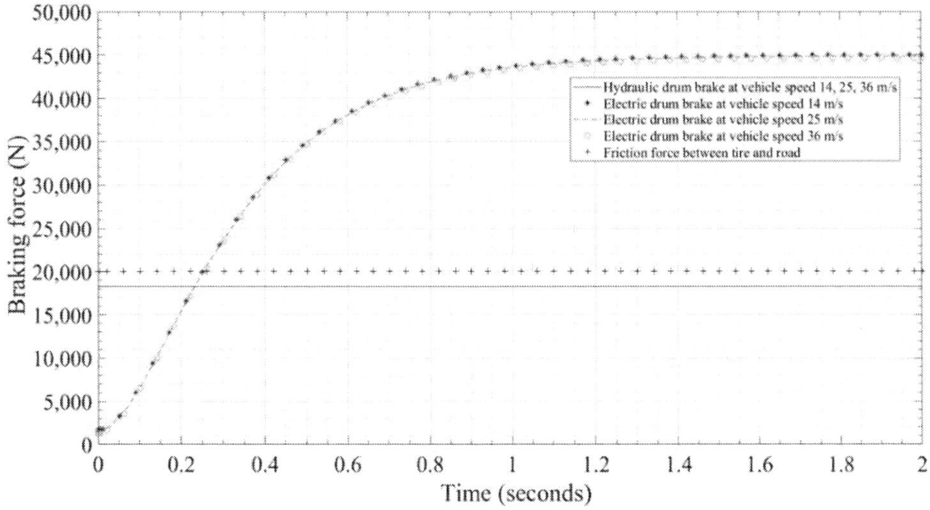

Figure 7. The braking force of hydraulic drum brake vs electric drum brake

Figure 7 shows the simulated braking force as a result of the actuation force from section 4.1. The braking force from the hydraulic drum brake is 18,283 N at all vehicle speed. Though, the braking force from the electric drum brake is 45,032.99, 44,867.59, 44,678.83 N at a vehicle speed of 14, 25, 36 m/s respectively. These are higher than the hydraulic drum brake by 146.31, 145.41, 144.37 % respectively. The electric drum brake gains additional brake force from the frictional force between the magnet and the drum or the wheel hub (f_m). The magnet, which is powered by the electric circuit, tries to stick itself with the wheel hub, hence retards the wheel vehicle speed.

4.3 The braking performance discussion

While there are many ways to compare the brake performance of one braking system to another, due to limited resources and information of the electric drum brake from the literature, the braking force from the hydraulic drum brake and the electric drum brake are considered. From the simulation results, the braking force of the electric drum brake generates brake force two times more than that is of the hydraulic drum brake under the same input actuation force. In reality, the benefit of having higher braking force will have to be investigated further as the vehicle's wheel can be locked any time when the braking force is higher than the friction force between the tire and the road. Figure 7 shows the friction force between the tire and the road for the simulated case, which is 20,012.40 N.

5. Conclusions

This research investigates the development of the electric drum brake from the conventional drum brake system. The main reason is that the drum brake has a higher brake factor than the disk brake which seems appropriate for the electric vehicle as far as minimizing the energy usage is concerned. The development adopted from the already existing electromagnetic trailer brake system is commonly found in the commercial trailer truck. The model of both hydraulic drum brake and the electric drum brake are developed in this paper. The hydraulic drum brake model is validated using the brake testing data from the literature. The validation results show that the model can simulate the brake force with acceptable accuracy and hence it can be used as a reference for the electric drum brake development.

The simulation of the actuation force that pushes the brake shoe and the braking force at the wheel of both hydraulic drum brake and electric drum brake are shown in this paper. While the actuation force for both braking systems are about the same, the resulted braking force at the wheel for the case of electric drum brake is about two times higher than that is of the hydraulic drum brake. This is due to the fact that the electric drum brake gains additional braking effort from the frictional force acting between the magnet and the wheel hub. Moreover, the simulation does not take into account the wheel locking situation. Therefore, in reality, the benefit of having a higher braking force will have to be further investigated. The developed model will also be further refined for controller design and implementation.

References

[1] Day, A. J. (2014). *Braking of road vehicles*. Butterworth-Heinemann.

[2] Jerzy, B., & Mateusz, B. J. (2013). Influence of Temperature and Rotational Speed on the Properties of Magnetorheological Brake. In *Design and Modeling of Mechanical Systems* (pp. 143-149). Springer, Berlin, Heidelberg.

[3] Zheng, J., Li, Y., Wang, J., Shiju, E., & Li, X. (2018). Accelerated thermal aging of grease-based magnetorheological fluids and their lifetime prediction. *Materials Research Express*, *5*(8), 085702.

[4] Ahmad, F., Mazlan, S. A., Zamzuri, H., Hudha, K., & Jamaluddin, H. (2015). Study on the potential application of electronic wedge brake for vehicle brake system. *International Journal of Modelling, Identification and Control*, *23*(4), 306-315.

[5] Aparow, V. R., Hudha, K., Ahmad, F., & Jamaluddin, H. (2014). Development of Antilock Braking System using Electronic Wedge Brake Model. *Journal of Mechanical Engineering and Technology (JMET)*, *6*(1), 37-63.

[6] Hydraulic drum brake system, accessed 27 February 2019, https://www.freepik.com/premium-vector/hydraulic-brake-system_2195999.htm

[7] Ivan, F., Boran, P., & Davor, P. (2016). A Detailed Model for Simulation of Hydrodynamic Processes in Hydraulic Braking System with ABS on Motor Vehicles. *Journal of Scientific Research and Reports*, 1-12.

[8] Gajjar VY, Jaiveshkumar, Gandhi D 2011 *International Journal of Environment, Agriculture and Biotechnology* vol 2 p 301

[9] Geromel, N. (2014). Modelling and control of the braking system of the electric Polaris Ranger all-terrain-vehicle.

[10] Shigley, J. E. (2011). *Shigley's mechanical engineering design*. Tata McGraw-Hill Education.

[11] Khairnar, H. P., Phalle, V. M., & Mantha, S. S. (2015). Estimation of automotive brake drum-shoe interface friction coefficient under varying conditions of longitudinal forces using Simulink. *Friction*, *3*(3), 214-227.

[12] Braking system-electric, USA, accessed 19 September 2018, https://saubermfg.com/wp-content/uploads/2013/02/electric_brake_system_manual.pdf.

[13] Li, S., Guo, P., Jiang, W., Ding, H., & Yu, D. (2015, July). Research on response characteristics and parameters optimization of high-speed solenoid valve. In *2015 34th Chinese Control Conference (CCC)* (pp. 2327-2332). IEEE.

IOP Conf. Series: Materials Science and Engineering **886** (2020) 012032 doi:10.1088/1757-899X/886/1/012032

Detection of Contact-type Failure Based on Non-linear Wave Modulation Utilizing Self-Excited Ultrasonic Vibration: Evaluation of Failure Development Focusing on Frequency Modulation

T Tanaka[1,*], S Maeda[2], Y Oura[1], and Z Wu[1]

[1]Department of Mechanical Systems Engineering, School of Engineering, University of Shiga Prefecture, 2500 Hassaka, Hikone, Shiga, 522-8533, Japan
[2]Division of Mechanical Systems Engineering, Graduate School of Engineering, University of Shiga Prefecture, 2500 Hassaka, Hikone, Shiga, 522-8533, Japan

* Corresponding Author: tanaka.ta@mech.usp.ac.jp

Abstract
This research concerns the detection method of contact-type failure based on non-linear wave modulation utilizing ultrasonic vibration driven by self-excitation. When the structure is vibrating by environmental force or forced excitation, the contact area is fluctuated by vibration. In this condition, the vibration transfer characteristics of ultrasonic vibration have fluctuated in synchronization with vibration (non-linear wave modulation). The detection method based on non-linear wave modulation can detect contact-type failure. In this paper, the novel detection method based on non-linear wave modulation is presented. Firstly, the self-excitation method using local feedback control is introduced. Local feedback control can oscillate at the natural frequency automatically. Secondly, the concept of a novel detection method is introduced. When the structure with contact-type failure is vibrating, the frequency of the ultrasonic vibration excited by local feedback control will fluctuate in synchronization with natural frequency fluctuation. To explain the detection method, non-linear wave modulation is expressed 1 degree of freedom model. In this model, the local stiffness fluctuation is modelled by the spring element of which spring coefficient is fluctuated in synchronization with structural vibration. We regard the transfer function of this model as a linear time-varying system. The amplitude of the fluctuation of natural frequency caused by stiffness fluctuation can be the novel index of failure level. Lastly, the result of the experiment using the uniform beam specimen is shown. From the relationship between forcing pressure of the contact-type failure and the amplitude of the frequency modulation, it is proved that the novel index is associated with failure level.

Keyword: Detection, Contact-type Failure, Ultrasonic, Non-linear Wave Modulation, Self-excited Vibration.

IOP Conf. Series: Materials Science and Engineering **886** (2020) 012032 doi:10.1088/1757-899X/886/1/012032

1. Introduction

Ultrasonic nondestructive testing is effective to detect a small failure. However, the contact-type failure such as fatigue crack, delamination of composite material, bolt loosening and adhesive failure is difficult to detect by ultrasonic testing because of high transmissivity of an ultrasonic wave in the vicinity of failure. The non-linear ultrasonic method was proposed to solve this problem [1-3]. This method is difficult to use for structural health monitoring because of the necessity of large-scale equipment.

The authors study the detection method based on non-linear wave modulation (NWM) [4, 5]. While the structure with contact-type failure is vibrating by the low frequency vibration, e.g. environmental disturbance or vibration generated by force excitation, the contact condition is fluctuated by vibration. Under this condition, the ultrasonic vibration is modulated caused by fluctuation of contact-type failure. The method based on NWM can detect contact-type failure using ultrasonic vibration of small amplitude. In this method, the resonance of the structure in the frequency range of ultrasonic is used. One of the problems of this method is the necessity of the measurement of natural frequencies. Another one of the problems is the independence of the failure index based on a demodulation component of the amplitude and phase.

In this paper, the novel detection method of contact-type failure utilizing self-excited ultrasonic vibration to solve these problems is presented. The authors proposed the self-excitation method at the natural frequency automatically using local feedback control [6]. The frequency modulation of the excitation signal will be generated when the NWM can express as the linear and time-varying system. In this paper, the experimental verification is done. Firstly, the self-excitation circuit is designed. Secondly, the excitation experiment of ultrasonic natural vibration of the uniform beam structure is examined. The result is confirmed that the frequency of the self-excited ultrasonic vibration corresponds to the natural frequency. Finally, the detection experiment of contact-type failure is examined. The generation of frequency modulation of the excitation signal and the increase of the amplitude of frequency modulation in synchronization with the amplitude of vibration is confirmed.

2. Non-linear wave modulation (NWM) utilizing self-excited ultrasonic vibration

In this section, the overview of NWM and the self-excited ultrasonic vibration is introduced. First, the phenomenon theory of NWM based on the linear and time-varying model is expressed. From this model, the fluctuation of natural frequency is the essence of NWM. Secondarily, the mechanism of self-excitation method using local feedback control and the characteristics of the vibration excited by local feedback control is introduced. It is explained that the frequency modulation generates to the frequency of the self-excited ultrasonic vibration from these models.

2.1 Non-linear wave modulation

The NWM is the phenomenon caused by fluctuation of the contact condition. Figure 1 shows the illustration of NWM. When the structure with healthy condition is vibrating at a low frequency and an ultrasonic frequency, the amplitude and phase of the ultrasonic frequency component of the observation signal are constant. On the other hand, when the structure with contact-type failure is vibrating, the contact condition of the failure fluctuates in synchronization with a vibration frequency. In this case, the amplitude and phase of the ultrasonic wave through the vicinity of the failure is oscillated by the fluctuation of the contact condition. This inconsistency of the contact condition can be expressed as the fluctuation of local stiffness.

In the case of ultrasonic vibration excited at a natural frequency, the NWM can model with stiffness fluctuation. If there is a great difference between the frequency of the low frequency vibration and the ultrasonic vibration, this model can be regarded as the linear and time-varying model. The transfer characteristics of the time varying model are shown in Figure 2 (a). The natural frequency vibrates in synchronization with the stiffness fluctuation. As a result, the fluctuation of amplitude and phase occurs. Figure 2 (b) shows the characteristics with large damping. The peak of the gain becomes smaller caused by damping. The gradient of the phase delay is gentle. Therefore, the fluctuation width

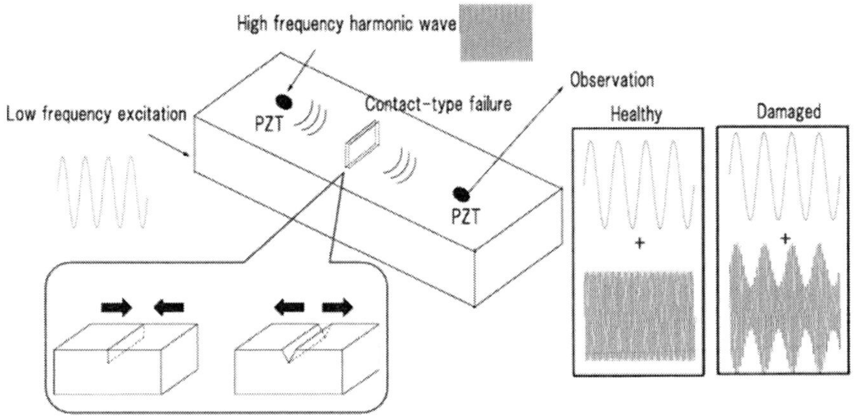

Figure 1. Illustration of NWM

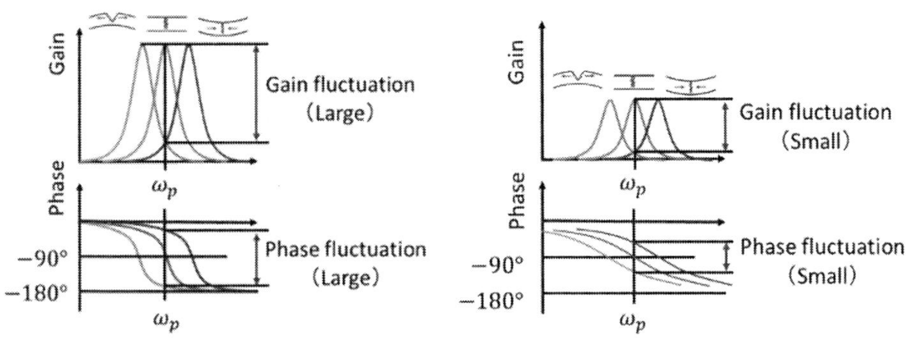

(a) In the case of small damping (b) In the case of large damping

Figure 2. Transfer function model of NWM [6]

of the amplitude and phase is smaller than Figure 2 (a). From these discussions, the failure index using the fluctuation width of the amplitude and phase is independent of damping.

The authors focus on the fluctuation of natural frequency, which is the essence of NWM. The natural frequency decides only the stiffness and the mass. Therefore, frequency modulation caused by NWM depends on only the fluctuation of the stiffness. The failure index using fluctuation of natural frequency will become the independent index of damping.

2.2 Self-excited ultrasonic vibration by local feedback control

The authors study the self-excitation method at the resonance point [6]. In this section, the mechanism of self-excitation method using local feedback is explained.

Firstly, the characteristics of the linear vibration system are introduced. Figure 3 shows the three degrees of freedom system model as an example and transfer characteristics when the actuator and sensor is the same location as this model. u is the excitation force and h is the relative displacement. Under this condition, the phase of all-natural frequencies equals a 90-degree delay in the transfer function.

Local feedback control is the control method using a 90-degree delay at the natural frequency. The condition of the oscillation of the self-excited vibration is a 180-degree delay on the open-loop transfer function. Thus, the self-excited vibration oscillates at the resonance point automatically when the phase of the feedback controller that the sensor and the actuator collocate designed a 90-degree delay at all frequencies.

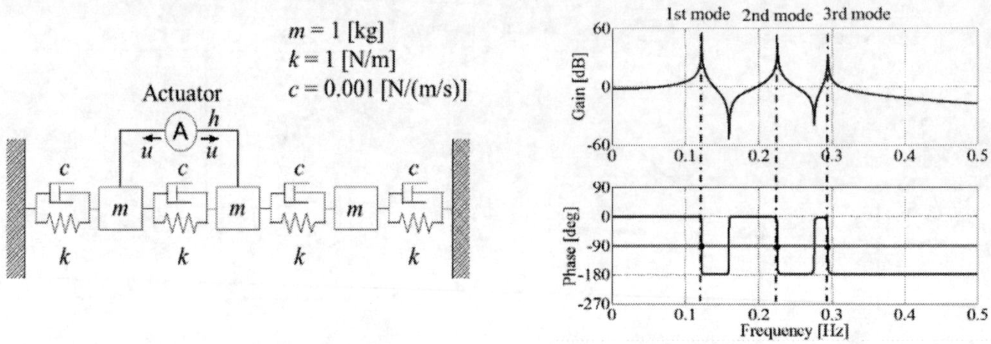

(a) Illustration of three degrees of freedom system (b) Transfer characteristics

Figure 3. Transfer characteristics of three degrees of freedom system [6]

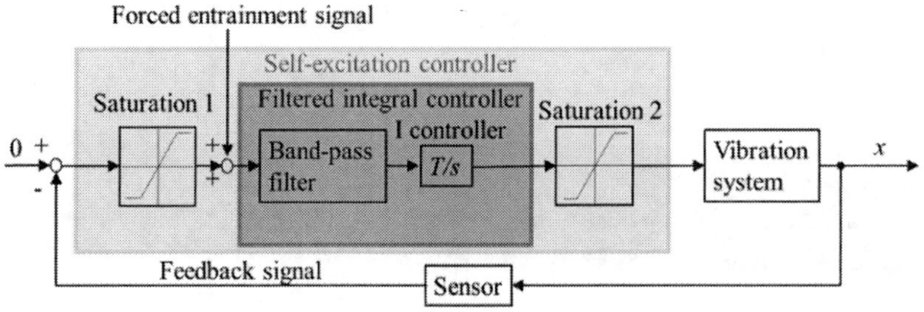

Figure 4. Block diagram of the self-excitation control using local feedback control

In this paper, the integral negative feedback control is used as the control that has the phase delayed by 90-degree. Figure 4 shows the block diagram of the local feedback control using the integral negative control. The saturation element 1 is the protection circuit for protecting the operational amplifier. The all saturation voltage of this self-excitation controller is -3V to 3V.

3. Detection of contact-type failure based on non-linear wave modulation utilizing self-excited ultrasonic vibration

In this section, the verification of excitation of self-excited ultrasonic vibration and detection of the contact-type failure are described. Firstly, the setup of the excitation experiment using the beam structure is expressed. Next, the self-excited vibration generated by the local feedback control is observed. Lastly, forced entrainment is done to change the oscillation frequency to the other natural frequencies.

3.1 Experimental setup

In this subsection, the experimental setup is described in detail. Figure 5 shows the experimental apparatus and setup. The beam structure whose length is 3000 mm, with a width of 50 mm. and a height of 6 mm. The boundary conditions of the heads of the specimen are fixed by vises. The piezo-electric patches for the actuator and the sensor are attached on both sides of the specimen of 600 mm. from the left edge. The additional piezo-electric patches are attached to confirm the vibration mode. The size of the piezo-electric patch is 30 mm× 20 mm The bipolar power amplifier to drive the piezo-electric patch amplifies the circuit output 10 times. The max voltage of circuit output is $30V_{p\text{-}p}$. The exciter is used to excite the specimen at low frequency. The oscilloscope is used to measure time response. The sampling frequency of measurement is 1 MHz. The measurement time is 1 s.

The open-loop transfer function is shown in Figure 6. The open loop transfer function is measured from the input signal the filtered integral controller to the output signal of the saturation element 1.

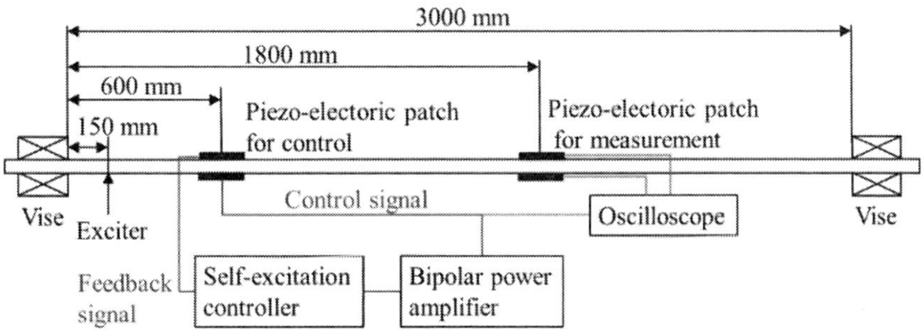

Figure 5. Experimental setup

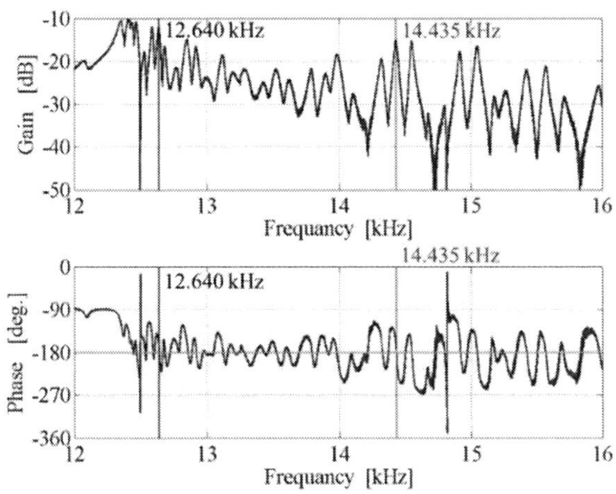

Figure 6. Open-loop transfer function: The black line shows the frequency of the self-excited vibration automaticaly in the experiment. The red line displays the entrainment frequency

It was confirmed that the phase delay at the natural frequencies is 180-degree. From this result, the self-excitation can only be generated at the natural frequencies.

3.2 Excitation of self-exited ultrasonic vibration by local feedback control

The result of the excitation experiment of self-excited ultrasonic vibration by local feedback control is described. Firstly, the self-excited ultrasonic vibration generated automatically is shown in Figure 7. Figure 7 (a) illustrates the feedback signal. Figure 7 (b) displays open-circuit voltages of the piezo-electric patch for measurement. The blue line and the red line are the output signals of the piezo-electric patch over and under the beam structure. The period is 0.0791 ms (12.64 kHz) that is the natural frequency of the black line of Figure 6. It can be seen that this natural vibration is the bending mode from Figure 7 (b).

Next, the generation of the self-excited ultrasonic vibration using forced entrainment attempts to use the various natural vibration for the proposed method. The result is shown in Figure 8. The period of Figure 8 (a) is 0.0639 ms (14.435 kHz) that is the natural frequency of the red line of Figure 6. It can be seen that this natural vibration is also the bending mode from Figure 8 (b).

3.3 Detection of contact-type failure using self-excited ultrasonic vibration

The detection experiment of the contact-type failure is described. Firstly, the frequency modulation of the feedback signal in the case of self-excitation automatically will be confirmed. In this experiment, the location of the contact-type failure is 2200 mm. from the left edge. The frequency of feedback signals are calculated by short-time Fourier transform when the voltage and frequency of exciter are 3.0 $V_{p\text{-}p}$ and 8.5 Hz (see Figure 9). Figure 9 (a) shows the feedback signal which is vibrating at 12.64 kHz. Figure 9 (b) displays the result of the short-time Fourier transform of feedback signal (Figure 9 (a)). From Figure 9 (b), the frequency modulation in synchronization with the low-frequency vibration is confirmed.

Finally, the variation of the amplitude of the frequency modulation to the failure development is investigated. The relationship between the amplitude of the frequency modulation and low frequency vibration is shown in Figure 10. The amplitude increases in synchronization with the voltage of the low frequency vibration which describes the failure level. This index is associated with failure level.

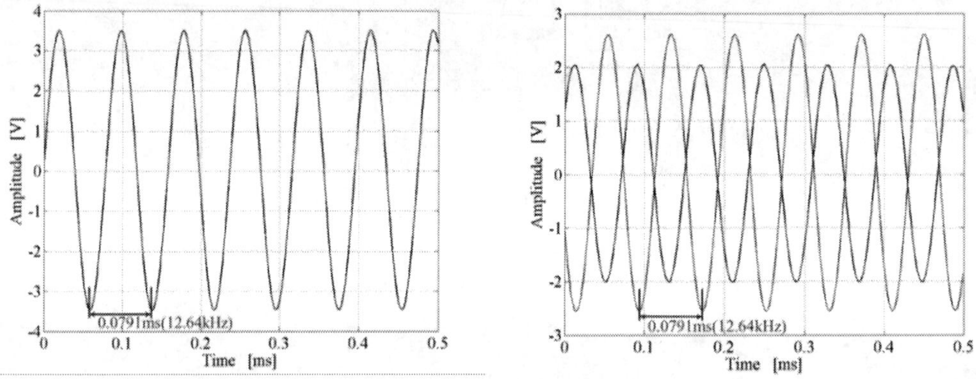

(a) Feedback signal (b) Output signal of the piezo-electric patch for measurement

Figure 7. Self-excited ultrasonic vibration generated automatically

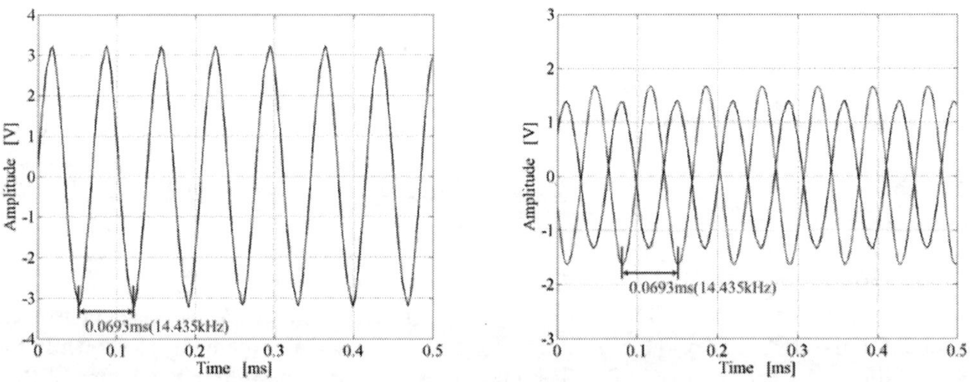

(a) Feedback signal (b) Output signal of the piezo-electric patch for measurement

Figure 8. Self-excited ultrasonic vibration generated by forced entrainment

4. Conclusion

This paper presented the detection method of the contact-type failure based on NWM using self-excited ultrasonic vibration. Especially, the verification experiment of the proposed method is done. As a result, the self-excited ultrasonic vibration can be generated at a natural frequency. Thus, the frequency of the self-excited ultrasonic vibration can change the other natural frequencies by forced entrainment. Therefore, the frequency fluctuation of the feedback signal occurs by the NWM. The amplitude of the frequency fluctuation increases in synchronization with the failure development. From this result, the amplitude of the frequency fluctuation is the novel failure index.

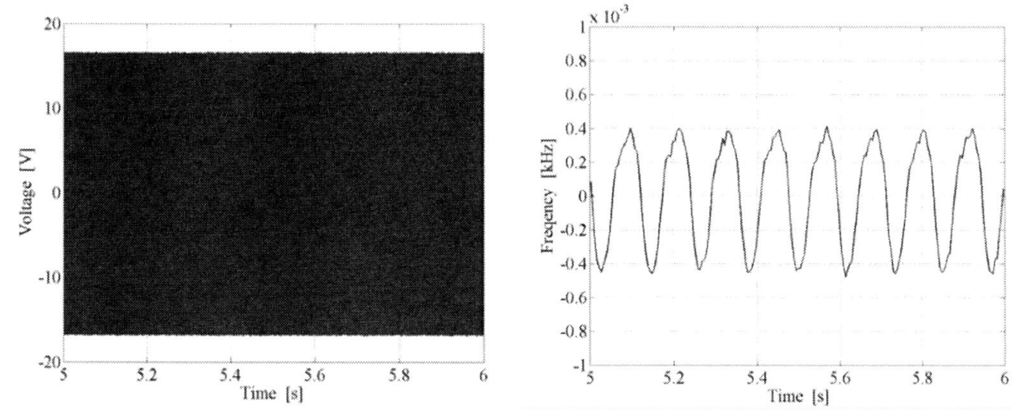

(a) Feedback signal (b) Frequency demodulation

Figure 9. Self-excited ultrasonic vibration when NWM is occurred

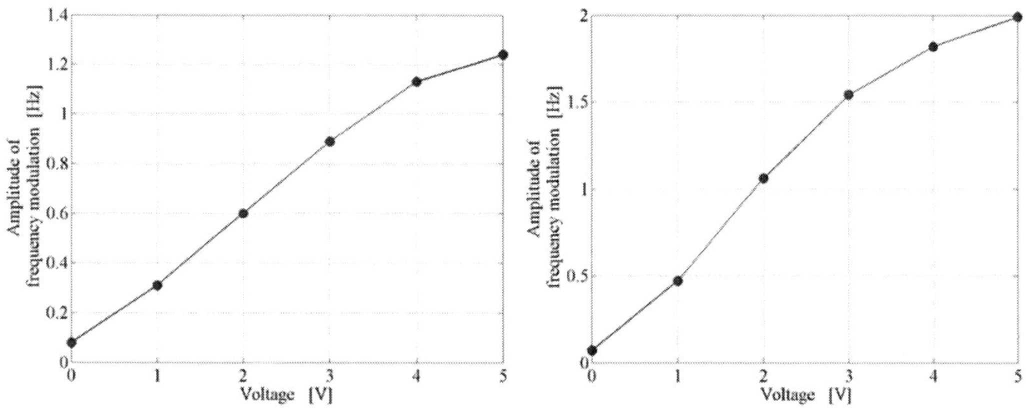

(a) In the case of 12.640 kHz (b) In the case of 14.435 kHz.

Figure 10. Relationship of the amplitude of the frequency fluctuation and low frequency vibration

Acknowledgement

This research was supported by JSPS KAKENHI Grant Number 18K13716. Additional fund to travel was supported by the Research Foundation for the Electrotechnology of Chubu.

IOP Conf. Series: Materials Science and Engineering **886** (2020) 012032 doi:10.1088/1757-899X/886/1/012032

References

[1] Solodov I Wackerl J Pfleiderer K and Busse G 2004 *Appl. Phys. Letters* **84** 26
[2] Ohara Y Kawashima K Yamada R and Horio H 2004 *Appl. Phys. Letters* **700** 944
[3] Ohara Y Mihara T Sasaki R Ogata T Yamamoto S Kishimoto Y and Yamanaka K 2007 *Appl. Phys. Letters* **90** 1
[4] Tanaka T Masuda A and Sone A 2014 *Proc. the 12th Int. Conf. on Motn. and Vib. Ctl. (Sapporo)* no2B22
[5] Tanaka T Masuda A and Sone A 2014 *Proc. the ASME 2014 Int. Des. Eng. Tech. Conf. and Comp. and Info. in Eng. Conf. (Buffalo)* **1** no DETC2014-34879
[6] Tanaka T Nakamura H Oura Y and Kurita Y 2018 *Proc. Int. Cong. on Sound and Vibration (Hiroshima)* T01-RS1-174

A Force-controlled Three-finger Prosthetic Hand via Three-Dimensional Printing

Wattana Chalong, Settanan Tanjaipet, Peeraphat Eurcherdkul, and Ittichote Chuckpaiwong*

[1]Department of Mechanical Engineering, Faculty of Engineering, Mahidol University, Thailand

* Corresponding Author: ittichote.chu@mahidol.ac.th

Abstract
This paper presents the development of a force-controlled 3D-printed prosthetic hand commanded by surface electromyography. The prosthetic hand used force control to pick up a 600-ml water bottle without any damages. Two experiments were carried out to determine the model of the system. The first one is to determine the relationship between the voltage of electromyography and the handgrip force. The second one is to determine the relationship between the current of DC motor and the water bottle grip force. Feedback control was used to control the gripping force. The prosthetic hand was tested for its gripping water bottles. Percentage of success in holding water bottle with a closed bottle cap is 90% of all brands. With opened bottles, the prosthetic hand cannot hold bottles from one brand, which is made of a soft plastic shell. The remaining brands of water bottle have a percentage of success above 90%.

Keywords: force-controlled, three-finger gripper, three-dimensional printer, surface electromyography

1. Introduction
The three-dimensional printing was widely applied in a variety of fields such as education, automotive industry, engineering, architecture, medical, etc. One of the medical applications is the creation of prosthetic hands to help disabled people. The survey of the National Statistical Office in 2007 found that 43.7 percent of the disabled in Thailand do not have access to the aiding equipment [1]. While the number of disabled people in Thailand increases (from 2017 to 2018). The three-dimensional (3D) printing promises an ability to customize the size to fit with the body of a disabled person. However, the current designs of 3D printed prosthetic hands do not function sufficiently for their needs. This work focuses on the grasping function to help the disabled to hold things in daily life.

In order to gain information from experts in the field of prosthetic hands, we contacted the head of Prosthesis and Orthosis division, Sirindhorn National Medical Rehabilitation Institute to provide the requirements of the prosthetic hand. From his experience, a prosthetic hand using strings and elastics [2] cannot firmly grasp an object, cannot control the grasping force and has a short lifetime.

In addition, objects that the disabled people want to grasp in their daily lives include spoon, fork and plastic water bottle.

The low-cost prosthetic hands using 3D printing were designed with various requirements [3-4]. This research focuses on the design of a prosthetic hand with minimal mechanical movement. The prosthetic hand must handle a 600-ml. plastic water bottle with a firm grip, which prevents damage on the plastic water bottle. In addition, the grasping force must be controlled without extra sensors, but using the signal from the motor itself.

2. Prosthetic hand design

The prosthetic hand was designed specifically to hold water bottles. Therefore, the inside of the finger was designed to be a circular shape (as shown in Figure 1). The gripper has three fingers – two fingers on one side, and one finger on the other side, which moves between the other two fingers in a closing position.

2.1 Grasping Force and Motor Selection

A 12-V worm-gear motor is chosen with a speed of 10 rpm. At such speed, the motor can provide a torque of 722 N-mm. The grasping force places on a plastic water bottle. At 50 mm. away from the motor's rotation point, as shown in Figure 1. The forces are the same on both sides of the gripper as shown by the free body diagram in Figure 2. Note that the diagram shows only one finger on each side for simplicity, but the summing forces are the same either for one or two fingers on one side.

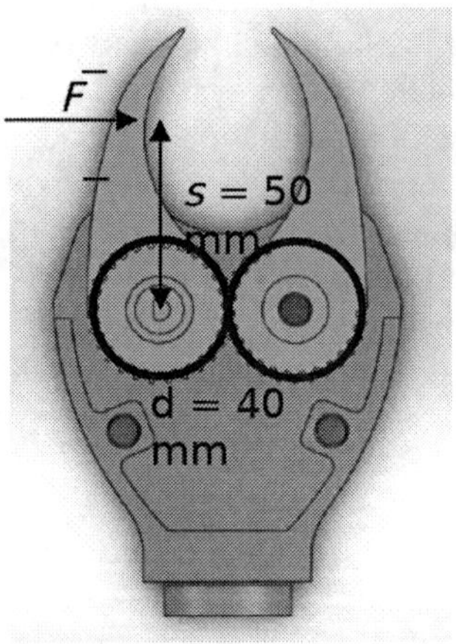

Figure 1. The Free-body diagram of the gripping force

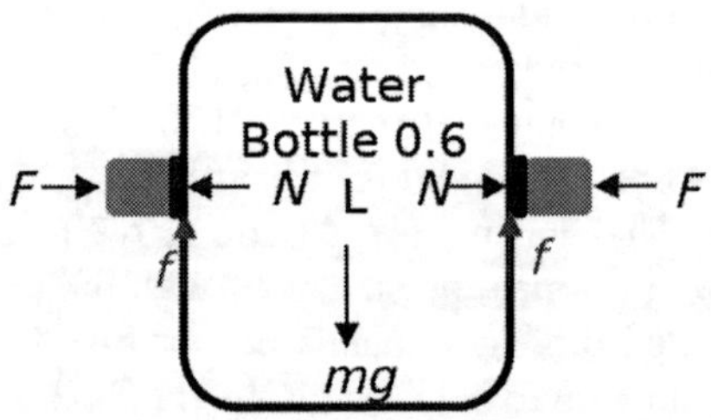

Figure 2. The Free-body diagram of force acting on a 600-ml water bottle

The grasping force F can be calculated by the motor torque as follows:

$$F = \frac{T}{s} = \frac{722}{50} = 14.44\ \text{N}$$

The maximum weight, which the gripper can lift up at the maximum grasping force, can be computed by the following equations,

$$2f = 2\mu N = mg$$

$$2(0.85)(14.44) = m(9.81)$$

$$m = 2.5\ \text{kg}$$

Therefore, the gripper has sufficient force to carry the 600-ml water bottle.

2.2 The Prosthetic hand prototype

A prosthetic hand prototype was designed as shown in Figure 3. It connected with the control devices including Arduino Uno R3, ACS712 current sensor, Myo-Ware Muscle Sensor, L298N dual H-bridge motor controller and Li-Po battery. The control operation will be described in the next section.

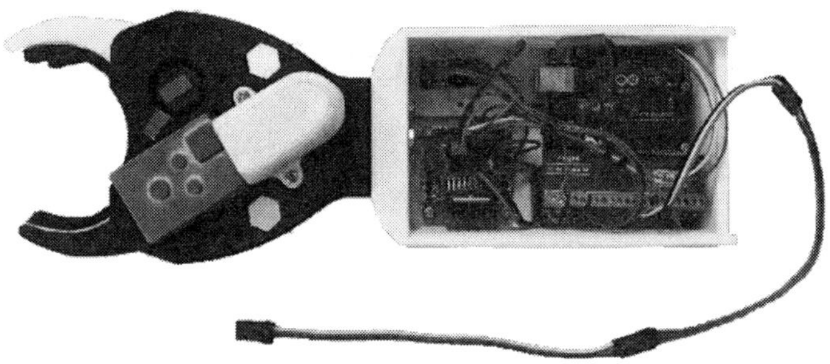

Figure 3. Prosthetic hand prototype

3. The Control System

3.1 The Control Scheme

The control objective is to control the grasping force according to the user's desired force measured by a Myo-Ware Muscle Sensor to detect the electromyography signal. When the motor starts working and the gripper starts squeezing the water bottle, the value of the current that applies to the motor increases. The automatic control system relies on a current sensor and converts the motor feedback current to the actual force against the water bottle.

Then the microcontroller compares the actual force and the desired force and then feeds the difference to drive the motor as shown in Figure 4.

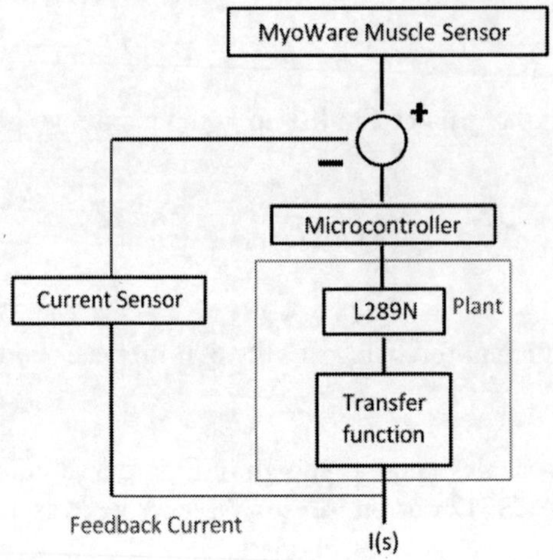

Figure 4. The force-feedback control diagram

3.2 Relationship between the Myo-Ware Muscle Sensor and the desired force

To find the relationship between electromyography and the desired gripping force, an electromyography signals and corresponding gripping force were measured and plotted as shown in Figure 5. The electromyography signal was measured by attaching the Myo-Ware Muscle Sensor at a forearm. The gripping force was measured by squeezing a hand dynamometer. The relationship equation between the electromyography signal (x) and gripping force (y) is $y = 198.67x - 17.995$.

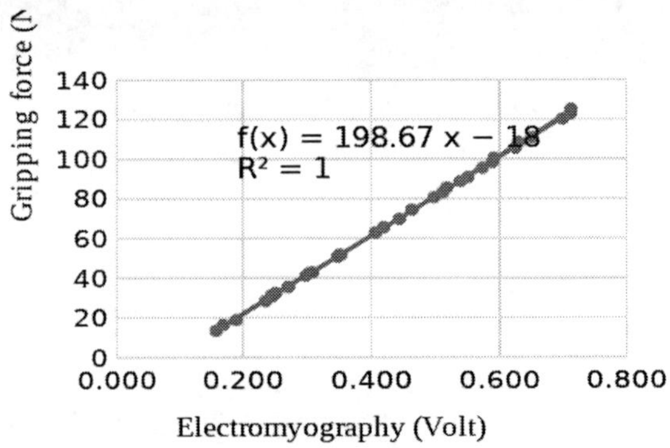

Figure 5. The relationship between electromyography and the desired gripping force

3.3 Relationship between the Motor Current and the Actual Gripping Force

To find the relationship between DC motor current and the actual gripping force of the prosthetic hand, test weights, range 0-600 gram, were hung at the middle of the finger, 50 mm. away from the motor axis. The motor currents were measured with respect to the test weights and plotted on a graph as shown in Figure 6. The relationship equation between the motor current (x) and the gripping force of the prosthetic hand (y) is $y = 29.715x - 0.0435$.

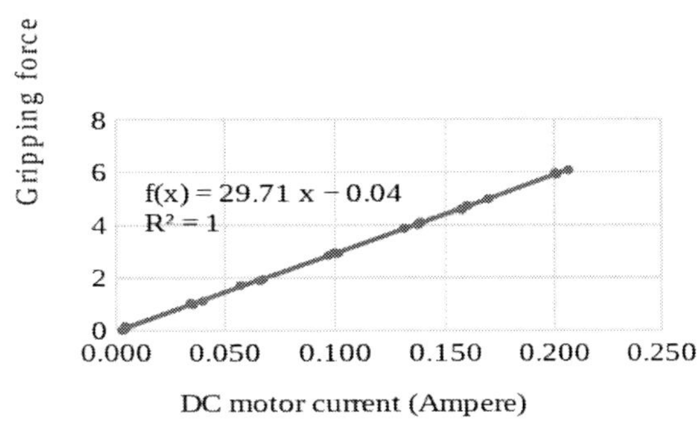

Figure 6. The relationship between motor currents and actual gripping forces of the prosthetic hand

3.4 Programming

The control system of the prosthetic hand was implemented on Arduino Uno, which is a low-cost microcontroller using C programming language. The control diagram can be shown by the block diagram in Figure 7.

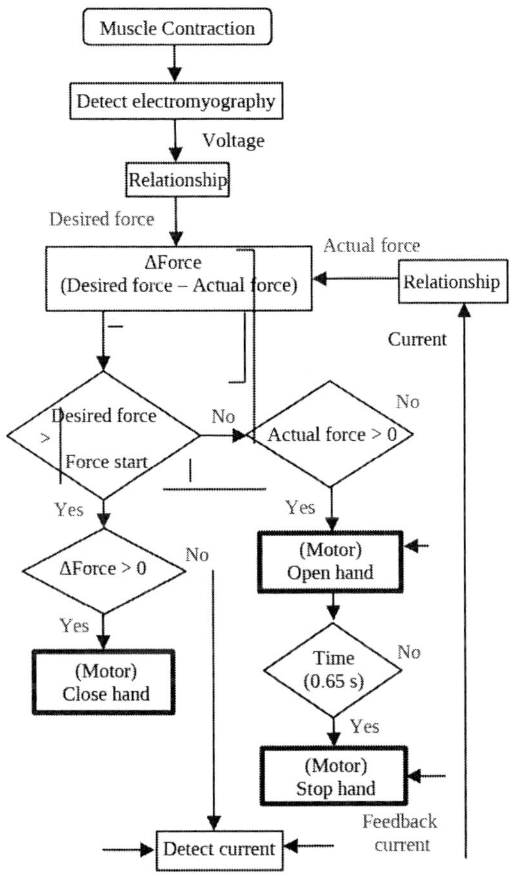

Figure 7. Block diagram of the control system programming

IOP Conf. Series: Materials Science and Engineering **886** (2020) 012033 doi:10.1088/1757-899X/886/1/012033

4. The prosthetic hand grasping test

The prosthetic hand was used to grasp plastic water bottles from 9 manufacturers (represented by brands A to I). The dimensions of the plastic water bottles are described in Figure 8 and Table 1.

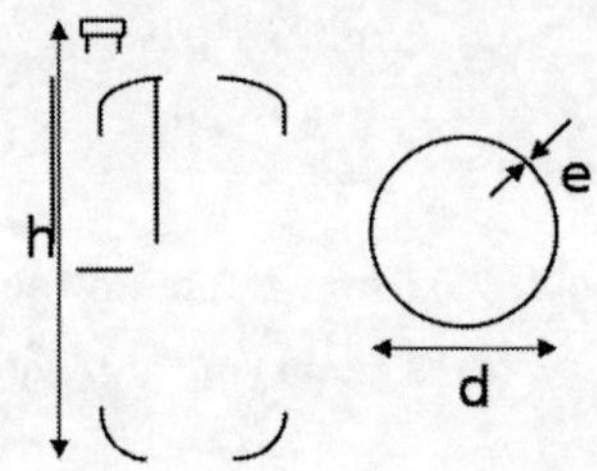

Figure 8. Dimensions of plastic water bottle

Table 1. Dimensions of 600-ml plastic water bottle

Brand	h (mm)	e (mm)	d (mm)
A	225.83	0.19	63.30
B	233.21	0.22	63.06
C	237.10	0.29	61.06
D	233.36	0.25	63.14
E	236.26	0.26	61.00
F	233.45	0.24	63.10
G	232.81	0.28	64.20
H	237.83	0.25	62.89
I	240.20	0.23	60.69

The prosthetic hand was tested to hold the opening water bottles from 9 brands, with 20 repetitions for each brand. The test was considered/regarded as successful if the water bottle was held still and no water was spilt.

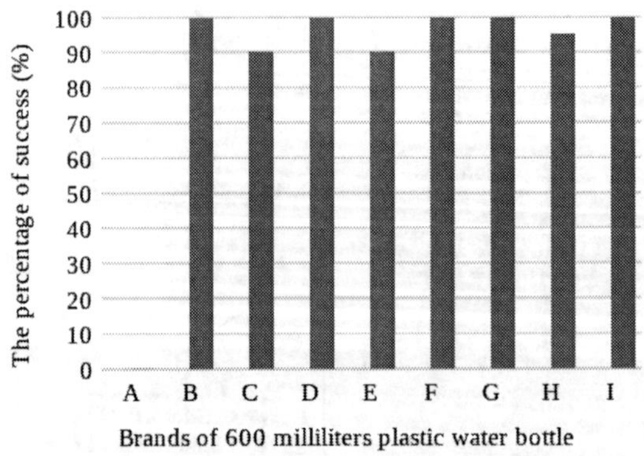

Figure 9. Percentage of success from holding water bottles from 9 brands

From the test, the prosthetic hand can hold 8 brands of opening water bottles with a percentage of more than 90% success in holding water bottles. Brand A failed the test because of the small thickness of the bottle. However, brand A passed the test more than 90% when the water bottle was closed.

5. Conclusion

The prototype of the prosthetic hand-worked according to the objective. It can hold the opening 600-ml plastic water bottles with 8 out of 9 brands. A problem from the force control using the electromyography is that the accumulated fatigue from the contraction of the muscle directly affected the function of the prosthetic hand. While the prosthetic hand holding a water bottle but the value of electromyography is reduced, the prosthetic hand spread the finger immediately causing the failure of holding a water bottle.

References

[1] Institute SNMR 2019 *Accessibility in assistance equipment of Thai's disabled 2007 http://www.snmri.go.th/snmri/download/disability/assistive%20_devices/accphy1.pdf*
[2] Kate T Jelle A Smit G and Breedveld P 2017 *Disability and Rehabilitation. Assistive Technology* **12**(3) 300–314
[3] Telegenov K Tlegenov Y and Shintemirov A 2015 *IEEE Access* **3** 638-647
[4] Ariyanto M Munadi A Haryadi G D Ismail R Pakpahan J A and MustaqimK A 2016 *Proc. 3rd International Conference on Information Technology, Computer, and Electrical Engineering* 42–46

IOP Conf. Series: Materials Science and Engineering **886** (2020) 012034 doi:10.1088/1757-899X/886/1/012034

LQR-Based Stabilization and Position Control of a Mobile Double Inverted Pendulum

D T Ratnayake[1,*], and M Parnichkun[1]

[1]Industrial Systems Engineering, Asian Institute of Technology, Khlong Luang, Pathum Thani 12120, Thailand.

* Corresponding Author: dhanika.1004@gmail.com

Abstract
The double inverted pendulum is a system that can be used efficiently to test and validate the performance of control algorithms because of its instability and the many degrees of freedom system. The paper presents the design and development of a unique mobile double inverted pendulum system. LQR is applied for the balancing control of both pendulums and position control of the mobile platform of the system. Both simulations and experiments are conducted to evaluate the system performance.

1. Introduction

The inverted pendulum is a system that can be used to evaluate the control performance of control algorithms. This experimental apparatus has been brought into the tests since the mid-1960s [1]. There are several versions of this system such as double inverted pendulum, self-balancing robot, and so forth. The double inverted pendulum is unstable with many degrees of freedom system whereas the self-balancing robot can be considered as a mobile inverted pendulum.

There have been many controllers used to control the inverted pendulum systems which can be categorized as linear and non-linear controllers. The single inverted pendulum is simple. Linear control algorithms such as the Proportional-Integral-Derivative (PID) control and the state-space control have been successfully applied for the single inverted pendulum system [1]. Neural network (NN) [2] and the nonlinear controllers such as, an energy-based control technique [3] have been applied to control the system. A double inverted pendulum system is more complicated and has many degrees of freedom. Both linear controllers [4] and NN [5] have been applied to control the double inverted pendulum systems. The swinging up of this system was successfully conducted using an energy-based controller [6] and a sliding mode controller [7]. In addition to the inverted pendulum systems, self-balancing robots have been developed and applied to evaluate the performance of control algorithms as shown in [8]. Several control algorithms; such as PID, Linear Quadratic Regulator (LQR) [9] and Fuzzy Logic Controllers (FLC) [10] have been used to determine the self-balancing robots. Their control performances have been compared with the self-balancing robots as shown in [5].

There have been many controllers used to influence the inverted pendulum systems which can be categorized as linear and non-linear controllers. The single inverted pendulum is simple. Linear control algorithms such as the Proportional-Integral-Derivative (PID) control and the state-space control have been successfully applied for the single inverted pendulum system [1]. Neural network (NN) [2] and nonlinear controllers such as an energy-based control technique [3] have been applied to control the system. A double inverted pendulum system is more complicated and has many degrees of

IOP Conf. Series: Materials Science and Engineering **886** (2020) 012034 doi:10.1088/1757-899X/886/1/012034

freedom. Both linear controllers [4] and NN [5] have been applied to control the double inverted pendulum systems. The swinging up of this system was successfully conducted using an energy-based controller [6] and a sliding mode controller [7]. In addition to the inverted pendulum systems, self-balancing robots have been developed and applied to evaluate the performance of control algorithms as shown in [8]. Several control algorithms such as PID, Linear Quadratic Regulator (LQR) [9] and Fuzzy Logic Controllers (FLC) [10] have been used to control the self-balancing robots. Their control performances have been compared with the self-balancing robots as shown in [5].

2. System Design

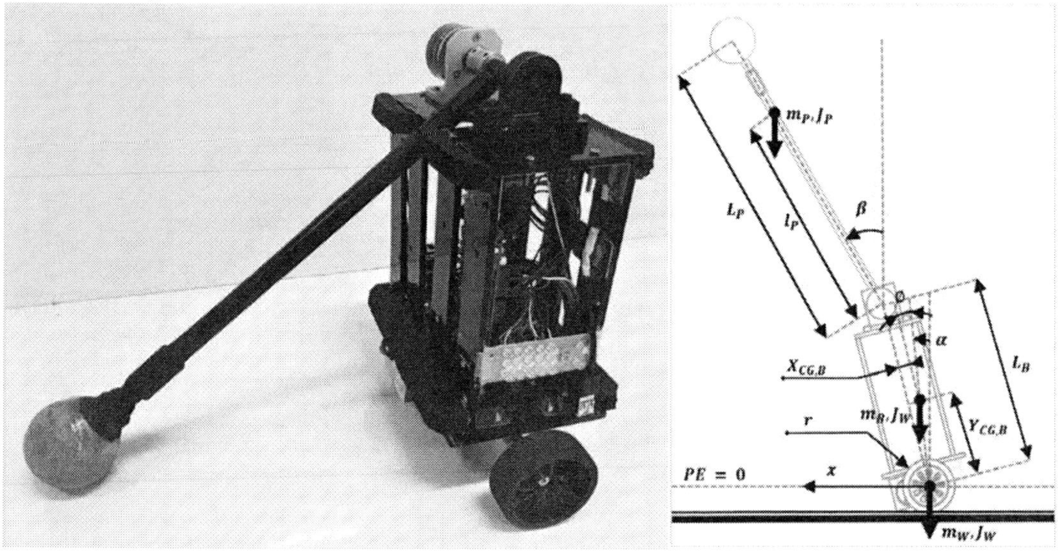

a. **b.**

Figure 1. System architecture

Figure 1a shows the mobile double inverted pendulum system developed. The frame of the self-balancing robot is fabricated from black acrylic. The wheels are mounted on the bottom plate of the frame by means of two rotary bearings (with friction coefficient less than 0.02 each) and a steel shaft. A single $28.9W$ DC motor is used to drive the system which is coupled to the wheels using two spur gears of similar diameter. Only the forward and backward motions are tested, thus, only a single actuator is sufficient for the system to drive. An incremental encoder with a resolution 0.18° is coupled with the rear end of the DC motor. An IMU unit with the output frequency of $100Hz$ is mounted on the body (frame) of the self-balancing robot to measure the leaning angle of the robot. The pendulum is fabricated using an aluminum rod with a rubber ball attached to the top of the rod. The bottom end of the pendulum is mounted on the self-balancing robot by means of a rotary bearing (with friction coefficient of less than 0.02) and an incremental encoder. This encoder is used to measure the angle between the pendulum and the body of the self-balancing robot. The motor is driven using a motor driver with a 12-bit resolution and $20kHz$ PWM signal. An Arduino due microcontroller is used to control the whole system. A 4-cell $2200mAh$ lithium-polymer battery is used to power the whole robot.

3. System identification

The schematic of the system used to derive the dynamics model is shown in Figure 1b. The subscripts W, B, and P denote the wheels, body of the balancing robot and the pendulum, and α, β, and γ represent the rotation/leaning angles, respectively. Note that ∅ is the angle formed between the vertical and the radial axis in addition to ∅ when the body of the robot is tilted (This is due to the offset between the radial axis and the point of action of CG of the robot body, $X_{CG,B}$). Notations $m, J, L, l,$ and r stands for the mass, inertia, total length, length to the CG from the axial axis and radius of the wheel, respectively. The derivation is initiated using the Euler-Lagrange equation, whereas the negligibly small friction components at bearings are disused in the calculation. Initially, the kinetic and potential energies acting on the three generalized coordinates (α, β, and γ) of the system are obtained as expressed in Equations (1) and (2).

$$T_W = \frac{1}{2}m_W\dot{x}^2 + \frac{1}{2}J_W\frac{\dot{x}^2}{r^2}$$
$$T_B = \frac{1}{2}m_B((\dot{x} + l_B\dot{\alpha}c_\alpha)^2 + (-l_B\dot{\alpha}s_\alpha)^2) + \frac{1}{2}J_B\dot{\alpha}^2 \tag{1}$$
$$T_P = \frac{1}{2}m_P\left(\left(\dot{x} + L_B\dot{\alpha}c_{\alpha+\emptyset} + l_B\dot{\beta}c_\beta\right)^2 + \left(L_B\dot{\alpha}s_{\alpha+\emptyset} + l_B\dot{\beta}c_\beta\right)^2\right) + \frac{1}{2}J_P\dot{\beta}^2$$

Where $l_B = \sqrt{X_{CG,B}^2 + Y_{CG,B}^2}$ and $c_\emptyset = \frac{Y_{CG,B}}{l_B}$.

$$V_W = 0, V_B = m_Bgl_Bc_\alpha, V_P = m_Pg\left(L_Bc_{\alpha+\emptyset} + l_Pc_\beta\right) \tag{2}$$

Afterwards, the total potential energy is deducted from the total kinetic energy to form the Lagrangian. Consequently, the Euler-Lagrange equation outputs the nonlinear dynamics model of the system which is expressed in Equation (3).

$$a_1\ddot{x} + (a_2c_\alpha + a_3c_{\alpha+\emptyset})\ddot{\alpha} + a_4c_\beta\ddot{\beta} - (a_2s_\alpha + a_3s_{\alpha+\emptyset})\dot{\alpha}^2 - a_4c_\beta\ddot{\beta} = \frac{\tau}{r}$$
$$(a_2c_\alpha + a_3c_{\alpha+\emptyset})\ddot{x} + a_5\ddot{\alpha} + a_6c_{\alpha+\emptyset-\beta}\ddot{\beta} + a_6s_{\alpha+\emptyset-\beta}\dot{\beta}^2 - (a_2s_\alpha + a_3s_{\alpha+\emptyset})g = 0 \tag{3}$$
$$a_4c_\beta\ddot{x} + a_6c_{\alpha+\emptyset-\beta}\ddot{\alpha} + a_7\ddot{\beta} - a_6s_{\alpha+\emptyset-\beta}\dot{\alpha}^2 - a_4s_\beta g = 0$$

Where,

$a_1 = m_W + m_B + m_P + (J_W/r^2)$
$a_2 = m_Bl_B$
$a_3 = m_PL_B$
$a_4 = m_Pl_P$
$a_5 = m_Bl_B^2 + m_PL_B^2 + J_B$
$a_6 = m_PL_Bl_P$
$a_7 = m_Pl_P^2 + J_P$
τ/r is the input force in the direction of x

Subsequently, the nonlinear model is linearized by assuming that $q \approx 0$, $\sin(q) \approx 0$, $\cos(q) \approx 0$ and $\dot{q}^2 \approx 0$. The DC motor dynamics are rearranged neglecting the armature inductance and friction in the motor shaft to obtain the form is expressed in Equation (4).

$$\tau = \frac{K_T}{R}\left(V - K_e\frac{\dot{x}}{r}\right) \tag{4}$$

Finally, motor input torque in equation (4) is substituted in equation (3) to obtain the linearized model that is shown in Equation (5).

$$\begin{bmatrix} a_1 & a_2 + a_3c_\emptyset & a_4 \\ a_2 + a_3c_\emptyset & a_5 & a_6c_\emptyset \\ a_4 & a_6c_\emptyset & a_7 \end{bmatrix}\begin{bmatrix} \ddot{x} \\ \ddot{\alpha} \\ \ddot{\beta} \end{bmatrix} + \begin{bmatrix} a_8\dot{x} \\ \dot{\alpha} \\ \dot{\beta} \end{bmatrix} + \begin{bmatrix} x \\ -(a_2 + a_3)g \\ a_4g\beta \end{bmatrix} = \begin{bmatrix} b_1V \\ 0 \\ 0 \end{bmatrix} \tag{5}$$

Where,

$a_8 = K_eK_T/Rr^2$ $\quad$ $b_1 = K_T/Rr$

The parameters of the experimental system are measured and calculated as shown in Table 1. The state-space model of the system is formed as expressed in Equation (6) by substituting parameters to Equation (5).

Table 1. Parameter Identification

Coefficient	Description	Coefficient	Description	Coefficient	Description
m_W	150g	l_B	100mm	$\emptyset$	0.859°
J_W	0.000Nm	L_B	480mm	L_P	420mm
r	35mm	m_P	150g	K_T	0.155Nm/A
m_B	2200g	J_P	0.016Nm	K_e	0.327V/rads^{-1}
J_B	0.022Nm	l_P	330mm	R	2.5Ω

Since $\emptyset$ is negligibly small, the offset in CG of the body in the tangential axis is approximated to be zero in the derivation of the linearized model.

$$X = [x \quad \alpha \quad \beta \quad \dot{x} \quad \dot{\alpha} \quad \dot{\beta}]^T, A = \begin{bmatrix} 0 & 0 & 0 & 1 & 0 & 0 \\ 0 & 0 & 0 & 0 & 1 & 0 \\ 0 & 0 & 0 & 0 & 0 & 1 \\ 0 & -8.151 & 0.504 & -8.170 & 0 & 0 \\ 0 & 80.782 & -7.864 & 34.127 & 0 & 0 \\ 0 & -46.395 & 19.819 & -12.438 & 0 & 0 \end{bmatrix}, B = \begin{bmatrix} 0 \\ 0 \\ 0 \\ 1.207 \\ -5.093 \\ 1.837 \end{bmatrix} \tag{6}$$

Matlab is used to identify the open-loop poles of the system, which deduce that the system is unstable. In addition, the controllability and observability confirm that the system is both controllable and observable.

4. Control Architecture

4.1. Input signal filtering for experimental system

The Signal from the gyro has noise in the angular velocity reading. Hence, a simple moving average filter is applied as expressed in the first line of the Equation (7). In this case, $\dot{\alpha}_n$ represents original signal and $\dot{\alpha}_T$ represents filtered output. Likewise, the differentiated output from the encoder is conditioned with a simple low pass filter with the gain (K_{lp}) of 1 and cut-off frequency (T_{lp}) at $10Hz$. The second line of Equation (7) shows the filter transfer function for the two encoders connected to the motor and the pendulum. In this case, $\dot{\theta}$ represents the original signal and $\dot{\theta}_{filtered}$ represents the filtered output.

$$\dot{\alpha}_T = \sum_{n=T-29}^{T} \frac{\dot{\alpha}_n}{30} \qquad \frac{\dot{\theta}_{filtered}}{\dot{\theta}} = K_{lp} \frac{1}{T_{lp}s + 1} \tag{7}$$

4.2. LQR

LQR is used to control the robot position, the leaning angle of the body of the self-balancing robot, and the pendulum at the upright position. LQR is an optimal controller that minimizes the cost function expressed in equation (8). The state depending matrix Q and input weighting matrix R have to be adjusted according to the desired cost function. The control law of the LQR follows $u = -Kx(t)$ where K is the optimal gain matrix. K can be obtained from the Algebraic Ricatti Equation (ARE).

$$J = \int_{\tau}^{T} (x'(t)Qx(t) + u'(t)Ru(t))\, dt \tag{8}$$

4.3. Position control of the system

Figure 2 shows the complete control block diagram of the system. The input u that is obtained using the control law is fed to the system to obtain the outputs x, α, and β. α and β are used to estimate $\dot{\alpha}$ and $\dot{\beta}$ while $\dot{x}$ is also estimated. The estimations are followed up by a series of filters.

The external step input applied to change the position of the robot (x_{set}) is converted to a linear input using a linear trajectory generator (LTG), summed with the x, and the six states are inputted to obtain the control law. This loop is repeated at $250Hz$ to maintain the stability and the position of the system.

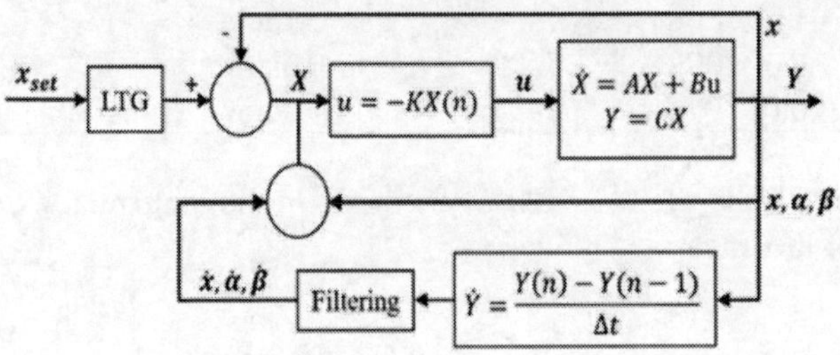

Figure 2. Control block diagram

5. Results

5.1 Simulation results

Simulation is conducted on Matlab/Simulink to evaluate the performance of the controller. Q and R matrices are defined to achieve the desired control conditions. Q matrix is defined in such a way that higher weight is given on the position states rather than the velocity states. The input weighting matrix is defined as such that the input does not exceed the limitation of the driving DC motor. Firstly, the gains for balancing the robot and the pendulum without robot position control are obtained by defining lower weight for the wheel position (Case I). Then, Q matrix is defined as $diag(10000, 5000, 5000, 5, 5, 5)$ and R as [1.0] which focuses on both stabilizing and position maintaining.

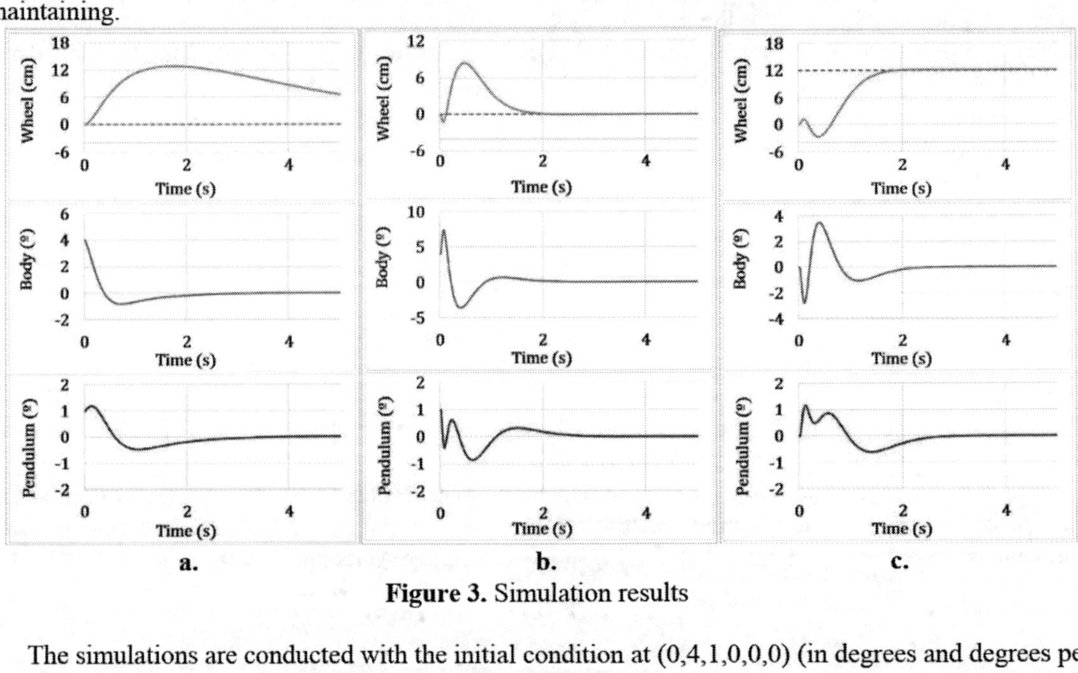

Figure 3. Simulation results

The simulations are conducted with the initial condition at (0,4,1,0,0,0) (in degrees and degrees per second) and the optimal gain obtained as (100.00, −525.87,527.09,99.05,64.24,133.09). Figure 3 shows the simulation results (in generalized coordinates) for Case I, Case II (the robot position is controlled at zero position), and Case III (the robot position is controlled by altering the target position with a $12cm$ distance). Figure 3a shows that the self-balancing robot and the pendulum reach the upright position within $3s$, however, the robot position does not stay at zero position. In Figure 3 b, the position is controlled at zero position within $2.4s$ while the pendulum takes $2.7s$ to be balanced at the upright position. The result of tracking a step position command of $12cm$ distance is shown in Figure 3 c. This results in the system reaching the new position within $3s$.

5.2 Experimental results

Figures 4a and 4c show the signal after the low pass filter from the two encoders. It is observed that the differentiation of the motor encoder contains the noise of $750°s^{-1}$ the amplitude at $3Hz$. This noise is removed by a low pass filter. However, this filtering causes a delay of $16ms$ to the original signal. A similar result is observed from the pendulum encoder signal with a delay of $14ms$ to the original signal. The averaging filter acting on the gyro signal removes the amplitude noise and causes an $8ms$ delay to the signal, as depicted in figure 4b, the plot in grey contained the actual signals whereas the colored plots represent the filtered signals.

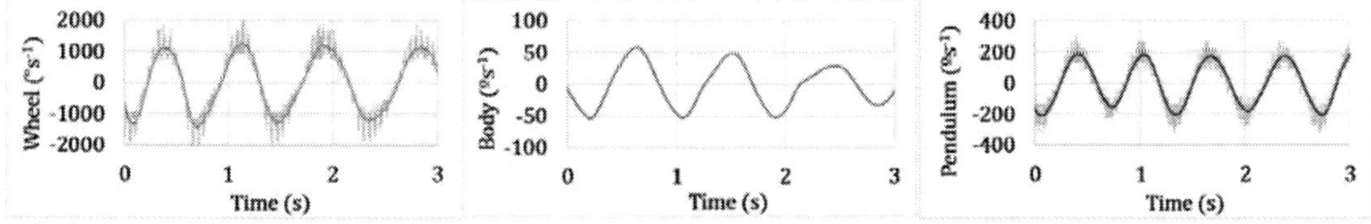

a. b. c.

Figure 4. Filtering outcomes for the three velocity estimations

Next, three experiments are conducted. The first experiment is conducted to balance the robot and the pendulum without the robot positioning control (Case IV). The weight matrix is defined so that higher priority is given for the leaning angle of the self-balancing robot and the pendulum, than the robot position. Then, optimal gain matrix is obtained as $(105.0, -525.0, 540.0, 60.9, 60.4, 39.8)$, which corresponds to both balancing and robot positioning control. Thereafter, all the positions are controlled at zero (Case V). In the final experiment, both leaning angles of the robot and the pendulum are controlled at zero and the robot position is controlled at different set positions (Case VI).

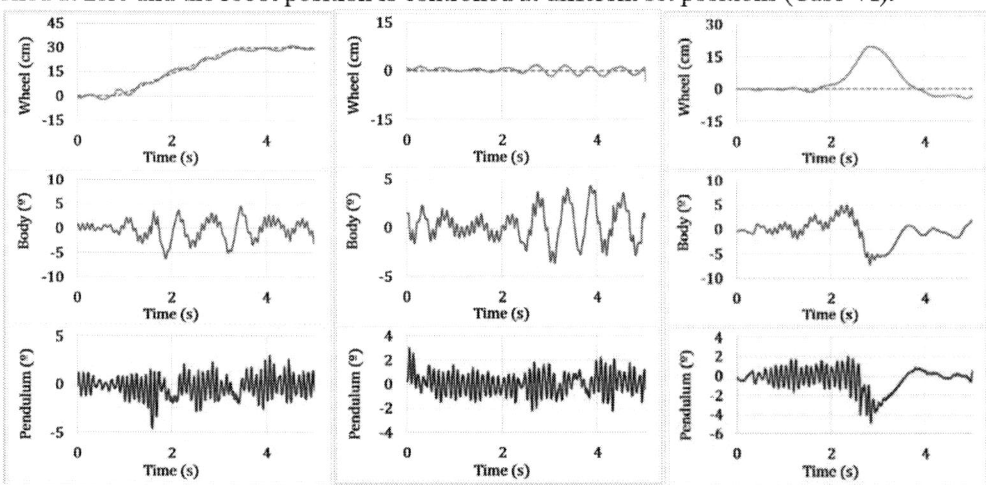

a. b. c.

Figure 5. Experimental Results

Figure 5 shows the experimental results (in generalized coordinates) from the three cases. Figure 5a shows the result of Case IV. It can be seen that the robot position does not stay at zero position while the body and the pendulum leaning angles are controlled near zero degrees. However, high-frequency noises can be observed from both measurement signals. The body leaning angle has a standard deviation of 4.58° whereas the pendulum shows a standard deviation of 2.29° from the upright position. It also can be seen that when the robot position is not controlled, the robot continuously moves to balance the body and the pendulum. In this experiment, oscillation is less compared with the other two cases. Figure 5b shows the result from Case V. It can be observed that the robot position is controlled at the starting position (zero position) with oscillation. The oscillation has a standard deviation of $0.84cm$ while the mean is $0.14cm$ away from the desired position. The standard deviations of the leaning angle of the self-balancing robot and the pendulum are reduced to $2.28°$ and $0.71°$, respectively. The result of a single position change from Case VI is shown in Figure 5c. The off-set position is applied as shown in the dotted lines, whereas the actual desired position is shown in a solid line. The position is changed for $30cm$ at the speed of $10cms^{-1}$. The robot position oscillates when the offset position is changed linearly. It can be seen that the robot moves to the new position at the end of the ramp, however, it still oscillates. The oscillation has a standard deviation of $0.74cm$ with a shift of $0.78cm$ as observed from the $30cm$ mark. The result from the step change of $75cm$ shows a similar result. The oscillation is seen after reaching the new position with a standard deviation of $2.05cm$ and a shift of $3.12cm$ from the desired position.

5.3 Comparison between the Two Test Methods
Firstly, the gain matrices from simulations and experiments show a similarity in each gain except for those corresponding to wheel and pendulum angular velocities. This is due to the undiminished noises at low amplitudes from the low pass filters. This phenomenon results in constraining the two individual gains to a lower value than those from the simulations. The minor differences are mainly caused due to the negligibly small friction parameters uncompensated during simulations. The result of the analysis conducted to obtain the maximum recovery angles shows that the angles recovered by the body of the balancing robot are comparatively higher in simulations, where the recovery angles of the pendulum are higher in experiments.

6. Conclusion
This paper described the development and control of a mobile double inverted pendulum. Both simulations and experiments were conducted to evaluate the performance of the system in balancing and position control. The results showed successful balancing of body and pendulum at the upright position and the position control at the desired positions, using LQR.

References
[1] Phelps F and Hunter J 1965 *American Journal of Physics* **33** 285295
[2] Williams V and Matsuko K 1991 *Proc. 1991 IEEE International Joint Conference on Neural Networks* (Singapore) 18-21
[3] Maeba T Deng M Yanou A and Henmi T 2010 *Proc. 2010 International Conference on Modelling, Identification and Control* (Okayania)
[4] Cheng F Zhong G Li Y and Xu Z 1996 *Fuzzy Sets and Systems* **79** 315-321
[5] Bogdanav A 2004 *Oreg. Heal. Scie. Univ* CSE-04 006
[6] Yamakita M Iwashiro M Sugahara Y and Futura K 2002 *Proc.1995 American Control Conference* (Seattle)
[7] Henmi T Deng M and Inoue A 2004 *Proc. 2004 American Control Conference* (Boston)
[8] Chee O Y Shukri M and Abidin Z 2006 *Proc. 2006 4th Student Conference on Research and Development, IEEE 2007* (Selangor)
[9] Xu C Li M and Pan F 2011 *Proc. 2011 International Conference on Electrical and Control Engineering, IEEE 2011* (Yichang)
[10] Bature A A Buyamin S Ahmed M N and Muhammad M 2014 *International Journal of Mechanical Mechatronics Engineering* **14**(3) 62-68

IOP Conf. Series: Materials Science and Engineering **886** (2020) 012035 doi:10.1088/1757-899X/886/1/012035

Development of Vertical Articulated Robot Deburring System by Using Sensor Feedback

Hiroki Kakoi[1*], Kiyoshi Yanagihara[2], Koji Akashi[2] and Kensuke Tsuchiya[3,]

[1]Department of Creative Engineering, National Institute of Technology, Ariake College, 150 Higashihagio-Machi, Fukuoka, 836-8585, Japan
[1]Faculty of Creative Engineering, National Institute of Technology, Ariake College, 150 Higashihagio-Machi, Fukuoka, 836-8585, Japan
[2]Institute of Industrial Science, The University of Tokyo, Komaba4-6-1 Megur-ku, Tokyo, 153-8505, Japan

* Corresponding Author: kiyoshi@ariake-nct.ac.jp

Abstract
Demand of customer for higher quality of product is increasing. Number of skilled engineers, however, is chronically lacking in worldwide. Thus, the change of environment of production drives robotization of hand-finishing process recently. This study is about Robotization of deburring by means of a hand grinder, which is one of the hand finishing tools. Basically, in robotization of the hand-finishing process, teaching operation is selected for the programming of the tool path of the robotic arm. In this study, however, a vertically articulated robot equipped with various sensors is developed for the robotic deburring system, the feasibility of fully automated robotic deburring by using hand grinder is investigated.

Keywords: industrial robot, computer vision, depth camera

1. Introduction

At the early stage of robot deburring study, Proctor F. M., et. al. (1989) developed the hand-finishing by utilizing industrial robot [1]. In their study, the technology which is a robot splaying and trimming is tried. Such technologies are being studied and evolved continuously until now. In the studies, teaching process which one of the robot programing methods is used to input move-path to the robot.

Regarding an industrial robot, its moves and positionings are important because they affect the accuracy directly [2] [3]. In order to teach tool path, there are various studies [4] [5] [6] including offline teachings, such as [7] [8] [9] [10] [16] concerning auto programing system using Computer Aided Design (CAD) data of workpiece, and Thomessen T, et.al . developed a method[14] that the workpiece geometry is pre-scanned by using force censor which is installed in the end of robot hand to generate grinding paths automatically.

The move-path calculation by employing one of these methods needs compensation of machining error when processing workpiece. Thus, some studies developed the methodology to compensate for the work path. They are roughly divided into the following two categories. The first one is the method that utilizes camera and image processing technology; the other is the one, which utilizes force sensor [14-26].

Verl, et al. (2019) reviewed trends and problems of utilization of industrial robot in recent years [26]. This review predicts that industrial robot will be applied not only for transferring of the workpiece but also machining including deburring here after, the application must increase. Needless to say, that the goal of development of robot deburring system is a fully automated one that not only enables removing burrs but also recognizes burrs without any teaching for the robot.

Along the trend, Naoki et al. had been developed a deburring robot system since 1980s [7]. They integrated the state of art components of those days into a robot deburring system, which included the sensor, the computer, the software, and so on. The applied components, however, was relatively expensive in those days. Thirty years on the environment of the development of robot, deburring system has become lowering the cost by providing high-speed network and sensors such as camera technology.

In this study, we develop a new deburring robot system using image processing technology and newly developed displacement sensor, which is using optical technology. The system, however, is different from previous studies. It relies on data not from CAD, but from image processing and displacement sensors for a deburring simple bar of a square in cross-section.

In the designing of robotic deburring system, our surveillance about previous study found that those studies are categorized by 4 technical components as shown in Table 1. Teaching means a necessity to program arm trajectory manually for a robot controller. Sensor feedback means whether the robot system equipped with sensors to detect the state of the deburring process. Target application means the production process to which robot deburring is applied.

Table 1. components listed from previous studies

Category of Componen	Selected	unselected
Teaching	Not Necessary	Necessary
Sensor feedback	Equipped	Not equipped
Handling object by End effector	Tool	Work
Target application	Edge burr by milling	Bead by welding, Fin by casting

Since our consideration from table1. We select teaching less method. The method do not need for human hands. It is important to use a computer for complete automation. A sensor that makes decisions on behalf of humans is needed.

Regarding the workpiece and tools installation, there are two ways. Either mounting the workpiece on the robot hand or mounting the deburring tool on the robot hand must be selected. For mounting the workpiece on the robot hand, it is too difficult to hold a heavy and large workpiece because of the ability of utilized robot. On the other hand, mounting deburring tool such as hand grinder is easy because the tool is light and the size is regulated. Therefore setting the deburring tool on the robot hand as the end effector is selected.

In the experiment, burr is produced on the edge of workpiece with milling. This burr, what we call "edge burr", is dealt as a target.

There are several similar robot systems. [27][28] In these researches, a camera is used as a sensor for the sensor fusion. We used a depth camera and a force sensor instead of a camera.

In this paper, an outline of the developed robotized deburring system and experimental result of detection of tool contact is reported by using the sensing system, which is composed of image processing and force feedback.

2. System configuration

2.1 set-ups

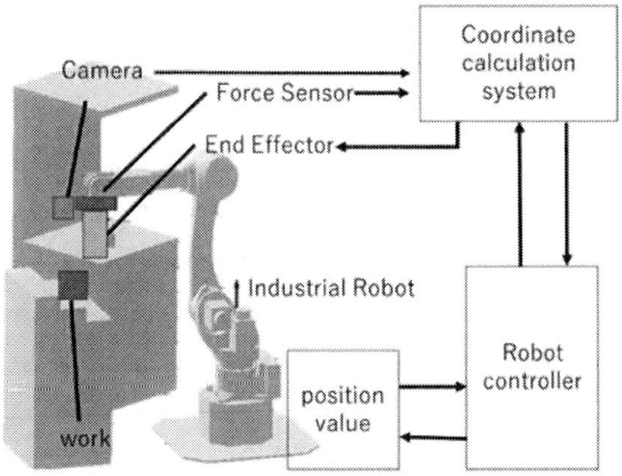

Figure 1. System configuration

The developed system is shown in Figure 1. The system consists of an arm robot (Kawasaki Heavy Industry Co. Ltd. RS20N), a workpiece recognition system by using image capturing, and a coordinate calculation system to generate toolpath. A router with a fixed abrasive tool is attached to the robot for deburring. For the recognition system, a depth camera (Intel Realsense D435) is installed on the tip of the arm of the robot.

The installation of various sensors and routers at the end of the robot and the coordinate system of the sensors are shown in Figure 2.

The value of the force sensor increases when a force is applied in each axis of the arrow direction. It can measure not only the forces of x, y, and z-axis, but also the torque around those three axes. The obtained torque of x, y, and z axis is labeled rx, ry, and rz, respectively. Clockwise and left-hand rules are applied to view from the front of each axis.

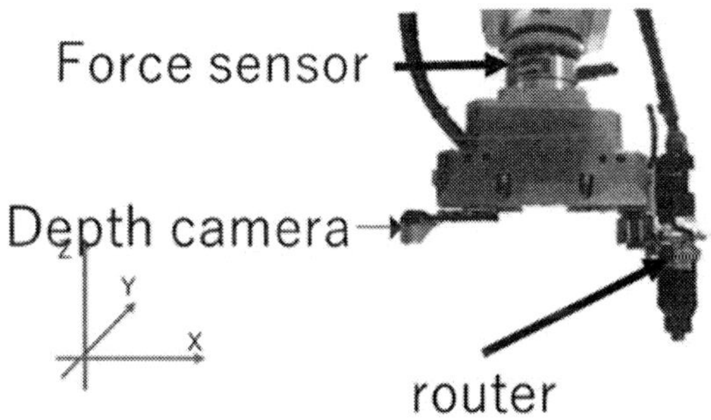

Figure 2. Hand tools and origin of work

2.2 function of system

When the workpiece is mounted on a table, the depth camera captures the image while measuring the distances from the camera, which is mounted at the end of the arm of the robot to the workpiece. The image data are sent to coordinate calculating system. The work is recognized by means of comparing the image with CAD data, and the tool path is calculated according to the work shape based on the image. The calculated path data is sent to the robot controller, the controller commands the robot to move arm along to the programmed path. The coordinate calculation system also monitors not only distance data from the depth camera but also the value of force sensor in the deburring process. If the value excesses the programmed threshold level, the coordinate calculation system corrects the tool path in processing.

3. Experiment

3.1 Measurement of shape and position of the workpiece by using a depth camera

The goal of development of this deburring system is to produce tool paths automatically by using sensor feedback that consists of the depth camera and the force sensor without any teaching of the robot controller.

Regarding tool motion in deburring, it is considered that the robot arm is programmed to trace the outside edge of a workpiece with a fixed abrasive tool. In order to realize the idea, it is necessary to obtain the distance between the tool and the workpiece, and the top shape of its surface.

Both left and right images in Figure 3 are camera images each of which shows, the distance output of depth sensors, respectively. In this experiment, the square top shape of the work piece is selected, and it is fixed in keeping parallel to the position of the outside shape of the table. Those images demonstrate that it is possible to obtain the outer shape of the work and the distances between the depth sensor and the top surface of the workpiece. Distance value measured by depth camera has potentially 2% of error. Therefore, for precision deburring, more accurate detection of tool position is needed to determine the original position of the tool.

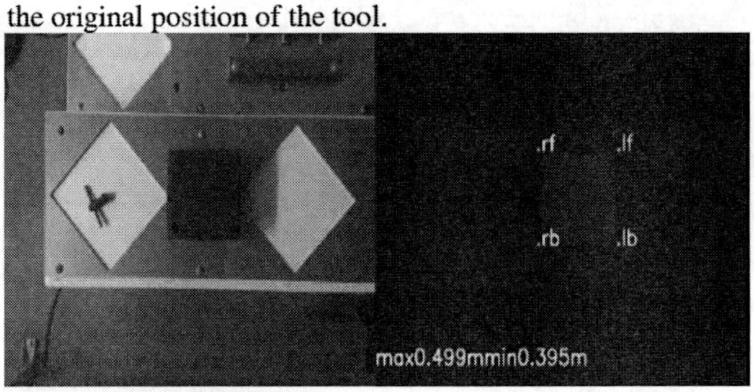

Figure 3. image and depth value by depth camera

3.2 The Influence of vibration of router rotation on force sensor value

In order to aid the limitation of accuracy of the depth camera, the utilization of the force sensor is mentioned in the previous section. The output of the force sensor, however, can be influenced by the rotational vibration of the router. Thus, the influence is investigated by recording the output of the sensor with or without rotation of the router. Table 2 shows the dispersion of the force sensor output. The dispersion of rotation y-axis is 13.7 when the router turns off. Meanwhile, it is 265 when the router turns on. From the results, it is determined that the torque value of ry is utilized for monitoring because it is relatively easy to distinguish the difference between on and off.

Table 2. Distribution of force sensor values by rotation of the router

	x	y	z	rx	ry	rz
on	165	217	180	235	265	425
off	138	76	135	8.79	13.7	27.6

3.3 The experiment detecting touch-down of the tool by utilising sensor fusion
In order to detect touch-down of the tool onto the workpiece, following the experimental sequences, as shown in figure 4, is developed for an experiment. In this sequence, firstly the shape of the workpiece is recognized by using the depth camera. Then the tool is rapidly approached to upward position of 5 mm. from the surface of the work. Thirdly, the tool is fed at relatively low speed of feed while referring ry value of torque of y-axis of the force sensor.

In the sequence, the torque threshold must be programmed to the robot controller before the operation.

After the command is given to the robot arm to move tool, the robot controller gets the value of the force sensor with the move of every 0.03 mm. If the output value becomes equal to or greater than the set threshold value, the robot arm is held and then the original tool position is programmed on the controller.

Figure 5 shows sensor output values of ry when the touch down of tool is recognised and the origin is programmed according to the sequence. In this experiment, the distance of free run of the tool from detection to stop is 0.03 mm.

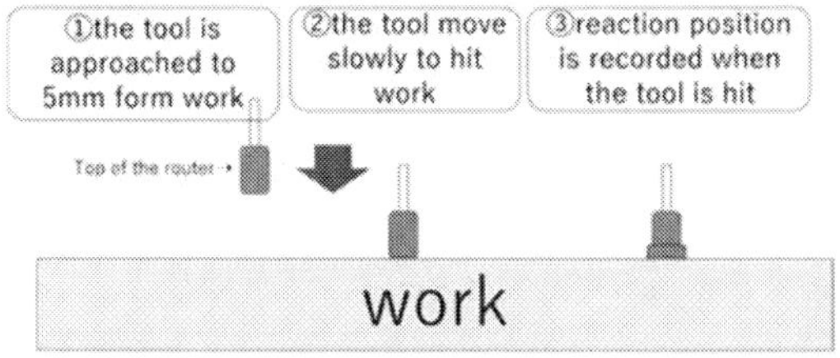

Figure 4. Touch detection sequence

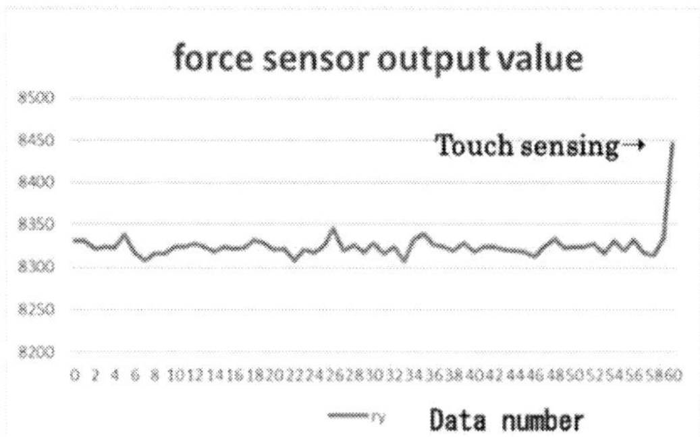

Figure 5. Change in the output shaft torque value ry

4. Conclusion
The feasibility of fully automated robotic deburring by using hand grinder with the sensor feedback system was investigated in this study. The goal of development of robotic deburring system not only can automatically program tool movement without any motion teaching of robot, but also can provide appropriate modification to the path in processing. A work recognition system and a tool path generation system was developed by using a depth camera, also a router with a fixed abrasive was created too as an end effector was utilized for deburring. The results obtained from the system-verification experiments are as follows:

(1) To realise fully automated teaching-less robot deburring system, a combination of the depth camera and the force sensor is employed for sensor feedback controlling system.
(2) By utilizing the depth camera, not only capturing the front shape of the workpiece but also measuring the distance to the edge of the workpiece object are available. The acquired image of work shape and the measured distance can also be useful for path generation of the robot movement.
(3) In the experiment to detect touchdown of the tool by using the distance data measured by the depth camera and the feedback from the force sensor's results demonstrate that the system can detect the distance between the tool and the surface of the workpiece within less than 0.30 mm.

How to distinguish the burr from the edge of the workpiece by using the captured image and positioning data of the edge, and how to make a tool to approach to the edge are subjects for further studies.

Acknowledgments

The use of vertically- articulated robot for this study is provided by Prof. Akashi of NITA, and the helpful assistance of Takaya Matsumura, our lab colleague, is greatly appreciated..
A part of this work was supported by a research grant of AMADA FOUNDATION.

References

[1] Proctor F M Norcross R J and Murphy K N 1989 *Technical Paper SME* 89–138
[2] Song H C Kim B S and Song J B 2012 *Proc. IEEE/ASME Int Conf on Adv Intelligent Mechatronics* 890–895
[3] Dietz T Schneider U Barho M Oberer-Treitz S Drust M and Hollmann R and Hägele M 2012 *Proc. VDE 7th German Conf on Robotics Proceedings of ROBOTIK* 1–6
[4] Solvang B Sziebig G and Korondi P 2008 *Proc. IEEE Int Conf on Networking, Sensing and Control* 949–954
[5] Bi Z and Kang B 2014 *Enterprise Information Systems* **8**(2) 225–245
[6] Park S C and Chung Y C 2003 *Computer-Aided Design* **35**(5) 467–475
[7] Asakawa N Itoh K and Takeuchi Y 1998 *Journal of the Japan Society for Precision Engineering* **64** (5)
[8] Toda K Asakawa N and Takeuchi Y 1999 *JSME* (**65-631**) 1288–1294
[9] Gary M B and Mohamed A E 1994 *IEEE Trans. On Industrial Electronics* (**41-2**) 137–146
[10] Nonaka Y Sakaue S Yanagihara Y and Yokoshima K 1995 *Proc. IEEE Int. Conf. Robotics Autom* 2084–2090
[11] Asada H and Izumi H 1989 *IEEE Trans on Robotics and Automation* **5**(2) 166–173
[12] Zhang H Chen H Xi N Zhang G and He J 2006 *Intelligent Robots and Systems* 9–15
[13] Shimokura K Asada H and Liu S 1996 *Robotics Society of Japan* **14**(1) 144
[14] Thomessen T Elle O J Larsen J L Andersen T Pedersen T Pedersen J E and Lien T K 1992 *Proc. of the 23th Int. Symposium on Industrial Robots* 457–464
[15] Jinno M Ozaki F Yoshimi T Tatsuno K Takahashi M Kanda M Tamada Y and Nagataki S 1995 *Proc. 1995 IEEE Int. Conf. on Robotics and Automation*1455–1460
[16] Kuss A Drust M and Verl A 2016 *Proc. 49th CIRP Conference on Manufacturing Systems* **57** 8–13
[17] Wang X Wang Y and Xue Y 2006 *Proc. IEEE Int Conf on Industrial Informatics* 921–925
[18]] Robertsson A Olsson T Johansson R Blomdell A Nilsson K and Haage M 2006 *Proc. IEEE/RSJ Int Conf on Intelligent Robots and Systems* 2743–2748
[19] Pagilla P R and Yu B 2001 *Proc. IEEE American Control Conference* **1** 630–635
[20] Elbestawi M A Tuen K M Srivasava A K and Dai H 1991 *Annals of the CIRP* **40**(1) 391–394
[21] Kashiwagi K 1991 *J. Robotics Society of Japan* **9**(3) 94–98
[22] Kazerooni H Bausch J J and Kramer B M 1986 *ASME J. Dynamic Systems Measurement Control* **108** 354–359

[23] Her M G and Kazerooni H 1991 *ASME J. of Dynamic Systems Measurement and Control* **113** 60–66
[24] Ling H C and Li C F 1997 *Proc. IEEE Int. Conf. Robotics Autumn meeting* 2844–2849
[25] Muto S and Shimokura K 1997 *J. Robotics Society of Japan* **15**(5) 744–751
[26] Verl A Valente A Melkote S Brecher C Ozturk E and Tunc L T 2019 *CIRP Annals-Manufacturing Technology* **68** 799–822
[27] Aasakawa N and Kiriyama Y 2011 *JSME* **77** 782
[28] Kiriyama Y Noguchi T and Asakawa N 2011 *Proc. of JSPE autumn Meeting* 335–336

IOP Conf. Series: Materials Science and Engineering **886** (2020) 012036 doi:10.1088/1757-899X/886/1/012036

Evaluation of Outer Structure Solar Reflection Characteristics of Highly Reflective Material in Consideration of Human Thermal Sensation

A. Yoshida[1], S. Kinoshita[1,*] and Y. Shimazaki[2]

[1]Department of Mechanical Engineering, Osaka Prefecture University, 1-1 Gakuen-cho, Naka-ku, Sakai, 599-8531, Japan
[2]Department of Architecture and Civil Engineering, Toyohashi University of Technology, 1-1 Hibarigaoka, Tempaku-cho, Toyohashi, Aichi, 441-8580, Japan

* Corresponding Author: kinosita@me.osakafu-u.ac.jp

Abstract
As a solution to the influence of solar radiation on thermal comfort in urban streets, constructing outer surfaces with highly reflective materials is considered to be controlling the directional reflection of the materials, and to reflect most of the radiation in the retroreflective direction—the same direction as that of the incident radiation. This study experimentally and numerically evaluates the directional solar reflectance of the outer cover material having an uneven surface. There is found to be a strong impact of radiation on human thermal comfort, and reduction of this radiation is the key to improving the thermal environment. Accordingly, the use of highly reflective materials is a good strategy for enabling this improvement, and retro reflection is a promising concept to consider in this regard. The simulation results show that human thermal comfort is determined by retro reflection if solar reflectance is the same.

Keywords: Heat island, Solar radiation, Building outer skin, Retro reflection, Human thermal load

1. Introduction

Recently, the thermal environment in urban areas has worsened with the aggravation of the urban "heat island" phenomenon. As a countermeasure to this problem, it has been suggested that the absorption of solar radiation be restrained by covering the surfaces of buildings and streets with materials exhibiting high solar reflectance. When the solar reflection from the surface is radiated onto the surroundings of the urban space, however, it is a concern that the absorption of solar radiation by other surfaces of structures increases and thermal comfort in urban streets deteriorates. As a solution, it has been suggested that directional reflection of materials be controlled and most of the solar radiation be forced to reflect in the recursion direction, i.e., the same as the direction of insolation, which is called "retro reflection."

Sakai et al. (date) showed experimentally that retroreflective materials reduce the heat generated by reflected insolation using a miniature model realized in a thermostatic oven, and they evaluated the solar reflective performance of several retroreflective materials [1, 2]. Iyota et al. (date) proposed a method for measuring retro reflectance using an apparatus containing an emitting–receiving optical fiber and spectrometers for both the visible and the infrared band. The retro reflectance and angular dependency of several sheet-type retroreflective materials were measured using the apparatus [3]. The development of surface materials with a retroreflective function was also promoted, and their performance evaluation was performed by Rossi et al. [4] and Morini et al. [5]. A method for performance testing of retroreflective material was studied by Harima et al. [6]. The same group also developed a film-shaped retroreflective material [7]. The thermal effect when using retroreflective material was investigated by numerical calculation [8, 9].

Human-centered technology has received considerable attention. For the thermal environment, feeling temperature or thermal comfort is an important factor in the study of human-oriented improvement. In general, feeling temperature or thermal comfort is determined by six dominant factors: air temperature, humidity, radiant temperature, wind speed, metabolism, and clothing. However, the present urban thermal environment cannot be improved completely merely by a single idea, and large-scale and complex urban planning projects are required.

In this study, the control of the directional reflection by a sheet material with an unevenly shaped surface was considered, and the characteristics of directional reflection were evaluated with measurement in the laboratory and the field and with numerical computation. The effect of highly reflective materials on human thermal comfort was considered for what?.

2. Measurement on reflection directivity and solar reflectance of a sheet with an uneven surface

Reflection directivity measurements of sheets of polyvinyl chloride resin were made. Two kinds of samples with flat and structurally uneven surfaces were prepared for the measurement. The size of each sample was 150 mm^2, and the materials were the same kind. Figure 1 shows the surface pattern of the uneven sample. The unit pattern size was 10 mm^2, and a wider surface element existed in the unit. Figure 2 shows a schematic drawing of the apparatus for measurement of directional reflection strength. A high-intensity discharge lamp was used as a light source. The lamp was fixed above the measuring apparatus, and the light was collimated with a Fresnel lens. The Zenith angle θ_i and azimuth angle ϕ of the incident ray onto the sample surface could be specified by rotating an arm table fixed at the end with a hinge and another table on the arm of the table for fixing the sample. Directional reflection strength was measured with a photodiode module. As shown in Figure 2, reflection from the direction of angle θ_r was measured by changing the angle by moving an arm with a photodiode module installed at the end. The field of view of the photodiode was 8.3°, and the distance between the diode and measuring sample was 250 mm. The observed area of the module was a circle with a diameter of 37 mm and $\theta_r = 0°$, and it spread wider in an elliptical shape with an increase of θ_r.

Figures 3 and 4 show directional reflection strength profiles concerning reflection angle θ_r and azimuth angle ϕ for the samples in the cases wherein the incident angle θ_i was fixed at 0° or 30°. The ordinate in each figure shows the value of the photodiode module output divided by $\cos\theta_r$ to eliminate the influence of the increase of the observed surface by θ_r, and the value is defined as the directional reflection strength. As shown in Figure 3, the directional reflection strength profile in the azimuth direction was almost uniform in the case of $\theta_i = 0°$. However, that of $\theta_i = 30°$ changed with the azimuth angle. In Fig. 4, the left side of the figure corresponds to the result of the flat-surface sample, and the right side to that of the uneven-surface sample. It was found that the reflection directivity in reflection zenith angle θ_r existed for both incident angle conditions. As shown in the left side of the figure, maximal directional reflection strength is observed near the specular reflection angle for each incident angle θ_i. However, at the right side of the figure, maximal strength is observed in the direction of the specular reflection angle. The directional reflection strength of the uneven surface was smaller than that of the flat surface.

Field measurement of the solar reflectance of two kinds of sheet sample with flat and uneven surfaces was performed in an open space on the terrace of Bldg. A5 at Osaka Prefecture University. The roof surface was covered with tiles. The sizes of the flat and uneven samples were 700 mm^2 and 600 mm^2, respectively. The net radiometer was located above the sample sheet, and global solar radiation and reflected solar radiation were measured every 5 s. The apparent solar reflectance $\rho_{m\text{-}app}$ calculated with the ratio of global and reflected solar radiations including the influence of reflection from surroundings. To eliminate the effect, the solar reflectance of the roof surface without sample sheet ρ_r was measured. This value was considered to be true reflectance because the roof surface was comparatively wide and uniform. The true solar reflectance of sample sheet ρ_m could be calculated with the roof reflectance and geometrical factor of the sample sheet from radiometer F with the following Equation.

$$\rho_{m} = \{\rho_{m\text{-}app} - (1-F)\rho_r\}/F \tag{1}$$

The measured solar reflectance of the flat and uneven sheets were 0.319 and 0.311, respectively. In the solar radiation condition of the day, whose maximum altitude was 34.2°, there was almost no difference of reflectance between the two sheets.

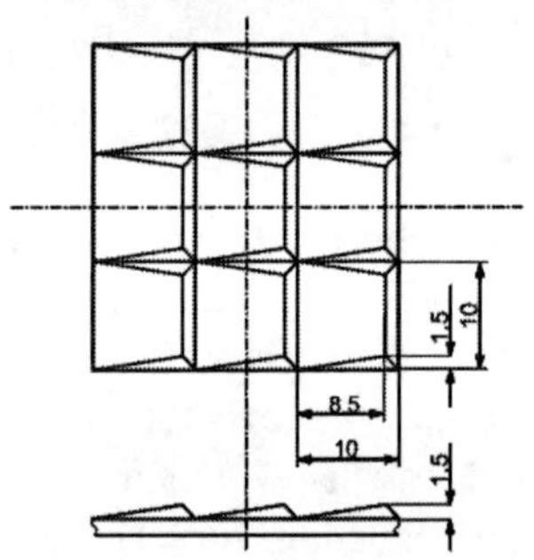

Figure 1. The surface pattern of uneven sample

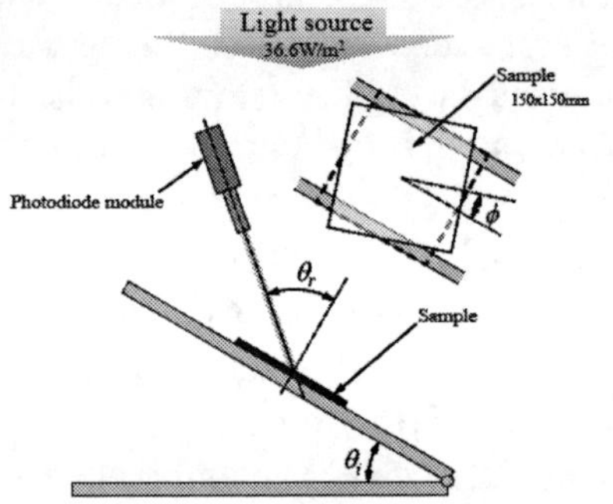

Figure 2. Schematic of the apparatus for measurement of directional reflection strength

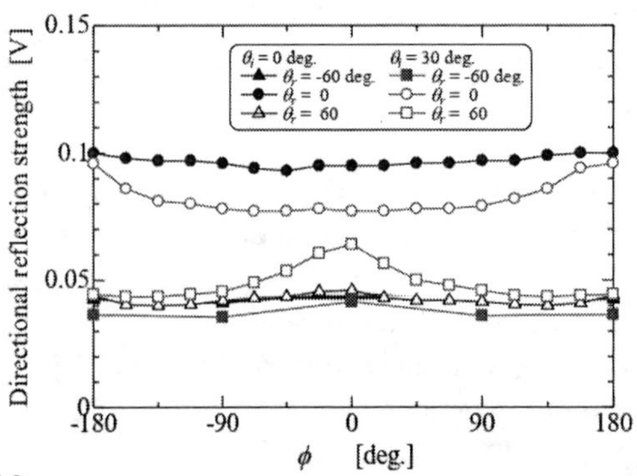

Figure 3. Directional reflection strength profile in azimuth angle

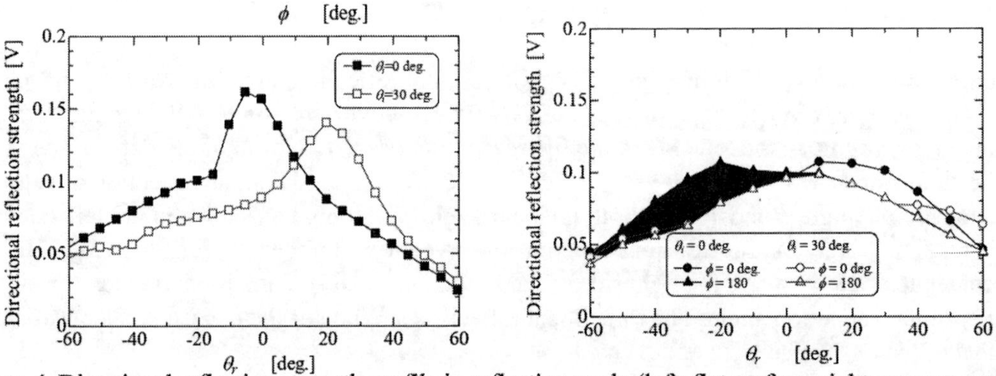

Figure 4. Directional reflection strength profile in reflection angle (left: flat surface, right: uneven surface)

3. Numerical analysis for reflection directivity of uneven surface

Numerical analysis based on a ray-tracing method was performed for evaluation of reflection directivity on the sample with an uneven surface. The conditions of solar radiation, incident angle, total radiation, and spectrum were specified, and the conditions of the sample surface, spectral absorptance, and the ratio of specular and diffuse reflection to the total reflection were specified. In the analysis, multireflection between the surface elements constructing the uneven surface was also considered. The incident position and wavelength of the beam element, reflection/absorption on the surface element, and direction of reflection in the case of diffused reflection were decided according to the Monte Carlo method.

Numerical analysis was performed while assuming the solar radiation at Osaka on August 1. The solar altitude and azimuth and the spectrum of direct solar radiation were calculated every 1 h employing the Bird model [10]. The shape of the sheet surface was assumed to be the same as that of Fig. 1 used in the measurement. The analysis was performed for the conditions of 0- and 2-mm height. The condition of 0-mm height corresponded to that of the flat surface. In the analysis, the sample was located such that the widest surface element faced south. The spectral absorptance was assumed to be gray, and the value was considered to be 0.7 uniformly in wavelength.

The results were evaluated for hemispherical reflectance and the ratio of specular reflection or retro reflection to solar reflection every 1 hour. Each ratio was defined as the total energy of the reflected beam within the cone whose axis represented the exact direction of reflection based on the horizontal surface, and the opening angle of 60° was divided by the total energy reflected in the hemispherical direction. In the overlapping area between two cones, the beam reflected into half the area of the incidence side was counted as retro reflection and the rest as specular reflection. Figure 5 shows the results at the condition wherein all surface elements were perfectly diffuse reflective surfaces. This figure shows the effect of the unevenness of the surfaces on the hemispherical reflectance and reflection directivity. In the figure, the left ordinate corresponds to hemispherical reflectance and the right ordinate to the ratio of specular reflection or retroreflection. As shown in the figure, hemispherical reflectance tended to be smaller as the unevenness increased, which is similar to the experimental result in Figure 4. Almost no difference between the ratios of specular and retro reflection on a flat surface existed. A slight difference on an uneven surface existed between the morning and evening. Figure 6 shows the results of changing the ratio of specular reflection to total reflection for each surface element of an uneven surface, which was set to be 0, 5, 20, and 50%.

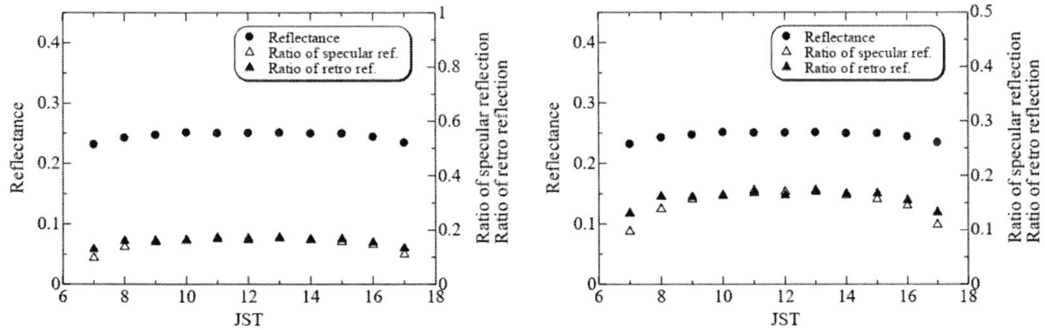

Figure 5. Effect of unevenness on hemispherical reflectance and reflection directivity (left: flat surface, right: uneven surface): all surface elements are a perfectly-diffused reflection

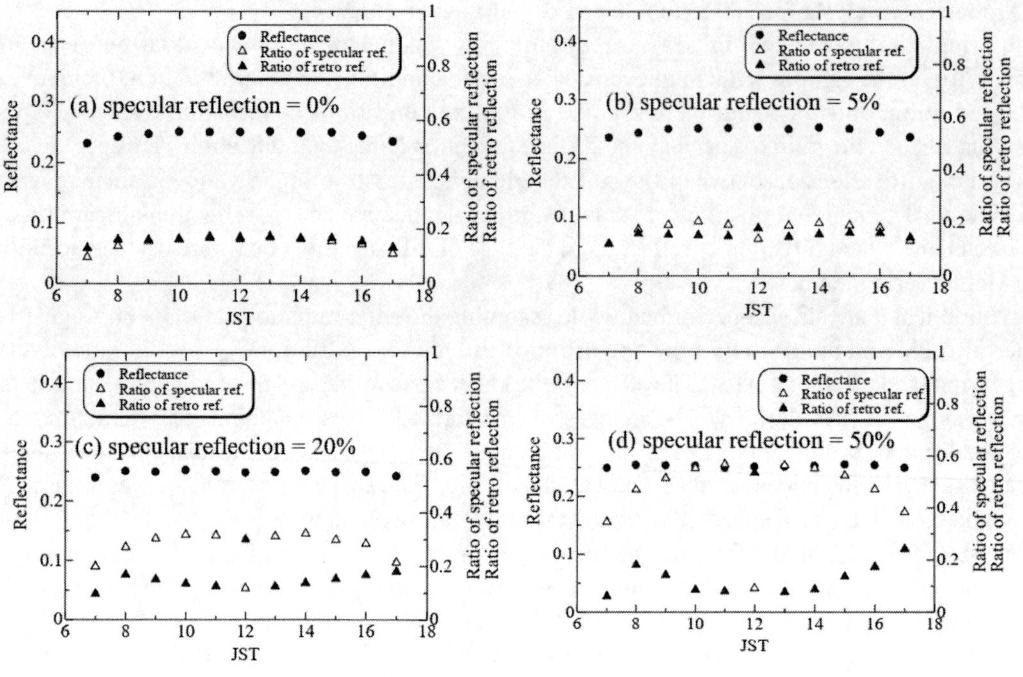

Figure 6. Effect of specular reflection of an uneven surface on reflection directivity

The effect of the ratio of specular reflection on hemispherical reflection was not observed. At noon, when the incident direction almost agreed with the normal direction of the widest surface element facing south, the ratio of retroreflection became larger with the increase of specular reflection of the surface element. These results indicate that it may be possible to control the reflection directivity by changing the unevenness of the sheet surface and specular reflection of the surface elements constructing the uneven surface.

4. Effect of surface treatment on human thermal comfort

An attempt was made to describe the thermal environment in human living spaces by combining the characteristics of highly reflective materials with a human thermal-comfort index. Any heat gain or loss from a thermally neutral state was considered to be a hot or cold sensation of discomfort. The human thermal load Q was calculated from Equation (2) as the remaining amount of each energy balance item [11, 12]. It is objective and is the basis of the energy balance formula for the human body incorporating the amount of physiology.

$$Q = M - W + R_{net} - E - C, \quad (2)$$

where M is metabolism [W/m^2], W is workload [W/m^2], R_{net} is net radiation [W/m^2], E is latent heat loss [W/m^2], and C is sensible heat loss [W/m^2]. The relationship between human thermal load and thermal sensation based on the American Society of Heating, Refrigerating and Air Conditioning Engineers (ASHRAE) standard was expressed using the least mean square approximation for the numerical analysis as:

$$(\textit{Thermal sensation}) = 0.0146 \times (\textit{Human thermal load}) - 0.748 \quad (3)$$

For the human model, the 65MN model [13] was modified for outdoor use [11]. The 65MN model is a thermoregulation model and was used for estimating skin temperatures. This is because the human thermal load was calculated from skin temperatures, clothing insulation, and environmental conditions.

It was assumed that various materials of interest were placed horizontally on the ground surface in a wide and open space. A human was assumed to be standing in the south. One-dimensional heat transport was used for thermal environmental analysis [14]. The earth's surface heat was focused on, and surface temperature, air temperature, and humidity were calculated. The wind speed was constant at 0.5 m/s. For the urban space, typical surface treatments, such as water-retentive material and asphalt concrete, were also considered. The details of each case are shown in Table 1. The results of time average values of air temperature, human perception based on human thermal load, thermal sensation, sensible heat flux, and latent heat at 12:00-2:00 p.m. JST at Osaka on August 1 are shown in Table 2. The average values of thermal sensation shown were obtained using Equaton (3). The human considered was 1.72 m in height and 56 kg in weight. The human was standing ($M = 80\ W/m^2$) with 0.68 clo of clothing insulation, and he did not perform external work.

Because water-retentive material holds water and this water vaporizes slowly, the temperature of the material remained relatively low for a long time. The degree of heat storage decreased, and this material contributed to thermal environment improvement. In fact, the surface and air temperatures were lower than that of asphalt concrete, which had the same reflectance because of the latent heat component. Because the material exhibited relatively high reflectance, the amount of heat in terms of radiation that was absorbed on the earth's surface was small, and this material could reduce the environmental thermal load. The primary cause of human thermal load can be seen in direct and reflected solar radiation represented in Figure 7. As the reflectance became higher, the amount of reflected solar radiation became larger, and the human was subjected to more heat. This is why the human felt hot, even when the temperature levels around him were lower. Because the highly reflective material's effect of reducing environmental thermal load was known, the best way to achieve a human- and the eco-friendly surface was to cut the solar radiation reflected into the human. Cases 3–7 represent the investigation based on this idea. When the material had the same reflectance, human thermal comfort was determined by retroreflection (Cases 3–5).

Table 1. Properties of material

Case	Material	Evaporation ratio	Solar reflectance	Retro reflectance
1	Asphalt	0	0.10	0
2	Water retentive	0.15	0.10	0
3	Highly reflective	0	0.30	0
4	Highly reflective	0	0.30	0.2
5	Highly reflective	0	0.30	0.5
6	Highly reflective	0	0.25	0
7	Highly reflective	0	0.20	0

Table 2. Results of each case (*thermal sensation is normally expressed from -3 to 3)

Case	Air Temperature °C	Surface Temperature °C	Thermal Sensation N.D.	Sensible heat W/m²	Latent heat W/m²
1	31.6	45.9	2.7	146	0
2	30.9	44.4	2.6	135	53
3	30.6	41.8	<<3 (3.4)*	114	0
4	30.6	41.8	<<3 (3.2)*	114	0
5	30.6	41.8	2.7	114	0
6	30.8	42.8	<<3 (3.2)*	122	0
7	31.1	43.8	≈3 (3.0)*	130	0

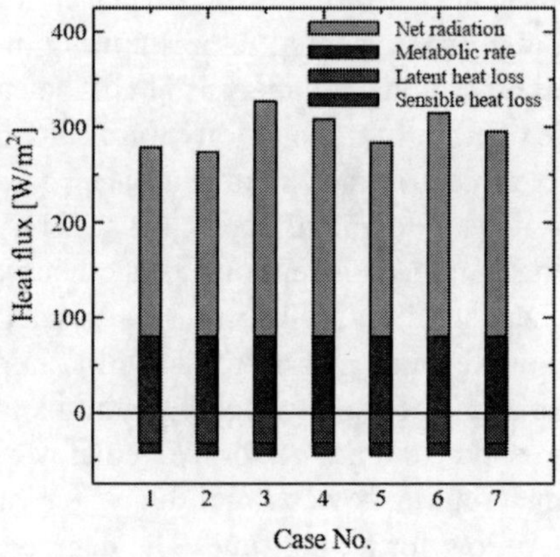

Figure 7. Heat flux components of the human body

The improvement effect in the case of 50% retro reflection (30% reflectance, case 5) was larger than that in the case of 20% reflectance (case 7) based on human thermal sensation.

5. Conclusions

Measurements and numerical analysis for sheet materials having uneven surfaces were performed to evaluate reflection directivity. As a result, it was possible to measure the reflection directivity given to the sheet material surface by constructing an uneven surface and changing the reflection directivity of each surface element.

The effect of surface treatment was considered based on human thermal comfort, and the following results were obtained. (1) Radiation has a strong impact on human thermal comfort, and reduction of radiation is a method to establish a better thermal environment. (2) The use of highly reflective materials can enable improvement, and retroreflection is an important concept to be considered.

References

[1] Sakai H Emura K Igawa N and Iyota I 2012 *J. Heat Island Inst. Int.* **7-2** 218
[2] Sakai H and Iyota H 2017 *Int. J. Thermophys.* **38** 131
[3] Iyota H Sakai H Emura K Igawa N Shimada H and Nishimura N 2012 *J. Heat Island Inst. Int.* **7-2** 213
[4] Rossi F Castellani B Presciutti A Morini E Filipponi M Nicolini A and Santamouris M 2015 *Applied Energy* **145** 8
[5] Morini E Castellani B Presciutti A Filipponi M Nicolini A and Rossi F 2017 *Energy & Build.* **139** 690
[6] Harima T and Nagahama T 2017 *Solar Energy* **148** 177
[7] Harima T and Nagahama T 2017 *Solar Energy* **148** 164
[8] Mauri L Battista G Vollaro E and Vollaro R 2018 *Solar Energy* **171** 150
[9] Qin Y Liang J Tan K and Li F 2016 *Solar Energy* **133** 172
[10] Bird R E 1984 *Solar Energy* **32** 461
[11] Shimazaki Y Yoshida A and Kinoshita S 2009 *Tran. Japan Soc. Ref. Air Cond. Eng.* **26** 113
[12] Yoshida A Hayashi D Shimazaki Y and Kinoshita S 2019 *Architectural Science Review* DOI: 10.1080/00038628.2019.1597678
[13] Tanabe S Kimura K Nakao J Ozeki Y and Konishi M 2002 *Energy & Build.* **34** 637
[14] Shimazaki Y Yoshida A Suzuki R and Kinoshita S 2009 *Proc. 7th Int.l Conf. Urban Climate* 4-35

IOP Conf. Series: Materials Science and Engineering **886** (2020) 012037 doi:10.1088/1757-899X/886/1/012037

Improvement of Background Oriented Schlieren Method Focused on Amplitude of Wavelet Transform

Miyaoku Koki[1], Fukuoka Hiroshi[1]*, Nakamura Shigeto[1], Yao Masanori[1] and Hiro Kazuki[1]

[1]National institute of technology, Nara college, 22 Yamatokoriyama, Nara, 639-1058, Japan

* Corresponding Author: fukuoka@mech.nara-k.ac.jp

Abstract
The BOS method is a visualization method of density variation in the flow field. This method is simpler and cheaper to visualize than the Schlieren method. The primary purpose of this study is to propose WA-BOS (a simpler BOS method focusing on the amplitude of the Wavelet transform). Output images by WA-BOS and WP-BOS (BOS method focused on the phase of the Wavelet transform) and the Schlieren image were compared to verify the validity of the output images. Also, output images by the captured image lost periodic information partially were verified, in order to confirm the validity of the image processing for the deviation out of the domain of phase. The targets of visualization are the supersonic jet and the shock wave ejected from the shock tube. The measurement system consists of an LED, a sine pattern background image and a high-speed camera. As a result, the supersonic jet and the shock wave ejected from the shock tube are visible in the visualized image that was obtained by WA-BOS and WP-BOS. Moreover, it is found that WA-BOS can visualize a supersonic phenomenon without containing an error.

1. Introduction

In recent years, many types of researches focusing on compressible fluid dynamics have been conducted in various fields such as aerospace mechanics [1]-[3], nuclear engineering [4], medical physics[5,6], and materials sciences[7,8], which are developing rapidly. The visualization technology is the most important technology to understand these phenomena. Representative examples are Schlieren and interferometry. Unlike these, the BOS (Background Oriented Schlieren) method, which is a unique visualization method, is proposed by Meire[9]. The visualizing principle of the BOS method is similar to the Schlieren method, which utilizes the change of the fluid density. One of the characteristics of the BOS is the ability to visualize by using straightforward experimental apparatus. The BOS system consists of a background image, a light and a camera. This is a straightforward system compared with the Schlieren method. The schlieren method requires a lot of complex components, for example, concave mirrors, pin-hole and a knife-edge. In the Schlieren method, much cost is required to expand the area of visualizing. While in the BOS, if the background image can be taken, there is no limitation on the physical size of the visualization target. This is because the BOS method depends on the area of the background image. Many experiments for visualizing have been performed on a real scale [10]-[12]. Thus, to visualize with a simple system without depending on the size of the experimental apparatus is a major characteristic of the BOS. In the BOS, image processing is required to obtain a visualized image. This image processing is said to be a disadvantage of the BOS, as it is not required in other visualization methods like the Schlieren method.

Various image-processing methods have been proposed to increase the accuracy and efficiency of the BOS method. An output method of the visualized image by the BOS method shown in Figure 1. A positional deviation is calculated by subtracting an image having no density change from an image for measurement in which light is refracted due to the density change. The visualized image is a result obtained by reconstructing the deviation of an image. Random dot patterns were used for background images in the calculation of displacement of the background image, and PIV is used for image processing. Ota et al. [13,14] proposed CGBOS method using a color grid pattern as a background image. CGBOS method does not require a reference image and can be visualized only with captured images. Therefore, it is possible to visualize even when the position of the model is largely shifted before and after the experiment. Akatsuka[15] proposed the Wavelet-based BOS method. This technique provides a Schlieren image using the Continuous Wavelet transform for the periodic background pattern. Our studies of WA-BOS (BOS method focused on the amplitude of the Wavelet transform) have proven that it can be obtained a visualized image. However, the verification of the characteristics of the visualized image obtained by WA-BOS and the comparison of the characteristics of WA-BOS and WP-BOS (BOS method focused on the phase of the Wavelet transform) were not sufficient.

The purpose of this study is to clarify the characteristic of WA-BOS. In order to clarify the characteristics, the output images by WA-BOS and WP-BOS were compared with the Schlieren image and considered. Targets of visualizing are the supersonic jet and the shock wave that is ejected from the shock tube [16,17]. The experimental apparatus consists of the background image that has the periodic pattern, the LED and the high-speed camera.

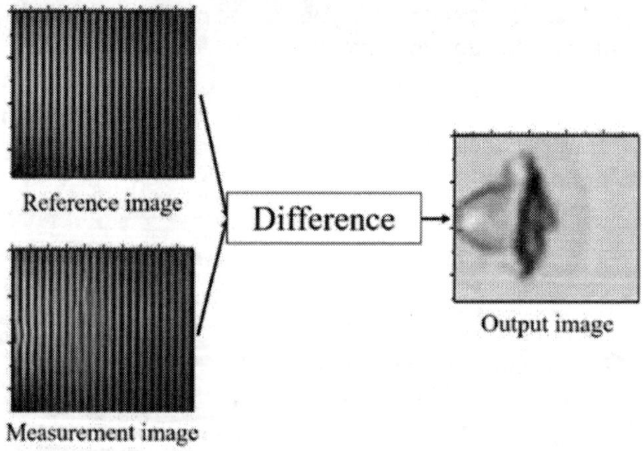

Figure 1. Schematic image of the BOS method

2. Principle

2.1 Principle of the background-oriented Schlieren method

Figure 2, shows the principle of the BOS method and measurement system. The measurement system consists of a background image, light and a camera. The red line and the blue line in the figure are light paths from the background image. The red line is the light path in the case of no density gradient. The blue line is the refracted light path by the density gradient. In the BOS method, Δh is obtained by using image processing applications a captured image that is taken by a camera. In this paper, the image processing method is the main focus. The equation of change in refractive index by density change is given as

$$\int_0^{\Delta r} \frac{\partial n}{\partial z} dx = n_0 \frac{\Delta h}{L_B - \Delta r/2} \quad (1)$$

where Δh is the positional deviation amount, $\partial n/\partial z$ is the refractive index gradient, n_0 is the refractive index of air, L_B, L_C and f are the distance from the background image to the density change area, from the measurement target to the camera and distance of focal length. Δr and ε are the width and refraction angle of the density change area. Also, the equation of the relationship between refractive index and density is given as

$$n = G\rho + 1 \quad (2)$$

where n, ρ and G are the refractive index, the density of air and the Gladstone-Dale constant respectively. The density information can be determined quantitatively from these relational equations.

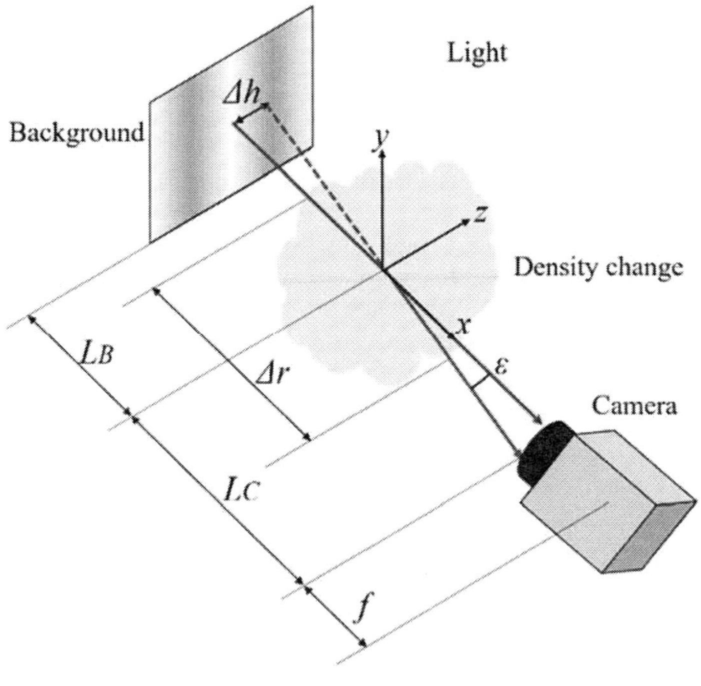

Figure 2. Optical set up of the BOS method

2.2 The WA-BOS method and the WP-BOS method

In this study, the Wavelet transform is applied to the BOS method. The Wavelet transform is spatial frequency analysis. The most major method of frequency analysis is FFT. The Fourier transform maps the target signal to the basis of sine or cosine. In this analysis, the periodic information is obtained clearly, but time information is lost. That is because of using the signal on the assumption that it repeats infinitely. On the other hand, the Wavelet transform uses a Mother wavelet that is temporally localized. Thus, the Wavelet transform can obtain the frequency band that has information of a specific time. Many Mother wavelet function is proposed. Gabor wavelet that is a good balance between the resolution of time and frequency is used in this study. The amplitude and phase can be obtained by applying the Wavelet transform for the target signal. The Wavelet transform is applied to the BOS method uses this information to visualize the flow field. In the BOS method, a positional deviation is needed to visualize quantitatively. That deviation can be calculated by a phase. The visualized image from the phase of the Wavelet transform is called the WP-BOS. In the WP-BOS, an image equivalent to the Schlieren image can be obtained. Also, the visualized image from the amplitude of the Wavelet transform is called the WA-BOS. Thus, In WA-BOS, an image equivalent to the Shadow Graph image can be obtained.

3. The experimental setup and image processing method

Figure 3 shows the experimental apparatus which consists of the shock tube and the BOS system. The jet flowing towards the test section from the shock tube is measured by the BOS system. The measurement system of the BOS method consists of the background image, light and high-speed camera. The background image has the periodic pattern that the period is 2 mm. The LED light is SELMIC SLS-75-08W and the high-speed camera is Vision Research Phantom v7.3.

Figure 4 shows the shock tube which is an experimental device that is suitable to study for a characteristic of the supersonic jet and the shock wave. The shock tube has a long tube of constant cross-section, which is divided into the high-pressure chamber and low-pressure chamber by diaphragm. In Figure 4, the x-axis is parallel to the jet flow, the shock tube diameter D = 10mm, the high-pressure chamber length L_h = 100mm and the low-pressure chamber length L_l = 10 mm. P_h and P_b are the pressure of the high-pressure chamber and the low-pressure chamber. By breaking the diaphragm, the high-pressure air was discharged. Then, the supersonic jet and the shockwave are formed by discharged air. Both supersonic phenomena are visualized by the BOS system.

Figure 5 shows the image processing procedure to visualize the supersonic phenomena. Time t = 0μs is one frame before immediately after the jet is ejected. Figure 5(a) is the reference image at t = 0μs. Figure 5(a') is the measurement image at t = 195μs. Figures 5 (b) and (b') show the luminance on y/D = 0 in Figures 5 (a) and (a'). Figures 5 (c) and (c') show the amplitude that is obtained by using the Wavelet transform for the luminance information of (b) and (b'). The y-axis shows the amplitude of the Wavelet transform. Figure 5 (d) shows the value obtained by subtracting (c) and (c'). The y-axis shows the amplitude difference. This image processing has performed all rows of images and is reconstructed to obtain the output image. Figure 5 (e) is the output image that has finished a process of image processing.

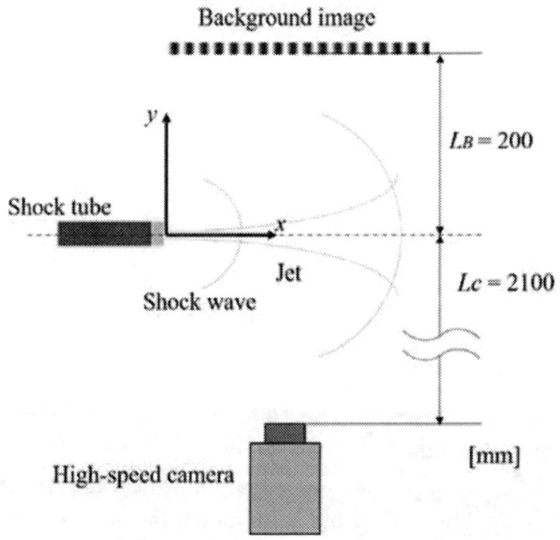

Figure 3. The experimental apparatus

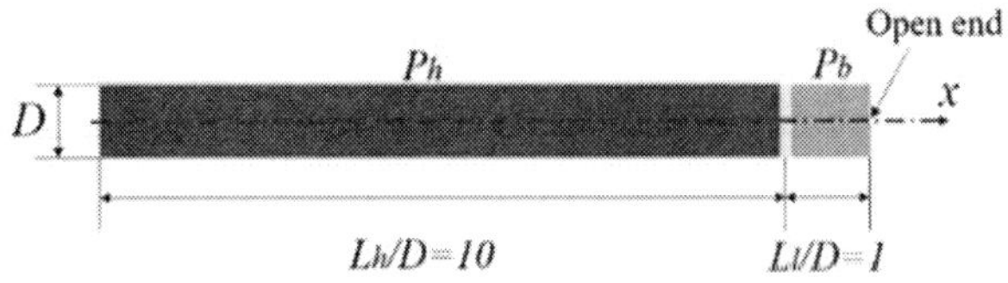

Figure 4. The schematic image of the Shock tube

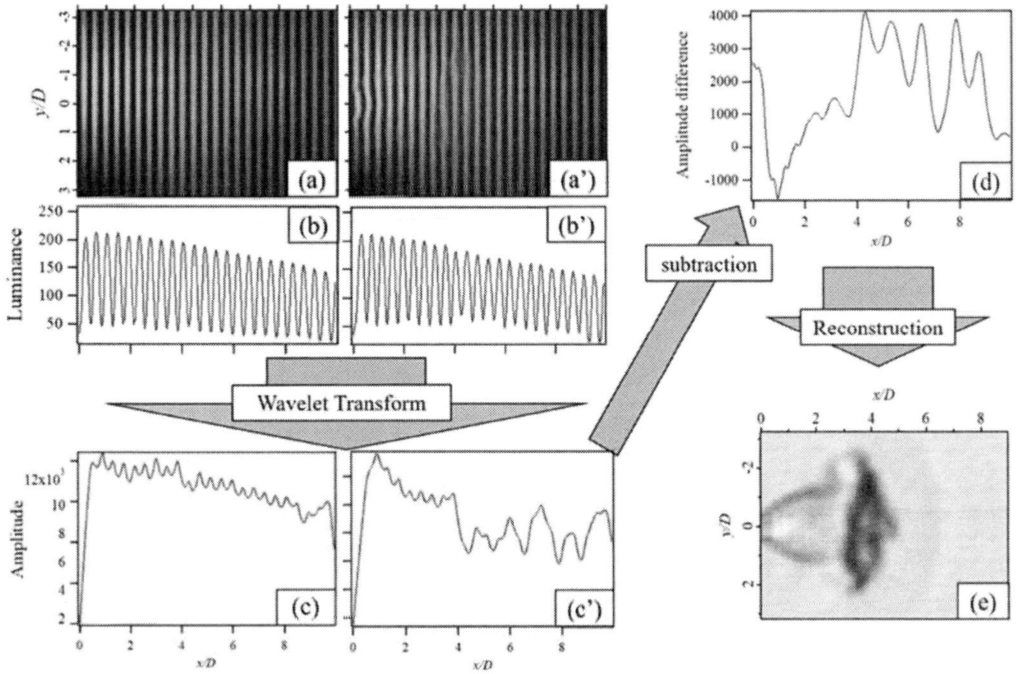

Figure 5. The visualizing process for WA-BOS and $P_h/P_b = 48.6$

4. Results and Discussions

Figure 6 shows a captured image of the supersonic jet ejected from the shock tube and the output image. The thickness of the diaphragm is 100μm and P_h/P_b is 50.2. Figures 6(a), (b) and (c) are the reference images at t = 0μs, the measurement image at t = 145μs and output image by WA-BOS respectively. In Figure 6(b), there is a large distortion of the background image at x/D = 0~1.5 and y/D = -0.6~0.6. The distortion amount of the background image confirms to the density gradient that is formed by the supersonic jet from the shock tube. In Figure 6(c), there is the head of the shock wave at x/D = 6.8, y/D = 0 and the head of the supersonic jet on x/D = 5.0, y/D = 0. While applying the image processing of this study to the captured images, the shock wave and the supersonic jet become visible.

The output images by WA-BOS and WP-BOS and the Schlieren image are compared to verify the validity of the output images. Figures 7 (a), (b) and (c) are the output images by WA-BOS and WP-BOS for P_h/P_b = 50.2 and the Schlieren image for P_h/P_b = 40.8, respectively. In Figures 7 (a) and (b), there is the head of the shock wave at x/D = 6.8, y/D=0 and the head of the supersonic jet at x/D = 5.0, y/D = 0. In Figures 7(c), there is the head of the shock wave and the head of the supersonic jet at the same points of (a) and (b). Therefore, the output images of WA-BOS and WP-BOS are valid. The shock wave in Figure 7 (a) is visualized more clearly than that in (b). This could be because of the differences in image processing between WA-BOS and WP-BOS. The phase is the position of a point in time on a periodic cycle. In the case of visualizing by using phase, that deals with the displacement of the background image as the vector that has luminance and information of position. While the amplitude is maximum displacement in a periodic signal and non-negative scalar. The vector deals with different values as a gradient. Yet, the scalar cannot do it. Therefore, it is obvious that the output image from the phase has less noise than the output image from the amplitude.

The wavelet transform deals with the argument of a complex function as a phase. The domain of the argument is limited ± π; it becomes discontinuous bordering on ± π. To calculate a phase difference requires phase unwrapping, in case of the deviation of background image exceeds ± π, it is not calculated correctly. Output images by the captured image lost periodic information partially are shown in Figure 8, in order to confirm the validity for the image processing for the deviation out of the domain of phase. The thickness of the diaphragm is 50μm and P_h/P_b is 23.5. Figures 8 (a), (b) and (c) are the measurement image at t = 145μs, the output image by WP-BOS and WA-BOS. In Figure 8 (a), the background image is lost on the range of x/D = 0~3 and y/D = 0.2~0.5. In Figure 8 (b), the pattern is confirmed in the same range; in (c) it is filled in black. The zebra pattern in Figure 8(b) shows the density gradient that does not exist the actual phenomenon. This is because the deviation exceeds the domain of phase. While the amplitude needs to calculate only the luminance at that pixel and does not depend on the periodic information, therefore, the advantage of WA-BOS is error resistance in image processing and able to visualize the captured image that does not depend on the background image.

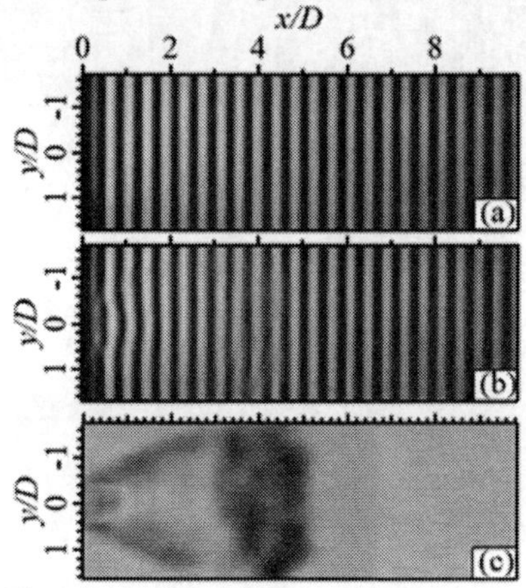

Figure 6. Visualization of captured images for P_h/P_b = 50.2.
(a) Reference image t = 0μs, (b) Measurement image t = 145μs, and (c) WA-BOS image.

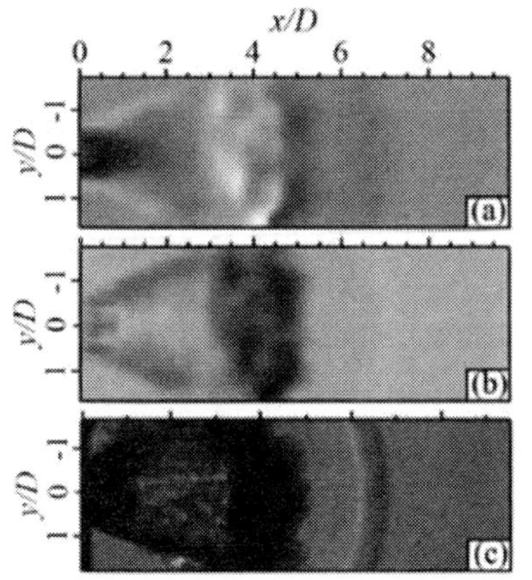

Figure 7. Output images (P_h/P_b = 50.2) compared with the Schlieren images (a) WP-BOS image, (b)WA-BOS image, and (c) Schlieren image (P_h/P_b = 48.3)

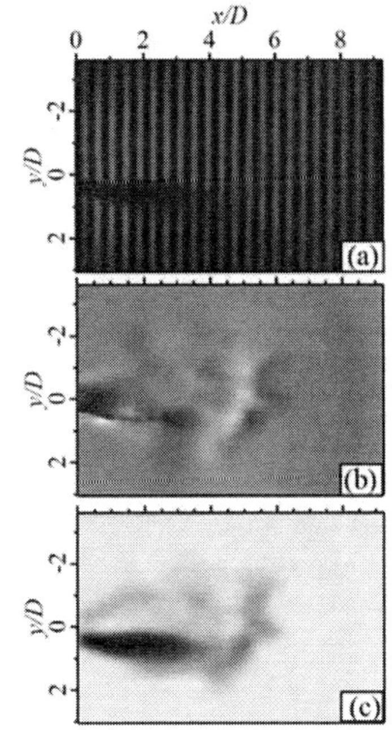

Figure 8. Visualization of defect images for P_h/P_b =23.5. (a) Measurement image with a lost background, WP-BOS image, and (c) WA-BOS image

5. Conclusions

The purpose of this study is to clarify the characteristic of WA-BOS (BOS method focused on the amplitude of the Wavelet transform). The visualized image was obtained by applying the Wavelet transform as an image processing. Targets of visualizing are the supersonic jet and the shock wave ejected from the shock tube. Output images by WA-BOS and WP-BOS (BOS method focused on the phase of the Wavelet transform) and the Schlieren image were compared in order to verify the validity of the output images. Also, the output images by the captured image lost periodic information partially were verified, in order to confirm the validity for the image processing for the deviation out of the domain of phase. The results are as follows.

1. WA-BOS and WP-BOS can obtain the visualized image of the supersonic jet and the shock wave ejected from the shock tube.
2. The head position of the supersonic jet and the shock wave in the output image by WA-BOS and WP-BOS correspond to the Schlieren image. Thus, visualizing by WA-BOS and WP-BOS is valid.
3. The output image by WP-BOS has less noise than the output image by WA-BOS, which is visualized more clearly.
4. In WP-BOS, an error occurs in the case of the periodic information of the background image which is lost,
5. In WA-BOS, no error was detected in the amplitude calculation. Therefore, WA-BOS can obtain a visualized image faithful to the actual phenomenon.

IOP Conf. Series: Materials Science and Engineering **886** (2020) 012037 doi:10.1088/1757-899X/886/1/012037

References

[1] Okubo M Toyota A Yamashita H Ogawa T Obayashi S Shimizu K Suzuki K Matsuda A and Sasoh A 2011 *JSASS* **59**(688) 119
[2] Toyota A 2009 *FDC/ANSS* **41** 71
[3] Peixue J Zhiyuan L Zheng H Yanbing X and Yinhai Z 2019 *Int. J. Heat Mass Transf.* **129** 965
[4] Kawashima T Shigemori K Hironaka Y Kato H Sakaiya T Batani D Jakubowska K Trela J Sakaki T Le E Bel and Maheut Y 2016 *Meeting abstracts of JPS* **71**(1) 21 CB-7
[5] Nakagawa K Tsukamoto A Arafune T Liao H Kobayashi E Ushida T and Sakuma I 2013 *JBSE* **8**(1) 17
[6] Tachikawa R Tsukamoto K Monden Y Nakagawa K Ando T Kobayashi E Ushida T and Sakuma I 2015 *Adv. Biomed. Eng.* **4** 96
[7] Kinoshita T Fukuoka H and Umezu I 2018 *Mater. Sci. Forum* (Trans Tech Publications) **910**
[8] Katayama K Horai Y Fukuoka H Kinoshita T Yoshida T Aoki T and Umezu I 2018 *APA* **124**(2) 150
[9] Meier G 2002 *Exp. Fluids* **33**(1) 181
[10] Kolja K Erik G Friedrich L and Markus R 2007 *Exp. Fluids* **43**(2-3) 233
[11] Kudo M Kazama M and Abe I 2009 *JAFST* **14**(2)93
[12] Michael A. Hill, and Edward A. Haering 2017 *Exp. Fluids* **58.1** 4
[13] Ota M Hamada K Kato H and Maeno K *J. Vis. Soc. Jpn.* **31**(9) 51
[14] Ota M Kato H and Maeno K 2012 *Int. J. Aero. Inn.* **4** 1
[15] Akatsuka J Nagai S and Honami S 2011 *Trans. JSME* B **77**(784) 2391
[16] Fukuoka H Yaga M and Takiya T 2008 *JFST* **3**(7) 881
[17] Takemura R Fukuoka H Enoki S Nakamura S and Hiro K 2018 *Mater. Sci. Forum* (Trans Tech Publications) **910**

IOP Conf. Series: Materials Science and Engineering **886** (2020) 012038 doi:10.1088/1757-899X/886/1/012038

Numerical Simulation of Flow past Glove-shaped Formers in Latex Dipping to Consider Tip Effect to Free Surface Flow

W Koranuntachai[1,2], T Chantrasmi[1,2] and U Nontakaew[1,2]

[1]Development of Machinery and Industrial Equipment Research Center (DMIE), King Mongkut's University of Technology North Bangkok, 1518 Pracharat1 Road, Wongsawang, Bangsue, Bangkok 10800 Thailand.
[2]Department of Mechanical & Aerospace Engineering, Faculty of Engineering, King Mongkut's University of Technology North Bangkok, 1518 Pracharat1 Road, Wongsawang, Bangsue, Bangkok 10800 Thailand.

Abstract
Latex-dipping process in a medical-gloves production line is one of the most critical components that affect the final product quality. In this process, glove-shaped formers are moved through an open-channel latex-dipping tank. Since the gloves have long arms and the latex compound is opaque, it is difficult or impossible to directly observe the flow beneath the free surface. On the other hand, the surface wave and vortex shedding patterns are easily observed or measured on-site. Therefore, it will be beneficial to identify and establish a relationship between the observable surface quantities and the flow underneath. In this work, Computational Fluid Dynamics (CFD) is used to simulate a free-surface flow past a glove-shaped former in comparison to that over a vertical cylinder of the same diameter. The computational results showed that the vortex generation in the near wake is inhibited by the effect of the free surface, and reducing the vortex shedding and vorticity. The shedding from the fingertips has little to no effect on the surface. Most surprisingly, the free surface flow past a glove-shaped former in this setting does not exhibit a transient vortex-shedding pattern unlike the case with a vertical cylinder. This disparity among others makes a vertical cylinder a poor simplified surrogate model for the glove-shaped former.

Keywords: Latex Dipping, Free-Surface Flow, Computational Fluid Dynamics

1. Introduction

In a production line of the rubber medical gloves, the final products are the solid elastic gloves made from the liquid latex compound. In a continuous production line, the long conveyor chain carries an array of glove-shaped ceramic formers (molds) through a variety of processes such as former cleaning, former drying, coagulant dipping, latex dipping, gelling, vulcanizing, beading, and stripping [1–3]. One of the most important processes is the latex dipping (see Figure 1), the formers are dipped into latex compound in the latex dipping tank to create the thin coating layer of latex film around the formers through a chemical process called "Latex Film Formation" [4]. This liquid film would then be turned into rubber medical gloves in a later process.

During the latex dipping process, the formers are lowered into a flow of latex compound. Usually, there is one dipping tank on each side of the conveyor chain. Each tank usually has a slow-rotating propeller to drive the latex compound in a loop circulation with one long straight section into which the formers will be lowered. The latex compound flow is generally in the same direction as the movement of the formers. In this way, the relative velocity between the formers and the fluid is reduced. This is believed to improve the quality of the products and reduce the number of rejects.

Final product quality and rejection rate are affected by several fluid flow phenomena that could occur in the straight dipping section of a latex dipping tank. Certain realistic combinations of the properties of latex compound, geometrical dimensions, and the conveyor chain speed might result in turbulent flow in the tank. On the other hand, some other combinations might result in vortex shedding from the flow past an array of formers. Note that the driving propeller can control the speed of the latex flow. However, the turbulent channel flow happens due to high relative velocity between the latex and the stationary tank walls, while the vortex shedding is dependent on the relative velocity between the latex and the moving formers. Since the formers move at the conveyor speed, which in turn dictates the production rate of a continuous manufacturing line, it is not an easy problem to predict and optimize the latex dipping process.

Figure 1. Latex dipping process in a continuous production line

Even with an operating latex-gloves production line, the flow phenomena are difficult to observe due to the latex compound being opaque. In practice, only the free surface can be easily observed. Thus, this work is an attempt towards understanding the behaviors of the fluid flow in the latex dipping tank with an emphasis on observable free surface flow.

Historically, the free surface analysis was used to study the surface wave influencing the offshore structures (such as an oil drill). In the old days, this was done by using the mathematical theory [5–7]. In recent years, computational power and simulation software has been developed to be powerful enough to simulate such problems with more accuracy via numerical methods. In particular, Computational Fluid Dynamics (CFD) simulations are frequently employed to model complex multi-physics problems. Flow simulation with interfaces and free surface flows are more of a prevalent topic [8,9]. In many works, vertical cylinders were used as a representative solid structure within the free surface flow [10–12]. The transient free surface can clearly be shown and monitored [13,14].

This work presents CFD simulation of the free surface flow past one latex-gloves former. A similar flow past a vertical cylinder of the same diameter as the former was also carried out as a comparison reference. Note that flow in a real latex dipping tank will be more complicated due to its being a flow past an array of formers instead of a single one. Nonetheless, the present work can serve as a basis of understandings of the real flow. It will also answer the question of whether it is reasonable to estimate the geometry of a former with that of a cylinder.

This paper is divided into five sections, including this one. The next section formulates the model problems and describes the case studies. The numerical setup is described in the third section. The simulation results are presented and analyzed in the following section. Lastly, the conclusions are made in the final section.

2. Model problem

Figure 2 shows a simplified drawing of a latex dipping tank unit. It is designed for both sides of the conveyor chain (so only one unit is needed for one production line). On each side of the conveyor chain, an array of formers will be moving lengthwise in a single file. This type of former array configuration is generally called "single-former."

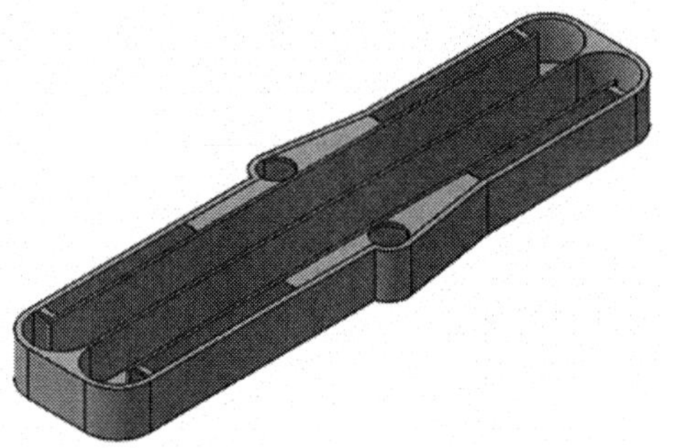

Figure 2. An example of a simple latex dipping tank design

The conveyor speed *(U)* and the dwell time (dipping time) are the main variables that determine the length of the latex dipping channel (L_x). In general, faster conveyor speed is desired for a higher production rate. A latex-dipping tank must be designed accordingly to work suitably at such a high speed, and the dwell time is the parameter that is directly related to the thickness of the rubber gloves, the dwell time between 8-12 seconds is generally enough to pass the rigorous testing requirements of the industry standards [15–16].

Conveyor speed of 0.35 m/s, which is typical of this type of production line, is used in this work. The latex-dipping channel is 40 cm wide, and the latex height is 50 cm from the channel bottom. The pressure gradient is calculated to be approximately 1.0 Pa/m.

Three material properties of the latex compound are needed for the analysis in this work: density (ρ), viscosity (μ) and surface tension (σ_s). Their values depend on the compound formula. In this work, nominal values of 1,000 kg/m^3, 200 cP and 44.2 mN/m are used, respectively. Note that the fluid is assumed to be Newtonian for the sake of simplicity. The air above the free surface is assumed to be at 25^{oC} and 1 atm.

For this work/paper, the former landing and departing regions are not considered and the flow is approximately as fully developed away from these regions. Figure 3(a) shows a schematic diagram of the simplified model problem and the flow direction. (Note that this is a positive flow direction. In certain scenarios, the latex compound can flow the other way in some regions.) As such, the left boundary will be called an inlet and the right boundary will be called an outlet.

3. Numerical setup

This section describes the second step in the proposed analysis. The CFD simulations based on the Finite Volume Method [17–20] were performed. The commercial software ANSYS Fluent was used. The Volume of Fluid (VOF) model [18] was selected for solving the interaction between two fluid phases. Only one former was drawn in the domain. The side and bottom walls were set as moving walls (with constant velocity) relative to the stationary former. The latex flow was expected to be laminar (possibly with laminar vortex shedding) at this Re of 122.

Two cases were simulated in this work (1) latex dipping flow past a vertical cylinder, and (2) latex dipping flow past a glove-shaped former. The diameter of the former's forearm and that of the cylinder are the same. The length of the former resembles the actual piece (3D-scanned CAD was performed from a real former) while the vertical cylinder spans the entire depth of the fluid domain.

Figure 4 shows the mesh for the first case (cylinder). The mesh is refined near the former surface as well as the interface between the two fluids (latex compound and air) in order to capture potential vortex shedding and the free surface accurately. The hexahedron mesh is used for the computational domain and the total number of cells is around 440,000 cells.

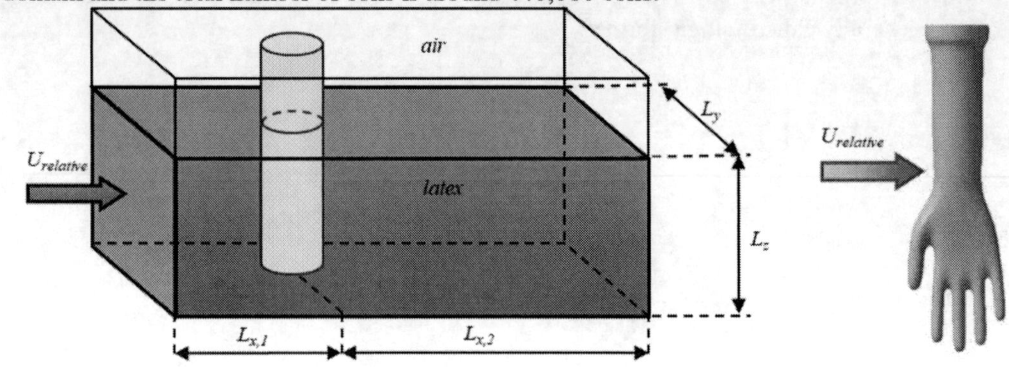

(a) (b)

Figure 3. A schematic diagram of the model problem

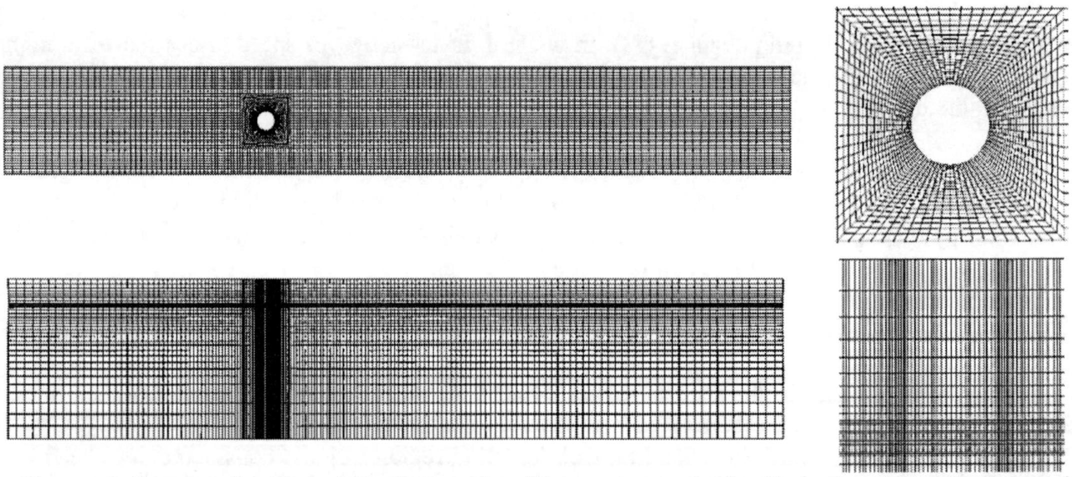

Figure 4. Computational mesh for free surface flow past a vertical cylinder in a latex dipping tank

Figure 5 shows the mesh for the second case (glove-shaped former). The meshing method far from the former is similar to the first case. However, the meshing of a latex-gloves former is much more difficult due to the complex shape of the former (and the tolerance of the 3D scan in the CAD file). Once again, the hexahedron mesh is used and the total number of cells is around 1,046,000 cells.

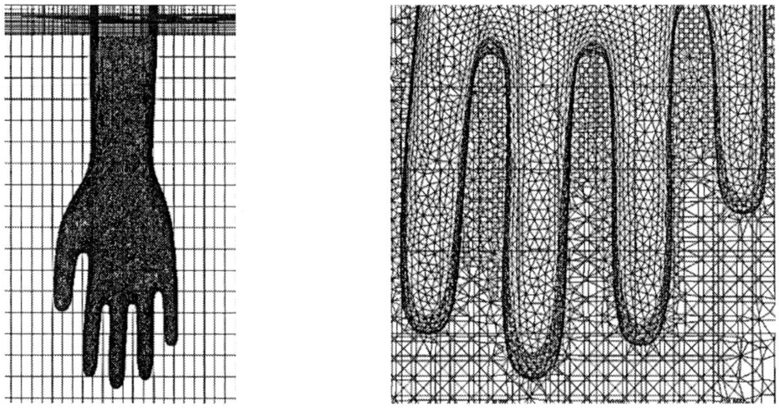

Figure 5. Detailed sections of the computational mesh for free surface flow past a glove-shaped former in a latex dipping tank

The main driving force for the flow comes from the propeller installed as part of the dipping tank. It creates the driving pressure difference which can be simplified to pressure gradient, $\partial P/\partial x = -\Delta P/L_x$ since it is assumed that the flow in this section is fully developed. When the moving frame of reference (fixed former) is considered, the velocity profile at the inlet is then:

$$u_{rel}(y,z) = U - u(y,z) \qquad (1)$$

The fully developed velocity profile $u(y,z)$ of an open channel flow can be obtained by solving a Poisson's equation with appropriate boundary conditions (three zero Dirichlet boundaries and one zero Neumann boundaries).

$$\nabla^2 u = -\frac{1}{\mu}\frac{\partial P}{\partial x} \qquad (2)$$

The fully developed velocity profile $u(y,z)$ of an open channel flow for latex compound can be obtained by solving a Poisson's equation with appropriate boundary conditions (three zero Dirichlet boundaries and one zero Neumann boundaries). The gauge pressure at the liquid inlet is set as $-L_x$ $\partial P/\partial x$. As for the air, the uniform velocity equal to the conveyor speed is used at the inlet with atmospheric pressure. No-slip boundary conditions were applied to the surface of the cylinder and former. The side and bottom boundaries are set as moving wall with (negative) conveyor speed, while the pressure outlet boundary conditions are all set to atmospheric (zero gauge pressure).

4. Results and discussion

It turned out that the flow over a vertical cylinder exhibits laminar vortex shedding as could be theoretically predicted at this Reynolds number, $\boldsymbol{Re_D = 122}$ (without considering the free surface and wall effects). Surprisingly, however, the flow over a glove-shaped former was steady without the transient vortex shedding. It would seem that the two flows were in different flow regimes and thus many quantitative comparisons would be quite meaningless. Nevertheless, the authors expect that when an array of formers are considered in a future, laminar vortex shedding in a more complex fashion would be present as observed from an actual latex dipping tank. Thus, in this work, the two cases were compared as if they both had transient laminar vortex shedding.

The measures for comparison of the two cases were (4.1) free surface visualization, (4.2) velocity field in the middle vertical plane, (4.3) vertical vorticity field at various depths, (4.4) free surface elevation vs. streamwise coordinate, and (4.5) Strouhal number and maximum sideway lift coefficient. For the first three measures, instantaneous fields will be shown. The details of the comparison are as follows.

41. Free surface visualization
Figure 6 shows the instantaneous free surface visualization of the two cases. Since the VOF method uses a continuous volume fraction field as a way to keep track of the interface between the two fluids, the volume fraction of 0.5 was selected as isosurface value to render this free surface. In the case of flow past a vertical cylinder, the wave profile exhibits a bow wave pattern in front of the cylinder and a clear vortex-shedding pattern in the wake region. In the case of flow past a glove-shaped former, bow wave pattern in the front and steady circulation regions in the wake region could be observed but the elevation levels were much less pronounced.

4.2 Velocity field in the middle vertical plane
Figure 7 shows interior flow structures in the fluid domain by way of velocity vector plots in the middle vertical plane. In case of the flow past a vertical cylinder, a vortex-shedding pattern can be seen all the way down to the tank bottom. It could also be observed that the pattern was less pronounced near the free surface due to the surface tension and air resistance. In case of the flow past the former, a large single recirculation region was formed. In this view, the recirculation region appeared as a triangle with the last fingertip as one vertex. No other visible flow structure could be observed.

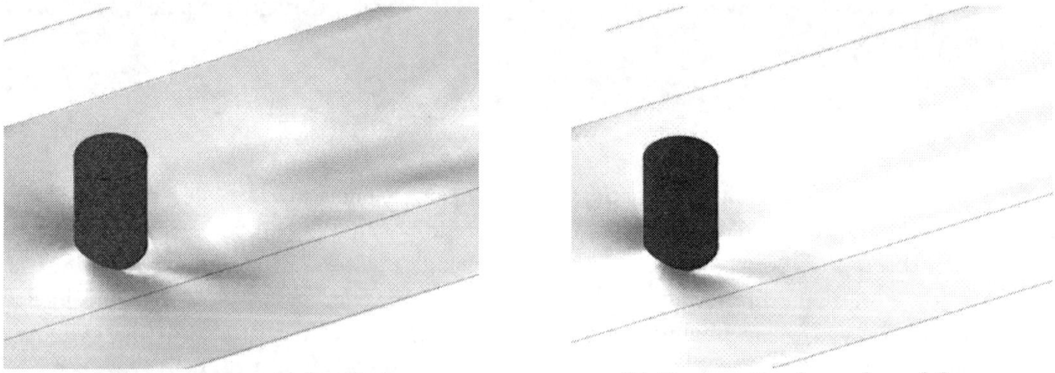

(a) flow past a vertical cylinder (b) flow past a glove-shaped former

Figure 6. The instantaneous free surface of the two cases

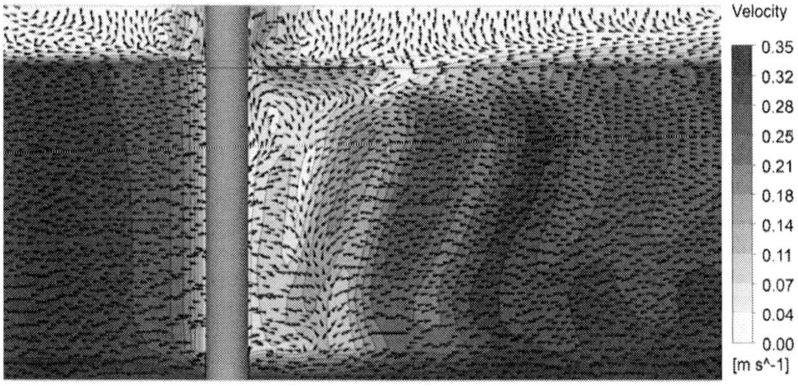

(a) flow past a vertical cylinder

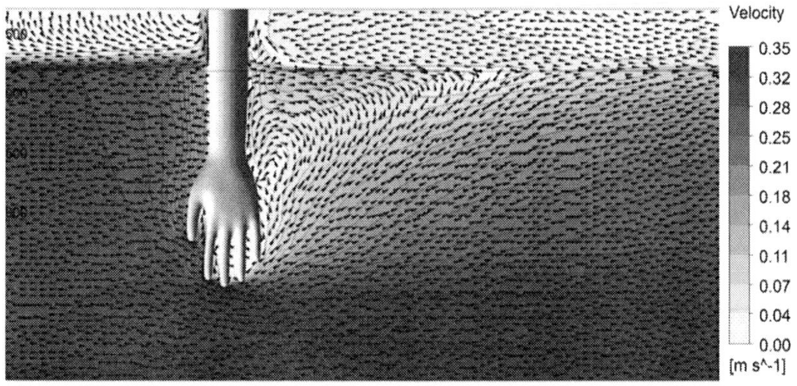

(b) flow past a glove-shaped former

Figure 7. The velocity field in the middle vertical plane of the two cases

4.3 Vertical vorticity field at various depths

Figures 8 and 9 show instantaneous vertical vorticity (rotation around z-axis) magnitude at four depth levels – (a) the original interface level, (b) 5 mm. deep, (c) 200 mm. deep and (d) 300 mm. deep – of the flow past a cylinder and that past a former, respectively. In the first case, Kármán vortex street pattern can be observed, especially near the interface level. At the lower depth levels, it could still be seen but the vortices dissipated much more quickly downstream.

In case of the flow past a glove-shaped former, asymmetric twin vortices were formed likely due to the asymmetry in the hand section of the former. At the interface level, the vortex strengths were much weaker than those in the first case. However, at the deeper levels, the strengths were comparable.

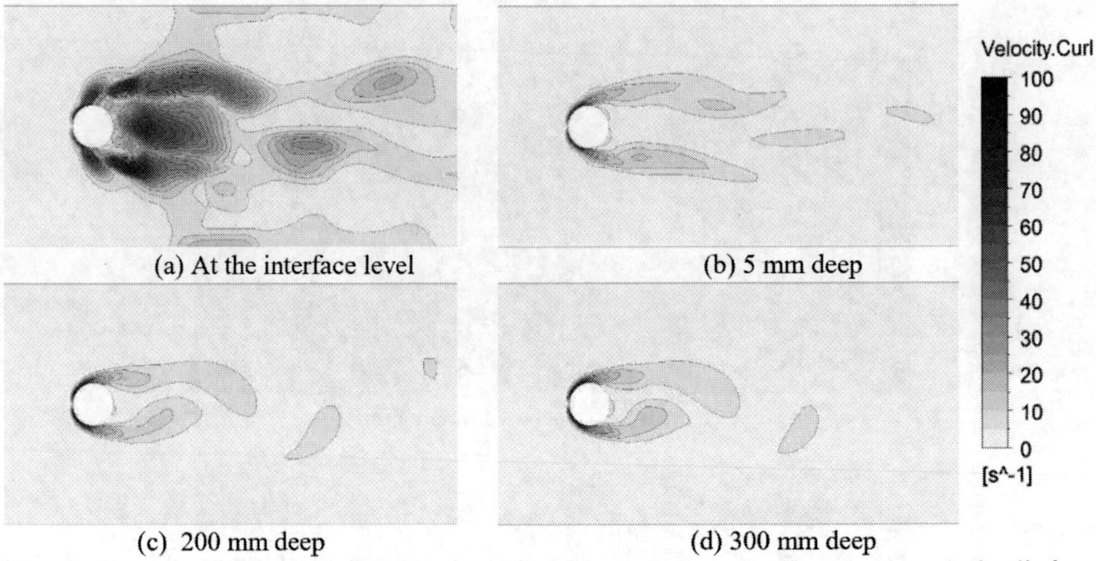

Figure 8. Instantaneous vertical vorticity contour in horizontal planes for flow past a vertical cylinder

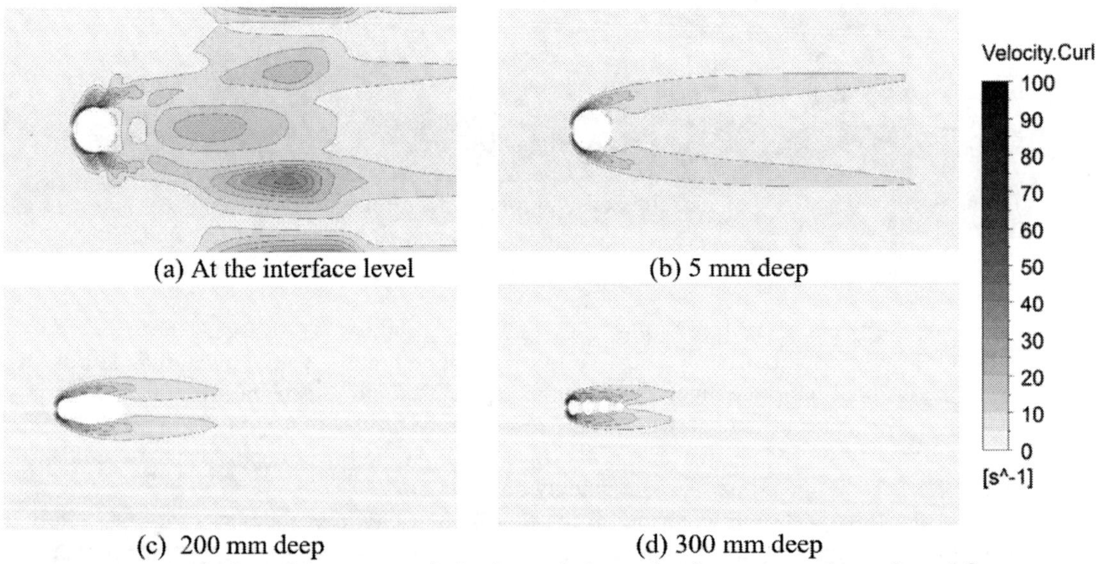

Figure 9. Vertical vorticity contour in horizontal planes for flow past a glove-shaped former

4.4 Free surface elevation vs. Streamwise co-ordinate

A more quantitative measure of the free surface level could be visualized by taking the spatial average of the free surface elevation field across the transverse coordinate (y). Similar to the visualization of the free surface, the VOF volume fraction value of 0.5 was used to identify the surface elevation level. For the first case with vortex shedding, additional time averaging was performed. This resulted in a plot of the free surface elevation vs. streamwise coordinate as shown in Figure 10. Note that the streamwise coordinate (x) was set to zero at the center of the cylinder and the former.

It could be seen that the crest and valley of the first case were more pronounced than those of the second case. This could be directly attributed to the fact that the cylinder has more frontal area perpendicular to the free stream.

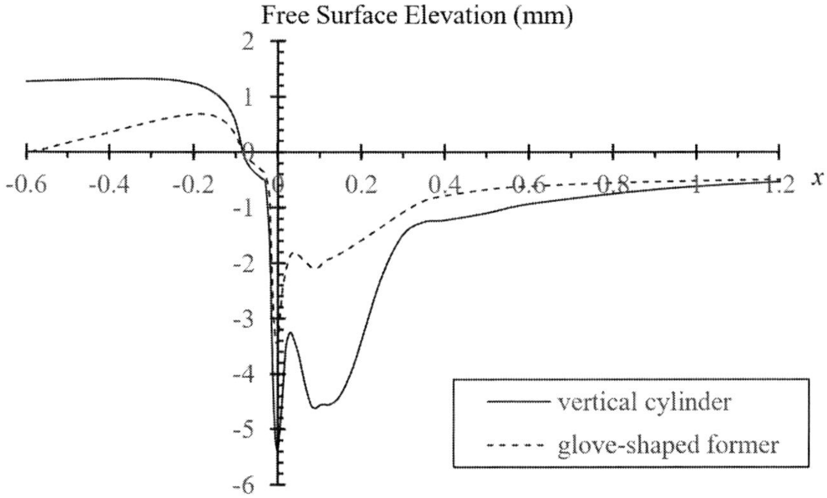

Figure 10. The plot of averaged free surface elevations of the two cases

4.5 Strouhal number and maximum sideway lift coefficient

More quantitative measures for characterizing the vortex shedding were Strouhal number (St) and the (maximum) sideway lift coefficient (C_L). The former measure can be computed by measuring the vortex shedding frequency (e.g. by monitoring C_L over time) and non-dimensionalizing it with the conveyor speed and cylinder/former diameter. The latter measure can be calculated by measuring the total resultant force acting on the solid surface in the transverse direction (F_y) then normalizing it with the dynamic pressure $\left(\frac{1}{2}\rho U^2\right)$ times the solid body's projected area in the xz-plane (A), i.e.

$$C_L = \frac{F_y}{\frac{1}{2}\rho U^2 A}. \qquad (3)$$

In case of the flow over a vertical cylinder, the Strouhal number was calculated to be 0.185, which was very close to the value of 0.18 from a case of 2D flow past a circular cylinder between two parallel walls with the same Reynolds number and blockage ratio [21]. The lift coefficient as a function of time oscillates symmetrically around zero with the highest magnitude of 0.518.

In case of the flow over a glove-shaped former, no transient vortex shedding occurred, therefore the Strouhal number could not be calculated. However, the flow was slightly asymmetrical due to the shape of the former, so the (steady) lift coefficient had a nonzero value of 0.017.

5. Conclusions

In this study, a three-dimensional CFD simulation with VOF technique of free surface flow representing a latex-dipping process was carried out with realistic parameters taken from an actual latex-gloves production line. The glove-shaped former was physically scanned and its geometry was used in the simulation. Another free surface flow simulation with a simple vertical cylinder of the same diameter (and the same dipping tank) was also carried out as a reference.

Many qualitative and quantitative comparisons were made between the two simulations. Notably, and quite surprisingly, the simulation case with the glove-shaped former did not exhibit a transient vortex-shedding pattern while the other case did as expected. In case of the flow past a glove-shaped former, a single circulation region was formed behind the former instead.

The free surface elevation was compared between the two cases. With a more streamlined shape and a smaller frontal area, the glove-shaped former generated less pronounced wave in front and in the wake. A quantitative calculation shows that the difference was about two times in elevation level measured from the original unperturbed baseline.

Even though the vortex structures from the forearm were less pronounced on the free surface due to air resistance and surface tension effect, they could still be visually observed from the free surface as one would inspect an actual dipping tank. The effect from fingertips could not be observed at all on the free surface as the vorticity quickly dissipated at that depth.

Due to various dissimilarities presented in this work, a vertical cylinder would be an inaccurate substitute model for the glove-shaped former. This presents some challenges in the future work when an array of formers should be considered.

References

[1] Blackley D C 1997 *Polymer Latices Science and Technology Volume 3: Applications of latices* (New York: Springer Science)
[2] Cacioli P 1997 *Rev. Fr. d'Allergologie d'Immunologie Clin.* **37**(8) 1173–1176
[3] Yip E and Cacioli P 2002 *J. Allergy Clin. Immunol.* **110**(2) S3–S14
[4] Hill D 2018 *The Science and Technology of Latex Dipping* (Smithers Rapra)
[5] Morison J R Johnson J W and Schaaf S A 2012 *J. Pet. Technol.* **2**(5) 149–154
[6] Demirbilek Z and Gaston J D 1985 *Ocean Eng.* **12**(5) 375–385
[7] Faltinsen T V O M and Newman J V 1995 *J. Fluid Mech.* **289** 179–198
[8] Scardovelli R and Zaleski S 1999 *Annual Review of Fluid Mechanics* **31**(1) 567–603
[9] Reichl P Hourigan K and Thompson M C 2005 *J. Fluid Mech.* **533** 269–296
[10] Bai W and Taylor R E 2006 *Appl. Ocean Res.* **28**(4) 247–265
[11] Kang A Lin P Lee, Y J and Zhu B 2014 *Comput. Fluids* **106** 41–53
[12] Wu Y L 2017 *Ocean Eng.* **141** 477–492
[13] Zhang, J Liang D Fan X and Liu H 2019 *Adv. Water Resour.* **123**(800) 96–108
[14] Yu G Avital E J and Williams J J R 2008 *J. Fluids Eng.* **130**(10) 101304
[15] Sasidharan K K Joseph R Rajammal G Viswanatha P and Gopalakrishnan K S 2001 *J. Appl. Polym. Sci.* **81**(13) 3141–3148
[16] Gorton A D T 1967 *Journal of the Rubber Research Institute of Malaya* **20** 1966.
[17] Quarteroni A 2014 *Numerical Models for Differential Problems* (Springer Publishing)
[18] Eymard R Normale E and De-Provence U 2000 *Handb. Numer. Anal.* **7**(3) 2000
[19] Versteeg H K and Malalasekera W 2007 *An Introduction to Computational Fluid Dynamics* (Pearson Education Limited)
[20] Nichols C W and Hirt B D 1979 *J. Comput. Phys.* **42**(3) 201–225
[21] Singha S and Sinhamahapatra K P 2010 *Ocean Eng.* **37**(8–9) 757–769

IOP Conf. Series: Materials Science and Engineering **886** (2020) 012039 doi:10.1088/1757-899X/886/1/012039

Minimization of Energy and Water Cost for the Main Building of Suranaree University of Technology Hospital (SUTH)

A Prasetyadi[1,2], and A Koonsrisuk[1,*]
[1]Suranaree University of Technology, 111 University Avenue, Muang District, Nakhon Ratchasima, 30000, Thailand
[2]Universitas Sanata Dharma, Jl. Affandi, Mrican, Yogyakarta, 55281, Indonesia

* Corresponding Author: atit@sut.ac.th

Abstract
Hospitals consume a significant amount of energy and water. Energy and water are intertwined in a way that an improvement of one can worsen the other one. However, as the water price was relatively cheaper in comparison to energy, several studies paid their attention to the minimization of energy usage of the building only. In this research, a study of energy and water for the main building of Suranaree University of Technology Hospital (SUTH) was conducted in order to find the minimum cost of energy and water of the building using proposed a new method namely variables pair. The investigation showed that a single meter of electricity usage as the only energy consumption of the building was applied. Meanwhile, water usage was composed of tap water, flush water, and distilled water. Therefore, a method to determine the cost of energy and water was proposed. A variables pair in a form of (water usage, electricity consumption) was used in the proposed method to determine the cost of energy and water for the main building. Several water-saving options were proposed and examined. It was found that the water and electricity consumptions in the form of variables pair were (82.9 m^3/day, 26.3 kWh/day) for a current system, while they were (82.9 m^3/day, 21.2 kWh/day) for an option without water regeneration with minimum energy. Meanwhile, they were (82.9 m^3/day, 24.4 kWh/day) for a system with water supply tanks allocated at the rooftop and 8th floor of the main building, instead of using only one supply tank installed at the rooftop as in the current system. A system that collects the condensed vapor of the HVAC system and then uses it, had consumptions of (77.3 m^3/day, 16.3 kWh/day). Applying the current cost of energy and water at 4.5 Baht/kWh and 10 Baht/m^3 respectively it was found that the cost of water could be reduced up to 10.2% with water regeneration. The water cost contributed 7.1% cost reduction, while the energy part was 38.2%. It shows water volume saving is more significant than the energy of water processing saving. The proposed method could separate process and cost parts, therefore it is easy to handle and provides a broader perspective of the saving strategy.

Keywords: Energy and water nexus, energy cost, water cost, energy and water conservation

1. Introduction

Energy and water become the main factors in building conservation. Intensities of energy and water usage portray the building conservation level [1] in addition to the cost. The more efficient a building is, it consumes less energy or water and has less metric intensity. Energy and water usage over space and over inhabitants are two main metrics that people usually use for evaluating the building. Codes [2, 3] and benchmarking [4] are applied to set a standard of intensity. The codes also relied on planning the energy needs of an area or a country [3].

Energy and water conservation were investigated separately in many cases. The discussions only focused on conservation of energy or water neglecting the other. Domestic hot water was the main issue of energy-related to water [4] where the energy consumption of the hot water preparation was the single object of concern [5]. The system consuming a lot of water generating abundant energy requirement of the cycle was still far from discussing the water aspect of the energy. Even, about a system consuming abundant water such as a cooling tower, the discussion just focused on energy conservation [6]. Effectiveness of the evaporation cooling became the main issue of the proposal as minimum energy consumption and exergy destruction. Highly efficient energy consumption can aggravate the water usage of the building. In a broader view, the condition may imply the disadvantage of energy conservation.

Some studies about energy and water in a hospital have been performed. Some organizations and governments conducted a study about energy and water in hospital and produced benchmark [2,5,7]. Nevertheless, a deeper analysis of the measure for optimizing energy and water is still rare. Neural network method was applied for energy optimization in operational through forecasting the consumption [8]. Improvement of the intensity consumption by evaluation of energy and water usage was proposed in facilities audit by merit system approach [9]. A multi-objective approach using genetic algorithm was elaborated for optimizing CHP system in a hospital. It showed that a primary energy saving should be companied with payback analysis for economical optimum [10].

Strategy for optimizing two different materials can be categorized into 2 main approaches. The first method is projecting to one of the forming material and the second is projecting to another material. The 1st category includes mapping water to energy and vice versa. In analyzing the energy and water of the wind power generation, water was nominated into energy [11]. Then analysis was conducted in the energy unit. The reason for the method was that the water cycle needs energy. Therefore, the energy can be used as water metric. The economical unit [12] and CO_2 [13] evaluation belong to the 2nd type of the method. The economical unit was determined in fixed cost unit or its derivations. CO_2 was applied in order to trace the green-house impact which is mostly presented in CO_2 accumulation equivalence[14]. The 1st method can be applied if the interrelation between both materials can be determined clearly. In other words, it can be said that the interrelation has fixed relations. The advantage of this method is its adaptability for another comparison and parameters existence. The contra of the 1st type is the inability to determine the policy direction while the price of the material is unstable, the material is irrelevant, or the amount is insignificant. The 2nd type method cannot be adapted when the economic states are changing and may mislead in term of non-economic purposes. However, the second method is very practical and useful.

Purpose determines the way of energy and water interconnected. LCA (*life cycle analysis*) was applied for calculating the intensity of wind power [11]. NEA (*network environmental analysis*) was applied to show the interdependency [11] as an alternative to input-output analysis methods. The Input- output analysis was applied to understand the impact of the broader system by inter-sectoral interdependence [15]. It implies that there is not any single strategy being able to depict both economic and environmental basis thoroughly. Optimization on economy often has to cross the environmental issue that an adjustment of the economic structure should be conducted to save the environment [16]. Integrating a heat exchanger network can neglect the economical aspect of investment [17]. The optimization in the environmental aspect sacrifices cost of the system. According to extreme value theorem, it can be guessed that the optimum condition range boundaries are the environmental condition and the cost minimization.

Some methods have been proposed for integrating energy and water in industries. Wang and Smith [18] proposed a method advancing pinch method for water allocation problem of the contaminant system. They searched for optimum wastewater treatments number and capacities. This work has been improved by Kuo and Smith [19] for multiple treatment case. Both methods were based on the graphical approach and had assumptions of a single contaminant system and steady-state system [20]. The advancement of water and heat problem of integration does not yet cover the problem of mechanical which comes from a combination of static and dynamic conditions as mechanical generates a driving force of the mass flow. The driving force problem is more dominant in low quality thermal flow as the thermal aspect becoming less significant.

The aims of this article are to propose a method of energy conservation considering water factors in a building and to propose a method for evaluating the energy of water system using paired variables consisting of energy consumption and water usage. The proposed method fits the hospital main building problem categorized as low-quality thermal system. A modification of a commonly used composite curve is proposed and called as pair variables. Energy evaluation of water was calculated as an important part of energy optimization. Therefore, mapping of energy and water flow and its energy projection became the main part of this work. A new method separating the process and material parts were developed. Some options of reconfiguring water streams were presented. Finally, the cost of water was provided as an evaluation.

To meet its aims, the article is composed of 4 sections. It is started with setting the problem in the introduction as the first section, then it is followed by the method as the 2nd. The method has two main parts. The first part is about presenting the energy and water pair variables model. The second focuses on applying the variables pair approach for energy and water in SUTH main building. This part consists of strategy in estimating water usage in hospital, energy estimation of the system and streams of energy and water. Evaluation of energy consumption of the water flow is proposed as the third section followed by some advancement options. The last section shows the conclusion of the work.

2. Methods

The method is presented in two main parts. The first part discusses the variables pair method. This method consists of steps to represent the problem into energy and water variables pair model until the minimum energy condition reached as an ideal condition without any reuse and regeneration. The problem as mentioned in Kuo and Smith [19] is used to show the case method. Comparison of the result is presented to show the fitness of the proposed model into the existing method. Following the first, the application of the energy and water at the water network of the SUTH main building is presented.

2.1. Energy and water variables pair model

A description of the water allocation problem can be done using requirement diagram a shown in Table 1. The diagram consists of a stream number, flow-rate, and concentration. The stream number designates the streams of the problems. The flow-rate indicates the amount of each stream. The concentration shows the amount of the contaminant over the mass of water at the respective inlet stream. The concentration of the contaminant released to the environment was 30 ppm [19].

Table 1. The requirement diagram of the allocation problem [19].

stream number	flow-rate (t/h)	concentration (ppm)
1	20	800
2	30	400

3	50	200

To find minimum concentration removing in water allocation problem, a composite curve is generated to complete the requirement diagram. The concept of composite curve bases on the assumption that the overlapping processes can reduce the total water streams. The composite curve is managed to find a minimum. The gradient of the slope represents the rate of a contaminant over the water flow rate. The higher the gradient means less water need for each contaminant rate. The optimum condition can be found as a line connecting output requirement and pinch point. This line should only cross in single point or equal to any line without crossing the rest. The more complex a system is, the more difficult to find the pinch point is. A thorough examination has to be done to find the minimum. Various algorithms can be applied to compare the gradients in order to find the minimum. An example of a composite curve of the problem from Kuo and Smith [19] is shown in Figure 1.a.

There are two main metrics of the composite curve method showing that conservation of contaminant is the base of requirement. The first is the capacity of the wastewater treatment that is represented by the flow rate of contaminant (X-axis). The second is the removal range (Y-axis). Higher capacity and range represent a higher cost [18]. Therefore, minimization is an effort to reduce cost in term of the capacity of wastewater treatment and the removal range. The ideal range shows a minimum single range of wastewater treatment that can be used for the system.

Based on these metrics, the variables pair composed of energy and water flow rate is proposed to exchange the removal range and flow rate. The removal range limits the contaminant mass removed in a specific flow rate. The difference in the contaminant between input and output determines cost linearly. The flow rate in addition to limiting the capacity of the system also determines cost in a linear way. Then, it can be inferred that cost constraint can be exchanged with another function that is linear to the cost. The relations can be formed mathematically as shown in (1).

$$\forall C(x,y)\, ^\wedge\, E(x,y) \rightarrow \exists C(x,E),\ x,y \mid x,y \in R \tag{1}$$

In order to minimize the wastewater treatment, the variables pair uses energy and flow rate as its components as shown in Figure 1 b. The X-axis is accumulative stream flow rates and Y-axis is energy consumption over cost factor. The requirement diagram presenting the streams can be changed easily. The slopes of the requirement diagram representing the contaminant flow rate that should be removed are used for calculating the flow of the water. Accumulation of the contaminant water needs can be plotted to Y-axis. The X-axis of the variables pair diagram is calculated from the stream of contaminant.

Energy represents the cost of the system for specific flow rates. These variables pair directly shows cost represented by energy need and the stream to handle at the same time. The energy is directly proportional to the contaminant flow rate at a composite curve. The stream is an additional feature that shows the input of the system. Minimum of the system can be found trough accumulating the stream and its required energy. This number becomes the limit of the system. The gradient shows the capacity of the treatment to remove the contaminant. The higher is the gradient, the more its ability to remove the contaminant.

Minimization procedure can be done by selecting a probable sequence of the streams and its energy requirement. The strategy of the minimizing is applied for reducing the flow rate as much as possible to minimize the overcapacity of the removal contaminant. Combination of some streams can be done as long as the combination of specific streams should be less than the gradient of the maximum targeted. The maximum stream should also be considered.

Adjustment of the variables should be done for a different case. Accumulative water flow rate can be changed to be accumulative water usage. The accumulative energy for the system can be transformed to be accumulative energy.

2.2. SUTH Energy and Water System

Suranaree University of Technology Hospital (SUTH) main building is the center of the SUTH activities which has 11 stories plus 1 ground level. This building is functioned for clinics, wards, emergency, operation rooms, office and support system. It becomes the center for employees and visitors. The building is equipped by centralized HVAC with 3 x 250 TR chillers, 2 x 1.25 MW electricity system substations, and 1.10 MW back up power generator and 30 minutes batteries. The building has a tap water system operating 2 x 30 kW pump, and provides water heaters with total power electricity power need 234 kW.

Energy and water usage of the SUTH main building was recorded daily for maintenance. The data consisted of building daily water and electricity consumption as the only energy type of the building. The electricity energy usage was measured by a meter at the electricity control cubicle on the ground floor. The water meter for the building was located before the groundwater tank supplying the system the tap water for the system.

To calculate the energy of water usage, estimation of water usage of the building activities were conducted. The activities of the building involving water usage are sanitary, medical activities, cafeteria and HVAC. Gardening and food preparation are excluded from the counting as they were held by other building and had its own meter. Sanitary system water usage was estimated by combining a proportion of employees and patient water usage. The medical activities were estimated from some medical treatments and their patients. The HVAC water usage was estimated from the simulation of the centralized HVAC system of SUTH main building. The cafeteria usage was assumed as the rest of the water usage of the building.

Energy intensity of the water was calculated from the ideal condition that becomes the benchmark (named: *benchmark*), current condition operation (named: *current*) and 2 options as an enhancement (named: *option 1* and *option 2*). The ideal condition (*benchmark*) takes place when every system uses water without any regeneration or reuse in minimum energy processing. The *current* was the operation condition as the data were taken. It used a single water tap tank on the rooftop. *Option 1* is a proposed condition when there were 2 tap water tanks. They were the current water tap tank for supplying floor 5th – 11th and additional water tap tank in floor 8th for supplying the ground floor until floor 4th. *Option 2* is proposed retrofit by installing a system to collect condensed water from HVAC.

As the building is multistory one, the heights of the level determine the ideal potential energy of water. Additional energy intensity was added as a specific quality of water needed. The HVAC needed to soften water for chiller water and medical activities needed RO water. Domestic hot water is excluded as a limited number of inpatients which are the main hot water consumers.

2.2.1. HVAC water

HVAC water usage was determined from the make-up water needed for centralized HVAC of SUTH main building with a temperature setting 24°C and relative humidity 50% using simulation in TRNSYS as shown in previous work. The temperature setting was chosen as it provided reasonable result compared to sampled data condition in 2017. The simulation also reported that the HVAC system consumed 1.8 GWh of electricity in a year. The result represents 56% of the electricity consumption of the building in the same year. Total evaporated make up water for the HVAC in a year was 8,506 m^3, which is equal to 23.3 m^3 water per day on average.

Water usage of the HVAC consists of make-up water and blow-down water. The make-up water is defined as the amount of water compensating evaporated water at cooling tower. Therefore, the make-up water keeps the condenser water is always equal. Amount of make-up water is equal to the amount of evaporated water for cooling. In addition, keeping an equal amount of water, the cooling tower also needs to maintain the mineral concentration of water is acceptable. Assuming that soften water and hard water have the conductivity 50 μS/cm and 300 μS/cm respectively, the blow-down water can be calculated as Equation (2). Therefore, the HVAC water usage is presented in Equation (3). It shows that the water of HVAC is a function of the water evaporated at simulation.

$$V_{bw} = \frac{V_m}{5} V_{cdw} \tag{2}$$

$$V_w = V_m + V_{bw} \approx \frac{6}{5} V_m \tag{3}$$

2.2.2. Sanitary water

Sanitary water consists of water for flushing, water for showering and water of hand washing. According to the type of water contacts with a human, it is clear that water of flushing does not have contact with a human. The water for showering and hand washing is considered to have indirect contact with a human because it does not go directly into the human body system. The water for flushing can be relatively low quality in term of medical requirement. The water for showering and hand washing can be tap water quality with bacteria colony less than 150 fcu/cm^3 [21].

Sanitary water for flushing was determined from a number of inhabitants and their times staying at the hospital in a day. The inhabitants of the hospitals are employees, outpatients and inpatients. The employees were assumed to stay in the hospital for 8 hours. The outpatient stays at a hospital on an average of 2 hours. The inpatient category lived 24 hours in hospitals. The flushing water equation is shown in Equation (4).

$$W_f = \sum C_i N_i t_i \tag{4}$$

In a similar way, water for handwashing and showering was calculated using Equation (4). The difference was the coefficients. The coefficients for flushing water, hand washing, and showering were 50.4 l/day, 29.8 l/day, and 69.6 l/day respectively [22]. The number of daily inhabitants is provided in Table 2.

2.2.3 Medical activities water

Medical activities water is the water used for medical activities such as lubricating at the dentistry clinic, operation and hemodialysis. The other water types were neglected due to the minimum amount. Medical activities water is determined by a coefficient of water usage and patient number assumed as activities. The equation of the medical activities water is presented in Equation (5).

$$W_m = \sum C_m N_m \tag{5}$$

There were some clinics which had significant water usage per patients. They were the operation room, delivery room, dental clinics, and hemodialysis unit. The coefficients of those clinics were 100 l/patient, 100 l/patient, 2 l/patient, and 200 l/patient in respective sequence for the preceding units. The water of hemodialysis and dental clinics was distilled water. While the other clinics used tap water mostly for the hand washing.

2.3. The energy of water usage

The energy of water usage is depended on the type of water usage and clinic position. The clinic position implies a minimum energy need for pumping. The types of water usage determine the energy intensity of water processing. The energy of tap water usage was assumed to consists of water distribution only which was depended on story. Soften water was assumed to have single-phase RO water processing. The intensity of this activity is 0.36 kWh/m^3 [23]. The RO water was assumed as double stage RO water processing which has 0.47 kWh/m^3 [23] of energy intensity. The hot water is assumed to be proportional to the number of patients and became part of water showering or hand washing. The hot water of consumption per patient per day was assumed to be 11 liters. The energy needed for the hot water was assumed to heat to 65°C from 25°C as the average of the water temperature with 4.2 kJ/kg°C heat capacity. Therefore, the energy of the water according to the type can be calculated according to Equation (6).

$$E_{wt} = \sum C_t W_t \tag{6}$$

The water usage can be broken down into story energy-water requirement. Every story has a different function and water usage. Some stories only need tap water; other stories might need all type of water. It also has a different level that determined the energy of tap water. Therefore, the water usage of stories can be calculated by Equation (7).

$$W_s = W_{hs} + W_{ds} + W_{ms} \tag{7}$$

The energy of the water at a story was composed of energy for distribution which was proportional to the amount of story tap water consumption, and energy according to the function. Therefore, energy for water in every story could be calculated by Equation (8).

$$E_{ws} = C_s W_s + C_{ms} W_{ms} + C_{hs} W_{hs} \tag{8}$$

The HVAC water was calculated as additional water. It was settled at the rooftop, therefore the energy of the HVAC was assumed as much as the tap water energy for 11th story. The processing of the HVAC water was in accordance with Equation (6) with energy intensity as a single step RO. Energy for HVAC was calculated as energy for make-up water and blow-down and calculated as Equation (9).

$$E_{HVAC} = (C_{11} + C_{RO})(V_m + V_{bw}) \tag{9}$$

After usage, the water was sent to the wastewater treatment before going to the environment. The intensity of water treatment at the building is 0.344 kWh/m^3. Except for the water evaporated at a cooling tower, the water from the hospital was treated at the facility. Then the water treatment needed energy as shown in Equation (10).

$$E_{WT} = C_T (W - V_m) \tag{10}$$

The energy of the water was calculated as an accumulation of energy of water processing for water, energy for distribution and energy for wastewater treatment. Accordingly, energy necessity of the water at the building was calculated using Equation (11).

$$E_W = \sum E_{ws} + E_{HVAC} + E_{WT} \tag{11}$$

2.4. Energy and water cost

Energy usage and water amount could be presented as variables pair as there were energy and water at the same time in every process cycles. Amount of water in a specific time (could be assumed as flow rate) was projected into X-axis. The Y-axis projection was used for energy usage of the water. The ordered pair variables can be done in every step of the system. The cumulative becomes the system of water and energy variables pair.

The variables pair can be transformed for other variables such as cost or CO_2 emission. The cost of energy and water can be calculated as a dot product of energy water paired variables and the respective cost prices. The total cost of energy and water can be considered as the summation of the water cost and energy cost. Equation (12) shows the cost of energy and water.

$$C_{EW} = \begin{bmatrix} f_{EC} & f_{WC} \end{bmatrix} \begin{bmatrix} E_W \\ W_s \end{bmatrix} \tag{12}$$

3. Results and discussion

Comparing the proposed method and the composite curve on the problem of wastewater treatment [18] provided a similar result. The composite curve gave 35 t/h contaminant removed capacity with 437.5 ppm contaminant rate at the water as the optimum condition in a single water treatment system with an

IOP Conf. Series: Materials Science and Engineering **886** (2020) 012039 doi:10.1088/1757-899X/886/1/012039

unspecific removal factor. The new method in the other hand mentioned that the energy needed for removing the contaminant should be at least 35F; with F is the conversion factor connecting removed contaminant and the cost. The rate of contaminant removal shows the cost of the system [18]. Accordingly, F is a function that shows energy necessity for removing every ton contaminant. Assuming energy as the basic cost of the removal, the new method also easily shows the condition to reach that.

Specified remove factor of wastewater treatment means that the constraint depends on the structure. Basically, the established method and the proposed one can be applied in a similar way. The established method minimizes the area on the left of the graph. The X-axis shows the capacity which is related to the cost of the system. The present method minimizes the area under the graph with energy as the main factor. Comparison of the methods can be seen in Figure 1.

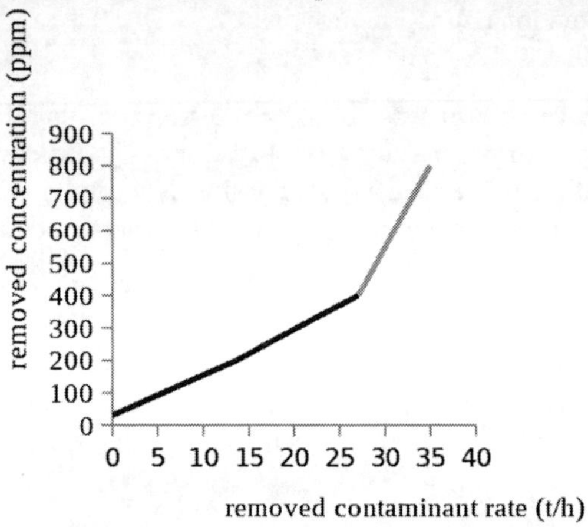

(a)

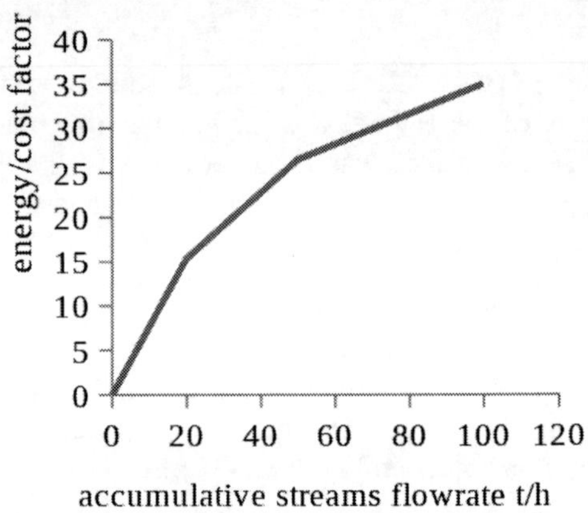

(b)

Figure 1. a) The composite curve of Kuo and Smith problem [19], minimization can be considered as integration to the Y-axis. b) The ordered pair variables of the same problem, the minimization process has an area of integration to the Xaxis.

The hospital water usage of the SUTH main building was composed of tap water, RO water in exchange for distilled water. The RO choice was based on assuming a practical process. The water should not be kept for a long time in order to avoid contamination. This composition basically represents the quality of the water which also shows the energy necessity of water processing. In addition to the quality, as the building is multi-story, division of the water into floors requirements is also presented. Table 3 shows the story and its water requirement. Outpatients were assumed to stay 2 hours in the hospital during their visit. The employees were assumed to stay for 8 hours per day for 1 shift work.

Table 2. The daily building inhabitants (in person) and water usage (m^3) of the building.

Floor	Inhabitants			Tap	Flush	RO
	OP	IP	Staff	Water		
Ground			53	14.53	0.89	
1	356.4		77	8.24	2.79	
2	141.2		34	3.76	1.16	0.32
3	126.6		38	3.88	1.17	0.25
4			20	0.99	0.34	
5	91.4		16	2.27	0.65	
6	0.5		16	0.96	0.27	

7			43	2.13	0.72	
8	62.5	7.6	13	7.55	0.86	3
9	4.0	18	21	3.77	1.28	
10	1.3	21.8	21	4.30	1.46	
11		13.6	11	2.57	0.87	
HVAC						27.96

The energy of water processing consisted of energy for tap water distribution, water processing to be RO or softening. The minimum pressure of the floor was 20 psi as compensation of the network pressure loss. Therefore floors 8t^h^, 9th, 10th and 11th needed booster pumps. The additional energy for those floors should be provided. Table 2 shows the distribution of the energy coefficient for the floor and energy usage for water on every floor.

Table 3. The coefficients of energy water processing and energy necessity for the main building usage.

Floor	Ct (Tap)	Ct (RO)	Tap (kWh/d)	RO` (kWh/ d)	Booster (kWh/d)
Ground	0.013		0.191		
1	0.026		0.217		
2	0.040	0.47	0.137	0.164	
3	0.053	0.47	0.180	0.133	
4	0.067		0.066		
5	0.080		0.182		
6	0.094		0.090		
7	0.107		0.229		
8	0.121	0.47	0.187	1.772	0.030
9	0.134		0.566		0.058
10	0.148		0.636		0.116
11	0.161		0.414		0.098
HVAC	0.173	0.36	5.705	10.044	

Accumulation of the energy for the water processing as the minimum without any reuse or regeneration can be shown in Figure 2. This condition is represented as an ideal condition as there is not any water reuse or regeneration and the water resource. At this condition, the energy for distribution of the water is considered as a minimum. The final water usage of this condition was 82.9 m^3/day consuming 21.2 kWh/day of electricity.

The current operation condition (*current*) has energy necessity higher than the ideal condition without regeneration (*benchmark*). The *benchmark* assuming the energy distribution is different for every story, while the *current* need single energy distribution as much as the eleventh story. The single water tower at the rooftop is the reason. The tap water is distributed from the rooftop tank. The current operation (*current*) has variables pair as (26.3 kWh/d, 82.9 m^3). *Option 1* of proposing a new water tower on the 8th floor to supply the ground up to 4th floors need less energy intensity for an equal volume of water. Regeneration strategy by collecting condensed water from the HVAC and use it can decrease the necessity for water by 8.6% and the energy for water usage by 39%. The ordered pair variables of the *option 1* and *option 2* are (16.3 kWh/d, 77.3 m^3/d). The reduction of energy comes from the collected condensed water from HVAC operation which is used for flushing.

This final water usage and energy necessity can be depicted into a vector-like of energy and water ordered pair variables as shown in Figure 3. The X-axis is water amount which is withdrawn or water usage and the Y-axis is energy usage. Figure 3 shows that water *option 2* have less water energy

IOP Conf. Series: Materials Science and Engineering **886** (2020) 012039 doi:10.1088/1757-899X/886/1/012039

intensities than the current operation condition but higher energy intensity than the *benchmark*. It also mentions that *option 3* has less water and water energy intensity compared with *current* and *benchmark*. *Option 3* is more preferable for conserving energy and water.

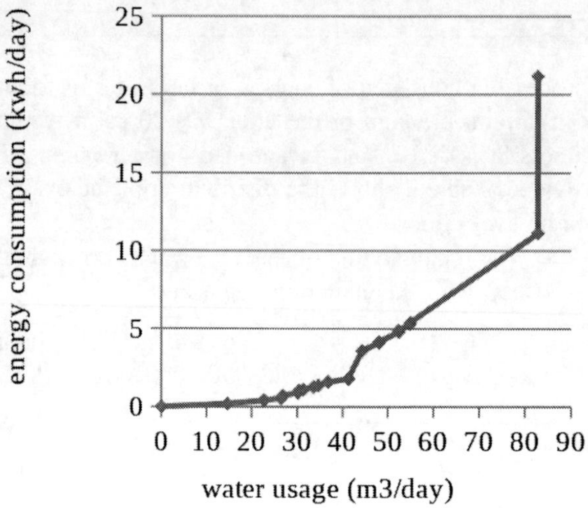

Figure 2. The accumulation of water usage and its energy for preparation is presented in variables pair approach.

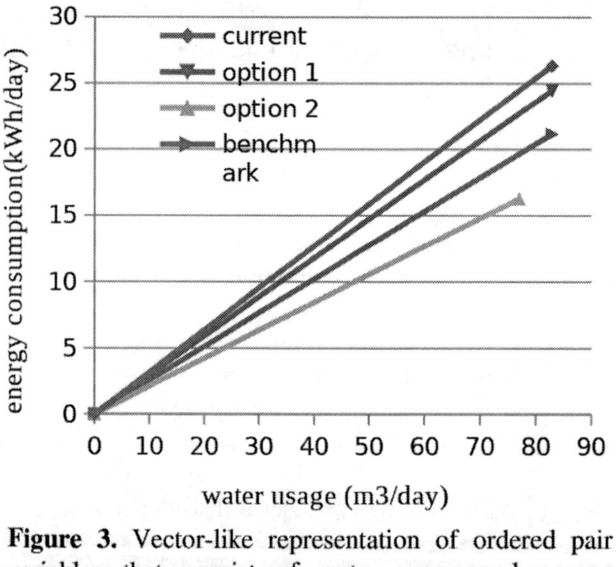

Figure 3. Vector-like representation of ordered pair variables that consist of water usage and energy consumption. The method also shows the energy intensity of the water as the gradient.

Cost of the water and energy can be calculated using Equation (12). As long as the price function of the water and energy is linear or exponentially progressive, cost of energy and water is proportional to the daily water usage and energy consumption. It can be inferred easily that *option 3* is the best. At the current prices, the daily energy and water consumption concerning water in SUTH main building can be seen in Table 4.

Tabel 4. Comparison of daily electricity and water cost (in Baht)

	cost of electricity	cost of water	total cost
current	829	118	947
option 1	829	110	939
option 2	770	73	844
benchmark	829	95	924

4. Conclusion

Managing water flow in SUTH main building can be used for conserving water and energy at the same time. Proposed option 1st, where the additional water tank should be built for supplying the ground until 4th floor, can reduce the energy of water by 7.1% without water usage reduction. The second proposed option which includes regeneration of the water from HVAC, can reduce the water usage by 7.1%. It also reduces the energy for water reduction by 38.2%. A similar pattern took place for water and electricity for water cost.

The variables pair can be used for minimizing the energy and water of the water usage at the SUTH main building. It shows the compatibility with composite curve method in water allocation problem and provides two variables that can easily be connected to the cost as the main variable for conservation.

Acknowledgments

The authors would like to express gratitude to SUT for providing the Postgraduate Scholarships for Foreign Students to the first author and the support for the conference. The authors also express their sincere thanks to the SUTH director, Dr. Niwatchai Namvichaisirikul, and the staff of the Engineering and Maintenance Department at SUTH, particularly Mr. Watcharin Singtana, Mr. Songkrit Jitpakdeebordin, Mr Chaleomkiat Kaewkurn, Mr. Prashya Suranachit, and Ms. Naphatthira Decheerawas.

Abbreviation and Variables

SUTH	Suranaree University of Technology Hospital	HVAC	Heating Ventilation Air Conditioning
LCA	Life Cycle Analysis	NEA	Network Environmental Analysis
RO	Reverse Osmosis	IOA	Input Output Analysis
GWh	giga watt hour	kWh	kilo watt hour
ppm	part per million	µS	micro siemens
V_{bw}	The volume of blow-down water	V_m	The volume of evaporated make-up water
V_{cdw}	The volume of condenser water	V_w	The volume of water for HVAC
W_f	water sanitary function	C_i	coefficient of function
N_i	number of people	t_i	time of staying at the hospital
W_m	water of medical function	C_m	coefficient of medical usage
N_m	number of patients	E_{wt}	energy for type of water processing
C_t	energy intensity of water processing	W_t	type of water volumetric flow rate
W_s	water usage of story	W_{fs}	sanitary water usage of story
W_{ds}	domestic water usage of story	W_{ms}	medical water usage of story
E_{ws}	The energy of water of story	C_s	The distribution coefficient of story
C_{ms}	medical water coefficient of story	W_{ms}	medical water of story
C_{hs}	coefficient of HVAC water condensed of story	W_{hs}	The volume HVAC water condensed of story
E_{HVAC}	The energy of HVAC water processing	C_{11}	coefficient of distribution for 11th floor
C_{SO}	coefficient of single-step RO	E_{WT}	energy for treating wastewater
C_T	coefficient of wastewater treatment	W	total water usage of the building
E_W	energy of water	C_{EW}	cost of energy and water
f_{EC}	The function of energy cost	f_{WC}	The function of water cost

References

[1] Saidur R Hasanuzzaman M Yogeswaran S Mohammed H A and Hossain M S 2010 *Energy* **35** 4780–4785
[2] Chirarattananon S Chaiwiwatworakul P Hien V D Rugkwamsuk P and Kubaha K 2007 *Proc. Jt. Int. Conf. Sustain. Energy Environ.* **2** 1–10
[3] Chirarattananon S Chaiwiwatworakul P Hien V D Rakkwamsuk P and Kubaha K 2010 *Energy* **35** 1741–1753
[4] Singer B C Mathew P Greenberg S Tschudi W and Sartor D 2009 *Hospital Energy Benchmarking Guidance - Version 1.0* (San Fransisco: Lawrence Berkeley National Laboratory LBNL) Report LBNL-2738
[5] Bujak J 2010 *Energy Build.* **42** 1047–1055
[6] Al-Hadban Y Sreekanth K J Al-Taqi H and Alasseri R 2018 *JERT* **140** 1–11
[7] González A G García-Sanz-Calcedo J and Salgado D R 2018 *Sustain. Cities Soc.* **36** 169–175

[8] Bagnasco A Fresi F Saviozzi M Silvestro F and Vinci A 2015 *Energy Build.* **103** 261–270
[9] Buffoli M Capolongo S di Noia M Gherardi G and Gola M Healthcare Sustainability Evaluation Systems *Improving Sustainability During Hospital Design and Operation* vol. 218 ed Capolongo S, Bottero M C, Buffoli M and Lettieri E (New York: Springer) chapter 3 pp 23-30
[10] Gimelli A and Muccillo M 2013 *Appl. Energy* **104** 910–923
[11] Yang J and Chen B 2016 *Appl. Energy* **169** 1–13
[12] Sanders K T Blackhurst M F King C W and Webber M E 2014 *Environ. Sci. Technol.* **48**
[13] Ackerman F and Fisher J 2013 *Energy Policy* **59** 235–241
[14] Racoviceanu A I Karney B W Kennedy C A and Colombo A F 2007 *J. Infrastruct. Syst* **13** 261– 270
[15] Fang D and Chen B 2017 *Appl. Energy* **189** 770–779
[16] Shi C and Zhan J 2015 *Phys. Chem. Earth* **79–82** 47–53
[17] Liu F Ma J Feng X and Wang Y 2018 *Appl. Therm. Eng.* **128** 1510–1519
[18] Wang Y P and Smith R 1994 *Chem. Eng. Sci.* **18** 3127–3145
[19] Kuo W C J and Smith R 1997 *Chem. Eng. Sci.* **52** 4273–4290
[20] Ataei A Panjeshi M H and Karbassian S 2009 *Res. J. Environ. Sci.* **3** 604–618
[21] Judy A 2006 Guide to Infection Control in the Hospital *Hospital Water* (International Society for Infectious Diseases)
[22] BIS and Cranfield University 2009 *Study on water performance of buildings* **33** 14–53
[23] Wakeel M Chen B Hayat T Alsaedi A and Ahmad B 2016 *Appl. Energy* **178** 868–885

IOP Conf. Series: Materials Science and Engineering **886** (2020) 012040 doi:10.1088/1757-899X/886/1/012040

Investigations of Stack Ventilation Operations Using An Energy Modelling And The Bas System

D Woradechjumroen[1,*], and T. Tongshoob[1]

[1]School of Engineering, Sripatum University, 2410/2 Jatujak, Bangkok, 10900, Thailand

* Corresponding Author: denchai.wora@gmail.com, denchai.wo@spu.ac.th

Abstract
Green buildings have been intensively constructed due to energy crisis and damaged environmental issues. Additionally, a green building focuses on efficient resource utilization and reducing environmental damage. Mechanical designers emphasize solving energy problem issues in large commercial buildings via a stack ventilation application; it can be used to reduce cooling load and mechanical fan power consumptions. However, stack ventilation installation has been problematic due to Thailand climate, because heat is absorbed on the lower floor of a building causing higher temperature than upper floors; this issue obstructs buoyancy force. Without consideration of the concern, mechanical designers may misunderstand stack ventilation operations from its theory. To avoid the issue, this research applies the three software for assuring the performance composing of (1) LoopDA for stack ventilation simulation to acquire different temperatures for buoyancy force; (2) SketchUp for developing a simplified building configuration; and (3) TRNSYS used for full energy modelling with and without stack ventilation. The energy modelling is validated via the existing building automation system (BAS) data in the example building with the accuracy within ±10 %. The verified building model can be further used to test the stack ventilation operations via the potential simulation instead of sensor installations.

Keywords: Energy Modelling; Green Building; LoopDA; Stack Ventilation; TRNSYS

1. Introduction
Due to population growth, the number of buildings has been increasing continuously, causing inefficient energy use and affecting the more severe environment. That said, green building construction is, therefore increasingly expansive in developing countries because this building type is mainly designed to use efficient resources, reducing wastes, pollution and environmental destruction. One of the requirements in green buildings is the application of sustainable energy, which mechanical designers focus on reducing electrical energy consumptions of heating, ventilation and air conditioning (HVAC) systems, approximately 70 percent of the total electrical energy in large Thai buildings [1]. Natural ventilation is widely utilized to reduce the workload of air conditioning systems

IOP Conf. Series: Materials Science and Engineering **886** (2020) 012040 doi:10.1088/1757-899X/886/1/012040

and mechanical fan ventilation called "stack ventilation"; it is an alternative way to reduce inside temperature of the building and to increase the effective electrical energy greatly, which uses the principle airlifting to create natural movement due to temperature differences [2, 3].

Since stack ventilation is a vertical airflow resulting from the differences between outdoor air temperatures and building interior temperatures, this system typically occurs when higher temperature difference causes lower air density; entering the air in the building will lift up and leave at any openings on the top of a building. However, in reality, it is unable to produce theoretical airflow, especially in tropical countries because a basement floor absorbs heat causing the temperature of a lower level to be higher than the top level of the building. Thus, the air cannot be lifted without buoyancy forces. In the design process, if designers neglect to consider this point, it may cause design errors. To reduce errors, energy simulation programs [4] and real-time measurement systems called building automation system (BAS) can be used to track the system performance from the design phase to the construction phase and throughout the building use. One of the programs used is called "LoopDA" that is not complicated and can be used to enhance and recheck typical design processes. This program can calculate pressure variations according to the height of a building as well as calculating the minimum temperature difference between the top and lowest levels of the building.

In order to solve the aforementioned problems, this article investigates the performance and behavior of a stack ventilation system in an example building. The methodology includes the following three steps: (1) simplified pressure and temperate differences are estimated by LoopDA program (NIST) [5]; this step is able to analyze the temperature differences between inside and outside the building. In addition, the optimum airflow rate is computed based on the possible size of the opening at the bottom and the top of the building (2) TRNSYS simulation studio is an energy modeling program used to create a building model including simplified zones, chiller plant system and building chimney. The model can simulate the airflow in the chimney, and (3) BAS of the building is used to obtain the actual zone temperature and humidity to validate the accuracy of the building simulation. The findings can be further used to assure the stack ventilation concept before construction.

2. Background

2.1 Stack Ventilation

The temperature differences between inside and outside of the building cause pressure differences; this causes natural airflow movements into the building called "Stack Effect" as depicted in Figure 1. This airflow can be used to reduce the temperature inside the building and can be used to reduce the electrical energy used in air conditioning systems [6]. The air pressure inside the building depends on a building height corresponding to the correlation as Equation (1):

$$\frac{dP}{dz} = -\rho g \tag{1}$$

Where z is the vertical height, ρ is the density and g is the gravity acceleration. Due to a relatively small change of density versus the height of the building, the density used in the equation is constant and p_0 is defined as the pressure at the ground level.

$$p = p_0 - \rho g z \tag{2}$$

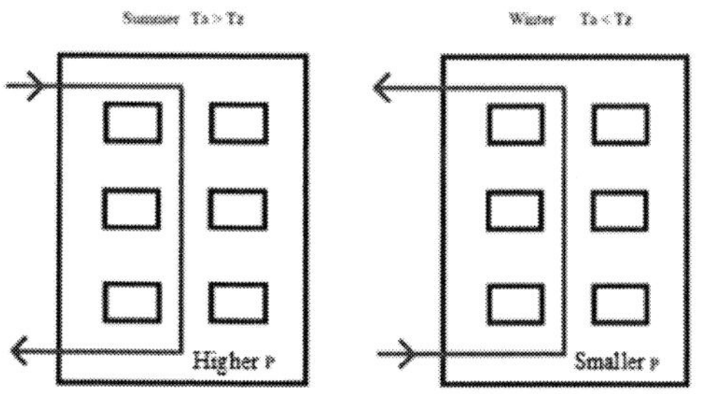

Figure 1. The airflow in summer and winter

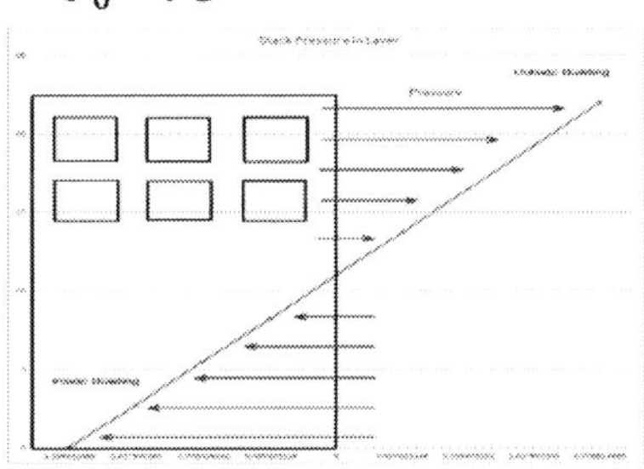

Figure 2. Pressure direction versus height in winter

In the winter, the inside air on the ground floor of the building is typically higher than the outside. Thus, indoor air density is lower than the density at the higher-level resulting indifferent pressure between inside and outside. As a result, the airflow enters at the bottom and then leaves at the top of the building. In the summer, the flow direction is opposite to the winter because the inside air temperature on the ground floor is lower than outside temperatures leading to higher density and pressure than the outside conditions. Therefore, the outside air flows into the top of the building, and then leaves at the bottom of the building as shown in Figure 1.

From Figure 2, the pressure is varied according to the height of a building in which h is the neutral height, and is equal to half the height of the building; it can find the differential pressure between inside and outside building from Equation (3).

$$\Delta p_s = p_A - p_z = \left(\rho_A - \rho_z\right) g(h - z) \tag{3}$$

In this equation, the pressure difference (Δp_s) of the opening is used to calculate the equilibrium balance pressure, and ρ is the density of air passing through openings. The pressure difference is caused by external wind when z is the height varied from the floor. The stack ventilation can be estimated by the following Equation:

$$\Delta p_s = \rho_A \left(\frac{T_A - T_z}{T_z}\right) g(h - z) \tag{4}$$

When T_A is outside air temperature and T_Z is indoor air temperature, the computed pressure is used to determine the airflow direction of the building as shown in Figures 1 and 2.

2.2 LoopDa Program

LoopDA is a program used to design natural airflow developed by the National Institute of Standards and Technology (Dols et al. 2012). This program can determine the size of air outlets to achieve the desired airflow rate for stack ventilation. By using the method of creating air flow loops, an opening size depends on the pressure of the wind and the pressure of the stack effect. The steps are as follows:

1. Sketch a simplified building into the program in front or side view of the building.
2. Specify the outside temperature, air velocity and data in each zone including the type of opening in each zone and the details of any openings.
3. Draw airflow lines in each loop and consider the airflow rate in each loop based on the minimum required ventilation flow rate for and heat flow rate.
4. Consider the opening size with the graph obtained from the LoopDa simulation, and then check the flow rate from the simulation; the computed/calculated value must be close to the design value. A minimum desired airflow rate equation is related to the temperature of the air inlet and outlet as follows:

$$Q_{cool} = \frac{UA_s\left(T_{ambt} - T_{des}\right) + \dot{q}_{int} A_f + \dot{q}_{other}}{\rho C_p\left(T_{des} - T_{ambt}\right)} \tag{5}$$

From Equation (5), the heat balance equation is generated between the air temperature of entering opening and exiting opening. The variables are as follows:

Q_{cool} is the airflow rate (m^3/s); U is the heat transfer coefficient for overall heat transfer through the surface; A_s is an envelope surface area (m^2); A_f is the outside surface area; T_{ambt} is the ambient temperature of a building, T_{des} is an opening design temperature (K); $\dot{q}_{int}$ is the internal heat gain per area (W/m^2), and $\dot{q}_{other}$ is another internal heat.

In addition, the equation of the opening airflow rate (Q_d) can be computed by Equation (6) with differential pressure (ΔP_{wind}) caused by the outside wind which can be determined by Equation (7).

$$Q_d = C_d A \sqrt{2\Delta P/\rho} \tag{6}$$

$$\Delta P_{wind} = \frac{1}{2}\rho C_p u^2 \tag{7}$$

where C_p is the pressure coefficient of the outside air passing through an opening, and u is the wind velocity.

Based on the aforementioned equations, all variables of stack ventilation can be related in terms of the energy balance equation at steady state condition shown in Equation (8).

$$\sum_{inlets} \dot{m}_j C_p T_j - \sum_{outlets} \dot{m}_i C_p T_i + UA_s \left(T_{ambt} - T_i\right) + \sum_{heat\ sources} \dot{q}_i = 0 \tag{8}$$

However, in real cases of the design process, one of the significant equations that designers practically forget to re-check is the equilibrium equation of the pressure balance, the total pressure in the zone which must be equal to the total released pressure:

$$\sum_{unsized} \Delta P_i = \Delta P_{stack} + \Delta P_{wind} - \sum_{sized} \Delta P_j \tag{9}$$

From Equation (6) to (10), C_d is the discharge coefficient (m/s); $\dot{m}$ is the mass flow rate through openings; ΔP_i is the unsized opening pressure (Pa); ΔP_j is differential pressure of specified openings (Pa); ΔP_{stack} is the pressure difference from the stack effect, and ΔP_{wind} is the pressure difference from the wind.

3. Methodology

3.1 LoopDA building model

The building example in Figure 5a) can be simplified into four zones (Figure 3 b) including low-level office; high-level office; basement and the building chimney. This building can be created as a simple model on the SketchUp program as shown in Figure 3 b) so as to study stack ventilation and wind through the chimney for reducing activity area load without using the air-conditioning system.

Figure 3a. Example building

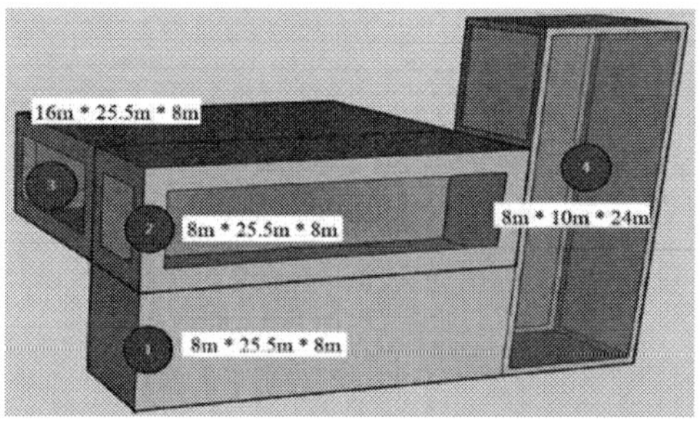

Figure 3b. SketchUp model

IOP Conf. Series: Materials Science and Engineering **886** (2020) 012040 doi:10.1088/1757-899X/886/1/012040

3.2 SketchUp building model

To simulate zone conditions of the example building, SketchUp model is required for assigning the connecting zones in TRNSYS program. The simple model with the dimensions consists of the two office zones, basement floor and the chimney as shown in Figure 3 b.

3.3 TRNSYS simulation model

The actual air-conditioning system components of the example building are simulated with TRNSYS program using the measured weather data from BAS ranged from 24°C to 35°C for the outdoor air temperature and the humidity range between 50 and 75% relative humidity as shown in Figure 4. This simulation model consists of (1) a cooling tower, three water-cool and normal water flow chiller units (2) two simplified cooling coils and constant air volume fans for the two office zones (3) three constant condensers and primary pumps (4) SketchUp model embedded into TRNSYS program, and (5) The measured weather data from the BAS of the building.

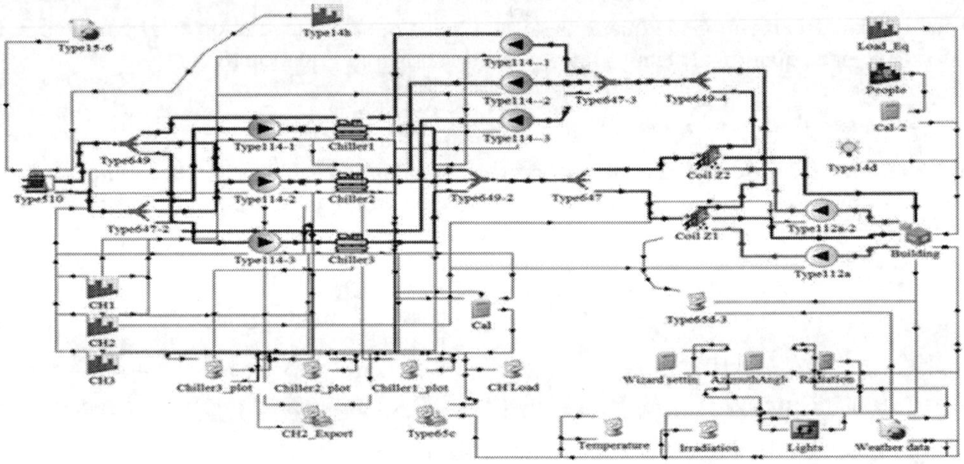

Figure 4. TRNSYS model for the water-cool chiller system of the example building

3.4 Validation of the example building model

The building simulation model using TRNSYS is validated via the data obtained the BAS of the building before installing the stack ventilation system. The test is scheduled Monday through to Friday, using the operation period and the actual opening and closing building information. Then, the validation of the model is conducted via the comparisons between the zone temperature and relative humidity of the BAS data and TRNSYS simulation.

3.5 Stack effect investigations

From the verified model, it can be used to further simulate airflow inside the chimney by using air flow data from LoopDA to investigate the stack effect instead of physical sensor installations due to cost-prohibitive and installation difficulty issues.

4. Results

4.1. LoopDa Program

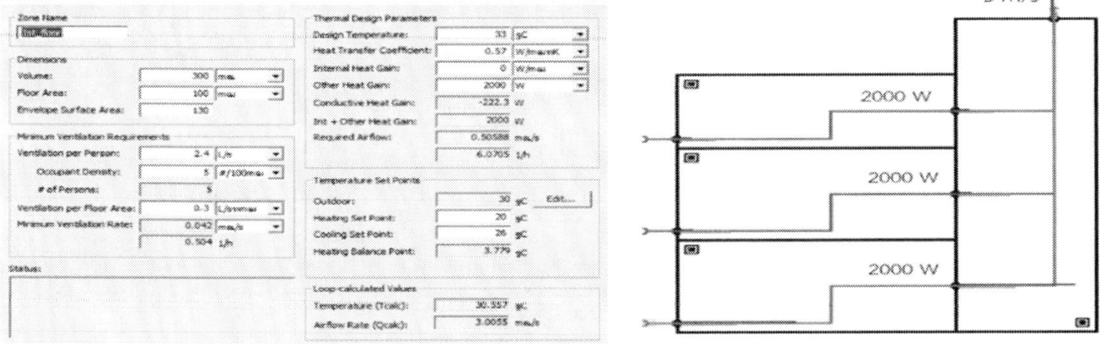

Figure 5. LoopDa Results

Air-flow through the chimney is simulated with the wind speed at 3 m/s according to the design condition. Also, internal heat gain is 6000 W added (recommended value from a designer) to the top of the chimney in order to create sufficient stack pressure. By inserting the internal heat into each floor equally, heat flow at 2000W can create a layer of temperature in the chimney as shown in Figure 5.

4.2. BAS data

The measurement data of the example building can be downloaded from the BAS (Figure 6) using the web browser control. Indoor relative humidity and temperature of the tow zone offices including outdoor air data are from 121/0 2018/to 121/62018/.

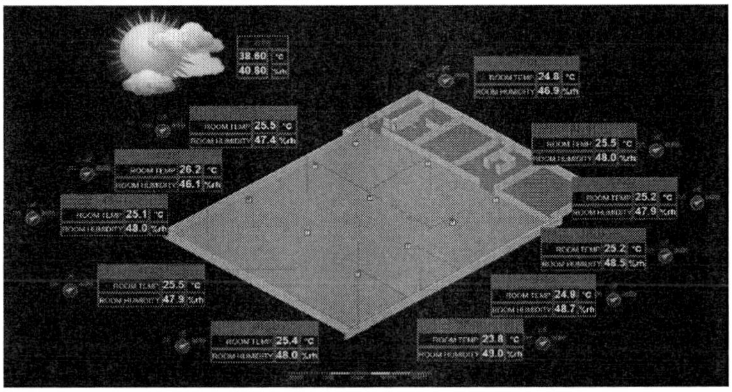

Figure 6. BAS graphical user interface of the example building

4.3. TRNSYS model validation

In Figure 7, the temperature of the simulated model is compared to the actual BAS data; the maximum temperature difference is around 3.6°C and the minimum temperature difference is -1.8°C for the whole periods. However, if the work periods are considered only, the temperature differences are within ±5% errors, while relative humidity differences are within ±10%, which is accurate enough within ±15% according to ASHRAE (American Society of Heating, Refrigeration and Air-conditioning Engineers) Guideline 14 [7] for measurement of energy saving applications.

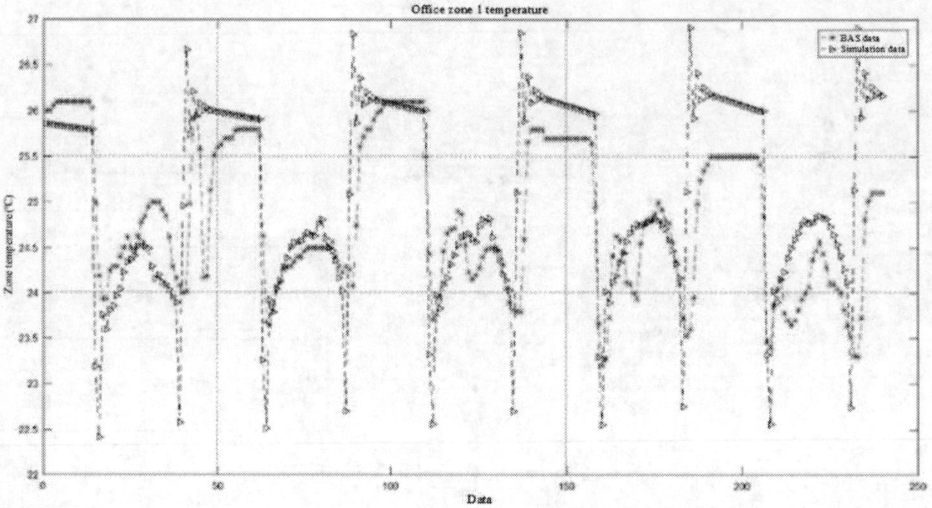

Figure 7. Comparison results between BAS data (red line) and simulated data (black line)

4.4. Stack ventilation effect

The data of LoopData are used as the inputs of stack ventilation simulation in the buiding's chimney. When comparing the zone temperature next to the chimney with wind velocity 3 m/s, there is the air flowing through the chimney causing the average office temperature to decrease by 0.2% from the original temperature (red-line in Figure 8). In terms of energy based on the equations in Ref [8], the enthalpy is 6900W, which is close to the energy obtained from the loopDA model on 6000W. The findings are the heat sources from 5500 to 6900 W required to incur natural stack ventilation in the chimney of the building.

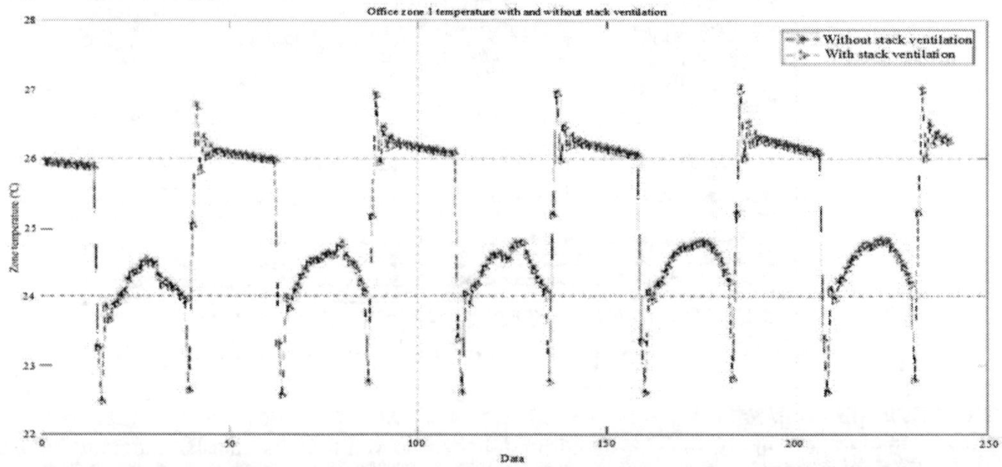

Figure 8. Results between without chimney (red line) and with chimney (black line)

5. Conclusion

The temperature layer in the chimney causes sufficient stack pressure to increase airflow according to the stack ventilation theory. The system development process in this research can be used to test the design of a stack ventilation system efficiently before the actual construction begins. However, real air-flow stack ventilation should be measured by physical sensors for leveraging building simulation platform for larger building systems.

Acknowledgments
The authors are deeply grateful to MyEnergy Domain (Thailand) Co., Ltd and EEC Engineering Network Co., Ltd for financial support.

References

[1] Department of Alternative Energy Development and Efficiency, Ministry of Energy 2014 *Guidelines, Best Practices for Energy Conservation in Designed Building and Designed Factory* (Thailand)
[2] Papadakis G Mermier M, Meneses J Boulard T 1996 *J. Agric. Eng.* **63** 219–228
[3] Brown G Z and Kay M D 2001 *Sun, Wind & Light: Architectural design strategies* (New York:John Wiley & Sons)
[4] Fiedoruk K G Zhelykh V and Pushchinskyi A 2019 *Energies* **12** 28–45
[5] Dols W S Emmerich S J Polidoro BJ 2012 *NIST Technical Note 1735 LoopDA 3.0 – Natural Ventilation Design and Analysis Software User Guide* (NIST Pub Series)
[6] Mitchell J W and Braun J E 2012 *Heating, Ventilation, and Air Conditioning in Buildings* (New York:Wiley–Interscience)
[7] Liu G and Liu M 2011 *Building Environ* **46**(2) 409–420
[8] Promwattanapakdee T and Woradechjumroen D 2016 *Proc. Int. Conf. on Engineering Science and Innovative Technology* (Phuket:Thailand)

IOP Conf. Series: Materials Science and Engineering **886** (2020) 012041 doi:10.1088/1757-899X/886/1/012041

Influence on the Electrical Efficiency of a Hybrid MGT-SOFC-System by µ-fogging in a-Two-Staged Compressor System

R Dückershoff[1*], H P Berg[2], A Himmelberg[2], M Lehmann[2] and M Kleissl[2]

[1]University of Applied Sciences Mittelhessen (THM), Friedberg, Department of Mechanical Engineering, Mechatronics and Material Technology, Wilhelm-Leuschner-Str. 13, 61169 Friedberg, Germany
[2]Brandenburg University of Technology Cottbus-Senftenberg, Chair of Combustion Engines and Flight Propulsion, Siemens-Halske-Ring 14, 03046 Cottbus, Germany

* Corresponding Author: roland.dueckershoff@m.thm.de

Abstract
Hybrid combinations of solid oxide fuel cell and recuperated micro gas turbines can convert the chemical energy of hydrocarbon-based fuels in electrical energy with high electrical efficiency. With an integrated and improved cycle management, more than 70% of the energy content of the fuel could be converted. Therefore, the systems are highly suitable for the Power-To-Gas conversion. In particular, a pressure charging of the SOFC fuel cell leads to an increase in stack performance. By a downstream turbo set, after residual fuels are intentionally oxidized with an afterburner, additional electrical energy can be gained from the expansion of the hot exhaust gas stream and the overall efficiency can be increased. In order to increase the electrical efficiency of the system, it is proposed, to ensure the required compression of the process air in particular by a-two-staged turbo compressor with an intermediate cooling system. By thus achievable reduction of the dissipation of the compressor and by targeted condensation of finest drops in front of the second compressor stage affected by intermediate cooling, an increase in efficiency of the system is possible. This is achieved by targeted cooling of the process air behind a low pressure compression, so that it is saturated over 100% relative air humidity. As a result, a slightly supersaturated airflow is available for the second compressor stage, which enters the compressor after heat removal via an intermediate cooling having a small number of microdroplets. Therefore, the condensed water evaporates again by the heat of compression in the second stage and the compressed flow ultimately enters the recuperation at a lower temperature than during normal compression. Thus, more heat can be recovered within the recuperation system. Therefore, the electrical energy of the system can be produced having higher efficiency, because the heat dissipation of the overall system decreases. In this article it is presented, how such a process is thermodynamically modelled and how a technical realization can be built after optimization by simulations. Finally, in this study, the process-influencing factors are analyzed to show the highest possible electrical yield of such a system.

1. Introduction

The hybrid power plants from Micro Gas Turbines (MGT) and high-temperature fuel cells (Solid Oxid Fuel Cells / SOFC) promise high electrical efficiency in energetic recovering Power-To-Gas fuels from renewable energy systems [1], [2], [7], [7]. The residual fuel released by an incomplete fuel conversion in the SOFC is treated by an afterburning. So, it provides additional energy in the gas turbine component.

In particular, the efficiency of the micro gas turbine can be increased by using the so-called Fogging and thus, its specific output of system performance can be raised [3], [4], [5]. The investigation is intended to show that the intermediate appearance of liquid water in the compression system leads to an increase in the efficiency of the MGT part and thereby to an increase in the overall efficiency of the system [10].

For this purpose, the development of a-two-staged compression system with intermediate cooling down to the entrance temperature is being presented. By taking out heat after the first stage, it is expected that microdroplets will be generated in the airflow by condensation. Then, the adiabatic cooling by the evaporation effect helps to run the second compression stage more efficient. Now, the air has been compressed to the operating pressure but having a lower temperature than without intermediate cooling. The air is led through a recuperator, wherein, regarding the lower entrance temperature, more heat within the system can be recovered. Thereafter, additional heat required will be supplied in one case by a combustion chamber system and in the other case using a hybrid system of SOFC fuel cell and the afterburner.

As a result, it will obtain an increase in the performance of the overall machine. It also requires less energy during the compression process. The heat rate shows shows a significant increase. Inlet fogging also reduces emissions of oxides of nitrogen (NOx) because the additional water vapour quenches hot spots in the combustors of the gas turbine [10].

All tasks that concern the mechanical stress regarding the thermal impact on blading, rotor and casing of the compressor have to be kept in mind. Therefore, we have to avoid large drops and agglomerations of water and still ensure a sufficient evaporation rate should be established. Furthermore, in a hybrid system with SOFC fuel cells, the operating temperature must be ensured via recuperator and a high-temperature heat exchanger. Through a targeted recovery of combustion heat and waste heat of the SOFC fuel cells, the overall process can be made highly efficient. In this study, the processes MGT and MGT-SOFC are presented with a thermodynamic comparison, both with and without intermediate cooling within the compressor. It will be shown that an increase in efficiency is possible having the appropriate use of micro-fogging by condensation.

The reduction in heating during compression shows an energetic advantage of the procedure

2. Thermodynamic comparison of the processes MGT, SOFC-MGT with or without intermediate cooling

A micro gas turbine process with double compression and intermediate cooling will be presented, in this section. Initially, air with a defined humidity, e.g. moistened by a wetted packed bed, is sucked in by a supercharger, which is used to be the first stage and thereby is precompressed to an intermediate pressure. Then an intermediate cooling by an intercooler follows. Since the airflow is now cooled down to the level of ambient temperature, its vapour pressure is lowered to the level of the ambient conditions, while the pressure is increased to the intermediate pressure level. Depending on the inlet conditions, the airflow is now supersaturated with moisture. As a result, microdroplets condense. Then, in a second stage, the actual compression to the operating pressure of the micro gas turbine system is performed. Due to the lower inlet temperature and the evaporation of the microdroplets during compression within the second stage, the airflow from the compressor occurs at a lower temperature compared to the normal process. In addition, having a lower level of temperature, the compression done in the second stage can be done with less workload. Now, the airflow is going to be heated in a recuperator system up to the combustion chamber inlet temperature, resp. the supply temperature of the SOFC stacks. Due to the lower inlet temperature of compressed air in the recuperator, more heat can be recuperated from the flue gas from turbine exhaust.

In the combustion chamber, the heat is generated by gas combustion at a working pressure of around 5 bars. This is followed by the expansion of the working gas in a high-pressure turbine to almost the pre-pressure of the low-pressure turbine. Further, on, the exhaust gas is passed through the recuperator and is used for preheating the combustion air. As a final step, the exhaust gas is driving the supercharger by passing its low-pressure turbine.

In the modified hybrid process, the combustion chamber is replaced by a combination of a SOFC fuel cell cluster and an afterburner system, as shown in Fig. 1. It is accepted that the compressed air flow on the airside into the SOFC stack system is with higher moisture content, but sub saturated.

An airflow with a higher vapour content has a higher heat capacity. As a result, smaller mass flow can dissipate the same amount of heat produced during the fuel cell conversion. However, the partial pressure of the water vapour is so low on the airside that only a marginal change can be expected there.

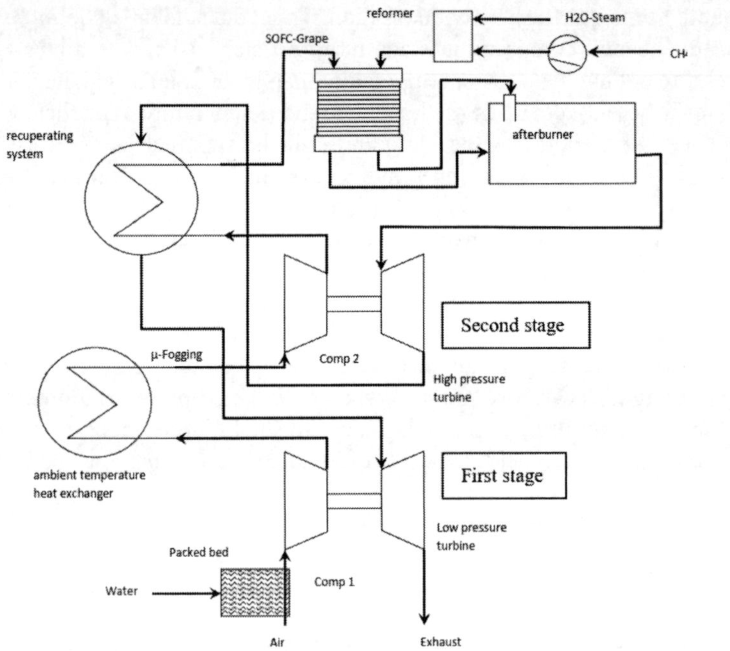

Figure 1: Connection scheme of MGT-SOFC-Process with intermediate cooling and µ-fogging installation between the first and the

The heat diagram (Fig. 2) shows the MGT-Process and the advantage of micro-fogging in the second stage. The micro-fogging causes a shift of the point D (to D') by adiabatic cooling and thus, a shift of

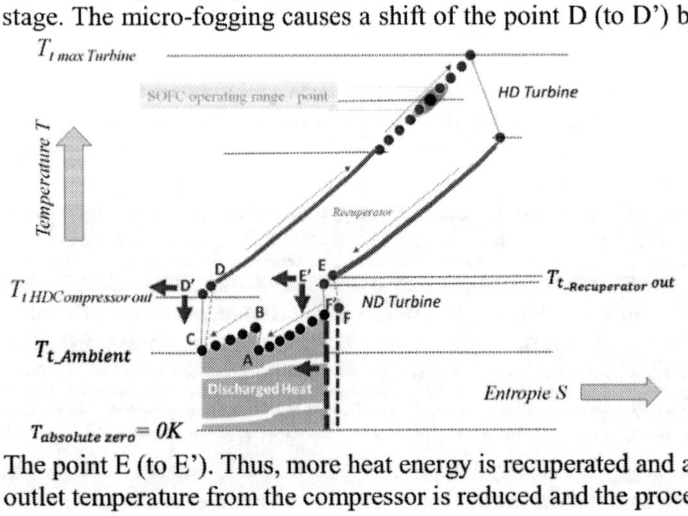

Figure 2: TS-Heat diagram of the MGT-Process with intermediate cooling and µ-fogging

The point E (to E'). Thus, more heat energy is recuperated and as a result, exergy losses are lower. The outlet temperature from the compressor is reduced and the process step needs less energy.

IOP Conf. Series: Materials Science and Engineering **886** (2020) 012041 doi:10.1088/1757-899X/886/1/012041

The efficiency of the overall process can be determined from the recovered and usable energy. It is to be distinguished between the electrical efficiency and the overall efficiency, which includes subsequent waste heat utilization. Since both the fuel cell and a micro gas turbine jointly supply electrical energy, more electrical energy can be obtained from the energy content of the gas than is possible with the individual components. The prerequisite for an increase in efficiency is that heat integration can be successfully achieved.

The energetic behaviour of the humid air depends on its thermal and caloric state equation variables. Density an specific heat capacity depend directly on the relative humidity resp. its degree of saturation, as shown e.g. in [9], 0. The air mass flow is directly related to the density of air and the flow of volume at the entrance of the system.

$$\dot{m} = \rho \cdot \dot{V}$$

The density of the air depends on the relative humidity, pressure and temperature. See also [10]:

$$\rho_{humid\ air} = \frac{p_d}{R_d \cdot T} + \frac{p_v}{R_v \cdot T} = \frac{p_d \cdot M_d + p_v \cdot M_v}{R \cdot T}$$

with index *d* for dry air, *v* for water vapour and R as Universal Gas Constant.

Typically, thermodynamic relationships of humid air are given by its relative humidity φ and its water content *x*.

$$\varphi = \frac{p_v}{p_s} \quad and \quad x = \frac{m_v}{m_d}$$

The relation between both is given by:

$$x = \frac{R_d}{R_v} \cdot \frac{p_s}{\frac{p}{\varphi} - p_s}$$

Thereby, p_s is the pressure of saturation, which is mainly depending on the temperature as long as the overall pressure of the humid air is less than 10 bars. Exceeds the vapour pressure p_v the pressure of saturation p_s, the relative humidity φ is above 100% and water will start to condense.

The overall pressure is to be calculated by the partial pressures:

$$p = p_d + p_v$$

By pressurizing the humid air, its partial pressures will increase by the same ratio as the overall pressure does:

$$\frac{p_{out}}{p_{in}} = \frac{p_{d,out}}{p_{d,in}} = \frac{p_{v,out}}{p_{v,in}}$$

Therefore, the relative humidity φ will increase at a higher pressure having the same temperature.

Caloric state variables of the humid air are to be calculated by its water content *x*, as in the example shown for the specific heat c_p.

$$c_p = \frac{c_{p,d} + x \cdot c_{p,v}}{1 + x}$$

The needed chemical energy flow of the fuel can be determined by its mass flow and the lower calorific value:

$$E_{chem.} = \dot{m}_{Fuel} \cdot H_{U,Fuel}$$

In a micro gas turbine system, the electrical efficiency can be determined by the ratio of the electric energy yield to the chemical energy flow:

$$\eta_{el} = \frac{E_{el.,MGT}}{E_{chem.}}$$

To determine the electrical efficiency, the electrical energy yields of fuel cell fraction and micro gas turbine fraction in a hybrid system have to be taken into account. In order to consider the compression

IOP Conf. Series: Materials Science and Engineering **886** (2020) 012041 doi:10.1088/1757-899X/886/1/012041

work of the fuel gas and the steam, needed by the internal steam reformer, the energy expenditure for these sub-processes have to be subtracted from the total energy yield:

$$\eta_{el} = \frac{E_{el.,SOFC} + E_{el.,MGT} - E_{comp.,fuel} - E_{comp.,steam}}{E_{chem.}}$$

The energy effort for the evaporation of the water of the reformer system and the gas heating up to working temperature is available internally by heat exchanger systems and other heat integration methods. In addition, if, heat energy is to be used via an exhaust gas heat exchanger, the overall efficiency will be higher and can be calculated as:

$$\eta_{ges} = \frac{E_{el.,SOFC} + E_{el.,MGT} - E_{comp.,gas} - E_{comp.,steam} + E_{therm.}}{E_{chem.}}$$

By use of humid air and an inter-cooling system, an increase in efficiency can be achieved in a two-staged compressor system. The lower intermediate inlet temperature in the compressor requires less energy to compress an equal mass flow. By cooling the airflow after a first compression stage in a heat exchanger to ambient temperature, the inlet temperature of the second stage is reduced. In this case, the cooling process must be controlled so that only microdroplets condense, which then evaporate during the second compression stage. This action reduces the expected outlet temperature. The entire airflow can be compressed to the working pressure of the combustion chamber or fuel cell system with less energy expenditure. Next, after expansion in the turbine, the system can recover more energy in the recuperator from the exhaust stream. This is mainly possible due to the lower inlet temperature of the recuperator rear of the compressor exit side.

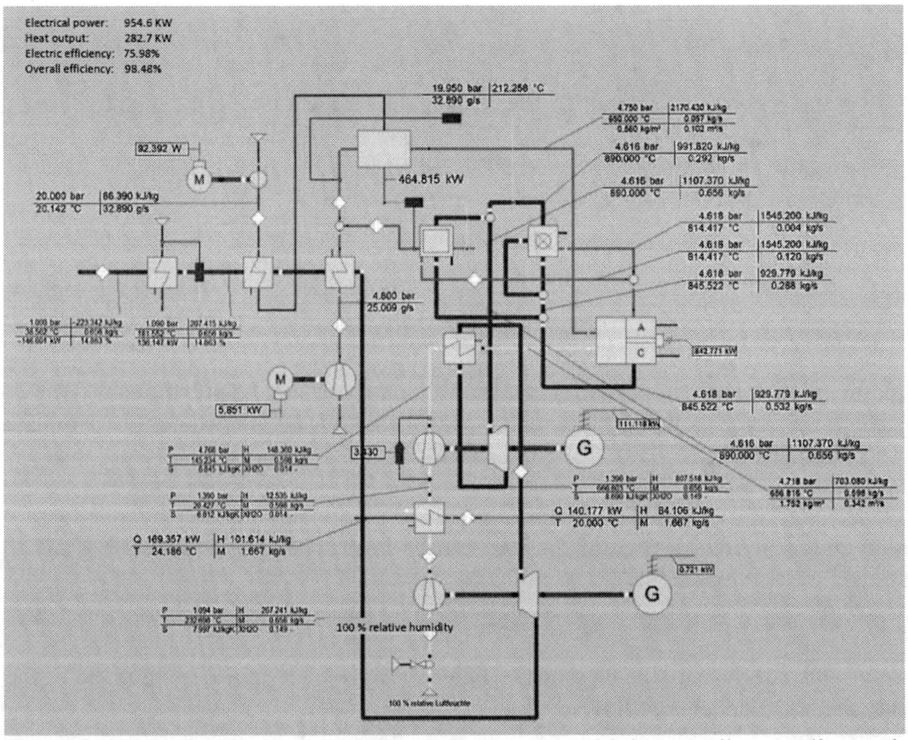

Figure 3. Simulation scheme for MGT-SOFC-Process with intermediate cooling and µ-fogging with EBSILON Professional

In hybrid power plants made of MGT and SOFC systems, an increase in the efficiency of the MGT fraction also leads to an increase in the efficiency of the complete system. Moreover, the higher water content due to the increased air humidity at systems entry can improve the conditions for the SOFC conversion on its airside, since here the heat capacity increases slightly.

The simulation is done with the program system Ebsilon 0. There, a 1D model of the considered processes was mapped. Fig. 3 shows a simulation schema for the MGT-SOFC-Process. First, the system simulation has been running with MGT fogging, therein variations of the air humidity at the inlet of the first compressor stage have been done. Subsequently, the simulation of MGT-SOFC hybrid system has been conducted. The fogging took place in both systems in front of the first compressor stage. Therefore, the heat of compression has been used to vaporize the mist so that the temperature rise is less than in no fogging systems. At lower temperatures, the required compression energy is also less.

In order to make the different simulations comparable, the same inlet mass flow into the high-pressure turbine was defined as the boundary condition. For this purpose, compression pressure and temperature level for the combustion chamber area resp. of the SOFC fuel cells area were specified as further fixed boundary conditions. The variation parameter used has been the specific water input using the fogging at the system entrance. Therefore, humidity at entry into the second compressor stage as well as regarding the overall process has been increased.

However, by the simulation program, only a limited water content, which exceeds a saturated atmosphere, could be simulated. Thus, the maximum relative humidity at ambient temperature and under pressure conditions has been limited to 120%. Required for comparison, the constant mass flow has been assumed.

3. Efficiency Analysis

A comparison of the thermodynamic efficiencies has been carried out in particular for a one-stage and an intercooler two-stage compressor MGT system. Therefore, the relative humidity at the entrance of each system has been used for the variation parameter. A sensitivity study regarding performance and efficiency has been done. Fig. 4 shows the impact of the relative humidity on the electrical efficiency in both cases.

In all simulations, it has been found that an increase in air humidity leads to an increase in the overall efficiency of the MGT system. In the case of having a single-stage compressor within the MGT system, it has been found that the system's electric efficiency does not increase if the air at the entrance of the system is supersaturated. Furthermore, it can be shown that the efficiency of the investigated MGT system can reach 32.7% at maximum level.

In contrast, the electric efficiency of the MGT system having an intercooled two-staged compressor is a much higher, at least 36.5%. Moreover, is shown, using fogging for increasing the relative humidity at the system's entrance increases the overall efficiency. A special point of inflexion is shown. There the air humidity reaches 70%. From here, the overall system's efficiency increases sharply regarding the air humidity at the system's entrance. At this point, the relative humidity at the entrance of the second compressor stage has reached 100% and water will start to condense. Here the micro-fogging process within the compressor has taken place. The simulations have been done up to a relative humidity of 120% based on ambient pressure and temperature at the inlet of the system. At higher water content level, convergence problems occurred in the simulation system, which did not allow for overall results of the calculation.

At a relative humidity of the entrance airflow by 120% in the simulations regarding the two-stage process, the maximum electrical efficiency of approximately 40% has been achieved for the MGT process. Thus, this variant offers a good opportunity to increase the overall efficiency of the MGT system. If by exhaust heat exchanger systems, additional usage of heat could be done, e.g. regarding heating purposes, the overall efficiency can be increased up to over 99%, based on the lower heating value.

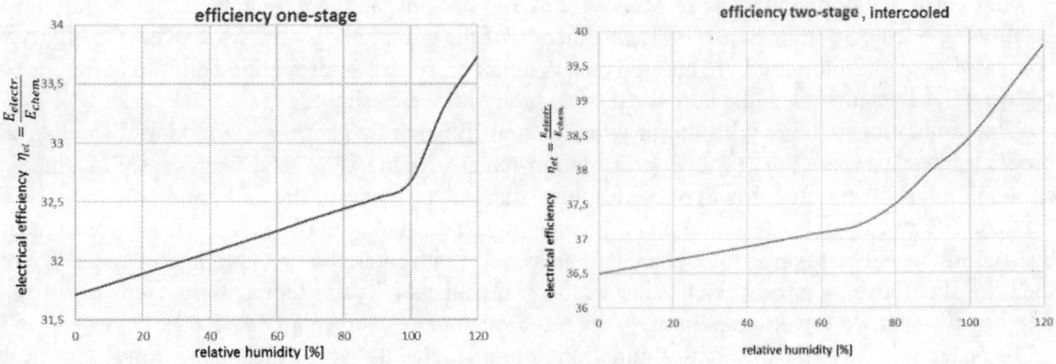

Figure 4. Electrical efficiency of the simulated MGT systems with a one-stage and a cooled two-stage compressor process in relation to the relative humidity at the system entrance

Moreover, placing instead of the combustion chamber a combination of SOFC and afterburner, as shown by Fig. 1, the so-called hybrid MGT-SOFC system has been simulated. The comparative simulation by Ebsilon 0, even much more complex, as shown by Fig. 3, shows that the electrical efficiency of 74% can increase by 2% using an intercooled compressor having micro-fogging internally. At least, the simulation shows an electrical efficiency of around 76% (see Fig. 3).

4. Discussion

In this publication, a two-stage compression system with a fog system in front of the first compressor stage in combination with an intercooler was presented. Overall, the simulations show that with such a system, the electrical efficiency of a MGT system can be significantly increased. A benefit in electric efficiency of 8% seems to be possible. Regarding the MGT-SOFC system the benefit in electric efficiency seems to be lower, being 2%. However, this is not surprising, since the electric efficiency of the MGT-SOFC process already is at a high level having more than 70% and the MGT part of this system, which has the gain by intercooling and micro-fogging, has a minor share of around 1 to 7 regarding its overall performance [1].

On one hand, such an increase in efficiency can be explained by the reduction of the workload required for compression. Intercooling generates lower inlet temperatures into the second compression stage, as well as supersaturation of the inlet air with microdroplets. This reduces the energy expenditure for the compression and reduces the outlet temperature from this subsystem. On the other hand, it is achieved by the lower inlet temperature into the recuperator, that from the hot gas side after the first turbine stage more energy can be taken out. This allows remaining more exhaust heat in the process. Regarding the MGT-SOFC system, it has to be pointed out that an increase in water content, due to higher heat capacity of the water vapour, may increase slightly the heat capacity of the air at the entrance of the SOFC stack. A positive effect can be here, that heat conduction within the SOFC fuel cells will be improved. However, this has to be investigated more in detail. However, higher water vapour content in the overall system ultimately may entail risks concerning the risk of fuel cell de-integration.

Not at least, increasing the operating pressure of the SOFC fuel cells by use of a two-stage compressor will have a positive effect. An increase of the output by the SOFC system can certainly be expected up to a system pressure of 8 bar [6]. Future investigations have to be done in particular to determine an optimal setting of the pressure level between the first and second compressor stage. Overall, the balancing by optimization tasks has to be done, so that a maximum in efficiency gain can be achieved.

Appropriate technical process management should also prevent problems regarding agglomeration of water or too large droplets, e.g. to protect the second stage of the compressor against damage by erosion and thermal stress. Thus, the generation of fine microdroplet formation by water condenses in the intercooler, the so-called micro-fogging, is required as a control variable.

However, significant is the risk that, due to the lower process temperature and the higher vapour content at the exhaust gas side of the recuperator water droplets due to condensation will occur and will reach the second turbine stage. In the waste heat recovery system, a sufficient distance must be maintained.

5. Conclusion

The concept and simulations regarding the determination of efficiency for a two-staged system using humid air for compression within MGT and hybrid MGT-SOFC processes have been presented in this study. Through the simulations and analysis, efficiency advantages have been proven useful for both systems. The MGT system will be able to increase electrical efficiency up to 40% having a two-staged compressor with intercooler and a fogging system at the entrance of the compressor. In a hybrid MGT-SOFC system, this may result in an increase of the electrical efficiency by up to 2% and therefore up to an electrical efficiency of about 76%.

Further investigations will be carried out. To study process variants with a relative humidity more than 120% at the entry into the main compression stage, an adaptation of the thermodynamic calculation models handling a higher water content are required. A realization of the intercooled two-stage compression having micro-fogging on a laboratory scale and further investigations, e.g. regarding the problem of erosion, corrosion and material stress, are planned.

In an update to the hybrid system, Turbo-Fuel-Cell intercooling which has micro-fogging within the system should have to be taken into account. For this purpose, assembly capability and handling the heat management of the overall system are to be optimized. However, as a first step, the technical system should be implemented as MGT-SOFC Turbo-Fuel-Cell 1.0 without micro-fogging, enhancements to do this process to increase efficiency by micro-fogging systems are conceivable following their implementation phase (to be in service in 2026).

References

[1] Berg H P Himmelberg A Lehmann M Dückershoff R Neumann M 2018 *IOP Conf. Series: Materials Science and Engineering* **297** 012004

[2] Berg H P Kleissl M Himmelberg A Lehmann M Prechavut N Vorpahl M 2019 *IOP Conf. Series:Materials Science and Engineering* **501** 012007

[3] Bracco S Pierfederici A Trucco A 2007 *Applied Thermal Engineering* **27** 699-704

[4] El-Shazly A A Elhelw M Sorour M M El-Maghlany W M 2016 *Alexandria Engineering Journal* **55** 1903

[5] Tahani M Masdari M Salehi M Ahmadi N 2018 *Applied Thermal Engineering* **143** 955

[6] Henke M Willich C Westner C Leucht F Leibinger R Kallo J Friedrich K A 2012 *Electrochimica Acta* **66** 158-163

[7] Berg H P Krienke C 2015 *Magdeburger Maschinenbautage* CD-ROM, UB Magdeburg *12*

[8] Berg H P Dückershoff R Lehmann M Prechavut N 2017 *Proc. 12th European Conference on Turbomachinery Fluid dynamics & Thermodynamics* (Stockholm:Sweden)

[9] Baehr H D Kabelac S 2016 Thermodynamik (Springer)

[10] Marek R Nitzsche K 2015 *Praxis der Wärmeübertragung* (München:Hanser)

[11] STEAG System Technologies 2018 *EBSILON®Professional-User Manual* (Zwingenberg, STEAG Energy Services GmbH)

[10] Sanjeev Jolly, https://www.powermag.com/inlet-air-cooling-systems-improve-gas-turbine-performance/, accessed 30 Apr 2017.

IOP Conf. Series: Materials Science and Engineering **886** (2020) 012042 doi:10.1088/1757-899X/886/1/012042

Development of a Solar Water Distiller with a Receiver and Condenser

Hieu Tri Le[1], Tawit Chitsomboon[1], and Atit Koonsrisuk[1*]

[1]School of Mechanical Engineering, Institute of Engineering, Suranaree University of Technology, Muang District, Nakhon Ratchasima 30000, Thailand

* Corresponding Author: atit@sut.ac.th

Abstract

The solar water distillation distills brackish water, which can mitigate the demand of pure water for the human beings. This study proposes a solar water distiller with a condenser and collector made from a tube bank. The receiver and condenser are connected by the rubber tubes. The brackish water from a tank flows along with condenser. The water temperature is relatively low, so this tube bank acts as a condenser for the water vapor that evaporates from a water basin of the distiller. It is believed that this can increase the condensation rate of the system. After passing through the condenser, the brackish water flows up along the receiver and its temperature is increased. Then the water continuously flows to the basin. According to the experimental results, the water's temperature in the proposed system is higher than that of the conventional distiller. This leads to the higher productivity of the proposed system (approximately 3.3 L/m2 day). In addition, it was found that the temperature of the condenser of the proposed system was lower than that of the top cover of the conventional system.

Keywords: solar still, distillation, water production, solar collector.

1. Introduction

Water is an important substance on our planet, which is vital for a sustainable future. Although water covers approximately three-quarters of the Earth, 0.014 percentage of water is used directly for the human being and other demands [1]. Global climate change and population explosion cause the depletion and deterioration of potable water and create water-stressed areas. When water scarcity become the challenges in the worlds, seawater plays an important role as an alternative source for drinking water as well. In order to use this water, it has to be desalinated. There are various advanced techniques for seawater distillation, which among them, solar still is a promising and environmentally friendly device. The performance of a solar distillation of water based on the principle of the hydrologic cycle, which refer to two processes namely evaporation and condensation. Distillation is always taking place in a closed system. The sun's energy goes through the glass to evaporate brackish water to vapor. Then water vapor condenses on the surface of condenser cover. Phadatare et al. [2] study the impact of water level on the performance of a single slope plastic solar still. Experimental results showed that the maximum yield is 2.1 L/m2/day at a water level of 2 cm. Meanwhile, double slope solar still was performed by Kulandaivel and Karuppaih [3]. The experiment indicated that the average yield of the still was 2.1 L/m^2/day however, Improving the productivity of solar still has been remained a challenge to scientists. Vishwanath Kumar et al. [4] summarized various types of passive and active solar still.

The improvement of the temperature difference between water and the glass cover (which acts as the condenser) enhance the solar still's productivity. This temperature difference could be raised by the increase of feed water and/or the decrease of condenser's temperature. The former can be done by heating the feed water through the solar collector. As a result, solar still coupled with different solar receivers, e.g. evacuated tube collector, heat pipe collectors, parabolic trough collector and flat plate solar collector, have been investigated and reviewed [5-9]. The latter can be achieved by using air or water to cool condenser surface. Tiwari et al. [10] use water to reduce the condenser temperature of passive solar still and the yield of solar still is almost double. However, increasing water flow rate reduced the productivity of solar still. Arunkumar et al. [11] used water and air to improve the performance of tubular solar still. The distillate production of tubular solar still was 2.05 l/day without the cooling flow and it was improved to 3.05 l/day and to 5 l/day with the flow of cooling air and water respectively.

Furthermore, reuse of latent heat of vaporization is one of the interesting methods to enhance the efficiency of solar water distillation, however, few experiments have conducted to utilize it. In the solar still, most of the heat energy is lost to ambient through the top cover. Some authors were performed to reuse this heat loss to heat the water in the additional distillate, which known as a multi- stage or multi basin or multi-effect distillation. Both experiment and theoretical study of a double basin of solar still was carried out by Sodha et al. [12]. The result showed that the daily yield of a conventional solar still is 36 per cent lower than that of such a solar stiller.

The present work is to create a new technique for solar water distillation. A tubes system, which is used as collector and condenser, replace glass cover in conventional still. The water vapour is transformed into fresh water on the surface of tubes and heats up the water in the tube transfer its latent heat in the process. In this way, it is hoped that an energy-saving will be achieved resulting in an efficient solar distillation.

2. System design and experimental performance

2.1. Solar still description

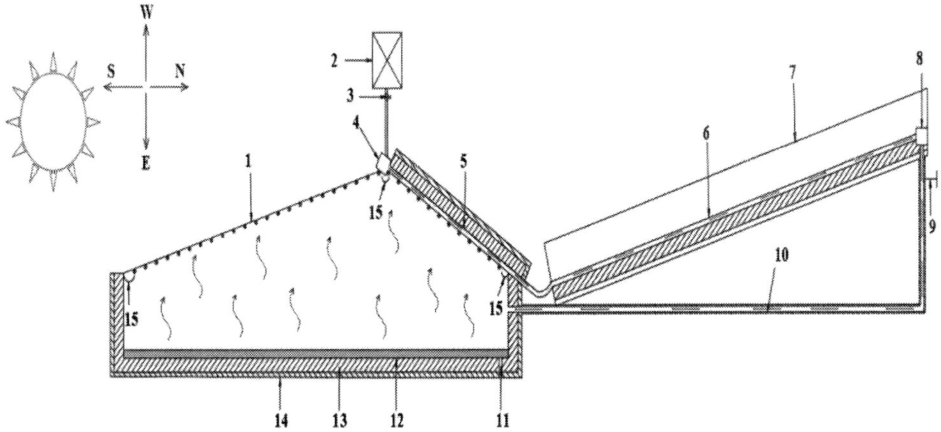

Figure 1. A cross-sectional view of a schematic diagram of solar still model. 1 - Glass cover of a solar still; 2 - water tank; 3 - valve of the water tank; 4 - condenser trough; 5 - condenser tubes; 6 - receiver tubes; 7 - glass cover of receiver tubes; 8 - receiver trough; 9 - valve; 10 - connecting tube; 11 - Drainage hold; 12 -Basin liner; 13- foam insulator; 14-wood cover ; 15- freshwater collector.

The design of solar still is described in Fig. 1. The number of condenser tubes equal to that of collector tubes (40 tubes). The tubes are made of copper of 8.5 mm inner diameter. The collector tubes are painted black. The receiver and condenser tubes connect to rectangular troughs respectively. Singh and Tiwari [13] observed that in order to maximize the annual productivity of solar still, the angle of glass should be the latitude of the experimental location. Hence, the tilt angle of the collector tubes and glass cover was chosen to equals the latitude of the location (15°) while the condenser tubes incline 45°. The basin liner is made of stainless steel, 0.4 mm thickness. It was insulated through the four sides and the bottom by aero foam, 20 mm thickness. The basin liner and insulation are fabricated inside the wooden box. The silicon is used to prevent any leakage between the wooden box and top cover. The materials selection and design specification of the solar still is shown in **Table 1.**

Table 1. The material of the different parts of the solar still

Glass cover	Material: glass Thickness: 5 mm
Condenser and receiver tubes	Material: Cooper Inner diameter: 8.5 mm Length of receiver tubes: 1000 mm Length of condenser tubes: 450 mm
Rectangular trough of receiver tubes	Material: Copper Size: 400x30x30 mm
Rectangular trough of condenser tubes	Material: acrylic Size: 400x30x30 mm
Basin liner	Material: stainless steel Size: 1000x200x380 mm Thickness: 4 mm
Insulation	Material: Aero foam Thickness: 25 mm
Wood cover	Thickness: 10 mm

Figure 2. Pictorial view of solar water distiller

IOP Conf. Series: Materials Science and Engineering **886** (2020) 012042 doi:10.1088/1757-899X/886/1/012042

2.2. Operation principle

The experimental work of this study was conducted on the campus of Suranaree University of Technology, Nakhon Ratchasima, Thailand (Latitude 14.9738°N and longitude 102.0837°E).

The brackish water was supplied from a water tank and flow through condenser and collector tubes respectively. After preheating in receiver tubes, feed water that is controlled the mass flow rate by a valve flow to the solar still through the connecting pipe. The mass flow rate of water flowing to solar still is 1 L/h and the water level is kept constant with 1 cm. Water continuously is heated by solar energy in solar still to evaporate and water vapour condense on the surface of glass and condenser tubes. The condensed water is collected through the distillation trough. During the experiment, the parameters were measured and recorded by data - logger: ambient temperature (T_a), the collector temperature (T_{cl1} and T_{cl2}), condenser tubes temperature (T_{cd}), the temperature of inlet water (T_{wi} is measured in the water tank), the temperature of water out collector (T_{wo}), basin water temperature (T_{ws}), basin liner temperature (T_b), humid air temperature (T_{abs}) and relative humidity (RH). A K-type digital thermometer measures the glass's temperature. The solar radiation was measured and recorded by a Kipp and Zonen pyranometer. Pictorial view of present work is illustrated in figure 2. All measurement points are illustrated in Figure 3.

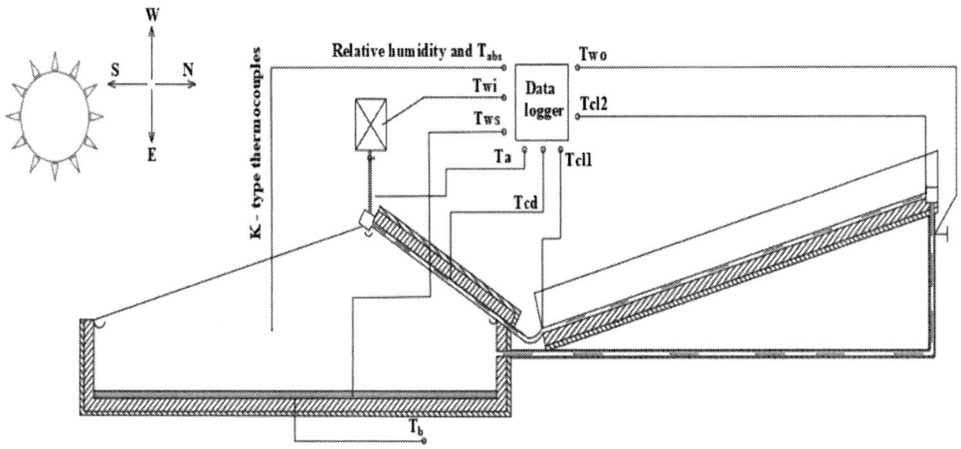

Figure 3. Measurement points of solar still

3. Result and discussion

Figure 4 depicts the solar radiation with respect to time on an experimental day (21/06/2019). It is observed that solar irradiation increased to the highest value at 12.30 o'clock (approximately 700 W/m^2) and decreased after that gradually. During the test, it was cloudy. This was the reason for the fluctuation appeared in the figure.

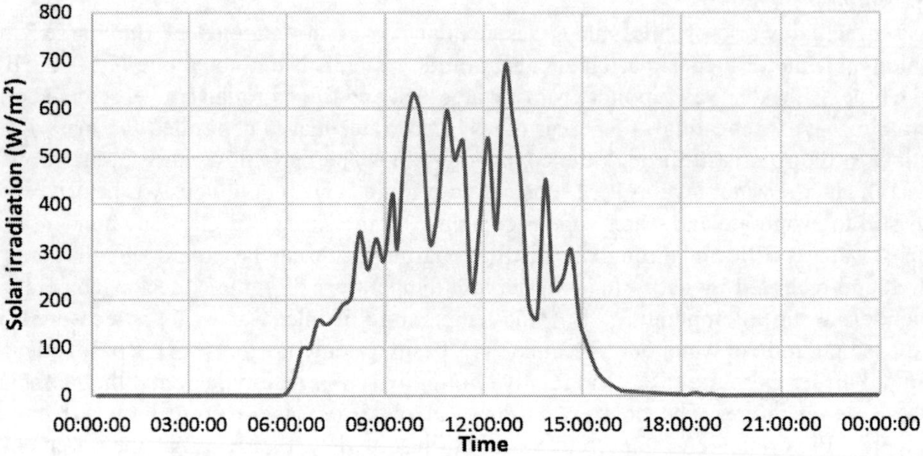

Figure 4. Variation of solar intensity with time of the day

The temperature of ambient air, basin water, glass and condenser tubes according to time for solar still are described in figure 5. The graph presented that the temperature of basin water reaches a peak at 67°C at 13 o'clock, while the temperature of the glass cover and condenser tubes is in the range of 35–54°C and 34 – 48.1°C, respectively. Additionally, the temperature of the condenser tubes is lower than that of the glass cover. This is because of the cooling effect of the water flowing in the tubes. The different temperature between water in basin and glass cover range from 1°C to 14°C, while with condenser tubes range from 2°C to 18.9°C. The temperature of the other parts of the present model is shown in table 2. There is a significant rise in water temperature after flowing through collectors. The collector and the water out collector got the highest temperature at 13 o'clock at 78.97°C and 76.1°C. Relative humidity remarkably is very high during the experimental time (from 86.75 to 92.4%).

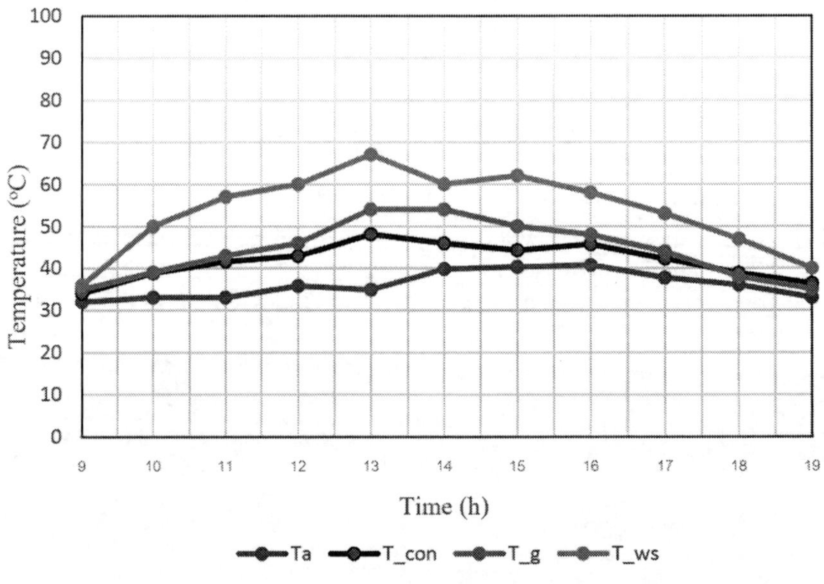

Figure 5. Hourly variations for the temperature of ambient (T_a), water in basin (T_{wb}), glass cover (T_g) and condenser tubes (T_{con})

Table 2. The temperature variation in the different parts of solar still

Time	T_a	T_{wi}	T_{cl1}	T_{cl2}	T_{wo}	T_{abs}	RH
9	32	28	51	55	38.42	43.2	86.175
10	33.06	28.85	65.67	62.9	47.39	47.457	89.92
11	33.02	29.1	62.83	67.34	63.66	49.14	92.375
12	35.76	29.39	63.52	70.17	66.34	52.848	92.855
13	34.88	30.95	62.82	78.97	76.1	63.02	91.5
14	39.82	30.36	60.68	70.84	68.21	51.57	91.535
15	40.28	31.17	69.16	74.16	69.05	51.138	89.535
16	40.75	32.35	67.2	71.37	66.71	49.536	86.755
17	37.72	33.49	54.5	63.25	61.17	44.208	88.595
18	36	33.78	38.69	44.67	41.7	40.761	90.175
19	33	32.01	33.33	37.39	37	37.296	91.5

Figure 6 shows the hourly freshwater productivity achieved from glass cover, condenser tubes and rectangular trough of condenser tubes. The distillate water increases to maximum values at 13 o'clock because the water evaporation rate and condensation rate are increased. The hourly accumulated yield peaked at 1 pm at 0.61 L/m^2 h. According to Figure 6, the freshwater gets from glass cover is higher than that of the others, because of its larger area. Figure 7 describes the accumulated productivity of solar still. The accumulated distillate freshwater is up to almost 3.3 L/m^2 day.

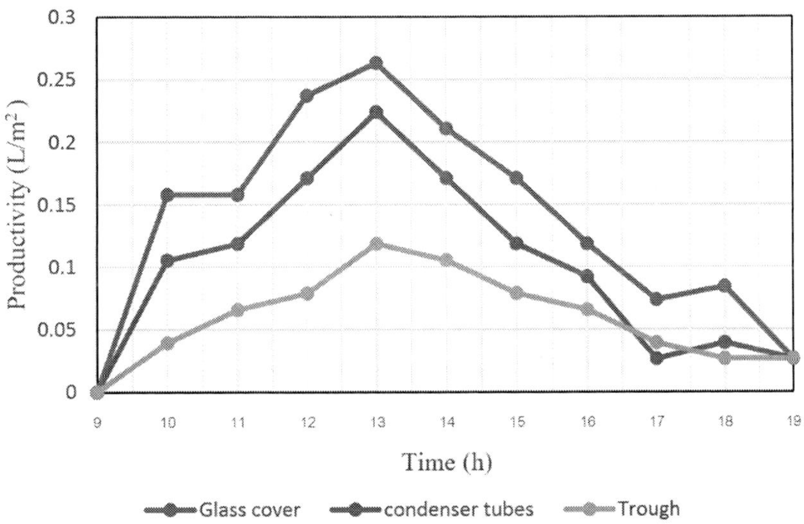

Figure 6. Variation of hourly productivity for solar still collected from glass cover, condenser tubes and rectangular trough

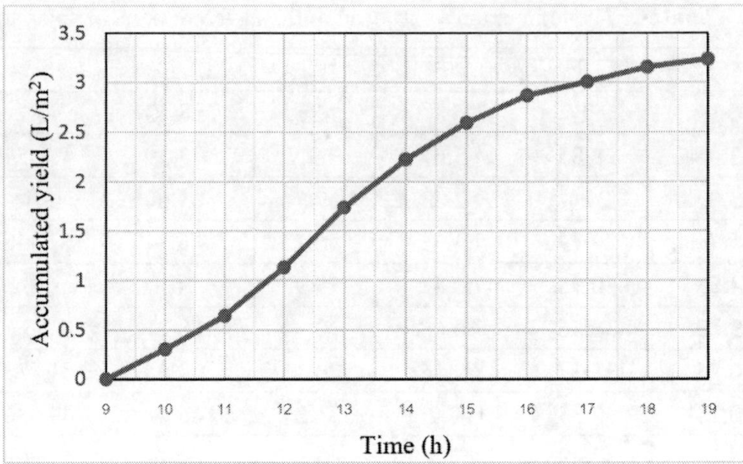

Figure 7. The accumulated yield solar still

4. Conclusions

This paper introduces a new model of the basin type solar still and improve its yield by using condenser and receiver tubes. The experiment was conducted with a mass flow rate of water 1 liter per hour and a constant water level 1 cm. The results of the experiment have underlined the following conclusion:

- The daily yield of the proposed solar still is about 3.3 L/m^2 day under cloudy condition.
- The temperature difference between water in basin and glass cover, water in a basin and condenser tubes ranged from 1°C to 14°C and from 2°C to 18.9°C respectively.
- The tube system is good for preheating feed water. After flowing through collector tubes, the temperature of the water was 76.1°C at 13 o'clock
- The performance of the studied system was improved compared with conventional solar still.

Acknowledgement

This work was supported by the SUT-Ph.D. Scholarship Program for ASEAN of Suranaree University of Technology (SUT), Thailand.

Reference

[1] Kabeel A E 2009 *Energy* **34** 10
[2] Phadatare M K Verma S K 2007 *Desalination* **217** 1–3
[3] Kulandaivel M K Karuppaih S 2014 *Therm. Sci.* **18**(Suppl. 2) S429–S438
[4] Kumar P V Kumar A Prakash O Kaviti A K 2015 *Renew. Sust. Energ. Rev.* **51**(C) 153–181
[5] Rai S N Tiwari G N 1983 *Energ. Convers. Manag.* **23** 3
[6] Singh S K Bhatnagar V P Tiwari G N 1996 *Energ. Convers. Manag.* **37** 2
[7] Singh R V Kumar S Hasan M M Khan M E Tiwari G N 2013 *Desalination* **318** 25–33
[8] Tanaka H Nakatake Y 2004 *Desalination* **160** 2
[9] Kumar P V Kumar A Prakash O Kaviti A K 2015 *Renew. Sust. Energ. Rev.* **51** 1038–1054
[10] Tiwari G N Bapeshwara Rao V S V 1984 *Desalination* **49**(3) 231-241
[11] Arunkumar T Jayaprakash R Amimul A Denkenberger D Okundamiya M S 2013 *Appl. Energ* **103** 109–115
[12] Sodha M S Nayak J K Tiwari G N Kumar A 1980 *Energ. Convers. Manag.* **20** 1
[13] Singh H N Tiwari G N 2004 *Desalination* **168** 145–150

Effects of a Microporous Layer on Water Transport Behaviour in PEM Fuel Cell Gas Diffusion Layers

F Jinuntuya* and W Kamsanam
Department of Mechanical Engineering, School of Engineering, University of Phayao, Phayao, 56000, Thailand

* Corresponding Author: fontipj@gmail.com

Abstract
This paper studies water transport behaviours in proton exchange membrane (PEM) fuel cell gas diffusion layers (GDLs) under the effects of a microporous layer (MPL) and some GDL structure parameters, namely thickness and pore size. Different paper GDL samples with and without MPL treatment were used in this study. The breakthrough pressure of liquid water and the amount of water retention in the GDL were measured. The results indicate that applying MPL on the GDL substrate has a greater impact on water transport behaviours in the GDL than changing the structure parameters of the GDL substrate. Compared to the GDL without MPL, the results show that applying MPL on the GDL surface considerably increases breakthrough pressure up to 4.3 times, while it greatly decreases water retention in the GDL by up to 13.7 times. For the GDL thickness, the results indicate that thicker GDL of the same structure requires up to 60% higher pressure to break through the GDL, while it can retain up to 4.8 times more water in its structure than thinner versions. Besides, the results indicate that twice-larger mean pore size GDL requires about 1.4 times lower breakthrough pressure, while it retains approximately twice more water in the GDL.

Keywords: Gas diffusion layer, Microporous layer, Water transport behaviour, Breakthrough pressure, Water retention

1. Introduction

The gas diffusion layer (GDL) plays an important role in the performance of a proton exchange membrane fuel cell (PEMFC) in the form of facilitating gas transport from a gas channel to a catalyst layer (CL), and product water removal in the opposite direction. An excessive presence of liquid water in the GDL limits the amounts of gaseous reactants accessing the reaction areas of the CL, causing a decrease in the cell performance of the PEMFC. Therefore, GDLs are usually wetproofed with a hydrophobic substance, e.g. PTFE (polytetrafluoroethylene), in order to alter their fibres towards becoming more hydrophobic, which eases gas and water transport across the GDL. Several studies [1–8] have demonstrated experimentally and numerically a decrease in water saturation level in the GDL with an increasing degree of hydrophobicity by applying PTFE. Benziger et al. [1, 2] illustrated that PTFE treatment made the pores more hydrophobic limiting the penetration of water into the GDL, thus providing free void space for gas transport. Also, a fine microporous layer (MPL), generally consisting of carbon black and PTFE particles is usually added on one side of the GDL between the GDL and the

CL in order to improve water management and cell performance. In recent studies, Lu et al. [9, 10] demonstrated the saturation reduction in GDL with MPL, compared to GDL without MPL. They attributed this reduction to the role of MPL in limiting the number of locations where liquid water can penetrate to the GDL. In the presence of MPL, Gostick et al. [11] also reported a considerable decrease in water saturation in the GDL, which might benefit gas transport in the PEMFC.
Similarly, in [12], the present author studied the impact of MPL on water transport in paper and cloth GDLs, and its effect was found to be much greater than applying PTFE, both in terms of increasing breakthrough pressure and limiting the amounts of water retention in the GDL. The samples used in this work, however, are limited to a pair of paper and cloth GDLs with and without MPL, which might not be enough to comprehend the role of MPL on water transport in the GDL. In the present work, different commercially available paper GDLs, which are the most commonly used GDLs, with and without MPL, was investigated in order to evaluate the effect of MPL on water transport behaviour in the GDL. By varying the GDL substrate, the effects of its structure on water transport in the GDLs can also be examined. This could contribute to the understanding of liquid water transport characteristics in the GDLs, which is crucial for the design and development of PEMFC.

2. Experimental

2.1. Materials

The GDL samples tested are listed in table 1. They are commercially available carbon paper GDLs with and without MPL. The selected GDL samples cover a range of GDLs commonly used in various PEM fuel cell applications. The thickness and pore diameter of these GDL samples range from about 103 μm to 355 μm, and from less than 1 μm to about 60 μm respectively. The micrographs of the tested samples are presented in figure 1. In this study, similar GDLs with different thicknesses were also examined to evaluate the effect of thickness on water transport in the GDL. In addition, different GDLs with almost similar pore sizes were investigated.

Table 1. Carbon paper GDLs used in this study.

GDL sample	MPL	Thickness (μm)	Mean pore diameter (μm)
Sigracet SGL29AA	No	180	59.5
Sigracet SGL29BC	Yes	220	< 1
Sigracet SGL39AA	No	277	47.9
Sigracet SGL39BC	Yes	322	< 1
AvCarb P75	No	209	54.6
AvCarb GDS2230	Yes	235	< 1
Toray TGPH030	No	103	30.5
Toray TGPH060	No	184	27.7
Toray TGPH090	No	277	30.4
Toray TGPH120	No	355	27.6

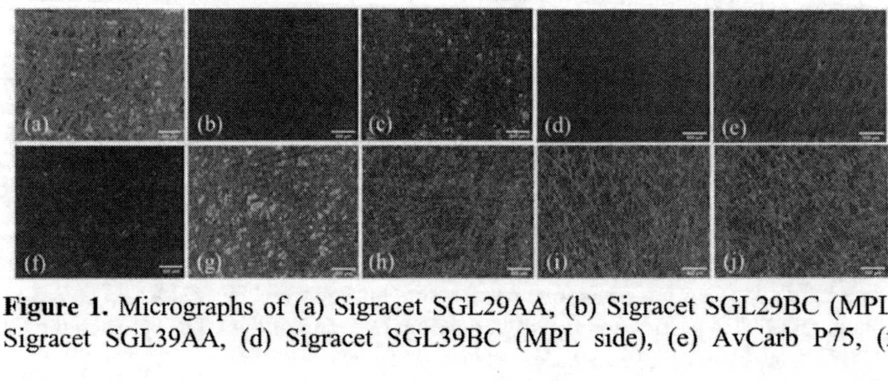

Figure 1. Micrographs of (a) Sigracet SGL29AA, (b) Sigracet SGL29BC (MPL side), (c) Sigracet SGL39AA, (d) Sigracet SGL39BC (MPL side), (e) AvCarb P75, (f) AvCarb

IOP Conf. Series: Materials Science and Engineering **886** (2020) 012043 doi:10.1088/1757-899X/886/1/012043

GDS2230 (MPL side), (g) Toray TGPH030, (h) Toray TGPH060, (i) Toray TGPH090, and (j) Toray TGPH120

2.2. Experiments
In this work, a test apparatus, as shown schematically in figure 2, was built to examine the breakthrough pressure and water retention in the GDL samples. This apparatus is composed of a 2-m high clear acrylic tube of 10 mm in diameter, which is located on top of the two acrylic plates where the GDL sample is placed. A water column is allowed to build up in the acrylic tube creating hydrostatic pressure according to the height of the water above the sample.

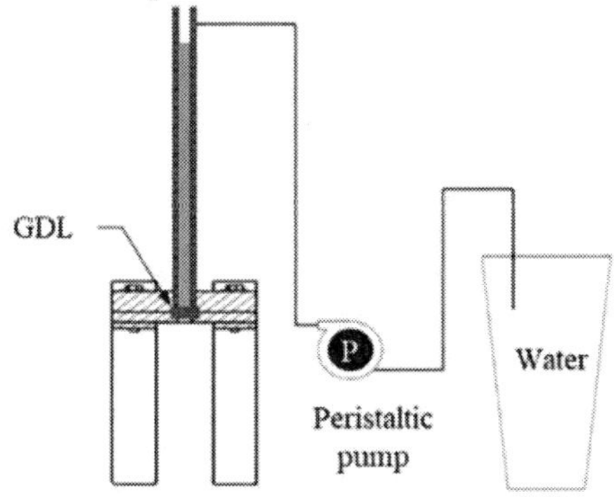

Figure 2. Schematic of the test apparatus

For the breakthrough experiment, the required minimum pressure to force liquid water through the GDL was measured. Liquid water was added into the acrylic tube by a peristaltic pump. As the water started to pass through the sample thickness, the height of the built-up water column above the tested sample corresponding to hydrostatic pressure was measured. Also, each sample was weighed before and after the experiment in order to determine the amount of water retained in it.

3. Results and discussion

3.1. Effects of MPL on breakthrough pressure
In order to examine the effect of MPL on the pressure required to force liquid water through the GDL, different GDL samples with and without MPL were used in this study. These include the Sigracet series, i.e. SGL29AA/BC, SGL39AA/BC, and the AvCarb series, i.e. P75/GDS2230. As shown in figure 3, a striped bar represents GDL without MPL, while the adjacent solid bar represents the same GDL substrate but with MPL. The figure shows that MPL has a great impact on the pressure required to force water through the GDL samples. By applying MPL, the minimum pressure required has increased by about 3.7, 4.3 and 2.0 times for the SGL29, SGL39 and AvCarb series respectively. This is because the MPL has diminutive pore sizes (less than 1 μm), which creates large capillary resistance to water flow. Consequently, much greater pressure is needed to push liquid water through the GDL with MPL. This agrees with the results in [9, 10], which report a considerable increase in the required breakthrough pressure for the GDL with MPL.

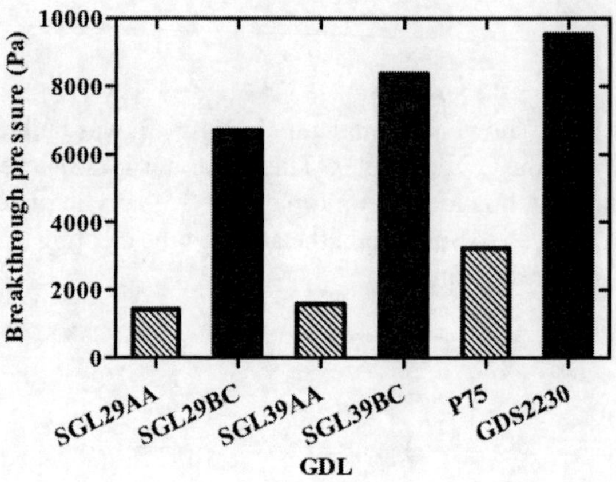

Figure 3. Effect of MPL on the breakthrough pressure

3.2. Effects of MPL on water retention

Figure 4 compares the amount of water retention in the GDL samples with and without MPL. A striped bar represents the GDL without MPL, while a solid bar represents the GDL with MPL. It is observed that applying MPL greatly decreases the amount of water retention in the GDL samples, amounting to over 5.4, 12.8 and 13.7 times reduction in water retention for the SGL29BC, SGL39BC and GDS2230 respectively, compared to their corresponding GDL without MPL, namely SGL29AA, SGL39AA and P75 respectively. This indicates that the MPL can limit liquid water from entering the GDL through its diminutive pore sizes, as seen in figure 1. These small pore sizes create much larger resistance to water flow than those of the GDL without MPL and, thus, lower water saturation in the GDL, offering more available pores for gas transport. This, again, corresponds to the results in [9, 10].

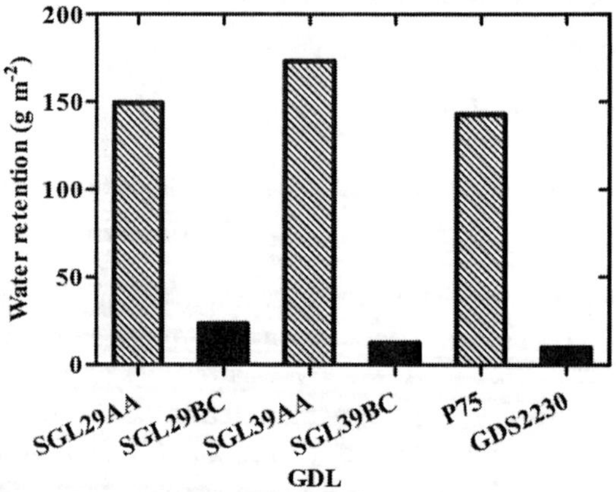

Figure 4. Effect of MPL on water retention in the GDL.

3.3. Effects of thickness on breakthrough pressure

Figure 5 shows the minimum pressure required to push liquid water through GDL samples with different thicknesses. Only the GDL samples without MPL were used in this study, which includes the Toray TGPH series, represented by the solid bar, with a thickness ranging from about 103 μm to 355 μm, and the Sigracet AA series, represented by the striped bar, with a thickness of 180 μm to 277 μm. For both the Toray and Sigracet series, it is observed that the required breakthrough pressure increases alongside the thickness. For example, the Toray TGPH030 requires around 2,449 Pa to push liquid water through its structure, while the thick TGPH120, the thickest in this series, requires about 60% greater pressure at around 3,911 Pa. The increase in breakthrough pressure with increasing GDL thickness agrees with the results reported by Tamayol and Bahrami [13], and Mortazavi and Tajiri [14], who compared similar carbon paper GDLs with different thicknesses. Likewise, in this study, a comparison can be made only among GDLs in the same series, which have a similar structure, and not the entire range of GDLs. For example, the Toray TGPH090 and Sigracet SGL39AA have a similar thickness at around 277 μm, but the Toray requires about 1.3 times larger pressure to push water through it than the Sigracet.

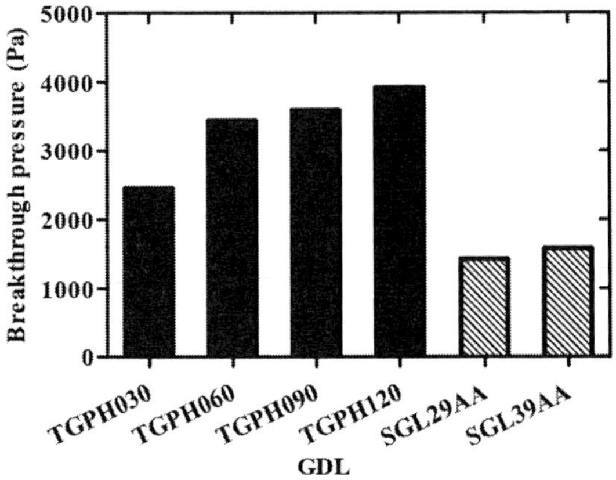

Figure 5. Effect of thickness on the breakthrough pressure

3.4. Effects of thickness on water retention

In order to examine the effect of thickness on water retention in the GDL, the GDL samples without MPL with different thicknesses were used. Figure 6 shows the amount of water retention in the GDLs, which include the Toray TGPH series with a thickness ranging from about 103 μm to 355 μm, represented by a solid bar, and the Sigracet AA series with a thickness of about 180 μm to 277 μm, represented by a striped bar. For both the Toray and Sigracet, it is observed that water retention in the GDLs of the same series increases as the GDL thickness increases. The Toray TGPH120, the thickest GDL in this series, retains about 4.8 times more water than the TGPH030, the thinnest one. This indicates the higher water retention capability of a thick GDL over its thinner counterpart.

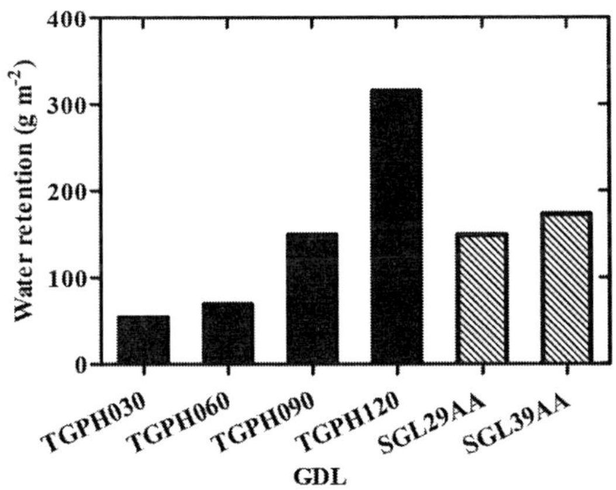

Figure 6. Effect of thickness on water retention in the GDL

3.5. Effects of pore size on breakthrough pressure

Figure 7 shows the required breakthrough pressure to push water through the GDL as a function of mean pore size. In this study, only the GDL samples without MPL were used. The solid line represents the breakthrough pressure required by the GDLs with a thickness of about 180 μm, while the dashed line corresponds to GDLs with a thickness of about 277 μm. It is observed that the breakthrough pressure decreases as the mean pore size increases. For example, the 180-μm-thick GDL with a mean pore size of about 27.7 μm requires about 1.4 times larger pressure than that of the 59.5-μm-pore-size GDL to break through the other side. According to the Young-Laplace equation, this is due to the decrease in capillary resistance at the increased pore size of the GDL. As a result, smaller pressure is required to overcome the decreased resistance of large pore sizes.

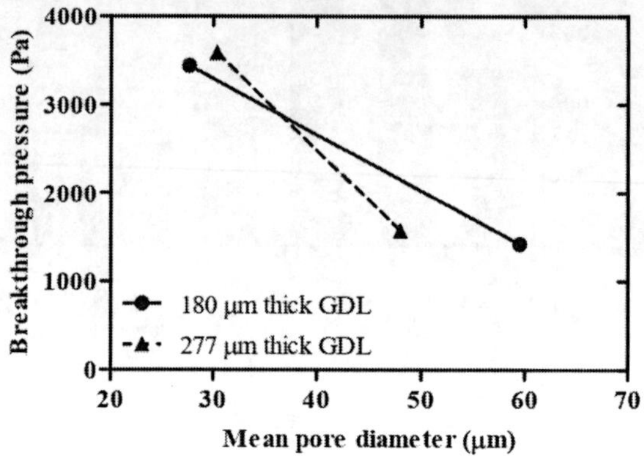

Figure 7. Breakthrough pressure as a function of mean pore diameter at different GDL thicknesses

3.6. *Effects of pore size on water retention*

Figure 8 shows the amount of water retention in the GDL samples as a function of mean pore size. The solid line represents the amount of water retention in the 180-μm-thick GDLs, whilst the dashed line represents that in the 277-μm-thick GDLs. For both thicknesses, it is observed that the amount of water retention increases with increasing mean pore size. For example, the amount of water retention in the thin GDL increases about twofold from 48.2 g m^{-2} to 149.3 g m^{-2}, as the mean pore size increases from 27.7 μm to 59.5 μm. This is because the GDL with larger pore sizes creates smaller capillary resistance to water flow according to the Young-Laplace equation, which allows more water entering the GDL and being retained by it. The results also show that the thick GDL absorbs more water than the thin one of about similar pore size. For example, the thick GDL with a pore size of about 30.4 μm can hold about twice more water than the thin one of about a similar pore size of 27.7 μm.

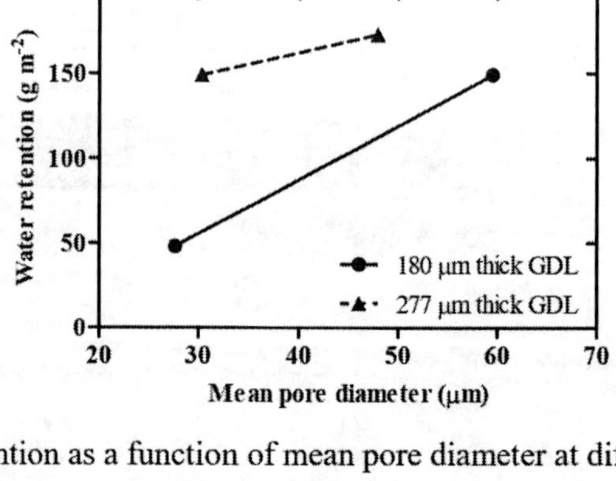

Figure 8. Water retention as a function of mean pore diameter at different GDL thicknesses

4. Conclusion

The effects of MPL and some GDL structure parameters on breakthrough pressure and water retention in GDLs were investigated. In applying MPL on GDL substrates, it was found that larger pressure is needed to force liquid water through the GDL due to its diminutive pore sizes creating greater resistance to water flow, while water retention in GDLs with MPL was greatly decreased compared to GDLs without MPL. Applying MPL on the GDL substrate, therefore, could effectively control water saturation levels in the GDL of a PEMFC operating at high current densities, where product water is rapidly produced. This could help in providing more available pores for gas transport during high water production rates. For the GDL structure parameters, which are thickness and pore size, it was shown that they have significant effects on both breakthrough pressure and water retention in the GDL, although their influence is much less than the MPL. It was observed that thicker GDLs require larger pressure for water to travel through the GDL, and they can hold more of it in their structure than thinner ones. As regards pore size, it was found that GDLs with larger mean pore size require smaller pressure to break through the GDL, while they allow more water accumulation within their structure than GDLs with smaller mean pore size.

References

[1] Benziger J Nehlsen J Blackwell D Brennan T Itescu J 2005 *J. Membrane Sci.* **261** 98-106
[2] Gauthier E Duan Q Hellstern T Benziger J 2012 *Fuel Cells* **12** 835-847
[3] Bevers D Rogers R von Bradke M 1996 *J. Power Sources* **63** 193-201
[4] Chapuis O Prat M Quintard M Chane-Kane E Guillot O and Mayer N 2008 *J. Power Sources* **178** 258-268
[5] Hao L Cheng P 2010 *J. Power Sources* **195** 3870-3881
[6] Rama P Liu Y Chen R Ostadi H Jiang K Gao Y Zhang X X Fisher R Jeschke M 2012 *Aiche J.* **58** 646-655
[7] Gao Y Zhang X X Rama P Chen R Ostadi H and Jiang K 2013 *Comput. Math Appl.* **65** 891-900
[8] Jinuntuya F Whiteley M Chen R Fly A 2018 *J. Power Sources* **378** 53-65
[9] Lu Z J Daino M M Rath C Kandlikar S G 2010 *Int. J. Hydrogen Energ.* **35** 4222-4233
[10] Kandlikar S G Garofalo M L Lu Z 2011 *Fuel Cells* **11** 814-823
[11] Gostick J T Ioannidis M A Fowler M W Pritzker M D 2009 *Electrochem Commun.* **11** 576-579
[12] Jinuntuya F Kamsanam W 2019 *IOP Conf. Ser., Mater. Sci. Eng.* **501** 012051
[13] Tamayol A Bahrami M 2011 *J. Power Sources* **196** 6356-6361
[14] Mortazavi M Tajiri K 2014 *Int. J. Hydrogen Energ.* **39** 9409-9419

IOP Conf. Series: Materials Science and Engineering **886** (2020) 012044 doi:10.1088/1757-899X/886/1/012044

The Prototype of Assistive Cane for Climbing Up the Stair

Pipatpong Boonthaworn[1], Panya Aroonjarattham[1,*] and Chompunut Somtua[1]

[1]Department of Mechanical Engineering, Faculty of Engineering, Mahidol University, Nakornpathom Thailand 73170

* Corresponding Author: panya.aro@mahidol.ac.th

Abstract
The elder and patient with osteoarthritis knee had the problem to climb up the stair. The purpose of this research was to design and construct the prototype of assistive cane for climb up the stair. The assistive cane use Arduino to control the motor, send the power to the gear set to lift the user up the stair. The lifting plate help to reduce the burden at the knee joint while climbing up the stair. The foot force sensing device used to measure and collect the force act on the insole while walking on the floor, climbing up the stairs and climbing up the stairs using the assistive cane. The assistive cane could be help to reduce the burden at the knee joint and the force on the insole when climbing up the stair.

1. Introduction

At present, Thailand is entering the society of the elderly. According to the Ministry of Social Development and Human Security in 2018, the number of elderly people (over 60 years old) was 16.06 percentage of the total population. The elderly population distribute in Thailand as shown in Figure 1.

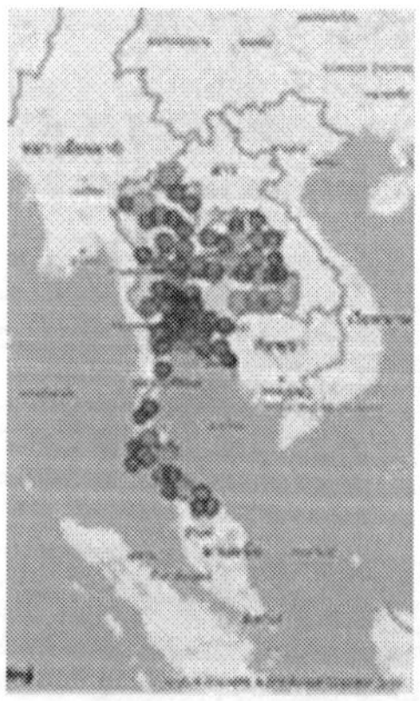

Figure 1. The elderly population distribute in Thailand in 2018 [1].

The common problems of the elderly due to knee degeneration include muscle weakness that occurs with the elderly or patients with osteoarthritis who receive physical therapy after undergoing knee surgery. Although there are many currently facilities that helped in the daily life such as crutches and wheelchairs but these devices does not support to climb up the stair. This study aims to design and

IOP Conf. Series: Materials Science and Engineering **886** (2020) 012044 doi:10.1088/1757-899X/886/1/012044

construct the prototype of assistive cane to climb up the stair using the lifting plate controlled by Arduino to reduce the burden at the knee joint while climbing up the stair.

2. Materials and Methods

2.1 The conditions and limitations
The main condition is the height and width of each step that has 16.67 cm. height and 26.27 cm. width as shown in figure 2. The symptoms of the tester have a problem with left leg, so the tester can stand by two legs but cannot stand by one leg which painful.

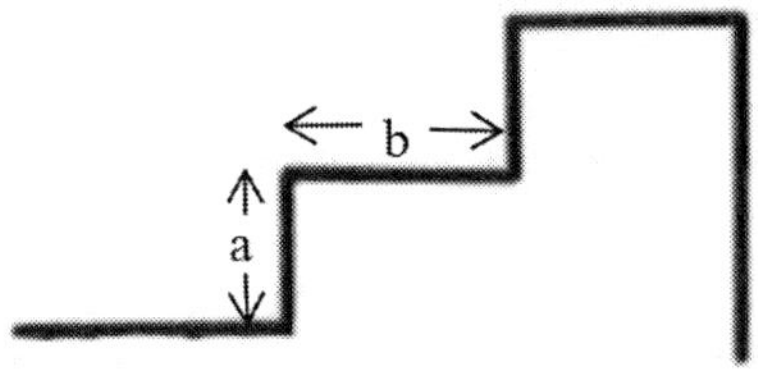

Figure 2. The height and width of stair: a = 16.67 cm. and b = 26.27 cm.

2.2 The prototype of assistive cane for climbing up the stair
Three-dimensional model of the assistive cane, which was designed by SolidWorks software, shown in figure 3 and the prototype shown in Figure 4.

Figure 3. Three-dimensional model of the assistive cane for climbing up the stair.

Figure 4. The prototype of the assistive cane for climbing up the stair.

The usability of assistive cane for climbing up the stair has 5 steps as follows:

- First, place the right foot on the plate.
- Second, step left foot up the stairs.
- Next, turn on the prototype of assistive cane to lift up to the same level as the upper stair.
- Then, step right foot on the stairs.
- Finally, drag the prototype of assistive cane onto the stairs.

2.3 DC Motor and Gear

Conditions of the motor's load are the tester weight 50 kilograms, the height of each step was 17 cm. and the time to move the tester foot to next step was 5 seconds. Three parameters used to calculate the power of motor in equation (1).

$$P = \frac{Fs}{t} \quad (1)$$

Torque of motor calculates by equation (2). The selected motor, 36 watts used to find the size of the rack and pinion.

$$P = T\omega \quad (2)$$

Table 1. Spur gear's radius calculation

rpm	ω (rad/s)	a (m/s^2)	F (N)	T (N.m)	R (cm)
5	0.524	0.003	522.138	68.755	13.17
10	1.047	0.006	522.298	34.377	6.58
30	3.142	0.018	522.934	11.459	2.19
50	5.236	0.030	523.571	6.875	1.31
80	8.378	0.048	524.526	4.297	0.82
100	10.472	0.060	525.163	3.438	0.65

From Table 1 at various speeds, the selected motor has an axis size 10 mm. The choosing motor with a speed of 10 rpm has the smallest radius of 6.58 cm. Equation (3) and (4) is used to find the spur gear size [2].

$$S_F = \frac{S_t Y_N}{\sigma Y_\theta Y_Z} \quad (3)$$

$$\sigma = W^t K_o K_v K_s \frac{1}{bm_t} \frac{K_H K_B}{Y_J} \quad (4)$$

S_t is Allowable bending stress equal to 267.66 MPa.
Y_N is Stress cycle factor for bending strength equal to 1.08.
Y_θ is Temperature factor for bending strength equal to 1.
Y_Z is Reliability factor for bending strength equal to 0.99.
W^t is Tangential transmitted load equal to 522 N.
K_o is Overload factor equal to 1.25.
K_v is Dynamic factor equal to 1.69.
K_s is Size factor used equal to 1.
m_t is Transverse metric module.
K_H is Load distribution factor equal to 1.453.
K_B is Rim thickness factor equal to 1.
Y_J is Geometry factor for bending strength equal to 0.42.

The safety factor of the gear that is commonly used is 1.75. Instead, take the safety factor in equation (3) to find the face width.

$$S_F = \frac{S_t Y_N}{\sigma Y_\theta Y_Z}$$

$$1.75 = \frac{267.66(1.08)}{\sigma(1)(0.99)}$$

$$\sigma = 166.853\ MPa$$

form $\sigma = 166.863$

$$\sigma = W' K_o K_v K_s \frac{1}{bm_t} \frac{K_H K_B}{Y_J}$$

$$166.863 = 0.522(1.25)(1.69)(1)\frac{1}{b}\frac{(1.453)(1)}{0.42}$$

$$b = 0.022\ m$$

$$b = 22\ mm$$

Finally, the appropriate module for rack and pinion is 2.5, which has a face width of 25 mm by the pinion or spur gear with diameter 12 cm.

2.4 Tri-star wheel

From Figure 5, it has five parameters to design a tri-star wheel. Two known parameters and three unknown parameters calculated by Equation (5), (6), (7), (8) and (9) [3].

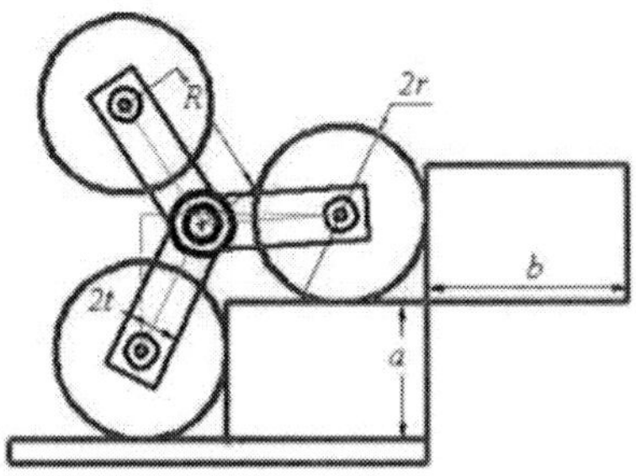

Figure 5. Five parameters on tri-star wheel

$$R = \sqrt{\frac{a^2 + b^2}{3}} \tag{5}$$

$$r_{max} = \frac{\sqrt{a^2 + b^2}}{2} \tag{6}$$

$$r_{min} = \frac{6Rt + a\left(3b - \sqrt{3}a\right)}{\left(3 - \sqrt{3}\right)a + \left(3 + \sqrt{3}\right)b} \tag{7}$$

$$t_{1max} = \frac{3r(a + b) - \sqrt{3}r(a - b) + a\left(\sqrt{3} - 3b\right)}{6R} \tag{8}$$

$$t_{2max} = \frac{r\left(r + \sqrt{3 \times \left(a^2 + b^2 - r^2\right)}\right)}{2\sqrt{a^2 + b^2}} \tag{9}$$

2.5 Electrical circuit

The circuit including on-off switch, motor, li-on battery, two-ways and limit switch and battery monitor shown in Figure 6.

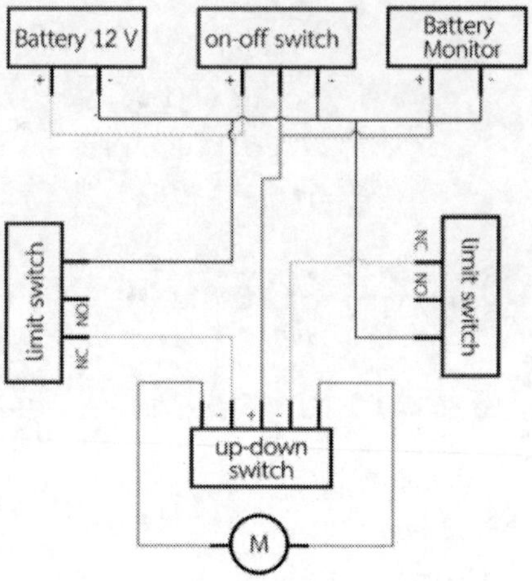

Figure 6. The assistive device's system circuit

When switching on, the on-off switch will change from NC to NO as shown in Figure 7 (a). Therefore, the NO of the power switch connects to the battery and supply power to the device system.

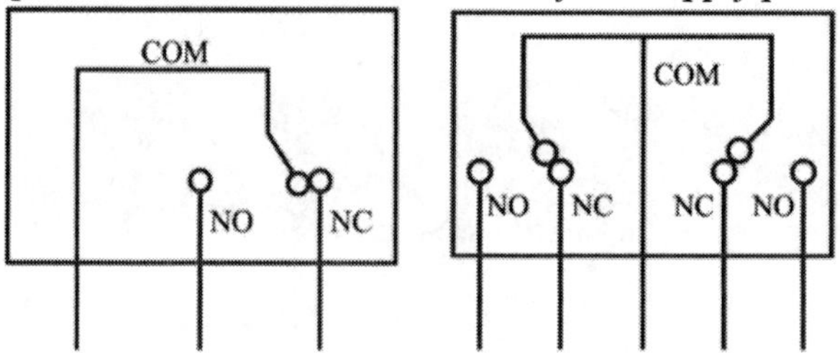

Figure 7. The switch circuit: (a) On-off and (b) Two-ways.

For two-ways switch as shown in Figure 7 (b), the motor rotate clockwise when press the button down. The contact on the left will change from NC to NO, which connect to the motor as well. When the motor rotate counterclockwise, the contact on the left will change from NO to NC and the contact on the right will change from NC to NO, which connect to the motor as well. The limit switch connects in the series to cut off the power when moving to the specified distance.

2.6 Foot force sensing device

The foot force sensing device used to measure and collect the force act on the insole while the user walking on the floor, climbing up the stairs and climbing up the stairs using the assistive cane. Six force sensitive resistors installed on the insole as shown in Figure 8.

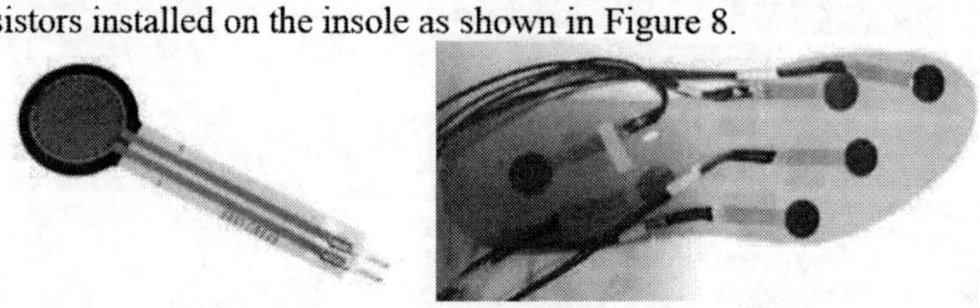

(a) (b)

Figure 8. (a) Force sensitive resistor and (b) Six sensors install on the insole.

3. Results and Discussion

The prototype of the assistive cane used to help the people, who had the knee joint problem to climb up the stair. The foot force-sensing device was installed on both insole as shown in figure 9 to measure and collect the force act on the insole.

Figure 9. The foot force sensing device measure and collect the force act on the insole.

The data from foot force sensing device used to compare between the normal knee and osteoarthritis knee when walking on the floor, climbing up the stairs without assistive cane and climbing up the stairs with assistive cane as shown in Figure 10.

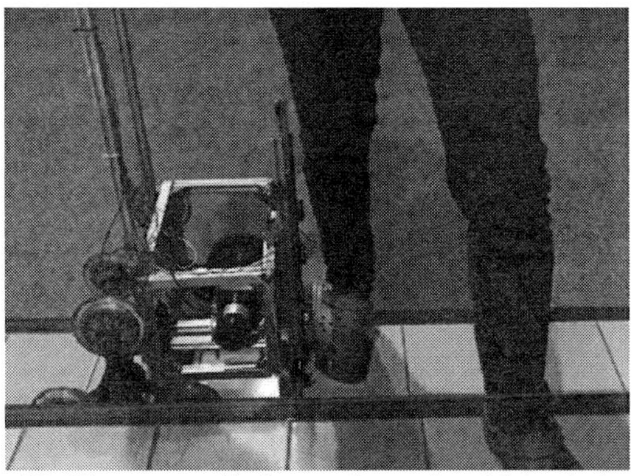

Figure 10. The tester climbs up the stair with assistive cane.

The tester have problem at the left knee joint was tested under three condition. The forces on insole under walking on the floor were shown in Figure 11 (a) and 11 (b) for left and right foot respectively.

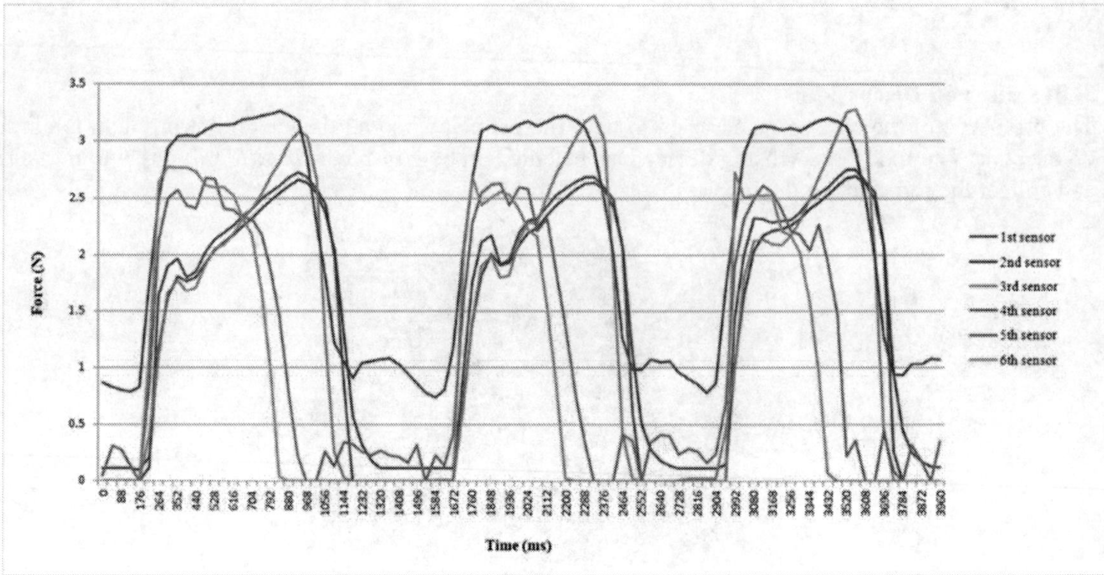

(a)

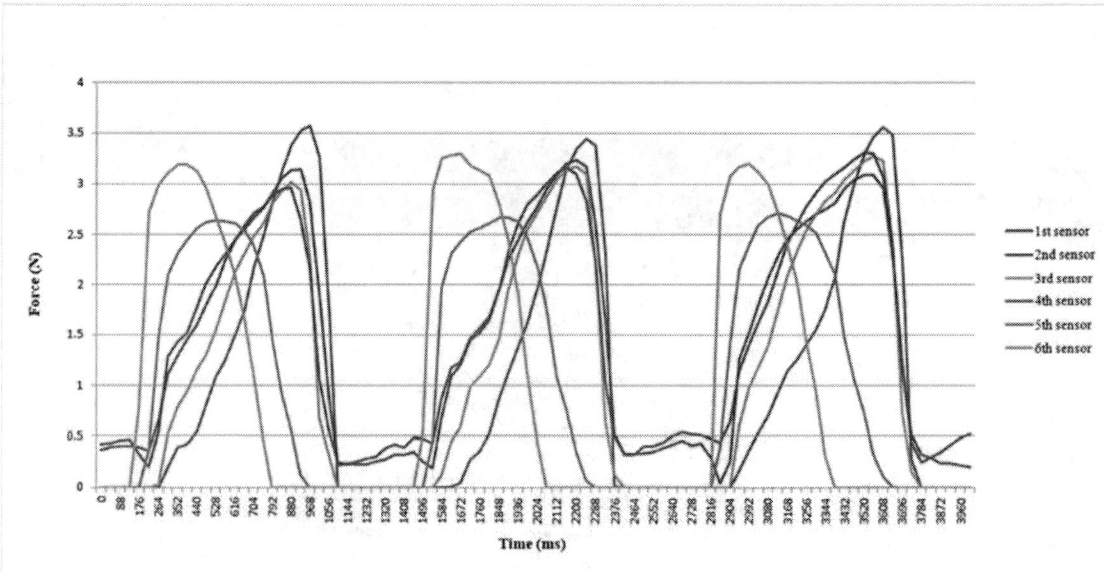

(b)

Figure 11. The force on insole under walking on the floor: (a) Left foot and (b) Right foot.

The left foot will receive less force than the right under walking on the floor condition. All sensors show more force acting on the right compared between six sensors on both insole. The maximum force on the left foot occurs at the middle of insole because the tester avoids putting the load on the left and transferring the load to the right. The forces on insole under climbing up the stairs without assistive cane were shown in Figure 12 (a) and 12 (b) for left and right foot respectively.

IOP Conf. Series: Materials Science and Engineering **886** (2020) 012044 doi:10.1088/1757-899X/886/1/012044

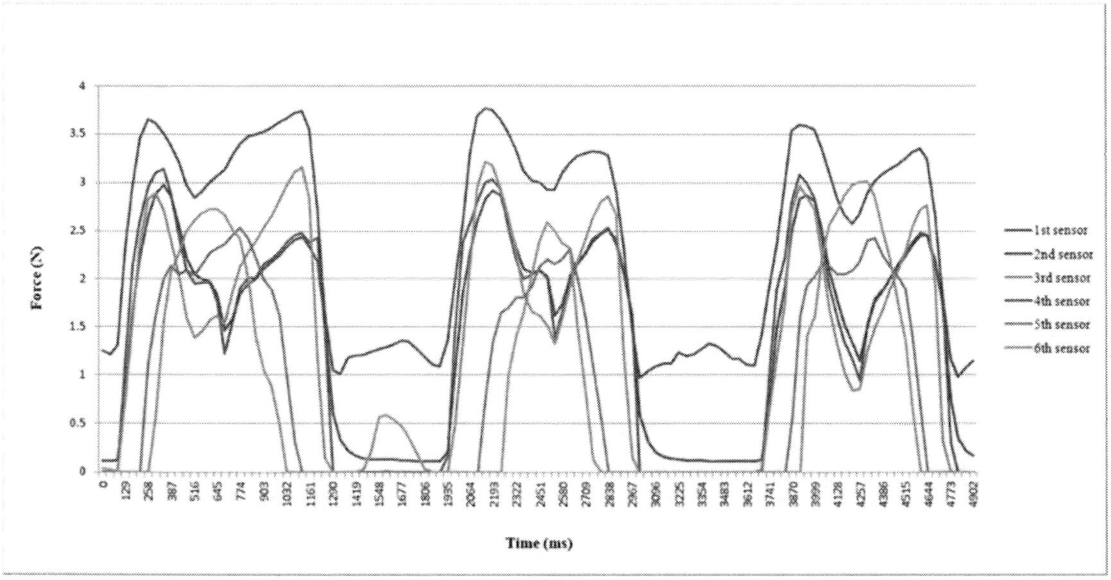

(a)

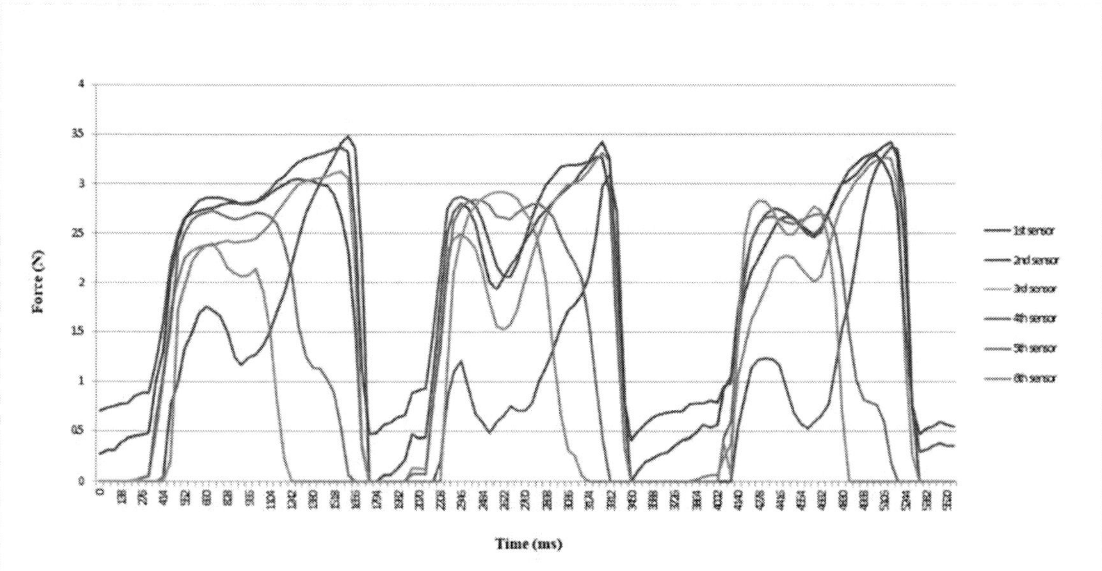

(b)

Figure 12. The force on insole under climbing up the stairs without assistive cane: (a) Left foot and (b) Right foot.

The left foot must bear the load similar to the right under climbing up the stairs without assistive cane condition. The occurring force on the left insole is closely to the right as shown in the graph. The force on insole under climbing up the stairs with assistive cane were shown in Figure 13 (a) and 13 (b) for left and right foot respectively.

IOP Conf. Series: Materials Science and Engineering **886** (2020) 012044 doi:10.1088/1757-899X/886/1/012044

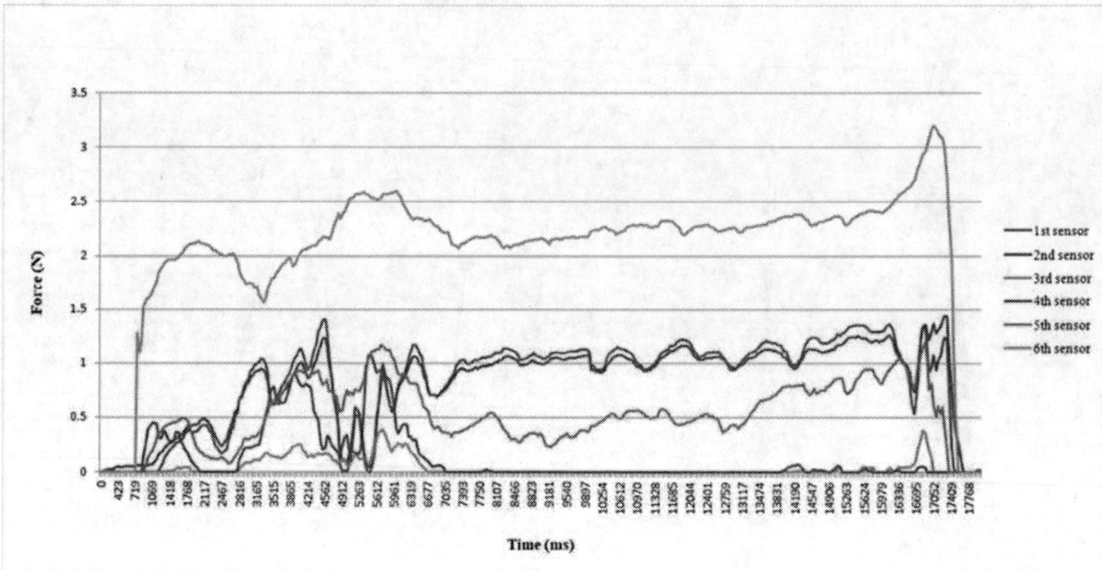

(a)

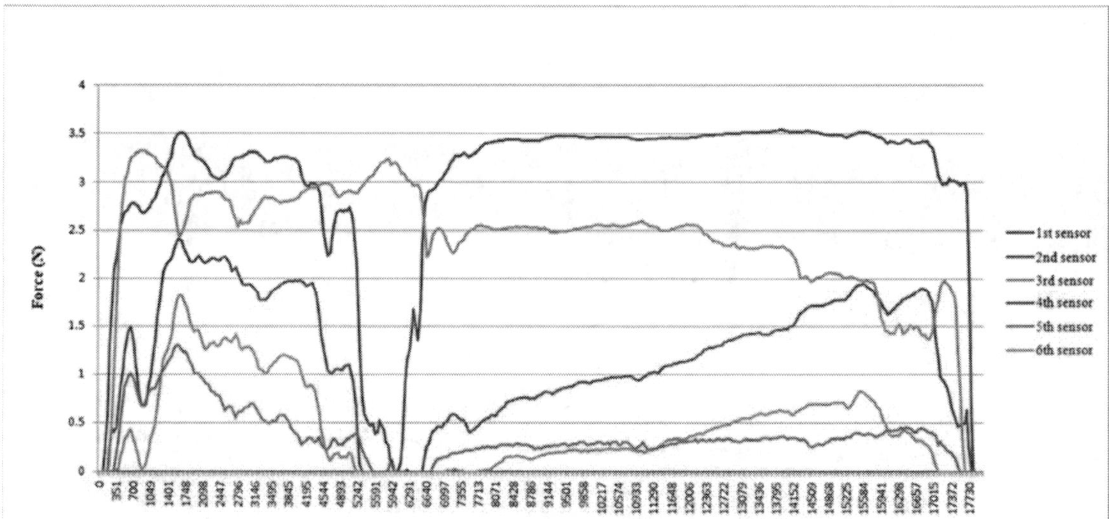

(b)

Figure 13. The force on insole for climb up the stairs with assistive cane: (a) Left foot and (b) Right foot.

While climbing up the stairs with assistive cane, the force on the left insole reduce from the Figure 12 (a). All sensors show the lower force on the left than the right insole that will be able to reduce the burden on the left knee joint. The sensor that receives the most force is located at the heel and middle of the foot as shown in the graph, which shows that the body weight does not transfer to the front, so in summary, the burden on the knees has been reduced.

In summary, the burden on the knees has been reduced.

4. Conclusion

The prototype of assistive cane can be help the user with the osteoarthritis knee to climb up the stair with the less burden on the knee joint. This prototype device could be used with ladder distance not exceed 17 cm. The design of the mechanism of tri-star wheels can help the users easily drag the underlying device up and down the stairs.

5. Acknowledgments

The authors wish to thank Ms.Ariyo Boonchasri, Ms.Patteera Kunkrisada and Mr.Vichya Wattana for construct the device, Biomechanics Analysis and Orthopedic Device Design Laboratory (B-AODD LAB), Department of Mechanical Engineering, Faculty of Engineering, Mahidol University for their support with facilities.

References

[1] Dop.go.th. Inc. 2018 *http://www.dop.go.th/th/know/1/153* accessed 20 March 2019
[2] Richard G Budynas J Nisbett K and Shigley J E 2015 *Shigley's mechanical engineering design.* (New York, NY: McGraw-Hill Education)
[3] Iqbal M 2012 *Mechanical design and development of tri-star wheel system for stair-climbing robot https://www.researchgate.net/publication/256 256942* accessed 1 April 2019

IOP Conf. Series: Materials Science and Engineering **886** (2020) 012045 doi:10.1088/1757-899X/886/1/012045

Analysis of Tympanic Membrane Reconstruction with Various Thickness Cartilage Plates

Zhiqiang Wu[1,*], Takashi Tanaka[1], Kazuki Kamitani[1], Yasunori Oura[1], and Hidayat Hidayat[2]

[1]The University of Shiga Prefecture, 2500 Hassaka, Hikone, Shiga, 522-8533, Japan
[2]Politeknik Negeri Samarinda, Jl. Ciptomangunkusumo Gn. Lipan Samarinda, Indonesia

* Corresponding Author: wu.z@mech.usp.ac.jp

Abstract
The purpose of this research is to investigate an appropriate design of the sliced cartilage used for the reconstruction of the tympanic membrane (TM). A great number of people are suffering from TM perforation caused by diseases or accidents. A surgical procedure, called cartilage myringoplasty, can save the people from hearing loss by replacing the damaged part of the membrane with thin-sliced cartilage. In this research, as the first step of designing the sliced cartilage, we established a three-dimensional finite element model of the human ear for numerical analysis. The ear model consists of the middle ear (TM, ossicular chain, ligaments, muscle), and inner ear cochlea. The joints between ear bones were also modelled with softer tissues. The analysis result of the frequency response shows a good correspondence with measurement results reported by other researchers. Then, using the numerical model, we investigated the optimal thickness of the cartilage plate. The TM was cut at the bottom position with a size of about 40% of TM and filled by cartilage plate with the thickness of 0.1mm to 0.7mm. Comparing with the frequency response of the healthy ear, we found that the optimal thickness of the plate is different in a various frequency range. The reason is considered as the difference of material properties between the membrane and the cartilage. Finally, to evaluate the performance of repaired models in all frequency range, we proposed an equation of matching rate as the evaluation equation. The results show that 0.5mm cartilage plate model shows a better performance at peak response area (600Hz-1kHz), and 0.3mm. cartilage plate model shows better average performance in the total frequency range.

1. Introduction

The human ear is divided into three sections, namely the outer ear, middle ear, and inner ear. The middle ear system is a complicated mechanism, which consists of the TM, three tiny bones called ossicular chain (malleus, incus, and stapes), the ligaments and the muscle. The TM is a thin and semi-transparent membrane located between the outer and middle ear that vibrates after it is hit by sound pressure. Many people have conductive hearing loss due to a problem on the TM such as eardrum perforation. Cartilage myringoplasty is a surgical procedure commonly performed by an otolaryngologist for closing TM perforation and improving hearing ability.

In recent years, a number of studies on TM perforation closing by cartilage myringoplasty have been reported by many researchers. Some reported the dynamic behaviour and the numerical modelling of the TM using a numerical and experimental method [1-5]. An experimental and theoretical approach to TM perforation which causes hearing loss in the human middle ear, has been reported [6]. Other papers investigated the experimental result of cartilage in varying thicknesses to determine the acoustics transfer characteristics of cartilage used in the reconstruction of the TM [7-8]. Furthermore, a few studies demonstrated a proper thicknesses of the use of sliced cartilage utilized for myringoplasty to repair the TM perforation [9, 10]. The numerical models they used are simplified to some extent, especially in regards to the treatment of boundary condition of TM, the joints of ear bones, and the ligaments.

In this study, the investigation of the optimal thicknesses of the sliced cartilage closing the TM perforation is carried out. First, a three-dimensional finite element model of human ear consisting middle ear (TM, ossicular chain, ligaments, and muscle) and cochlea in the inner ear is generated by using finite element analysis software. As for the connection of the ossicular chain, the softer tissue is modelled as joints between the ear bones. Then, a part of the TM is replaced by sliced cartilage. The values from 0.1mm to 0.7mm are used as thicknesses of the sliced cartilages to perform frequency response analysis. In the frequency response analysis, the optimal thickness of sliced cartilage used in myringoplasty is calculated by comparing with the healthy ear in the frequency range of 100Hz to 10kHz.

2. FE Modelling of Middle Ear System

To perform the numerical analysis and to investigate the dynamic characteristics of the middle ear, a 3D FE model was established. The geometric data is downloaded from the website, which is a CT scan data obtained from a 75 years old male, and provided by De Greef et al. [11].

2.1. FE Model of middle ear system

Geometric data of each part of the ear was downloaded with STL format, and imported to NX-Nastran, a general-purpose FEM software, to be discretized from the geometrical data into solid elements. After assembling each part, the FE model of the total ear system was built up as shown in Figure 1. It consists of the TM, ossicular chain (malleus, incus, and stapes), ligaments, joints, stapedius muscle, manubrial fold and others. The TM is separated into 3 parts, TM tension, TM relaxation, TM annular ring, and given with different material properties. The part names corresponding to the numbers are shown in Table 1.

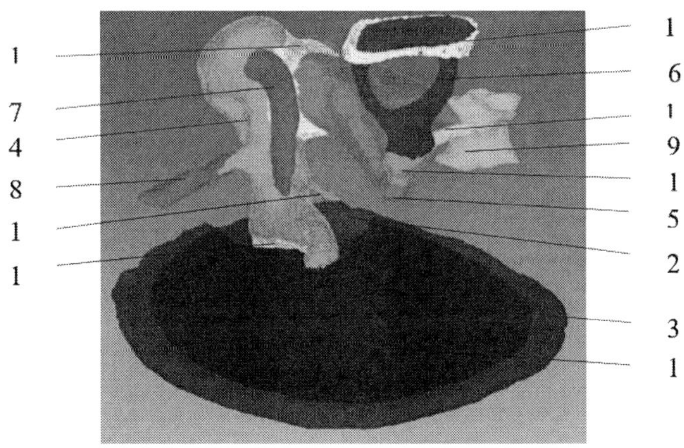

Figure 1. FE-model of total middle ear system.

2.2. Material properties of the middle ear
Table 1 shows the material properties of all components in the middle ear model. The values in Table 1 were decided by considering the references [3, 5].

Table 1. Material properties of components of middle ear

No.	Anatomical name	Young's modulus (MPa)	Poisson's ratio	Density(kg/m³)
1	TM tension	33.4	0.3	1200
2	TM relaxation	11.1	0.3	1200
3	Tympanic annular ring	0.6	0.3	1200
4	Malleus	14000	0.3	2390
5	Incus	14100	0.3	2150
6	Stapes	14000	0.3	2200
7	Tensor tympani	5	0.3	2500
8	Anterior malleolar ligament	21	0.3	2500
9	Posterior incudal ligament	4.8	0.3	2500
10	Lateral mallear ligament	6.7	0.3	2500
11	Stapedius muscle	0.38	0.3	2500
12	Annular ligament	0.15	0.3	2500
13	Incudomallear joint	7	0.3	2500
14	Incudostapedial joint	6	0.3	2500
15	Manubrial fold	3.34	0.49	1200

2.3. Boundary conditions
Some researches modelled the boundary conditions of the edge of TM with torsion springs or combination springs [1, 3]. In this study, the nodes at the circumference of the tympanic annular ring were fixed. Notice that the Young's modulus of the ring is quite smaller than that of the TM tension. This boundary condition is more suitable to express the translation and bending motions at the edge area of the TM. The ends of tensor tympani, anterior malleolar ligament, posterior incudal ligament, lateral mallear ligament and stapedius muscle were also fixed. As for the load condition, the pressure, converted from the sound pressure was applied at the normal direction of the TM surface.

3. Analysis and Verification of Numerical Model
A dynamic analysis was performed, and the result was compared to the measurement results reported from other researchers to examine the validity of the numerical model.

3.1. Sound pressure
The frequency response analysis of the middle ear model was performed in the frequency range from 100 Hz to 10 kHz. NX Nastran was used as the solver of the analysis. For the convenience of comparing to the measurement results, the sound pressure was set at 90 dB, and it was converted into pressure using equation (1).

$$L_p = 20\log_{10}(p/p_0) \qquad (1)$$

where $L_p = 90$ dB is the relative sound pressure, and $p_0 = 20 \times 10^{-6}$ Pa is the reference sound pressure. The pressure of $p = 0.632$ Pa was obtained and applied on the surface of the TM.

IOP Conf. Series: Materials Science and Engineering **886** (2020) 012045 doi:10.1088/1757-899X/886/1/012045

3.2. Damping
Some researchers have suggested using Rayleigh's damping for the analysis of the middle ear [1, 3]. Since the software we use cannot select the Rayleigh's damping in frequency response analysis, structural damping was used. Then, the equation of motion of the finite element model can be written as follows:

$$[M]\{\ddot{x}\}+(1+jG)[K]\{x\}=\{F\}e^{j\omega t} \tag{2}$$

As for the structural damping coefficient G, it is assumed as a function of frequency f, and determined as below:

$$G=0.00021f+0.4 \tag{3}$$

The initial value G_0= 0.4 refers to the research of Hidayat et al. [4], and the coefficient 0.00021 is determined by try-and-error to fit the analyses frequency response curve to the measured one.

3.3. Result and Consideration
Figure 2 shows a frequency response curve obtained by the finite element analysis. It presents the relation between the normal displacement of stapes-footplate and frequency. The normal displacement of stapes-footplate, of which the direction is towards to the cochlea, is considered as an intensity index of the sound wave hitting the cochlea. As a comparison, a measurement curve, which is a mean curve measured from 17 subjects by Gan et al. [2], is also shown in Figure 2. It is concluded that the analyses curve is roughly matched to the measurement curve. Especially, the peak position and peak value of the two curves are almost the same. Therefore, it is reasonable to consider this numerical model can reflect the dynamic behaviours of a healthy ear. This model was used as a reference model to evaluate the repaired model later.

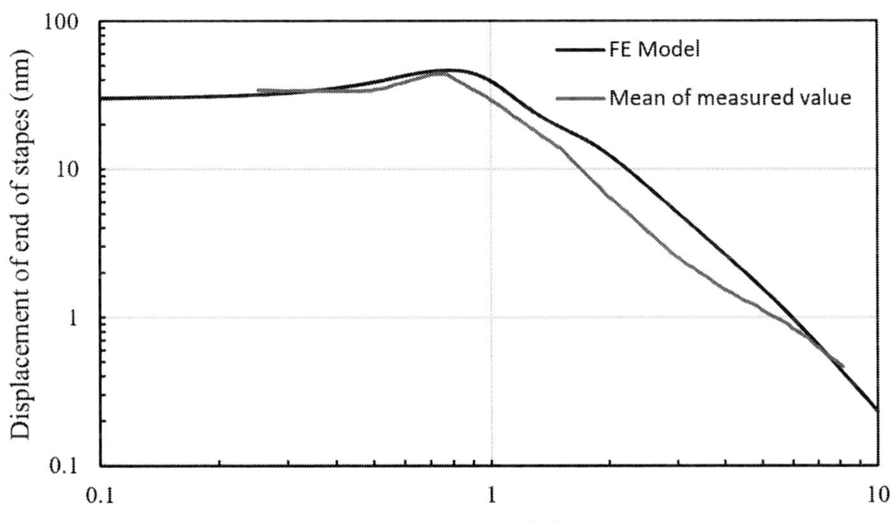

Figure 2. Comparison of analysis result and experimental result [2].

4. Repairing TM by Cartilages of Various Thickness
Cartilage myringoplasty is an operative procedure that fixes the TM perforation by thin-sliced cartilage. In this study, as a fundamental research for the optimization of this surgery, we investigated the optimal thickness of the sliced cartilage for closing the TM perforation.

4.1. Perforated TM model and sliced cartilage model

As an example, a perforated TM model was created by cutting about 40% area of the TM at the bottom area, as shown in Figure 3. Replacing the perforated part with sliced cartilage, the repaired model was created as can be seen in Figure 4. The red area shows the sliced cartilage plate attached to the TM. The values from 0.1mm to 0.7mm were used as thicknesses of the sliced cartilages to perform frequency response analysis. We attempted to find the optimal thickness of sliced cartilage used in myringoplasty by comparing the frequency response curve with that of the healthy ear in the frequency range of 100Hz to 10kHz. The Young's modulus and density of the cartilage are 2.8Mpa and $1.3\times10^3 \text{kg/m}^3$, respectively.

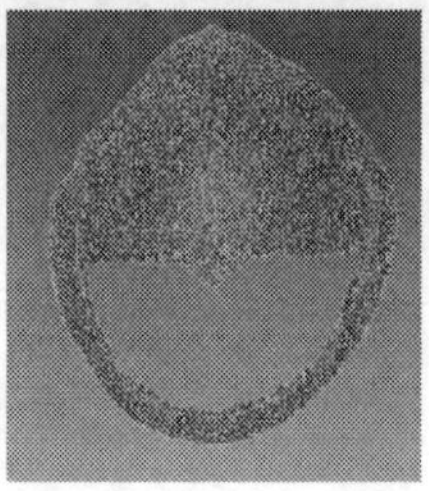

Figure 3. Perforated TM

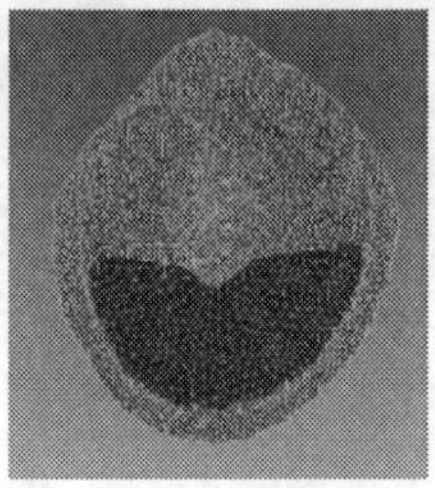

Figure 4. A Repaired model with sliced cartilage.

4.2. Analysis results

The frequency response curves of repaired models of the thickness of 0.3mm/0.4mm/0.5mm/0.6mm /0.7mm cartilage plates are shown in Figure 5. Curves of a thickness of 0.1mm/0.2mm are not shown here to avoid confusion, and for other reasons which will be mentioned in Section 5. The frequency responses of the healthy ear model and the perforated model are also shown in Figure 5 as references. The following results are found from the analysis: a) As a natural result, the response of the healthy ear is higher than that of others in almost all frequency ranges. b) There is only a small difference between the responses of various thickness cartilage plates. c) And among them, the peak position and value of the response of 0.5mm plate model shows a better matching to those of the healthy ear model. d) The response of the 0.3mm plate has a better performance than others except for the range of about 600Hz to 1kHz, the neighborhood of the peak. e) The response of the perforated model shows a large difference to the healthy ear model at the low-frequency range, which is considered to be a reasonable result. However, it shows a better performance at the range of 1kHz to 5kHz. It is interesting to examine further reasons for such results.

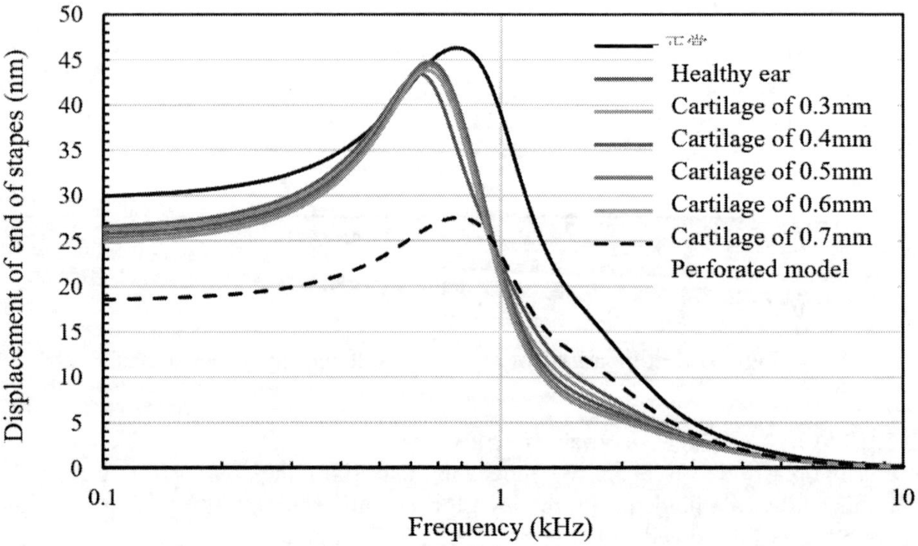

Figure 5. Analysis results of models with different cartilage thickness.

5. Considerations

Only from the mechanical viewpoint, the thickness larger than 0.5mm is not a choice since its frequency response has no advantage in the peak position, peak value and response performance in all frequency range. The thickness of 0.1mm/0.2mm are also not considered because of the larger gap of the peak position, and peak value to those of healthy ear. Another reason is the increased risk of the fracture using such thin-sliced cartilage in a relatively larger area.

5.1. Evaluation of the dynamic characteristics at the peak area

The frequency response curve of the healthy ear shows a peak point at about 800Hz, which is almost the same as the measurement curve. Around the peak point, there is a plateau in the frequency range of about 500Hz to 1kHz. This frequency range is important to the discrimination of the vowels since the first formants of vowels are in this range.

As for the healthy ear model, the eigen-modes mainly contributing to the plateau area are identified as shown in Figure 6 through the modal analysis.

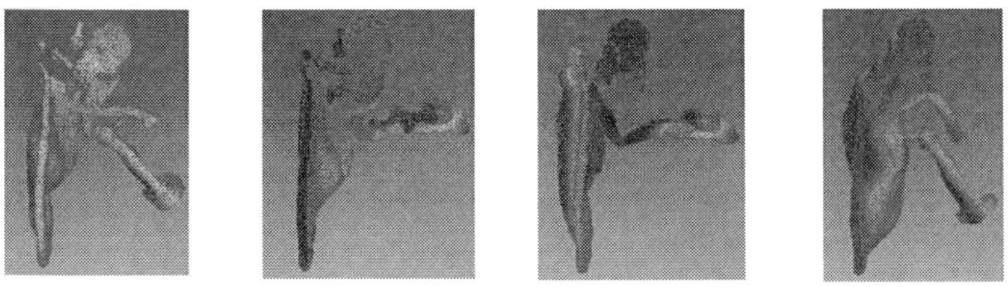

Mode 1: 723.0Hz   Mode 2: 906.5Hz   Mode 3: 962.4Hz   Mode 4: 1096.4Hz

Figure 6. Eigen-modes and frequencies of healthy ear model in the plateau area.

Concerning the repaired model with 0.5mm cartilage, the eigen-frequencies in this area are obtained as mode 1: 706.9Hz, mode 2: 817.1 Hz, mode 3: 932.6Hz, mode 4: 1065.1Hz, respectively.

As for the repaired model with 0.3mm cartilage, the eigen-frequencies in this area are obtained as mode 1: 691.4Hz, mode 2: 770.0 Hz, mode 3: 915.3Hz, mode 4: 975.8Hz, mode 5: 1075.8 Hz, respectively. Comparing with 0.3mm cartilage model, the 0.5mm cartilage model shows closer performance to the healthy model in both eigen-frequencies and mode shapes in the plateau area.

5.2. Evaluation of response performance in all frequency range

In order to do a quantitative evaluation of the effect of the repaired models, we proposed an equation of *matching rate* defined as follows:

$$MR = \left(1 - \sum_{i=0}^{N} w_i \left| X_i - X_{0i} \right| / X_{0i} \middle/ \sum_{i=0}^{N} w_i \right) \times 100 \tag{4}$$

where X_i is the frequency response of a repaired model at the frequency f_i of the sampling point i, and X_{0i} is the frequency response of the healthy ear model at the same sampling point. The frequencies of sampling points are decided as the formula below:

$$f_i = 100 \times 1.1^N \,(\text{Hz}), \qquad N = 0,1,2,\cdots,48 \tag{5}$$

The weight coefficients w_i of sampling point i should be decided according to the importance of the frequency for daily conversation. As an example, here we provided the weight coefficients as, 2 in the frequency range of 300Hz to 3kHz, and 1 in the frequency range of others.

Using the proposed equation, the matching rate of each thickness model is calculated as shown in Table 2.

Table 2. Matching rate of each thickness model

Thickness (mm)	0.3	0.4	0.5	0.6	0.7	Hole
Matching rate (%)	76.2	74.7	73.1	71.3	69.6	66.8

As a reference, the index of the perforated model is also listed in this table. Notice that the matching rate of the healthy ear model is 100%. By comparing the values of matching rate to that of the healthy ear model, we can find that 0.3mm model is the best one for the recovery of hearing ability, while perforated model is the worst one.

6. Conclusions

In this research, the proper thickness of the sliced cartilage used for the reconstruction of the TM was investigated numerically. Firstly, a detailed 3D FE model of the human ear was established from the CT scan data. The edge of the TM, ligaments, muscle and joints were faithfully modeled. After adjusting the damping coefficient, the frequency response curve shows a good correspondence with the measurement curve. Then, the optimal thickness of the cartilage plate using for closing the TM perforation was investigated. Comparing with the frequency response of healthy ear, it is found that: 1) in the case of a hole created at the bottom of the TM with a size of about 40% of TM, thickness larger the 0.5mm is not suggestable from the mechanical viewpoint; 2) 0.5mm cartilage plate model shows a better performance at the peak response area (600Hz-1kHz); 3) evaluating with the matching rate equation we proposed, 0.3mm cartilage plate model shows a better average performance at the recovery of hearing ability in total frequency range. To decide the specific dimension of optimal thickness, further investigation is required.

References

[1] Koike T and Wada H 2002 *J. Acoust. Soc. Am.* **111**(3) 43-64
[2] Gan R Z Sun Q Dyer R K Chang K H and Dormer K J 2002 *Otology & Neurology.* **23** 271-280
[3] Sun Q Gan R Z Chang K H and Dormer K J 2002 *Biomechanics Modeling Mechanobiology.* **1** 109-122
[4] Hidayat H Okamoto S Lee J H Hato N Yamada H and Takagi D 2016 *J. Biomed. Sci. Eng. Online*
[5] Greef D De Pires F and Dirckx J J J 2017 *Hear. Res.* **344** 195-206
[6] Mehta R P Rosowski J J Voss S E O'Neil E and Merchant S N 2006 *Otol. Neurotol.* **27**(2) 136-143
[7] Yetiser S and Hidir Y 2009 *Ann. Otol. Rhinol. Laryngol.* **118**(8) 570-574
[8] Lee C F Hsu L P Chen P R Chou Y F Chen J H and Liu T C 2006 *Audiol. Neurotol.* **11**(6) 380-388
[9] Lee C F Chen J H Chou Y F Hsu L P Chen P R and Liu T C 2007 *Laryngoscope,* **117**(4) 725-730
[10] Hidayat H Okamoto S Lee J H Hato N Yamada H and Takagi D 2017 *Int. Sympo. Eng. Appl. Sci.* 94- 106
[11] Greef De Buytaert J A Aerts J R Hoorebeke L Van Dierick M and Dirckx J 2015 *J. Morphol.* **276**(9) 1025-1046

Rapid Fabrication of Close-Typed Electrowetting on Dielectric Devices

K Sukthang[1], E Pengwang[2], A Tuantranont[3] and W Wechsatol[1,*]

[1]Department of Mechanical Engineering, King Mongkut's University of Technology Thonburi, 126 Prachautid Rd, Bandmod, Thungkru, Bangkok 10140, THAILAND
[2]Institute of Field Robotics, King Mongkut's University of Technology Thonburi, 126 Prachautid Rd, Bandmod, Thungkru, Bangkok 10140, THAILAND
[3]Nanoelectronics and MEMS Laboratory, National Electronics and Computer Technology Center (NECTEC), 112 Thailand Science Park, Paholyothin Rd., Klong Luang, Pathumthani 12120, THAILAND

* Corresponding Author: wishsanuruk.wec@kmutt.ac.th

Abstract
Electrowetting on Dielectric (EWOD) devices have becomes a common device for manipulating liquid droplets in chemical, electrochemical, disease diagnosis and biomaterial detection processes. EWOD devices are prefect platform for Lab-on-a-Chip devices due to their simplicity with no moving parts and high manoeuvrability precision. In this paper, the rapid fabrication procedure of close-type EWOD devices is proposed. The EWOD electrodes was designed to manipulate droplets as small as 8 microlitre. The experimental test reveals that the droplet velocity increases with the magnitude of applied voltage. The further experiment also confirms the capability of the fabricated EWOD on droplet dispersion, which is common requirement for Lab-on-a-chip applications.

1. Introduction
Electrowetting on dielectric devices are a microfluidic device that can be designed specifically for transport, disperse and mixing of liquid droplets. EWOD devices can be used to manipulate small droplets in the vicinity of microlitre [1] to nanolitre [2]. With their precise manipulation of small droplets, EWOD devices are widely used in biomaterial detection such as electrochemical property detection [3], and polymerase chain reaction PCR [4]. Recently, EWOD devices have become common platform for Lab-on-a-Chip devices [5]–[7]. EWOD devices utilize the induced electromotive force to disturb the equilibrium of droplet surface tension, thus causing the droplet to move in order to sustain the equilibrium state.

In this paper, the focus is on proposing the rapid fabrication technique of closed-type EWOD devices. The configuration of typical closed-type EWOD devices in Figure 1 are composed of bottom plate, lower electro layer, hydrophobic layers, dielectric layer, and top electrode layer. The bottom plates or substrates must be rigid, stiff, durable to heat, non-electroconductive, and well-adhesive to metal such as SiO2 plate [7] and glass slide [8]. The lower electrode must be fabricated with high conductive materials. The high conductive electrode allows precise manipulation of droplet movement on EWOD devices [9]. The common materials for electrode layer are gold [10], silver [10], and copper [12]. The dielectric layer must be fabricated with high dielectric constant materials such as PDMS [13], and Parylene C [14]. Both hydrophobic layers are important to the magnitude of surface tension and droplet contact angles. The common hydrophobic layers are fabricated by Teflon spin coating [15]. The top layer is normally made of IOT glasses or FOT glasses to allow visual possibility, rigidity and electricity conduction. The top layer then also acts as the top electrode layer. Medium layer sandwiched between the hydrophobic layers are left for material tested. Sometimes the medium layer is filled with mineral oils in order to control the droplet size and to prevent the outside chemical interference on tested droplets.

Normally, fabrication of EWOD devices are complex and requires sophisticated equipment in order to create precise layers of EWOD devices. In this paper, the rapid fabrication technique is proposed with the objective of providing common accessibility to microfluidic and biomaterial detection research on EWOD devices to the Lab-on-a-chip platform. The responsive velocity of droplet movement in EWOD devices fabricated by proposed rapid technique due to applied voltage is also reported in this paper. Further dispersion behaviour of droplets in EWOD device is also mentioned in this paper.

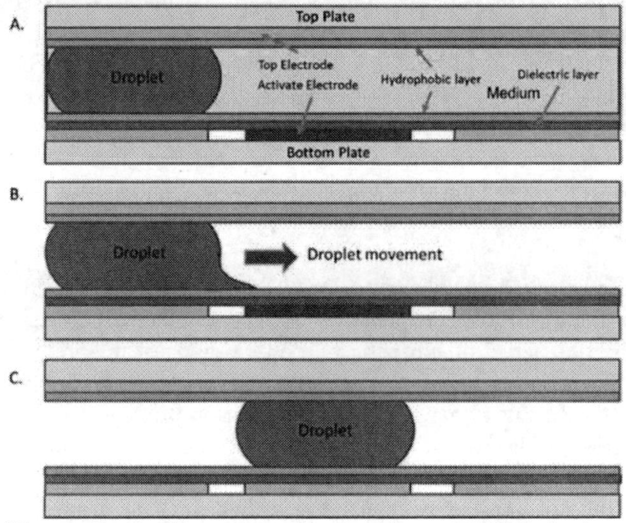

Figure 1. Configuration of a closed type EWOD device

2. Bottom Electrode Design

The fabrication of closed type electrowetting on dielectric devices started from imprinting lower electrode layer on the substrates. The size of each electrode is important and directly related to the volume of controlled droplets. The smaller droplet requires the smaller size of electrode. The relation of droplet volume and the geometry of electrodes and void space in Figure 2 [16] is described in Equation (1)

$$V_{full} = \frac{2L}{3}\pi(3w^2 + 2L^2) \tag{1}$$

where 2w is the width of each electrode [m]
2L is the width of the void space of EWOD device [m]

In this paper, the size of bottom electrodes is fixed at 3 mm. x 3 mm. and the width of the void space of medium space is 1 mm. When considering the corresponding droplet volume according to relation in Equation (1) and the width of medium space at 1 mm, the suitable volume of droplets for

3 mm. x 3 mm. electrodes is about 7.59 microlitres. In this design in Figure 3, the gap space between two adjacent electrodes is 0.15 mm.

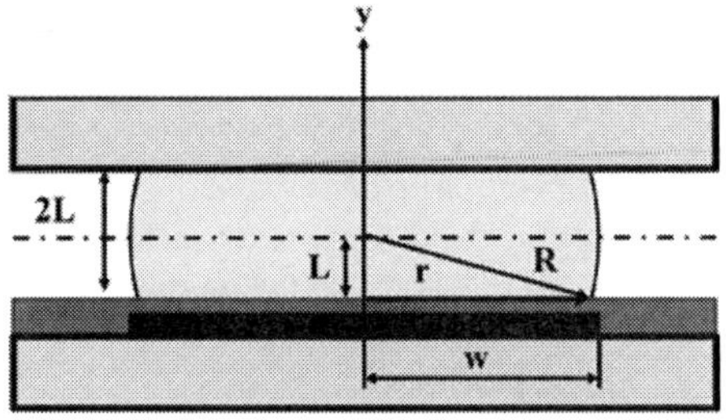

Figure 2. Geometry of Droplets and Closed Type EWOD devices

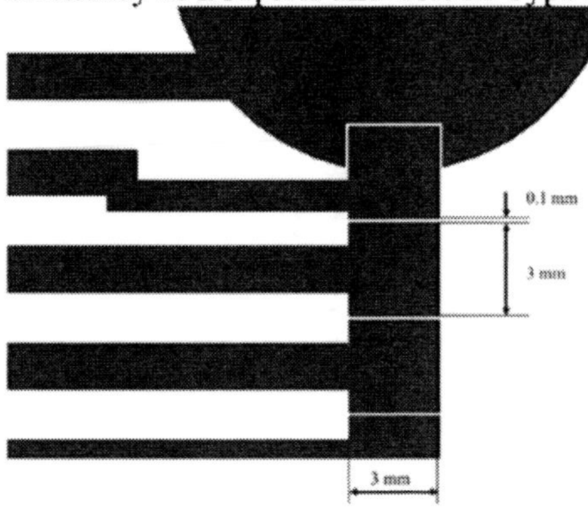

Figure 3. Bottom electrode geometry

Table 1 The droplet volume according to electrode size at the width of medium space equal to 1 mm

w (mm)	Volume of Droplet (microliter)
1.00	3.67
1.25	5.43
1.50	7.59
1.75	10.14
2.00	13.09

3. Proposed Rapid Fabrication Procedure

This paper is aimed to propose the rapid fabrication procedure of EWOD devices with low fabrication cost and requiring less complex equipment. This work will be benefit in providing the accessibility to public and educational institutes for acquiring EWOD and Lab-on-a-chip technology. Materials was carefully selected in this work to provide common usage in fabricating each EWOD layer. The main materials in this work are included of:

1. Printed Circuit Board (PCB), which is used as substrate and imprinted bottom electrode layer,
2. Teflon Film, which can be used to fabricate both dielectric and hydrophobic layers simultaneously,

IOP Conf. Series: Materials Science and Engineering **886** (2020) 012046 doi:10.1088/1757-899X/886/1/012046

3. Mineral Oil, which is used to fill the void or medium space for controlling droplet size and contact angles of testing droplets,
4. Conductive coat glass (FTO), which is used as the top electrode layer and rid of the EWOD devices.

3.1 Fabrication of the bottom plate and bottom electrode layer by utilizing PCB
The photolithography technique using photo-responsive dried film in Figure 4 is applied to imprint electrode layout on Printed Circuit Board (PCB). The fabrication in this step is included of

3.1.1 Preparation of photo-responsive dried film
The photo-responsive dried film must be cut with similar size of printed circuit board (PCB). The hot roll process with the rolling temperature of 150°C by heat plate is used to strict the dried film on to the PCB as in Figure 5.

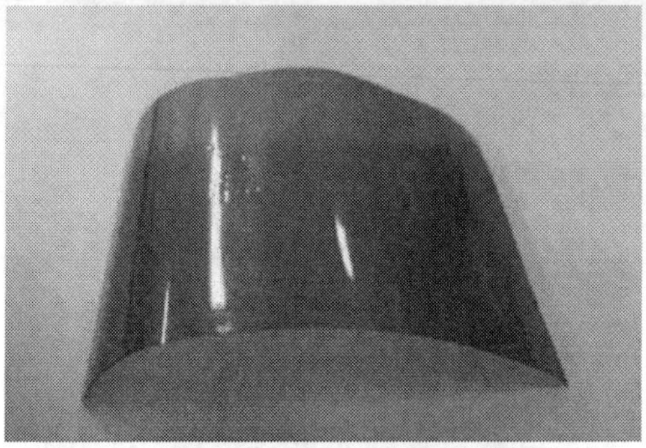

Figure 4. The photo-responsive dried film

Figure 5. The photo-responsive dried film on PCB after hot rolling process

3.1.2 Fabrication of negative film mask
Negative dried film mask is used to create the electrode layout on the photo-responsive film. For simplicity, the transparent paper was used to create the negative film mask. The typical laser print with high dot per inch (dpi) is used to print the electrode layout on the transparent paper as in Figure 6.

Figure 6. The negative film mask made from laser printing on a transparent paper

3.1.3 .Imprinting Electrode Layout on PCB by Applying UV Light

The chemical substance coated on the photo-responsive film will be solidified under ultraviolet light. After covering the photo-responsive dried film rolled on PCB in Figure 5 with the negative film mask in Figure 6, the typical UV lighting as in Figure 7 was then applied for 8 seconds to form the electrode layout on the PCB.

Figure 7. the common UV light lamp

3.1.3 Etching process

This etching process is a common electronic circuit fabrication technique. The part of copper coated on PCB that is not covered with the solidified film on top of the PCB can be removed under etching process. During the etching process, the PCB was soaked with aqueous acid solution at 60–70°C for 15 minutes. After the uncovered copper was removed from PCB in Figure 8 (a), the PCB was then soaked in aqueous sodium hydroxide to remove the photo-responsive chemical substance. In the end, the clear bottom electrode layer was successfully imprinted on PCB as in Figure 8 (b).

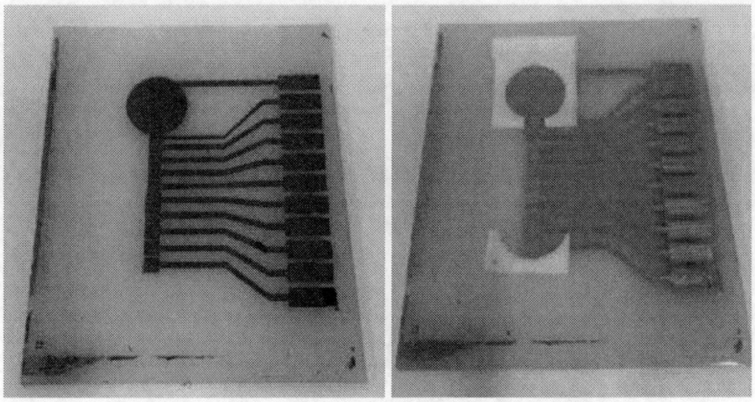

(a) (b)

Figure 8. (a) the PCB after etching process, and (b) the PCB after removing the photo-responsive substance

3.2 Fabrication of Both Dielectric and Hydrophobic Layers by Applying Teflon Film

Instead of coating the dielectric materials on PCB to create the dielectric layer and then coating Teflon-AF on the dielectric layer to create the hydrophobic layer, Teflon film with thickness of 40 micron as in Fig. 9 was applied to create directly on the PCB to create both dielectric and hydrophobic layers. The mineral oil film was applied on the PCB to prevent air bubbles prior to applying the Teflon film. The surface tension of mineral oil also adheres the Teflon film to the PCB and prevent the peeling off between the film and the PCB surface.

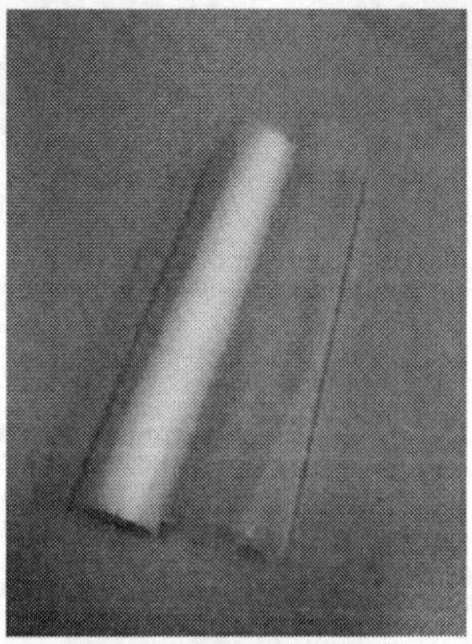

Figure 9. Commercial Teflon Tape

3.3 Mineral Oil Filling For Droplet Protection

Mineral Oil has very low chemical activity and low viscosity comparing to other oils. Mineral oil was applied to fill the void space or the medium space between the top and bottom plates of closed-type EWOD devices for protecting the tested droplets from excessive heat that may cause evaporation, for improving droplet contact angles, and for protecting droplets from chemical contamination.

3.4 Conductive FTO Glass As the Top Plate

FTO glasses normally coated with conductive materials such as Fluorine-doped Tin Oxide was used as the top plate of the closed-type EWOD device in this work. The light transparent characteristics of FTO glasses also allows the visibility of droplet movement in the medium layer. Thus, the FTO glasses are perfect for fabricating EWOD devices for Lab-on-a-chip application.

4. Experimental Set-Up

The experimental test was set-up to study the droplet movement in the EWOD devices fabricated by the proposed method. The equipment for controlling experimental test conditions in Fig. 10 are composed of

1. A step-up transformer with the discharge voltages in the range of 450 – 750 Volts,
2. A function generator for controlling the frequency of applied electrical field,
3. A power amplifier for controlling applied voltages,
4. A sequential controller for controlling the sequence of applying voltage on each electrode.

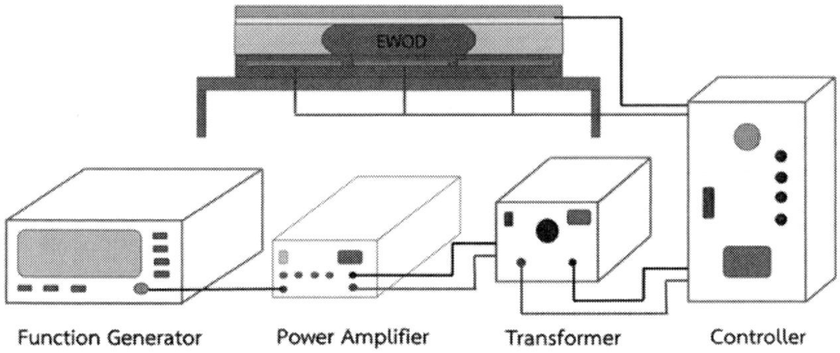

Figure 10. Diagram of control equipment set-up and power supply for EWOD devices

5. Experimental Results

The experimental test was done of the droplet with the controlled volume of 8 microlitre, which is slightly bigger than the 7.56 mm limit reported in Table 1. The electrical control was connected to the EWOD device fabricated following the guideline in this paper as in Figure 10.

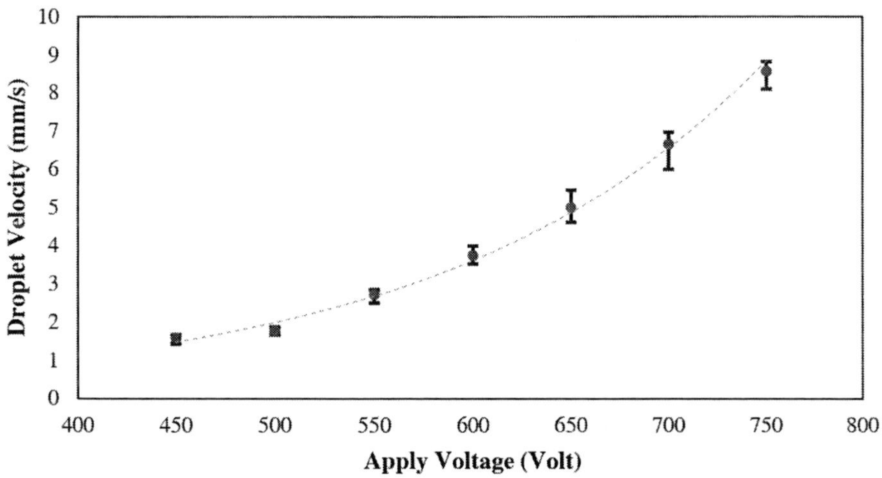

Figure 11. the responsive velocity according the magnitude of applied voltages at 1,000 Hz

The frequency of the applied electrical field was fixed at 1,000 Hz, while the applied voltage was in the range between 450 –750 Volts. The responsive velocity of droplets was determined based on the minimum of switching time for applying voltage on each EWOD electrode. The responsive velocity is reported in Figure 11, in which the droplets move faster as the voltage increases.

Further study on dispersion behaviour of droplets in the EWOD devices is shown in Figure 12, where the droplet volume was successfully split into half on the EWOD device fabricated following the guideline in this paper.

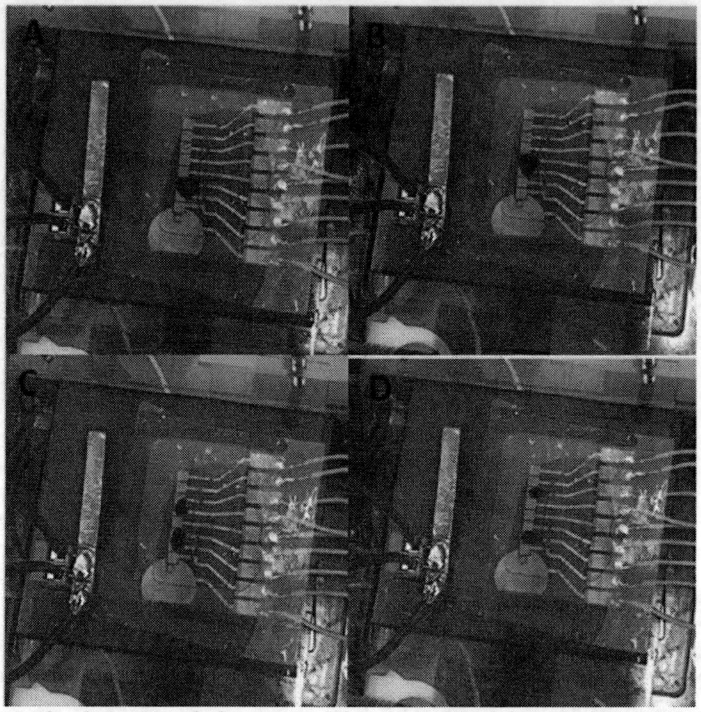

Figure 12. the dispersion behaviour of droplet in the fabricated EWOD devices

6. Conclusion

In this paper, the new rapid fabrication procedure of EWOD was proposed in order to provide accessibility of such devices to public and educational institute, who are interested on microfluidic and Lab-on-a-chip technology. The responsive velocity on EWOD fabricated following the guideline of this paper are collected and reported, in which the responsive velocity increases with parabolic relation with the magnitude of applied voltages. In this case, the empirical relation between the responsive velocity (v) and the applied voltage (V) at fixed frequency of 1,000 Hz can be described as Eq. (2) for the applied voltage in the range between 450 -750 Hz. The maximum responsive velocity is 8.57 mm/s at the applied voltage of 750 Volt.

$$v = 6E\text{-}5V^2 - 00482V + 11.078 \quad (2)$$

The further experiment also confirms the capability of such devices for droplet manipulation, -i.e. in this case, the dispersion capability, thus allowing such devices as the perfect platform for microfluidic and Lab-on-a-chip technology.

Acknowledgments

Authors would like to show sincere gratitude to the financial supports from Thailand Graduate Institute of Science and Technology (TGIST) and Faculty of Mechanical Engineering King Mongkut's University of Technology Thonburi. Also we would like to thank a Technical and Machine support for fabrication from Nanoelectronics and MEMS laboratory at Thai National Electronics and Computer Technology Center (NECTEC).

References

[1] Shah G J Ding H J Sadeghi S Chen S Kim C J and van Dam R M 2011 *Proc. IEEE* Transducers'11 (Beijing:China) 1260–1263
[2] Uwe T Gabriel S M Abhishek S and John S M 2015 *Biomicrofluidics* **9** 04410
[3] Chanpen K Kreeta S Wishsanuruk W and Adisorn T 2011 *Talanta* **84** 1384–1389
[4] Beatriz C Bruno V and Pedro V B 2017 *Sensors* **17** 1495
[5] Fair R B 2007*Microfluid Nanofluid* **3** 245–281
[6] Fouillet Y Jary D Achard J L and Peponnet C 2006 *Proc. 4th International Conference on Nanochannels, Microchannels and Minichannels* (Limerick:Ireland)
[7] Yuhua Y Jianfeng C and Jia Z 2014 *Journal of Micromechanics and Microengineering* **24**(015020) 1–7
[8] Andreas T Herbert E and Bernhard J 2016 *Procedia Engineering* **168** 1061–1065
[9] Mun M N Jagath B N Seung M Y and Hyejin M 2016 *Micromachines* **7** 71
[10] Xuebin T Zhixian Z and Mark M C 2012*Nanotechnology* **23**(375501) 1–8
[11] Veasna S Yunpyo K Oh-Sun K and Kwanwoo S 2019 *Micromachines* **10** 109
[12] Vandana J T Pravin R Raghvendra D and Rajendra P 2015 *Microsystem Technologies Micro- and Nanosystems Information Storage and Processing Systems* (New York:Springer)
[13] Shiraz S Debanjan D Soumen D and Karabi B 2014 *Physics of Semiconductor Devices, Environmental Science and Engineering* (Switzerland:Springer International Publishing)
[14] Chenhui L and Hongrui J 2014 *Micromachines* **5**(3) 432–441
[15] Carlos E Clement D J Si K T and Sung Y P 2016 *Materials* **10** 41
[16] Sukthang S Pongsomboon C Jiartaweesin P Wechsatol W and Kreeta S Wishsanuruk W and Tuantranont A 2016 *Maharakham Int. J. Eng. Tech.* **2**(1) 1–5

IOP Conf. Series: Materials Science and Engineering **886** (2020) 012047 doi:10.1088/1757-899X/886/1/012047

The Comparison of Strain Distribution on Thai Scoliosis and Thai Scoliosis Adjust by Screw Fixation System

Chompunut Somtua[1], Panya Aroonjarattham[1], and Kitti Aroonjarattham[2,*]
[1]Department of Mechanical Engineering, Faculty of Engineering, Mahidol University, Nakorn Pathom Thailand 73170
[2]Department of Orthopaedics, Faculty of Medicine, Burapha University, Chonburi Thailand 20131

* Corresponding Author: kittaroon@gmail.com

Abstract
The secondary curve (S-shape) is one of the scoliosis deformity types. Scoliosis usually occurs in the thoracolumbar region of the adolescents, adulthood and the elderly. If the patient is not advised and treated by the surgeon, the patient will be in more pain and the curvature of scoliosis will be increased. The screw fixation system or pedicle screw system is popular in the treatment of spinal deformity. Because the surgical treatment is highly considered as successful, the postoperative of the pedicle screw system will become virtual homogeneous vertebral and movement together; however, there are few studies reporting the post-operative results in Thai scoliosis surgery. Therefore, this study aims to analyse the strain of Thai scoliosis (S-shape) as compared with normal Thai spine and Thai scoliosis adjusted by screw fixation system under the compressive load with a finite element method. The results showed the maximum strain occurred on Thai scoliosis, normal Thai spine and Thai scoliosis adjusted by screw fixation system respectively. It appeared that the strain occurred on the model of scoliosis adjusted by screw fixation system were more decreased/reduced than the scoliosis model and it is reasonable if compared with the normal model, due to the pedicle screw fixation system could be absorbed the strain occurred on the spinal very well. The center of scoliosis pre-operative was at T7 and L1 levels which had the maximum displacement before being adjusted by screw fixation system. It affects the strain occurred on the models by the reaction force. As mentioned above, the patient should be advised and treated by the surgeon as quick as possible in order to return to daily activities better than before the surgery.

Keywords: Strain, Scoliosis, Thoracic to Thoracolumbar, Screw fixation system and Finite element analysis.

1. Introduction
The most deformity of spinal cord is scoliosis, which is one of the spine curvature disorders, and usually occurs in an adolescent. The deformation is a disorder between and within vertebrae, too must curvature in the frontal plane [1]. Scoliosis can be divided into two types as a single primary curve (C-shape) and secondary curve (S-shape) [2] as shown in figure 1.

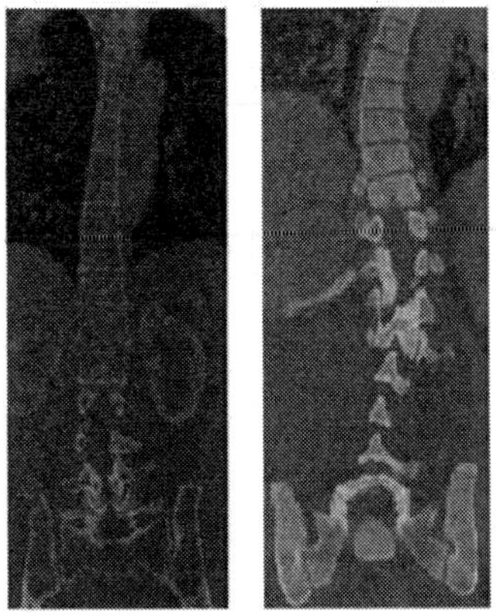

(a) (b)

Figure 1. The types of scoliosis: (a) C-shape and (b) S-shape

This research aims to analyze the strain dispersion on Thai scoliosis (S-shape) compared with normal Thai spine and Thai scoliosis adjusted by screw fixation system using finite element analysis. The scoliosis model measured the curvature of deformity by the Cobb method.

2. Materials and Methods

The three-dimensional models in this research were Thai scoliosis, normal Thai spine and Thai scoliosis adjusted by screw fixation system.

2.1 The three-dimensional models of Thai middle of thoracic to thoracolumbar (T5-L5)

Three-dimensional models of Thai spine recreated by ITK-SNAP program from the computerized tomography (CT) data. The middle of thoracic to thoracolumbar models was scoliosis and normal models as shown in figure 2.

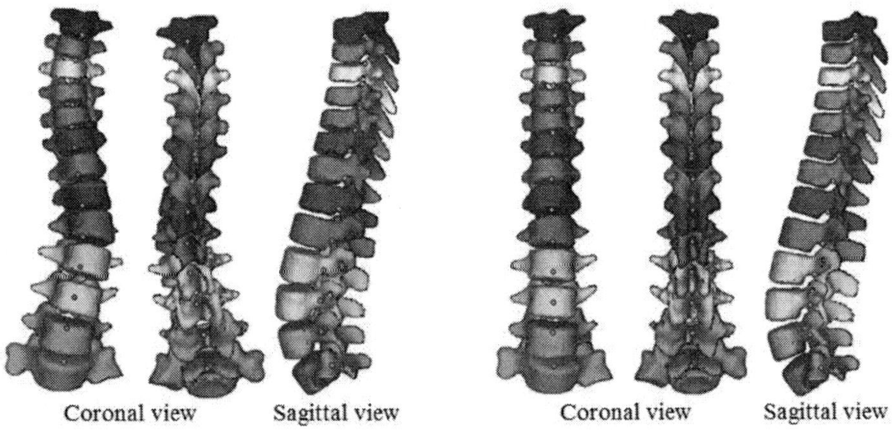

(a) (b)

Figure 2. Three-dimensional models of Thai spine: (a) Scoliosis model and (b) Normal model

2.2 The three-dimensional model of an intervertebral disc

The three-dimensional model of an intervertebral disc were created from SolidWorks software with the actual anatomical position. The intervertebral disc models consisted of nucleus pulposus and annulus fibrosus as shown in Figure 3.

2.3 The three-dimensional model of the screw fixation system

This model was created from SolidWorks software. The model of pedicle screw was the mono-axial type with Moss Miami. The diameter of the pedicle screw was 5 mm and 30 mm length as shown in figure 3.

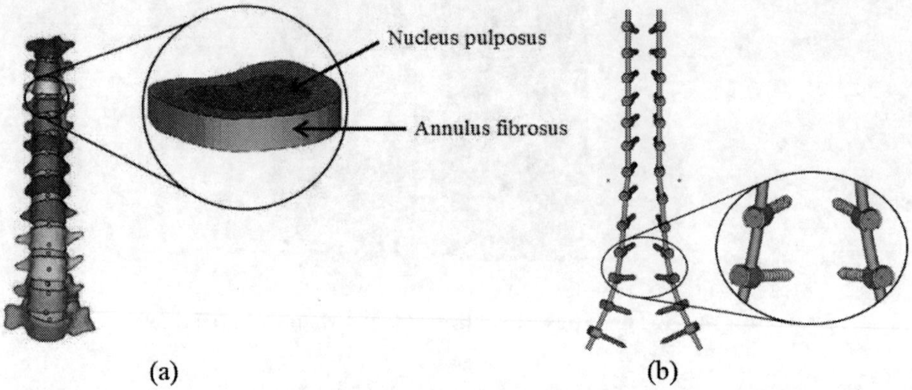

Figure 3. Three-dimensional models: (a) The intervertebral disc and (b) The screw fixation system

2.4 Measurement of scoliosis

Cobb method was used to measure scoliosis curvature angle. Scoliosis in this study was the secondary curve (S shape) and the Cobb angle of the middle thoracic (T5-T10) were 20 degrees and 21 degrees of the thoracolumbar (T10-L5) as shown in figure 4.

2.5 Virtual Simulation

The virtual simulation was used to correct and adjust the scoliosis to a normal model, according to the anatomical human spine. In this process, the models have inserted the intervertebral disc and adjusted by screw fixation system as shown in figure 4.

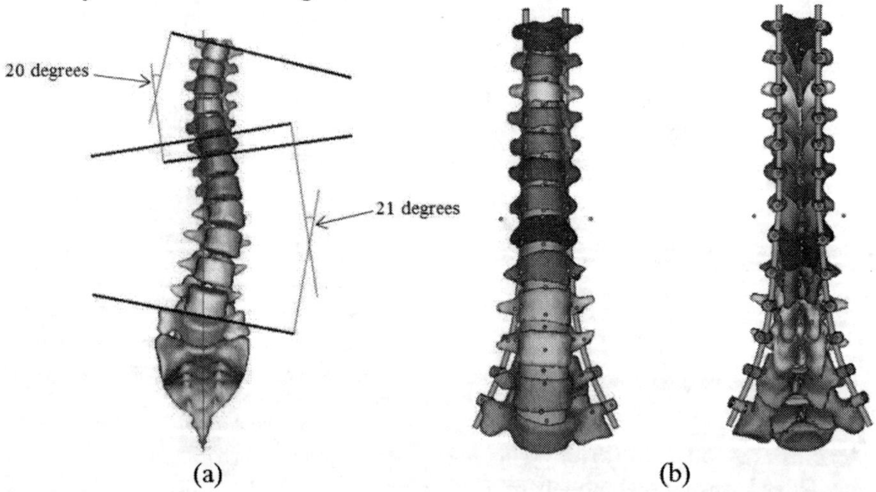

Figure 4. (a) The measurement of scoliosis (Cobb angle) and (b) The virtual simulation model

2.6 Material properties

Material properties of cortical bone, annulus, nucleus and titanium alloy assume as homogeneous, isotropic material and linearly elastic. Anterior longitudinal ligament, posterior longitudinal ligament, ligamentum falvum, capsular ligament, interspinous ligament and supraspinous ligament were nonlinear properties. The specifications of all material properties were shown in table 1.

Table 1. The materials properties of the spine and ligament [3, 4, 5]

Materials	Elastic modulus (MPa)	Poisson's ratio
Cortical bone	14,000	0.30
Annulus	8.4	0.45
Nucleus	1.0	0.499
Titanium alloy	110,000	0.30
Anterior Longitudinal Ligament	Nonlinear	-
Posterior Longitudinal Ligament	Nonlinear	-
Ligamentum Falvum	Nonlinear	-
Capsular Ligament	Nonlinear	-
Interspinous Ligament	Nonlinear	-
Supraspinous Ligament	Nonlinear	-

2.7 Loading and boundary condition

The finite element (FE) models have represented an average sized Asian woman with the height of 150 cm and weight 50 kg. The models were compressed with axial load from the weight of the head 35.80 N at the top of the fifth thoracic vertebrae (T5) [6]. The End of the fifth lumbar vertebrae (L5) was fixed as shown in figure 5.

2.8 Mesh Generation

The element type used to generate the mesh model was the four-node tetrahedral element. All mesh models were generated by MSC Marc software. The scoliosis model had a total of 72,584 nodes and 277,038 elements, the normal model had a total of 73,939 nodes and 281,780 elements and the scoliosis adjusted by screw fixation system had a total of 1,087,751 nodes and 4,892,176 elements. The finite element models are shown in figure 5.

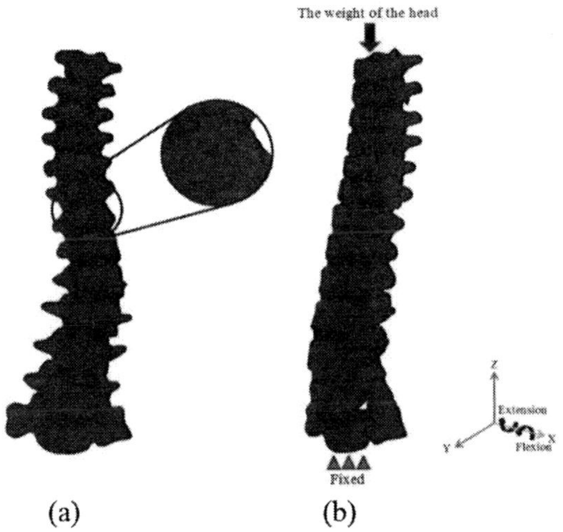

Figure 5. The position of loading and boundary condition act on the middle of thoracic to thoracolumbar mesh model: (a) Scoliosis model and (b) Normal model

3. Validation Model

The validation model constructed the experimental test kit by 3D printing machine to determine the mechanical properties for validating the finite element models. The test kit was tested with Universal Testing Machine (UTM) for evaluating stress and strain distribution compared with finite element models as shown in figure 6 and the reliability of the validation was 90.83%. Therefore, the percentage error obtained in this study is "acceptable" and the finite element model is reliable [7].

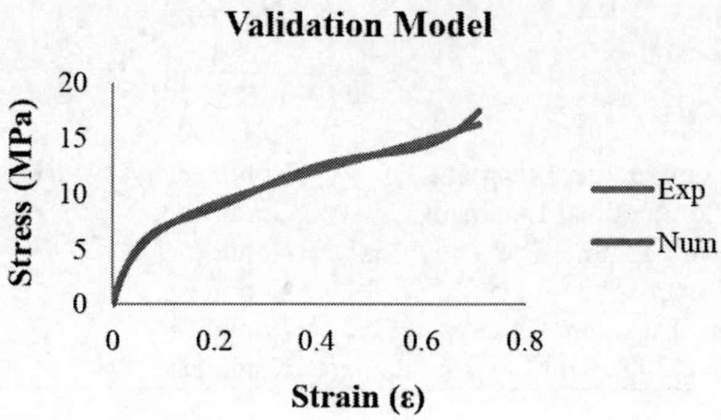

Figure 6. Graph of stress-strain relationship between experimental testing and numerical simulation

4. Results and Discussion

Based on the results, we make the following observations:
The maximum equivalent of total strain on Thai scoliosis, normal Thai spine and Thai scoliosis adjusted by screw fixation system under compressive load are shown in table 2.

Table 2. The maximum equivalent of total strain on Thai the middle of thoracic to thoracolumbar

Thoracic to Thoracolumbar	The Maximum equivalent of total strain (με)		
	Normal	Scoliosis	Scoliosis adjusted by screw fixation system
T5	47.163	157.229	26.657
T6	36.099	265.898	37.676
T7	39.061	185.619	74.100
T8	49.743	158.349	43.753
T9	34.503	181.880	31.633
T10	52.649	70.393	44.104
T11	53.838	103.081	35.572
T12	87.918	165.563	8.943
L1	85.516	50.505	83.456
L2	111.181	70.133	53.312
L3	80.168	67.148	64.323
L4	60.147	92.830	26.566
L5	107.979	78.784	79.503

The maximum equivalent of total strain occurred on scoliosis adjusted by screw fixation system tended to be less than scoliosis and normal spine models as shown in figure 7. The screw fixation system could be absorbed the strain occurred on the middle thoracic to thoracolumbar. In addition, the results show that the maximum equivalent of total strain of scoliosis adjusted by the screw fixation system model is less than the normal and scoliosis models at 27.94% and 63.00%, respectively.

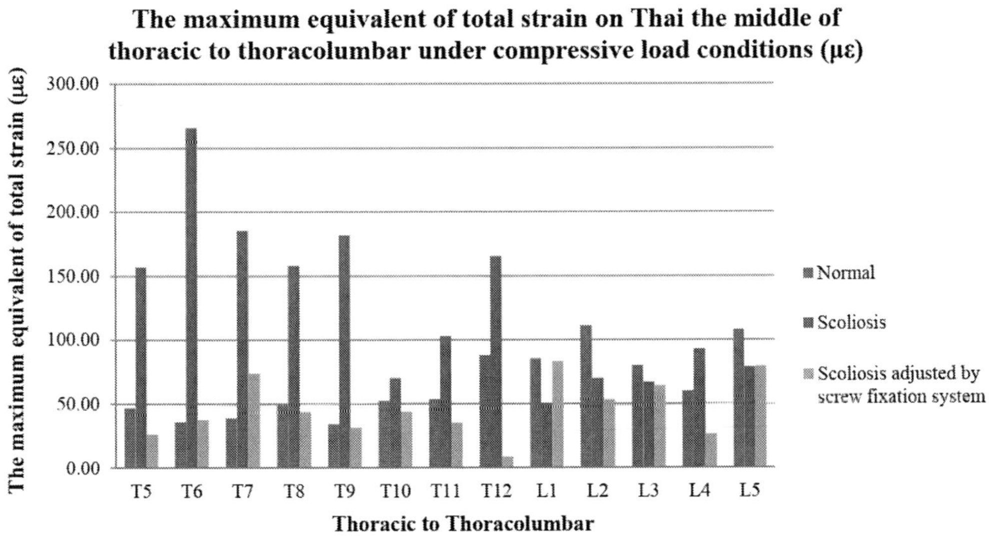

Figure 7. The maximum equivalent of total strain on Thai the middle of thoracic to thoracolumbar under compressive load conditions (με)

Figure 8, illustrates the maximum equivalent of a total strain occurred at the center of scoliosis (T7 and L1) of normal, scoliosis and scoliosis adjusted by screw fixation system models. From figure 8(a) and 8(d), the maximum equivalent of total strain occurred around the body to spinous process, according to the axial force transmission from the weight of the head. The figure 8(b) and 8(e) show the maximum equivalent of total strain occurred around the body to the spinous process more than figure 8(a) and 8(d). As the result of the scoliosis deformity, cause the axial force not transmission along the axis. The last one, the strain concentration occurred at the area around the hole of a pedicle screw, especially when T7 and L1 levels were the center of scoliosis. Because the area was the connection between the vertebrae to the screw fixation.

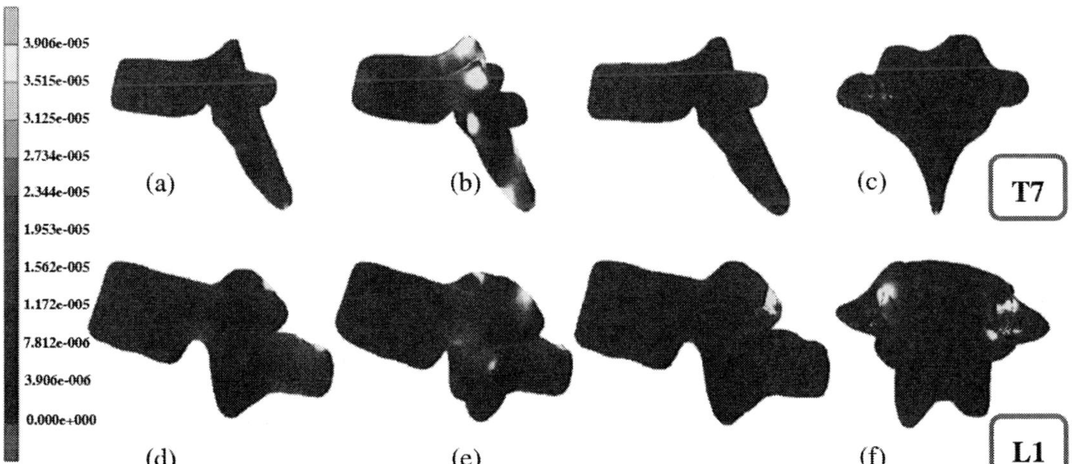

Figure 8. The maximum equivalent of total strain occurred at T7 and L1: (a) T7 of normal, (b) T7 of scoliosis, (c) T7 of scoliosis adjusted by screw fixation system, (d) L1 of normal, (e) L1 of scoliosis and (f) L1 of scoliosis adjusted by screw fixation system.

5. Conclusion

In conclusion, the findings of this study show that the center of scoliosis pre-operative was T7 and L1 levels which had the maximum displacement before being adjusted by screw fixation system. It affects the strain occurred around the models by the reaction force. However, the patient had the Cobb angle less than 40 degrees and the maximum equivalent of the total strain of scoliosis adjusted by implant less than normal spine lightly. So, the surgeon should advise the patient to find a suitable treatment, such as wearing "a brace" [8, 9].

Acknowledgments

The authors wish to thank the Faculty of Medicine, Burapha University for their kind support of CT data and wish to thank Biomechanics Analysis and Orthopaedic Device Design Laboratory (B-AODD Lab), Faculty of Engineering, Mahidol University for their kind support of letting us use the facilities.

References

[1] White A A Panjabi M M 1990 *Clinical biomechanics of the spine* (Philadelphia:Lippincott) **2** 128-155
[2] Janicki J A Alman B 2007 *Paediatrics & child health* **12**(9) 771-776
[3] Niinomi M Liu Y Nakai M Liu H Li H 2016 *Regenerative biomaterials* **3**(3) 173-185
[4] Shin D S Lee K Kim D 2007 *Computer-Aided Design* **39**(7) 559-567
[5] White A A Panjabi M M 1990 *Clinical biomechanics of the spine* (Philadelphia:Lippincott) **2** 22
[6] Okamoto Y Murakami H Demura S Kato S Yoshioka K Hayashi H Sakamoto J Kawahara N Tsuchiya H 2015 *The Spine Journal* **15**(4) 713-720
[7] Lin C L Chang Y H Liu P R 2008 *Journal of dentistry* **36**(3) 194-203
[8] Weiss H R 2008 *Disability and rehabilitation* **30**(10) 799-807
[9] Kuroki H 2018 *Journal of clinical medicine* **7**(6) 136

An Analysis of Temperature Distribution and Ablated Volume in the 3-D FEM Tissue Model with Blood Vessel during Radio-Frequency Ablation

P. Keangin[1,*] and P. Rattanadecho[2]

[1]Department of Mechanical Engineering, Faculty of Engineering, Mahidol University, 25/25 Phutthamonthon 4 Road, Salaya, Nakhon Pathom 73170, Thailand.
[2]Department of Mechanical Engineering, Faculty of Engineering, Thammasat University (Rangsit Campus), 99 mu 18, Paholyothin Road, Klong Nueng, Klong, Pathumthani, 12120, Thailand.

* Corresponding Author: pornthip.kea@mahidol.ac.th

Abstract
The current treatment for a brain tumor has many methods such as surgery or chemotherapy but often the treatments may affect the patient or treatment is still limited in some aspects. A promising technique for brain tumor treatment is radio-frequency ablation. Radio-frequency ablation utilizes alternating current (AC), typically at about 500 kHz, to destroy unwanted tissues by heating to temperatures exceed 45°C to 50 °C. The objective of this research is to study the treatment of brain tumor during radio-frequency ablation with computer simulation through finite element method (FEM) for solving the problem. A three-dimensional FEM model of brain tissue with a brain tumor and blood vessel is considered. Three types of blood vessels i.e. aorta, main veins and terminal veins are studied. The electric current equation and time-dependent bioheat transfer equation coupled with time-dependent convective blood vessel heat transfer equation are solved to predict temperature distribution and ablated volume within brain tissue model. The simulation results are compared with the simulation results from previous work to verify the accuracy of the presented model. The effects of the blood vessel are included and not included in the brain tissue model, types of blood vessel and treatment time during radio-frequency ablation on the temperature distribution and ablated volume are investigated. The outcomes indicated that size of blood vessel results in heat sink effects which increase with the velocity of blood flow. The aorta causes more heat sink effects than main veins and terminal veins, respectively; thereby the ablated area and ablated volume in case of the aorta is higher than one in case of main veins and terminal veins, respectively. In addition, the temperature value increases with greater treatment time. The results from this study provide the basis for planning the radio-frequency ablation of brain tumor, leading to approaches of medical practice process improvement.

Keywords: Ablated volume, Brain tissue, Radio-frequency ablation, Finite element analysis

1. Introduction

The human brain is the command center for the human nervous system. It is responsible for controlling activities of body, motion, homeostasis such as heartbeat, blood pressure or balancing temperature, and so forth related to cognition, thought, memory, emotion, even decision making. A Brain tumor is one of the diseases concerned with the brain in which abnormal cells uncontrollably form in the tissue of the brain. It can be either benign tumors or cancerous tumors and can start within the brain or spread from elsewhere which is cancerous [1]. The Risk factor of brain tumor includes family history, head injury, infections, environmental risks such as exposure, radiation or electromagnetic field and metastases that spread from any types of cancer such as lung cancer, breast cancer, colon cancer, kidney cancer etc. [2-3]. A Brain tumor can cause many symptoms regardless of whether a benign tumor or cancerous tumor it is such as headache which is an early symptom of brain tumor and reflects increased intracranial pressure, a seizure which causes loss of consciousness and some symptoms reflect the location, size or growth of tumor such as weakness, loss of vision, loss of hearing or impairment of language [1]. Diagnosis of brain tumor is usually by magnetic resonance imaging (MRI). Brain tumor whether benign or cancerous, start within the brain or spread from elsewhere is usually treated by surgery, radiation and chemotherapy – alone or combination based on a case-by-case basis and depend on the number, size and location of a brain tumor [4]. There are risks, side effects and limitations for each type of treatment. The alternative treatment option is radio-frequency ablation.

Radio-frequency ablation is one of the thermal treatment techniques that are employed in the destruction of the tumor. Radio-frequency ablation can be used to treat tumors in the liver, the lungs, kidneys, bones, breasts, the brain as well as other organs [5-9]. During radio-frequency ablation, alternating current (AC) with high frequency (about 500 kHz) is delivered into tumorous tissue through an electrode inserted percutaneously under image guidance into the tissue. The alternating electrical currents then move across the inserted electrode to the ground pad. The movement of ion against friction results in Joule heating, so the tissue temperature is raised [8-9]. The goal of this technique is to elevate the tissue temperature beyond 50 °C to cause the irreversible death of tumorous cells and coagulative necrosis [8-11]. Slightly increasing in temperature in approximately 1°C in the human brain can affect properties of single neurons and neural network function [12]. An increase of approximately 1 – 5 °C in human organs temperature can cause brain lesions. In addition, increases to approximately 40 °C for an extended period of time may lead to tissue damage [12-13]. The study examined the effect of radio-frequency by experiment such as the experimental treatment in the human brain will encounter the problem due to research ethics. Therefore, the computer simulation will be an alternative study which will help reduce the limitations from the experimental methods.

The Pennes' bioheat model is widely used for modeling the heat transfer in human tissues during radio-frequency ablation to investigate the interaction of radio-frequency and heat transfer. In 2002, Tungjitkusolmun et al. [14] determined tissue temperature distribution during radio-frequency ablation using the finite element method (FEM). A three-dimension model of a four-tine radio-frequency probe, liver tissue and a large blood vessel was constructed. The effects of the presence of a blood vessel and its distance to probe were studied. The results showed that the blood vessel acted as a heat sink and the adjacent tissue was cooled down when it was near the probe. Zhu et al. [15] investigated a numerical model coupling the transport of electrical current, heat and the evaporation process in the tissue. The findings showed that the largest specific absorption rate happens at the edge of the circular surface around the electrode. In addition, the time duration between the beginning of the evaporation was 65 s and the cooling situations of the inside agent greatly affect the water loss pattern. Huang [16] computational study focused on the effects of blood vessels on the formation of the thermal lesion during radio-frequency ablation. The finite difference method was used to solve both Pennes' bioheat transfer equation and convection-diffusion equation for a three-dimensional model of liver under several conditions of blood flow, the dimension of blood vessel and location of the blood vessel. The results showed that small blood vessel produced thermal lesion region like a tail, which could hurt the normal tissues. Zorbas et al [17] simulated radio-frequency ablation in computational models of human liver, lung and kidney in order to study the effect of realistic geometry in the treatment outcome. It was shown that the ablation with the conventional single electrode inside the tumor must be careful in voltage and time setting in order to achieve effective treatment because of the difference in an ablated zone in various body sites. Singh et al. [18] studied on three-dimensional FEM models of breast cancer ablated by radio-frequency by varying applied voltages, tissue thermogenic capacities and sizes of a blood vessel. Integrating the bioheat equation with blood flow conservation was solved. The results showed that the increase in thermogenic capacity due to an increase in tumor volume caused an increase in treatment time.

Although a bulk amount of research has developed models to study on radio-frequency ablation, it was found that most of them have no regard for the complexity of organs such as including of different types of the blood vessel. In addition, most studies have mainly focused on the modeling of heat transfer within homogeneous tissue but do not include a tumor in a model. In fact, having different types of blood vessel as well as blood flow velocity and embedded brain tumor are affected by the treatment outcome. Therefore, in order to provide adequate information on the appropriate treatment of radio-frequency ablation process, it is essential to consider all of the previously mentioned parameters in the analysis. This research aims to study the treatment of brain tumor during radio-frequency ablation with computer simulation through FEM for solving the electrical-thermal problem. A cylindrical geometry of brain tissue which has spherical tumor included and Y-shape blood vessel included with a four-tine radio-frequency probe is considered. There are three different types of blood vessel considered, all of which have different diameters as well as blood flow velocity. The coupling of electrical current equation, time-dependent bioheat transfer equation and time-dependent convective blood vessel heat transfer equation is explained to predict temperature distribution and ablated volume within brain tissue model. The simulation results are compared with the simulation results from previous work to verify the accuracy of the presented model. The effects of a blood vessel which is included and not included in the brain tissue model, types of blood vessel and treatment time on the temperature distribution and ablated volume are studied. The results from this study provide the knowledge of heat transfer within brain tissue with blood vessel during radio-frequency ablation. It can be used as a guideline to treat brain tumor during radio-frequency ablation.

2. The formulation of the mathematical model

In this research, a numerical analysis of the electrical current equation, time-dependent bioheat transfer equation and time-dependent convective blood vessel heat transfer equation within the heterogeneous brain tissue is studied.

2.1. The simulation models

The three-dimensional finite element model is used in this study to analyze electrical-thermal problem within human brain tumorous tissue which blood vessel included and not included during radio-frequency ablation. The model of a problem can be divided into four sections: brain tissue, brain tumor, Y-shape blood vessel and radio-frequency probe (RF probe) as shown in Figure 1. The model of brain tissue is a cylindrical shape and has a diameter of 100 mm and a height of 120 mm [14] as shown in Figure 1(a). For another section, the brain tumor is modeled to be 50-mm diameter spherical shape within the brain tissue model as shown in Figure 1(b). The effects of a blood vessel which is included and not included in human brain tissue during radio-frequency ablation are considered. The blood vessel is modeled to be Y-shape looking alike branches of the human blood vessel. The angle between two branches of the Y-shape blood vessel is 30 degree. The length of the blood vessel is 100 mm. The diameter of the blood vessel depends on the types of blood vessel. The blood vessel will be considered in the case that includes blood vessel effect as shown in Figure 1(c). The study uses a four-tine electrode RF probe to transfer electrical current into a brain tumor. Figure 1(d) shows that the RF probe consists of two parts: a trocar and four-tine electrode. The trocar is made from stainless steel and has a diameter of 1.83 mm and a length of 60 mm. The four-tine electrode is made from Nickle-Titanium (Ni-Ti) and has a diameter of 0.53 mm as shown in Figure 1(c). The distance from the distal trocar to the distal electrode is 14.47 mm. The overall width of the electrode is 31.53 mm as shown in Figure 1(e). The RF probe is inserted into the brain tissue with 60 mm depth. The distal trocar is at the center of the brain tumor and parallelizes 31.38 mm away to the centerline of Y-shape blood vessel as shown in Figure 2. Table 1 illustrates the dimensions of model

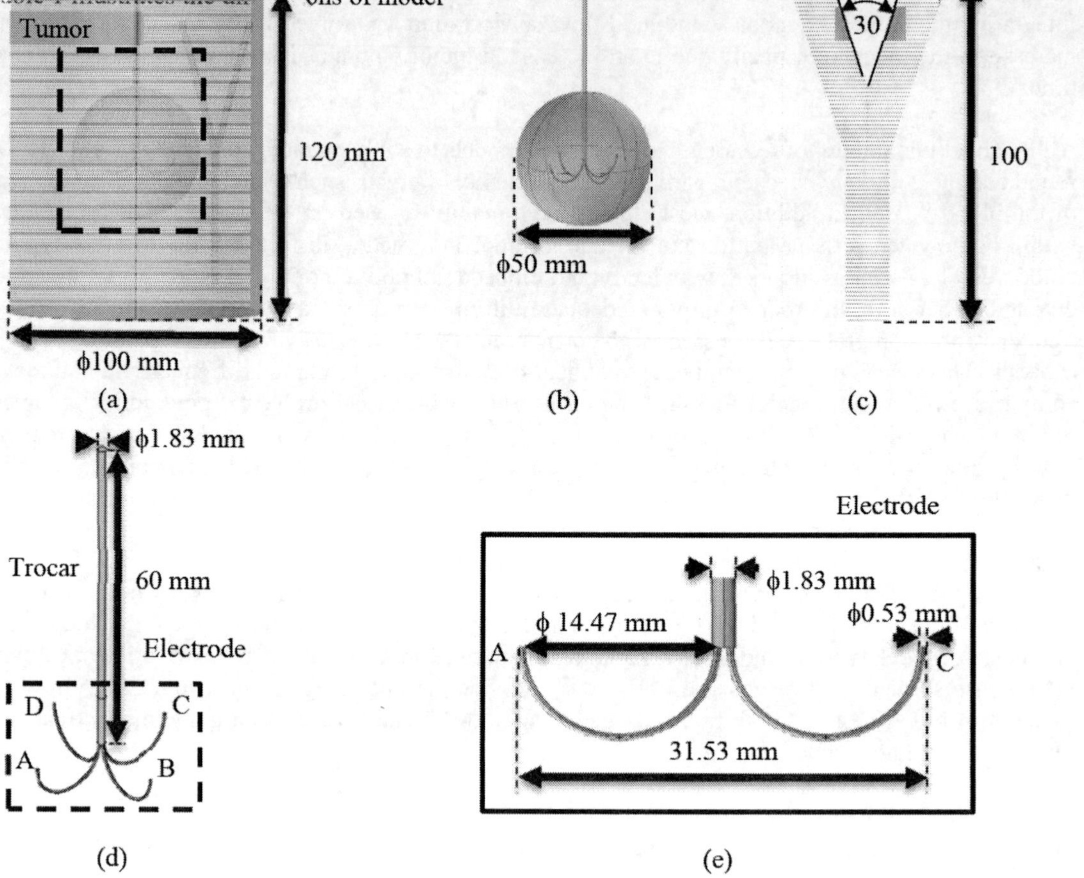

Figure 1. The physical domains of the problem: (a) overall model for human brain tissue during radio-frequency ablation (b) brain tumor (c) Y-shape blood vessel (d) Four-tine electrode of RF probe and (e) electrode

2.2. Governing equation
Mathematic analysis being relevant to this research consists of the electric current equation, heat transfer and blood flow. The system of governing equation is solved numerically using FEM in order to predict temperature distribution and ablated volume in human brain tissue during radio-frequency ablation, as follows;

2.2.1. Electric current
In radio-frequency ablation, an electric current through RF probe causes the electric field in the tissue. The nearby region from the RF probe has the strongest field and causes resistive heating around the electrode. Thus, the electric field is solved by the generalized Laplace equation [12]:

$$-\nabla \cdot (\sigma \nabla V) = Q_j \quad (1)$$

where σ is the electrical conductivity (S/m), V is potential (V) and Q_j is the current source (A/m^3).

The boundary conditions for the electric current analysis as follows: the outer surface of brain tissue is assumed to be ground ($V = 0$), the potential at electrode surfaces is V_0 ($V = V_0$) and the surface between the brain tissue and the brain tumor is considered as a continuity boundary condition, as shown in Figure 3.

2.2.2. Heat transfer

The Pennes' bioheat equation with the metabolic heat source and the external heat source applied to the brain tissue can be written as [12]:

$$(\rho c_p)\frac{\partial T}{\partial t} = \nabla \cdot (k\nabla T) + (\rho c_p)_b \omega_b (T_b - T) + Q_{met} + Q_j \quad (2)$$

where subscript b represents the blood, ρ is the density (kg/m^3), c_p is the specific heat (J/kg·°C), k is the thermal conductivity (W/m·°C), ω_b is the blood perfusion (1/s), Q_{met} is the metabolic heat source (W/m^3), is neglected since it was small [14], Q_j is the external heat source generated by electric current (W/m^3) and have relations as Equation (1), T is the tissue temperature (°C) and T_b is the blood temperature (assumed to be uniform throughout the tissue at 37 °C).

The Pennes' bioheat equation is frequently used to describe heat transfer and the temperature distribution in biological tissue.

The convective heat transfer equation applied to the blood vessel can be written as:

$$\rho c_p \frac{\partial T}{\partial t} + \rho c_p u \cdot \nabla T = \nabla \cdot (k\nabla T) \quad (3)$$

where ρ is the density of blood (kg/m^3), c_p is the specific heat of blood (J/kg·°C), u is the velocity of blood (cm/s), k is the thermal conductivity of blood (W/m·°C) and T is the blood temperature (°C).

The boundary conditions for the heat transfer analysis as follows: the outer surface of brain tissue is set to be body temperature of 37 °C and the velocity field of blood in Z-axis is constant ($u_z = u_0$). Corresponds to the electric current analysis, the surface between the brain tissue and the brain tumor and the surface between the brain tissue and the blood vessel is considered as a continuity boundary condition, as shown in Figure 3.

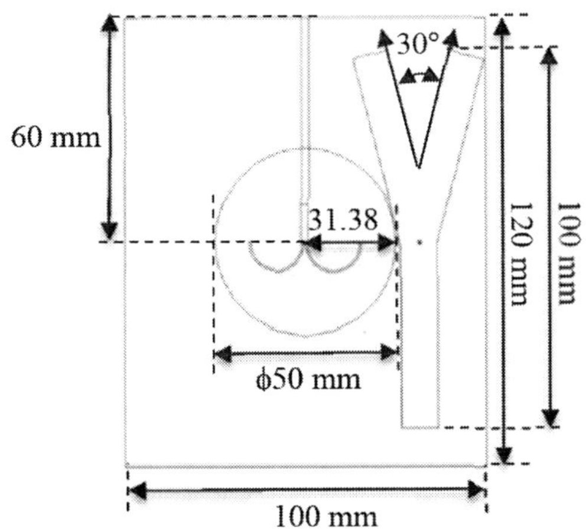

Figure 2. Dimensions and position of physical domains of the problem

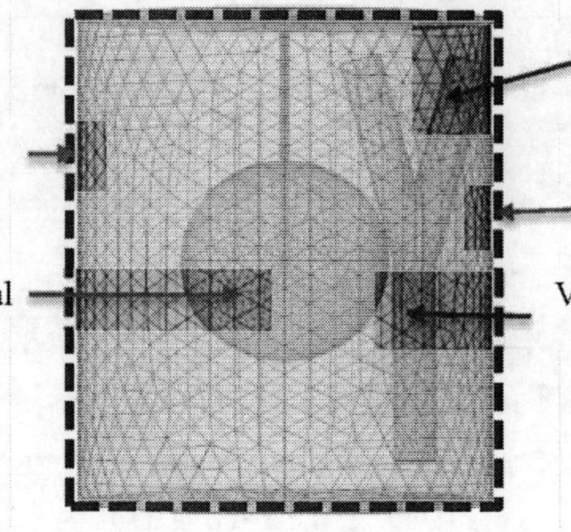

Figure 3. Boundary conditions for analysis

Table 1. Dimensions of model components [14,19].

Components	Materials	Dimensions	
		Diameter (cm)	Length (cm)
Electrode	Ni-Ti	0.053	-
Trocar	Stainless Steel	0.183	6.0
Blood Vessel	Blood	depend on types of blood vessel	10.0
Tissue (Brain)	Brain	10.0	12.0
Tissue (Tumor)	Tumor	5.0	-

2.3. Material and tissue properties

The electrical and thermal properties of the components and tissues in this study i.e. electrode, trocar, blood vessel, brain tissue and brain tumor are shown in Table 2.

Table 2. The electrical and thermal properties of the model components in this study [14,19].

Components	aterials	ρ (kg/m^3)	c_p (J/kg·°C)	k (W/m·°C)	σ (S/m)
Electrode	Ni-Ti	6450	840	18	1×10^8
Trocar	Stainless steel	21500	132	71	4×10^6
Blood vessel	Blood	1000	4180	0.543	2.54
Tissue (Brain)	Brain	1038	3650	0.53	0.76
Tissue (Tumor)	Tumor	1040	3960	0.57	2.09

In addition, there are three types of blood vessel included in this study, namely, aorta, main veins and terminal veins. The parameters such as the diameter of the blood vessel and velocity of blood depend on types of blood vessel. Blood vessel parameters used in the calculation are shown in Table 3.

Table 3. Blood vessel parameters for the calculation [20].

Types of blood vessel	Diameter (mm)	Velocity (cm/s)
Aorta	10	50
Main veins	2.4	1.5
Terminal veins	0.28	0.8

3. Calculation procedure

The temperature distribution and the ablated volume in a human brain during radio-frequency ablation are investigated through a finite element model. The triangular elements and the Lagrange quadratic elements are used to estimate the temperature through each element in the three-dimension model. This study uses an element-independent formulation to solve the problem. To obtain a reasonable estimate, a fine mesh is specifically identified in a sensitive area. The matrix of governing equations with the boundary and initial conditions are solved numerically by the FEM via COMSOL™ Multiphysics program to demonstrate the phenomenon that occurs in the human brain treated by radio-frequency ablation. A grid independence test is carried out to identify the appropriate number of elements that are required. This mesh independence test led to a mesh with approximately 400,000 elements. It is reasonable to suppose that, with this element number, the accuracy of the results will be independent of the number of elements.

This study will systematically investigate the impacts of blood flow and types of a blood vessel on temperature distribution and ablated volume in tumorous human brain tissue during radio-frequency ablation. There are effects of blood vessel included and not included in the model and three types of blood vessels, namely aorta, main veins and terminal veins to be considered.

4. Results and discussion

In the present study, the three-dimensional model is used to analyze coupling of electrical-bioheat transfer and blood flow within tumorous human brain tissue which blood vessel included and not included during radio-frequency ablation.

4.1. Verification of the model

To verify the accuracy of the present model, the result in case of the blood vessel is not included in human brain tissue is validated against the computer simulation results obtained by Tungjitkusolmun et al., 2002 [14]. The validation model is a three-dimensional model having a cylindrical shape (100-mm diameter and 120-mm length). The four-tine RF probe is fully deployed and situated in the middle of the hepatic tissue model. The average applied power is 60 W for 8 min, the average voltage is 30 V and impedance is 60 Ω. Table 4 shows the validation results which are the ablated zone dimensions (at a temperature of 50 °C) in the same positions with the results of validation model when a blood vessel is not included in tissue. The diagram of a typical lesion and the parameters measured is shown in Figure 4. It is found that the simulated results correspond closely with the Tungjitkusolmun et al study. [14]. Some certain mis-match involved in the simulation is caused by some difference in the numerical scheme.

Table 4. Validation of the present simulation results against the previous simulation results [14]

Case	W1 (mm)	W2 (mm)	D1 (mm)	D2 (mm)	R1 (mm)	R2 (mm)
Tungjitkusolmun et al., 2002	21.0	21.0	17.6	18.5	11.8	11.8
Present Study	22.02	21.90	17.00	16.45	13.29	13.29

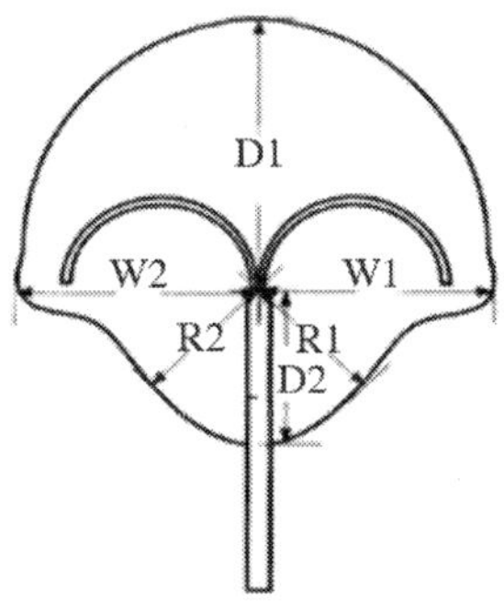

Figure 4. The diagram of a typical lesion and the parameters measured of validation model from Tungjitkusolmun et al., 2002 [14]

4.2. Effects of blood vessel

In this study, the models of three different types of blood vessel included in brain tissue, namely, aorta, main veins and terminal veins which have different diameters and blood flow velocity as shown in Table 3 undergone radio-frequency ablation are investigated to discuss the effects of blood vessel included and not included in the model and effects of the difference between each type of blood vessel. The computer simulation of radio-frequency ablation is performed based on the power of 15 W for a treatment time of 480 s via a four-tine electrode RF probe. Brain tumor has 50 mm diameter and distance between the center of brain tumor and the centerline of blood vessel included in the brain tissue is 31.38 cm. Figure 5 (a) - (d) illustrate temperature distribution within brain tissue on the x-z plane which various types of blood vessel included, i.e. aorta, main veins, terminal veins and no blood vessel included, respectively, undergone radio-frequency ablation. The red bound in temperature distribution identifies the margin in which temperature is 50 °C isotherm. The 50 °C isotherm correlates with coagulation zone boundary within acceptable accuracy and the region inside the red bound is called an ablated zone. The hot spot zone with the highest temperature occurs at the four-tine electrode tips and decreases with distance. In cases that blood vessel included in brain tissue, the results show that the temperature distributions and ablated zones are asymmetric around the electrode because they converge to the blood vessel. The asymmetry in temperature distributions and ablated zones occurs more in case of the aorta, main veins and terminal veins, respectively. Considering cases of blood vessel included, the temperature distribution and ablated zone within brain tissue are affected by blood flow due to heat sink effect. Aorta causes the larger temperature distribution and ablated zone with a higher temperature than main veins and terminal veins, respectively, because its higher blood flow velocity results in more influential to heat transfer within brain tissue during radio-frequency ablation. In case of no blood vessel included, radio-frequency ablation creates symmetric larger temperature distribution and ablated zone within brain tissue than cases of blood vessel included because it is not disturbed by a blood vessel.

The simulation results from this study illustrate that the temperature distribution and ablated zone from radio-frequency ablation are influenced by heat sink effects due to convective heat transfer by blood flow. The electric field is created between two electrodes, one is percutaneously inserted into the tumor and another is a ground pad. The high-frequency AC is delivered into the tissue and induces Joule heating that leads to irreversible destruction of the tumor cell at a higher temperature above 50 °C. When the bulk motion of blood flow occurs near the region in which radio-frequency ablation is conducted, the heat generated by Joule heating from radio-frequency ablation is carried and dispersed. Heat is first transferred from ablated zone to surface of a blood vessel by conduction and then carried away from the surface by convection. This is by combined effects of conduction within blood flow due to randomness of blood flow and the bulk of blood flow that removes heated blood which is adjacent to the surface of a blood vessel and replaces with cooler blood. The blood flow enhances heat transfer, since it carries cooler bulk of fluid into contact, initiating rate of heat conduction at a number of sites in the fluid. The rate of heat transfer by convection strongly depends on fluid (blood) velocity. The higher fluid velocity leads to a higher rate of heat transfer. Considering the blood vessels included in this study, the aorta has a larger diameter with higher blood flow velocity than main veins and terminal veins, respectively, as shown in Table 3. Therefore, the rate of heat transfer due to heat sink effects in the aorta is higher than the main veins and terminal veins, respectively. It results in more hot spot zone with higher temperature in the aorta than main veins and terminal veins, respectively. In addition, heat sink effects result in the region between near electrode and blood vessel has a higher temperature and leads to asymmetric ablated zone compared with the case that no blood vessel included. The asymmetric ablated zone with a high temperature between the near electrode and blood vessel occurs more in case of the aorta and less in case of main veins and terminal veins, respectively, corresponding to the rate of heat transfer resulted by heat sink effects. Furthermore, the radio-frequency ablation raises the temperature in tissue by Joule heating, being the process where the energy of electric current flows through a conductor and converted into heat due to the passage of it through resistance. The blood has higher electrical conductivity than brain tissue as shown in Table 2, AC delivered into the tissue is consequently diverted toward blood flow. Therefore, more Joule heating occurs in the region between the near electrode and the blood vessel along with the blood vessels, especially the nearby site to the electrode, results in the asymmetric ablated zone with a high temperature in these regions.

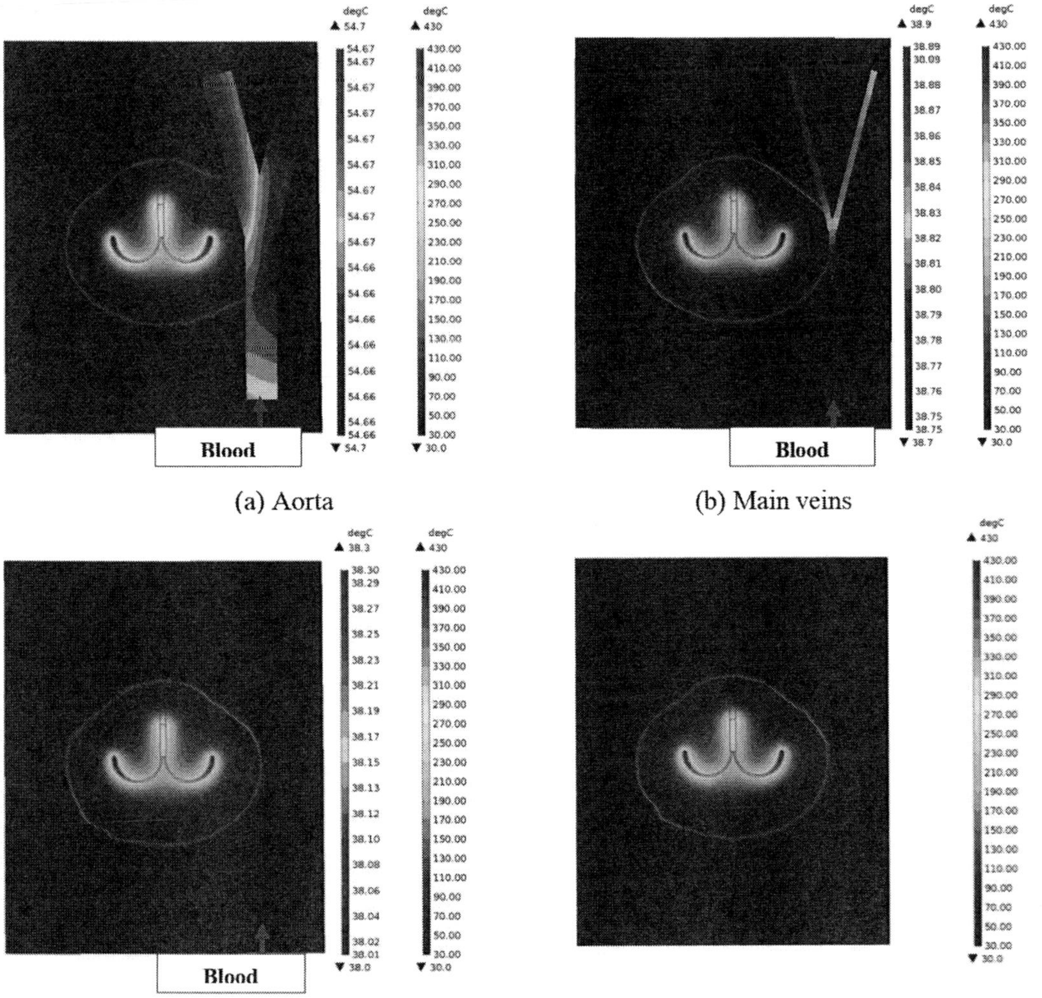

(a) Aorta

(b) Main veins

(c) Terminal veins

(d) No blood vessel

Figure 5. Temperature distribution within brain tissue on an x-z plane based on the power of 15 W for a treatment time of 480 s via four-tine electrode radio-frequency probe for each type of blood vessel included (a-c) or not included (d)

Temperature trends obtained at three points around the tumor during radio-frequency ablation based on the power of 15 W for a treatment time of 480 s via four-tine electrode RF probe are exhibited in Figure 6(a) – (d) for the case of the aorta, main veins, terminal veins included and no blood vessel included, respectively. The three points i.e. R1, R2 and R3 are exhibited in Figure 7. Temperature trends show that the temperature at all points increases along the treatment time for all cases. The temperature distributions in cases of blood vessels included are asymmetric as they converge to the blood vessel. In case of aorta included, the temperature trend at R3 is higher than R1 and R2, respectively, while in cases of main veins and terminal veins, temperature trends at R3 are slightly higher than the R1 but significantly higher than R2. This result is consistent with the temperature distribution and ablated zone in Figure 5. The temperature trend at the point which near the blood vessel (R3) is the highest as a result of the effects of Joule heating caused by the diverted passage of radio-frequency electric current toward blood flow due to higher electrical conductivity of blood compared with brain tissue as well as heat sink effects due to convective heat transfer by blood flow. The aorta which has a larger diameter with a higher velocity of blood flow results in more heat sink effects than main veins and terminal veins, respectively. Thus, the temperature trend at R3 in case of the aorta is higher than the main veins and terminal veins, respectively. Considering two other points (R1 and R2), since the point R1 is nearer to the electrode tip in which the highest temperature occurs as shown in Figure 5, the temperature trend at the point R1 is higher than one at R2 in all case. It is because blood flow is included and effects in these cases that the temperature trends at R1 and R2 are also in agreement with the temperature trend at R3. For the case of no blood vessel included, the temperature trend at R3 is about the same high as one at R1 because of no heat sink effects from blood flow, whereas the temperature trend at R2 is lower than the others because of the further distance from the electrode tips. The result agrees with the temperature distribution in Figure 5. The temperature distributions in case of no blood vessels included are symmetric as there is no heat sink effect by blood flow. The temperature trend at R3 is lower than one in the cases of blood vessels included while the temperature trend at R1 and R2 to be close to the temperature trend at R1 and R2 in the cases of blood vessels included. This is because there are no heat sink effects due to convective heat transfer by blood flow as well.

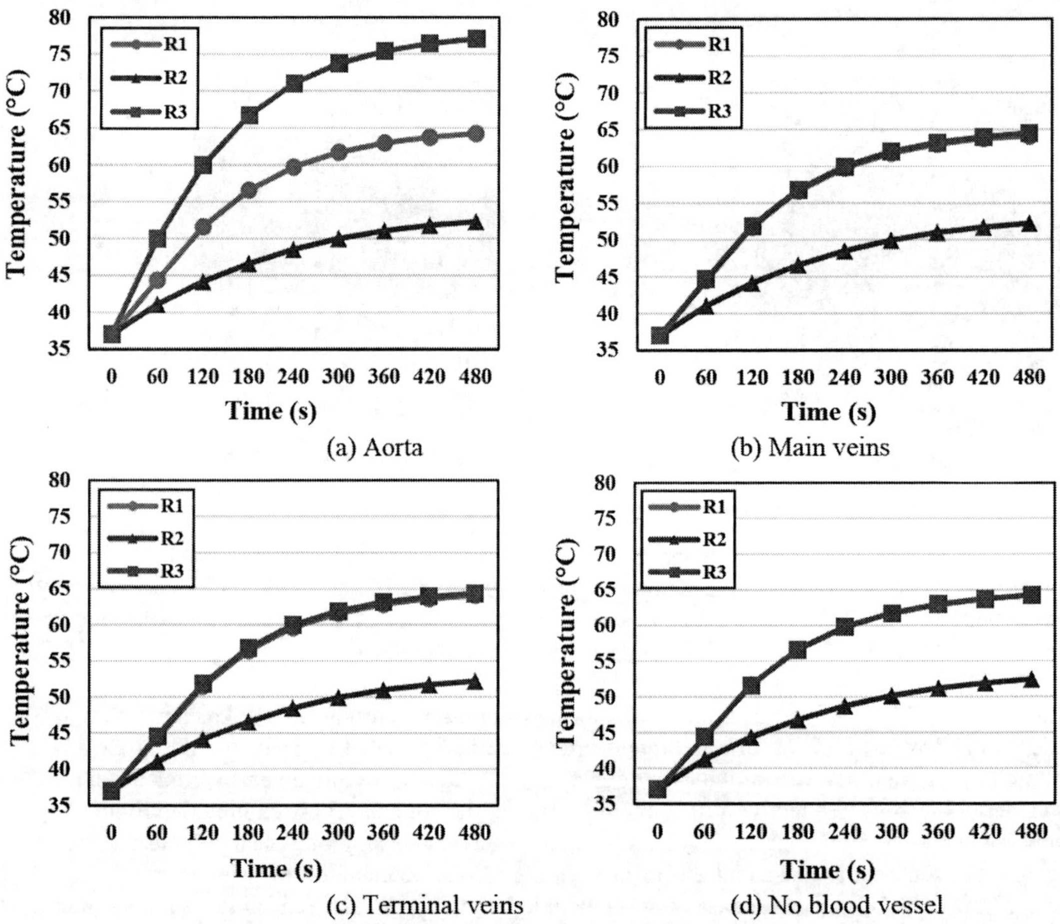

Figure 6. The temperature trends evaluated at three points (R1, R2 and R3) within brain tumor based on the power of 15 W for a treatment time of 480 s via four-tine electrode radio-frequency probe for each type of blood vessel included (a-c) or not included (d)

Table 5 shows a comparison of the ablated area and ablated volume from a computer simulation of radio-frequency ablation in cases of the aorta, main veins, terminal veins and no blood vessel included. Figure 8 illustrates a diagram of the measured dimension of the ablated zone. The results show that heat sink effects due to convective heat transfer by blood flow dominate restricted ablated area and ablated volume. Because 50 °C isotherm (as the boundary of the ablated zone) is limited by blood vessel and heat dissipates in the directions perpendicular to the electrode, W1 is used to calculated ablated area and ablated volume in this study. Considering cases of blood vessel included, the results show that ablated area and ablated volume in cases of the aorta is larger than main veins and terminal veins, respectively, because aorta produces a broader area of temperature distribution due to its higher blood flow velocity. However, the ablated area and ablated volume in case of no blood vessel included are higher than cases of blood vessel included because it is not disturbed by a blood vessel. This result corresponds to the temperature distribution and ablated zone in Figure 5.

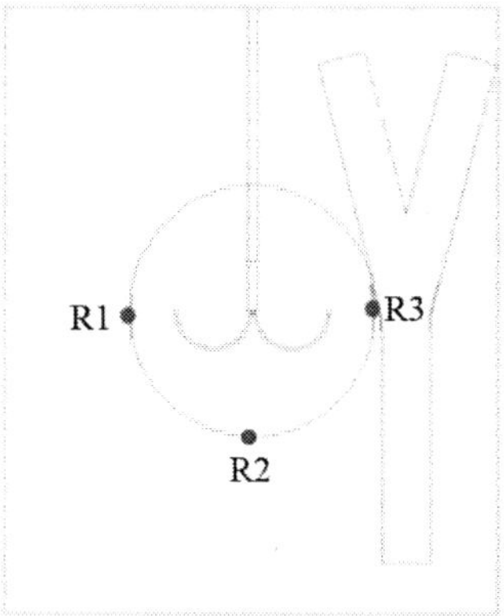

Figure 7. The points where temperature trends are evaluated within the brain tumor (R1 - R3).

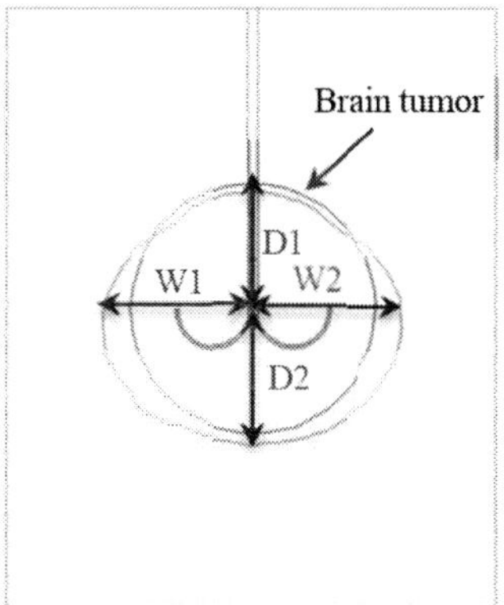

Figure 8. Diagram of a measured dimension of the ablated zone

Table 5. The ablated area and ablated volume from a computer simulation of radio-frequency ablation in cases of the aorta, main veins, terminal veins and no blood vessel included.

Blood Vessel	Diameter of tumor (mm)	Power (W)	Time (s)	W1 (mm)	W2 (mm)	D1 (mm)	D2 (mm)	Ablated area (cm^2)	Ablated volume (cm^3)
Aorta	50	15	480	30.436	26.309	23.388	26.315	29.10215	118.1004
Main veins	50	15	480	30.411	30.113	23.638	27.050	29.05436	117.8096
Terminal Veins	50	15	480	30.209	30.722	23.657	27.086	28.66966	115.4776
No vessel	50	15	480	30.680	30.619	23.358	27.056	29.57063	120.9636

5. Conclusions

This numerical study presents the three-dimensional finite element analysis of radio-frequency ablation in the brain tumor using a four-tine electrode. The model of cylindrical brain tissue which spherical brain tumor embedded is constructed. The effects of blood vessel included and not included on temperature distribution and ablated volume within brain tissue are investigated. There are three types of blood vessel including aorta, main veins and terminal veins. The results reveal that blood vessel results in heat sink effects that dissipate heat from surrounding tissue. The aorta causes more heat sink effects than main veins and terminal veins, respectively, because the higher velocity of blood flow leads to a higher rate of convective heat transfer by blood flow. Therefore, the temperature distribution within aorta is higher than main veins and terminal veins, respectively. In addition, the results indicated that ablated area and ablated volume in cases of the aorta is larger than the main veins and terminal veins, respectively, because aorta produces a broader area of temperature distribution due to its higher blood flow velocity. Nevertheless, the ablated area and ablated volume in case of no blood vessel included are higher than cases of blood vessel included since it is not disturbed by a blood vessel. Even though it is found that temperature distribution and ablated zone in case of no blood vessel included are symmetric and broader than cases of blood vessel included, consideration of blood vessel and its influences is very essential because it is consistent with radio-frequency ablation in clinical treatment. This simulation can be applied to use in clinical brain tumor treatments by radio-frequency ablation as a guideline for treatment planning.

Acknowledgment
The authors would like to gratefully acknowledge the Thailand Research Fund under the project no. MRG6180295 for supporting this research.

References

[1] DeAngelis L M 2001 *N Engl J Med.* **344**(2) 114–123
[2] Schlehofer B Blettner M Becker N Martinsohn C Wahrendorf J 1992 *Cancer* **69**(10) 2541–2547
[3] Connelly J M Malkin M G 2007 *Curr Neurol Neurosci Rep.* **7**(3) 208-214
[4] Lin X DeAngelis L M 2015 *J Clin Oncol.* **33**(30) 3475-3484
[5] Zhu J C Yan T D Morris D L 2008 *Ann Surg Oncol* **15** 1765–1774.
[6] Kanagaratnam L Lee A Whalley D Figtree G A 2013 *Pace* **36**(1) 1–3
[7] Givehchi S Wong Y H Yeong C H Abdullah B J J 2018 *Minim Invasiv Ther* **27**(2) 81–89
[8] Singh S Repaka R 2017 *Int J Hyperthermia.* **33**(2) 122–34
[9] Singh S Repaka R 2018 *ASME J of Medical Diagnostics* **1**(1) 011002–011011
[10] Chang I A 2010 *Open Biomed Eng J.* **4** 3–12
[11] Zhang B Moser M Zhang E Zhang W J 2013 *J Med Eng Technol.* **37**(2) 150–159
[12] Elwassif M M Kong Q Vazquez M Bikson M 2006 *Proc. IEEE Eng Med Biol Soc.* 1 (New York City: USA) 3580–3583
[13] Datta A Elwassif M Bikson M 2009 *Proc. IEEE Eng Med Biol Soc.* (Minneapolis, Minnesota: USA) 670–673
[14] Tungjitkusolmun S Staelin S T Haemmerich D Tsai J Z Webster J G Lee F T Jr Mahvi D M Vorperian V R 2002 *IEEE Trans Biomed Eng.* **49**(1) 3–9
[15] Zhu Q Shen Y Zhang A Xu L X 2013 *Biomed Eng Online.* **12**(1) 127
[16] Huang H W 2013 *Med Phys.* **40**(7) 073303
[17] Zorbas G Samaras T 2014 *Int J Hyperthermia.* **30**(8) 570–8
[18] Singh S Repaka R Bhowmik A 2015 *Proc. of the 23rd National Heat and Mass Transfer Conference and 1st International ISHMT-ASTFE Heat and Mass Transfer Conference* (Thiruvananthapuram: India)
[19] Chaichanyut M Tungjitkusolmun S 2016 *Comput Math Methods Med.* **2016** 1–14
[20] Crezee J Lagendijk J J 1992 *Phys Med Biol.* **37**(6) 1321–1337

IOP Conf. Series: Materials Science and Engineering **886** (2020) 012049 doi:10.1088/1757-899X/886/1/012049

The Contact Patch Analysis of Solid Tire on Drum Testing by Finite Element Method

J Phromjan[1,2], and C Suvanjumrat[1,2,*]

[1]Department of Mechanical Engineering, Faculty of Engineering, Mahidol University, Nakhon Pathom, 73170, Thailand
[2]Laboratory of Computer Mechanics for Design (LCMD), Department of Mechanical Engineering, Faculty of Engineering, Mahidol University, Nakhon Pathom, 73170, Thailand

* Corresponding Author: chakrit.suv@mahidol.ac.th

Abstract
The drum testing method is an effective method to study the tire-rotating characteristic. The rolling resistance and contact force are carried out for tire performance development. The tire was tested by the cylindrical drum indoor laboratory. Thus, the contact patch of rolling solid tire on the drum surface should provide the same results as the tire testing on a flat surface. This research uses a finite element method to develop the drum testing of solid tire model and to investigate the contact patch. The compressed contact patches of solid tire on drum were compared with the validated contact patches of solid tire on the flat surface. The different results showed the effect of the drum curvature. Consequently, the diameter of the drum wheel was optimized for solid tire testing. It is interesting to note that the suitable ratio between drum diameter and solid tire diameter was 1.81.

Keywords: Solid Tire, Contact Patch, Drum Testing, Finite Element Method

1. Introduction

The solid tire is important in the support of the heavy load and plays a key role in the industry. The purpose of this investigation is to develop solid tire performance. This study uses the performance of the tire testing to provide the physical behavior of tire. Evidence suggests that the drum testing is among the most important method for testing a rolling tire in the laboratory [1]. To mimic the flat road, a large diameter drum was selected to provide the contact patch characteristic between tire and drum. The diameter of the steel drum was 1.7 m. and was presumed the suitable drum diameter for study contact force response of rolling solid and pneumatic tires [2]. It could indicate the peak-to-peak of the contact force between solid tire and drum. It was also found that it was higher than the contact force of a pneumatic tire. Moreover, the drum testing machine was performed to study the noise and vibration behavior of a rolling mechanical system by using tire to represent the flexible ring model. The Laser Doppler Vibrometer (LDV) was installed with drum testing machine to measure the vibration velocity of rolling tire [3]. Previous research findings into the drum testing method have been consistent to investigate the air-related tire noise generation mechanisms (e.g. Eisenblaetter et al.,

IOP Conf. Series: Materials Science and Engineering **886** (2020) 012049 doi:10.1088/1757-899X/886/1/012049

2010). Researchers attempted to evaluate the impact of the drum testing on the small tire with the ratio between drum and tire diameter of 41.67. The experimental results from the drum testing were compared well with an existing theoretical model. Nowadays, the computer-aided design (CAD) and computer-aided engineering (CAE) have been employed popularly to develop the tire performance. Prior to the work of Palanivelu et al., (2015), the role of the explicit finite element analysis was largely used to simulate a tire rolling for the development of tire noise generation mechanisms model. The finite element tire model with OMA technique was compared with the experiment. The results indicate that the established procedure can be used to study the dynamic behavior of tire and road interaction. Wei and Olatunsun (2014) [6] studied and validated the effect of different height obstacle on pneumatic tire behavior traversing by the finite element method (2.44 m. drum diameter). Besides, the transient dynamic responses of a pneumatic tire rolling over road obstacles were investigated. Using this approach, researchers have been able to simulate the speed and height of obstacle, which shows a significant influence on reaction force between tire and road. Using the resonant amplitude of vertical force, we were able to increase the speed of the tire and the tire deformation when the tire impacted the higher obstacle. Moreover, the drum testing method was created to develop the accurate simulation of a pneumatic tire rolling over cleat by 3-D finite element method [7]. Besides, the drum was modelled with a rigid element. The tire rolling speed gave a remarkable influence while the change of the tire inflation pressure only produced a remarkable variation in vertical dynamic force. Unfortunately, the large drum, which was used to represent the flat road, has to use a lot of resources for finite element analysis. The diameter of the steel drum is one of the most common characteristics, which might affect the rolling solid tire was studied by 3D finite element analysis. The 3D finite element analysis was developed in this research. Also, the diameter of the drum wheel was optimized to reduce the computing time for simulation the rolling solid tire.

2. Solid Tire Testing on Flat Surface

The solid tire sizes of 6.00-9.00 inch KOMACHI series II was used in this study. As can be seen in Figure 1 the KOMACHI series II consists of three main rubber layers and they were assembled comprising internal, middle and tread layer. To obtain the rigid the internal layer, the four steel rings were inserted. The tire stiffness tester (EKTRON TEK model: PL-2003) was the testing equipment and used to study the solid tire stiffness and the solid tire contact patch. The testing was performed by mounting the solid tire with the mounting arms and raising the measurement table to press the mounted tire (see Figure 2). Using the deformation and looking at the compression force of solid tire, it was possible to identify press the mounted tire. Meanwhile, the contact patch of the solid tire also occurred on a pressure measurement film, which was placed between a solid tire and measurement table. This was the pigment of color on the film which illustrated the contact patch characteristics comprising footprint area, contact area and contact pressure. Figure 3 describes the contact patch characteristics of the KOMACHI solid tire. The red contour was a contact area (CA) while the footprint area (FA) was calculated by multiplying width (a) and height (b) of the red contour. Subsequently, this contour has been transformed into the contact pressure and contoured by the image processing method.

3. Finite Element Analysis

Analysis of a large strain was based on a hybrid formulation, which is effective for the analysis of rubberlike material [8]. The finite element analysis (FEA) of the solid tire testing was governed by the strain energy density per unit volume, ${}_0^t\bar{U}$, and is written by the following equation (1).

$$d\,{}_0^t\bar{U} = {}_0^t\bar{S}d\,{}_0^t\varepsilon_{ij} \tag{1}$$

(a) (b)

Figure 1. The KOMACHI series II solid tire (a) overall dimension and (b) the inside structure.

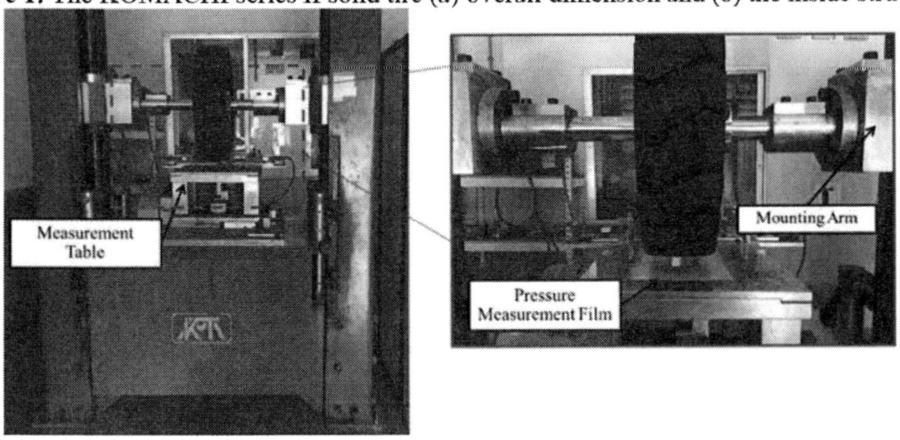

Figure 2. The tire stiffness tester, EKTRON TEK model: PL-2003.

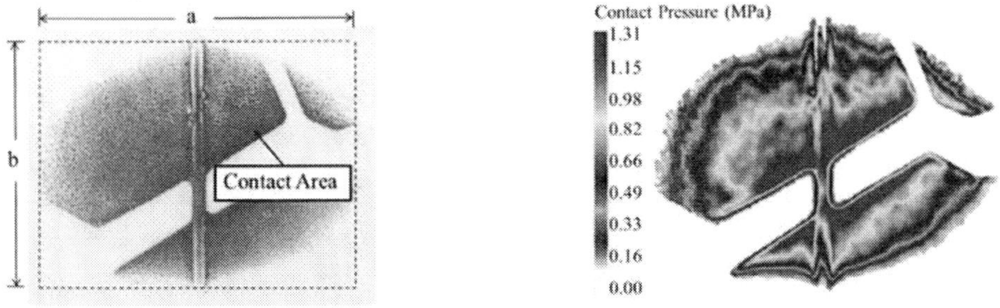

(a) (b)

Figure 3. The contact patch result by (a) a pressure measurement film and (b) image processing translation.

where $d\,{}_0^t\bar{U}$ and ${}_0^t\bar{S}$ are incremental potential energy and Piola-Kirchhoff stress which was computed only from the displace field. The integral form of a total Lagrangian formulation is shown in equation (2).

$$\int_{V_0} {}_0^tS_{ij}\,\delta\,{}_0\varepsilon_{ij}\,dV_0 = R \tag{2}$$

The general form of a principle of virtual work, which was written in the total Lagrangian formulation, is given by equation (3).

$$\delta\left(\int_{V_0} {}_0^tU\,dV_0\right) = R \tag{3}$$

where ${}_0^tU$ is the incremental potential which can be modified to include the effect of interpolated pressure by adding to the term of the displacement based on the total element pressure.

3.1 Solid tire model
The rubber layers of KOMACHI solid tire were model by using a different material formula of rubber compounds. The internal layer (M058) was the hardest, the middle (M047) and the tread layer were softer. The material property of each rubber layer was carried out using the tensile testing, which is based on the ASTM D412 standard. To obtain a material model from the tensile testing results, the linear regression method was employed in order to fit with the stress-strain curves. The Ogden hyperelastic constitutive model (see Equation 4) which is a strain-energy density function. It is found that it is the most suitable material model to represent the solid tire deformation in all rubber layers [9]. The constant values of Ogden constitutive model are presented in Table 1.

$$U=\sum_{i=1}^{n}\frac{\mu_i}{\alpha_i}(\bar{\lambda}_1^{\alpha_i}+\bar{\lambda}_2^{\alpha_i}+\bar{\lambda}_3^{\alpha_i}-3)+4.5K(J^{1/3}-1)^2 \tag{4}$$

$$\text{and } \bar{\lambda}_i = J^{(-1/3)}\lambda_i,\ J=\lambda_1\lambda_2\lambda_3$$

where λ_i is the deviatoric principle stretches, J is the Jacobean determinant, K is the initial bulk modulus, and μ_i, α_i are constants.

Table 1. The constant values of Ogden constitutive model.

Rubber layer	μ_1	μ_2	α_1	α_2	K
M067	1.075	154.616	2.927	1.14e-7	7,866.61
M047	0.362	510.855	3.233	0.0056	10,078
M058	3,615.31	2,915.36	0.0028	0.0024	42,449

The 3-D solid tire model was created using miscellaneous methods the 3D scanner and Computer-Aided Design (CAD) in response to the geometry of KOMACHI series II solid tire, which has 6.00-9.00 inch in size. Each component of the solid tire was divided by hexahedral elements. The rubber layer elements are presented by different colors, as shown in Figure 4(a). The internal, middle, tread layer and steel wires were assembled by 16,623, 8,448, 14,366 and 7,040 hexahedral elements, respectively. In Figure 4(b) there are finite element pouts that are connected by conjunct nodes regardless of the connection between the outer tread and the inner tread layer. Both of them were defined in the glue contact condition.

3.2 Compression testing of solid tire on flat surface
The purpose of the experiment was to set up the finite element model of the solid tire to compress on a rigid flat plate according to the physical tire stiffness testing. The multi-point constraint (MPC) was defined on nodes at the inner surface to link an axis of solid tire model and to model the steel wheel, which was installed by the solid tire. The fixed boundary is assigned at a wheel fixed point to mimic the solid tire that mounts on the mounting arm of the stiffness tester (see Figure 5a). The contact boundary condition between the solid tire model and the rigid flat plate was specified with the static friction coefficient of 0.8. Thus, the rigid flat plate was assigned to lift in a vertical direction to press the solid tire model with three different loads, 400, 800 and 1,200 (kgs), respectively.

3.3 Compression testing of solid tire on drum
To compress a solid tire model on the drum, the cylindrical steel drum was created using the curvature rigid element. The diameter of the rigid drum was adjusted to study the effect on the contact patch. It was varied in a range of 0.2 to 2.45 m., which was the feasible curvature of the drum surface.

The solid tire model was fixed at the wheel fixed point in the same method as the solid tire model and was compressed on a flat surface. The vertical load was assigned on a rigid drum to move and press the solid tire model at 400 kgs. The finite element model of the solid tire, which was compressed by the steel drum, was used to study the impact of curvature surface on a contact patch, as shown in Figure 5 b.

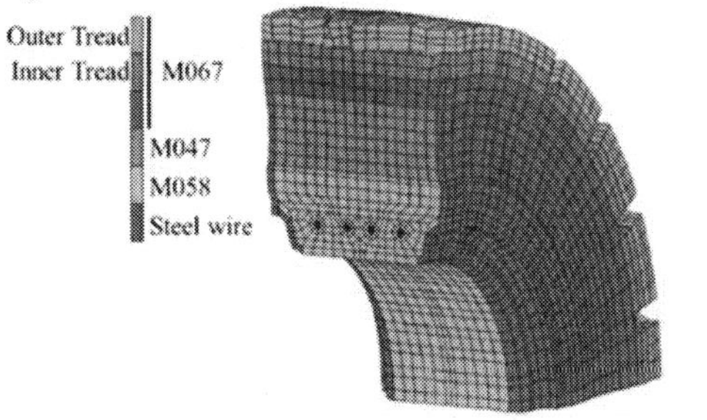

(a)

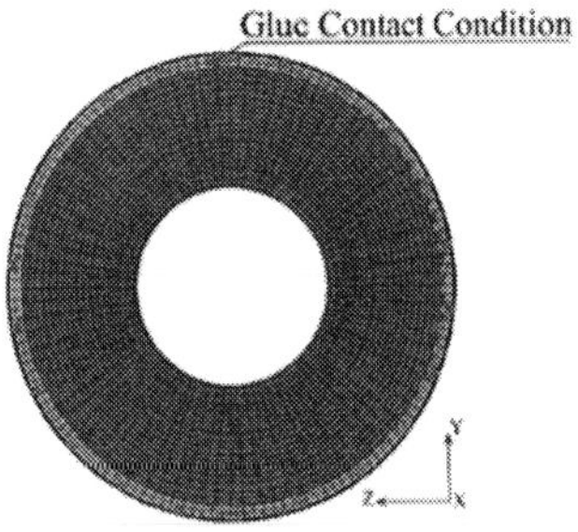

(b)

Figure 4. The finite element model of (a) solid tire components and (b) contact boundary conditions.

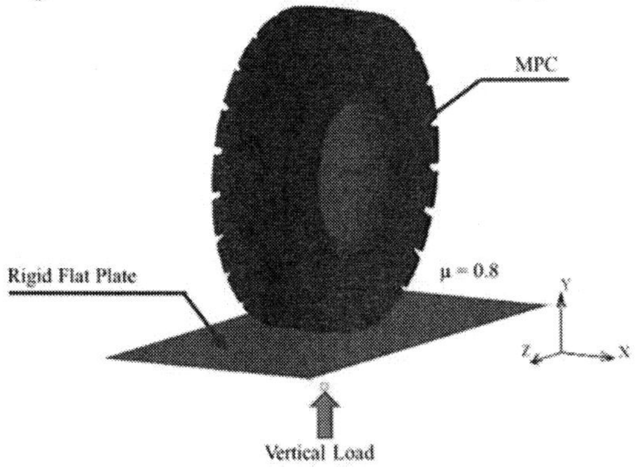

(a)

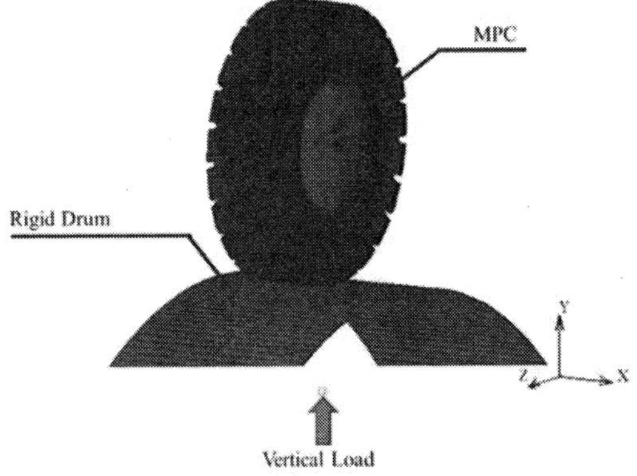

(b)

Figure 5. The finite element model of solid tire compressing by: (a) flat surface and (b) drum surface.

4. Results and Discussions

4.1. Validation

To assess the accuracy simulation, the contact patches of the experiment were carried out to validate the finite element model. There was comprised the contact patched by the vertical load of 400, 800 and 1,200 kgs. The contact coefficient of the solid tire model, which is the ratio of the contact area to footprint area (CA/FA), was used to be a benchmark. Figure 6 provides a contact patch obtained by the experiment and simulation. The results revealed that the contact patch of the solid tire model showed an agreement with the pressure measurement film. From the data, it shows a similar contact coefficient at 0.4. It can thus be suggested that the finite element model of the solid tire could simulate the contact patch on the curvature surface and obtain the accuracy results.

4.2. The effect of drum curvature

The mechanical behaviours of solid tire FE model i.e. vertical deformation, contact patch and contact pressure were used to investigate the effect of drum curvature. The finite element model of the solid tire,

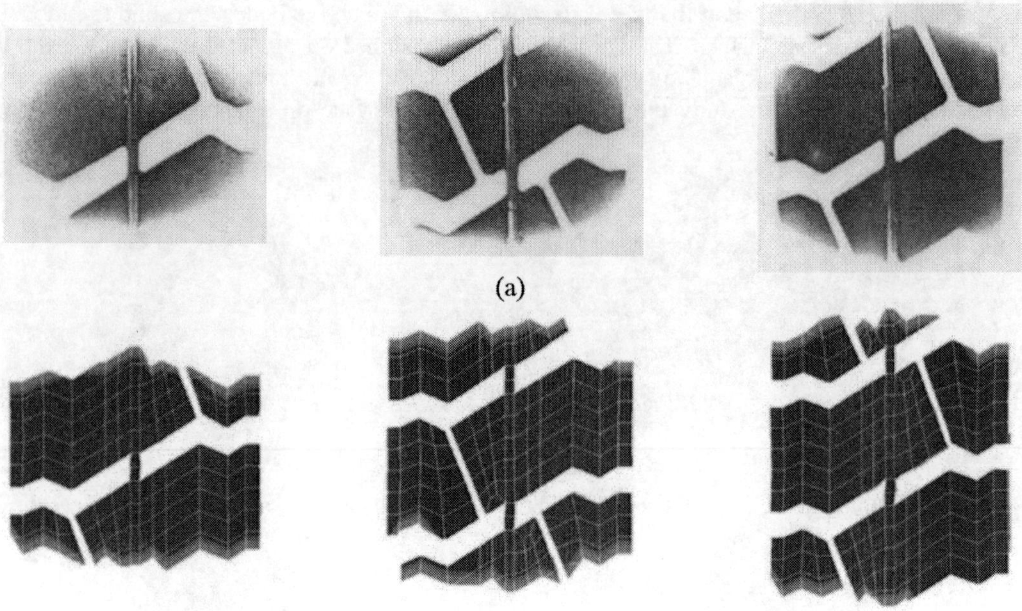

(a)

(b)

400 kg 800 kg 1,200 kg

Figure 6. The contact patch of solid tire on flat surface of (a) experimental and (b) finite element analysis.

which was validated on a flat surface, was carried out to compress on rigid drums by the compression load of 400 kgs. The small radius of the drum surface was intense to the contact pressure on the solid tire. It is possible to hypothesise that the bigger the vertical deformation increased by the small road, the smaller the footprint area was. Figure 7 shows the results of the simulation, the vertical deformation and the contact pressure of the compression testing obtained by the drum surface. The colour contour represents the results from the intense value of the simulation. It was found that the contact patch distribution was better at the larger drum. Particularly, the footprint area as found to be increased to constant values after the size of the diameter of the drum was larger than 0.95 m. The contact coefficient at the constant footprint indicated a good agreement on a flat surface. The experiment found an error, which is less than 4.90%. The vertical deformation and the contact pressure are decayed in the exponential form. At the drum diameter of 0.95 m, the vertical deformation and contact pressure were constant. As can be seen in Figure 8, the graphs from the simulation results regarding the diameter of the drum was varied.

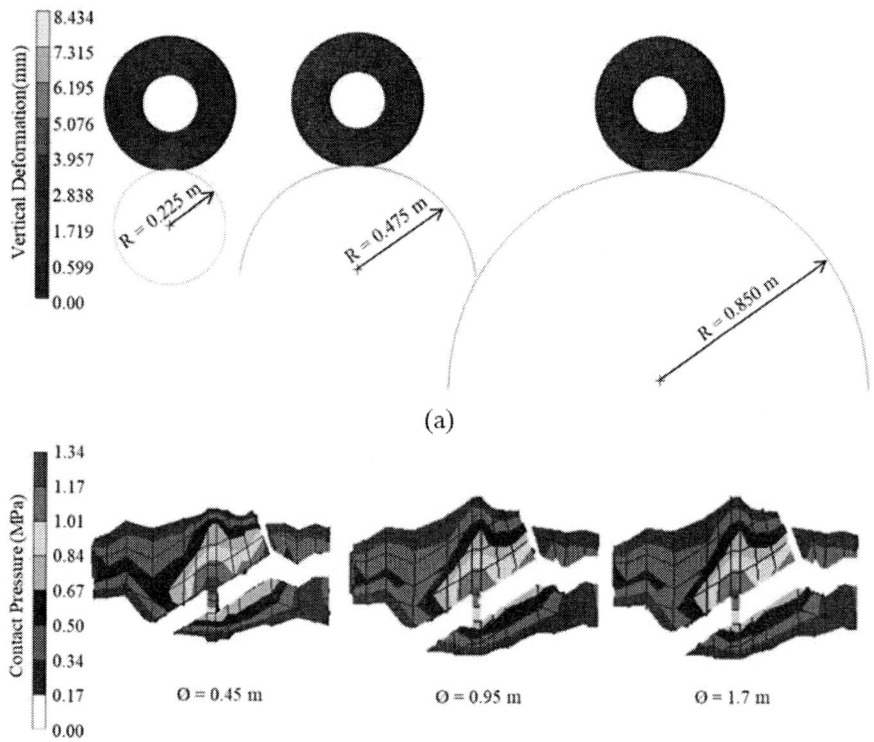

(b)

Figure 7. The finite element analysis result of: (a) vertical deformation and (b) contact pressure under compressing of solid tire by various drum diameter.

5. Conclusion

The main goal of the current study was to develop the drum testing of solid tire using the finite element method. The contact patch by different compressing surface, flatness and curvature, were investigated. Concerning the benchmark, the contact patches of solid tire modelled by the flat surface compression were validated. The second aim of this study was to investigate the effects of that the finite element model was error less than 3.87%. The current study was designed to determine the effect of curvature surface by testing the solid tire with drum. This study has found that generally the diameter of rigid of the drum was in a range of 0.2 to 2.45 m. Also, it was found out that the small drum causes the vertical deformation. In addition, the higher the contact pressure, the smaller the footprint area was. The results of the distribution of the contact patch was satisfied when the solid tire was compressed by the larger drum. In regards of the analysis of the finite element, the results indicated that the contact patch characteristic of the solid tire model was constant at the drum diameter ranged from 0.95 m to 2.45 m. The contact patch at those state showed a good agreement with the compressing solid tire of the flat surface. The research has also shown that the experiment of the optimum drum diameter of the solid tire suggested a ratio of the tire diameter 1.81. These findings have significant implications for the understanding of how to reduce cost of the drum testing machine and to reduce the time-consuming process of the tire testing simulation.

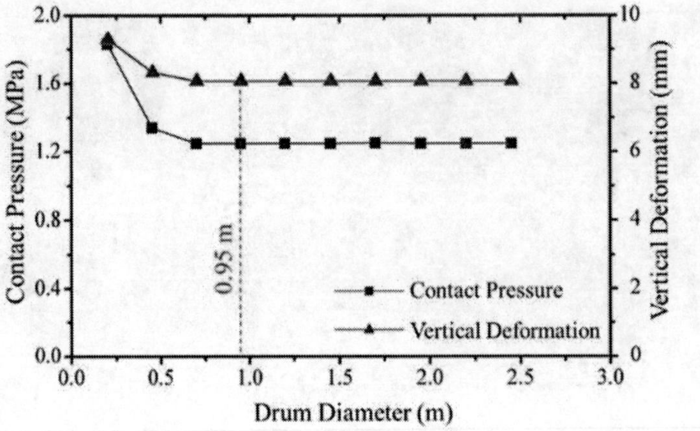

Figure 8 .The vertical deformation and contact pressure of solid tire compressing on various drum.

References

[1] Andersen L G Larsen J K Fraser E S Schmidt B and Dyre J C 2015 *J. Transp. Eng.* **141**(2) 1-10
[2] Phromjan J and Suvanjumrat C 2018 *J. Mech. Sci. Technol.* **32**(4) 1539-1548
[3] Lee J Wang S Pluymers B Desmet W and Kind P 2015 *Mech. Syst. Signal Proc.* **60-61** 604-618
[4] Eisenblaetter J Walsh S J and Krylov V V 2010 *Appl. Acoust.* **71**(9) 854-860
[5] Palanivelu S Rao N and Ramarathnam K K 2015 *Mech. Syst. Signal Proc.* **64-65** 385-402
[6] Wei C and Olatunbosun A O 2014 *J. Terramechanics* **56** 1-16
[7] Cho J R Kim K W Leon D H and Yoo W S 2005 *Eur. J. Mech. A-Solid.* **24** 519-531
[8] Bathe K J 1996 Finite element nonlinear analysis in solid and structural mechanics *Finite Element Procedure* (London: Pearson Education Limited) chapter 6 pp 485–641
[9] Phromjan J and Suvanjumrat C 2018 *Key Eng. Mater.* **777** 416-420

Numerical Validation of Heat Conduction in 2D Binary Granular Mixtures under Mechanical Loading

K Jongchansitto[1], P Jongchansitto[1,*], and I Preechawuttipong[1]

[1]Department of Mechanical Engineering, Faculty of Engineering, Chiang Mai University, 239 Huay Kaew Rd., Chiang Mai, 50200, Thailand

* Corresponding Author: pawarut.j@cmu.ac.th

Abstract
This study set out to investigate numerical simulations of heat conduction to validate existing experimental results of two-dimensional binary granular mixtures subjected to mechanical loading. Data for this study were collected using molecular dynamics (henceforth MD) method. Three different configurations of the granular composite samples were systematically prepared under similar experimental conditions. A confined mechanical loading was applied to the granular samples. The fields of normalized temperature change of each particle were plotted for an individual sample. The results were statistically analyzed under a static equilibrium condition. The results indicate that simulation is in good correlation with the experiments in terms of statistical analysis via the probability of the distributions of the normalized temperature change. It also revealed that the normalized temperature changes which are greater than the average temperature distributes as an exponential decreasing for all tested samples. This study is in line with other studies that are related to force distribution of law. Less than 50% of particle numbers that have the normalized temperature changes, which are greater than the average value, is also explored. Besides, localizations of the temperature were found in the individual sample.

1. Introduction

Granular materials are one of the most widely used materials in our daily life and many industrial fields. For example, sand, rocks and soil are important components used in civil engineering or industrial field to prepare and build roads. Rice, crops and beans are fundamental to argo-food and our daily life. Some natural phenomenon is also involved in a granular movement like landslides and avalanches. In this study, granular materials are defined as a collection of solid particles whose macroscopic mechanical behavior is controlled by interparticle forces. They are composed of grains in various sizes, shapes and materials. This is a reason why behaviors of granular materials are generally complicated and it completely differs from the common solids and fluids [1, 2]. The distribution of the interaction forces has a significant consequence in transport phenomena, especially in heat transfer. Besides, heat transfer plays a vital role in many industries regarding handling and processing of granular media, such as mixing, drying, granulation, and coating. Nowadays, numerical simulations based on the discrete element method (henceforth DEM) are an effective and useful tool for studying the effect of various parameters such as particle sizes, shapes [3-5]. Studies of DEM show the importance of using DEM to simulate heat transfer processes in granular materials [6-9]. Previous research on the applications of granular flow i.e. rotating drum [7-9], fluidized systems [10-12] has been proposed. Previous studies of DEM have not dealt with the heat transfer phenomena in two-dimensional granular composites under mechanical loading. Therefore, it enables us to validate our thermoelastic stress analysis (TSA) experiments recently performed in 2014 [13]. An objective of this present study is to statistically compare the simulation results with the existing TSA experimental results. Simulation configurations are prepared as the similar to what we done in the experiments. Molecular dynamics method is carried out for the method of simulations.

2. Molecular dynamic (MD) simulation

One study by Cundall (1971) examined the discrete element method (DEM) for rock mechanics [14]. Using this approach, researchers have been able to study the mechanical behavior of granular media. The MD method belongs to the DEM and relies on an explicit algorithm. Moreover, this method considers the particles as rigid bodies with non-conforming surfaces [15]. The MD method provides a mean of simplicity and flexibility because it applies only Newton's second law for calculating kinematical quantities:

$$m_i \frac{d^2 x_i}{dt^2} = \boldsymbol{F}_i \ , \qquad i = 1,2,3,\ldots,N \tag{1}$$

where N is the number of particles in the simulation, m_i is the mass of particles i, x_i is the position of particles, and $\boldsymbol{F}_i$ is the force exerted on that particle. The method consists of calculating force $\boldsymbol{F}_i$ and solving ordinary differential equations (ODE) in equation (1). Integration of Newton's equations of motion can be achieved by applying a predictor-corrector scheme with Gear's set of corrector coefficients [16]. By using this method, positions, velocities, accelerations, the third-order time derivative of particles are provided.

2.1 Numerical models

The contact force between two dry particles can be separated into two components: a normal force and a tangential force. In the MD method, the deformation is simply modeled by using a virtual overlap δ. Due to the unchanged particle shapes, the normal force can be calculated by a function of the virtual overlap, which is expressed in terms of vector position of the particle centers x_i and x_j and radius r_i and r_j:

$$\delta = \left(r_i - r_j\right) - \left|x_i - x_j\right| \tag{2}$$

In this study, the "linear spring-dashpot" [17, 18] is a common model to calculate the normal force between two particles. The normal force F_n is calculated by:

$$F_n = k_{eff}\delta + \alpha_n v_n \tag{3}$$

where k_{eff} is the effective contact stiffness, α_n is the normal damping coefficient, and v_n is the normal velocity (the first-order time derivative of the virtual overlap). The integration of Newton's equations of motion requires a smooth friction law (mono-valued) i.e. the tangential force must be expressed as a linear function of the sliding velocity v_t, while the classical Coulomb's law friction is a non-smooth function. Hence, the tangential force F_t in this study can be commonly determined by using a regularized form of the exact Coulomb's law, which can be calculated by this expression:

$$F_t = \min\left\{\left|\gamma_t v_t\right|, \mu F_n\right\} \cdot \mathrm{sign}\left(v_t\right) \tag{4}$$

where γ_t is the tangential viscosity coefficient and μ is the coefficient of friction. In this study, the rotational motion that occurs from the tangential force is assumed to be free because its value is very small compared to the normal force. The phenomena of heat transfer consist of heat conduction, heat convection, heat radiation, and heat generation, which are complex and depend on many variables. Therefore, the phenomena of heat transfer in the present study are considered only effects of heat conduction and heat generation. When the particles inside the granular system move and two bodies are in contact, their relative motion can be taken into account as a source of heat generation, as denoted as Q_{ij}^s. The heat generation can be computed in many different methods depending on the nature of the considered system. The heat generation model used in this study is directly based on the first principle of thermodynamics, accounting for normal motions as well as tangential motions. The heat generation of each particle, which is related to an energy dissipation at the contact level [19] can be expressed as:

$$Q_i^s = \sum_{i=1}^{N} Q_{ij}^s;\quad Q_{ij}^s = \frac{1}{2}F_n^{ij}v_n^{ij} + \frac{1}{2}F_t^{ij}v_t^{ij} \tag{5}$$

Then, the heat conduction of particle i is the sum of all heat exchanges of contacting particles [6] which can be expressed as:

$$Q_i^c = \sum_{i=1}^{N} Q_{ij}^c;\quad Q_{ij}^c = H_{ij}^c\left(T_i - T_j\right) \tag{6}$$

where H_{ij}^c is the heat conductance of contact and the term of $T_i - T_j$ is the temperature difference between the particles which relies on two assumptions. The first assumption is no temperature distribution inside the particle i.e. each particle has only one temperature for the whole body. The second assumption is that the temperature exchange between the particles is slow enough during each time step. The coefficient H_{ij}^c, which is a function of compression force, depends on the particle properties. In this study, the particle is assumed as a cylinder similar to the experiments. The solution was then assayed for coefficient H_{ij}^c c using Hertz theory, as shown in [18]:

$$\frac{H_{ij}^c}{k_{eff}^s} = 4\left(\frac{F_n^{ij}R_{eff}}{\pi L E_{eff}}\right)^{\frac{1}{4}} \tag{7}$$

where R_{eff}, E_{eff}, k_{eff}^s and L represent the effective contact radius, the effective Young's modulus, the effective thermal conductivity and the length of the cylinder, respectively. For the evolution of the thermal system, the Euler scheme is applied to discretize the differential equation characterizing the evolution of the temperature of each particle. Thus, the temperature at each time step *dt* can be discretized explicitly using a □-method over time *t* to time *t+dt* [19], which can be expressed as:

$$T_i^{t+dt} = T_i^t + \left(1-\theta\right)\dot{T}_i^t + \theta\dot{T}_i^{t+dt} \tag{8}$$

where □ indicates a value between 0.5 to 1 for stability reasons. In order to reach the conservation of the simulation, the value of □ equals to 0.5 was used in this study. The term $\dot{T}_i^{t+dt}$ is a function of the local thermal phenomena of particle i by accumulating from multi-contacts [6]. It can be thus calculated from:

$$\dot{T}_i^{t+dt} = \frac{dT_i^{t+dt}}{dt} = \frac{Q_i^{t+dt}}{\left(\rho_i c_i V_i\right)^{t+dt}} \tag{9}$$

$$\dot{T}_i^{t+dt} = \frac{dT_i^{t+dt}}{dt} = \frac{Q_i^{t+dt}}{(\rho_i c_i V_i)^{t+dt}} \tag{9}$$

where ρ_i, c_i and V_i is the bulk density, the specific heat capacity and the volume of the particle. In 2D simulations, the particle volume is replaced by an area of the particle. It must be noted that the quantities ρ_i, c_i and V_i is constant during the simulations. The term Q_i^{t+dt} in equation (9) implies that the amount of heat is accumulated from all thermal phenomena of particle i at time $t+dt$. The total amount of heat in each particle can be thus separated into two components:

$$Q_i = Q_i^c + Q_i^s \tag{10}$$

where Q_i^c represents the heat exchange (heat conduction) between particle i and its neighbors, Q_i^s corresponds to the heat generation by the relative motion between particle i and its neighbors.

2.2 *Numerical preparation*

Three granular composite configurations in the TSA experiments [13] were chosen to be prepared as the two-dimensional numerical composite samples. The granular systems were made of two different constitutive materials: one is polyoxymethylene (POM) called a "stiff" particle. Another one is high-density polyethylene (HDPE) called a "soft" particle. It must be noted that the stiff particles are approximately four times stiffer than the soft particles [13]. Table 1 presents configurations of all tested samples. Sample #1 is a monodisperse system that has a diameter ratio is equal to one Whereas others are bidisperse systems.

Table 1. All tested samples used in this study.

	Diameter ratio D_{stiff}/D_{soft}	Number of particles $N_{stiff}:N_{soft}$	Total number of particles
#1	1.0	638 : 597	1235
#2	1.6	334 : 371	705
#3	3.0	86 : 466	552

Prior to the experiment, the locations of particles for each configuration were extracted from an optical image of a real composite granular system. After extracting the locations, the initial states were prepared for numerical simulations. Then, each particle was deposited inside a box consisting of four rigid walls. quasi-static conditions, a vertical compression force of 60 kN was incrementally applied to granular samples by the bottom wall. Whilst the others were stationary during the simulations. The gravitational force was also taken into account, although it can be negligible compared to the magnitude of the external loading. The normal and tangential forces between the particles are calculated by using Eq. (3) and Eq. (4), respectively. The contact force between particle and wall is also computed by these expressions. The effective contact stiffness (k_{eff}) depends on the type of contacting particles: the case of interparticle contact is defined by $(k_{stiff} \times k_{soft})/(k_{stiff}+k_{soft})$, while the case of particle-wall contacts is defined by the stiffness of such particle that contact to the wall.

After the contact force calculation, the contact forces and relative velocities were then employed to compute the heat generation inside the contacting particle. The particle which has a different temperature from its neighbors will exchange heat to one another. This phenomenon can be expressed in terms of the heat conduction and it depends mostly on the contact area and the effective thermal conductivity. Additionally, the initial temperature of all particles in all configurations is defined as 25°C. After the calculation of the local heat effect, this local heat effect was used to update the temperature of each particle in each time step. Finally, the simulations were accomplished when the granular systems reached to static equilibrium conditions.

3. Results and discussion

In order to improve data consistency within the database, normalized temperature change rather than the temperature is applied for the analysis. It is defined as the temperature change in the particle divided by the average temperature change of all the particles inside the granular sample.

3.1 Field of normalized temperature change

The normalized temperature change ΔT_{norm} of all the particles was plotted with their positions to create a field of normalized temperature change. The fields of normalized temperature change obtained from the simulations and experiments for all configurations are illustrated in Figure 1.

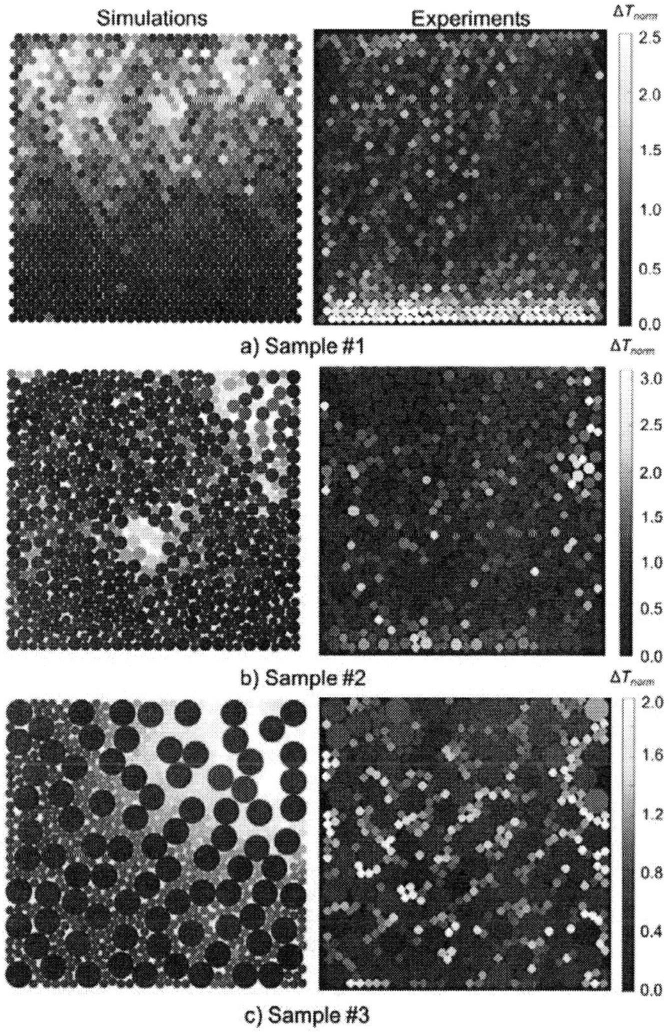

Figure 1. Fields of normalized temperature change generated by simulations and experiments in the cases of: a) Sample #1, b) Sample #2, and c) Sample #3.

It can be seen that the field of normalized temperature change for sample #1 exhibits the triangular network (see Figure 1a). It is well-known for the monodisperse granular system [13]. This behavior is similar to what we have observed from the force or stress fields [13, 20]. Considering samples #2 and #3 in Figures 1b and 1c, most of particles which change in their temperature has a lower temperature than the average temperature is stiff particles. This can be explained by the fact that both the mass and

IOP Conf. Series: Materials Science and Engineering **886** (2020) 012050 doi:10.1088/1757-899X/886/1/012050

the specific heat capacity of stiff particles are larger than those of the soft particles. It must be noted that this change in temperature does not imply to the amount of the heat storage inside the particles. Indeed, it is not possible to obtain precisely the same temperature pattern between simulations and experiments at the particle scale. This might be due to an intrinsic variation of parameters in the experiment, such as variations in terms of diameter size, roundness, physical material properties, etc. A technical problem regarding particle locations might be also included. For this reason, a statistical analysis of the experimental and simulation results is performed in the next section.

3.2 Probability distribution function of normalized temperature change $P(\Delta T_{norm})$

In this section, a probability distribution function of the normalized temperature change was determined for each configuration. It must be noted that five additional simulations with randomly changes the particle locations were systematically performed for each sample: simulations A, B, C, D, and E, respectively. Figure 2 shows the $P(\Delta T_{norm})$ of each configuration obtained from the simulations and experiments.

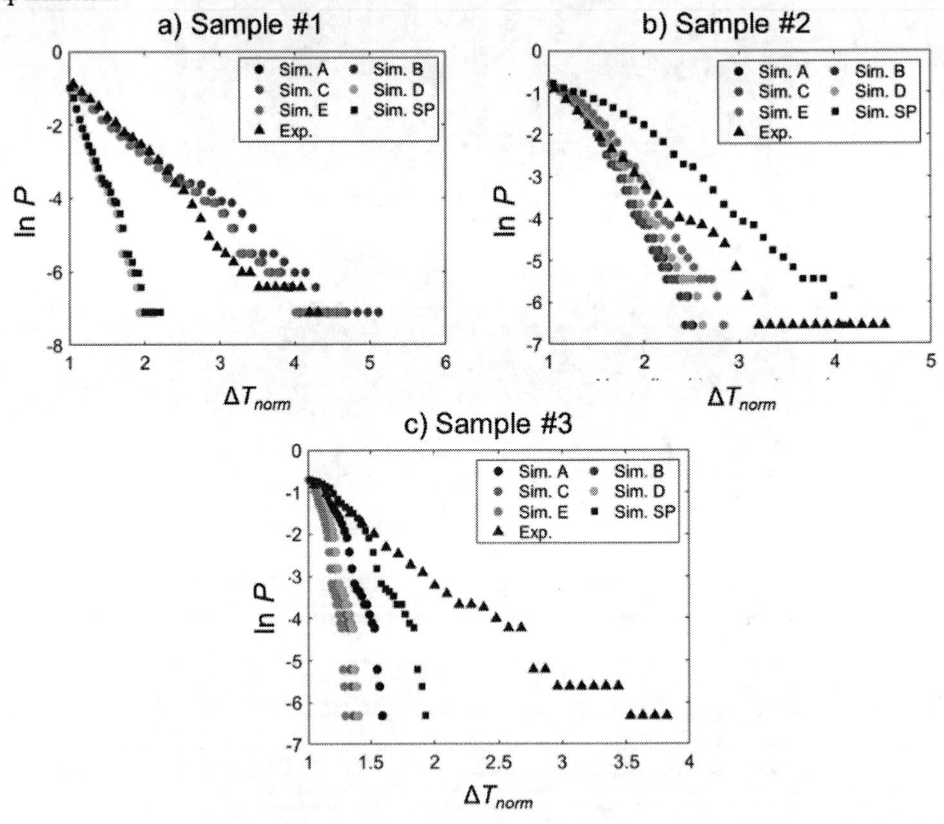

Figure 2. Probability distribution function of normalized temperature change obtained from both simulations and experiments for: a) Sample #1, b) Sample #2, and c) Sample #3. Note that simulations "A", "B", "C", "D", and "E" were performed under the same experimental configuration, excluding the particle locations, while the simulation "SP" was conducted under all the same as in the experiments.

The results from the distributions of $P(\Delta T_{norm})$ obtained by both simulations and experiments are in the same trend for all the samples, which exhibit as an exponential decay. This distribution is characterized by the coefficient β, which can be written as:

$$P \propto e^{\beta(1-\Delta T_{norm})} \tag{11}$$

These results are consistent with data obtained in force and stress distributions [13, 20]. Moreover, the function of exponential decreasing appears when only stiff particles and only soft particles are considered. It is worth noting that the temperature of the experimental and simulation percentages of the particles having the temperature larger than the average temperature appear to be less than 50% ($P(1) < 0.5$). This value is also found for both experiments and simulations when only stiff particles and only soft particles are considered.

4. Conclusion

The present study was designed to validate the heat conduction in two-dimensional granular composites under a confined mechanical loading. The discrete element simulations using the MD method was employed for this purpose. This study selected three configurations of the existing TSA experiments [13] to prepare numerical samples. Each sample was subjected to a compressive loading. Only heat conduction and heat generation were taken into account for the simulations. The normalized temperature change, which can be determined from the temperature change in the particle divided by the average temperature change of all the particles inside the granular sample, was adopt in the statistical analysis through the field of normalized temperature change and the probability distribution function of normalized temperature change. The investigation show that the experiments and simulations results follow the same trends in the overview. The evidence from this study suggests that the distribution of the inhomogeneous temperature was clearly observed in all the samples for both simulations and experiments, as shown in the plot of T_{norm} fields. It also showed that some localizations were found and the exponential decreasing distribution of T_{norm} was clearly observed by both experiments and simulations for all tested samples. This study appears to be positively related to previous studies in terms of force distribution law. In addition, less than 50% of the particle numbers that have T_{norm} was greater than the average value, was explored in the experimental and simulation results for all samples.

References

[1] Jaeger H M 2005 *Phys. World.* **18** 34
[2] Jaeger H M and Nagel S R 1996 *Rev. Mod. Phys.* **68** 1259
[3] Nguyen D H Azéma E Radjai F and Sornay P 2014 *Phys. Rev. E.* **90** 012202
[4] Azéma E Preechawuttipong I and Radjai F 2016 *Phys. Rev. E.* **94** 042901
[5] Binaree T Preechawuttipong I and Azéma E 2019 *Phys. Rev. E.* **100** 012904
[6] Vargas W L and McCarthy J J 2001 *AIChE J.* **47** 1052
[7] Chaudhuri B Muzzio F J and Tomassone M S 2010 *Powder Technol.* **198** 6
[8] Xie Q Chen Z Hou Q Yu A B and Yang R 2017 *Powder Technol.* **314** 175
[9] Yazdani E and Hashemabadi S H 2019 *Granul. Matter* **21** 1
[10] Gomes J L M A Pain CC Oliveira C R E and Goddard A J H 2007 *Heat Transfer Eng.* **28** 576
[11] Wahyudi H Chu K and Yu A 2016 *Int. J. Heat Mass Tran.* **97** 521
[12] Al-Arkawazi S 2018 *Appl. Therm. Eng.* **128** 696
[13] Jongchansitto P Balandraud X Grédiac M Beitone C and Preechawuttipong I 2014 *Soft Matter.* **10** 8603
[14] Cundall P A 1971 *Proc. the International Symposium on Rock Fracture (Nancy)* 47–65
[15] Johnson K L 1985 *Contact Mechanics* (New York:Cambridge University Press)
[16] Allen M P and Tildesley D J 1987 *Computer Simulation of Liquids* (New York:Oxford Science Publications)
[17] Schäfer J Dippel S and Wolf D E 1996 *J.Phys. I.* **6** 5
[18] Matuttis H G Luding S and Herrmann H J 2000 *Powder Technol.* **109** 278
[19] Renouf M and Berthier Y 2011 *Discrete-element modeling of granular materials*, ed F Radjai and F Dubois (London: ISTE Ltd) 347-366
[20] Radjai F Jean M Moreau J J and Roux S 1996 *Phys. Rev. Lett.* **77** 274

Augmented Reality for Dies Alignment in Machine Setup

T. Tothong[1], S. Vongbunyong[1,*], W. Chinthammit[2], A. Piyasinchart[3]

[1]Innovation and Advanced Manufacturing Research Group,
Institute of Field Robotics, King Mongkut's University of Technology Thonburi.
126 Pracha Uthit Rd, Bang Mot, Thung Khru, Bangkok, Thailand 10140
[2]Human Interface Technology Laboratory Australia (HIT Lab AU),
Information and Communication Technology, University of Tasmania. 2 Invermay Rd, Launceston, Australia, TAS 7248
[3]Asahi-Thai Alloy Co., Ltd. 9/23,
Suk Sawat 78, Bang Chak, Phra Pradaeng District, Samut Prakan 10130

* Corresponding Author: supachai.von@kmutt.ac.th

Abstract
Forging is one of the material deformation processing techniques. While the operation cycle is short and efficient, the machine setup process is generally long and time consuming. One of the difficulties in operation is position alignment procedure of the equipment during the die changing process. Due to the size and shape complexity of the machine and equipment, the operators are not able to clearly see the important parts – e.g. guided pins, holes, and sub-plate which need to be moved to finely adjust the alignment. Augmented Reality (henceforth AR) is used to solve this problem by rendering virtual objects and guides associated to those parts over the real equipment. Thus, the operators can monitor the position of the parts during the alignment process through the virtual objects. This AR system is developed based on ARcore SDK together with Unity on the Android platform. The experiment was conducted in real factory environment.

Keywords: Machine Setup, Forging, Augmented Reality, Die changing

1. Introduction

1.1 Forging Processes and Machine Setup

Forging has emerged as powerful platforms for bulk material deformation processing and is also able to produce high-strength workpieces. Traditionally, a metal billet has been inserted between an upper and a lower dies which is pressed by a hydraulic press machine to form the workpieces according to the shape of the cavity [1, 2].

Machine setup is a process of preparing the system that includes a workstation, machine, and operator to complete the work. The machine setup for the forging process generally consists of several complicated steps. The main challenge faced by many researchers is the setup time, which is considered time-consuming. Thus, recent evidence suggests that the setup time should be reduced [3].

Shingo (1985) [4] developed a well-known machine setup improvement method, *Single Minute Exchange of Dies* (SMED). The setup activities are categorized into four groups. The "measurement" and "adjustment" may take up to 15% of a whole process of time. It has been demonstrated that using special tools can reduce the time and difficulties in individual operations.

Regrading the forging process in the factory used in this current study, the original setup process for dies changing between the two models (i.e. 80 mm. ball valve: MB80 and BV80, in ENOMOTO 700GFH hydraulic press machine) take approximately 90-120 minutes by the two operators. Vongbunyong et. al. (2019) [5] improved the setup process by developing a semi-automatic system. The setup time is reduced to around 25 minutes, which are reduced by 65%. In a study conducted by Vongbunyong et. al. (2019), it was shown that an additional AR tool developed could assist the operators in adjustment and measurement operations during the IED activities.

[20] Radjai F Jean M Moreau J J and Roux S 1996 *Phys. Rev. Lett.* **77** 274

1.2 Augmented Reality (AR)

More recent attention has focused on the provision of AR which is combined between a real-world and a virtual-world. The computer graphical models are created and rendered on the top of the real-world image for showing desired contents. AR can be classified into two categories. The first group is called a marker-based AR. It is simple and has been widely used in most applications. The program can locate and track the objects and environment in reality. In this regards, the marker that is a part of an object in computer vision tracking e.g. 2D image, 3D object, and sound. Second, the makerless-based AR is also considered to be more complicated than the marker-based AR. Concerning development of the technique, It does not require markers to operate but it can calculate the position directly from the environment. The localization can be done from an environment around the object by using related information, which is acquired, from many sources, e.g. computer vision, GPS location, and accelerometer.

Early examples of research into AR technology include industrial applications of AR to develop [develop an AR system for planning and installing pipes in large ship assemblies (Olbrich *et al.*, 2011). AR created virtual images from the CAD model and render over the real world image. The user can plan the route, customize the shape of the pipe and check the accuracy of the position immediately before the actual building and installation. Oliver Wasenmuller et al. (2016) [11] developed an AR system for the inspection of manufacturing errors which are discrepancy between 3D models and the real objects. RGB-D Camera was used and resulted in less than 0.01m error.

1.3 Summary of SDKs

Recent advances in developing information technology methods have facilitated the investigation of the software development kit (henceforth SDK). AR applications were carried out using SDK which is based on our existing platform, Android and Unity. Criteria for selecting the SDK were shown in Table 1 below.

Table 1. Comparison between SDKs

Factor	Vuforia	Wikitude	ARcore
(1) License	Free for developer	Free for academic license	Free API
(2) Unity3d plugin	Yes	Yes	Yes
(3)Supported platform	Android, IOS, Windows	Android, IOS	Android
(4) 2D tracking	★★★★★	★★★★	★★
(5) 3D tracking	★★★	★★★★	-
(6) SLAM tracking	★★	★★★	★★★★★

(a) *Vuforia [6]* is the most popular SDK for new developers, especially for Marker-based AR. It can work on a variety of platforms. Simple AR programs can be created with minimal knowledge of programming. Tracking of high-speed objects is possible but less accurate than ARcore. It works well with 3D object tracking. It offers many features e.g. extend tracking, instant tracking, etc.

(b) *Wikitude [7]* can work on a variety of platforms. Intermediate level of programming skills is required for developing moderate tasks. It is compatible with Unity, Cordova, Xamarin and Android studio. For high-speed 2D/3D SLAM tracking, the performance is very high in comparison to Vuforia. It offers special features that are unavailable in most SDKs, i.e. a 3D marker for large scale object.

(c) *ARcore [8]* as an SDK working on Google's platform for developing AR on Android devices. Both marker-based and markerless-based AR can be developed. It can learn from the environment so that the detection and measurement of objects can be done precisely. Extended tracking is more efficient in comparison to (a) and (b), in which the main problem is to keep track of the objects. The limitation of ARcore is the speed of the objects that need to be tracked. However, from the observation of this machine setup system, ARcore is capable. As a result, ARcore is the most suitable SDK for this research.

Regarding the related technique, simultaneous localization and mapping (SLAM) [9] is one of the important localization concepts used in robotics. SLAM consists of two primary functions: creating a map from the surrounding environment and finding a location of the device in the map relative to the position and gesture data from various sensors. Initially, SLAM is often used in machine vision with 3D sensors. It is then developed to use with 2D imaging, e.g. 2D phone camera. This concept is Monocular SLAM [9], which is a starting point of various AR systems on current mobile devices.

2. Problems overview

2.1. Related components in dies changing process

The machine setup, in this case, is to exchange the dies between two models, MB80 and BV80. For each model, the dies consist of 3 parts which are *upper die, lower die*, and *insert*. (a) The upper die is attached with the upper plate, which is mounted on the ram. (b) The lower die is mounted on the sub-plate, which is mounted on the base. On the sub-plate, there are 4 holes for the guideposts and 4 holes for the cushion posts, where all the posts are on the base (see
Figure **1** a). These posts are to align the sub-plate and lower die. The detail operations are presented in [5].

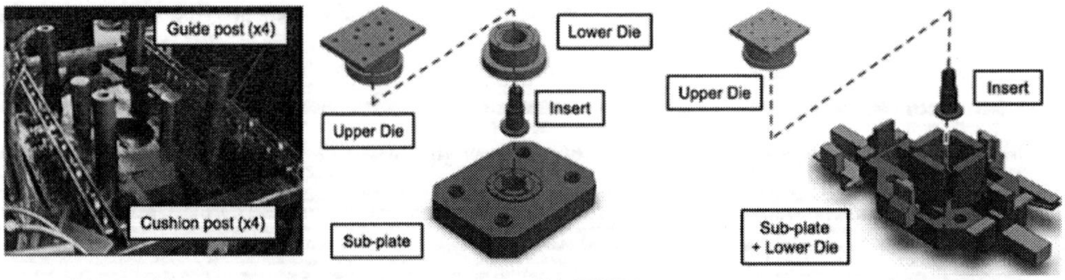

(a) Cushion and guide posts (b) BV80 (c) MB80

Figure 1. Components in the machine and dies set [5]

2.2. Problem and solution

This research focuses on the operation that moves the sub-plate and the lower die into the machine where the alignment is crucial. The sub-plate and the lower die maximum size is 910 mm. x 1,921 mm. x 430 mm. and around 1,100 kg weights is very heavy for manual operation. These components are placed on the guide rails and manually slid into the machine on top of the cushion and guideposts.

As the alignment of the posts is essential, the problem arises when the operators need to see the components clearly for the best alignment. The discrepancy between the holes on the sub-plate and the posts on the machine needs to be observed while sliding the components. As can be seen in Figure 2, this is impossible according to the workspace and parts of the machine that obstruct the view of the operators.

Thus, this work aims to implement an AR technology to help the operator as a role of a decision tool for localization of the components. *Markerless-based AR* is considered because of two main reasons (a) attaching a marker in the workspace is impractical due to the rugged condition of the shop floor. The surface of components is dirty and coated by lubricants so that the marker will be occluded. (b) Direct object recognition of the components is impractical and complicated due to the limited number of feature points. To rule out the possibility that Markerless-based AR and SLAM will occur, the algorithm can be used to memorize the position and orientation of the markers. It is also attached out of a workspace area to create the virtual model of the machine. Two AR concepts are adopted to resolve this problem.

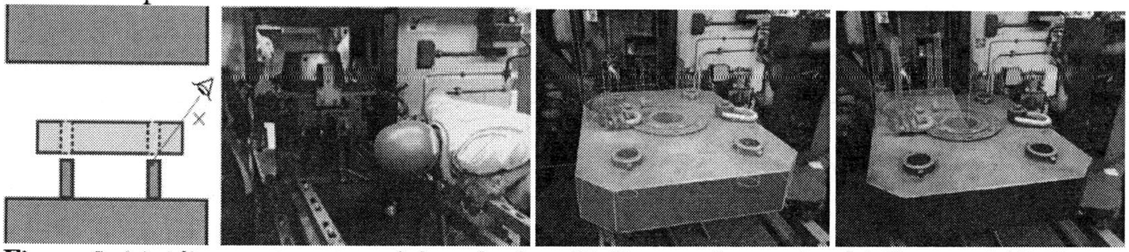

Figure 2. (a) Alignment: posts and the sub-plate (b) Current observation (c) Concept 1 (d) Concept 2

2.2.1. Concept-1: using 2 AR markers to notify a discrepancy of 2 models

The two markers were chosen to show two virtual models in this concept. There is one at the sub-plate and one another at the machine so it allows the system to calculate the distance and orientation between the two markers. The user interface can inform the operator how to move the sub-plate to the right position directly, such as showing a direction arrow, showing the top view, showing distance numbers as shown in Figure 2 c.

A major problem with the experimental method is that the technical issues and usability is becoming problematic. For example, the process of overlaying of many virtual models can be confusing than being assisted the user. As a result, the appropriate content display management is considerably complicated. Besides, the accumulated error in SLAM affects the discrepancy calculation. Therefore, this problem is related to SLAM algorithm within the SDK, which cannot be modified at this stage.

2.2.2. Concept-2: using only 1 marker to generate the final position model of sub-plate

This concept is likely to assist the user and is similar to the previous concept. The difference is that only a marker is used to construct the virtual model of the sub-plate at the final position. Then, the user needs to move the sub-plate to that position. By attaching a marker to the machine, the virtual guidance posts can be elongated from the actual posts so that they are observable directly from above the sub-plate (see Figure 2 d). This guidance model would be tested to investigate the accumulated error and the practicality of the AR system. This can be further developed to implement the procedure in concept-1 in the future.

3. Methodology

In this research, concept-2 is selected to develop the AR system for the die alignment process. Also, the purpose of the study is to test the position error of monocular SLAM technology whether it has enough level of accuracy and precision to be used in the high precision industry.

3AR system design

As shown in Figure 3a, an Android mobile device was used to communicate between the operator and the AR system. The AR system consists of a marker attached on the hydraulic press machine for recalling the virtual model of the machine and rendering the operator.

The position and the orientation of the virtual model are calculated by SLAM's map, which obtains from an environment scanned with a device camera. During the process, the SLAM's map will find many reference planes with respect to the feature point clusters. The default plane should be manually selected as the main reference plane. Next, the marker is scanned to obtain the actual machine location and recall the virtual model. After that, a little distance error between the virtual model and the real object may be presented. Manually offsetting is required for the first time. After the virtual model has been rendered, the virtual guide posts can be seen by the operator. These posts will be the target position that the sub-plate will be removed.

The AR program is run according to the flowchart presented in Figure 3b. The working of this program can be divided into three parallel parts: (a) point cloud generator for finding feature point, (b) plane generator for finding feature point cluster that probably is the plane, and (c) the main process. In the main process, there are three operations involved: environment scanning, selecting the default plane, and marker scanning. These operations will be conducted as described above.

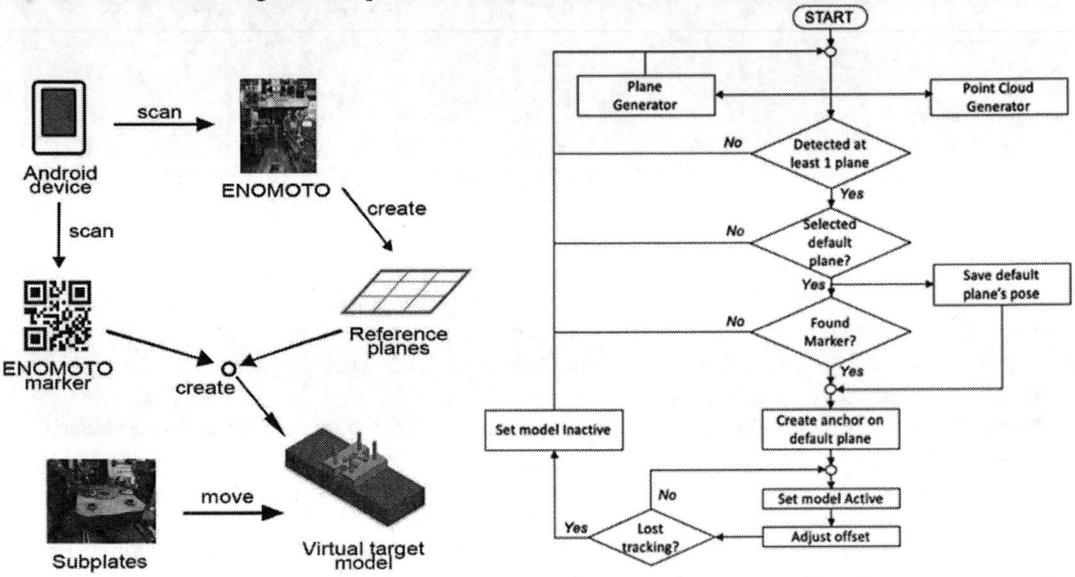

Figure 3. System overview (a) system diagram (b) program flowchart

3.2. System configuration and setup

AR core is used to develop the markerless-based AR with SLAM in this research due to its robustness, speed, stability, and accuracy, which are better than other reviewed SDKs (see Table 1). It runs on the Android platform with the Unity environment. Some limitations are found. One of these is that it cannot track a fast object and detect a 3D marker. However, they do not affect the requirements of this research.

The Android phone is connected to the computer via a USB cable. The development environment includes: (a) Unity 2017.4.15f1, (b) ARcore SDK for Unity 1.4.0 and (c) Android SDK 7.0 API Level 24. After the application has been built, it will deploy on the phone to be used in the experiment.

The development and preliminary experiment were conducted in the lab. Significant changes in the environment occurred when the experiment was conducted on the shop floor. The setup can be described in the following issues:

(a) Workspace refers to the area that the sub-plate can be moved, which is 950 mm. x 950 mm. This area is limited inaccessibility due to the structure of the machine so that the operator can see the sub-plate from only two sides.

(b) Marker is 100 mm. x 100 mm. attached on the hydraulic-press machine.

(c) Environment, the factory floor is metal and oily surface while the lab floor is a carpet with some pattern. This difference affects the performance for defining the reference plane. The patterned carpet has better features in term of the number, clarity, and consistency over time.

(d) Guideposts are mocked-up with two tubes for the lab version to prove the concept, while there are four guideposts on the actual machine. The diameters are 70 mm.

(e) Sub-plate is mocked-up with a cardboard box with guide holes for the lab version for easy moving. For the actual sub-plate, it is a heavy part which can be mobilized with assisted tools only.

4. Experiment

4.1. Objectives and Procedure

The purpose of the experiment was to verify if this AR system can achieve sufficient accuracy and precision of the position and orientation of the virtual model concerning the real objects. An acceptable tolerance between the guidepost and the guide hole is 7 mm. The system performance needs to satisfy this value.

The experiment was conducted in the lab and on the shop floor. The lab experiment was used to predict the maximum performance with minimal effects from the external environment. On the other hand, the shop floor experiment aimed to investigate the factors related to the actual working conditions that might affect the performance of the system. The experiment procedure is set out to:

(1) Prepare the model and the guideposts (see Figure 4 a-b);
(2) Start scanning the marker (see Figure 4 c);
(3) Scan the area associated with the machine and select the default plane (see Figure 4 d-e);
(4) Manually adjust the offset of the model to match the machine (see Figure 4 f); and,
(5) Move the camera angle from the original position and observe the changes.

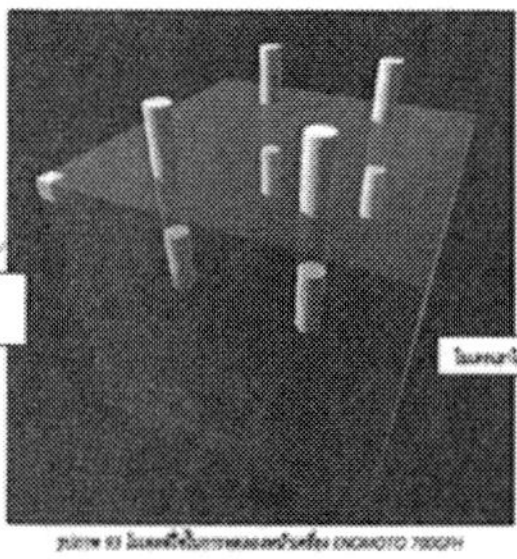

(a) Virtual model of extended posts

(b) Actual posts on the machine

(c) Marker on the machine.

(d) Scanning for the reference plane

(e) Render the virtual model

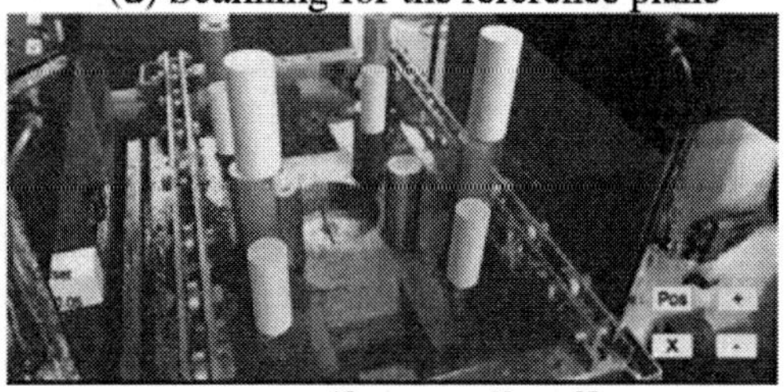

(f) manually offset

Figure 4. Experiment on shop floor

4.2. Results and Discussion

From the lab experiment, the experiment shows an error of the AR system. In an attempt to keep all factor to reach the highest performance, the average error is between 80-50mm. The test last 10 times.

Concerning the shop floor experiment, the program was quite unreliable because the number of the feature points to be detected is decreased in the actual working area. The current study found that it is due to the surface that is coated with oil stains and similar color.

Regarding the error of ARcore, the movement of the model is similar to the experiment in the lab environment. Within the 800 mm around the front side of the machine, the system has an error between 50–80 mm. The results of this study indicate that the maximum error occurred when observed from the side view. Returning to the first position of the offset adjustment, the virtual model moved slightly from the original location. A possible explanation for this might be that there is an appearance of the ARcore due to the discrepancy ARcore SLAM's world and Unity's world. This issues should be taken in to account.

Due to the uncertainty of SLAM map, in every model reconstruction, the virtual guidance model cannot be rendered at the same position with respect to the fixed offset from the marker. Therefore, manually offsetting is allowed to ensure that the model position is coincident with the real object. In regard to the change of the camera angle, the errors are associated with the SLAM mapping and data management of ARcore. These relationships may partly be explained by manual offsetting is allowed for fine positioning of the virtual model in two dimensions. Whereas the z-axis (height direction) is attached to the reference plane obtained from the ARcore algorithm. There are, however, other possible explanations. This result may be explained by the fact that the plane simplification algorithm of ARcore alleviated the virtual plane in order to compensate the roughness of SLAM map. Also, the error is caused by uncertainty data management algorithm in relation to internal problem of ARcore and general RGB-camera. Future studies on the current topic are therefore recommended.

5. Conclusion

The present study was designed to determine the effect of AR technology to solve the problem of wasting time from aligning the die's subplate. The results of this investigation show that creating of a virtual image model of the final position of the subplate can be replaced by the operator i.e. moving the subplate to that position. During the development of this program, problems were found in the process of adjusting models offset, due to three main problems: human error, plane error, and ARcore error. The study contributes to our understanding of a system that can help adjust the alignment. This study appears to be the first study to use a system to eliminate the human error. The major limitation of this study is that the other errors cannot be resolved because of an internal problem caused by the SLAM algorithm of the ARcore SDK.

Further work needs to be done to establish whether other errors can be resolved and clearly identified. Further research should be undertaken to explore how error values occur from those problems. There is, therefore, a definite need for conducting an experiment to collect the error values that occur from those problems. These findings suggest using the test with simulated environments that control various factors in the condition than the program will be the most effective. Also, the result of the experiment showed that the maximum error value occurred was 50-80 mm. which is worse than the acceptable error value of 7 mm. SLAM algorithm must be improved to resolve this problem.

6. References

[1] Groove M P 2010 *Fundamentals of Modern Manufacturing: Materials, Processes, and Systems* (John Wiley & Sons) 405-420

[2] Vongbunyong S Archeewawanich P Akaratham P Martchan C and Thamrongaphichartkul K 2019 *IOP Conference Series: Materials Science and Engineering* **501** 012046

[3] Andrade F D 2019 *Machine Set Up Definition* *https://www.academia.edu/10073499/Machine_Set_Up_Definition* accessed 2019
[4] Shingo S 1985 *A Revolution in Manufacturing: The SMED System* (Productivity Press)
[5] Vongbunyong S Wongpakdee A Suwanarawat N Phansaeng S and Piyasinchart A *Proc. the 10th TSME International Conference on Mechanical Engineering* (Pattaya:Thailand)
[6] Vuforia developer 2019 *Getting Started with Vuforia Engine in Unity* *https://library.vuforia.com/articles/Training/getting-started-with-vuforia-in-unity.html* accessed 2019
[7] Wikitude developer 2019 *Image Recognition* *https://www.wikitude.com/external/doc/documentation/latest/unity/imagerecognitionnative.html#introduction* accessed 2019
[8] Googles Developers 2019 *Fundamental Concepts https:// developers. google.com/ar/discover/concepts* accessed 2019
[9] Imperial College London 2019 *Mono SLAM https://www.doc.ic .ac.uk/~ab9515/monoslam.html* accessed 2019
[10] Olbrich M Wuest H Riess P and Bockholt U 2011 *Fraunhofer Institute for Computer Graphics Research* (Germany)
[11] Wasenmuller O Meyer M and Stricker D 2016 *German Research Center for Artificial Intelligence* (Germany)

IOP Conf. Series: Materials Science and Engineering **886** (2020) 012052 doi:10.1088/1757-899X/886/1/012052

Experimental Investigation and Finite Element Modelling of the Influence of Hydrostatic Pressure on Adhesive Joint Failure

Kerati Suwanpakpraek[1,2*], Baramee Patamaprohm[1,2], Ekkarin Phongphinittana[1,2], and Arisara Chaikittiratana[1,2]

[1]Department of Mechanical and Aerospace Engineering, Faculty of Engineering,King Mongkut's University of Technology North Bangkok, 1518 Pracharat 1 Road, Wongsawang, Bangsue, Bangkok 10800, Thailand
[2]KMUTNB – TU Chemnitz Collaborative Center for Lightweight Structures Technologies (LiST), King Mongkut's University of Technology North Bangkok, 1518 Pracharat 1 Road,Wongsawang, Bangsue, Bangkok 10800, Thailand

* Corresponding Author: k.suwanpakpraek@gmail.com, Telephone: 0-2555-2000 ext 8315, Fax Number 0-2586-9541

Abstract

Structural adhesives have been increasingly used in automotive industries in conjunction with lightweight structural components to maximize weight reduction in modern car design. They are used to bond various kind of lightweight materials such as FRP composites or metallic alloys and provide a good distribution of force across joining area without damaging the substrates. For automotive applications, bonded components have been often used in complex loading environments which brings failure prediction of adhesive joints a key of success for this technology. As polymer-based materials, hydrostatic pressure can have a major impact on adhesive joint failure in addition to temperature and strain rate. Its influence can be investigated using a specific test fixture known as a modified-Arcan fixture. This article presents the characterization of epoxy-based adhesive using a series of modified-Arcan tests to apply different loading taking in account effect of hydrostatic pressure onto adhesive joints. A high-order exponential Drucker-Prager failure criterion is chosen to interpret the yield and failure surface. Finally, the non-linear behaviour of adhesive is identified via 0° and 90° Arcan tests which correspond respectively to tri-axial and shear modes and the validation of this behaviour is carried out by mix-mode 60° and 120° Arcan tests. The simulation shows a good agreement only with the result in direction 60° while in direction 120° the divergence is obvious due to the different in damage mechanism of this mode.

Keywords: adhesive bonding, modified-Arcan test, The high-order Drucker-Prager criteria, FEM simulation

1. Introduction

Structural adhesive joining technology is currently one of the attractive methods to assembly lightweight structural components such as composites in automotive and aerospace industrials. This method provides many advantages in comparison to conventional methods including a good distribution of force all over the bonded area, which noticeably decrease a local concentration of stress in components. It is also an available option for multi-material assembly and adds relatively low-weight onto structures, which is suitable for weight reduction philosophy to increase an efficiency of a car in automotive industrial design nowadays [1, 2]. For bonded multi-material structures, e.g. metal-composite joining, the optimize thickness of adhesives is suggested a certain structure (about 0.2-0.5 mm.) to ensure that the adhesive layer is bonded all over the overlap area [3-5] so that the thickness control of adhesive layer is also a topic in industrial applications where more sophisticated geometric components are bonded. The adhesive SikaPower-497 designed to bond metal-composite structure was investigated in this study. The mechanical properties in manufacturer's datasheet is insufficient to predict efficiently structural failures.

Since the adhesive presents a non-linear behavior under loading and its yield and failure domain also depend on hydrostatic pressure, a sophisticated experimental methodology is required to characterize its behavior. A modified-Arcan fixture is a specific test fixture that can be used to characterize an adhesive layer under hydrostatic pressure [6]. It can apply either tri-axial tension or compression and coupling with shear load to induce different state of hydrostatic pressure onto a specimen. In addition, a metallic substrate of modified-Arcan test has a special geometry to reduce the edge effects of adhesive layer [5]. With a good surface preparation, a cohesive failure, which is a failure in an adhesive layer, can be expected. This type of failure is preferable to ensure that the characterization is on an adhesive layer not on interfaces between adhesive and its substrate and confirm the good selection of adhesive. This paper investigates the effect of hydrostatic pressure on adhesive layer using a series of modified-Arcan tests.

An image correlation is used as a non-contact method to measure a strain field during the tests. The yield and failure surface are interpreted by a high-order exponential Drucker-Prager criterion. A non-linear behavior of adhesive is identified by tri-axial (0°) and shear (90°) mode of modified-Arcan test using FEM simulations and mix-mode direction (60° and 120°) are served as a model validation.

2. Material and Behavior model

An adhesive structural for composite-metal bonding in this study is a SikaPower-497 Epoxy adhesive manufactured by Sika. It is a one-part structural adhesive based on epoxy. Its cure requires 50 minutes at 160°C to form a high-performance thermoset adhesive [7]. As a polymer-based material, the adhesive mechanical behavior is a type of elasto-visco plasticity which depends on temperature, strain rate and also hydrostatic pressure [8]. In this study, the material is treated as elasto-plastic and only the effect of hydrostatic pressure is investigated. As can be seen in Figure 1, the effect of hydrostatic pressure on adhesive yield and failure function, $f(\sigma_v, p)$ can be explained by high-order exponential Drucker-Prager criterion for more accuracy at the state of negative hydrostatic pressure (Eq.1) [9]:

$$f(\sigma_v, p) = a\sigma_v^{\,b} - p - p_t = 0 \tag{1}$$

where σ_v is Von Mises equivalent stress (eq.2), p is hydrostatic pressure (Eq.3), p_t is hardening constant, a and b are material parameters that are independent to plastic deformation

$$\sigma_v = \sqrt{3J_2} = \sqrt{\left((\sigma_{11}-\sigma_{22})^2 + (\sigma_{22}-\sigma_{33})^2 + (\sigma_{33}-\sigma_{11})^2 + 6\left(\sigma_{12}^2 + \sigma_{13}^2 + \sigma_{23}^2\right)\right)/2} \tag{2}$$

$$p=-\frac{1}{3}I_1=-\frac{1}{3}(\sigma_{11}+\sigma_{22}+\sigma_{33}) \tag{3}$$

where σ_{11}, σ_{22} and σ_{33} are the normal stress components, σ_{12}, σ_{13} and σ_{23} are the shear stress components, I_1 and J_2 are the first invariant and the second deviatoric stress of the stress tensor respectively [10]. The functions for plastic flow follow the Equation 4 and 5.

$$d\varepsilon_p=d\lambda\frac{\partial g}{\partial\sigma} \tag{4}$$

$$g(\sigma_v,p)=\sqrt{(\xi\sigma_y\tan\beta)^2+(\sigma_v)^2}-p\tan\beta \tag{5}$$

where $d\lambda$ is the plastic multiplier, g is the flow potential, β is the dilation angle measured in the $p-\sigma_v$ plane, ξ is an eccentricity of hyperbolic function defined an approaching rate of function to its asymptotes. Since the flow function, $g(\sigma_v,p)$ is different from yield function, $f(\sigma_v,p)$, the plastic flow is non-associated $(g(\sigma_v,p)\neq f(\sigma_v,p))$. The flow direction will not be normal to the yield surface. To simplify the calculation, the constant ξ, which refers to the eccentricity of hyperbolic function in equation 5, is tuned to provide the flow direction as close as possible to the associated flow $(g(\sigma_v,p)=f(\sigma_v,p))$ which means that the flow direction is normal to yield surface. Then, the non-linear isotropic hardening from the theory of plasticity is used to interpret its non-linear behaviour [11]. Its response is described by the following equation:

$$\sigma_v=\sigma_y+q\left(1-e^{-c\varepsilon_{eq}^p}\right)+H\varepsilon_{eq}^p \tag{6}$$

where σ_v is the Von Mises equivalent stress, σ_y is the yield stress, ε_{eq}^p is the equivalent plastic strain and H, q and c are material parameters. All material parameters are identified using a series of modified-Arcan tests.

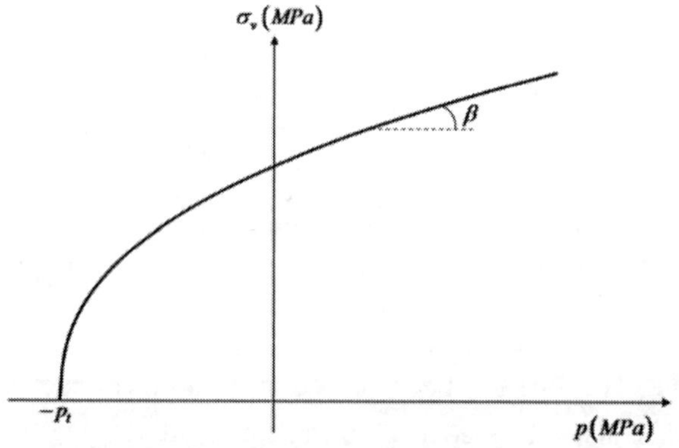

Figure 1. The high-order Drucker-Prager criteria

Figure 2. Modified-Arcan test

3. Modified-Arcan fixture

The modified-Arcan fixture is a specific fixture for testing adhesives. It was inspired by the work of Arcan [12] and followed by the modified version of Arcan-Mines testing fixture [13, 14]. The updated version for adhesive testing allows the fixture to vary the angle to impose a different state of stress onto the adhesive layer as shown in Figure 2 [15]. The tri-axial and pure shear state of stress can be obtained when testing at the direction 0° and 90°. The combination of the tensile-shear state of stress can be found at direction 30° and 60° and the compressive-shear state of stress occurs when testing at direction 120°. All stress combinations are calculated using Equation 7 [14].

$$\sigma_{11} = \frac{F}{S}\cos(\theta), \sigma_{22} = \sigma_{33} = \frac{\nu}{1-\nu}\frac{F}{S}\cos(\theta), \sigma_{12} = \frac{F}{S}\sin(\theta) \text{ and } \sigma_{13} = \sigma_{23} = 0 \quad (7)$$

where F is a force from the tensile test, S is a surface area of the substrate, θ is the angle of fixture and ν is Poisson's ratio of the material. The direction 0° and 90° are called the triaxial and shear mode respectively while the other directions are named as the mix modes, as shown in Figure 3. In addition, this fixture has the advantage of using the universal testing machine. The specimen's substrates for this test have a specific geometry called a beak, around their edge as shown in Figure 4. The adhesive layer bonds these two substrates together forming a specimen for testing. The 45° beak decreases remarkably the edge effect on the adhesive layer due to the substrate's geometry [5]. A stress field in the adhesive layer is more uniform especially in shear stress and maximum stress appear in the center of the joint.

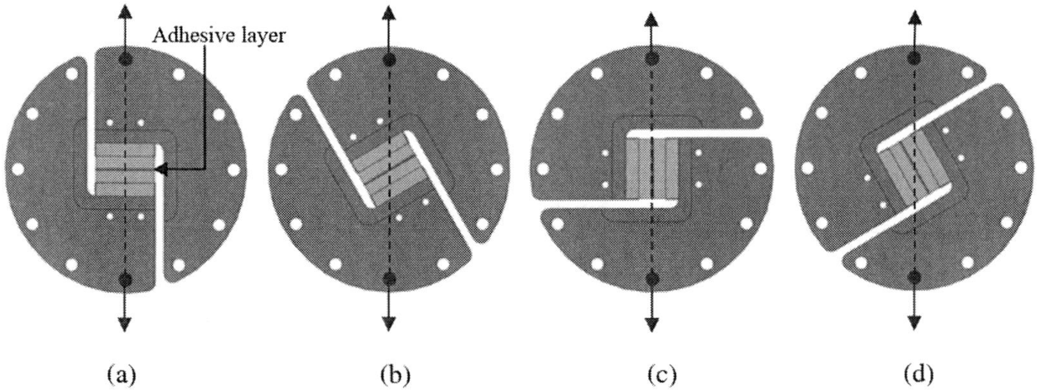

Figure 3. Modified-Arcan test mode: (a) triaxial mode, (b) mix modes: tensile-shear, (c) shear mode, (d) mix modes: compression-shear

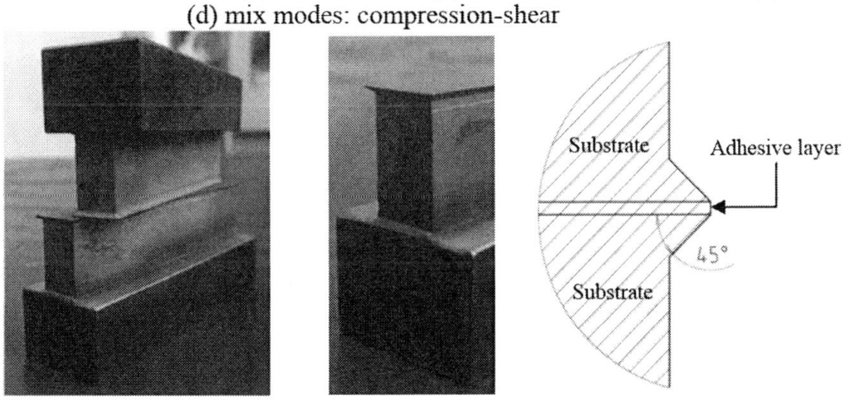

Figure 4. The substrates geometry with the beak

4. Experimental investigation

All tests have been carried out statically using an Instron 5567 universal testing machine with a load cell of 30 kN capacity (accuracy ± 0.5%) and a crosshead speed of 0.5 mm/min [16]. The strain measurement was done by an image correlation system, ARAMIS manufactured by GOM, which has a strain measurement ranged to 0.005 % - 2000 % [17]. All tests were performed at room temperature.

The substrates for adhesive bonding are made of steel with a rectangular cross-sectional bonding area of 70x10 mm. The bonding surface of substrates is sanded and degreased according to manufacturer's recommendations. The thickness of the adhesive layer and substrate's alignment are precisely controlled by the assembly fixture. This fixture consists of a shim plate for thickness control, four straight columns to control a substrate's alignment and coil spring to apply a clamping force, as shown in Figure 5. Specimens with a single component adhesive were cured at 160°C for 50 minutes to ensure a complete cure of adhesive layer. In order to utilize an image correlation strain measurement, a white-matte surface coating with a cloud of black points is required on the side of the specimen (see Figure 6). The deformation measured from point 1-10 was performed on steel substrates to capture the entire deformation field of the specimen. With a rigid substrate assumption, the average strain was calculated from the average deformations between points 1&6, 2&7, 3&8, 4&9, 5&10 and divided directly by the adhesive thickness (see Equation 8).

Figure 5. Specimen's assembly fixture

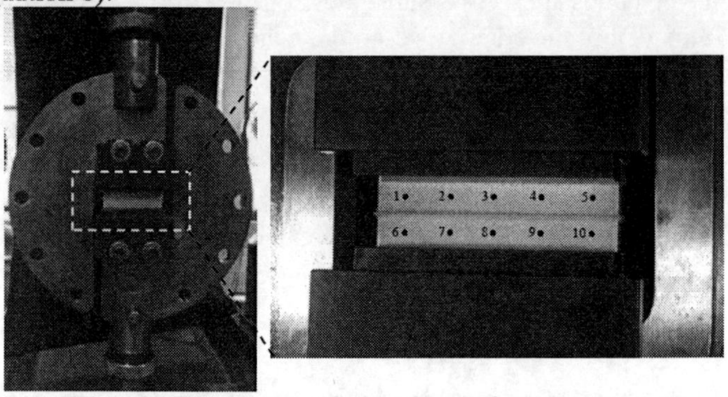

Figure 6. Image correlation for strain measurement

$$Average\ strain = \frac{\sum_{i=1}^{n}(\delta/t)_i}{n} \tag{8}$$

where δ is a deformation between the two points, t is an adhesive thickness, n is a number of measured deformation.

5. Experimental results

The results from the modified Arcan test show a non-linear behavior in all modes. The limit elastic was identified by loss of linearity of experimental curves. The recorded force from load cell was transformed to stress components using Equation 7 with relating to testing mode. Then, Von-Mises equivalent stress and hydrostatic pressure were obtained via Equations 2 and 3. The yield and failure surface plots obtained experimentally from Arcan tests were shown in Figure 7. It can be seen that the dependence of hydrostatic stress on yield and failure stress can be represented well with the higher-order Drucker-Prager. The corresponding criteria's parameters are summarized in Table 1. Table 2 and Table 3 summarize the parameters of the plastic model and plastic flow. They were identified using iteration method and FEM simulation to complete with experimental results (see Table 2 and Table 3). The results obtained from all testing mode show a cohesive failure on the specimen except in the direction 120° which is a combination of compression-shear mode. Regarding the direction 120°, the tested specimen was found out to be an adhesive failure. This can be explained by the nature of polymer behavior that becomes stronger under compressive loading (positive hydrostatic pressure) while the strength of interface remains the same (see Figure 8).

Table 1. Summary of identified parameters for the adhesive.

$F = a\sigma_v^{\,b} - p - p_t$	a	b	p_t(MPa)
Yield surface	1.3×10^{-2}	2	19.7
Failure surface	4.3×10^{-5}	3.2	32.8

Table 2. Summary of identified parameters for plastic model.

$\sigma_v = \sigma_y + q\left(1 - e^{-c\varepsilon_{eq}^p}\right) + H\varepsilon_{eq}^p$	σ_y (MPa)	q	c	H
Hardening (eq.6)	29.6	9.2	19.5	62.8

Table 3. Summary of identified parameters for plastic flow.

$g = \sqrt{(\xi\sigma_y \tan\beta)^2 + (\sigma_v)^2} - p\tan\beta$	σ_y (MPa)	ξ	$\beta(^\circ)$
Flow potential (eq.5)	29.6	18	14.6

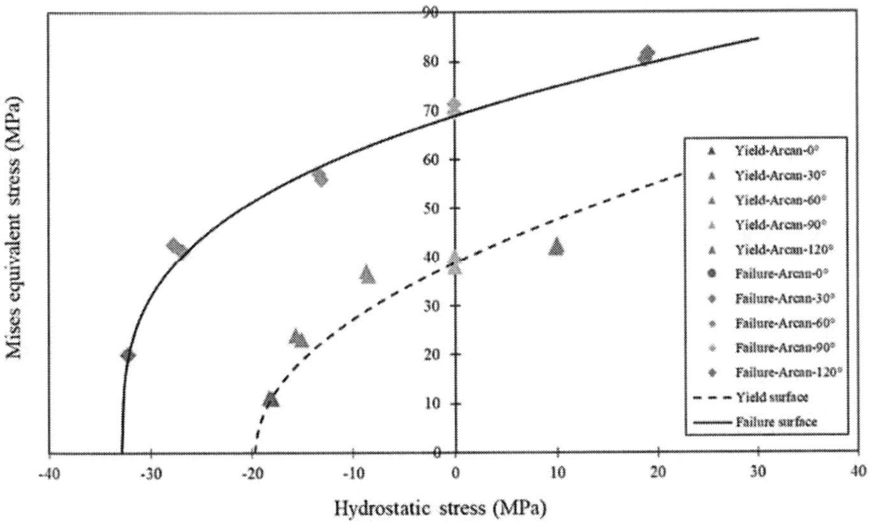

Figure 7. The Drucker-Prager yield and failure surface

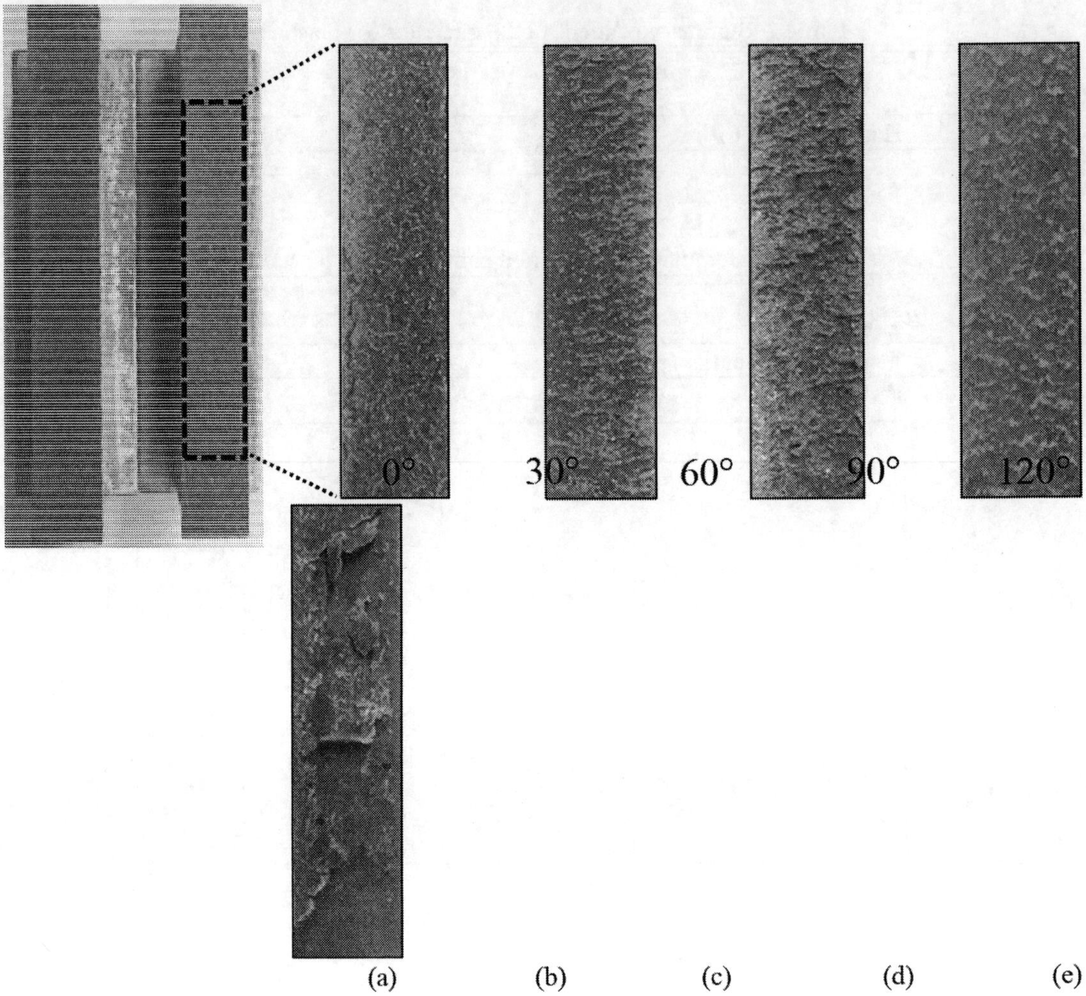

Figure 8. The failure of modified Arcan specimens in different directions: (a) direction 0°, (b) direction 30°, (c) direction 60°, (d) direction 90°, (e) direction 120°

6. FEM simulation

6.1. Structural analysis

The simulation of the modified-Arcan test was performed to identify and validate the Drucker-Prager model. Concerning the CAD model of the fixture, its boundary conditions and geometries of the specimen are shown in Figure 9. Since the fixture deforms under loading due to the strength of the adhesive layer, the simulation needs to be performed on the entire fixture taking into account of the deformation. However, the fixture can still be simplified to increase the efficiency of simulation since its symmetries in the x-y plane. A 3D 8-noded solid element with reduced integration type (C3D8R) was used to ensure the accuracy of the results. Taken together, the fixture model consists of 46,344 elements and 58,838 nodes. The adhesive layer has 1,600 elements and 2,525 nodes with the mesh refining zone at the edge of the adhesive layer. The mesh sizes are in the range of 9.375×10^{-3} to 4.688×10^{-2} mm^3. The FEM simulations were performed using ABAQUS software with Young's modulus (E) = 2120 MPa and Poisson's ratio (□) = 0.36 acquired from the previous work [11] and taken into account the nonlinear behavior of the material.

IOP Conf. Series: Materials Science and Engineering **886** (2020) 012052 doi:10.1088/1757-899X/886/1/012052

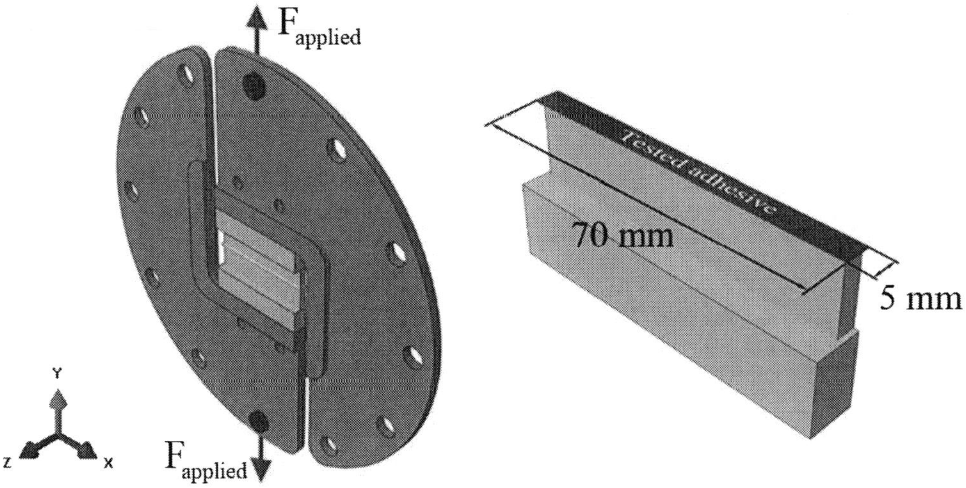

Figure 9. A modified-Arcan test simulation in direction 0°

6.2. Identify of model

The tests simulation in direction 0° and 90° were used to identify the plastic model's parameters for the non-linear behavior of adhesive. These two modes apply completely different state of stress onto the adhesive layer, which is suitable for the identification phase. Figure 10 demonstrates the comparison between experiments and simulation results using the parameters of the plastic model and plastic flow in Table 2 and Table 3.

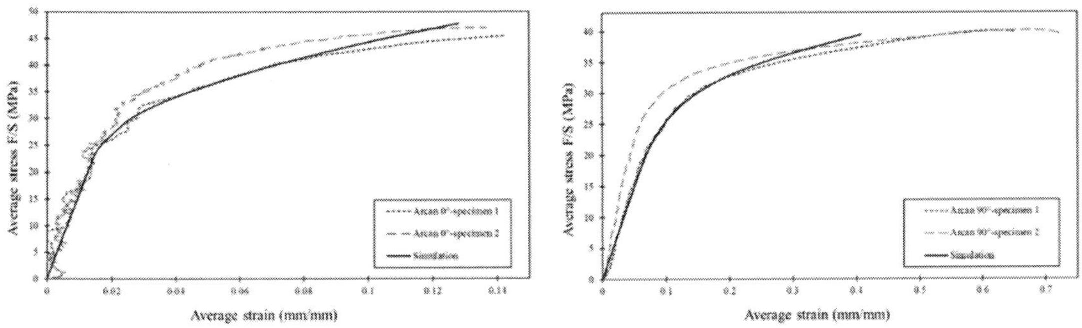

(a) (b)

Figure 10. Identify behaviour model by simulation: (a) modified-Arcan test in direction 0°, (b) modified-Arcan test in direction 90°

6.3. Validation of model

The model validation was carried out using the mode mix tests, which are the tests in the direction 60° and 120° corresponding to tensile-shear and compressive-shear, respectively. These complex states of stress are perfect for validation phase. Figure 11 provides the comparison between simulation results and experiments. The simulations agree well with experimental results in direction 60° while show an obvious divergence in direction 120°. This disagreement can be explained by the type of failure in this mode, which is an adhesive failure at the interface. As discussed previously that the adhesive layer becomes stronger under compressive loading and exceeds the interface strength between the adhesive layer and metallic substrates. With this scenario, the damage is localized at the interface rather than the adhesive layer, which is contrast from the other modes. Thus, the non-linear behavior observed from the experiment is a result of interfacial damage, which did not implement in the simulation. Since the effect of interface strength plays an important role in this mode, it needs to be analyzed by other approach [19].

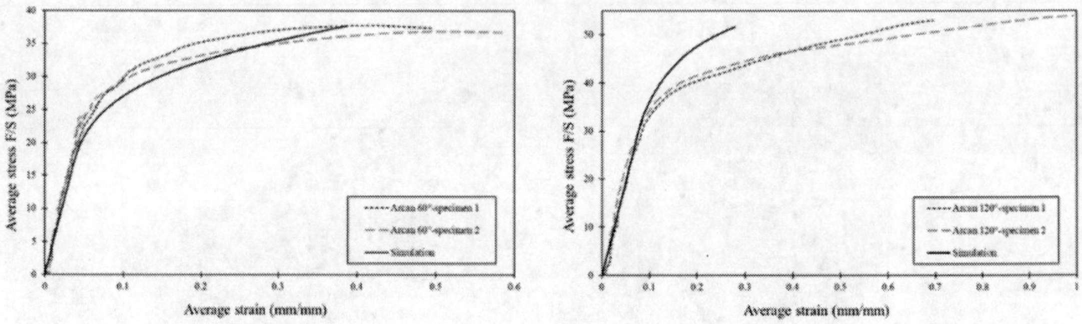

(a) (b)

Figure 11. Validation behaviour model by simulation: (a) modified-Arcan test in direction 60°, (b) modified-Arcan test in direction 120°

7. Conclusions

The investigation of hydrostatic pressure effect on structural adhesive SikaPower-497 clearly shows how it varies the yield and failure surface of the material. The strength of adhesive increases with hydrostatic pressure but greatly declines under negative hydrostatic pressure, which corresponds to tensile tri-axial state. The high-order exponential Drucker-Prager criterion is suitable to interpret the experimental results in both yield and failure scenarios in comparison to the original linear Drucker-Prager criterion. The numerical validation of non-linear behavior identified from tri-axial and shear modes show a good agreement with experiments only if damage takes place at the adhesive layer which relates to cohesive failure. For interfacial damage, the different simulation approach is required in order to predict its behavior more precisely.

Acknowledgments

This work was financially supported by the Faculty of Engineering at King Mongkut's University of Technology North Bangkok (ENG-NEW-62-27).

Special thanks are given to Sika (Thailand) Limited for the support of adhesive materials and STRI, KMUTNB for the support of DIC camera ARAMIS.

References

[1] Jahn J Weeber M Boehner J and Steinhilper R 2016 *Procedia CIRP* **50** 689–694
[2] Adams R D 2005 *Technology and Applications* (Bristol:Woodhead Publishing Ltd)
[3] Badulescu C Cognard J Y Créac'hcadec R and Vedrine P 2012 *International Journal of Adhesion and Adhesives* **36** 56–64
[4] Cognard J Y Davies P Sohier L and Créac'hcadec R 2006 *Composite Structures* **76**(1-2) 34–46
[5] Cognard J Y 2008 *Computers & Structures* **86**(17-18) 1704–1717
[6] Cognard J Y Davies P Gineste B and Sohier L2005 *Compos Sci Technol.* **65** 359–68
[7] Sika A G and Power S 2015 *Chemicals company* *https://zaf.sika.com/dms/getdocument.get/.../cmi-ms-bro- sikapower_sikaseal.pdf/,* accessed July 2015
[8] Mirkhalaf S M Pires F A and Simoes R 2016 *Computers & Structures* **166** 60–74.
[9] *Abaqus reference manuals version 6.12 Simulia* (Providence:USA)
[10] Suwanpakpraek K Patamaprohm B and Chaikittiratana A 2019 *IOP Conference Series: Materials Science and Engineering* (Phuket) **501**(1) 012023
[11] Jousset P and Rachik M 2014 *International Journal of Adhesion and Adhesives* **50** 107–118
[12] Arcan M Hashin Z A and Voloshin A 1978 *Experimental mechanics* **18**(4) 141–146

[13] Bassery J and Renard J 2012 *Fatigue Behaviour of Fiber Reinforced Polymers: Experiments and Simulations* 195–222
[14] Patamaprohm B Phongphinittana E Guinault C Gantchenko V Renard J Jeandrau J P Peyrac C and Lefebvre F 2016 *Revue des composites et des matériaux avancés* 26 63–85
[15] Arcan L Arcan M Daniel I 1987 *ASTM Tech Publ* **948** 41–67
[16] Instron Corporation *Instron Series 5500 Load Frames Including Series 5540, 5560, 5580* https://docplayer.net/35203019-Instron-series-5500-load-frames-including-series-5540-5560-5580.html accessed 15 July 2019
[17] Gom gmbh *Technical DataGOM https://www.gom.com/metrology-systems/aramis/aramis-adjustable.html* accessed 15 July 2019
[18] Da Silva L F Dillard D A Blackman B and Adams R D 2012 *Testing adhesive joints: best practices* (Springer Science & Business Media)
[19] Jousset P and Rachik M 2014 *Engineering Fracture Mechanics* **132** 48–69.

Investigation on Mechanical Properties and Fracture Mechanism of Carbon Fiber and Glass Fiber Composite at Different Curing Temperature

Ye Htet Lin[1*], T. Cuaiman[1], P. Karin[1], C. Charoenphonphanich[1], P. Larpsuriyakul[2] and N. Ohtake[3]

[1]Automotive Engineering Program, Faculty of Engineering, King Mongkut's Institute of Technology Ladkrabang, Bangkok 10520, Thailand.
[2]National Metal and Materials Technology Centre, National Science and Technology Development Agency, Pathumthani, Bangkok 12120, Thailand.
[3]Department of Mechanical Engineering, School of Engineering, Tokyo Institute of Technology, Tokyo, 152-8550, Japan.

* Corresponding Author: 61610005@kmitl.ac.th

Abstract
Weight reduction is one of the big challenges in the auto industry. In the aerospace industry, the high modulus of composite material makes weight savings to replace alloys such as aluminium and titanium. It is one way to reduce fuel consumption and reducing emissions in vehicles by using composite. Among them, carbon fiber and glass fiber are the most useful materials in composite to reduce the weight in vehicles not only in the automotive field but also in aircraft, marine, medicine, sport, etc. In this study, hand lay-up (wet lay-up) method is used to fabricate carbon fiber and glass fiber composites with different curing temperature 30°C (the equivalent of room temperature) until it is dry and 80°C for curing 6 hours. The strength of carbon fiber and glass fiber composites were increased over 10% and the hardness of carbon fiber composite was also increased two times at 80°C curing temperature. Furthermore, the fracture mechanism of carbon fiber and glass fiber composites were also observed by using Scanning Electron Microscopy (SEM) image processing method.

Keywords: Composites, Curing temperature, Mechanical Properties, Fracture Mechanism.

1. Introduction
Carbon fibers are fibers about 5–10 micrometers in one filament diameter and consist of at least 92 wt%, mostly more than 99 wt% of the carbon in the non-graphitic state [1]. Recently fiber composites are rapidly developing in the auto industry due to their unique characters and several advantages including low weight, high stiffness, high tensile strength, high chemical resistance, high-temperature tolerance, corrosion resistance, and rigidity [2, 3]. Their properties could be used not only in automotive parts, aerospace, marine, but also, they are used in civil engineering, medicine, military,

motorsports, and others [4]. There are several types of carbon fiber such as pitch, aramid, PAN and mesopitch carbon fiber. Among them, precursor polyacrylonitrile (PAN) is used as the most useful carbon fiber production because of its high strength and young modulus compared to other materials that contained PAN formula so-called C_3H_3N. And this formula is made from propylene and ammonia [5]. Nevertheless, they are proportionately expensive when compared with similar fibers like glass fibers or plastic fibers. Glass fiber is one of the most flexible materials that can be used in various fields. Most of them are based on silica (SiO_2) and combined with other chemical substances, such as oxides of calcium, boron, sodium, aluminum, iron. Glass comes from sand and it needs good care to get free of impurities material of process [6]. A composite material is a combination of two or more materials with significantly different in terms of physical or chemical properties. The two main functions in the composite are fiber and matrix [7, 8]. In composite, carbon fiber is the main substance and matrix is the one material, which can produce bonding strength between the fibers [9]. The polymer matrix resin can be divided into thermoset and thermoplastic resins. The overall influencing resin for composite structural is a thermoset, giving lower cost processability and high performance. For a thermoset, an irreversible cross-linking of the polymer chain consolidates the resin into an amorphous polymer forming the matrix. The main requirements for the resin are high ductility, high toughness, low viscosity, excellent processability, low shrinkage, and compatibility with fibers, sizing, and adhesives [10]. N. Ozsoy [11] observed 0, 90° and 0/90° orientation of both carbon fiber and glass fiber by using hand lay-up process and the composite was investigated in the tensile and flexural tests. In an analysis of nano-indentation technique, Hardiman (2015) [12] found that the technique of a nano-indentation can identify the hardness and modulus of carbon fiber polymer matrix composite with small pyramid shape spindle diamond knife. Singh (2018) [13] studied the effect of curing temperature ranging from 80°C to 130°C on natural jute fiber reinforced epoxy polymer matrix-based composites for various mechanical properties. While Cui et al. (2018) [14] observed carbon fibers coated with graphene reinforced TiAl alloys composite bulks sintered at 620°C for 10 minutes using the field emission scanning electron microscopy (FESEM S-4800, Hitachi, Japan). This study revealed that this scanning was equipped with Energy-Dispersive Spectrometry (henceforth EDS), and it leads to a characterization of the microstructure of the specimens. The objective of the study was to investigate the mechanical properties and microstructure fracture of composites when increasing the curing temperature. This study seeks to explain the fracture mechanism of carbon fiber and glass fiber composites that are the microstructures weakness using electron microscopy analysis with different curing temperature.

2. Material Properties and Experimental Setup

2.1 Material Properties

In this experimentation, two types of fibers have been used with epoxy resin YD 582 and hardener TH 7253 as reinforcement in this composite; carbon fiber plain weave based on poly-acrylonitrile (PAN) and E-glass fiber fabrics were obtained from a local market. The properties of the two fibers and matrix material are mentioned in Table 1.

Table 1. Material Properties

Material	Based	Tensile Strength (MPa)	Elongation (%)	Tensile Modulus (GPa)	Chemical Composition	Product name
Carbon Fiber Plain	PAN	152.70	1.83	17.253	C (95–99%)	PC6951500
Glass Fiber Fabrics	E-Glass	114.40	2.97	8.230	Si (32–37%)	Glass Cloth EW200
Epoxy/Hardener	Thermoset	60	4-6	2.850	Hydrocarbon	YD 582/ TH 7253

IOP Conf. Series: Materials Science and Engineering **886** (2020) 012053 doi:10.1088/1757-899X/886/1/012053

2.1.1 SEM with Energy-Dispersive Spectrometry (EDS) analysis on raw material

The carbon fiber plain (CFP) and glass fiber fabric (GFF) was checked in SEM/EDS to determine the elemental composition of fibers found, as shown in Figure 1. The cross-section of carbon fiber and glass fiber were cut by fiber cutter scissor. The surface of the carbon fiber plain consists of roughness, pitting sharp but there is no porous in the cross-section area of the fiber. The chemical composition of carbon fiber plain is carbon-containing 96 wt% and oxygen-containing 4 wt%, respectively.

Regarding glass fiber fabric, there are several defects on the fiber surface along the fiber axis, cracking area and porous and are found in the cross-section area. The chemical composition of glass fiber fabric contained Zn element in the range of 0.9-1.4% based on three filament measurement, as shown in Figure 5 (b). The chemical percentage of Al was not found in the fiber, some were from the holder of the sample while SEM analysis is performed. Table 2 provides a summary of the EDS chemical composite in each carbon fiber and glass fiber.

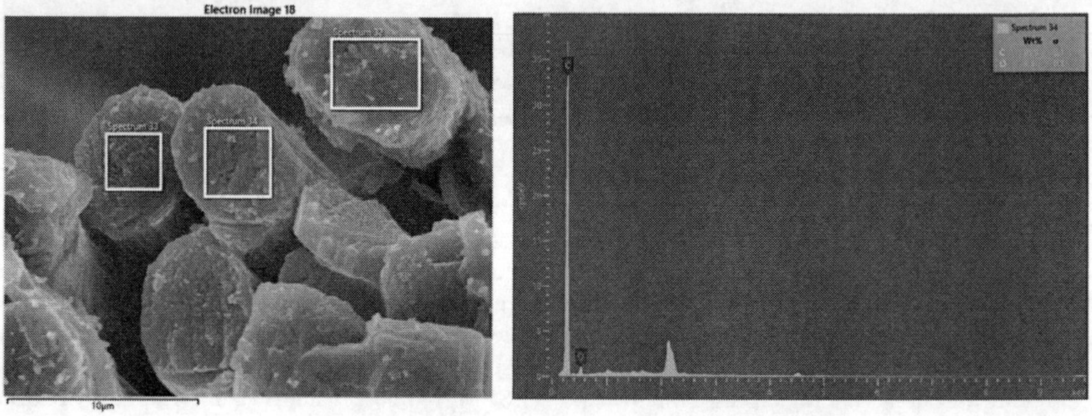

(a) Elemental composition of carbon fiber plain

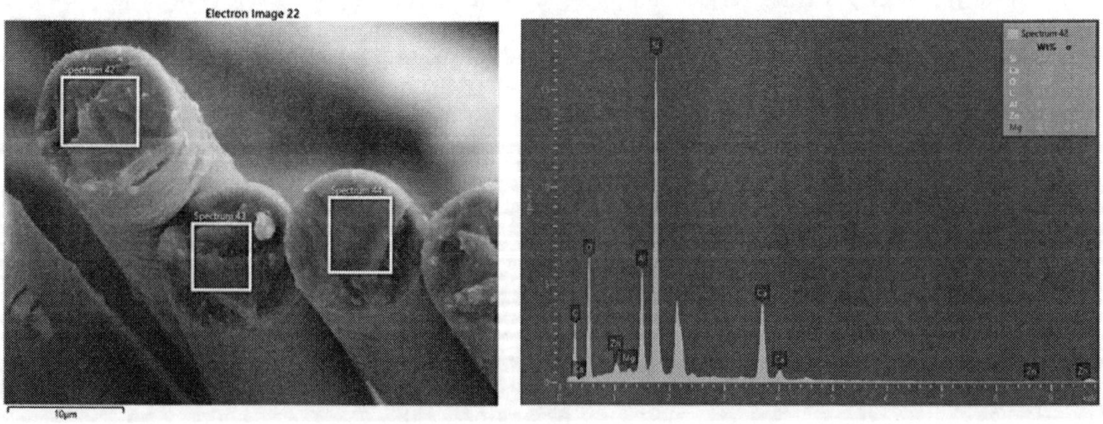

(b) Elemental composition of glass fiber fabric

Figure 1. Energy-Dispersive Spectrometry (EDS) analysis of fibers

Table 2. EDS result of CFP and GFF filaments

Fiber type		Element (Weight %)						
		Carbon	Oxygen	Silicone	Ca	Al	Zn	Mg
CFP	Spect:34	95.9	4.1	-	-	-	-	-
GFF	Spect:42	13.7	18.8	34.4	21.6	9.7	1.4	0.3

2.2 Fabrication of composites

All the reinforced polymer composites were fabricated using hand lay-up process and adjusted bidirectional 0/90 degree. Weave fibers were used with reinforcement and epoxy resin as matrix material. There are two parts of the epoxy resin mixing ratio such as epoxy and hardener mixing ratio (3:1) and they are calculated by equations (1-2). The weight of the composites was controlled based on the fabric weight of each layer. All the composites were cut by computer numerical control (CNC) machine Mazak FJV-20 with the parameter of spindle speed 150 mm/min and feed speed 50-100 mm/min according to ASTM specimens shape. Additionally, after cutting dried composites at 30 °C until it was dry, other specimens were cured at 80°C for 6 hours. Among them, the three specimens were tested twice in each mechanical testing and all the laid-up conditions (see Table 3).

Table 3. Composites Fabrication Procedure

No	Material	Fil Dia (μm)	Weight (g/m²)	Thickness (mm.)	Composite Size (mm.)	layers	Curing Condition	Name
1	Carbon Fiber Plain	5-7	200	0.29	240x280x3.5	12	(1) At 30°C (RT) until its dry	CF30
							(2) RT+ At 80°C for 6hrs	CF80
2	Glass Fiber Fabric	10-12	200	0.2	320x320x3.2	16	(1) At 30°C (RT) until its dry	GF30
							(2) RT+ At 80°C for 6hrs	GF80

$$4 \times \text{Hardener Weight (g)} = \text{Weight of fiber (g)} \quad (1)$$

$$3 \times \text{Hardener Weight (g)} = \text{Resin Weight (g)} \quad (2)$$

2.3. Mechanical testing

2.3.1. Tensile Test

The tensile test based on the ASTM D638 was used with the three samples of carbon fiber composite and glass fiber composite. After that, all the samples were tested INSTRON 8801 with the condition of 50 mm/min, 95 kN load cell.

2.3.2. Flexural Test

The flexural test was performed on 51% RH with INSTRON 55R4502 machine, 24°C temperature by three-point bending test. ASTM D790 method was used on 5.2 mm/min testing rate speed with cell load 10 kN for all the tests.

2.3.3. Indentation Test

Indentation test was used to observe the mechanical properties of the materials especially young modulus and hardness of composite that is cured at room temperature until it was dry and cured composite at 80°C for 6 hours. In this research, Elionix (ENT -1100a) indentation tester was used in

IOP Conf. Series: Materials Science and Engineering **886** (2020) 012053 doi:10.1088/1757-899X/886/1/012053

the operating condition at room temperature with max load 200 mN. The time set up was 10,000 sec and 500 steps (200 sec/ step).

2.3.4. Morphological analysis
Morphological analysis was investigated at the room temperature (RT) 30°C and RT + 80°C at a curing temperature to understand the fiber-matrix interface and mechanism of failure of the composites using SEM image analysis.

3. Result and Discussion

3.1. Tensile test results on carbon fiber and glass fiber composites with different curing temperature

In this research, tensile test for three specimens of carbon fiber plain and glass fiber fabrics epoxy-based composites were investigated with two different curing temperature such as (1) at room temperature until it was dry; (2) at room temperature + 80°C for 6 hours. As can be seen in Figure 2, the result shows the effect of curing temperature on tensile properties of carbon fiber and glass fiber composites. The ultimate tensile strength of CF80 was recorded as around 18% higher than CF30 while GF30 increased to 12% compared to GF80. The tensile strain of carbon fiber is two times shorter than glass fiber elongation. GF30 and GF80 are nearly similar in tensile strain around 2.9% elongation. The results from the CF80 slightly increased by about 13% than the CF30. Similarly, the results show a similarity of the effect of curing temperature 30°C and 80°C on the young modulus of the composites. The value of carbon fiber composite is just about two times higher than glass fiber composite of the young modulus. Higher tensile strain represents the higher ductility of composite that was found in glass fiber composite. It can be explained that glass fiber composite has ductile behaviour compared to carbon fiber. The two composites did not have the same thickness, but their thickness does not affect the young modulus of materials because the material properties are not influenced. In this experiment, the strain gauge was not used so that the elongation of composites is calculated from sample crosshead displacement. The effect of temperature on tensile strength, tensile strain, and tensile modulus of composites are shown in Figure 2 (c), Figure (d) and Figure (e).

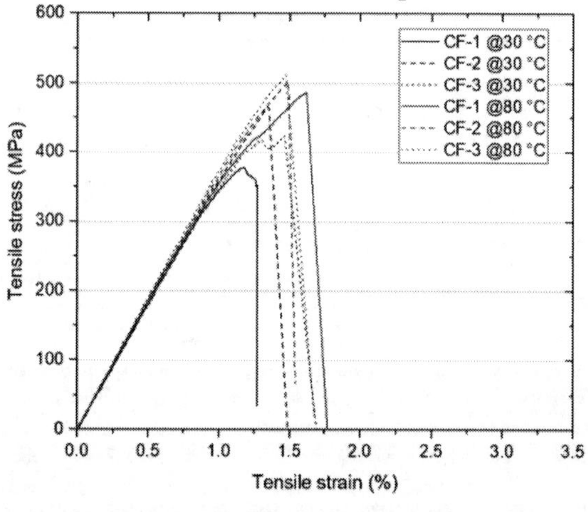

(a) Three specimens of carbon fiber composite

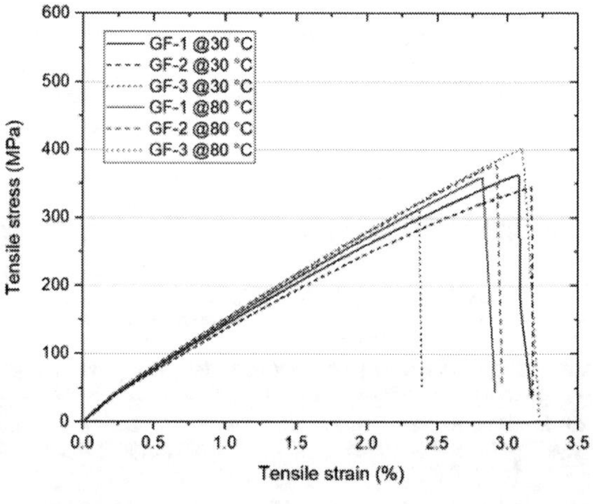

(b) Three specimens of glass fiber composite

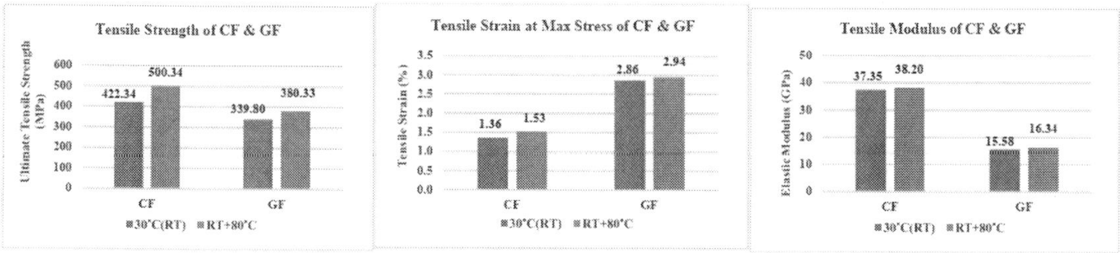

(c) Ultimate tensile strength (MPa) (d) Tensile strain (%) (e) Tensile young modulus (GPa)

Figure 2. Effect of curing temperature on tensile test results

3.2 Flexural test results on carbon fiber and glass fiber with different curing temperature

Regarding the flexural test condition, three specimens are used for flexural strength and flexural strain. They were calculated with the machine automatically, while the flexural modulus was calculated by the ratio of stress and strain. The mechanical properties of carbon fiber and glass fiber composites were observed at 30°C (room temperature) and RT+80°C for 6 hours. All the tests were performed with the strain rate 5.2 mm/min and all the stress-strain results of each specimen are shown in Figure 3 (a) and Figure (b). Besides, the two different curing temperature of carbon fiber composite, the high flexural strength of CF80 is higher than CF30 (almost 17%). The elongation of CF80 composite is slightly shorter than CF30 and it can be considered the heating at the oven for 6 hours with 80°C released all the moisture and evaporate. In contrast, the young modulus of CF80 is 28% higher than CF30. For glass fiber composite, its the flexural strength was 240 MPa for GF30 and 326 MPa for GF80 whereas the elongation rate also increases by 24% as shown in Figure 3 (c) and Figure (d). Young Modulus of both GF30 and GF80 did not show a significant change but it slightly increased later, as shown in Figure 3 (e).

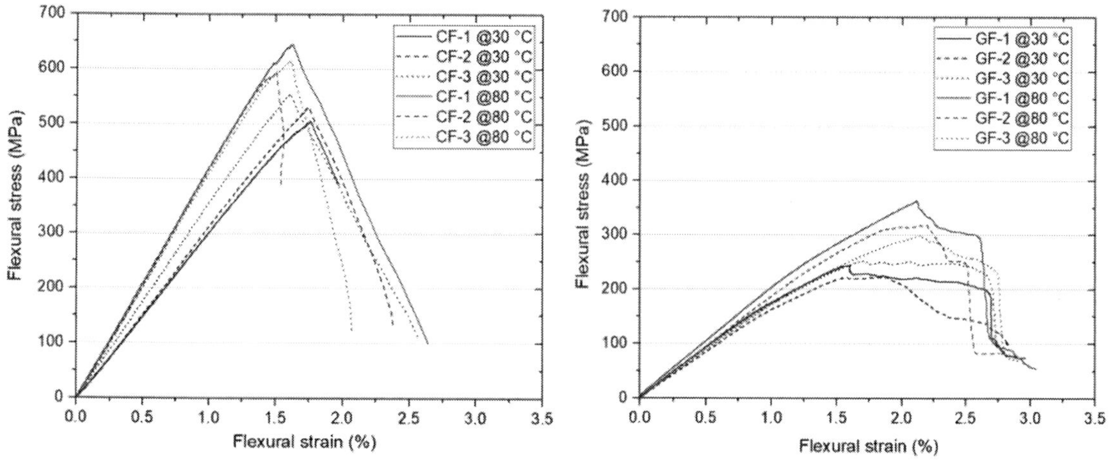

(a) Three specimens of carbon fiber composite (b) Three specimens of glass fiber composite

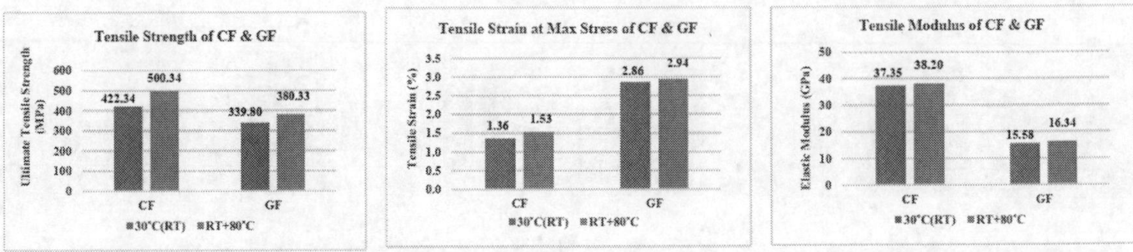

(c) Ultimate flexural strength (MPa) (d) Flexural strain (%) (e) Flexural young modulus (GPa)

Figure 3. Effect of curing temperature on flexural test results

3.3 Indentation test results on carbon fiber and glass fiber with different curing temperature

In the indentation test, 10mm^2 of composites were prepared by band saw (electric saw) and test 5 times in different location with 20 μm of the triangle shape spindle diameter. It was found that the hardness of composite and modulus of composites were investigated with two different curing temperature such as room temperature and 80°C for 6 hours. At 30°C cured result, glass fiber composite has the lowest depth at 4.5 μm where carbon fiber is the deepest depth at 6.5 μm. Glass fiber fabric contains a plain weave one by one bundle. The filament bundle thickness is much smaller, but the composite was built up by 16 layers of fabric. It will lead to difficulties for a spindle to penetrate through the composite. Concerning carbon fiber, the woven in plain weave and the layers were 12 layers fiber in composite and the filament bundle, and its amount is smaller than the glass fiber. Therefore, it will be easier to penetrate through composite.

Figure 4 (c), Figure (d) and Figure (e) show the hardness, max depth and modulus of all type of fibers composite. The results from the composite show that it is cured at 80°C for 6 hours, the depth of CF80 was less than 1.7 μm. compared to CF30. While the depth of GF80 slightly increased around 0.4 μm because the glass fiber composites increased in ductility based on the increase of curing temperature. In addition, the results of the higher curing temperature showed higher hardness. The results of the modulus in carbon fiber composite contrasts with the glass fiber composite result. The results remained unchanged when the curing temperature was increased. The main point of higher hardness and modulus might be higher adhesive cross-linking between fiber and epoxy. The spindle dent area image of both composites is shown in Figure 4 (f) with different curing temperature.

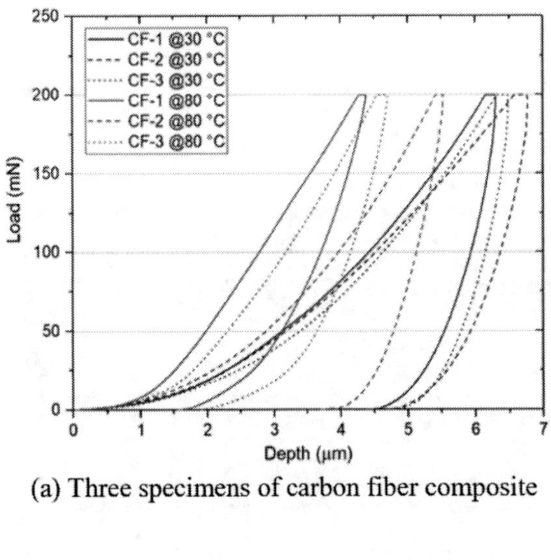

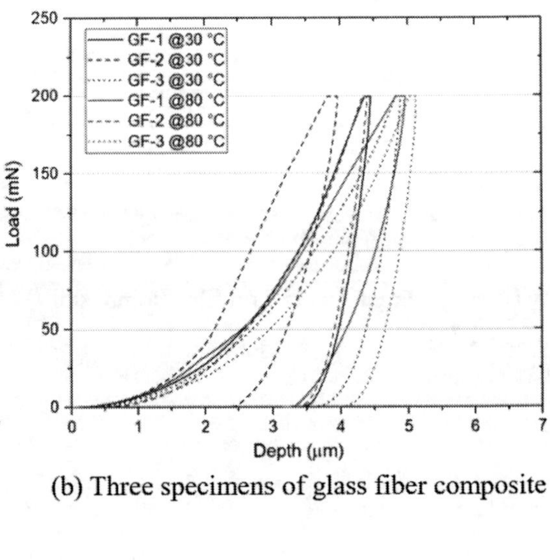

(a) Three specimens of carbon fiber composite (b) Three specimens of glass fiber composite

IOP Conf. Series: Materials Science and Engineering **886** (2020) 012053 doi:10.1088/1757-899X/886/1/012053

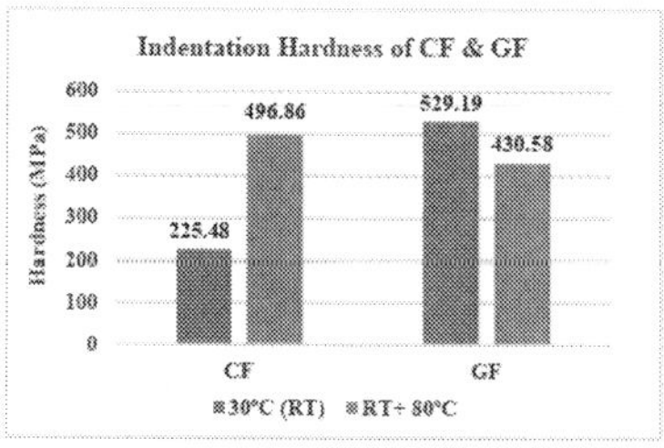

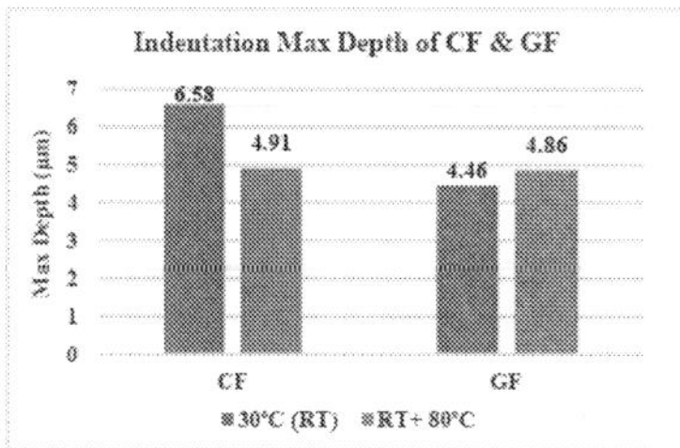

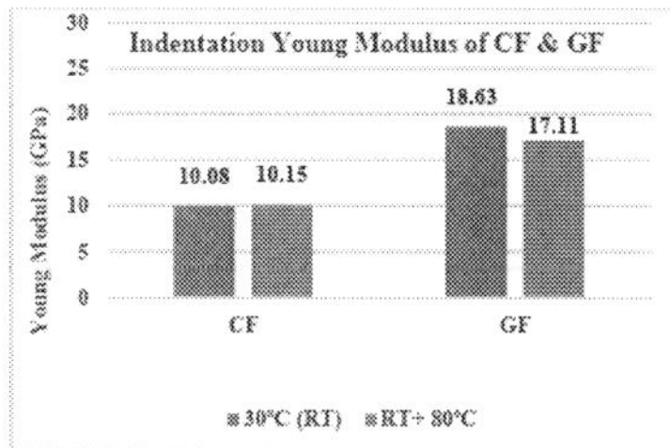

(c) Indentation hardness (MPa) (d) Indentation max depth (µm) (e) Indentation young modulus (GPa)

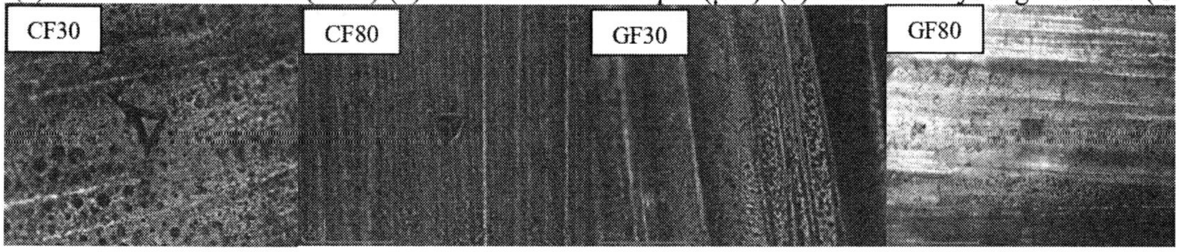

3.4. *Field-Emission Scanning Electron Microscope (FE-SEM) Analysis*

The fracture mechanism of carbon fiber and glass fiber composite was observed by using SEM image analysis at 30°C and 80°C cured fracture mechanism especially on tensile test failure, as shown in Figure 5. The fracture mechanism of carbon fiber composite was from vertical direction laid up fibers, it may start cracking when the load was applied the vertical axis of tensile force. Regarding the fracture mechanism of CF30, it indicated that several voids and less resin were attached between fibers and matrix. Whilst the CF80 is more compact than the CF30 because high temperature can remove the moisture and make brittle on composite, as shown in Figure 5 (b) and Figure (c). As can be seen in Figure 5 (a), carbon fiber was found to be a narrow channel around the fiber surface. A possible explanation for this might be that the mechanical bonding with a matrix can be a composite. Also, it was cured at a higher temperature for a long time so that it could be a brittle composite.

For glass fiber composite, their fracture area was quite large compared to carbon fiber composite due to the main extracted fibers from the matrix. Figure 5 (e) shows several defects and cracks on the cross-section of GF30 surface because the voids from the composites are one of the factors that allowed the cracking of the composite. In contrast, the results of the GF80 revealed a different failure condition compared to GF30 because the fiber and matrix are bonding tightly than the curing room temperature. And, the results were found to be less pull out fiber in this fracture in Figure 5 (f). The matrix holds the fiber tightly for not being pulled off from external force at high temperature. However, the results indicated that the fibers kept pulling out in GF80 due to the surface of the smooth condition of the glass fiber along the filament axis, as shown in Figure 5 (d). Therefore, the pull-off fibers are become breakage easily due to a very weak in the chemical reaction between the fiber and the matrix.

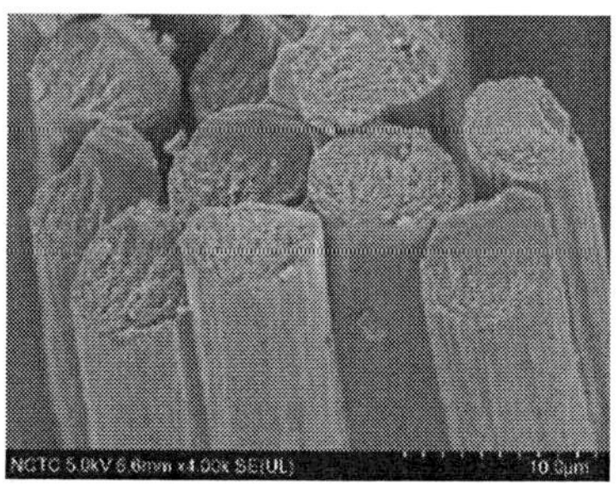

(a) Carbon fiber filament

(b) CF30 composite

(c) CF80 composite

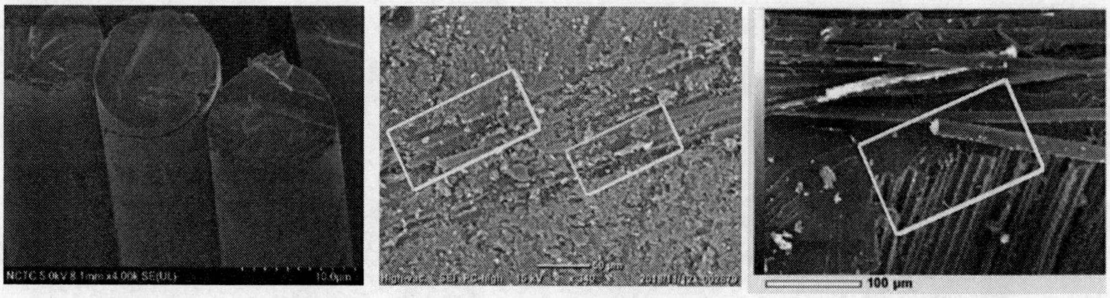

(d) Glass fiber filament (e) GF30 composite (f) GF80 composite

Figure 5. FE-SEM image analysis of raw fiber filaments and fracture of composites

4. Conclusion

The present study was designed to determine the effect of carbon fiber and glass fiber reinforced epoxy polymer matrix-based composites at different curing temperatures ranging from 30°C (RT) and 80°C. The SEM method was selected to identify the mechanical properties such as tensile, flexural, indentation, and morphology of the fracture mechanisms of the composites. The element composition of fibers was also observed using SEM-EDS.

During the tensile testing, carbon fiber and glass fiber composites are improved in tensile strength 18% and 12% with the increase of curing temperature. For the flexural test, both fiber composites are significantly increased in flexural strength 17% and 36% but it does not affect the modulus of flexural strength of glass fiber composite. In the indentation test, the hardness of CF80 was two times higher than CF30 while the hardness of GF80 was slightly 19% lower than CF30.

The fracture mechanism of composites and fiber filaments were observed by using SEM image analysis. The results from the carbon fiber composite at 30°C cured show that several voids could start cracking easily when the load is applied. The main failure mode is pulling out of the fiber due to the chemical bonding fiber and matrix were not well interlocked. At 80 °C temperature cured, the composite becomes brittle due to the increase of cross-linking in the matrix resulting in higher strength. However, it can create holes and cavities since fibers are shrunk at high temperature. For glass fiber composite, the fracture mechanism was quite different because glass fiber composite is broken gradually at load applied to point and break at the outer surface of the composite. This study set out to represent the ductile behavior of glass fiber composite with a large fracture area.

Acknowledgements
The authors would like to thank International College at King Mongkut's Institute of Technology Ladkrabang (KMITL) for granting a research fund. We also would like to thank the National Metal and Materials Technology Centre (MTEC), National Science and Technology Development Agency for providing facilities and their big support. We are thankful to be offered materials and poly plastics as well as being provided equipment from Somboon Advance Technology Company Limited.

References

[1] Huang X 2009 *Materials* **2**(4) 2369–2403
[2] Dhawan V Singh S and Singh I 2013 *Journal of Composites* **2013** 1–8
[3] Shinde S and Aniket S 2015 *International Journal of Engineering Research* **4** 446–449
[4] Pramendra K B Singh I and Madaan J 2012 *Journal of Thermoplastic Composite Materials* **27**(1) 52–81
[5] MorganP 2005 *The Forms of Carbon* (Toylor & Francis) 15–63
[6] Dwight D 2000 *Glass Fiber Reinforcements http://10.1016/B0-08-042993-9/00046-2*
[7] Fazeli M Jennifer F and Simão R A 2018 *Composites Part B: Engineering http://163. 10.1016/j.compositesb.2018.11.048*
[8] Rani E Valeria L S and Anastasia M 2017 *Smart Composites: Mechanics and Design* (Taylor & Francis)
[9] Enns J H 1993 *Creep and Mechanical Properties of Carbon Fibre Reinforced PEEK Composite Material* Thesis (Canada:Universitv of Manitoba)
[10] Bascom W D Yon K J Jensen R M and Cordner L 1991 *The Journal of Adhesion* **34**(1-4) 79–98
[11] Neslihan O Mimaroğlu A and Murat O 2015 *Acta Physica Polonica A* **127** 1032–1034
[12] Hardiman M and McCarthy C T 2015 *Proc. the 19th International Conference on Composite Material 2015* (Canada) 1–10
[13] Inder J Singh P Sehijpal S and Vikas D 2018 *Journal of Natural Fibers* **15**(5) 687–696
[14] Cui S Cui C Xie J Liu S and Shi J 2018 *Sci Rep* **8** 2364

IOP Conf. Series: Materials Science and Engineering **886** (2020) 012054 doi:10.1088/1757-899X/886/1/012054

Development of On-The-Machine Robotic Mold-Maintenance System by Utilizing Electro Discharge Deposition Technology

Takuma Mutaguchi[1], Kiyoshi Yanagihara[1], Kouji Akashi[1], and Kensuke Tsuchiya[2]

[1]National Institute of Technology, Ariake College, 150 Higashihagio Town, Omuta city, Fukuoka, 836-0097, Japan
[2]Institute of Industrial Science, The University of Tokyo, 4-1-6 Komaba, Meguro-ku, Tokyo, 153-8505, Japan

* Corresponding Author: m51137@ga.ariake-nct.ac.jp

Abstract
Maintenance of mold in a regular schedule has emerged as powerful platforms for quality assurance of the product. One of the greatest challenges is it may cause a dangerous situation for workers in the process of repairing mold using manual operation. This study set out to investigate the usefulness of a robotic system based on the on-the-machine repair of the worn mold without detaching and attaching the mold. Specifically, an electrodeposition applicator for repairing is attached to the hand of the work conveying robot which is installed near the machine tool. This study employed a vertical articulated robot, which is equipped with various sensors. Fundamental electrodeposition experiments were performed using the constructed robot system. The results from the experiments showed the feasibility of the developed system.

1. Introduction
Manufacturing methods are fast becoming a key instrument in the manufacturing processes. Some processes are considered important in mass production, such as pressing, injection molding, and casting. In particular, regular maintenance of the mold is prerequisite to keep the product qualified. In the case of mass production, spare molds are usually prepared for the regular maintenance of mold. On the contrary, if the mass production is not needed, an immediate repairing of mold wear and defects and modification of the design of the mold in the correction of processing errors are required. There is an urgent need to address the safety problems caused by maintenance of the mold which is installed on the machine tool. For example, workers are risky to have a burn by touching hot parts, receive injuries by falling on heavy parts or having cramps (see Figure 1).
This study uses a robot system to make awareness of the use of the on-the-machine maintenance of molds. The process begins by installing the robot system near a machine tool in the production line because the robot is used to transfer a work piece. Its duties involve pressing, injection molding, and die casting. In this process, we only focus on the process which the robot transfers workpiece in front of a machine. Recently, Verl et. al. (2019) studied the effects of utilization of industrial robot [4]. Considering all of this evidence, it seems that the industrial robot can be not only applied for transferring of the workpiece but it can also be used in machining and tooling. Using this approach, researchers have been able to know the concept of robot repairing system.

The goal of the robot system is that the robot itself replaces the end effector, recognizes mold defects such as wear, and performs on-machine repair of the mold. In the study, the robotic hand finishing was proposed in the paper [4]. Over the past decade, most research in technology has emphasized the use of the robot in trimming, splaying, and polishing, and it evolves today's technology.
On the other hand, in recent years, laser, electron beam, and micro-welding that enable us to control material deposition with low heat input. The mold repair options have been paid more attention [3]. Therefore, using installed Electro Discharge Deposition (EDD) technology in the robot can help perform a function of transferring and repairing mold.

Figure 1. The instance of mold repairing done by an engineer [1]

To use the mentioned concept, the first step in this process was to develop a robotic system that can build up metal on scratches using EDD technology and use in the experiment. The feasibility of on-the-machine mold-maintenance will be also discussed in this paper.

2. On-the-machine robotic mold-maintenance system

2.1. System Outline

Figure 2 shows a set-up experiment of a metal build-up using the vertical articulated robot. The vertical articulated robot used in this study is RS20N and it was made by Kawasaki Heavy Industries. EDD device, so-called Micro-depo, that was made by Techno-Coat Co. Ltd. was selected for deposition of metal build-up and is installed on the robot arm as the end effector. Micro-depo is a coating device that was developed to repair the mold. The control parameter of the Micro-depo are voltage, capacitance, and frequency. The stage of assuming mending scratch on mold, the SD-01 type of electrode was selected in the experiment of the study. This electrode can provide deposition of metal that is used to build-up on various materials, such as SCM, SC, SUS. The material of electrode is made of tungsten carbide and cobalt, and it can be applied to the process of mold repair.

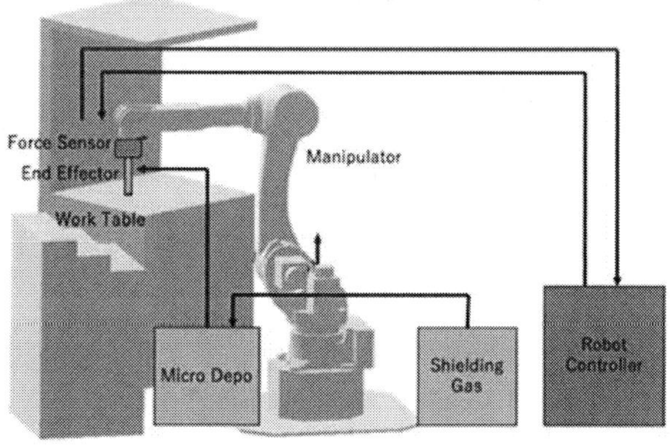

Figure 2. The process of system development

IOP Conf. Series: Materials Science and Engineering **886** (2020) 012054 doi:10.1088/1757-899X/886/1/012054

2.2. Method of Electro Discharge Deposition (EDD) Process

EDD is a process that has a composition of electrode and was melted by an electric spark generated between workpiece and electrode. It was also deposited on the surface of the workpiece.

The electrode material is heated to 8,000-25,000 [°C] at the contact portion. It functions by discharging the direct current for a very short time of 10^{-6} to 10^{-5} seconds in a cycle of 10^{-3} to 10^{-1} seconds. The heated electrode turns into a plasma and metallurgically transfers to the work surface. The transferred electrode material is alloyed and can be overlaid by depositing on the work surface (see Figure 3-a). In this stage, an appropriate gap is required because it can generate a spark between the workpiece and the electrode. This gap is called the discharge gap. Regarding the sparks, they do not occur when the workpiece and the electrode are in contact (the discharge gap is 0) [3]. When the method of Micro-depo EDD is used, the vertical position of the electrode is frequently moved up and down. Then, it continues to process. The electrodeposition is adjusted so that the discharge gap is suitable to generate a spark.

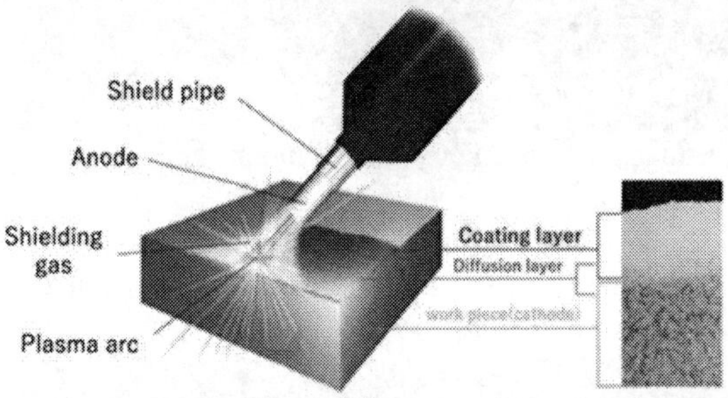

(a)

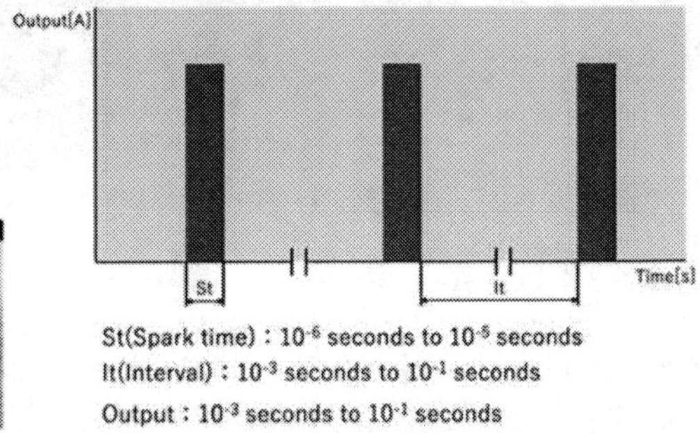

(b)

Figure 3. Illustration of replicator unit and discharge cycle[1]

There are two advantages of utilizing of Micro-depo EDD method. One is to investigate a low input of heat. The mold surface is less affected by heat i.e. thermal strain, cracks, sinks, etc. Because the spark time is very short (10^{-6} to 10^{-5} seconds) but the interval between discharges is (10^{-3} to 10^{-1} seconds) is long for the spark time (see Figure 2-b). It was found that other show a high adhesion that does not peel off after the build-up. The reason is that the electrode is alloyed and deposited on the workpiece. At the same time, it forms a strong diffusion layer like grown-up root below the surface (see Figure 2-a). In addition, the materials of plastic (S55C, SCM445, SCM430, etc) are generally adopted [2]. Therefore, there is an advantage of using electric conductors including those steel materials because they can be utilized as a workpiece by proper choice of electrodes.

2.3. Motion programming for die repair

Micro-depo EDD equipment is applied to repair defects of mold. To utilize it as an end effector of the robotic system, it is necessary to program the tool move as a human operator does. Figure 3 provides the programmed motion of the basic built-up motion of the electrode. To begin this process of mold repair operations, the electrode is contacted with the surface of the work, and the electrode is moved in the horizontal direction by the designated distance of the move. When a spark is generated from the tip of the electrode, the melted electrode is transferred to the surface of the workpiece. Once, the electrode leaves for upward in the vertical direction and returns to the start point of this sequence, the cycle is completed. Following the next cycle, after positioning of shortened electrodes is compensated, the sequence is repeated until the height correction of the deposited is in the vertical direction. Table 1 presents the parameters that can affect mold repair processing.

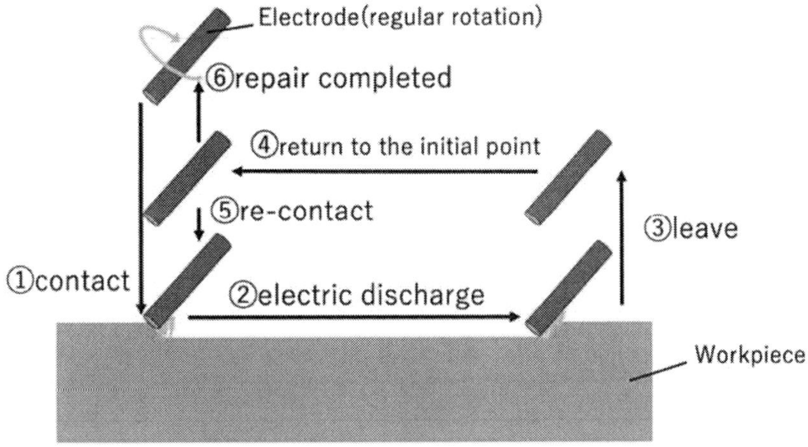

Figure 4. Illustration of repair sequence

Table 1. Control element

	control method	Control range
Cycle count[Times]	program	-
Discharge gap[mm]	program	-
Feeding speed[mm/s]	program	-
Height correction[mm]	program	-
Base material and electrode angle[°]	program	0 ~ 90
POWER[V]	Microdepo	50 , 100 , 150
Capacitor capacity[μF]	Microdepo	10 ~ 100
Spark frequency[Hz]	Microdepo	60 ~ 1600
Electrode rotation number[RPM]	measurement	2500 ~ 8000
Electrode diameter[mm]	-	1.6 , 2.4
Atmosphere	-	Argon gas:2~15[L/min]

3. Fundamental deposition experiment

To understand the mold repair cycle, the robotic system is used as a trial. Length error of deposition against to the programed length is investigated. In this process, it is assumed that there is one to repair a scratch of 10 mm length, 1 mm width, and about 0.5 mm depth on the surface of S35C, 10.0 [mm] of feed and this stage was programmed.

After the deposition experiment, the deposited line is obtained as shown in Figure 4. It is clear that the electrode material is deposited on the surface of the workpiece because it demonstrates the state of the weld bead.

The length of this deposited line was measured as shown in the red line (see Figure 5). As can be seen in Table 2, the length errors obtained under various conditions using tool feeding speed and capacitor capacity of EDD controller.

Figure 5. obtained surface in deposition experiment

Table 2. Length error in various control parameters

length errors[mm]

		capacitor capacity[μF]		
		10	50	100
feeding speed[mm/s]	1	+1.09	+1.21	+0.99
	5	+1.01	+1.09	+1.11
	10	+0.94	+0.95	+1.19

The results revealed that the feed rate and capacitor capacitance have less influence on the length of the obtained deposition length under this experimental condition. Meanwhile, the length was about 0.9 to 1.2 mm and was longer than the distance of the tool travel.

4. Cause of error in deposit length

To identify the error that affects factor of the deposit length, the generation of spark at the end of the electrode is observed by using a microscope. The results indicated that the melted electrode composition can transfer to the surface on the workpiece. The obtained microscope image is shown in Figure 6. It also shows that the sparks are generated inside from the acute angle side by the end of the electrode. A possible explanation for this might be that the length of deposition becomes longer than the length of the programmed tool feed in the X-direction. There are, however, other possible explanations. It can generate an extra spark of the electrode which can produce deposition of the melted droplet of the electrode.

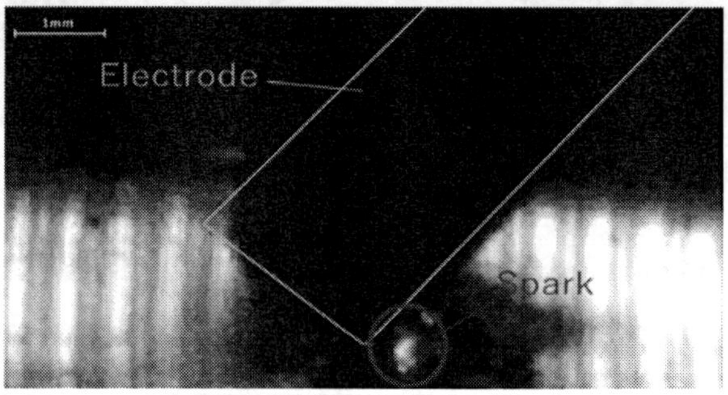

Figure 6. Spark generating point

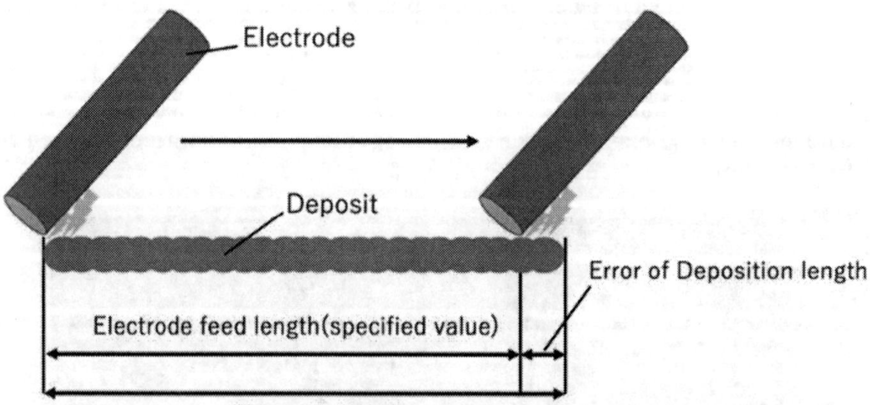

Figure 7. Illustration of mechanism about Occurrence of error

5. Conclusion

The aim of the present research was to examine the on-the-machine maintenance of molds. This study has shown that the robotic system can deposit metal on scratches by using the EDD device so-called Micro-Depo. The feasibility of the on-the-machine robotic mold-maintenance is also discussed. This study has found that generally, when moving the electrode in the horizontal direction by the specified value, the deposit length error is about 0.9 to 1.2 [mm]. It did not affect on the feeding speed and capacitor capacitance (1). The second major finding was that since the sparks are generated from the acute angle side of the tip of the electrode, extra spark is generated for the horizontal feed of the electrode resulting in the deposit length error (2). Further research could usefully explore how to build the on-the-machine maintenance of mold by developing of the mold-repair cycle made by repeating build-up deposition-processing of the straight beads. Further work could also be conducted using a deposition method to guarantee a sufficient hardness of deposited area in mold repairing.

Acknowledgments

The authors would like to acknowledge that some images and illustrations related to experimental equipment and molding technology were provided by Techno Coat Company Limited. We are thankful for the support from Mr. Takumi Tanaka, the 5th-year student in the department of mechanical engineering at NIT Ariake.

References

[1] Kenta A 2019 *On the build up repair of plastic molding dies http://www.technocoat.co.jp* accessed 15 June 2019
[2] Jhavar S Paul C P and Jain K 2013 *Engineering Failure Analysis* **34** 519-535
[3] Kobayashi T and Kunieda M 2002 *Japan Society of Electrical-Machining Engineers* **36** 11-17
[4] Okada Y 1983 Daido Steel Co., Ltd. **55** 68-73
[5] Verl A Valente A Melkote S Brecher C Ozturk E and Tunc L 2019 *CIRP Annals* **68**(2) 799-822
[6] Yongkun P and Hirata Y 2015 *Japan Welding Society* **33** 49-54
[7] Onori T Yanagihara K Nishijima R and Mitsunobu T 2017 *Proc. JSME annual meeting* 81-83
[8] Onori T Yanagihara K and Mitsunobu T 2018 *Proc. JSME annual meeting* 291-292

IOP Conf. Series: Materials Science and Engineering **886** (2020) 012055 doi:10.1088/1757-899X/886/1/012055

The Effect of Particle Shape on Magnetic Field-Induced Rubber-Like Behavior of Ni-Mn-Ga/Silicone Composites

P. Sratong-on[1,*], V.A. Chernenko[2-5], and H. Hosoda[2]

[1] Automotive Engineering Program, Faculty of Engineering, Thai-Nichi Institute of Technology, 1771/1 Pattanakarn Rd. Suanluang, Bangkok, 10250,Thailand
[2] Institute of Innovative Research (IIR), Tokyo Institute of Technology, 4259 R2-27 Nagatsuta Midori-ku, Yokohama, 252-8503, Japan
[3]Ikerbasque, Basque Foundation for Science, Bilbao, 48013, Spain
[4]BCMaterials, Basque Center for Materials, Applications and Nanostructures, UPV/EHU Science Park, Leioa, 48940, Spain
[5]University of Basque Country (UPV/EHU), Bilbao, 48080, Spain

* Corresponding Author: pimpet@tni.ac.th

Abstract
The single crystalline of the prototype Ni-Mn-Ga ferromagnetic shape memory alloy exhibits a huge magnetic field-induced strain of 6% due to the re-orientation of twin variants. Therefore, it is promising as the magneto-driven actuator material. However, the magnetic field or mechanical stress applied along the orthogonal direction is necessary to restore the deformed bulk single crystal into the original shape. In order to avoid an application of the extra fields, the Ni-Mn-Ga particles/polymer composite is an alternative solution supported by the concept of the accumulated stress in a polymer matrix which drives a deformation recovery of the composite during switching off of the magnetic field. The objective of this study is to clarify the effect of embedded particle shape on the deformation recovery of Ni-Mn-Ga particles/silicone composites by using the finite element method (FEM) and to investigate the local stress distribution in a polymer matrix between either spherical or rectangular shape particles pairs, whereby to provide the guidelines for design/optimization of the magneto-strain-active Ni-Mn-Ga particles/polymer composites. The case studies of the simulation are divided into an isolated particle and a pair of particles, the particles being positioned either parallel or perpendicular to the applied magnetic field direction. Particularly, the simulations reveal that in case of 200 μm of the inter-particle distance in the pair of spherical particles aligned perpendicularly to the applied field, the polymer layer between particles generates the compressive recovery stress of -0.32 MPa, which is insufficient to restore the deformation of the embedded Ni-Mn-Ga particles during removal of the magnetic field. By contrast, the strain recovery effect can be achieved for the rectangular particle pairs, generating the stress concentration in a matrix of about -0.5MPa, in a similar condition and arrangement.

Keywords: Ferromagnetic shape memory alloys, Ni-Mn-Ga/polymer composites, Finite element method

1. Introduction
The single crystalline Ni-Mn-Ga martensite, exhibiting a tetragonal structure with lattice parameters ratio of $c/a < 1$, is a prototype of the ferromagnetic shape memory alloys. It shows 6% of the magnetic field-induced strain due to the twin variant reorientation [1–3]. This deformation can be recovered by either stress or magnetic field applied along the orthogonal direction with regard to the initial direction of the magnetic field. Thus, unlike the conventional shape memory alloys, e.g., Ni-Ti alloys, which exhibit the strain recovery during reverse martensitic transformation [4,5], the actuation of Ni-Mn-Ga is not limited by the heat. As a result, Ni-Mn-Ga is a promising candidate for the magnetically driven high-frequency actuators. However, for such a cyclic actuation, an additional application of the magnetic field or stress along the orthogonal direction is needed to periodically restore the initial shape of Ni-Mn-Ga single crystal. This leads to more complex design and to the efficiency reduction in some applications. Besides, the fabrication of single crystal is a time-consuming and costly process. To avoid these problems, an idea of the fabrication of Ni-Mn-Ga particles from polycrystal [6,7] and their embedding into a polymer, whereby creating a particulate-filled polymer composite, was pursued by many researchers [8–10]. Especially, one of the main advantages of the composite is the storage of stress in the matrix during deformation of an embedded particle under applied magnetic field. The stress in the matrix could be turned into a force to propel shape recovery of individual particles under decreasing of an applied field.

An idea of stress accumulated in the matrix during a mechanical stressing of the Ni-Mn-Ga/silicone composite has been demonstrated in Ref.[11]. Recently, we have shown that elastic force in the matrix which exerted on particles during on/off of the applied magnetic field to the Ni-Mn-Ga/silicone composites depends on the inter-particle distance and particles orientation [11,12]. Therefore, it could be deduced that both the distance- and arrangement-dependent inter-particle interactions cause a non-uniform deformation. This implies the existence of the non-uniform stress fields in the local areas of the matrix, also giving rise to the change in the apparent stiffness of composite, as a global aspect. Hence, we can consider that the non-uniform stress fields in matrix contribute to the strain recovery of Ni-Mn-Ga/polymer composite during removal of the magnetic field. Such a reversible magnetic field induced straining of composite has been found experimentally and called "the magnetic field-induced rubber-like behavior" [12]. It has to be emphasized that this behavior was observed for the irregular shape particles contained in the studied composite [12]. On the other hand, it is known from the bulk of literature on other types of composites that the relationship between particles shape and internal stress considerably affects their mechanical properties and deformation behavior [13–15]. In the case of Ni-Mn-Ga/polymer composite, the influence of the particle shape on the internal stress field in a matrix is unknown and is needed to be clarified.

The objective of this study is to elucidate the effect of particle shape on both the local stress distribution in a polymer matrix and the reversible deformation of Ni-Mn-Ga particles/silicone composites by using a finite element method (FEM). The analysis is performed for the two representative shapes of particles, such as sphere and rectangular prism. The guidelines for a design/optimization of the magneto-strain active Ni-Mn-Ga particles/polymer composites are summarized.

2. Materials and Methods

2.1. Assumptions, modelling and case studies of simulations
The numerical model, used throughout this study, is based on the following assumptions and characteristics of the Ni-Mn-Ga single-crystalline particles [16].

i) The particles have a tetragonal martensitic structure with the lattice parameters ratio $c/a < 1$ at room temperature. The direction of an easy axis for magnetization is parallel to the c axis of the crystallographic unit cell.

ii) The principal axes of the cubic unit cell in the parent phase of a particle are parallel to x, y and z axes of the orthogonal coordinate system.

iii) Particles contain three different crystallographic directions of martensitic variants. When the magnetic field is applied, one variant, where *c*-axis is parallel to the direction of the applied field grows at the expense of the others.

iv) The deformation of Ni-Mn-Ga single crystal in the martensitic phase follows the plastic deformation criteria. Particularly, the total volume of a single crystal is conserved.

To satisfy the aforementioned assumptions in simulation modelling, the maximum crystallographic strains of Ni-Mn-Ga single crystal represent the strain input parameters along each of the x, y and z axes. That is, if the magnetic field is applied along the x-axis, -0.06 is the strain input along the x-axis, while +0.03 is the strain input along each of the two other orthogonal directions, to conserve the total volume.

In this study, the three representative configurations of the embedded Ni-Mn-Ga particles in the silicone matrix concerning the magnetic field direction are divided into an isolated particle and the pairs of particles that align either parallel or perpendicular directions to the applied magnetic field. The particle pair represents an elementary unit modelling the neighborhood of the close particle in the transversal to the particles chain direction. The spherical and rectangular shapes of particles are considered to investigate the influence of the particle shape on the stress distribution in a matrix. In the simulations, the composite contains a 30vol% of the Ni-Mn-Ga filler with the sizes of spherical and rectangular particles, equal to 100 μm and 150 μm, respectively. These particle sizes correspond to the minimum cross-sectional areas described in Refs.[6,12]. All simulation cases in this study are carried out using a commercial finite element software ANSYS 18.0.

The material constants of the Ni-Mn-Ga single crystal are 3100 MPa, 0.33, -0.06 and +0.03 for Young's modulus in martensitic phase, Poisson's ratio and crystallographically largest compressive and tensile strain, respectively. The Young's modulus of 1 MPa is a material constant for the silicone rubber. Since the model of the three configurations of particles is symmetrical, one-eighth of the model with the symmetry boundary conditions are taken into account, as shown in Figure 1(a) for the isolated spherical particle/silicone composite. The one-eighth model with a quadratic hexagonal mesh shape and symmetry boundary conditions in xy, yz and xz planes are shown in Figure 1(b). The minimum size of the mesh is 10 μm.

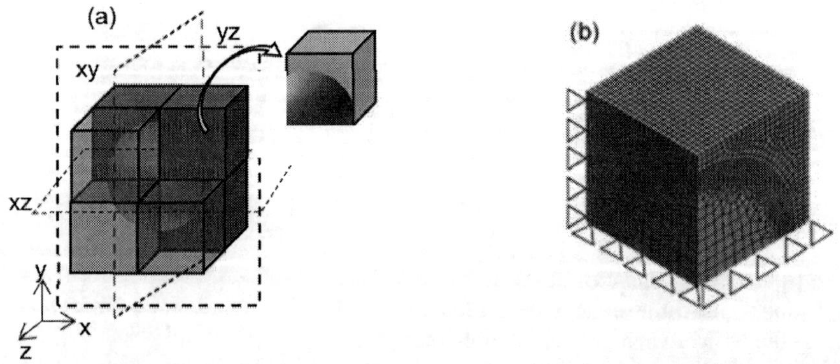

Figure 1. (a) The case of isolated particle/silicone composite with the one-eighth of model and symmetry boundary conditions. (b) The mesh shape of the one-eighth model of isolated particle/silicone composite. Fine mesh area at the boundary between the Ni-Mn-Ga particle and matrix is illustrated by the green color.

IOP Conf. Series: Materials Science and Engineering **886** (2020) 012055 doi:10.1088/1757-899X/886/1/012055

3. Results and Discussions

3.1. Results of simulation

3.1.1. Isolated particle/silicone composite

A 30 vol%-isolated particle/silicone composite, containing either spherical or rectangular shape particle (Figure 1), was simulated. Figure 2 illustrates the contour of stress distribution in a matrix of the isolated particle composite along the direction of the applied magnetic field. Isometric view of the model of spherical shape filler is shown in Figure 2(a). Since the matrix is unable to deform under a magnetic field, only particle contracts along the x-axis, in direction of the applied field, and elongates in the orthogonal direction. According to Figure 2(a), when particle contracts by 6%, a -0.45 MPa of compressive stress, illustrated by red colour, generates at the interface between the spherical particle and matrix. It slightly decreases from the interface to the edge of the composite. In the case of a rectangular particle, the largest compressive stress of -0.70 MPa, illustrated by orange colour, generates at the boundary between the rectangular particle and matrix. In addition, it reduces as the distance from the interface increases [16].

The composite can recover its initial shape due to the netback stress, which accumulates in the matrix during application of the magnetic field and releases during removal of the latter. This netback stress is defined by the difference between the absolute value of compressive stress at interface and stress at the edge of the composite. Therefore, netback stresses for one spherical particle (Figure 2a) and one rectangular particle (Figure 2b) in the silicone composite are -0.33 MPa and -0.40 MPa, respectively. It is known that the stress value close to -0.5 MPa, or more, is necessary to trigger a reorientation of the martensitic variants [2,17], hence, higher or equal bias stress for recovery deformation of the single crystal is necessary to be applied in the orthogonal direction. Therefore, the estimated values of the netback stress in the matrix influencing either isolated spherical or isolated rectangular particle is insufficient to recover the shape of the composite.

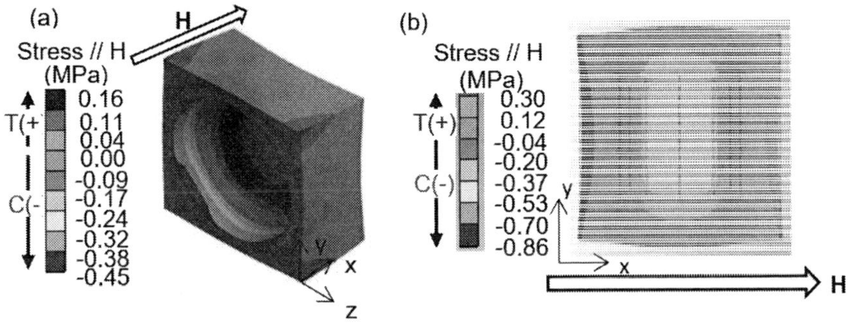

Figure 2. Stress distribution in the matrix of the 30vol%-isolated particle/silicone composite which is parallel to the direction of the applied magnetic field: (a) isometric view of the spherical particle; (b) front view of the isolated rectangular particle. T and C stand for the tensile and compressive stress, respectively.

3.1.2. Configurations of particles pair aligned parallel to the applied magnetic field

Figure 3 illustrates the stress distribution in a matrix containing a pair of particles; both two particles in the pair are either spherical or rectangular, with a 200 μm of the inter-particle distance aligned parallel to the direction of the applied magnetic field. Figure 3(a) shows that, as the particles contract along the applied field direction, this generates -0.33~-0.43 MPa of the compressive stress (yellow contour) at the interface of particles and matrix which spreads between the two nearby spherical particles. The maximum compressive stress of -0.64 MPa (the red color in Figure 3a) occurs at the circumference of a big circle of spheres. Figure 3(b) shows that -0.35~-0.61 MPa of the compressive stress (the orange color) generates at the interface between the rectangular particles and matrix. Unlike spherical particles, compressive stress does not spread between the two rectangular particles, it spreads to the right and left of the particles along the vertical direction (the yellow contour). Since the matrix is incapable to deform by the magnetic field, it tends to resist a contraction of the rectangular particles along the direction of the applied field. As a consequence, a 0.2~0.5 MPa of the tensile stress (the light green color in Figure 3b) occurs in a matrix. Thus, the tensile stress has no significant effect upon the deformation of rectangular particles. By comparison, the spherical particles aligned along the applied field are easier to deform than rectangular particles in a similar configuration, due to the spread of compressive stress field in a matrix along with the particle pairs.

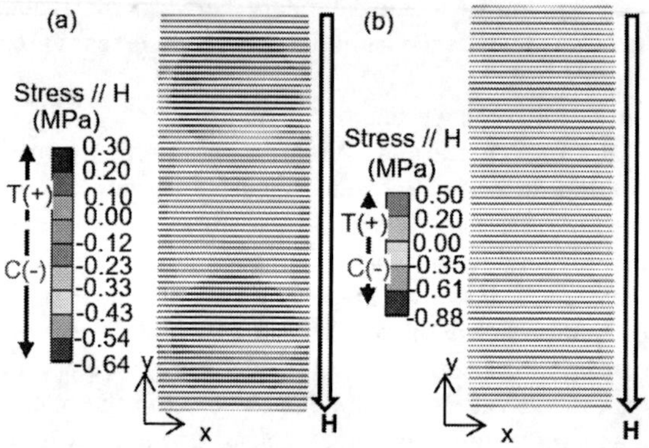

Figure 3. Stress distribution in the matrix in case of the configuration when the spherical particles pair (a) or rectangular particles pair (b) are parallel to the applied field. The inter-particle distance is 200 μm. T and C denote a tensile and compressive stress, respectively.

3.1.3 Configuration of particles pair aligned perpendicularly to applied magnetic field

Figure 4 shows the stress distribution in the silicone matrix when the spherical particles are aligned with an orthogonal direction to the applied field. The two inter-particle distances, 200 μm and 400 μm, are considered. Note that the direction of stress distribution shown in Figure 4 is perpendicular to the direction of the applied magnetic field. According to Figure 4, the particles contract along the y-axis and expand along the x-axis. As the horizontal neighbor particles are present, the particles are unable to elongate independently along the x-axis resulting in a -0.33~-0.43 MPa of the compressive stress concentration in the matrix between particles illustrated by the yellow and orange colors in Figure 4(a). This compressive stress concentration decreases as the inter-particle distance is greater indicated by the reduction of the yellow color area with a brighter shade as shown in Figure 4(b).

An occurrence of the compressive stress concentration can be explained by the following reasons. During the application of magnetic field in a perpendicular direction to the orientation of pair of particles, the particles contract along the applied field direction (y-axis in Figure 4) and simultaneously elongate along an orthogonal direction to the field (x-axis in Figure 4). In such a configuration, a presence of the horizontal neighbor particles causes difficulties to the expansion of particles. Hence, compressive stress concentration, which opposes the elongation generates in a matrix between the two particles. That is, not only the stiffness of the free matrix constrains the deformation of particles, but also an increase in the stress concentration in the matrix due to the presence of neighbor particles . As a result, the local effective stiffness of composite increases.

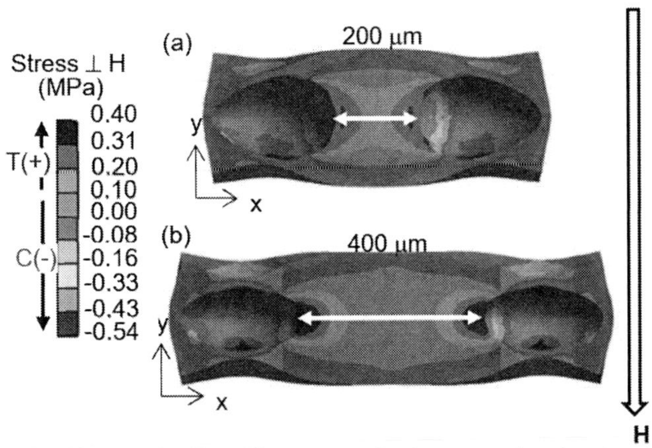

Figure 4. Stress distribution in the silicone matrix in case of pair of spherical particles aligned perpendicularly to the applied magnetic field. The direction of the stress distribution is perpendicular to the applied field. Two distances between the particles, 200 (a) and 400 μm (b), are considered. T and C denote the tensile and compressive stress, respectively.

Figure 5 shows the stress concentration in the silicone matrix of either rectangular or spherical particles pairs, oriented perpendicularly to the applied magnetic field, as a function of the reciprocal inter-particle distance. The inter-particle distance is shown along the upper horizontal axis. According to Figure 5, a compressive stress concentration in the matrix in the middle of two nearby particles decreases as the inter-particle distance becomes farther, in a hyperbolic-type manner, for both the rectangular and spherical particles pairs. The compressive stress concentration in a matrix for the rectangular particles pairs is almost two times higher than for the spherical one. The compressive stress concentration decreases to -0.36 MPa at 400 μm of inter-particle distance for the case of the rectangular particles. This value of stress concentration is close to the stress accumulated in a matrix for the isolated rectangular particle/silicone composite. Therefore, the rectangular particles pairs deform independently, similarly to the isolated particle/silicone composite. However, in case of the spherical particles, the maximum stress concentration of -0.32 MPa occurs at 200 μm of the inter-particle distance which is equal to the stress accumulated in a matrix of the isolated spherical particle/silicone composite (Figure 1a). Note that the stress restored in a matrix of the isolated particle/silicone composite is considered to be inadequate to reset the shape of the composite after shut down of the magnetic field. The statement is valid for the case of the spherical particle pairs. In sum, the Ni-Mn-Ga/silicone composites containing rectangular particles aligned perpendicularly to the applied magnetic field with a suitable inter-particle distance are superior to the case of the spherical particles in the same configuration when considering a recovery degree of the deformation of the composite during removal of the magnetic field.

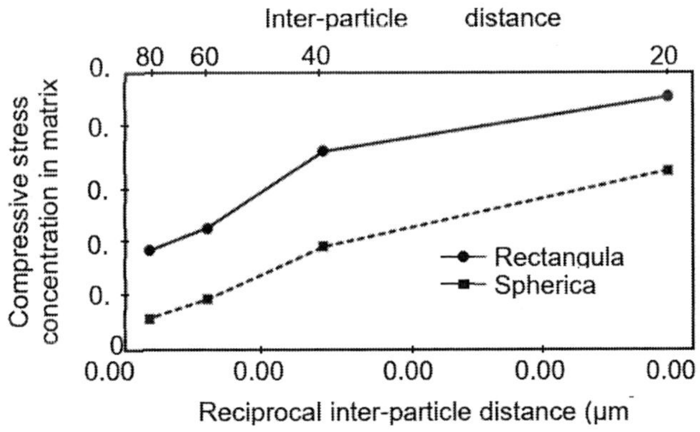

Figure 5. Compressive stress concentration in the matrix that generates due to the interaction of particles aligned perpendicularly to the applied magnetic field as a function of the reciprocal inter-particle distance. The data for spherical and rectangular particles are shown as the full squares and circles, respectively. The upper horizontal axis shows the inter-particle distance.

4. Conclusions

Recent findings of the giant and recoverable magneto-strain exhibited by the Ni-Mn-Ga particles/silicone composite in Ref. 12 were explained in that work by the outstanding role of the non-uniform internal stress in the matrix accumulated during magnetic field-induced particles straining and released as a driving force to recover the initial shapes of the particles during the switching off of the magnetic field. The stress distribution in the local areas was considered in Ref. 12 as inhomogeneous and highly dependent on the inter-particle interactions and mutual orientation of the particle chains and magnetic field direction. In the present work, by carrying out FEM simulations, we have analyzed the effect of the shape of particles on the internal stress fields generated in the matrix which facilitate the reversible magneto-strain behavior of Ni-Mn-Ga particles/silicone composite. In addition, we have found that the occurrence and change in stress in the matrix depends not only on the shape of neighboring particles, but also their configurations concerning the applied magnetic field. The following results can be highlighted in order to serve as a guide for design of the Ni-Mn-Ga/polymer composite with recoverable deformation under applied/removal magnetic field.

i) Regardless of the shape of the particles, the crystallographic easy magnetization axis (*c*-axis) in the particles should be aligned perpendicularly to the direction of the applied magnetic field.

ii) When the particles pairs are aligned parallel to the field, they can deform independently resulting in the compressive stress field spread in the matrix (the yellow area in Figure 3(a,b)). Compressive stress field spreads over a wider area of the matrix for the spherical particles than for the rectangular particles. Therefore, a composite made with the spherical particles can transfer the stress via the matrix to particles more efficiently than in the case of rectangular particles.

iii) In the case of complete magneto-strain reversibility, which is called the magnetic field-induced rubber-like behavior of composite [12], the rectangular particles aligned perpendicularly to the applied magnetic field with a suitable inter-particle distance are superior to the case of the spherical particles in the same configuration.

The fabrication method of the magneto-strain active spherical particles of Ni-Mn-Ga is under development. After elaboration of such particles and corresponding measurements, the simulation results will be compared with experiment ones.

Acknowledgments

This research project was supported by Grant-in-Aid of Scientific Research (Kiban S 26220907) from the Japanese Society for the Promotion of Science and by the Spanish Ministry MICINN & U (project RTI2018-094683-B-C53).

References

[1] O'Handley R C Allen S M 2002 *Encyclopedia of Smart Materials* (New Jersey:John Wiley & Sons, Inc.)
[2] Karaca H E Karaman I Basaran B Chumlyakov Y I Maier H J 2006 *Acta Mater.* **54** 233–245
[3] Pareti L Solzi M Albertini F Paoluzi A 2003 *Eur. Phys. J. B - Condens. Matter* **32** 303–307
[4] Krulevitch P Lee A P Ramsey P B Trevino J C Hamilton J and Northrup M A 1996 *J. Microelectromechanical Syst.* **5** 270–282
[5] Naresh C Bose P S C Rao C S P 2016 *Proc. IOP Conf. Ser. Mater. Sci. Eng.* **149** 012054
[6] Feuchtwanger J Richard M L Lázpita P Gutiérrez J Barandiarán J M Allen S M O'Handley R C 2008 *Mater. Sci. Forum* **583** 197–212
[7] Nilsén F Aaltio I Ge Y Lindroos T Hannula S P 2015 *Mater. Today Proc.* **2** S879–882
[8] Hosoda H Takeuchi S Inamura T Wakashima K 2004 *Sci. Technol. Adv. Mater.* **5** 503–509
[9] Glock S Zhang X X Kucza N J Müllner P Michaud V 2014 *Compos. Part A Appl. Sci. Manuf.* **63** 68–75
[10] Glock S Canal L P Grize C M Michaud V 2015 *Compos. Sci. Technol.* **114** 110–118
[11] Sratong-on P Tahara M Inamura T Chernenko V A Hosoda H 2018 *Smart Mater. Struct.* **27** 85024
[12] Sratong-on P Chernenko V A Feuchtwanger J Hosoda H 2019 *Sci. Rep.* 2019. **9**
[13] Lipetzky P Schmauder S 1994 *Int. J. Fract.* **65** 345–358
[14] Lee B J Mear M E 1999 *J. Mech. Phys. Solids.* **47** 1301–1336

[15] Rasool A Böhm H J 2012 *Int. J. Eng. Sci.* **58** 21–34
[16] Sratong-on P Chernenko V Hosoda H 2019 *Results Mater.* **2** 100037
[17] Straka L Hänninen H Soroka A Sozinov A 2011 *J. Phys. Conf. Ser.* **303** 012079

IOP Conf. Series: Materials Science and Engineering **886** (2020) 012056 doi:10.1088/1757-899X/886/1/012056

Development of a Transient Dynamic Finite Clement Model for the Drum Testing of a Non-Pneumatic Tire

R Rugsaj[1,2], and C Suvanjumrat[1,2,*]

[1]Department of Mechanical Engineering, Faculty of Engineering, Mahidol University, Salaya, Nakhon Pathom, 73170, Thailand
[2]Laboratory of Computer Mechanics for Design (LCMD), Department of Mechanical Engineering, Faculty of Engineering, Mahidol University, Salaya, Nakhon Pathom, 73170, Thailand

* Corresponding Author: chakrit.suv@mahidol.ac.th

Abstract
Non-pneumatic tire (NPT) or airless tire must be designed to meet the required performance such as weight, load carrying capacity and resistance to cyclic loading. The tire dynamic response is also needed to be observed since the tire are constantly subject to cyclic loading while rolling. Drum testing method is preferred over other rolling tire testing method since various factors from the vehicle suspension system and pavement structure are eliminated. This article aimed to develop the finite element (FE) model based on the transient dynamic approach for drum testing of NPT. The finite element model of NPT was created using Ogden hyperelastic material model. The FEM of NPT, which composed of Tread band and Spokes, was model using 3D hexagonal and 2D Quadrilateral elements respectively. The steel belt layers were modelled and embed using rebar elements and tying equation, respectively. The FE model of NPT was then assigned to contact with the rigid drum surface using Coulomb's friction model. The finite element analysis of drum testing method of NPT as the drum was assigned to be rolling at speed of 11 km/hr. The NPT spoke deformations at each position and the angle was compared to the experiment to prove the validity of the model. The analysis results of spoke deformation were shown to be in a good agreement with the experiment at the average error of 3.51%. The developed FE model of NPT based on transient dynamic approach can be used to evaluate the important dynamic response of NPT during rolling and can be useful in designing the NPT.

1. Introduction

Recently, the non-pneumatic tire (NPT) or airless tire was developed to overcome the disadvantage of a traditional pneumatic tire. The traditional pneumatic tire has several disadvantages such as the required maintenance of inflation pressure periodically, and a possibility to be damaged or flattened. The latter can cause catastrophic damage if the tire went flat while driving [1]. The NPT has been studied and developed over the past decade. The NPT can be designed to exhibit conventional pneumatic tires performance regardless of tire size and inflation pressure. The TWEEL, which was recently developed by Michelin, is the first commercially available NPT. The TWEEL is consisting of three main components which are 1) shear band, which can be described as a combination of tread and a solid sidewall inserted with steel belts, 2) polyurethane (PU) spoke structure, and 3) steel hub [2, 3].

The designing of NPT needs to be engineered carefully to meet the desired tire's performance such as load-carrying capacity, riding comfort, rolling resistance, etc. The optimum design of NPT to have required load-carrying capacity and vertical stiffness can be achieved easily by mean of static finite element method (hereafter, FEM) with appropriated material models [4-6]. However, the dynamic characteristics of the NPT while rolling, which including cyclic stress, strain, deformation, rolling resistance, and riding comfort, are still needed to be observed. Normally dynamic characteristic of the rolling tire can be obtained by using a drum testing method. The method is performed by pressing an inflated tire against a drum with the desired load. The drum is rotated and tire's interested properties can be measured and recorded. Drum testing method had an advantage when compared to the other rolling tire testing method since various factors from the vehicle suspension system and pavement structure are eliminated [7]. In recent years, finite element (FE) models of drum testing had been developed. The rotation system of drum and tire could be modelled by using transient dynamic analysis [8].

The tire, which is considered as a viscoelastic material, undergoes cyclic deformation while rolling. The stress and strain response under cyclic loading exhibit phase delay, which is the cause of hysteresis energy loss [9]. Thus, the development of the tire model with viscoelastic material is required to predict the dynamic characteristic of the rolling tire. The thermo-viscoelastic model of NPT with a lattice spoke was developed to predict rolling energy loss and the corresponding heat generation of NPT components. The Yeoh hyper-elastic model was obtained from the tension and compression tests, while viscoelastic material properties were obtained from dynamic mechanical analysis (henceforth, DMA) test. A 3-D stress analysis was performed using steady-state rolling analysis, then the cyclic strain energy was converted into heat to obtain temperature and compared with the experiments [10]. The effects of static and dynamic loading of NPT with honeycomb spoke was investigated using FEM. the three types of NPT with constant cell wall thickness but different cell geometries were numerically simulated to study the effects on deformation modes, stress distribution, load carrying capacity, and rolling resistance. While on dynamic loading, the effects of friction coefficient and angular velocity on rolling resistance were also studied. The maximum stress in spoke under dynamic loading was observed to be higher when compared to static loading [11].

This research aimed to develop the transient dynamic FE model for NPT. The Ogden hyperelastic model and generalized Maxwell's viscoelastic model were used to modelling the elastic and inelastic behavior of NPT components. The spoke deformation obtained from finite element analysis (hereafter, FEA) then was compared with the drum testing experiment to validate the model. Lastly, the displacement at the center of the NPT and impact force was observed and recorded from the FEA.

2. Finite Element Modelling of Non-Pneumatic Tire

The commercial NPT Tweel, which was developed by Michelin, is selected to study the mechanical behavior and dynamic characteristic of NPT in this research. The Tweel 12N16.5 SSL ALL-TERRAIN model is shown in figure 1(a). The FE software MSC. Patran was used to model the FE model of NPT. The components of FE model of NPT, which composed of 4 main components, can be concluded as follows: 1) tread, 2) shear band, 3) belt layers, and 4) PU spoke. The FE model of NPT, its overall dimension are shown in figure 1(b) and figure 1(c) respectively, while the NPT components are shown in figure 2. The hybrid formulation or Hermann element was used to model the rubber components such as tread and shear band elements of NPT. The formulation was selected to prevent the excessive stress due to volumetric locking that may occur during compression of rubber material elements by mean of separated integration of pressure and displacement field. The spoke was then modelled using thick shell approximation. The details of elements used for each NPT components can be summarized as shown in Table 1. The spoke thickness was measured from the real spoke and assigned to each element differently based on actual thickness distribution, which the average thickness was found to be 5.8 mm.
Additionally, the thicknesses at the outer and inner ring of the spokes were fixed at 6 and 7 mm, respectively corresponding to the real NPT.

The cross-section of shear band component including belt layers was obtained by mean of waterjet cutting technique, which the belt layers were found to consist of 4 main layers which were 1) the outer layer, 2) the 1st middle layer, 3) the 2nd middle layer, and 4) the inner layer. The outer layer was composed of three sublayers while the inner layer was composed of two sublayers. The reinforcing bar or rebar element was used to develop the FE model of belt layers, which was then embedded into the rubber element using tying equation [12]. The value of the degrees of freedom of the nodes in the host body element was tied based on their iso-parametric location in the elements. The approximated number of wires per unit length of 0.3582 mm^{-1} was measured and assigned to the model. The cross-section of the shear band of the FE model and real TWEEL along with the embed belt layer position is shown in Figure 3.

(a) (b) (c)

Figure 1. (a) The NPT Tweel 12N16.5 SSL ALL-TERRAIN model, (b) finite element model and (c) NPT's overall dimension.

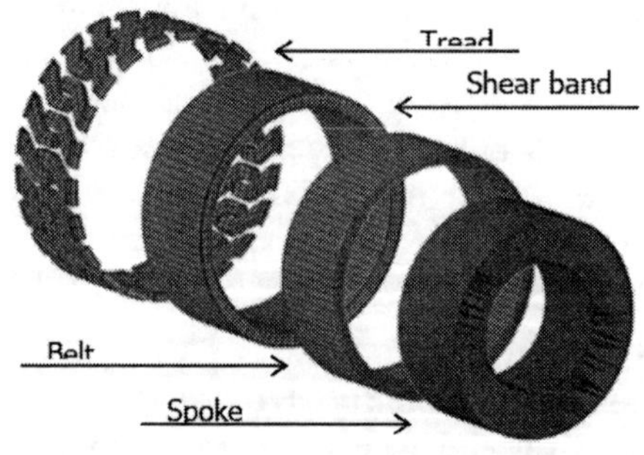

Figure 2. Finite element model of NPT's components.

Table 1. Details of elements used in a finite element model of NPT's component.

NPT's Components	Element type	Number of element	Averaged element length (mm)
Tread	Hexagonal	2,228	19.89
Shear band	Hexagonal	11,904	16.15
Belt layers	Quadrilateral	6,144	20.38
Spoke	Quadrilateral	35,500	8.74

(a) (b)

Figure 3. Cross-section of the shear band with embed belt layers of (a) Tweel and (b) finite element model.

3. Hyperviscoelastic Material Model

The Ogden hyper-elastic constitutive model was used to model nonlinear elastic behaviors of rubber tread, shear band, and polyurethane spoke. The general form of the Ogden hyperelastic model [13] can be expressed as shown in equation (1).

$$ {}_0^t\tilde{W} = \sum_{n=1}^{3} \frac{\mu_n}{\alpha_n}\left(\lambda_1^{\alpha_n} + \lambda_2^{\alpha_n} + \lambda_3^{\alpha_n} - 3\right) \tag{1} $$

Where λ_i is the principal value of the stretch tensor ${}_0^tU$, μ_i and α_i are hyperelastic material constants.

The inelastic behaviors of NPT components were modelled using the generalized Maxwell viscoelastic material model. The generalized Maxwell viscoelastic model can be express as follows:

$$ G(t) = G_0 - \sum_{i=1}^{n} G_i\left(1 - e^{-t/\tau_i}\right) \tag{2} $$

$$ \tau_i = \eta_i / E_i \tag{3} $$

where $G(t)$ is Shear Relaxation Modulus, G_0 is Shear Modulus at a time, $t=0$, G_i is i[th] term of Shear Modulus, τ_i is the ith term of Relaxation Time (sec), E_i is Modulus of Elasticity, and η_i is viscoelasticity.

The parameters of Ogden hyperelastic and generalized Maxwell material model of tread and PU spoke, which are obtained and validated from mechanical testing in the previous research, are shown in Table 2 and 3, respectively [4, 10]. The compressive testing and tensile testing are performed on the cylindrical shape and dumbbell shape specimens, which are prepared from NPT's tread and spoke respectively using waterjet cutting technique (figure 4) [5, 6]. In addition, linear elastic material with a modulus of elasticity (E) of 200 GPa and Poisson's ratio of 0.3 was used to model the steel belt layers of NPT.

(a) (b) (c) (d) (e)

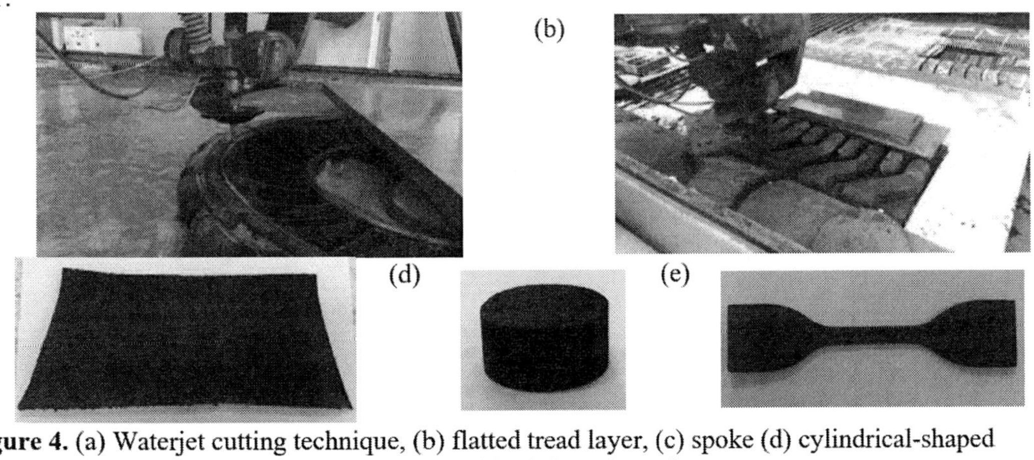

Figure 4. (a) Waterjet cutting technique, (b) flatted tread layer, (c) spoke (d) cylindrical-shaped specimen and (e) Dumbbell shaped specimen.

IOP Conf. Series: Materials Science and Engineering **886** (2020) 012056 doi:10.1088/1757-899X/886/1/012056

Table 2. Hyperelastic constants of NPT components [4].

Constant	Component	
	Spoke	Shear Band
μ_1	0.112983	1.15673
μ_2	-11.0664	1.06228
$\square_1$	3.1488	5.37146
$\square_2$	-1.75206	-2.31827

Table 3. Viscoelastic constants of NPT components [10].

i^{th}	τ_i	G_i	
		Spoke	Shear Band
1	0.2	0.125	0.2
2	0.02	0.125	0.2
3	0.002	0.125	0.2

4. Finite Element Modeling of Drum Testing on NPT

The FE model of NPT based on viscohyperelastic material model was brought into contact with the drum surface model. The drum surface model was defined as rigid, which means that the deformation did not occur. The Coulomb's contact friction model with a friction coefficient of 0.8 was used to model friction between tread elements and drum surface. The tire was assigned to be pressed against the drum surface with a constant force of 14 kN. After the load was fully applied and the NPT was deformed, the drum surface was started to rotate with a velocity of 11 km/hr while the load was maintained at 14 kN. The drum testing experiment is shown in figure 5(a). The boundary conditions of the FE model of drum testing on NPT is shown in figure 5(b). The linear velocity could be converted into corresponding angular velocity as 7.0159 rad/sec. It should be noted that the static solution FEA was used during the load applying state, while the transient dynamic solution was applied during the rolling state. In addition, the high-speed video camera is set up to capture the deformation of spokes while the NPT is rolling at high speed (figure 6).

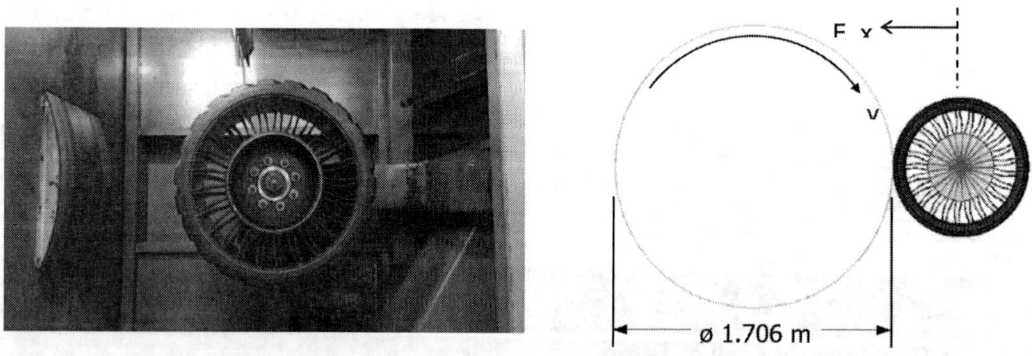

(a) (b)

Figure 5. (a) Drum testing experiment on NPT and (b) finite element model of drum testing on NPT with boundary conditions.

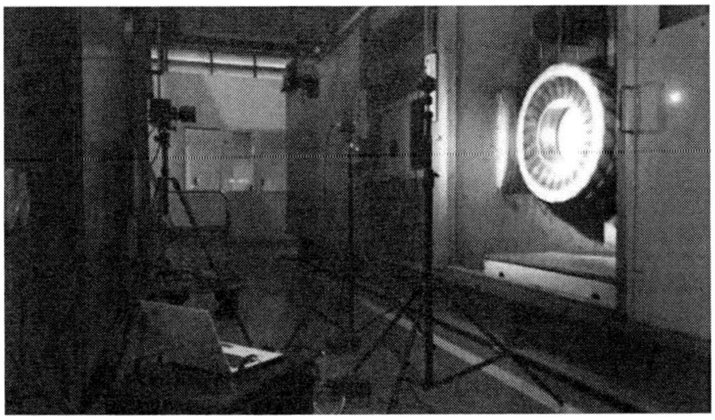

Figure 6. The setting up of high-speed video camera on drum testing of NPT

5. Finite Element Analysis of rolling NPT

The FEA of drum testing of NPT was carried out using the transient dynamic finite element solution. The spokes deformation at various analysis time and rolling angle were collected as nodal coordinates. The recorded rolling angle position can be shown as a schematic diagram as shown in figure 7(a). The analyzed local stress at spoke and NPT deformation is shown in figure 7(b), which the maximum local stress at the node is shown to be 1.20 MPa. The deformed spoke coordinates were compared with experimental recording spoke position, which was recorded using a high-speed video camera to validate the accuracy of the model. The comparison of spoke deformation between the FEA and experiment at various rolling angle i.e. 0°, 30°, 60°, 90°, 120°, 150°, 180°, is shown in figure 8. The average error of spoke deformation analysis is calculated to be 3.51% compared to the experiment. Thus, FE model of NPT was shown to be accurate and can capture the mechanical behavior of the NPT while rolling. The impact force and displacement of NPT's rolling center against the drum in the horizontal direction are plotted in figure 9. It should be noted that the displacement of the tire's center using load 14 kN is 18.97 mm at the start of the dynamic solution/end of the static solution. The value of displacement at the center of NPT was shown to fluctuate between 17.55 and 19.80 while the impact force was shown to fluctuate between the 12.18 and 16.23 kN, respectively.

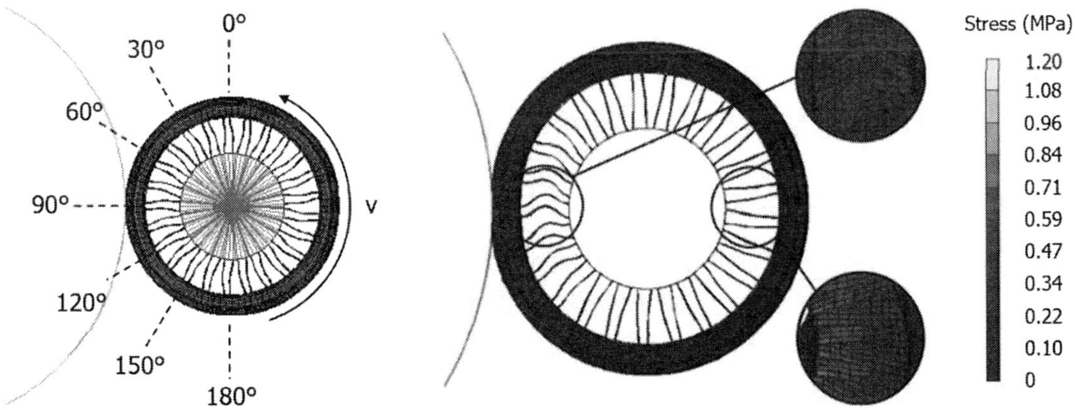

(a) (b)

Figure 7. (a) The schematic diagram of the angular position of the collected spoke position and (b) local stress at spoke with load 14 kN at time 1.2 sec.

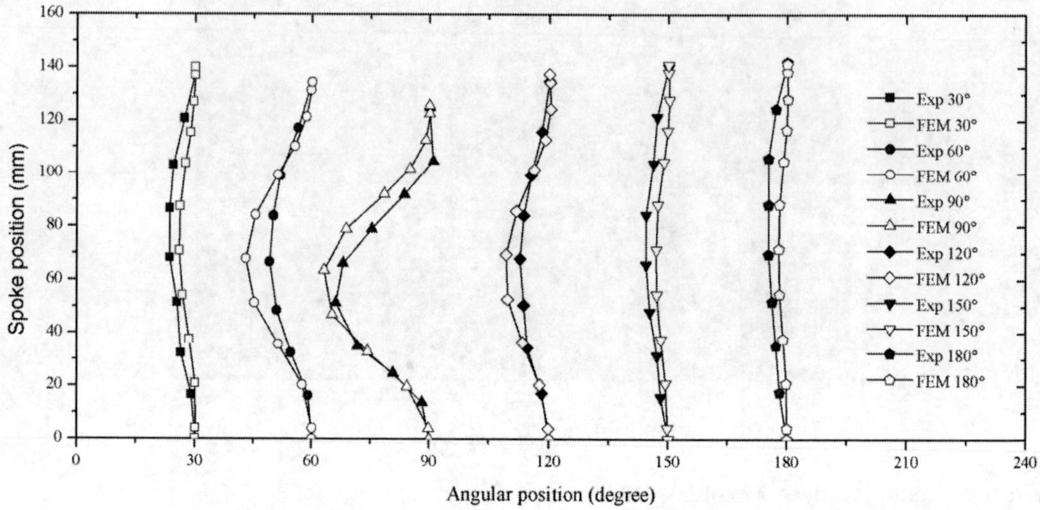

Figure 8. Collected spoke position at the various angular position

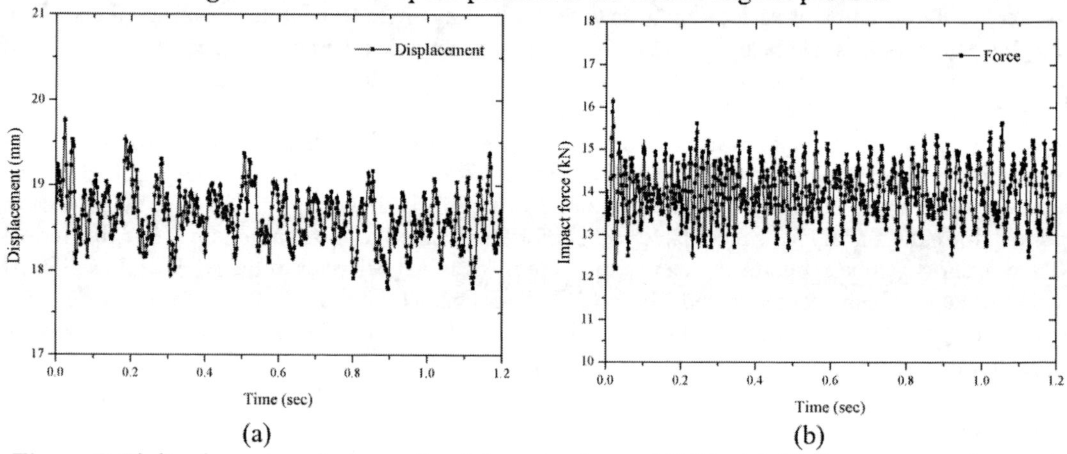

Figure 9. Finite element analysis results of drum testing on NPT by(a) displacement vs. time and (b) impact force vs. time.

6. Conclusion

The transient dynamic FE model based on viscohyperelastic material model for drum testing of NPT was developed. The Ogden hyperelastic material model and generalized Maxwell's viscoelastic was applied to model elastic and inelastic deformation of NPT, respectively. The FE model of NPT, which composed of Tread band and Spokes, was a model using 3D hexagonal and 2D Quadrilateral elements, respectively. The rebar elements and tying equation were used to model steel belt layers of the NPT. The FE model of NPT was then assigned to contact with the rigid drum surface using Coulomb's friction model. The finite element analysis of drum testing method of NPT as the drum was assigned to be pressed with a load of 14 kN while rolled at the speed of 11 km/hr. The NPT spoke deformations at each position and angle were compared to the experiment to prove the validity of the FE model. The analysis results of spoke deformation were shown to be in a good agreement with the experiment at the average error of 3.51%. The developed FE model of NPT based on transient dynamic approach can be used to evaluate the important dynamic response of NPT during rolling and advantageous to use in further designing of the NPT.

Acknowledgments

This work was financially supported by Rubber Technology Research Center (RTEC), Mahidol University and the Thailand Research Fund (TRF) under the TRF Research Grant No. RDG60T0140.

References

[1] Gent A Watler J 2006 *The pneumatic tire* (Washington DC: National Highway Traffic Safety Administration)
[2] Rhyne T Cron S 2006 *Tire Sci. Technol.* **34**(3) 150-169
[3] Gasmi A Joseph P Rhyne T Cron S 2012 *Int. J. Solids Struct.* **49**(13) 1723-1740
[4] Rugsaj R Suvanjumrat C 2019 *Int. J. Automot. Technol.* **20**(4) 801-812
[5] Rugsaj R Suvanjumrat C 2018 *Key Eng. Mater.* **775** 554-559
[6] Rugsaj R Suvanjumrat C 2018 *Key Eng. Mater.* **777** 411-415
[7] Phromjan J Suvanjumrat C 2018 *J. Mech. Sci. Technol.* **32**(4) 1539-1548
[8] Tönük E Ünlüsoy Y S 2001 *Comput. Struct.* **79**(13) 1219-1232
[9] Montgomery S MacKnight J 2005 *Introduction to Polymer Viscoelasticity* (John Wiley & Sons)
[10] Yoo S Uddin M Heo H Ju J Choi S J 2017 *P.I. Mech. Eng. D-J Aut.* **231**(2) 241-252
[11] Jin X Hou C Fan X Sun Y, Lu J, and Lu C 2018 *Compos. Struct.* **187** 27-35
[12] Helnwein P Liu H Meschke G Mang A 1993 *Finite. Elem. Anal. Des.* **14**(1) 1-16
[13] Bathe J 1997 *Finite Element Procedures* (London:Prentice Hall)

IOP Conf. Series: Materials Science and Engineering **886** (2020) 012057 doi:10.1088/1757-899X/886/1/012057

Texture Properties of a-2D Granular Mixture of an Angular Particle under Bi-axial Compression

Theechalit Binaree[1], Itthichai Preechawuttipong[1,*], and Emilien Azéma[2]

[1] Department of Mechanical Engineering, Faculty of Engineering, Chiang Mai University, 239 Huay Kaew Rd., Chiang Mai, 50200, Thailand.
[2] LMGC, Univ. de Montpellier, CNRS, Montpellier, France.

* Corresponding Author: itthichai.p@cmu.ac.th

Abstract
Utilizing Discrete Element Modeling (DEM), we investigate the shear strength properties and microstructure of 2D granular mixture composed of various regular polygonal particles (from triangle to icosagon and disk) under biaxial compression. Three cases were considered: 1) increase of proportion and angularity started from disk (S1); 2) increase of proportion by decreasing the angularity started from triangle (S2); 3) increase of proportion by random angularity started from disk (S3) and polygons (S4). The results show that the shear strength changes with the mean angularity. On the other hand, the packing fraction slightly varies with the mean angularity. At the microscopic scale, the granular texture in term of the force, the contacts transmitted the force higher than the average force (strong force) are oriented in the direction of externally applied loading but they represent only 40% of the total number of contacts. Besides, the tangential contact forces direction tend to be consistent with the maximum shear strength direction.

Keywords: Granular materials, Granular packing, Granular rheology, Particle shape effects Discrete Element Method

1. Introduction

Granular materials in nature are composed of a mixture of particles with different sizes, shapes and properties (such as density, friction, and elasticity). For example, sand, rock, plastic bead, a particle of pharmaceutical[1–3], etc. Exploring the effect of granular composition or polydispersity, the physical properties of granular media can help to analyze civil engineering structure and optimized industrial processes[4,5]. A well-known example is the particle size distribution of rock aggregates in concretes, which performances and workability were optimized by proportion[4]. In addition, the particle shape distribution of the granular mixture was studied, such as mechanical behavior and packing properties. Many kinds of particle shapes were found in everyday life and industry, for instance, angularity, roundedness, elongation or platy-ness and nonconvex shapes. These properties are used to characterize the particle shapes[6–11]. Binary mixtures of two shapes have been very investigation[12,13], such as segregation behavior or the strength properties, while the effect of multiple particle shapes remains still poorly investigated.

The aim of this paper is exploring/to explore the effect of angular particle shape polydispersity under biaxial compression by mean of the contact dynamic method simulation whereas analyzed in the steady-state in term of shear strength and packing fraction at the macroscopic scale, and force transmission at the microscopic scale.

2. Numerical Model

2.1. Contact dynamics methods.

In this study, our simulations were implemented by the contact dynamics (CD) method, which is appropriate with simulating mixture of rigid particles[14]. This method integrates the motion equation of collection of rigid objects with contact interactions. These interactions are defined by friction coefficient, normal restitution coefficient, and tangential restitution coefficient. A parallelized iterative research algorithm based on nonlinear Gauss-Seidel method. These method used to determine velocities and forces at the end of the time step, concurrently. The contact dynamics method is widely used both in 2D and 3D simulation[11,15]. LMGC90 was implemented by this method and is multipurpose software developed by the University of Montpellier, France.

2.2. Packing construction and bi-axial test

The Granular mixture is characterized by three parameters: 1) the number of species N_s, 2) shape properties of each species 3) the proportion of each species k_i. In this samples, we assume the particles in packing are generate with uniform distribution by particle number, $k_i = 1/N_s$. The total number of particles N_p= 10 000. Particles used to prepare the granular mixture are regular polygon with the number of sides $n_s \in [3,4,5,6,7,8,10,15,20]$ and disk. Shape properties were determined by angularity α is given by $2\pi/n_s$,where n_s is the number of sides of particle. The proportion of particle in a mixture is defined by the number of species, which varied from 1 to 10. There are three case considered 1) increase of angularity with N_s ($S1$) 2) decrease of angularity with N_s ($S2$) 3) random angularity with N_s ($S3$) and ($S4$) ;see Table 1. All samples considered in this paper are prepared according to the same protocol. First, Particles were geometrically placed into a rectangular box using a sequential layer by layer deposition[16]. In order to avoid long-range ordering, we introduced small size polydispersity by taking diameter in the range $[d_{min}, d_{max}]$ where $d_{max} = 1.25d_{min}$ with a uniform distribution in particle volume fractions.

Table 1. Numerical samples of geometrical characteristics (The numbers in brackets mention to the number of side of polygons of each species).

N_s	$S1$	$S2$	$S3$	$S4$
1	{disk}	{3}	{disk}	{7}
2	{disk,20}	{3,4}	{disk,3}	{7,5}
3	{disk,20,15}	{3,4,5}	{disk,3,20}	{7,5,15}
4	{disk,20,15,10}	{3,4,5,6}	{disk,3,20,4}	{7,5,15,10}
5	{disk,20,15,10,8}	{3,4,5,6,7}	{disk,3,20,4,15}	{7,5,15,10}
6	{disk,20,15,10,8,7}	{3,4,5,6,7,8}	{disk,3,20,4,15,5}	{7,5,15,10,disk}
7	{disk,20,15,10,8,7,6}	{3,4,5,6,7,8,10}	{disk,3,20,4,15,5,10}	{7,5,15,10,disk,8}
8	{disk,20,15,10,8,7,6,5 }	{3,4,5,6,7,8,10,15}	{disk,3,20,4,15,5,10,6 }	{7,5,15,10,disk,8,6}
9	{disk,20,15,10,8,7,6,5 ,4}	{3,4,5,6,7,8,10,15,20}	{disk,3,20,4,15,5,10,6 ,8}	{7,5,15,10,disk,8,6,2 0}
1 0	{disk,20,15,10,8,7,6,5 ,4,3}	{3,4,5,6,7,8,10,15,20, disk}	{disk,3,20,4,15,5,10,6 ,8,7}	{7,5,15,10,disk,8,6,2 0,3}

All packings are compacted by isotropic compression in a container (rectangle frame) which the right and top walls are applied by compressive stress σ, and the left and bottom walls are fixed. During the isotropic compression, the gravity acceleration g, the friction coefficients μ between particles, and the friction coefficients μ between particles and walls are set to zero. This procedure leads to a random close packing state. Dense samples obtained at the end of isotropic compression are then subject to biaxial compression by applied constant velocity v_y on the top wall and constant compressive stress σ_0 acting on the side walls. The friction coefficients μ between particles are set to 0.4 but wall friction kept to zero. Quasi-static (rate-independent) behavior was interested in this study which can be defined by the inertia number $I = \dot{\varepsilon}\sqrt{m/\sigma_0}$ where $\dot{\varepsilon} = \dot{y}/y$ is the strain rate, m are the mass of the particles[17]. In this simulation, I is set to 10^{-4}.

3. Numerical Result

3.1. Macroscopic behaviour

In the numerical simulation, the stress tensor can be determined from the geometrical configuration of packing and contact force. The stress tensor σ in a given volume V defined by[18]:

$$\sigma = \frac{1}{V}\square_{\omega \square V} M^{\omega} \quad (1)$$

Where M^{ω} is a tensorial moment given by $M_{ij}^{\omega} = \sum_{c \in \omega} f_i^c r_j^c$ which f_i^c is the force component on the particle ω at the contact c, r_j^c is the j the position vector component of the same contact c. In dry granular materials, the *shear strength* can be quantified by the macroscopic friction angle φ following:

$$\sin\varphi = \frac{q}{p} \quad (2)$$

Where $q = (\sigma_1 - \sigma_2)/2$ (deviator stress), and $p = (\sigma_1 + \sigma_2)/2$ (mean stress) which σ_1 and σ_2 are principal stress. For the bi-axial compression, the major principal direction is assumed to be vertical. The packing fraction is defined by $\rho = V_p/V$, where V_p is the volume occupied by particles, and V is the container volume.

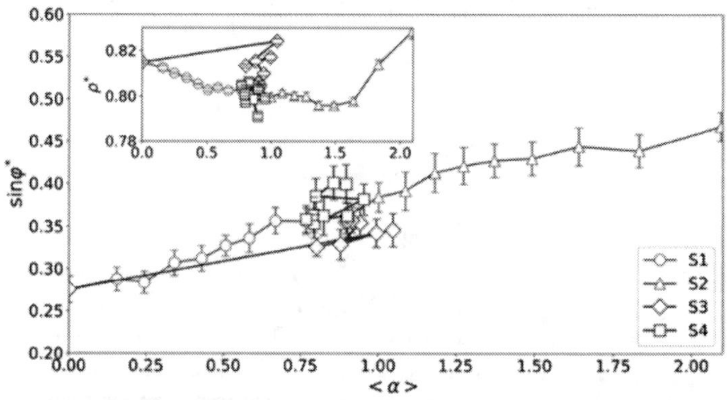

Figure 1. Averaged macroscopic friction angle $\sin\varphi^*$ and (inset) the packing fraction ρ^* in steady-state as a function of mean angularity $< \alpha >$

For the macroscopic friction angle $\sin\varphi^*$and the packing fraction ρ^*, both determined from the averaged of $\sin\varphi$ and ρ, respectively, in the steady-state. Figure 1 show $\sin\varphi^*$ and ρ^* as a function of mean angularity $< \alpha >$ of each sample. We observe the shear strength increase with the mean angularity, whereas the packing fraction slightly decrease. In other words, the packing becomes a little loose but with higher a shear strength. Furthermore, in random mixtures S3 and S4, at the identical value of mean angularity, the value of $\sin\varphi^*$ and ρ^*, no collapse data is observed.

3.2. Texture properties

In this section, texture properties are described by force networks and forces distributions at a microscopic scale. Figure 2 shows a map of normal in the steady-state for $N_S = 4$ in S1, and S2. The thickness of the segment joining the particle centers were represented the normal forces. The long force chains are observed from the upper wall to lower wall of samples and especially oriented near the compression direction which is 90 degrees. Although the thin forces chains distribute similar to the higher forces, it tends to be disconnected. Comparing with the polar representation of normal contact forces direction; see Figure 3(a). Mostly the probability of normal forces direction is in 90 degrees which be consistent with force network and direction of compression. Furthermore, the tangential contact forces tend to be aligned with the maximum shear strength angle (45 degrees) as shown in Figure 3(b). A large number of contacts were transmitted weak forces, whereas a small fraction of contacts carries strong forces which about 40% of active contacts; see Figure 4. Besides, the proportion of several strong force contacts with the total active contact nearly constant with mean angularity.

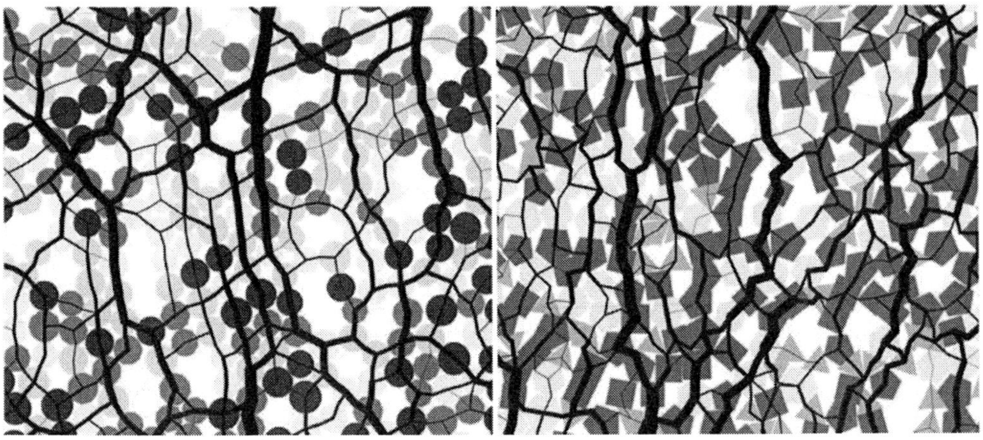

(a) (b)

Figure 2. The force network in S1 (a), and S2 (b) for $N_S = 4$

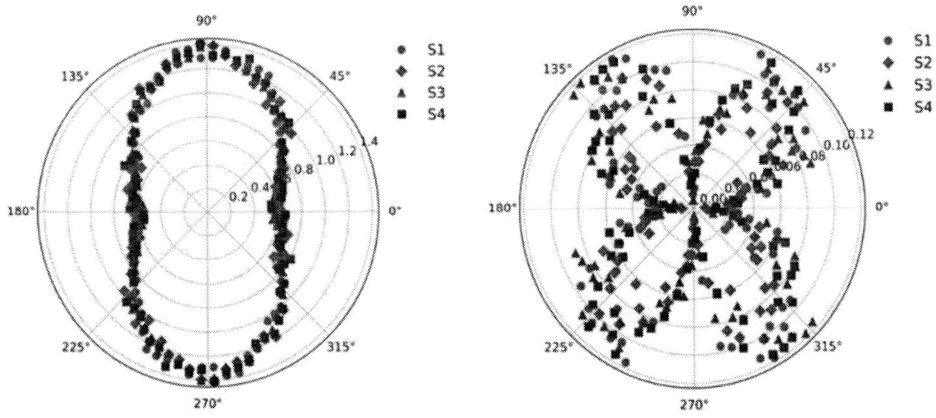

(a) (b)

Figure 3. Polar representation of the probability of normal contact forces direction (a), and tangential contact forces direction (b) in the steady-state for $N_S = 4$ in the steady-state.

Besides, the force transmission has been investigated by the probability distribution function (pdf) which separated into two networks following the weak network and the strong network. The strong network is a network of the contact carrying the force more than average force, while the weak network is a network of the contact transmitting the force less than average force. Figure 5 shows the probability distribution function (pdf) of normal contact force f_n normalized by average normal contact force in strong networks, we observed the normal force distribution is exponential decrease, whereas the normal force distribution in the weak networks is very flat with values close to 1 as shown in Figure 6, thus can be characterized by power-law $P(f) \propto (f_n/\langle f_n \rangle)^{\beta}$ where β is close to zero[16]. These distributions were found in all cases.

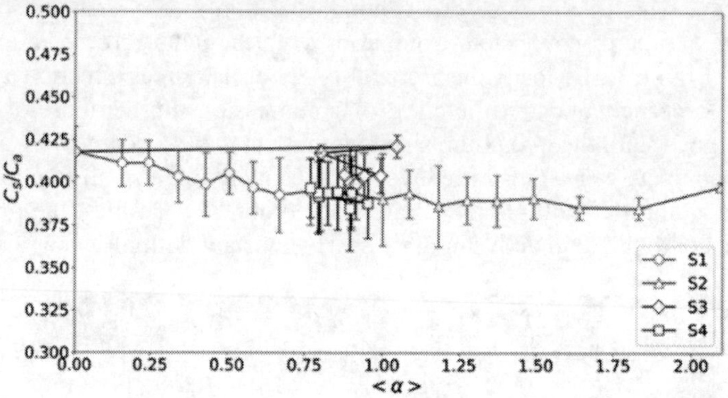

Figure 4. The Proportion of the number of strong force contacts C_s with a number of active contacts C_a in steady-state as a function of mean angularity $< \alpha >$

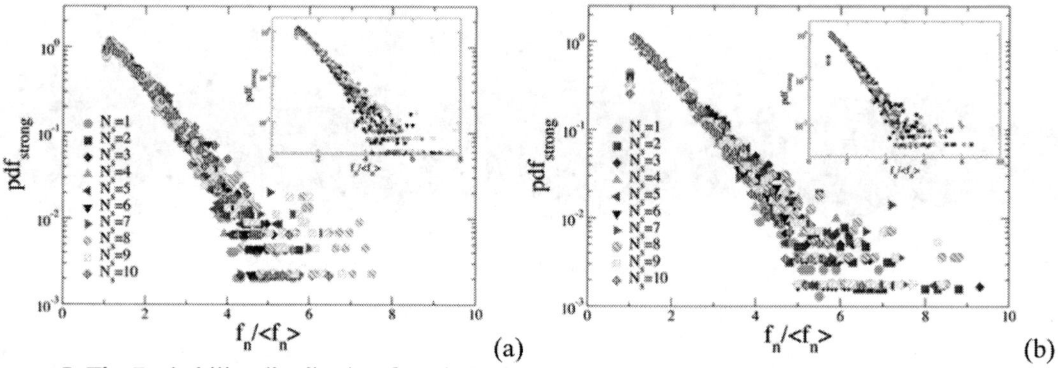

Figure 5. The Probability distribution function of normal force f_n normalized by average normal force $\langle f_n \rangle$ for the strong networks: (a) S1 and S3 (inset), and (b) S2 and S4 (inset), both represented in semi-logarithmic axes.

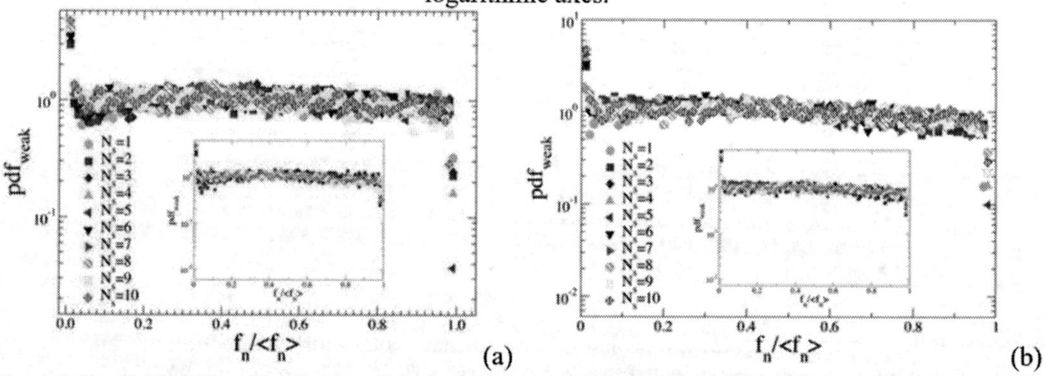

Figure 6. The Probability distribution function of normal force f_n normalized by average normal force $\langle f_n \rangle$ for the weak networks: (a) S1 and S3 (inset), and (b) S2 and S4 (inset), both plotted on semi-logarithmic axes.

Conclusion

The strength and texture properties of a 2D granular mixture of angular particles were investigated by the contact dynamic simulation. Each packing in this studied consists of 10000 particles which contained in rectangle frame under biaxial compression. The number of species is varied from 1 to 10 with various shapes (regular polygon and disk). The samples were built with three cases 1) increase of proportion of particle shapes and angularity by starting from disk, 2) increase of the proportion of particle shapes by decreasing the angularity started from the triangle; 3) increase of the proportion of particle shapes by random angularity started from disk and polygons.

At the macroscale, the mechanical behavior depends on geometrical properties of the granular mixture. The shear strength increases when mean angularity in packings increase, otherwise the shears strength decrease. On the other hand, the packing fraction slightly decreases, which nearly constant. These results represent that packing density does not affect the mechanical strength of the granular mixture.

At the microscopic scale, the forces were transmitted in direction of compression and the tangential contact force remain on maximum shear stress direction, which not depend on the mean angularity of packing. In all cases, only 40% of the number of contacts carry force higher than the average force whereas these forces are sorted to be a network from the upper boundary to lower boundary to support the system. Furthermore, the probability distribution function of normal force does not depend on mean angularity (particle shape). The force distribution of the strong network is regressed by exponential drop, while the weak network can be approximated by a power law.

Acknowledgments

This work was supported by Royal Golden Jubilee Ph.D. Program (Grant No. PFD/0062/2558) under the Thailand Research Fund (TRF) and the French Embassy in Thailand. The authors also acknowledge financial support from the Faculty of Engineering, Chiang Mai University.

References

[1] Vallejo L Lobo-Guerrero S 2012 *Int. J. Geotech. Eng.* **6** 371–380
[2] Lee J S Guimaraes M Santamarina J C 2007 *J. Geotech. Geoenvironmental Eng.* **133** 1136–1143
[3] Džiugys A Navakas R 2009 *Granul. Matter* **11** 403–416
[4] Brouwers H J H Radix H J 2005 *Cem. Concr. Res.* **35** 2116–2136
[5] Lee J Yun T Choi S U 2015 *Materials* **8** 3975–3991
[6] Mitchell J K Soga K *Fundamentals of Soil Behavior* (New York:John Wiley & Sons, Inc)
[7] Azéma E Radjaï F 2010 *Phys. Rev. E* **81**
[8] Nguyen D H Azéma E Sornay P Radjai F 2015 *Phys. Rev. E* **91**
[9] Nouguier-Lehon C Cambou B Vincens E 2003 *Int. J. Numer. Anal. Methods Geomech.* **27** 1207–1226
[10] Zhou W Xu K Yang L Ma G 2017 *Proc. the 7th International Conference on Discrete Element Methods* (vol 188) ed X Li Y Feng G Mustoe (Singapore:Springer Singapore) pp 245–252
[11] Azéma E Estrada N Radjaï F 2012 *Phys. Rev. E* **86**
[12] Abreu C R A Tavares F W and astier M 2003 *Powder Technol.* **134** 167–180
[13] Azéma E Preechawuttipong I Radjai F 2016 *Phys. Rev. E* **94**
[14] Jean M 1999 *Comput. Methods Appl. Mech. Eng.* **177** 235–257
[15] Azéma E Radjai F Dubois F 2013 *Phys. Rev. E* **87**
[16] Voivret C Radjaï F Delenne J Y El-Youssoufi M S 2007 *Phys. Rev. E* **76**
[17] Groupement de Recherche Milieux Divisés 2004 *Eur. Phys. J. E* **14** 341–365
[18] Staron L Radjai F 2005 *Phys. Rev. E* **72**
[19] Andreotti B Forterre Y Pouliquen O 2013 *Granular Media: Between Fluid and Solid* (Cambridge: Cambridge University Press)

IOP Conf. Series: Materials Science and Engineering **886** (2020) 012058 doi:10.1088/1757-899X/886/1/012058

Investigation of Joining Dissimilar Materials Using Hot Pressing Process

N. Wannapa[1], V. Tangwaradomnukun[2], V. Uthaisangsuk[1,*]

[1]Department of Mechanical Engineering,
[2]Department of Production Engineering,
Faculty of Engineering, King Mongkut's University of Technology Thonburi,
126 Pracha Uthit Rd., Bang Mod, Thung Khru, Bangkok 10140, Thailand,

* Corresponding Author: vitoon.uth@kmutt.ac.th

Abstract
Hot pressing has been increasingly applied as a joining technology for dissimilar materials because of its simple setup. However, various process parameters must be clearly defined here in order to obtain an optimum joint. In this study, an experimental hot pressing for joining an aluminium alloy sheet grade A5052 with a polybutylene terephthalate (PBT) plate was carried out. Different surface treatment on the aluminium samples were firstly employed by using both mechanical and chemical processes. Then, surface topographies of the treated samples were characterized by scanning electron microscope. The Fourier transform infrared spectroscopy analysis was carried out to examine the C=O in the carbonyl group of thermo-plastic at the joint. During the hot pressing, joining force and temperature were varied. Afterwards, tension shear tests were performed for the aluminium-PBT specimens to evaluate the resulted joint strength. It was found that the pressing force of 1-ton and a pressing temperature of 270°C were the optimum parameters for joining these investigated materials. The maximum joining strength was obtained by specimens subjected to sandblasting combined with chemical surface treatment, in which the surface roughness of aluminium sheet samples was in the range of 4-9 μm. The specimens after shear tests exhibited a cohesive failure. The depth of mechanical interlocking between surfaces of both materials was approximately 33 μm. No change of the molecular structure of PBT was observed by the direct bonding between aluminium and PBT using hot pressing.

Keywords: Dissimilar materials joining, Hot pressing, Surface treatment

1. Introduction

In recent years, joining of dissimilar materials has been increasingly required in the automotive industries due to the purpose of lightweight design, in which saving of fuel consumption and thus environmental protection can be ensured by road transportation. In order to achieve an effective weight reduction of the vehicle with improved structural performance, it is necessary to apply a wide variety of materials. As a result, materials for lightweight structures, for instance, various types of high strength steels, aluminium alloys, plastics and fiber-reinforced composites are more often combined in an automotive component. Plastics and composite materials can be even used to replace some structural metal parts such as a bumper, front grill, door panel and dashboard. Therefore, new joining technologies for different materials need to be strongly developed [1]. To establish a material joining process between metals and plastics, fundamental researches with regard to many techniques and mechanisms were carried out. Budhe et al. [2] investigated the effect of surface roughness of different adherend material on adhesive bond strength by adhesive bonding. Amend et al. [3] studied influences of laser structuring of metal surface on the joining of aluminium and thermoplastics by means of mono- and polychromatic radiation. Balasubramanian [4] established a relationship between the base material properties and friction stir welding process parameters. Kajihara et al. [5] developed the joining method for polymer-metal hybrid (PMH) by an injection molding process. Kondoh et al. and Hussein et al. [6, 7] studied a direct bonding process of dissimilar materials by using the conventional hot pressing. These processes showed some varying disadvantages depending on used materials and parameters. However, they could provide acceptable joining strength between metal sheets and thermoplastic materials. In most cases, before joining nanostructures or micro-sized holes had to be prepared on the metal surfaces by using either chemical, mechanical or heat-treating process. Most of the results [8, 9] noticed the importance of surface and its positive influence on the joining strength. Due to several types of materials and available specific processes, understanding of welding of dissimilar materials still needs to be enhanced. In this work, joining between aluminium sheet grade A5052 and PBT plastic was studied using a hot pressing technique. Different surface treatments based on both chemical and mechanical treatments were employed to increase the surface roughness of aluminium sheets. Additionally, joining conditions with regard to pressing force and joining temperature were investigated. The adhesion strengths of hot-pressed samples were characterized by performing lap shear tests. The Fourier transform infrared spectroscopy (FTIR) analysis was also carried out to examine the molecular structural changes of the plastic materials before and after joining to the aluminium sheet.

2. Experiment

In this work, an aluminium alloy sheet grade A5052 with a thickness of 0.8 mm and a polybutylene terephthalate plate including 30% glass fiber reinforcement (PBT-30GFR) with a thickness of 3.0 mm was used. Tensile testing of materials is specified according to European EN 10002 standard. The physical and mechanical properties of aluminium A5052 and PBT-30GFR are presented in table 1.

Table 1. Physical and mechanical properties of aluminium grade A5052 and PBT-30GFR plastic

Material	*Density at 25°C (g/cm^3)	*Heat distortion temperature (°C)	*Melting temperature (°C)	Tensile strength (MPa)	Yield strength (MPa)	Elongation (%)
Aluminium grade A5052	2.7	-	605	240	160	9.96
PBT-30GFR	1.5	150	260	96	61	3

* Supplier Data Sheet

Test specimens of the hot pressing experiments were prepared according to the standard ISO 19095-2:2015. The dimensions of aluminium and PBT samples were 55×20 mm^2 and 55×15 mm^2, respectively, as shown in figure 1. The joining area on both samples was 15x15 mm^2.

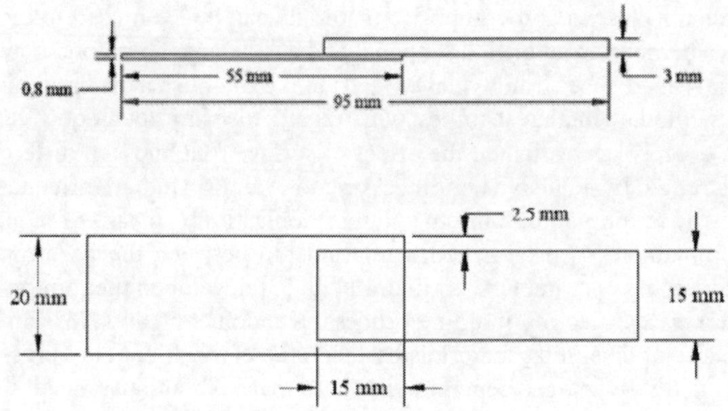

Figure 1. Dimensions of aluminium and plastic specimens for hot pressing process

2.1. Metal surface treatment

As the adhesion force between metal and polymer is significantly affected by the nanoscale surface roughness, a surface treatment process for preparing the metal surface is necessary. During such treatment, oil, oxide layers and corrosion layers are also removed and nanoscale surface morphology can be obtained simultaneously [10]. In this work, two approaches were used to generate nanoscale surface structures on aluminium tested samples. First, the nanoscale pattern was created on the aluminium sheet surface by different mechanical treatments, namely, using sandpaper, sandblasting and laser beam [11-13]. The details of each mechanical surface treatment are presented in table 2. Afterwards, chemical treatment [14] was conducted on all samples subjected to each individual mechanical processe. Furthermore, a group of samples was only treated by the chemical process.

Table 2. Mechanical surface treatment process on the aluminium surface

	Parameters used in the surface treatment process
Sandpaper	Grinding with P-240
Sandblasting	0.5 mm white aluminium oxide particles, pressures 0.3 MPa with a nozzle distance of 5 mm
Laser beam	Square texture with a length of 200 μm, laser power of 10 W and frequency of 50 Hz

By the chemical treatment [15], patterned sheet surfaces were cleaned with methanol for removing the aluminium oxide. The aluminium sheets were then immersed in a solution of 5 wt% NaOH and ultrasonically cleaned for approximately 5 mins. Subsequently, the sheets were dipped in acetone at 25°C for around 30 s to remove the residual NaOH. The aluminium sheets were afterwards immersed in a solution of 25 wt% NH_3, by which the immersing time was varied up to 30 mins and the temperatures up to 60°C were used. Finally, the aluminium sheets were rinsed with water and dried [16]. The aluminium sheet surfaces after treated by each procedure were characterized by scanning electron microscope (SEM). Figure 2(a) shows the surface structures generated by the chemical treatment, while figure 2(b), 2(c) and 2(d) depict those by the combinations of sandpaper, sandblast and laser texturing and the chemical treatment, respectively. It was clearly observed that the aluminium surfaces treated only by the chemical process exhibited randomly distributed shallow holes with the diameters ranging between 10-60 μm. Furthermore, the surface roughness of different samples was measured using a profilometer. The measurements were performed in the longitudinal direction and measuring length was 10 mm in all cases. Hereby, the average profile roughness (Ra) was used to evaluate the roughness of treated specimens. Note that the measuring range of the used profilometer was around 0.77–9.32 μm.

IOP Conf. Series: Materials Science and Engineering **886** (2020) 012058 doi:10.1088/1757-899X/886/1/012058

Table 3 provides the mean surface roughness values of prepared aluminium samples. It is seen that the only chemical treatment led to the lowest roughness, while the combination of sandblasting and chemical treatment provided the largest roughness value, which was larger than other combination methods several times.

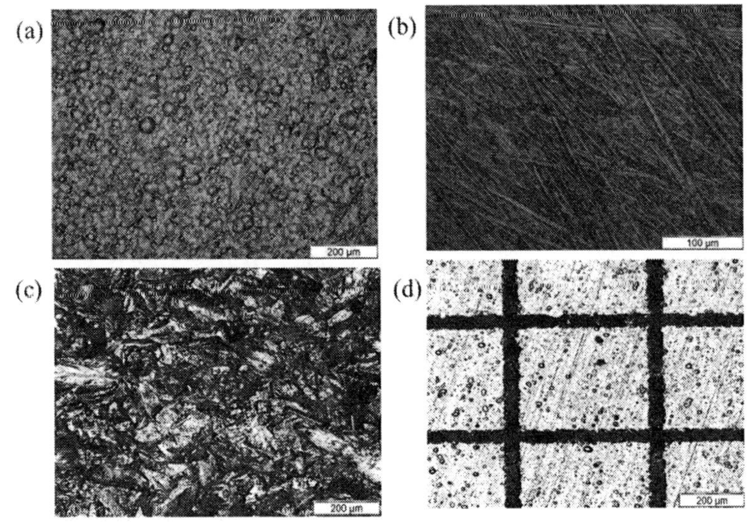

Figure 2. SEM images of the surface of aluminium samples prepared by (a) chemical treatment, (b) sandpaper, (c) sandblasting and (d) laser beam process

Table 3. The surface roughness of aluminium samples treated by different methods

Surface treatment process	Roughness surface Ra (μm)
Chemical	0.774
Sandpaper + chemical	1.962
Sandblasting + chemical	9.328
Laser beam + chemical	2.952

2.2 Hot pressing joining procedure

In this work, the aluminium alloy sheet with structured surface and PBT-30GFR sample were joined by a hot pressing process. The hot pressing setup consisted of a hydraulic machine (30 tons capacity), heater including temperature control (up to the maximum temperature of 300°C), punch and die, as shown in figure 3. The temperature at the interface between both samples (Al-PBT) in the middle within the joining area was measured by a thermo-couple until the beginning of pressing.

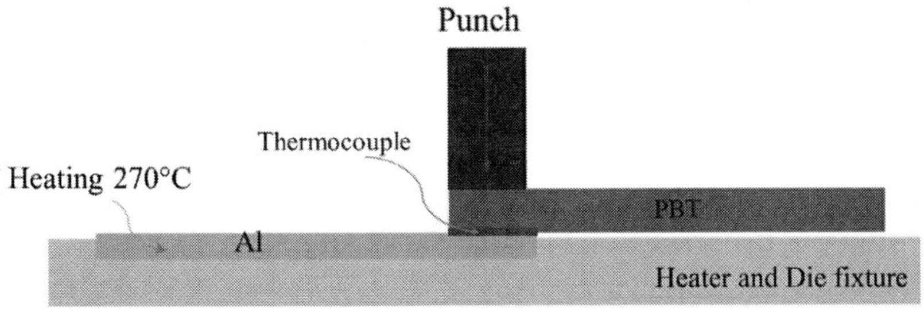

Figure 3. Schematic of direct bonding between aluminium and plastic using hot pressing

During the experiments, pressing force was varied between 0.5 and 5 tons, joining temperature was set between 240°C and 290°C and the holding time of 60 s was used. Totally, experiments with 36 different conditions were performed in order to investigate the effects of pressing force and temperature. Subsequently, tension shear tests of joined specimens were done on a universal testing machine at the crosshead speed of 2 mm/min. Note that the gripping length at both ends of the sample was 15 mm. Loads and displacements were recorded throughout the tests.

3. Results/findings and discussion

3.1 Effects of joining condition

The joining process parameters, namely, press force and temperature, were investigated with regard to the joining strengths obtained from the tensile shear tests. Note that only the aluminium samples prepared by the sandblasting treatment, which had the highest surface roughness, were here taken into account. The relationship between the pressing force and joining strength was determined and shown in figure 4(a). It can be seen that the maximum strength of the sample between aluminium A5052 and PBT-30GPR was reached at the pressing force of 1 ton. The pressing force less than 1-ton led to samples with insufficient bonding. However, by the press force higher than 1 ton it caused too large deformation of plastic specimens and thus weaker joining. Figure 4(b) presents the effect of joining temperature on the joining strength of samples. Obviously, lower temperatures provided joined samples between aluminium A5052 and PBT-30GPR with lower strengths, because the fluidity of plastic was likely too low to allow for a full penetration in aluminium surface texture. On the other hand, by setting high temperature over the melting point of plastic joined samples showed low strengths, since the plastic already decomposed. Therefore, the joining temperature of 270°C was most suitable for joining aluminium and PBT-30GFR samples using the hot pressing process. An example of a joined specimen prepared by the combination of sandblasting and chemical treatment is illustrated in figures 5.

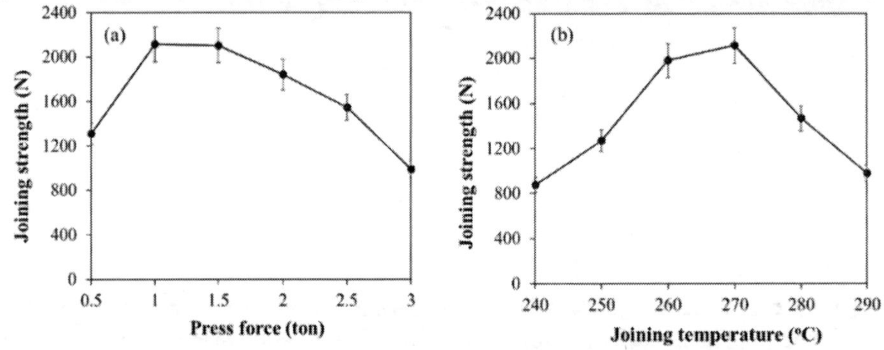

Figures 4. Effects of (a) press force and (b) joining temperature on the joining strength

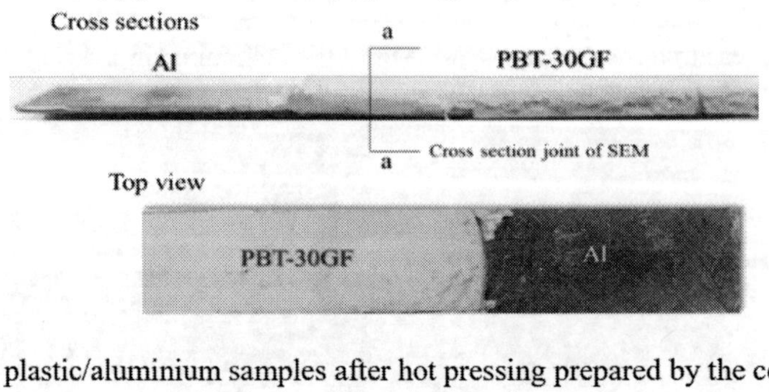

Figure 5. Joined plastic/aluminium samples after hot pressing prepared by the combination of mechanical and chemical treatment

IOP Conf. Series: Materials Science and Engineering **886** (2020) 012058 doi:10.1088/1757-899X/886/1/012058

In addition, the surfaces of aluminium sheet samples in the joining area were observed by SEM. It can be well recognized that the plastics were embedded in the textured grooves of an aluminium sheet prepared by the introduced method, as depicted in figure 6. The adhesion behaviour between both materials was due to the mechanical interlocking, in which the depths of observed interlocks between the surfaces of different samples were approximately 7-33 μm. This interlocking strongly depended on the surface preparation. The largest interlocking depth was observed in the case of sandblasting.

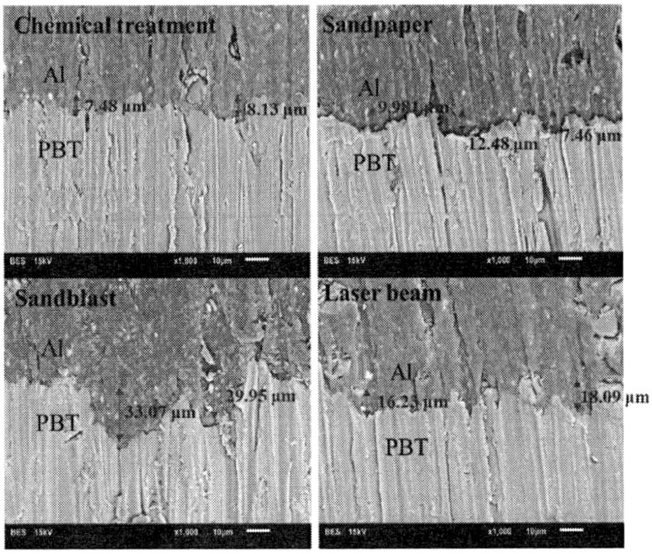

Figure 6. Surface morphologies of bonding areas of aluminium/plastic joined samples prepared by various methods

3.2 Joining strength vs. surface roughness

Figure 7 shows the resulted forces and displacement curves of joined samples between aluminum and plastic prepared by various surface treatments. All tensile shear tests were performed for the samples previously subjected to hot pressing under the temperature of 270°C, 1-ton force and holding time of 60s. It was found that the maximum joining strength of 2.20 kN could be achieved from the sandblasting combined with chemical surface treatment. It was found that this sandblasting treatment caused the highest surface roughness on the aluminium surface. On the other hand, the lowest joining strength of around 0.69 kN was obtained when only the chemical treatment was used. In addition, the correlation between joining strength and surface roughness of different samples was done, as shown in figure 8. The increased roughness of the aluminium surface obviously led to increased joining strengths in a non-linear manner. The surface roughness within the range of 4–9 μm is suggested for the aluminium surface in order to reach an acceptable bonding strength of joined sample in this work. A different trend of resulting shear strengths with respect to the surface roughness was reported by Li., et al. [17]. They showed that samples with higher surface roughness than 4.3 μm exhibited decreased joining strengths. This was likely due to the size of the particle in the sandblasting process was greatly larger than that used in this work. Consequently, joining dissimilar materials certainly needed an appropriate combination of joining process, process parameters and surface characteristic.

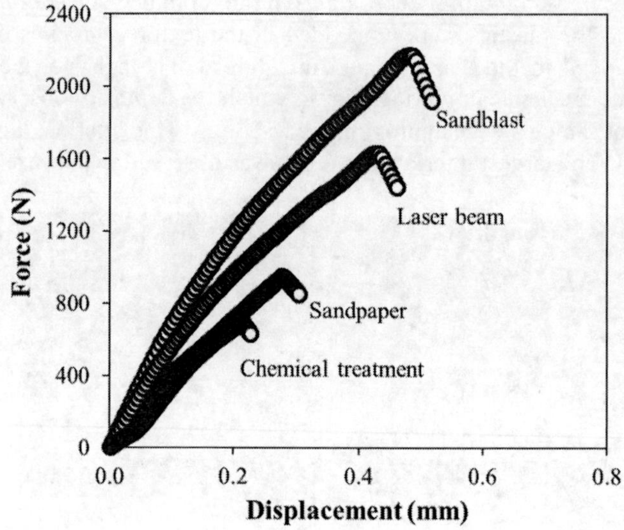

Figure 7. Determined force-displacement curves of joined samples between aluminium and plastic after different surface treatments

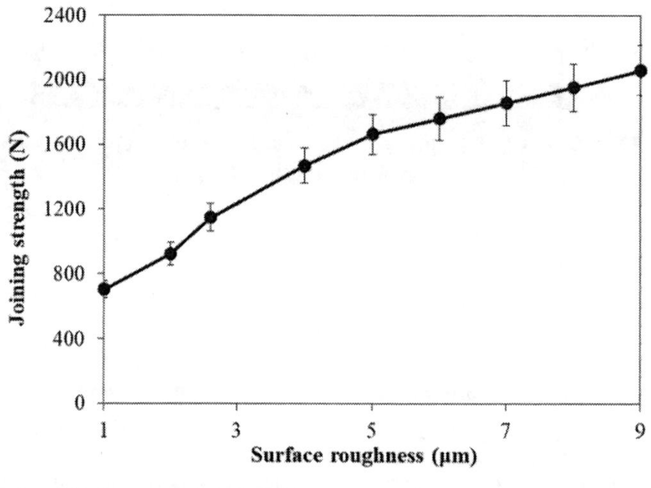

Figure 8. Relationship between joining strength and surface roughness of joined samples

3.3 failure characteristic

To identify the failure behaviour of joined samples after the shear test, the main rupture modes suggested by the ASTM D5573 standard [18] was referred to. Figure 9 shows the interface characteristic taken by SEM of the failed sample subjected to the mechanical and chemical treatments. It can be seen that the sample exhibited a mixed failure mode of interface and cohesive failure.

Figure 9. Failure surface of joined samples between aluminium and plastic

3.4 FTIR analysis on the plastic of aluminium and PBT-30GFR bonded material

It was reported in [6] that the C=O bond played an important role in the direct bonding of metal to plastics and thus resulting in shear joining strength. The FTIR analysis was thus conducted to detect the change of the C=O spectrum in the bonding area of investigated samples. Three conditions of specimens were hereby taken into account, which was as-received PBT-30GFR sample, PBT-30GFR sample heat-treated at 260°C and PBT-30GFR sample heat-treated at 260°C under 1-ton pressure. This temperature was the joining temperature used in the hot pressing. The results are illustrated in figure 10. The spectrum at the wavenumbers between 2800-3000 cm^{-1}, 1700-1750 cm^{-1}, 1000-1300 cm^{-1} and 700-990 cm^{-1} represented alkane C-H group (stretching), C=O bond, C-O and C-H (bending), respectively [19]. It can be seen that for the PBT-30GFR specimens after heat treatment and PBT-30GFR specimens after heat treatment combined with 1-ton pressure no molecular structural change of C=O bond was observed as compared to the spectrum of C=O in the as-received PBT specimen. It meant that the direct bonding between PBT-30GFR and aluminium A5052 by the hot pressing did not cause a change of molecular structure in the examined PBT-30GFR plastic.

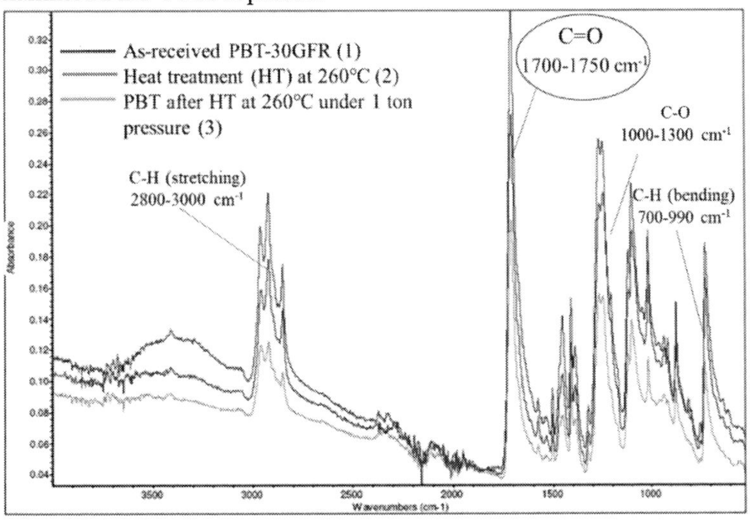

Figure 10. FTIR spectrum of (a) as-received PBT-30GFR sample, (b) PBT-30GFR sample heat-treated at 260°C and (c) PBT-30GFR sample heat-treated at 260°C under 1-ton pressure

IOP Conf. Series: Materials Science and Engineering **886** (2020) 012058 doi:10.1088/1757-899X/886/1/012058

4. Conclusion

In this work, the joining of aluminium alloy sheet grade A5052 and polybutylene terephthalate with 30% glass fiber reinforcement by means of the hot-pressing process was studied. Both mechanical and chemical surface treatment was applied to form a nanoscale surface structure on the aluminium sheet. Then, roughness measurement and SEM analysis were performed for the prepared sheet surfaces. The influences of joining process parameters resulted in joining strengths from tension shear tests were investigated. The conclusion can be drawn as follows.

- The surface roughness, joining temperature and press force showed significant effects on the joining strength of joined samples.
- The maximum joining strength could be achieved by using the pressing force of 1 ton, joining temperature of 270°C and holding time of 60 s.
- The joining strength of samples obviously increased with an increased surface roughness of the aluminium sheet. The sandblasting combined with chemical surface treatment led to the highest surface roughness and thus provided the largest joining strength. Hereby, the depth of mechanical interlocks between surfaces of both materials was approximately 33 μm.
- The observed failure mode of joined Al-PBT samples was a mixed-mode of interface and cohesive failure.
- No change of molecular structure in PBT-30GFR plastic after direct bonding with aluminium by the hot-pressing process.

Acknowledgments

The authors would like to acknowledge the National Research Council of Thailand and King Mongkut's University of Technology Thonburi (KMUTT) for financial supports.

References

[1] Stokes V K 1989 *Polymer Eng. Sci.* **29**(19) 1310
[2] Budhe S Ghumatkar A Birajdar N Banea M D 2015 *Applied Adhesive Sci.* **3** 20
[3] Amend P Pfindel S Schmidt M 2013 *Physics Procedia* **41** 98-105
[4] Balasubramanian V 2008 *Mater. Sci. Eng. A* **480** 397
[5] Kajihara Y Tamura Y Kimura F Suzuki G Nakura N Yamaguchi E 2018 *CIRP Annals* **67**(1) 591
[6] Kondoh K Umeda J 2018 *Mater. Letters* **211** 331
[7] Hussein S K Mhessan A N Alwan M A 2017 *Eng. J.* **21**(7) 157
[8] Li X P Xu D H Gong N N Xu Z Wang L Dong W 2018 *Mater. Design* **179** 107875
[9] Kimura F Kadoya S Kajihara Y 2016 *Precision Engineering* **45** 203
[10] Moseke C Gbureck U 2019 *Metals for Biomedical Devices* **355** 367
[11] Rodríguez-Vidal E Sanz C Soriano C Leunda J Verhaeghe G 2016 *Mater. Proc. Tech.* **229** 668–677
[12] Okadaa M Taketab H Toriib Y Iriea M Matsumotoa T 2019 *Dental Mater* **35** 169–175
[13] Arenas J M Alia C Narbon J J Ocana R Gonzalez C 2013 Composites **415** 423
[14] Fabrin P A Hoikkanen M E Vuorinen J E 2007 *Polymer Eng Sci* **47**(8) 1187
[15] Naritomi M Andoh N 2010 *US Patent App.* **12/669** 143
[16] Martinsen K Hu S J Carlson B E 2015 *CIRP Annals* **64** 679
[17] Li X P Gong N N Yang C Zeng S Fu S Zhang K 2018 *J. Mater. Proc. Tech.* **255** 635
[18] ASTM D5573-99 2019 *Standard Practice for Classifying Failure Modes in Fibre-Reinforced-Plastic (FRP) Joints* (West Conshohocken:ASTM International)
[19] Domininghaus H Elsner P Eyerer P Hirth T 2012 *J. Chem. Education* **537** 546

IOP Conf. Series: Materials Science and Engineering **886** (2020) 012059 doi:10.1088/1757-899X/886/1/012059

Numerical Investigation of the Effects of Particle Density and Dimensionless Acceleration on Segregation in 2D Vertically Vibrated Binary Granular Mixtures

P Jongchansitto[1,*], and I Preechawuttipong[1]

[1]Deparment of Mechanical Engineering, Faculty of Engineering, Chiang Mai University, 239 Huay Kaew Rd., Muang District, Chiang Mai, 50200, Thailand

* Corresponding Author: pawarut.j@cmu.ac.th

Abstract
The present work aims to investigate the influences of particle density and dimensionless acceleration on segregation phenomena in 2D binary granular mixtures by using contact dynamics method. The granular samples were prepared by randomly depositing two species of regular pentagonal particles inside a rectangular container. These two particle species are the same in size, but differ in terms of particle density. Different levels of vertical vibration were then applied to the mixtures. The simulation results were systematically analyzed by using segregation coefficient (H), segregation pattern, and velocity field related to convection mechanism. It was found that segregation phenomena are not observed in this study. This could be explained by the fact that a value of friction coefficient between interparticle and that value between wall and particle is set to be the same, that is why no segregation was found in the present study. Due to the friction, only convection rolls were observed for all the density ratios when dimensionless acceleration is greater than 2.

Keywords: Granular materials, Segregation, Density, Vertical vibrations, Contact dynamics method

1. Introduction

Granular materials are omnipresent around us in such a way that we can encounter in our daily life. For example, coffee beans, sugar, sands, rocks, etc. In this context, granular materials are defined as a collection of solid particles whose macroscopic mechanical behavior is governed by the contact forces between the particles. Due to a great diversity in grain properties, granular materials generally display complex behaviors, which cannot be easily categorized into ordinary solids, liquids, or gases [1]. These behaviors cause an occurrence of various phenomena that remains partially understood. One of the interesting phenomena is segregation, which is observed when a homogeneous bulk solid composed of various constituents or species evolves from an initially mixed state to a spatially non-uniform state due to the relative movement within the material [2]. The segregation is indeed an undesirable phenomenon observed in many industrial processes involving the processing and handling of granular mixtures such as pharmaceutical, powder metallurgy, agriculture, and food processing [2-4]. Such segregation often occurs during transportation and handling of granular mixtures due to shaken or vibrated systems [4-6].

It must be also noted that an occurrence of segregation phenomena is a major issue affecting low product yield and out-of-spec products. Consequently, this issue has motivated many researchers endeavoring to describe and understand the segregation phenomena in granular media for a long time. The particle size has been mainly employed to investigate such phenomena, compared to other particle properties such as density [2], shape [4, 6], or friction [7, 8]. It could be clearly seen that there is still a lack of knowledge and understanding in terms of the effects of other particle properties on the segregation phenomena.

Therefore, the influence of particle density on segregation phenomena is systematically investigated by changing densities of particles inside binary granular mixtures composed of regular pentagonal particles. These mixtures are subjected to various vertical vibrations characterized by dimensionless acceleration (Γ), in order to also understand the effect of driving parameters. The paper is organized as follows. Section 2 describes in brief about the fundamental of contact dynamics (CD) method, the details of numerical samples, and parameters used for segregation analysis. Section 3 is devoted to analyze and discuss the results obtained in the present work. Finally, the conclusions are then provided in Section 4.

2. Numerical Simulations

An open-source software so-called "LMGC90" [9, 10], which has been well-known and widely used for the simulations of granular materials, was employed in this study. It must be noted that LMGC90 is dedicated to the modelling of a large number of deformable or un-deformable interacting objects of any 2D or 3D shape with various mechanical behaviors. It also aims at considering interaction laws as complex as necessary, including multi-physics couplings such as thermal effects, fluids, etc.

2.1. Contact dynamics (CD) method

In this study, the simulations were performed by using the contact dynamics method or sometime called non-smooth contact dynamics (NSCD). This method is based on implicit numerical algorithms together with a non-smooth dynamic contact law and Coulomb's law of friction to integrate equations of motion of rigid bodies. It must be noted that elastic repulsive interactions and smooth Coulomb's law of friction are not required for the determination of contact forces. In the CD method, contact laws are employed to characterize the interactions between particles, instead of force laws. Only three parameters are required for the characterization of these interactions: the coefficient of friction when it is nonzero and the coefficients of normal and tangential restitution that control the dissipation rate. The unknown variables are only particle velocities and interaction forces. These variables are handled with an iterative method based on nonlinear Gauss-Seidel algorithm at each step for all potential contacts. A set of potential contacts is then identified or updated in each step. Note that the contact between two particles is detected by the overlaps of the space between them. In 2D simulations of the present study, the so-called "shadow overlap method" [11] was employed for the detection of contact between two convex polygonal bodies. For additional details in the implementation of the CD method, please see in refs. [12, 13].

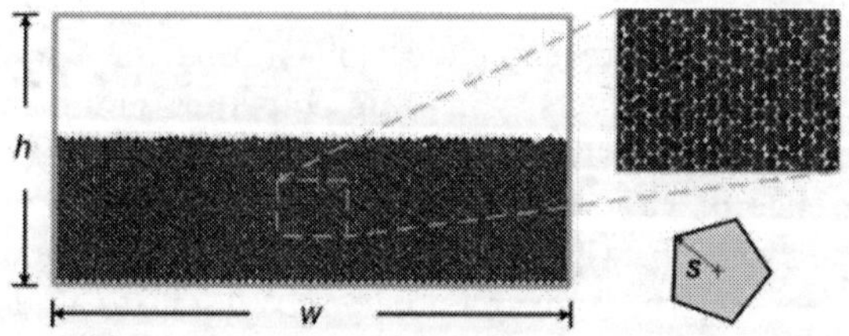

Figure 1. Example of preparation of granular samples in the present work. Dense particles are represented by "blue" color, while loose particles are in "red" color

2.2. Numerical preparations

Figure 1 shows an example of sample preparation in this study. Granular samples were prepared by two species of regular pentagonal particles with the same size, but different in terms of density. Five different density ratios consisting of 1.0, 1.7, 4.0, 8.0, and 12.0 were used in this study. Note that this density ratio is specified by the density between dense particle and loose particle (ρ_d / ρ_l). However, in order to prevent the effect of monodisperse granular materials, the particle sizes were necessary to be slightly different. Therefore, the particle sizes were generated by using a normal distribution between two values of the length measured from a centroid to a peak of the regular pentagon (s), i.e. s_{min} = 5.63 mm and s_{max} = 5.86 mm (see in Figure 1). Each particle species has the same number of particles, i.e. 2000 particles, which were randomly placed inside a rectangular container. It means that the total number of particles inside the granular sample is equal to 4000 particles. The aspect ratio between width and height of the container (w/h) was fixed at 2, see in Figure 1. The coefficient of friction between particles and that between the container wall and particle were set to be the same value of 0.3. Under the gravitational force, vertical vibrations in the manners of the sinusoidal wave were then applied to the granular samples. It must be also noted that the level of vibration was controlled by the dimensionless acceleration (Γ), which is defined by

$$\Gamma = \frac{A\omega^2}{g} = \frac{A(2\pi f)^2}{g} \quad (1)$$

where A is the vibrational amplitude, f is the vibrational frequency, and g is the gravity. In this study, the vibrational amplitude was fixed at 5.0 mm, while the vibrational frequency was progressively varied to reach the values of $\Gamma = 1 - 5$. During the simulations, the kinematical data were saved every 10 seconds until the simulation finished at 300 seconds.

2.3. Segregation analysis

After the simulations were completed, the kinematical data obtained during the simulations were used to methodically analyze the segregation. Segregation coefficient (H), segregation pattern, and velocity fields were employed for this purpose.

2.3.1. Segregation coefficient (H)

This parameter is applied to quantify the degree of segregation of the binary mixtures in a static state [14], which is defined as

$$H = 2\frac{h_d - h_l}{h_d + h_l} \quad (2)$$

where h_d is the average height of dense particles and h_l is the average height of loose particles. The height of each particle was measured at their centroid with respect to the container base. Factor 2 is a multiplication factor arising from the equal solid fraction of each particle species which limits the average height of particles in such species to $0.25 < h_\alpha < 0.75$ and thus $-1 < H < 1$. In this case, the dense particles initially rise to the top of the container when $H > 0$, while the dense particles start to migrate to the bottom of the container when $H < 0$. Note that the segregation coefficient at the initial configuration of each granular sample is equal or approach to zero.

2.3.2. Segregation pattern

Previously, no segregation indices or coefficients can be used to characterize the segregation pattern. Until in 2016, Coletto *et al.* [15] developed and proposed a new segregation index named "Three Thirds Segregation Indices Set (TTSIS)". This index cannot only describe the degree of segregation of the mixtures, but also the segregation pattern of interesting species. Hence, the TTSIS was applied for the analysis of the segregation pattern by modifying some conditions for the segregation patterns classification to consist of the conditions used in this study. In a classification of the segregation patterns, a total number of particle centroids of interested species along every three ranges inside the container were necessary to be quantified by three corresponding indicators (see in Table 1): a bottom indicator (p_B), Middle indicator (p_M), and top indicator (p_T). It must be noted that the species of interest in this study was taken into account only dense particles.

Table 1. Conditions used in this study for classification of segregation patterns

	p_B	p_M	p_T	and	Segregation pattern	
if	≥ 0.33	≥ 0.33	≥ 0.33	-	PU	Pure uniform
Else if	$0.3 \leq p_B \leq 0.4$	$0.3 \leq p_M \leq 0.4$	$0.3 \leq p_T \leq 0.4$	-	U	Uniform
Else if	$0.55 \leq p_B < 0.98$	-	-	$\lvert p_T - p_M\rvert < 0.14$	B	Bottom
Else if	-	$0.55 \leq p_M < 0.98$	-	$\lvert p_T - p_M\rvert < 0.14$	C	Central
Else if	-	-	$0.55 \leq p_T < 0.98$	$\lvert p_T - p_M\rvert < 0.14$	T	Top
Else if	$0.47 \leq p_B < 0.54$	-	$0.47 \leq p_T < 0.54$	-	V	Pure V
Else if	$> p_M$	-	$> p_M$ and $> p_B$	-	VT	V-top
Else if	$> p_M$ and $> p_T$	-	$> p_M$	-	VB	V-bottom
Else if	-	$p_B < p_M < p_T$	-	$p_M - p_B \geq 0.14$	TC	Top central
Else if	-	$p_B > p_M > p_T$	-	$p_M - p_T \geq 0.14$	BC	Bottom central
Else if	-	-	$p_M > p_T > p_B$	$p_T - p_B \geq 0.14$	CT	Central top
Else if	$p_M > p_B > p_T$	-	-	$p_B - p_T \geq 0.14$	CB	Central bottom
Else	-	-	-	-	TS	Transition

2.3.3. Velocity field

According to the results obtained from previous researchers, it was found that convection is one of the important mechanisms influencing strongly on segregation phenomena in granular materials under vibrations [6, 16]. In general, two symmetric convection rolls were observed along two sidewalls of the vibrated container. The particles migrate downward along the side walls and move upward in the central part of the container. This phenomenon is known as "normal convection" [17]. This convection can be represented by a plot of the velocity of each particle inside the vibrated container. Therefore, the velocity fields will be employed in this study to analyze the occurrence of convection in the granular sample.

3. Results and Discussions

In this section, the effects of particle density and level of vibration on segregation phenomena analyzed by the three parameters for segregation analysis as mentioned in the previous section are presented and discussed as follows.

3.1. Segregation coefficient (H)

Figure 2 shows the relationship between the segregation coefficient (H) and time. It is clearly observed that the values of the segregation coefficient are uniform and close to zero ($H \approx 0$) along all the time of vibration for any density ratios and any dimensionless accelerations. In this case, it can be said in the beginning that segregation phenomena do not occur in the vertically vibrated granular system, even though both density ratios and dimensionless accelerations were changed in this study. In order to be undoubtedly ensure that no segregation arises in the present work, the segregation pattern which can characterize the distribution of a collection of interested particle species inside the container is then employed for this purpose.

3.2. Segregation pattern

The segregation pattern of the vertically vibrated granular mixtures at each time is presented in Figure 3. Note that only a collection of dense particles was used for the analysis of the segregation pattern. It can be seen that only three segregation patterns are observed: Pure Uniform (PU), Uniform (U), and Transition (TS). Distributions of these patterns have a similar characteristic, i.e. the number of the centroids of interested particles is distributed uniformly in each range of the container, excluding TS pattern that is attempting to change their distribution to other segregation patterns. As a result, it is clearly found that there is no occurrence of segregation phenomena in this study, corresponding to the results analyzed by the segregation coefficient.

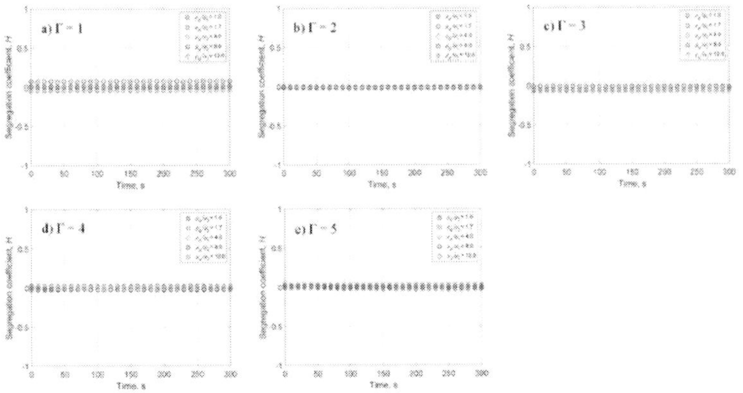

Figure 2. Segregation coefficient (*H*) as a function of time in the cases of ρ_d/ρ_l = 1.0, 1.7, 4.0, 8.0, and 12.0 under different dimensionless accelerations: a) Γ = 1, b) Γ = 2, c) Γ = 3, d) Γ = 4 and e) Γ = 5.

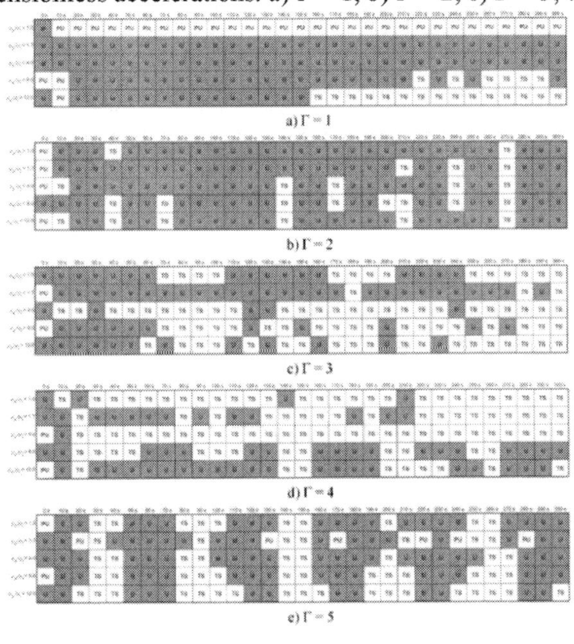

Figure 3. Segregation pattern as a function of time in the cases of ρ_d/ρ_l = 1.0, 1.7, 4.0, 8.0, and 12.0 under different dimensionless accelerations: a) Γ = 1, b) Γ = 2, c) Γ = 3, d) Γ = 4 and e) Γ = 5. Note that the segregation highlight with green color is a uniform pattern

3.3. Velocity field

Let us consider next the convection through a plot of velocity fields. As mentioned earlier, the convection is one of the main driving mechanisms for segregation phenomena. Therefore, the plotting of velocity fields enables us to clearly understand the role of the convection on the present work.

In the case of the dimensionless acceleration less or equal to 2, no convection rolls exist for any density ratios as shown in Figure 4a. It is evident that each particle inside the granular system just moves upward and downward alternately, according to a characteristic movement of the sinusoidal wave. In this manner, the granular system is more compacted by the vibration. In addition, it is interesting to note that some particles near the surface of the granular system are more excited in such a way that they move with high velocity and quite turbulence. In the case of the dimensionless acceleration greater than 2, the convection rolls are observed for all the density ratios used in this study as illustrated in Figure 4b. It is clearly seen that more than two convection rolls occurring in the granular system can be found in sometimes period. This observation corresponds to the experimental results reported in refs. [18, 19]. In addition, it is seemingly that a mixing occurrence of two types of convection is observed: normal and reverse convection.

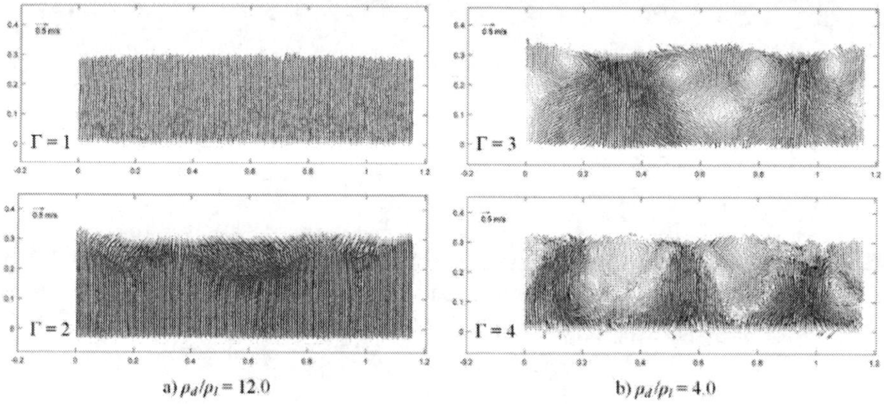

Figure 4. Examples of velocity fields of each particle in the granular mixtures with a) $\rho_d / \rho_l = 12.0$ at 140 seconds and b) $\rho_d / \rho_l = 4.0$ at 50 seconds under vertical vibrations

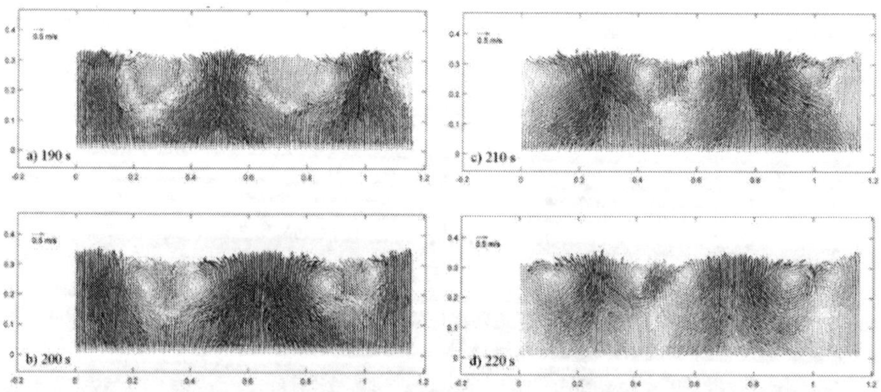

Figure 5. Examples of velocity fields of each particle in the granular mixtures with $\rho_d / \rho_l = 8.0$ under vertical vibrations with $\Gamma = 3$ at a) 190 seconds, b) 200 seconds, c) 210 seconds, and d) 220 seconds

Apart from an occurrence of more than two convection rolls, it is interesting to also note that there are some movements of the convection rolls inside the container as shown in Figure 5. This situation is in contrast to whatever reported in the previous studies [5, 16, and 17]. The number of convection rolls and their movement might be caused by the effect of friction. In this study, it can be said that the effect of convection due to friction dominates over the effect of particle re-organization due to density ratio and dimensionless acceleration. Significantly, a value of friction coefficient between interparticle and that value between the container wall and particle is set to be the same. This is a reason why segregation phenomena do not occur in this study.

4. Conclusion

The objective of this study is to methodically investigate the influences of particle density and dimensionless acceleration on segregation phenomena in two-dimensional granular media subjected to vertical vibrations. For this purpose, Binary granular systems composed of 4000 regular pentagonal particles with different density were systematically simulated by means of contact dynamics (CD) method via LMGC90 software. Under the gravitational force, vertical vibrations were then applied to the mixture. The level of vibration was controlled by a dimensionless acceleration (Γ). It was found that for all the density ratios under any dimensionless accelerations, segregation coefficient (H) is close to zero. In addition, only "Uniform" and "Transition" patterns are observed in terms of the distribution of a collection of all dense particles inside the granular system. These results clearly indicate that no segregation phenomena occur in the vibrated granular systems due to the effects of particle density and dimensionless acceleration. This situation could be explained by the fact that the coefficient of friction between interparticle was set to be the same as the coefficient of friction between particle and wall, leading to an occurrence of convection. The convection occurred here are mixing between normal and reverse convections, when the dimensionless acceleration is greater than 2 for any density ratios. It is interesting to note that the convection rolls do not occur only near the sidewalls of the container as found on the previous studies, but they also move to other areas inside the container. Moreover, more than 2 convection rolls are observed in some periods. In this case, it could be said that the effect of friction causing the convection dominates over the effects of density and dimensionless acceleration. That could be a reason why no segregation was found in this study.

References

[1] Jaeger H M Nagel S R1996 *Rev. Mod. Phys.* **68** 1259
[2] Rosato A D Lan Y Wang D T 1991 *Powder Technol.* **66** 149
[3] Menbari A Hashemnia K 2019 *Chem. Eng. Sci.* **207** 942
[4] Li L Wu P Zhang S Wang L 2019 *Powder Technol.* **342** 404
[5] Jain A Metzger M J Glasser B J 2013 *Powder Technol.* **237** 543
[6] Knight J B Jaeger H M Nagel S R 1993 *Phys. Rev. Lett.* **70** 3728
[7] Ulrich S Schröter M Swinney H L 2007 *Phys. Rev. E* **76** 042301
[8] Liao C C Hunt M L Hsiau S S Lu S H 2014 *Phys. Fluids.* **26** 073302
[9] Dubois F Jean M 2006 *Analysis and Simulation of Contact Problems* ed P Wriggers U Nackenhorst (Springer) pp 375-378
[10] Dubois F Jean M Renouf M Mozul R Martin A Bagnéris M 2011 *10e Colloque National en Calcul des Structures* (Giens:France)
[11] Saussine G Cholet C Gautier P E Dubois F Bohatier C Moreau J J 2006 *Comput. Method Appl.* **195** 2841
[12] Radjai F Richefeu V 2009 *Mech. Mater.* **41** 715
[13] Dubois F Acary V Jean M 2018 *CR. Mecanique.* **346** 247
[14] Ciamarra M P De Vizia M D Fierro A Tarzia M Coniglio A Nicodemi M 2006 *Phys. Rev. Lett.* **96** 058001
[15] Coletto M M Marchisio D L Barresi A A 2016 *Powder Technol.* **299** 77
[16] Gallas J A C Herrmann H J Sokolowski S 1992 *Phys. Rev. Lett.* **69** 1371
[17] Grossman E L 1997 *Phys. Rev. E.* **56** 3290
[18] Aoki K M Akiyama T Maki Y Watanabe T 1996 *Phys. Rev. E.* **54** 874
[19] Knight J B 1997 *Phys. Rev. E.* **55** 6016

IOP Conf. Series: Materials Science and Engineering **886** (2020) 012060 doi:10.1088/1757-899X/886/1/012060

Yield Parameters for Iron Powder Compaction on Different Particle Size

R Ogawa [1], J Fujii and Y Taniguchi [2, *]

[1]Student of National Institute of Technology (KOSEN), Nara College, Nara 639-1080, Japan
[2] Department of Mechanical Engineering, National Institute of Technology (KOSEN), Nara College, Nara 639-1080, Japan

* Corresponding Author: taniguchi@mech.nara-k.ac.jp

Abstract
Shear failure yield line in Drucker-Prager yield surface for iron powder on different particle size and amount of interparticle lubricant has been estimated. Critical stress state and uniaxial failure stress state have been determined by a single shear test and uniaxial failure test for a green compact. Compaction cap parameter has been estimated by the lateral force measurement test by instrumented die compaction system. By those estimations, materials input parameters for the Drucker-Prager Cap (DPC) model have been determined successfully, and differences in particle size and amount of lubricant can be recognized. Finite element method (FEM) analysis of simple closed-die compaction has been conducted. As a result, the materials input parameters for the DPC model on different powder properties have been successfully identified as the FEM analysis fairly predicts compaction pressure and density distribution on uniaxial compaction.

1. Introduction
Metal powder die compaction is widely applied as a forming method for several sintered parts such as gear, synchronizer hub and so on. It is necessary to control the compression tool velocity that accomplishes uniform consolidation until objective shape dimension of the green compact, and to avoid crack occurrences such as slip-crack [1, 2] and shear failure, but it is not easy to optimize especially in case of highly complex shape with high density. Therefore, it is effective to apply the numerical simulation for powder die compaction process to predict density distributions and shear failure occurrence [3-8]. The Constitutive equation of metal powder is complex because both consolidation yield properties and shear failure yield properties must be considered. The compaction yield function is expressed by elliptical function such as Shima-Oyane model [9]. On the other hand, the Drucker-Prager model [10] which is expressed by linear function is the simplest shear yield function for granular materials. Drucker-Prager Cap (*DPC*) model is defined as a combination of these functions which is indicated in the principal stress space as conical shape that the tip is closed with a consolidation cap, and its central axis is hydrostatic pressure axis. It is necessary to conduct a triaxial compaction test to estimate the cap parameters in the *DPC* model [11-14]. However, the testing device is not easy to construct due to the complicated tool moving mechanism. Also, more than two kinds of shear failure

test for green compact must be performed to estimate shear yield parameters in the model. Thus, it takes considerable effort to identify the materials parameters for the *DPC* model. So, we have suggested applying single shear test which is general strength-test in soil mechanics to the iron powder green compact [15] and succeeded in measuring Critical State Line (*CSL*) which is a stress state of *DPC* yield surface [16, 17]. Since it indicates a division point of consolidation stress state and shear failure stress state for granular materials, *CSL* provides both of consolidation cap parameters and Drucker-Prager failure parameters. Cap yield surface has been drawn by adding uniaxial compaction stress state that is easily determined by lateral force measurement during closed die compaction. Also, parameters of shear failure line have been identified by adding failure stress state which is measured by simple compression failure test for a green compact. By those examinations, we have succeeded in the estimation of materials parameters for iron powder in the *DPC* model [18] and prediction of failure occurrence in multi-step parts compaction [19]. Results of finite element method (FEM) analysis shows good corresponding with actual die compaction characteristics, but the influence of frictional force during compaction was not considered in the estimation of yield parameters. In this study, frictional force on the die wall in lateral force measurement test has been considered by improving testing device, thus its effect has been eliminated from the estimation of yield parameters. Then the differences in materials parameters on different particle size and different amount of interparticle lubricant have been investigated and its validity on compaction force with a magnitude of the density distribution of green compact has been confirmed by FEM analysis.

2. Definition of Drucker-Prager Cap model

2.1. Yield function

Yield criterion of the *DPC* model in *ANSYS* is defined by the relationship between the first invariant of stress I_1 and second invariant of deviatoric stress J'_2. The Compaction cap function Y_c, Shear failure function Y_s and Expansion cap function are given by equation (1) and (2). Moreover, these functions are combined with equation (3).

$$Y_c = 1 - H(K_o - I_1)\left(\frac{I_1 - K_o}{R_c Y_S(K_o, \sigma_o)}\right)^2 \tag{1}$$

$$Y_s = \sigma_o - Aexp(BI_1) - \alpha I_1 \tag{2}$$

$$F(I_1, J'_2, K_o, \sigma_o) = J'_2 - Y_c(I_1, K_o, \sigma_o) Y_s^2(I_1, \sigma_o) = 0 \tag{3}$$

Where Compaction cap parameter R_c means the ratio of elliptical I_1 to J'_2 axis. σ_o is the cohesion related hardening parameter with consolidation. Shear envelope exponential coefficient A, Shear envelope exponent B and Shear envelope linear coefficient α are material parameters which define shear yield stress respectively. H and K_o of equation (1) express Heaviside function and a key flag indicating the current transition point at which the compaction cap surface and shear failure surface intersect. Note that the Heaviside function roles as follow;

1. When I_1 is greater than K_o, the compaction cap takes no effect on yielding. In other words, only shear failure function Y_s is effective.
2. When I_1 is less than K_o, the yielding may only happen in the compaction cap portion, which is shaped by both the shear failure function Y_s and cap function Y_c as expressed by equation (1)

Equation (2) means shear failure envelope is simplified as the linear relationship with intercept σ_o and inclination α for $A = B = 0$. Therefore, shear yield surface is expressed as a conical shape having a central axis of mean stress σ_m in three principal stress space.

IOP Conf. Series: Materials Science and Engineering **886** (2020) 012060 doi:10.1088/1757-899X/886/1/012060

2.2. *Hardening function*

The cap hardening law is defined by describing the evolution of the parameter X_o, which is the current intersection point of the compaction cap and the I_1 axis. In other words, hardening is expressed by increasing volumetric plastic strain ε_{pv} with consolidation proceeding that is defined as:

$$\varepsilon_{pv} = W_1^c[exp\{D_1^c(X_o - X_i)\} - 1] \quad (4)$$

where: X_i =Initial value of compaction cap yield pressure X_o at which the cap function takes effects. W_1^c = Maximum possible volumetric plastic strain which is defined from densification curve of the powder material. D_1^c = Hardening parameter.

2.3. *DPC model in $I_1 - \sqrt{J'_2}$ plane*

According to equations (3), the yield surface of the plane *DPC* model in I_1 - $\sqrt{J'_2}$ can be drawn schematically as shown in figure 1. Subsequent yield surface can be defined as work hardening behavior which is obtained by variations in density ratio to ρ_0 from ρ_i = Initial yield surface by updating initial parameters X_i and σ_c. These parameters are not able to measure directly but can be estimated by interpolation from the stress states of several density ratios.

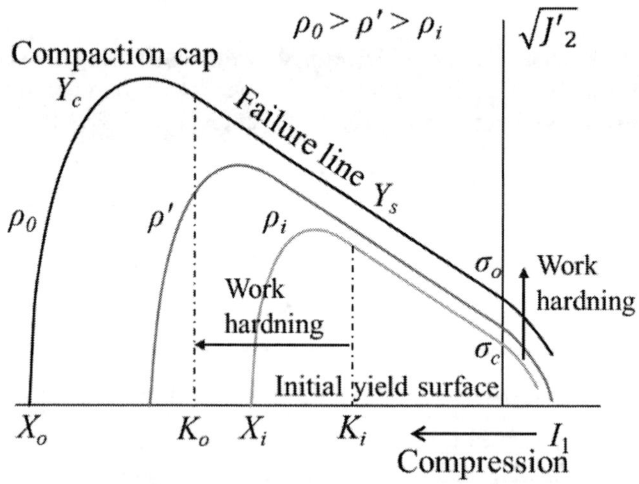

Figure 1. Schematic drawing of yield surface of the *DPC* model

3. Various examinations to identify the parameter of yield surface in the DPC model

3.1. *Uniaxial failure test and single shear test*

Uniaxial failure test of green compact provides a point of stress state in failure by uniaxial principal stress (σ_1, $\sigma_2 = \sigma_3 = 0$). The uniaxial failure stress state is obtained as:

$$\sqrt{J'_2} = \sigma_1/\sqrt{3} \quad (5)$$

$$I_1 = \sigma_1 \quad (6)$$

The shape of green compact for uniaxial failure test is cylindrical in 10 mm diameter with 15 mm height.

Single shear test of green compacts in several compact densities provides *CSL*. In previous work, *CSL* has identified [15]. When a green compact executes shear failure, vertical displacement on the shear failure plane indicates a positive value which means volume expansion. In the case of vertical force, which exceeds a certain value, vertical displacement indicates a negative value which means consolidation occurrence due to higher mean stress (hydrostatic stress) state. The stress states at which the vertical displacement was kept in zero during shear were estimated by a single shear test. Thus, we can identify K_o directly by the estimation of *CSL*. The slope of failure lines α and evolution of cohesion σ_0 with densification proceedings can be estimated by equation (2) by using stress states of *CSL* and uniaxial failure. The shape of green compact for a single shear test is cylindrical in 20 mm diameter with 8 mm height.

3.2. Lateral force measurement test

Lateral force measurement provides a point of stress state in densification by measuring forming force and lateral force during closed-die compaction. Principle of the lateral force measurement device is shown in figure 2. The device is composed of servo cylinder which gives compaction load as forming force F_{upper} and die set with load cells which measure lateral force F_L and transmitted force F_{lower} respectively. Frictional force occurs on the inner surface of the mold deduces applied forming force, therefore F_{lower} becomes lower than F_{upper} and influence of such frictional force should not be ignored in the identification of compaction cap parameters. The green compact which shape is a square column with 8 mm × 8 mm sectional dimension can be compacted.

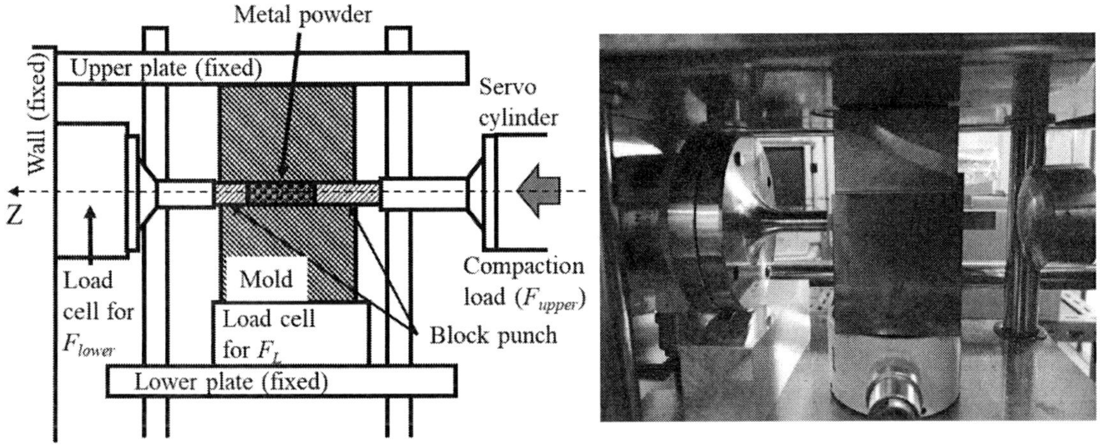

(a) Schematic view of a lateral force measurement device (b) Image of the die set in the system

Figure 2. Schematic view of a lateral force measurement device

σ_{upper} and σ_{lower} are derived from F_{upper} and F_{lower}. Therefore, σ_z is given by:

$$\sigma_z = \sigma_{lower}\left(\frac{\sigma_{upper}}{\sigma_{lower}}\right)^{\frac{Z}{H}} \tag{7}$$

where: H means the whole height of the powder part in the die, Z = representative value for the height of powder part. In this study, Z was considered as central in the powder. Therefore, H / Z was 1 / 2. So, equation (7) is rewritten as follows:

$$\sigma_z = \sigma_{lower}\sqrt{\frac{\sigma_{upper}}{\sigma_{lower}}} \tag{8}$$

Therefore, by using σ_z and lateral pressure σ_L, the stress state is given by:

$$I_1 = \sigma_1 + 2\sigma_L \tag{9}$$

$$\sqrt{J'_2} = |\sigma_1 - \sigma_L|/\sqrt{3} \tag{10}$$

where: $\sigma_1 = \sigma_z$, $\sigma_2 = \sigma_3 = \sigma_L$.

3.3. Experimental condition

Iron powder with zinc stearate, which is an additive as interparticle lubricant, has been used. Zinc stearate was spread on the inner surface of the mold in the lateral force measurement test. It is necessary to prepare green compact specimens with proper density for both uniaxial failure test and single shear test. The density ratios ρ, which are defined as the ratio of the compact density to material density, were chosen as 0.70, 0.75, 0.80 and 0.85. Atomized iron powders having approximately 100 µm mean particle size with 1.0 mass % zinc stearate (Fe 100 µm -1.0 %) and 0.5 mass % zinc stearate (Fe 100 µm-0.5 %) were tested. Besides, carbonyl iron powder having more fine particles size, which is approximately 10 µm with 1.0 mass % zinc stearate (Fe 10 µm-1.0 %) were also tested.

4. Results and discussion

4.1. Critical state line

Figure 3 shows the *CSL* as a relation between vertical stress σ and shear stress τ on the shear failure plane of powder compacts for several density ratio. It is well known that finer particle size shows fewer consolidation characteristics due to poor flowability, therefore the slope of *CSL* in 10 µm particle size becomes higher as it shows higher values of shear strength for each density ratio. Since less amount of interparticle lubricant decreases flow ability, also 0.5 mass% zinc stearate showed a higher slope of *CSL*.

4.2. Compaction yield characteristics

Figure 4 shows an example of the variations of punch stroke and applied forces in the lateral force measurement test. In loading, each force increases exponentially with increasing punch stroke. The friction coefficient on the inner surface in the loading stage is derived by:

$$\mu = \frac{F_{upper} - F_{lower}}{F_L} \tag{11}$$

The values of μ were approximately 0.10 for each tested powder. In the unloading stage, each force decreases almost linearly but lateral force remains significantly. The values of F_L / F_{upper} were 37% for atomized powder and 41% for finer carbonyl powder. It indicates that the finer particle size requires higher stress to densification because of too much point of contact which having frictional resistance.

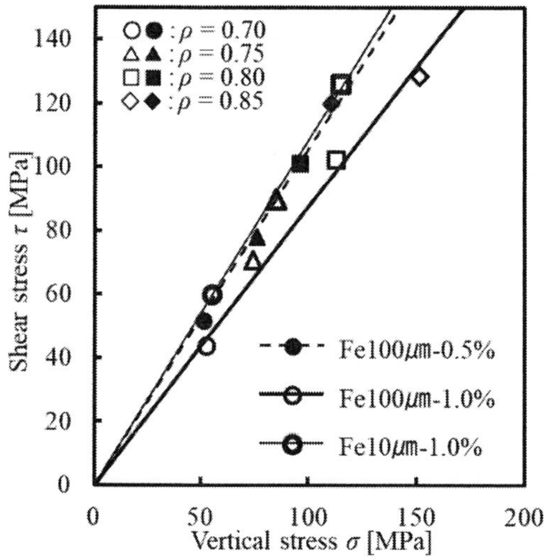

Figure 3. The critical state line of tested powders

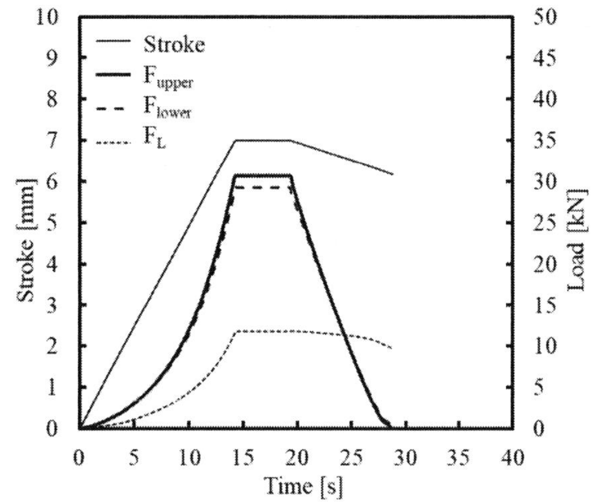

Figure 4. Variations of punch stroke and forces (Fe 100 μm-1.0%)

Figure 5 shows the loading pass in I_1 - $\sqrt{J'_2}$ plane for each tested powder. Poor compaction characteristics of Fe 10 μm-1.0 % powder is clearly recognized. Interparticle lubricant improves compaction characteristics slightly. However, the influence of I_1 to $\sqrt{J'_2}$ is comparatively negligible for each tested powder. It is considered that magnitude of work hardening increases significantly in finer particle size but hydrostatic pressure dependence of the deviatoric stress is not varied if substrate materials properties such as elastic modulus and tensile strength are same for the powder.

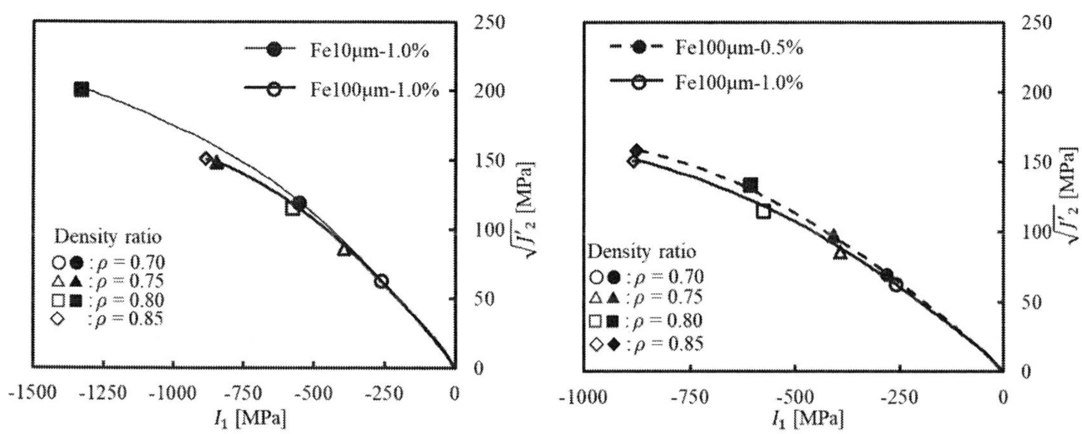

(a) Comparison in particle size (b) Comparison in lubricant amount

Figure 5. Loading pass in I_1 - $\sqrt{J'_2}$ plane

4.3. Estimation of yield surface of the DPC model

Figure 6 shows an estimated yield locus for each density ratio. The stress states which were investigated by different test, uniaxial failure, single shear and lateral force measurement were combined by *DPC* models. However, since $\sqrt{J'_2}$ values at the critical stress states of Fe 10 μm-1.0% powder are shown at lower than the compaction cap stress states, K_o value is not identified directly by a single shear test. It was considered that the shear failure occurs at different plane due to the clearance which is necessary to conduct single shear in case of finer particle compact. To avoid such crack occurrence at the different

IOP Conf. Series: Materials Science and Engineering **886** (2020) 012060 doi:10.1088/1757-899X/886/1/012060

plane with applied vertical force, modification of the shape of the specimen to cube shape having square shear plane will be effective.

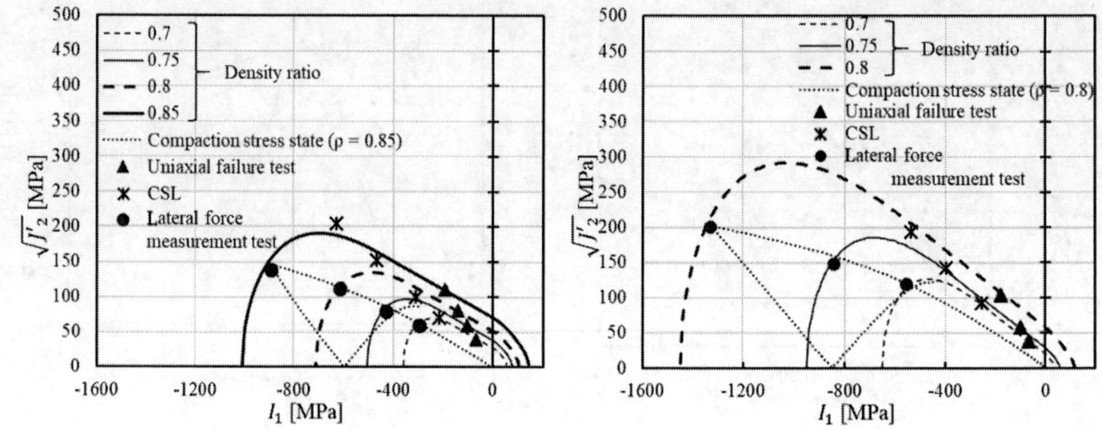

(a)Fe 100 μm-1.0% (b)Fe 10 μm-1.0%

Figure 6. Estimated yield surfaces of the *DPC* model for different particle size

Table 1 shows the yield parameters which have been identified to describe figure 6. The slope of shear failure envelope α was constant in each density ratio, and dependence exists on the particle size difference. It is also influenced by decreasing the amount of lubricant as it becomes 1.5 times higher in case of Fe 100 μm-0.5 % powder. Thus, the interparticle lubricant decreases the shear strength of the compact and its behaviour can be implemented by α in the *DPC* model. Poor compaction characteristics of Fe 10 μm-1.0% powder is implemented by lower hardening parameter D_1^c with higher compaction cap parameter R_c. It indicates that incrementation of densification in hydrostatic pressure rapidly decreases, thus the yield by deviatoric stress state becomes effective relatively in case of finer particle size. In contrast, little difference in W_1^c and D_1^c on amount of interparticle lubricant was recognized.

Table1. Yield parameters for the identification of *DPC* yield surfaces

Powder	Fe 100 μm -1.0 %					Fe 10 μm -1.0 %				
Density ratio	0.50	0.70	0.75	0.80	0.85	0.47	0.70	0.75	0.80	-
R_c	3.3					4.0				
X_0	-18	-360	-505	-710	-1010	-3.6	-650	-950	-1450	-
σ_c [MPa]	2	27.7	38.8	52.0	71.1	0.13	19.7	29.0	58.1	-
α	0.200					0.275				
W_1^c	0.693					0.755				
D_1^c	0.0022					0.00142				

4.4. FEM results and validation in densification curve

Figure 7 shows the relation between applied compaction force and density ratio in comparison of simulation and experimental result measured by lateral force measurement test. Despite measurement error in the single shear test for Fe 10 μm-1.0% powder, FEM results showed good correspondence since its yield locus was complemented by the other two testings with a probable slope of *CSL* as mentioned in figure 6(b). Insufficient density calculated at the bottom corner of the compact reflects adequately damage occurrence during ejection from mold in actual green compact in fine particle size as the photograph mentioned in figure 7(b). Fe 100 μm-0.5 % powder also shows good correspondence with the experimental result. Thus, it is concluded that proper yield parameters were estimated by the proposed procedure in this study.

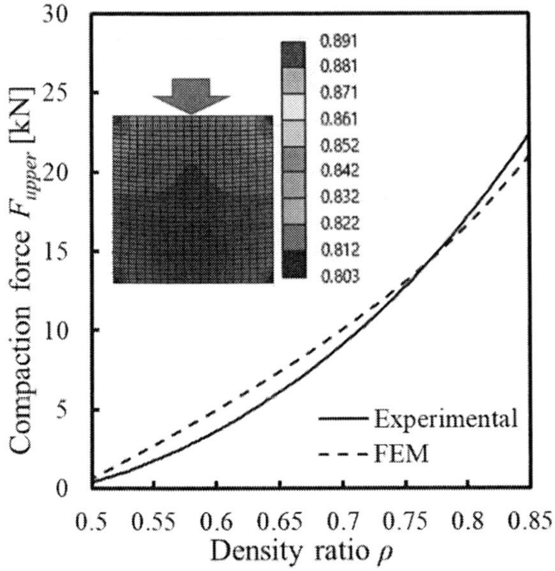

(a)Fe 100 μm-1.0%

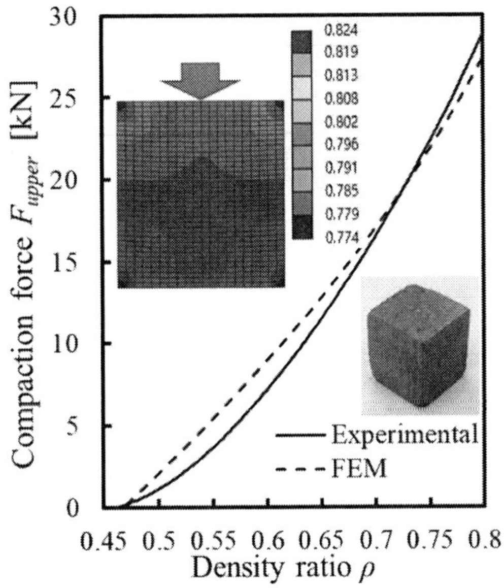

(b)Fe 10 μm-1.0%

Figure 7. Comparison of simulation and experimental result

5. Conclusions

Critical stress state and uniaxial failure stress state have been determined by a single shear test and uniaxial failure test for green compact for different particle size. Also, compaction cap parameter has been estimated by lateral force measurement test by instrumented die compaction system which can eliminate the influence of frictional force on the inner surface of compaction mold. By those estimations, materials input parameters for the *DPC* model have been identified successfully, and differences in particle size and amount of lubricant can be recognized adequately. FEM analysis of simple closed-die compaction has been conducted. As a result, the materials input parameters for the *DPC* model on different powder properties have been successfully identified as the FEM analysis fairly predicts compaction force and density distribution on uniaxial compaction.

References

[1] Takemoto S 1996 *Proc. Int. Conf. on Powder metallurgy & particulate Materials* **2** 87-100
[2] Takemoto S 2009 *J. Jpn. Soc. Powder & Powder Metallurgy* **56** 306-312 (in Japanese)
[3] Shima S Kotera H 1998 *J. High pressure Sci. and Tech.* **8** 260-267 (in Japanese)
[4] PM Modnet Computer Modeling Group 1999 *Powder Met.* **42** 301-311
[5] Mosbah P Shima S Habraken A Charlier R 1999 *J. Jpn. Soc. Powder Powder Metallurgy* **46**(7) 696-704
[6] Shima S Kotera H Yamamoto Y 2000 *J. Jpn. Soc. Powder & Powder Metallurgy* **47** 1312-1317 (in Japanese)
[7] Lewis R W Khoei A R 2001 *Int. J. Plasticity* **17** 1659-1692
[8] Shima S 2001 *Journal of the JSTP* **42** 316-320 (in Japanese)
[9] Shima S Inoue T Oyane M Okimoto K 1976 *J. Jpn. Soc. Powder Powder Metallurgy* **22** 257-263 (in Japanese)
[10] Drucker D C Prager W 1952 *Quarterly of Appl. Math.* **10** 157-165
[11] Shima S Mimura K 1986 *J. Jpn. Mech. Sci.* **28** 53-59 (in Japanese)
[12] Pavier E Doremus P 1999 *Powder Met.* **42** 345-352

[13] Gu C Kim M Anand L 2001 *Int. J. Plasticity* **17** 147-209
[14] Sinka I Cocks A Morrison C Lightfoot A 2000 *Powder Met.* **43**(3) 253-262
[15] Taniguchi Y Dohda K Wang Z 2004 *Trans. Japan Society of Mech. Eng.* **70**(696) 2555-2560 (in Japanese)
[16] Taniguchi Y Shiota K Kodama K 2015 *Proc. 6th TSME Int. Conf. on Mech. Eng. (Phetchaburi)* AMM020
[17] Taniguchi Y 2015 *Abstracts of Spring Meeting JSPM* 151 (in Japanese)
[18] Ogawa R Yoshimura T Yagi N Mizuta K Taniguchi Y 2018 *J. IOP Conf. Series Materials Science and Engineering* **501** 012021
[19] Mizuta K Taniguchi Y 2019 *J. Jpn. Soc. Powder & Powder Metallurgy* **66** 405-412 (in Japanese)

IOP Conf. Series: Materials Science and Engineering **886** (2020) 012061 doi:10.1088/1757-899X/886/1/012061

Effect of Chemical Solution on Tensile Strength of Bamboo Fiber

Jerachard Kaima[1], Itthichai Preechawuttipong[1,*], Pawarut Jongchansitto[1] and Nattaphat Charoenloe[2]
[1]Department of Mechanical Engineering, Faculty of Engineering, Chiang Mai University, Chiang Mai, Thailand
[2]Department of Civil Engineering, Faculty of Engineering, Chiang Mai University, Chiang Mai, Thailand

* Corresponding Author: itthichai.p@cmu.ac.th

Abstract
This study aims to investigate the effect of a chemical solution on the tensile strength of the bamboo fiber. The bamboo woods were cut into the size of small pieces 60 mm x 5 mm x 1 mm. Seven solutions were used in this work which is distilled water (control conditions), ASH and NaOH solution with concentrations of 20%, 10%, and 5% by weight. The bamboo pieces were soaked in the solution with 11 different soaking times except soaking in distilled water for 5 different soaking times. After that processed bamboo pieces were washed and dried at room temperature for 1 week. They were separated by hand to get processed bamboo fiber with cross-section size between 0.06-0.27 mm2. For all specimen, more than 5 pieces of processed bamboo fibre were tested following to ASTM D3039. The results show that the cross-section area in rang 0.06-0.27 mm2 from raw materials have similar tensile strength with 290 MPa [1]. The higher concentration of NaOH, the lower value of the tensile strength which decreases and reach below 100 MPa for 20% NaOH solution. For ASH solution, the values of the tensile strength as the function of soaking time are the same inclination with the different concentration. The tensile strength of processed bamboo fibres is pretty splattered between 200-400 MPa for the soaking time less than 48 hours. While tensile strengths have the range between 275-300 MPa for the soaking time more than 48 hours. Because some fungus grew on bamboo fibre surface soaked in distilled water, the tensile strength of processed bamboo fibres soaked in 20% ash solution is greater 100 MPa than the tensile strength of bamboo fiber soaked in distilled water for the soaking time more than 120 hours.

Keywords: Tensile Strength; Bamboo Fiber; Chemical Solution

1. Introduction
Thailand is one of the countries having the most distribute bamboo species in the world [2], i.e. more than 31 species of bamboo can be found in Thailand. In Thailand, the bamboo has been applied in many areas for a long time. Its applications can be found from kitchenware at home like chopsticks to the structure of buildings. This is because the bamboo is easy to find, rapid growth rate, and better

mechanical properties. It must be noted that the bamboo is ready to use within 3 years after planting, while other woods are available after planting more than 20 years [3]. Nowadays, bamboo becomes a popular raw material. However, a lifetime of bamboo applications must be concerned, because the lifetime can be reduced by major enemies: insects and funguses. In this case, a degradation of the lignin in bamboo by such enemies causes a reduction of the bamboo's strength [4, 5].
To obtain pure cellulose of the bamboo fiber consisting of xylem and phloem, the lignin inside the bamboo must be removed by mechanical or chemical treatments. This pure cellulose is also called an avascular bundle, which is an important structure related to the strength of the bamboo fiber [6, 7]. The widely- used chemical treatment to remove the lignin from the bamboo is the use of Sodium Hydroxide (NaOH) solution. Nevertheless, its base strength is very strong that can rapidly destroy the strength of the bamboo fiber [6]. In addition, this procedure is non-environmentally friendly. In the northern region of Thailand, an ash solution, which is local knowledge, was employed by Hmong peoples to wash dishes, clean clothes, and fabricate clothes. The authors expect that it would be useful for the removing of the lignin from the bamboo fiber. It is important to note that no previous studies have focused on the effect of concentration of ash solution on the tensile strength of the bamboo fiber. Therefore, the present study aims at experimentally investigating the influence of the concentration of ash and NaOH solutions on the tensile strength of the bamboo fiber.

2. Material and Method

2.1. Material Preparation

A bamboo tree can be generally separated into 3 parts: a base part (0-0.5 m from a ground), a middle part (0.5-2 m from) and a top part (more than 2 m). The middle part of the bamboo for the age of more than 3 years was used in this study [8], because this part consists of fewer bamboo joints and more wood compared with the other parts. The middle part of the bamboo was then dried out at the temperature of about 35-37°C for 7 days. After that, it was cut into a small piece with dimensions of 60 x 5 x 1 mm (length x width x thickness). The selected specimens are shown in Fig. 1. Then specimens were washed by distilled water for clean grime on bamboo skin and dried in control room temperature (25°C) for 7 days.

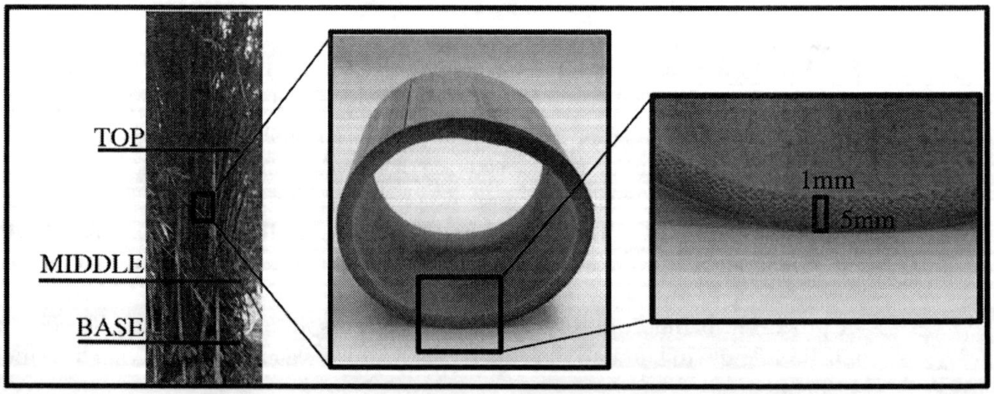

Figure 1. The selected part of the bamboo used as a specimen

Chemical solutions in this study were ash solution, NaOH solution. distilled water was used to mix the solution and be control condition. Ash solution was made by Longan wood ash that popular to use in local northern Thai kitchen. After Longan tree woods were burned in the burner, all the ash was cooled for 3 hours and collect ash powder in plastic bags to avoid moisture added. Ash powder was mixed with different concentration following equation (1) and measured by digital scales.

Then distilled water and ash powder were mixed by shaking the bottle. After that, the mixture was let to sit for 1 day. From the ash mixing process, water will get ash containing sediment lying on the bottom of the bottle. The ash solution used in this study will use the only solution with clear color above.

For the preparation of the NaOH solution, a commercial NaOH quartz 98% was used in this study. They were measured by digital scales. Then NaOH was separated in quite small 5 groups to avoid the heave chemical reaction. After that distilled water was added in each group and mixed them following equation (1).

$$\%Sol = \left(\frac{(M_{Sub}x\%Pure)}{(M_{Sub}x\%Pure)+M_{WATER}}\right)x100 \tag{1}$$

where *%Sol* is a concentration of a solution, M_{Sub} is a mass of a substance (g) *and* M_{WATER} is a mass of a distilled water (g), *%Pure is* a purity of a substance

Chemical solution mixture and number of soaking times in each specimen are shown in the Table 1. Dried bamboo strips were soaked in ash solution and NaOH solution in the cylinder bottles size 4 cm diameter 8 cm in height. Both chemical solutions have 3 concentrations which are 5%, 10%, 20%. The bottles were closed and leave it in control room temperature (25°C) for the 11 different soaking times which are 1, 3, 6, 9, 12, 18, 24, 48, 72, 120, 168 hours. The same method was applied with distilled water for the control group, but the soaking time change to be 5 different times which are 24, 48, 72, 120, 168 hours. After leaving the specimen from the solution with the different soaking time, the specimens were washed and dried in control room temperature for 7 days. After that specimens were packed in the plastic bag for more than 1 month. Each bamboo fiber was then separated by hand pullingout. All of the bamboo fiber was measured cross-section size in rang 0.06-0.27 mm^2.

Table 1. Chemical solution mixture in 1000 g and the number of soaking times in each specimen are show in table

			Mixture		
Substance	Purity of substance	Concentration (%)	Substance (g)	Distilled water (g)	Number of Soaking Times
		5	50.00	950.00	
Ash	100%	10	100.00	900.00	11
		20	200.00	800.00	
		5	51.02	948.98	
NaOH	98%	10	102.04	897.96	11
		20	204.08	795.92	
Distilled Water	100%	0	0	1000	5

2.2. Tensile strength test

Fibers destroy by the clamping is the problem in this part. So, paper grips and rubber glue were used to solve machine clamping destroy [9]. The paper grips method for the tensile test are shown in Fig. 2. Fibers were stuck in the centre line of paper grips by rubber glue. Top and bottom of paper grips were flipped to the centre. The paper grip with specimens was let in control room temperature more than 2 hours for dried glue. Following ASTM D3039 [10], a tensile loading was then applied to the specimen by using INSTRON 5566 testing machine. The loading was displacement-controlled of 1 mm/s. It must be noted that guideline areas beside the paper grips were cut before the machine start pulling out.

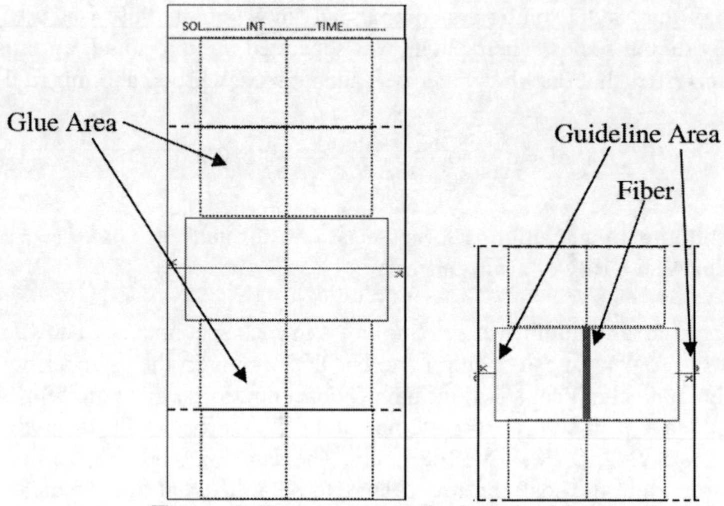

Figure 2. Paper grips method applied for tensile test

3. Result

3.1. Effect of the cross-sectional area of the fiber on tensile strength

Separating processed bamboo fibers by hand pull out cause fibers have varieties size of the cross-section area. The varieties size of cross section is effect to tensile test that are destroy by the clamping point while fibers are too thin and fibers are slip from clamping point when fibers are too big [11,12]. So raw bamboo fibers were separated by the same way to find the rang of fibers able to use in this method.

Fibers were classified in 5 groups by cross-section area size which is A, B, C, D, E. all of them were tested on tensile strength test. The result of this test is shown in Table 2.

Table 2. Effect of the cross-section on tensile strength

Group	Range of cross-section (mm^2)	Tensile stress (MPa)
A	Less than 0.06	Destroy by clamping
B	0.06-0.08	265.74
C	0.13-0.19	290.40
D	0.20-0.27	275.36
E	More than 0.27	Slip
G.Wang and F.Chen [1]	N/A	290.00

The result showed that the cross-sectional area of the fiber in group A is too thin. Fibers were destroyed by the clamping point when pulling out. Moreover, the cross-sectional area of the fiber in group E is too big for paper grip. The fiber is not torn apart. So only the cross-section in group B, C, D can use in this test. The values of tensile strength are the same as Wang and Chen's study [1,13].

3.2. Effect of chemical solutions on tensile strength

The results of the concentration of ash and NaOH solutions on tensile strength at the soaking time of 168 hours are shown in Table 3. For the NaOH solution, the tensile strength decreases with an increment of the solution concentration. This can be explained by the fact that both the lignin and the bamboo fiber are destroyed by the NaOH solution [14]. In the case of the ash solution, the tensile strength decreases with the concentration. The tensile strength of the processed bamboo fiber in 20% ash solution is higher than raw material. Because process bamboo fibers are soft and smooth than raw material. In the case of the distilled water, the tensile strength is lower than the raw bamboo fiber, because of the growth of fungus on processed bamboo fibers [15]. For bamboo fiber that soaked in a chemical solution, there is no fungus growth.

Table 3. Effect of a chemical solution on tensile strength at 168 hours.

Solution	Tensile strength (MPa)	Fungus growth
ASH 5%	244.21	No
ASH 10%	274.91	No
ASH 20%	368.24	No
NaOH 5%	199.628	No
NaOH 10%	165.82	No
NaOH 20%	25.27	No
Distilled water	264.90	Yes

3.3. Effect of soaking time on tensile strength

The result of soaking time on tensile strength is shown in figure 3. It is observed that the soaking time less than 48 hours causes a fluctuation of the tensile strength of the bamboo fiber. Tensile strength of processed bamboo fibers soaked in NaOH is decreased by the time. As same as ash solution at concentration 5%, 10% and distilled water tensile strength are decreased by the time. On the other hand, the tensile strength of process bamboo fibers soaked in 20% ash solution have a different tendency. The values are increased above distilled water after 120 hours. Our study recommends 20% ash solution as the best chemical solution to enhance the tensile strength of the bamboo fiber.

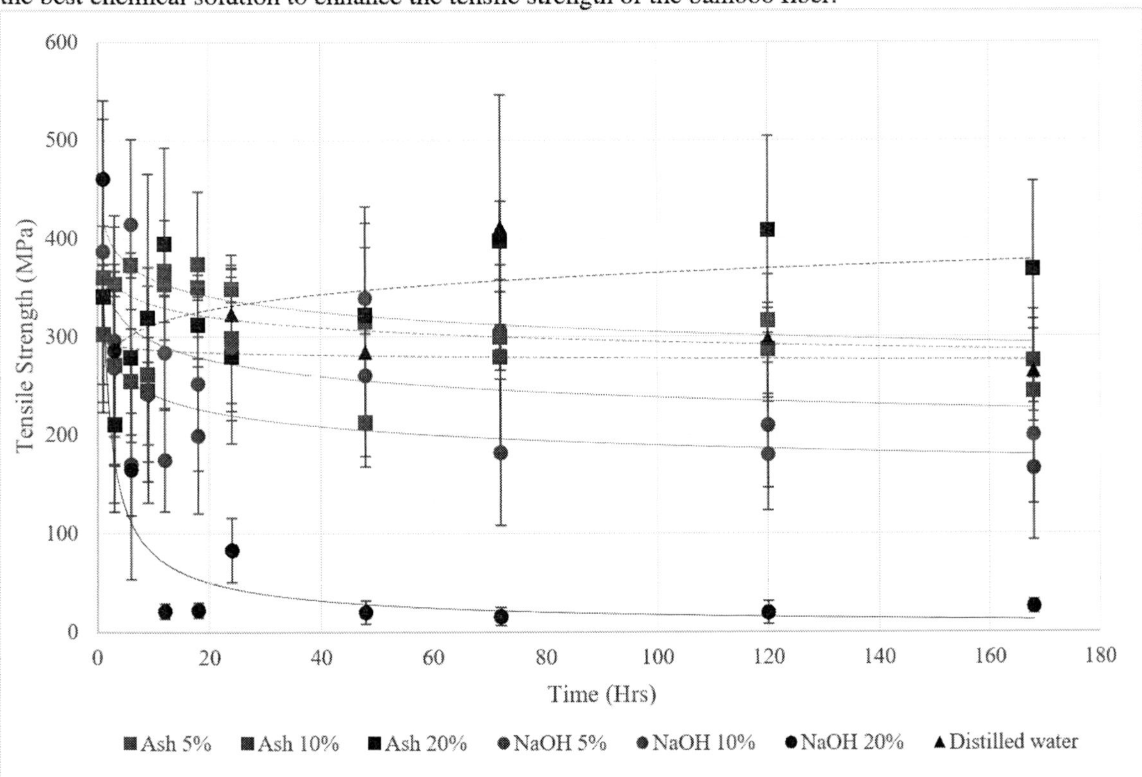

Figure 3. Effect of soaking time on tensile strength

4. Conclusion

This research has studied the effect of ash and NaOH solution on the tensile strength of bamboo fiber that extracts by hand pull out. For ash solution, the tensile strengths are decreased by the concentration. The best concentration recommended by this study is 20% ash solution. For NaOH not only lignin but also bamboo fibers were strongly destroyed so this chemical is not suggested by this study.

The effects of soaking time on tensile strength of bamboo fiber after 48 hours-soaked bamboos in chemical solution the tensile strength values change little. So, 48 hours is enough time for soaked bamboo in a chemical solution.

Acknowledgments

The authors would like to acknowledge Research and Researchers for Industries Scholarship (Grant No. PHD 60I0039) and Charoen Triphop Limited Partnership for the financial support during this research. The authors also would like to acknowledge Faculty of Engineering, Chiang Mai University for equipment support and scholarship for the conference at the 10th TSME International Conference on Mechanical Engineering.

References

[1] Wang G Chen F 2017 *Advanced High Strength Natural Fiber Composites in Construction* chapter 10 (Woodhead Publishing:United Kingdom) 235–253
[2] Dransfield S Widjaja, E.A. 1995 *Bamboos* **7** 189
[3] Chawla K K 2016 *Fibrous Materials* (Cambridge university press)
[4] Horrocks A R Price D 2001 *Fire retardant materials* (Woodhead Publishing:England)
[5] Liese W 1992 *Proc. 5th Int Conf on Forest Product* **2** (Nancy Publisher) 742–51
[6] Khan Z Yousif B F Islam M 2014 *Composites Part B* **116** 186–199
[7] Amada S Ichikawa Y Munekata T Nagase Y Shimizu H 1997 *Composites Part B: Engineering* **28** 13–20
[8] Zakikhani P Zahari R Sultan M T H Majid D L 2014 *Materials and Design* **63** 820–828
[9] Haameem J A M Majid M S A Afendi M Marzuki H F A Fahmi I Gibson A G 2016 *Composite Structures* **136** 1–10
[10] ASTM International 2017 *Standard Test Method for Tensile Properties of Polymer Matrix Composite Materials* (West Conshohocken)
[11] Pickering K L 2008 *Properties and performance of natural-fibre composites* (Cambridge:England)
[12] Chawla K K 2016 *Fibrous Materials* (Cambridge:England)
[13] Takaki H Ichihara Y 2004 *JSME International Journal* **47** 332–339
[14] Yong C Yi-Qiang W U 2008 *J. Cent. South Univ. Technol.* **15** 564–567
[15] Garcia C M Giron M Y Mabilangan L C 2000 *Proc. the BAMBOO 2000 International Symposium* (Chiang Mai:Thailand)

AUTHOR INDEX